Creo Parametric 8.0 for Designers

(8th Edition)

CADCIM Technologies
525 St. Andrews Drive
Schererville, IN 46375, USA
(www.cadcim.com)

Contributing Author
Sham Tickoo
Professor
Purdue University Northwest
Department of Mechanical Engineering Technology
Hammond, Indiana, USA

CADCIM Technologies

Creo Parametric 8.0 for Designers
Prof. Sham Tickoo

CADCIM Technologies
525 St Andrews Drive
Schererville, Indiana 46375, USA
www.cadcim.com

ISBN 978-1-64057-125-9

www.cadcim.com

DEDICATION

To teachers, who make it possible to disseminate knowledge
to enlighten the young and curious minds
of our future generations

To students, who are dedicated to learning new technologies
and making the world a better place to live in

THANKS

To the faculty and students of the MET department of
Purdue University Northwest for their cooperation

To employees of CADCIM Technologies for their valuable help

Online Training Program Offered by CADCIM Technologies

CADCIM Technologies provides effective and affordable virtual online training on various software packages including Computer Aided Design, Manufacturing, and Engineering (CAD/CAM/CAE), computer programming languages, animation, architecture, and GIS. The training is delivered 'live' via Internet at any time, any place, and at any pace to individuals as well as the students of colleges, universities, and CAD/CAM training centers. The main features of this program are:

Training for Students and Companies in a Classroom Setting

Highly experienced instructors and qualified engineers at CADCIM Technologies conduct the classes under the guidance of Prof. Sham Tickoo of Purdue University Northwest, USA. This team has authored several textbooks that are rated "one of the best" in their categories and are used in various colleges, universities, and training centers in North America, Europe, and in other parts of the world.

Training for Individuals

CADCIM Technologies with its cost effective and time saving initiative strives to deliver the training in the comfort of your home or work place, thereby relieving you from the hassles of traveling to training centers.

Training Offered on Software Packages

CADCIM provides basic and advanced training on the following software packages:

CAD/CAM/CAE: *CATIA, Pro/ENGINEER Wildfire, Creo Parametric, Creo Direct, SOLIDWORKS, Autodesk Inventor, Solid Edge, NX, AutoCAD, AutoCAD LT, AutoCAD Plant 3D, Customizing AutoCAD, EdgeCAM, and ANSYS*

Architecture and GIS: *Autodesk Revit (Architecture, Structure, MEP), AutoCAD Civil 3D, AutoCAD Map 3D, Navisworks, Oracle Primavera, and Bentley STAAD Pro*

Animation and Styling: *Autodesk 3ds Max, Autodesk Maya, Autodesk Alias, Foundry NukeX, and MAXON CINEMA 4D*

Computer Programming: *C++, VB.NET, Oracle, AJAX, and Java*

For more information, please visit the following link: ***https://www.cadcim.com***

Note

If you are a faculty member, you can register by clicking on the following link to access the teaching resources: ***https://www.cadcim.com/Registration.aspx***. The student resources are available at ***https://www.cadcim.com***. We also provide **Live Virtual Online Training** on various software packages. For more information, write us at ***sales@cadcim.com***.

Table of Contents

Chapter 3: Creating Sketches in the Sketch Mode-II

Chapter 5: Datums

Chapter 6: Options Aiding Construction of Parts-I

Chapter 7: Options Aiding Construction of Parts-II

Chapter 8: Options Aiding Construction of Parts-III

Chapter 9: Advanced Modeling Tools

Chapter 10: Assembly Modeling

Chapter 11: Generating, Editing, and Modifying the Drawing Views

Chapter 12: Dimensioning the Drawing Views

Chapter 13: Other Drawing Options

CHAPTERS AVAILABLE FOR FREE DOWNLOAD

In this textbook, four chapters have been given for free download. You can download these chapters from our website *www.cadcim.com*. To download these chapters, follow the given path: *Textbooks > CAD/CAM > PTC Creo Parametric > Creo Parametric 8.0 for Designers > Chapters for Free Download* and then select the chapter name from the **Chapters for Free Download** drop-down. Click the **Download** button to download the chapter in the PDF format.

Chapter 14: Working with Sheetmetal Components

Chapter 15: Surface Modeling

Chapter 16: Introduction to Mold Design

Chapter 17: Concepts of Geometric Dimensioning and Tolerancing

PROJECTS AVAILABLE FOR FREE DOWNLOAD

In this textbook, four projects are available for free download. You can download these projects from our website **www.cadcim.com**. To download these projects, follow the path: *Textbooks > CAD/CAM > PTC Creo Parametric > Creo Parametric 8.0 for Designers > Projects for Free Download* and then select the project name from the **Projects for Free Download** drop-down. Click the Download button to download the project in the PDF format.

Note

For additional projects, visit **www.cadcim.com** and follow the path: *Textbooks > CAD/CAM > Parametric Solid Modeling > Parametric Solid Modeling Projects*

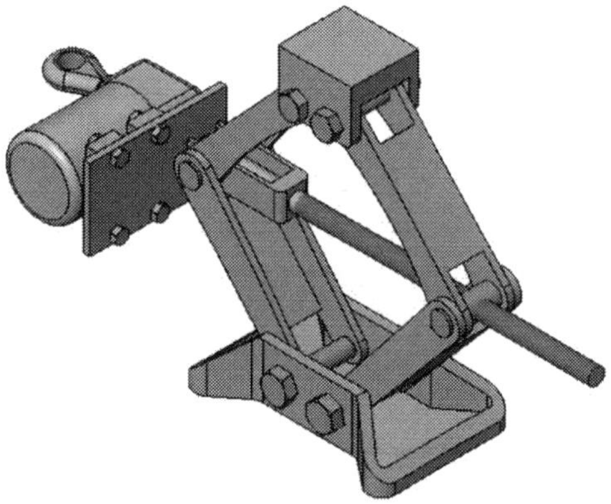

Project 1: *Car Jack Assembly*

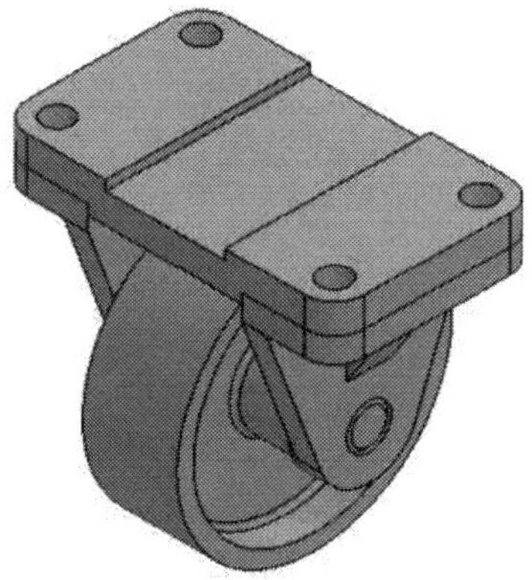

Project 2: *Wheel Assembly*

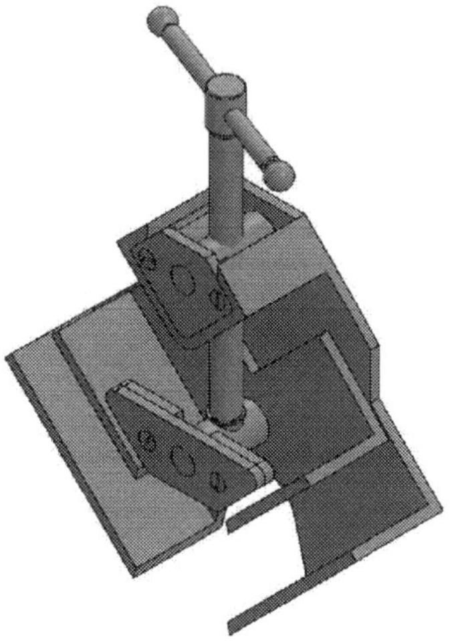

Project 3: *Angle Clamp Assembly*

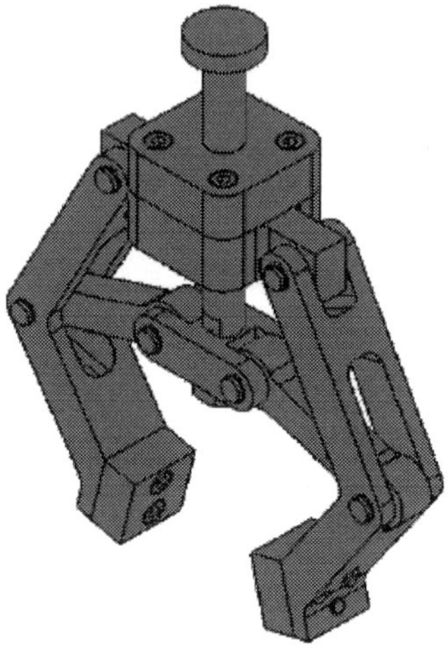

Project 4: *Pneumatic Gripper Assembly*

Preface

Creo Parametric 8.0

Creo Parametric is developed by Parametric Technology Corporation. It provides a broad range of powerful and flexible CAD capabilities that can address even the most tedious design challenges. Being a parametric feature-based solid modeling tool, it not only integrates the 3D parametric features with 2D tools, but also assists in every design-through-manufacturing process. This software is remarkably user-friendly.

This solid modeling software allows you to easily import the standard format files with an amazing compatibility. The 2D drawing views of the components are automatically generated in the Drawing mode. Using this software, you can generate detailed, orthographic, isometric, auxiliary, and section views. Additionally, you can use any predefined drawing standard files for generating the drawing views. You can display the model dimensions in the drawing views or add reference dimensions whenever you want. The bidirectionally associative nature of this software ensures that any modification made in the model is automatically reflected in the drawing views. Similarly, any modification made in the dimensions of the drawing views is automatically updated in the model.

The **Creo Parametric 8.0 for Designers** textbook has been written to enable the readers to use the modeling power of Creo Parametric 8.0 effectively. The latest surfacing techniques like Freestyle and Style are explained in detail in this book. The textbook also covers the Sheetmetal module with the help of relevant examples and illustrations. The mechanical engineering industry examples and tutorials are used in this textbook to ensure that the users can relate the knowledge of this book with the actual mechanical industry designs. In this edition, one chapter has been added to enable the readers to understand the concept of mold design. The salient features of this textbook are as follows:

- **Tutorial Approach**
  The author has adopted the tutorial point-of-view and the learn-by-doing approach throughout the textbook. This approach guides the users through the process of creating the models in the tutorials.

- **Real-World Projects as Tutorials**
  The author has used the real-world mechanical engineering projects as tutorials in this textbook so that the readers can correlate them with the real-time models in the mechanical engineering industry.

- **Tips and Notes**
  Additional information related to various topics is provided in the form of tips and notes.

- **Learning Objectives**
  The first page of every chapter summarizes the topics that will be covered in that chapter. This helps the users to easily refer to a topic.

- **Self-Evaluation Test, Review Questions, and Exercises**
  Every chapter ends with Self-Evaluation Test so that the users can assess their knowledge of the chapter. The answers to Self-Evaluation Test are given at the end of the chapter. Also, Review Questions and Exercises are given at the end of the chapters and they can be used by the instructor as test questions and exercises.

- **Heavily Illustrated Text**
  The text in this book is heavily illustrated with the help of around 1400 line diagrams and screen capture images that support the tools section and tutorials.

Symbols Used in the Textbook

Note
The author has provided additional information to the users about the topic being discussed in the form of notes.

Tip
Special information and techniques are provided in the form of tips that helps in increasing the efficiency of the users.

New
This symbol indicates that the command or tool being discussed is new.

Enhanced
This symbol indicates that the command or tool being discussed has been enhanced in Creo Parametric 8.0.

Formatting Conventions Used in the Textbook

Please refer to the following list for the formatting conventions used in this textbook.

- Names of tools, buttons, options, groups, tabs, slide-down panels, and Ribbon are written in boldface.

  Example: The **Extrude** tool, the **OK** button, the **Editing** group, the **Sketch** tab, and so on.

- Names of dialog boxes, drop-downs, drop-down lists, dashboards, areas, edit boxes, check boxes, and radio buttons are written in boldface.

  Example: The **Revolve** dashboard, the **Chamfer** drop-down of **Engineering** group in the **Model** tab, the **Thickness** drop-down of the **Shell** dashboard, the **Extended intersect surfaces** check box in the **Options** slide-down panel of the **Draft** dashboard, and so on.

- Values entered in edit boxes are written in boldface.

  Example: Enter **5** in the **Radius** edit box.

- Names and paths of the files are written in italics.

  Example: *C:\Creo-8.0\c03*, *c03tut03.prt*, and so on.

- Different options available for invoking a tool are given in a shaded command box.

  Ribbon: Model > Datum > Sketch

Naming Conventions Used in the Textbook

Tool

If you click on an item in a toolbar or a group of the **Ribbon** and a dashboard or dialog box is invoked to create/edit an object or perform some action, then that item is termed as **tool**.

For example:
Line tool, **Normal** tool, **Extrude** tool
Fillet tool, **Draft** tool, **Delete Segment** tool

If you click on an item in a toolbar or a group of the **Ribbon** and a dialog box is invoked wherein you can set the properties to create/edit an object, then that item is also termed as **tool**, refer to Figure 1.

For example:
To Create: **Extrude** tool, **Sweep** tool, **Round** tool
To Edit: **Extend** tool, **Trim** tool

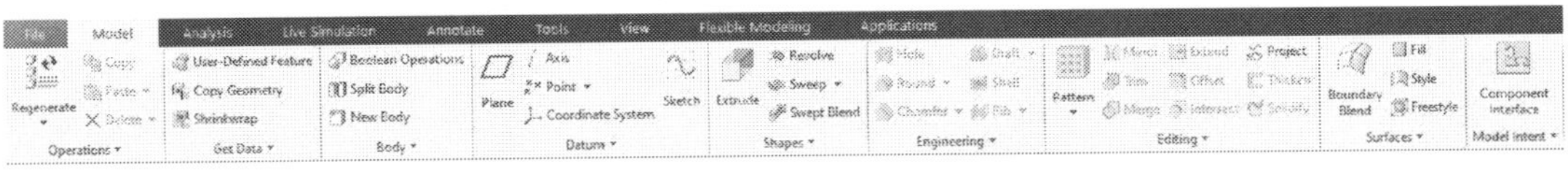

Figure 1** Various tools in the **Ribbon

Button

The item in a dialog box that has a 3D shape like a button is termed as **Button**. For example, **OK** button, **Cancel** button, **Apply** button, and so on.

Dialog Box

In this textbook, different terms are used for referring to the components of a dialog box. Refer to Figure 2 for the terminology used.

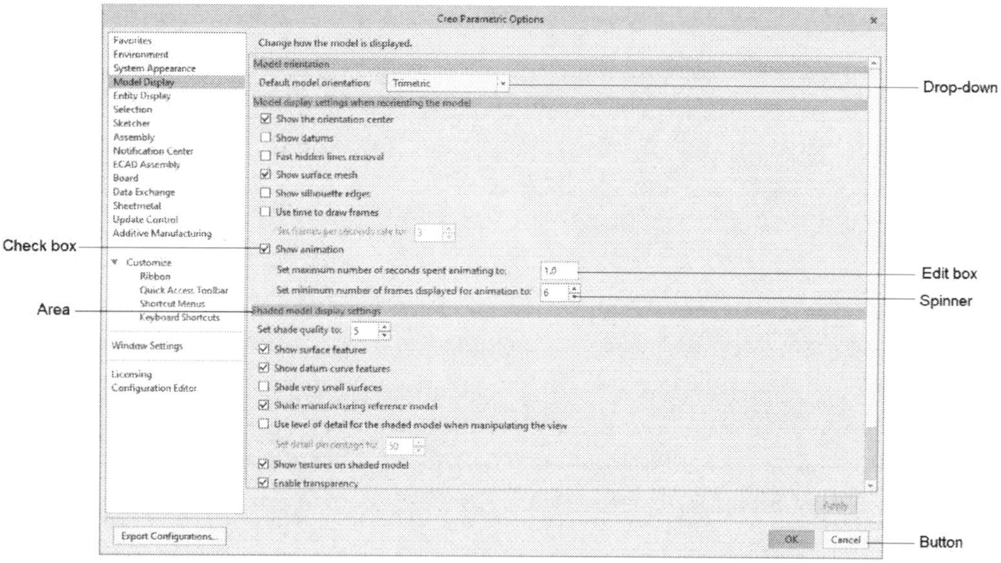

Figure 2 The components of a dialog box

Drop-down

A drop-down is the one in which a set of common tools are grouped together for creating an object. You can identify a drop-down with a down arrow on it. These drop-downs are given a name based on the tools grouped in them. For example, **Arc** drop-down (refer to Figure 3), **Chamfer** drop-down (refer to Figure 4), **Draft** drop-down (refer to Figure 5), and so on.

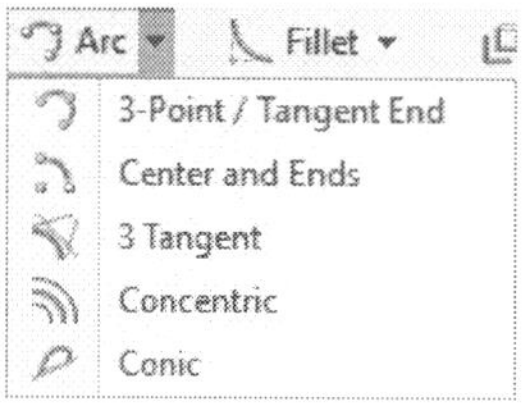

*Figure 3 The **Arc** drop-down*

*Figure 4 The **Chamfer** drop-down*

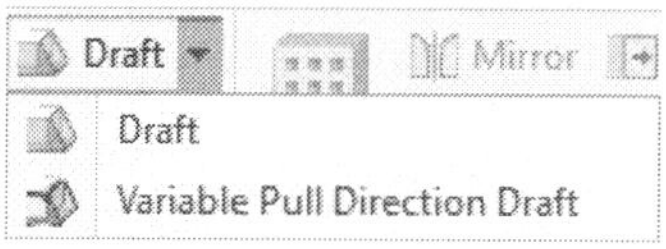

*Figure 5 The **Draft** drop-down*

Drop-down List

A drop-down list is the one in which a set of options are grouped together. You can set various parameters using these options. You can identify a drop-down list with a down arrow on it. For example, **Datum Display Filters** drop-down list, **Clear Appearance** drop-down list, and so on, refer to Figure 6.

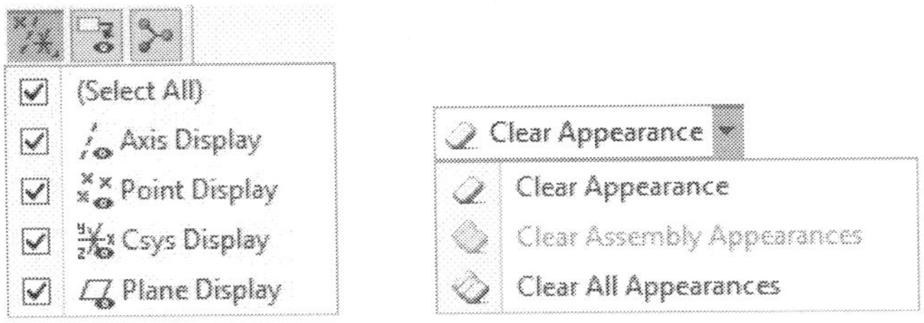

***Figure 6** The **Datum Display Filters** and **Clear Appearance** drop-down lists*

Options

Options are the items that are available in shortcut menu, drop-down list, dialog boxes, and so on. For example, choose the **Front** option from the **View Manager** dialog box, refer to Figure 7; choose the **New** option from the **File** menu, refer to Figure 8.

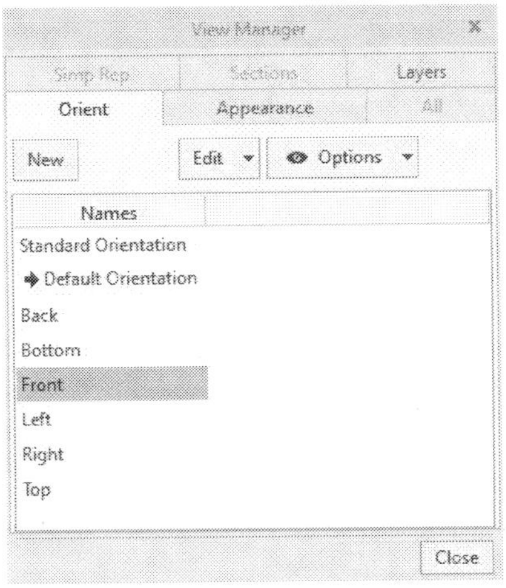

***Figure 7** The **Front** option in the **View Manager** dialog box*

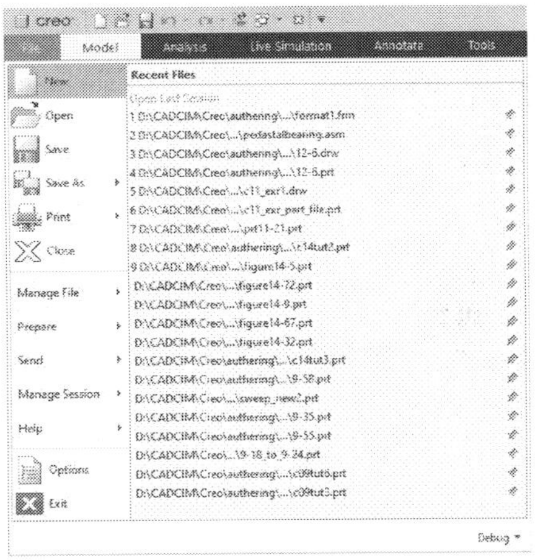

***Figure 8** The **New** option in the **File** menu*

Free Companion Website

It has been our constant endeavor to provide you the best textbooks and services at affordable price. In this endeavor, we have come out with a Free Companion website that will facilitate the process of teaching and learning of Creo Parametric 8.0. If you purchase this textbook, you will get access to the files on the Companion website.

The following resources are available for the faculty and students in this website:

Faculty Resources

- **Technical Support**
  You can get online technical support by contacting ***techsupport@cadcim.com***.

- **Instructor Guide**
  Solutions to all review questions and exercises in the textbook are provided in the Instructor guide to help the faculty members test the skills of the students.

- **Part Files**
  The part files used in illustrations, examples, and exercises are available for free download.

- **Input Files**
  The input files used in examples are available for free download.

- **Free Download Chapters**
  Chapters available for free download.

- **Free Download Projects**
  In this book, four projects are available for free download.

To access files, you need to register by visiting the **Resources** section at ***www.cadcim.com***.

Student Resources

- **Technical Support**
  You can get online technical support by contacting ***techsupport@cadcim.com***.

- **Part Files**
  The part files used in illustrations and examples are available for free download.

- **Input Files**
  The input files used in examples are available for free download.

- **Free Download Chapter**
  Chapters available for free download.

- **Free Download Projects**
  In this book, four projects are available for free download.

If you face any problem in accessing these files, please contact the publisher at ***sales@cadcim.com*** or the author at ***stickoo@pnw.edu*** or ***tickoo525@gmail.com***.

Video Courses

CADCIM offers video courses in CAD, CAE Simulation, BIM, Civil/GIS, and Animation domains on various e-Learning/Video platforms. To enroll for the video courses, please visit the CADCIM website using the link ***https://www.cadcim.com/video-courses***.

Stay Connected

You can now stay connected with us through Facebook and Twitter to get the latest information about our textbooks, videos, and teaching/learning resources. To stay informed of such updates, follow us on Facebook *(www.facebook.com/cadcim)* and Twitter (***@cadcimtech***). You can also subscribe to our YouTube channel *(www.youtube.com/cadcimtech)* to get the information about our latest video tutorials.

This page is intentionally left blank

Chapter 1

Introduction to Creo Parametric 8.0

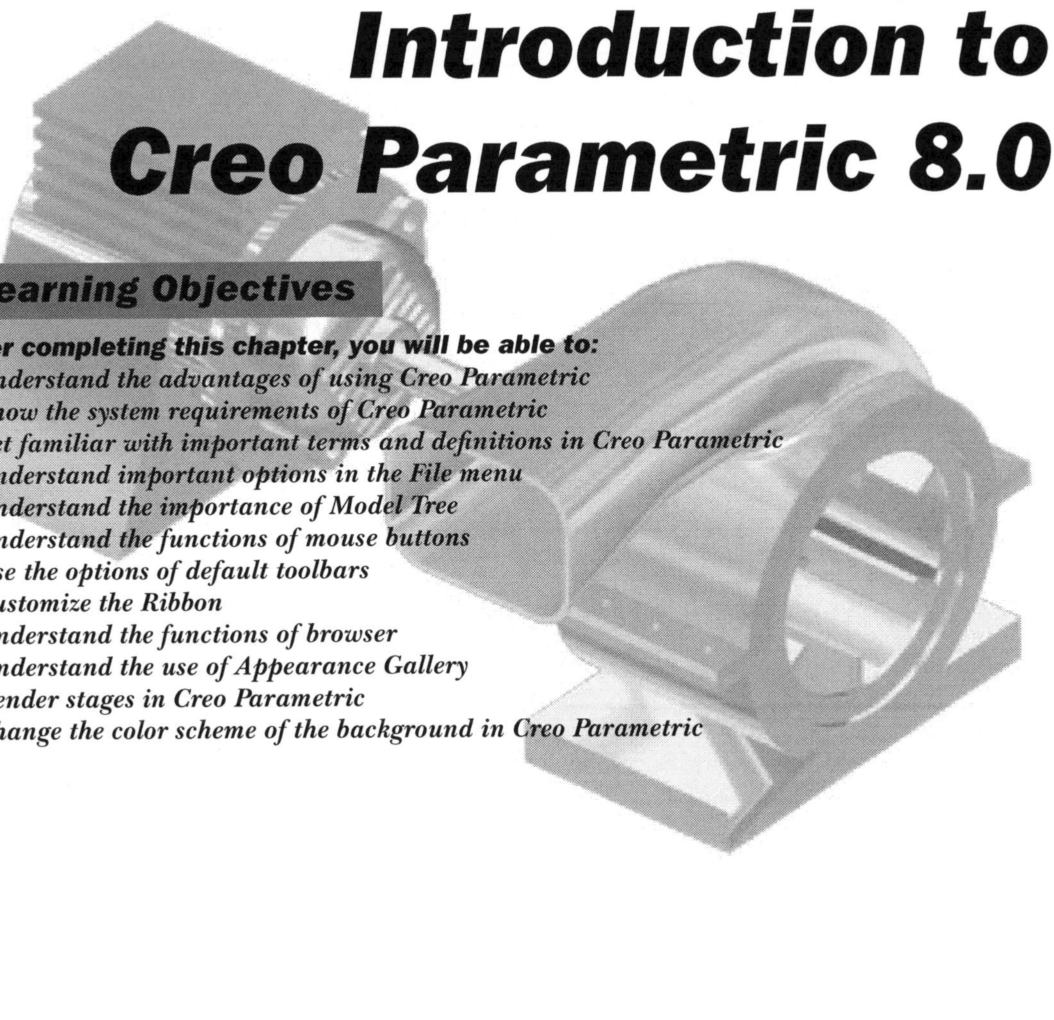

Learning Objectives

After completing this chapter, you will be able to:

- *Understand the advantages of using Creo Parametric*
- *Know the system requirements of Creo Parametric*
- *Get familiar with important terms and definitions in Creo Parametric*
- *Understand important options in the File menu*
- *Understand the importance of Model Tree*
- *Understand the functions of mouse buttons*
- *Use the options of default toolbars*
- *Customize the Ribbon*
- *Understand the functions of browser*
- *Understand the use of Appearance Gallery*
- *Render stages in Creo Parametric*
- *Change the color scheme of the background in Creo Parametric*

INTRODUCTION TO CREO PARAMETRIC 8.0

Creo Parametric is a powerful software used to create complex designs with great precision. The design intent of a three-dimensional (3D) model or an assembly is defined by its specification and its use. You can use the powerful tools of Creo Parametric to capture the design intent of a complex model by incorporating intelligence into the design. Once you understand the feature-based, associative, and parametric nature of Creo Parametric, you can appreciate its power as a solid modeling tool.

To make the designing process simple and quick, the designing process have been divided into different modules in this software package. This means each step of the designing is completed in a different module. For example, generally a design process consists of the following steps:

- Sketching using the basic sketch entities
- Converting the sketch into features and parts
- Assembling different parts and analyzing them
- Documenting parts and the assembly in terms of drawing views
- Manufacturing the final part and assembly

All these steps are divided into different modes of Creo Parametric namely, the **Sketch** mode, **Part** mode, **Assembly** mode, **Drawing** mode, and **Manufacturing** mode.

Despite making various modifications in a design, the parametric nature of this software helps preserve the design intent of a model with tremendous ease. Creo Parametric allows you to work in a 3D environment and calculates the mass properties directly from the created geometry. You can also switch to various display modes like wireframe, shaded, hidden, and no hidden at any time with ease as it does not affect the model but only changes its appearance.

FEATURES OF CREO PARAMETRIC

Different features of the software are discussed next.

Feature-Based Nature

Creo Parametric is a feature-based solid modeling tool. A feature is defined as the smallest building block and a solid model created in Creo Parametric is an integration of a number of these building blocks. Each feature can be edited individually to bring in the desired change in the solid model. The use of feature-based property provides greater flexibility to the parts created. For example, consider the part shown in Figure 1-1. It consists of one counterbore hole at the center and six counterbore holes around the Bolt Circle Diameter (BCD).

Now, consider a case where you need to change all the outer counterbore holes to drill holes keeping the central counterbore hole and the BCD for the outer holes same. Also, you need to change the number of holes from six to eight. In a non feature-based software package, you will have to delete the entire part and then create a new part based on the new specifications. Whereas, Creo Parametric allows you to make this modification by just modifying some values in the same part, see Figure 1-2. This shows that the solid parts created in Creo Parametric are a combination of various features that can be modified individually at any time.

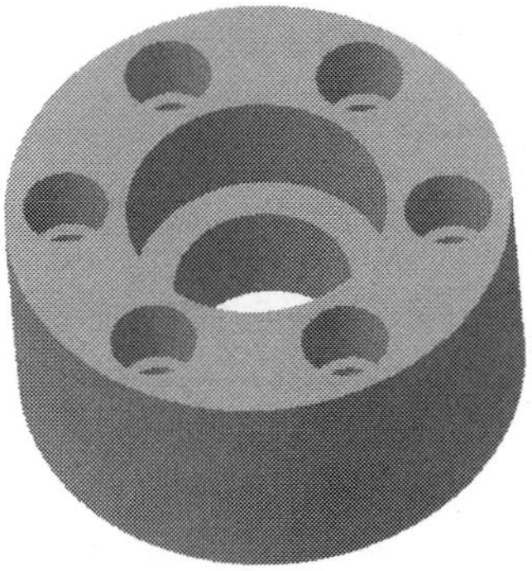

Figure 1-1 Model with counterbore holes

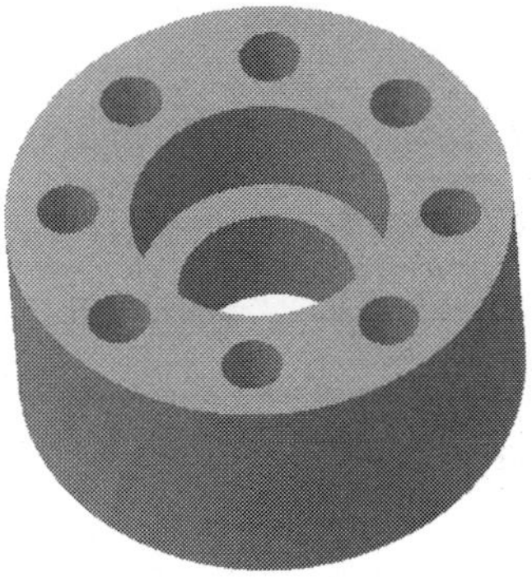

Figure 1-2 Model after making modifications

Bidirectional Associative Property

There is a bidirectional associativity between all modes of Creo Parametric. The bidirectional associative nature of a software package is defined as its ability to ensure that if any modifications are made in a particular model in one mode, then those modifications are also reflected in the same model in other modes. For example, if you make any change in a model in the **Part** mode and regenerate it, the changes will also be highlighted in the **Assembly** mode. Similarly, if you make a change in a part in the **Assembly** mode, after regeneration, the change will also be highlighted in the **Part** mode. This bidirectional associativity also correlates the two-dimensional (2D) drawing views generated in the **Drawing** mode and the solid model created in the **Part** mode of Creo Parametric. This means that if you modify the dimensions of the 2D drawing views in the **Drawing** mode, the change will be automatically reflected in the solid model and also in the assembly after regeneration. Likewise, if you modify the solid model in the **Part** mode, the changes will also be seen in the 2D drawing views of that model in the **Drawing** mode. Thus, bidirectional associativity means that if a modification is made to one mode, it changes the output of all the other modes related to the model. This bidirectional associative nature relates various modes in Creo Parametric.

Figure 1-3 shows the drawing views of the part shown in Figure 1-1 generated in the **Drawing** mode. The views show that the part consists of a counterbore hole at the center and six counterbore holes around it.

Now, when the part is modified in the **Part** mode, as shown in Figure 1-2, the modifications are automatically reflected in the **Drawing** mode, as shown in Figure 1-4. The views in this figure show that all outer counterbore holes are converted into drilled holes and the number of holes is increased from six to eight.

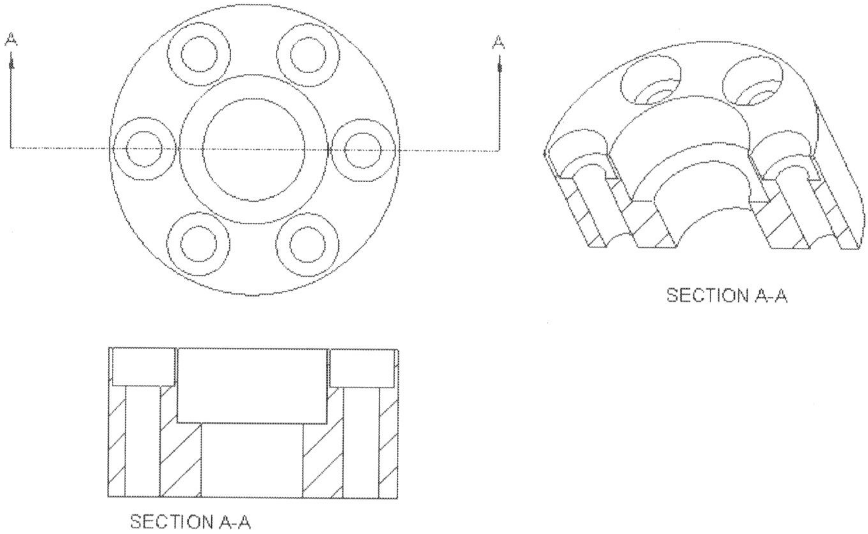

Figure 1-3 *Drawing views of the model before modifications*

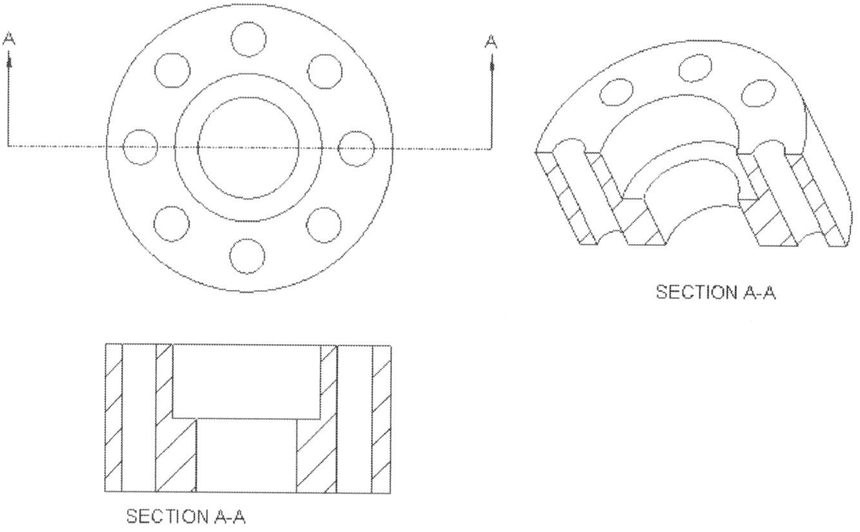

Figure 1-4 *Drawing views of the model after modifications*

Figure 1-5 shows the Crosshead assembly. It is clear from the assembly that the diameter of the hole is more than what is required (shown using black rounded rectangle). In an ideal case, the diameter of the hole should be equal to the diameter of the bolt.

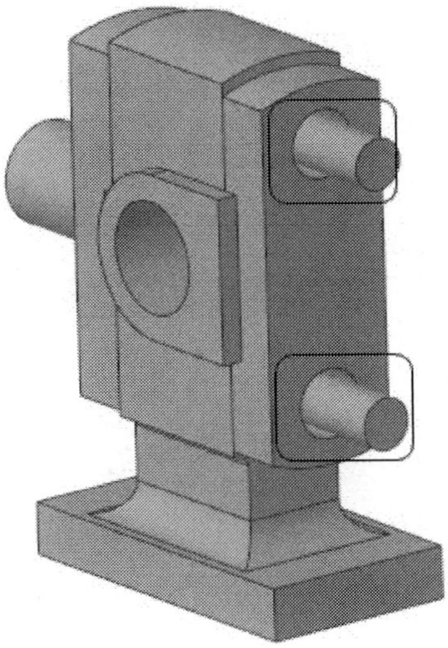

Figure 1-5 *Crosshead assembly illustrating difference in diameter of the hole and the bolt*

The diameter of the hole can be changed easily by opening the file in the **Part** mode and making the necessary modifications in the part. This modification is reflected in the assembly, as shown in Figure 1-6. This is due to the bidirectional associative nature of Creo Parametric.

Since all modes of Creo Parametric are interrelated, it becomes very easy to modify your model at any time.

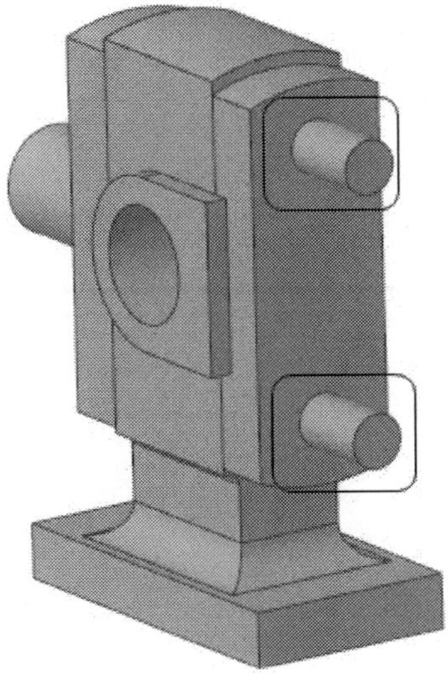

Figure 1-6 *Assembly after modifying the diameter of the hole*

Parametric Nature

Creo Parametric is parametric in nature, which means that the features of a part become interrelated if they are drawn by taking the reference of each other. You can redefine the dimensions or the attributes of a feature at any time. The changes will propagate automatically throughout the model. Thus, they develop a relationship among themselves. This relationship is known as the parent-child relationship. So if you want to change the placement of the child feature, you can make alterations in the dimensions of the references and hence change the design as per your requirement. The parent-child relationship will be discussed in detail while discussing the datums in later chapters.

SYSTEM REQUIREMENTS

The preferred system requirements for Creo Parametric are given below.

1. Operating System: Windows 7 64-bit Edition (Professional, Ultimate, and Enterprise), Windows 8.1 64-bit Edition or Windows 8.1 Pro 64-bit Edition, Windows 10 64-bit Edition (Professional, Enterprise).

2. Monitor: 1280 x 1024 (or higher) resolution support with 32-bit color.

3. Processor: 2.5 GHz minimum (Core i5 or higher).

4. Memory: 4 GB or higher.

5. Hard disk space: 5 GB minimum.

6. Microsoft TCP/IP Ethernet Network Adapter.

7. Microsoft approved 3-button mouse.

8. Microsoft Internet Explorer 6.0 or later.

9. A certified and supported graphics card.

GETTING STARTED WITH Creo Parametric

Install the Creo Parametric on your system; the shortcut icon of Creo Parametric 8.0.0.0 will automatically be created on the desktop. You can start it by double-clicking on its shortcut icon on the desktop.

Figure 1-7 shows the initial screen and Figure 1-8 shows the **Resource Centre** window of Creo Parametric. Both windows appear when you start Creo Parametric.

Figure 1-7 Initial screen of Creo Parametric

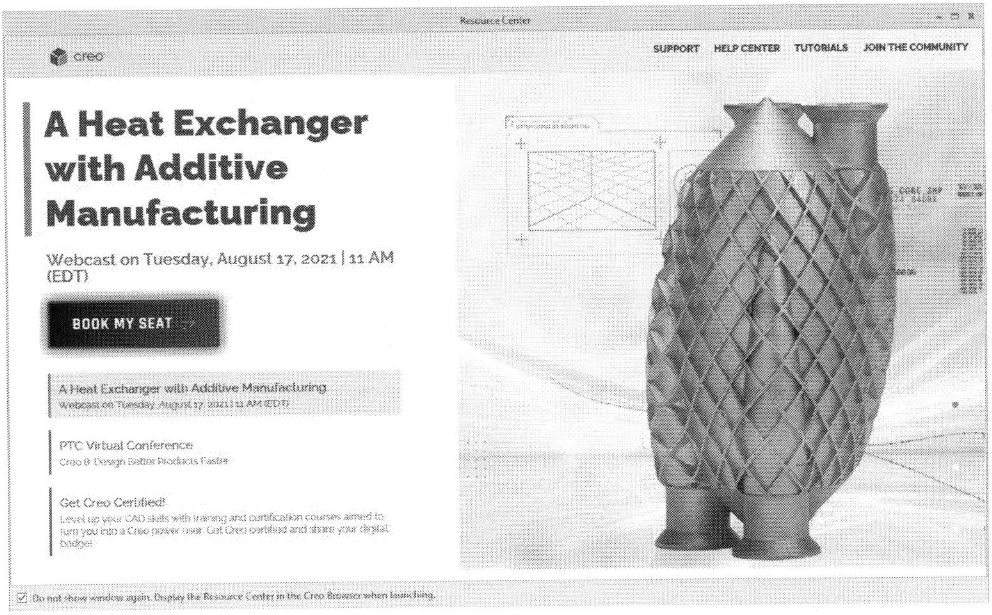

Figure 1-8 The ***Resource Center*** *window displayed on starting Creo Parametric*

IMPORTANT TERMS AND DEFINITIONS

Some important terms that will be used in this book while working with Creo Parametric are discussed next.

Entity

An element of the section geometry is called an entity. The entity can be an arc, line, circle, point, conic, coordinate system, and so on. When one entity is divided at a point, then the total number of entities are said to be two.

Dimension

It is the measurement of one or more entities.

Constraint

Constraints are logical operations that are performed on the selected geometry to make it more accurate in defining its position and size with respect to the other geometry.

Parameter

It is defined as a numeric value or a word that defines a feature. For example, all dimensions in a sketch are parameters. The parameters can be modified at any time.

Relation

A relation is an equation that relates two entities.

Weak Dimensions and Weak Constraints

Weak dimensions and weak constraints are temporary dimensions or constraints that appear in light blue color. These are automatically applied to the sketch. They are removed from the sketch without any confirmation from the user. The weak dimensions or the weak constraints should be changed to strong dimensions or constraints if they seem to be useful for the sketch. This only saves an extra step of dimensioning the sketch or applying constraints to it.

Strong Dimensions and Strong Constraints

Strong dimensions and strong constraints appear in dark blue color. These dimensions and constraints are not removed automatically. All dimensions added manually to a sketch are strong dimensions.

Note

In case there are conflicting dimensions and constraints, Creo Parametric makes the constraints and dimensions appear in blue box, and prompts you to delete one or more of them.

File MENU OPTIONS

The options that are displayed when you choose **File** from the menu bar are discussed next.

Select Working Directory

A working directory is a directory on your system where you can save the work done in the current session of Creo Parametric. You can set any directory existing on your system as the working directory. Before starting the work in Creo Parametric, it is important to specify the working directory. If the working directory is not selected before saving an object file, then the object file will be saved in a default directory. This default directory is set at the time of installing Creo Parametric. If the working directory is selected before saving the object files that you create, it becomes easy to organize them. In Creo Parametric, the working directory can be set in the following two ways:

Using the Navigator

When you start a Creo Parametric session, the navigator is displayed on the left of the drawing area. This navigator can be used to select a folder and set it as the working directory. To do so, click on the **Folder Tree** node displayed at the bottom of the navigator; the expanded **Folder Tree** area will be displayed. Browse to the required location using the nodes available next to the folders and then select the desired folder. The selected folder will become the working directory for the current session. Alternatively, right-click on the folder that you need to set as the working directory; a shortcut menu will be displayed, as shown in Figure 1-9. Choose the **Set Working Directory** option from this shortcut menu to set the selected folder as the working directory. To make a new folder, choose the **New Folder** option from the shortcut menu.

Figure 1-9 *Shortcut menu*

Using the Select Working Directory Dialog Box

To specify a working directory, choose **File > Manage Session > Select Working Directory** from the menu bar; the **Select Working Directory** dialog box will be displayed, as shown in Figure 1-10. Using this dialog box, you can set any directory as the working directory.

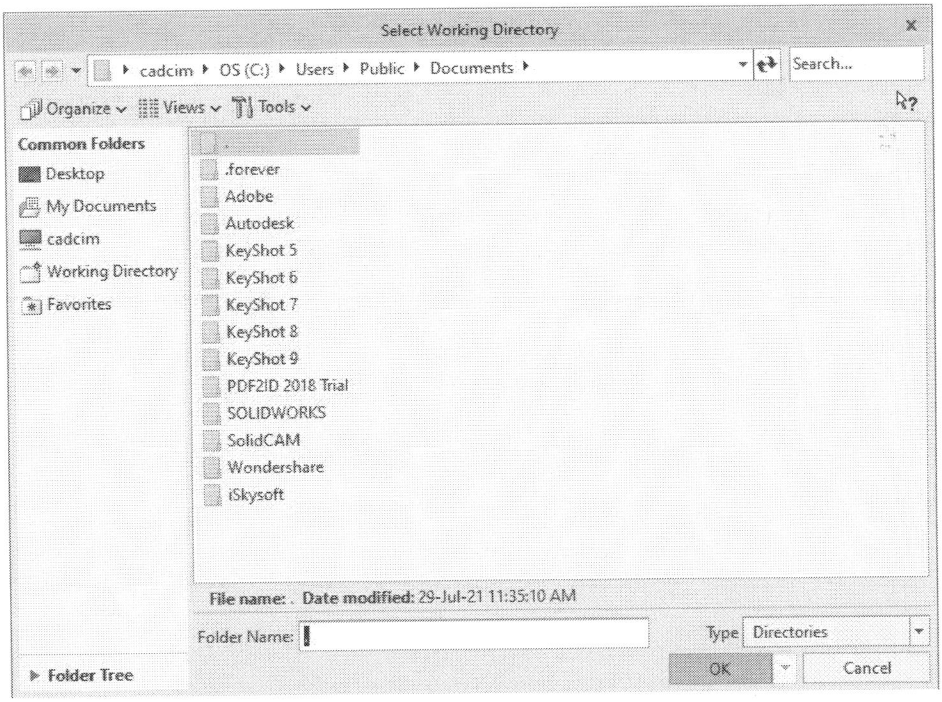

Figure 1-10 *The* ***Select Working Directory*** *dialog box*

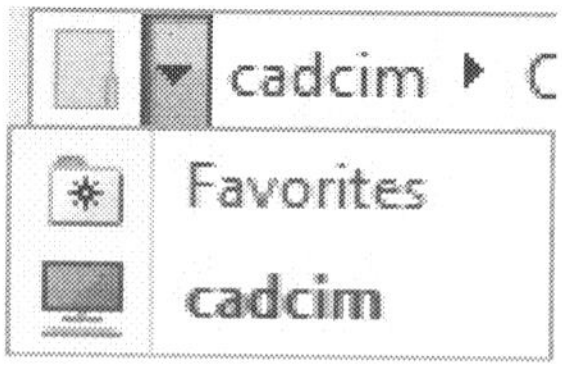

Figure 1-11 The flyout displayed

Choose the arrow at the upper left corner of the **Select Working Directory** dialog box; a flyout will be displayed, as shown in Figure 1-11. This flyout displays some of the drives present on your computer along with the **Favorites** folder. The **Favorites** folder contains all directories that you saved as favorites. The procedure to save the favorite directories will be discussed later. When the **Select Working Directory** dialog box is invoked by default, it displays the contents of the default directory. However, you can change the default directory that appears every time you open this dialog box. Various options in the **Select Working Directory** dialog box are discussed next.

Folder Name

This edit box displays the name of the directory selected in the **Select Working Directory** dialog box. You can select a directory using the flyout, as discussed earlier or by entering the path of any existing directory in this edit box.

Type

This drop-down list has two options, **Directories** and **All Files (*)**. If you select the **Directories** option, all directories present get listed, and if you select the **All Files (*)** option, then all files along with the directories are listed in the dialog box.

Organize

When you choose the **Organize** button from the **Select Working Directory** dialog box, a flyout will be displayed. The options in this flyout are used to create a new directory or rename an existing directory. You can also cut, copy, paste, and delete the existing folders using the options in the flyout. Moreover, you can add any existing folder in the Common Folders by using the **Add to common folders** option in this flyout, refer to Figure 1-12.

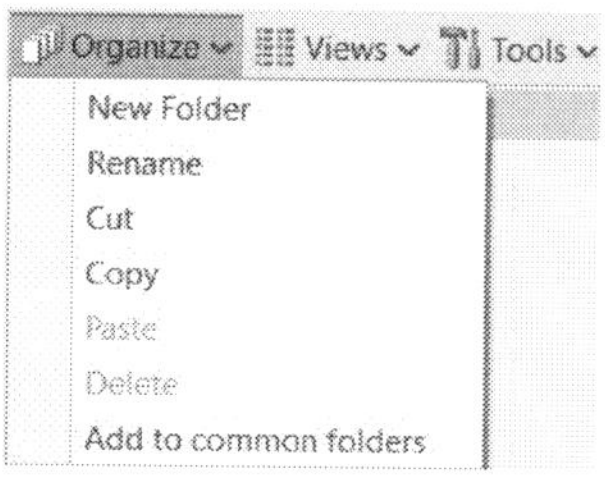

*Figure 1-12 The **Organize** flyout*

Views

When you choose the **Views** button from the **Select Working Directory** dialog box, a flyout will be displayed. The options in this flyout are discussed next.

List: The **List** radio button is used to view the contents of the current folder or drive. These include files and folders in the form of a list.

Details: The **Details** radio button is used to view the contents of the current folder or drive in the form of a table, which displays the name, size, and date on which it was last modified.

Tip

*The **Select Working Directory** dialog box has some of the properties of the Microsoft Windows operating system. You can set the working directory using this dialog box by browsing through directories and folders. You can also rename a file, directory, or a folder in this dialog box. Also, you can create a new directory using this dialog box.*

Tools

When you choose the **Tools** button from the **Select Working Directory** dialog box, a flyout will be displayed, as shown in Figure 1-13. The options in this flyout are discussed next.

*Figure 1-13 The **Tools** flyout*

Address Default: When you choose this option, the **'Look In' Default** dialog box will be displayed. Figure 1-14 shows this dialog box with the options in the drop-down list. If you select the **Default** option from the drop-down list and then invoke the **File Open** dialog box, it will display the directory that is set as default. If you select the **Working Directory** option from the drop-down list and then invoke the **File Open** dialog box, it will display the working directory that is set. If you select the **In Session** option and then invoke the **File Open** dialog box, the **File Open** dialog box will open with the **In Session** folder selected by default. Similarly, you can set the **Pro/Library** as the working directory.

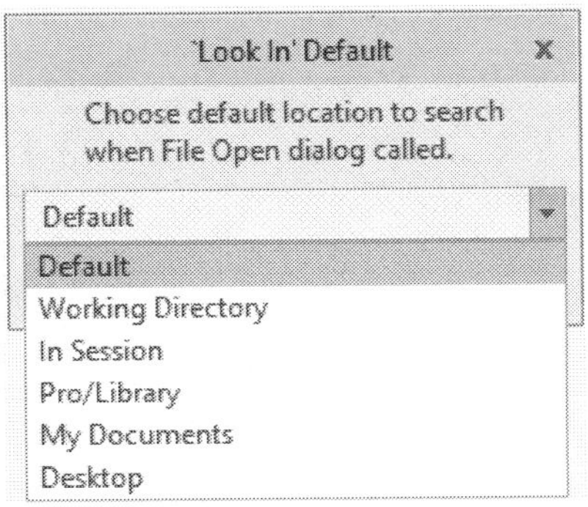

*Figure 1-14 The **'Look In' Default** dialog box with options in the drop-down list*

Up One Level: The **Up One Level** option allows you to move one level up in the directory. Choose this option; a directory that is one level above the current directory will be displayed. Alternatively, press ALT+UP arrow keys or **BACKSPACE** key to move one level up. You can also choose the arrow button on the left of the required directory in the address bar to display all folders in it.

Add to Favorites: The **Add to Favorites** option allows you to add the folders in the **Favorites** folder.

Remove from Favorites: The **Remove from Favorites** option allows you to remove the folders from the **Favorites** folder. This option is not enabled by default. To enable this option, you first need to add the folder in the **Favorites** folder using the **Add to Favorites** option.

Sort By: In the **Select Working Directory** dialog box, the **Directories** option in the **Type** drop-down list is displayed by default. From this drop-down list, if you select the

All Files option and then choose the **Tools** button, a flyout will be displayed with the **Sort By** option.

The **Sort By** option is used to list all files in the directory in an order to facilitate the process of searching a file. When you choose the **Sort By** option, a cascading menu is displayed. In the cascading menu, there are two radio buttons, **Model Name** and **Markup/Instance Name**. If you select the **Model Name** radio button, the file list will be sorted out alphabetically by the model name in the **Select Working Directory** dialog box. The **Markup/Instance Name** radio button sorts out the file list by specific markups or instance names in the **Select Working Directory** dialog box.

Common Folders and Folder Tree

The **Common Folders** and **Folder Tree** tabs are available on the left of the **Select Working Directory** dialog box. The Common Folders contains folders such as Desktop, My Documents, Computer, Working Directory, and Favorites. You can add more folders in the **Common Folders** by using the **Add to common folders** option available in the **Organize** flyout.

The **Folder Tree** contains all the drives available on your computer along with their contents. You can also set the working directory by using the **Folder Tree**. By default, the **Folder Tree** is in the collapsed state. To expand it, you need to click on the node that is available on the left of the **Folder Tree**. The **Working Directory** and **Favorites** folders available in the **Common Folders** are discussed next.

Working Directory: This folder is used when you have already set the working directory. You may browse through the directories in the **Select Working Directory** dialog box, but when you choose this folder, the directory selected previously as the working directory is displayed in the list box.

Favorites: This folder is used to save the location of the directories that are to be used frequently. You just need to specify the working directory to be used frequently and save its location by selecting the **Favorites** folder.

If you want to select one of the favorite working directories, then select the **Favorites** folder from the **Common Folders**; the list of all directories that were saved as favorites will be displayed in the list box. Select the required favorite directory and choose the **OK** button; the selected favorite directory will be set as the current working directory.

Note

*In Creo Parametric, an object can be created in different modes such as **Part**, **Drawing**, **Sketch**, and then saved as a file.*

New

To create a new object, choose the **New** tool from the **Data** group of the **Home** tab in the **Ribbon** or choose the **New** tool from the **Quick Access** toolbar; the **New** dialog box will be displayed, as shown in Figure 1-15. This dialog box displays various modes available in Creo Parametric. In this dialog box, by default, the **Part** mode radio button is selected and the default name of the object file is displayed in the **File name** edit box. You can also enter a new name for the object file. Note that the name must not contain any special character and a space between characters.

When you select the **Part**, **Assembly**, or **Manufacturing** radio button in this dialog box, the subtypes of the respective modes will be displayed under the **Sub-type** area of this dialog box.

Accept the default settings in the **New** dialog box by choosing the **OK** button; the default template will be loaded. To load a template other than the default one, clear the **Use default template** check box and then choose the **OK** button; the **New File Options** dialog box will be displayed, as shown in Figure 1-16. Using this dialog box, you can select the predefined templates or create a user-defined template. You can also open an empty template provided in the **New File Options** dialog box. In this case, you need to create the datum planes and the coordinate systems manually.

If the measuring units for creating models is millimeters, select **mmns_part_solid_abs** or **solid_part_mmks_abs** from the list in the **Template** area and then choose the **OK** button from the **New File Options** dialog box. On doing so, the three default datum planes and a coordinate system will be displayed in the drawing area. Also, the **Model Tree** will appear on the left of the screen, as shown in Figure 1-17.

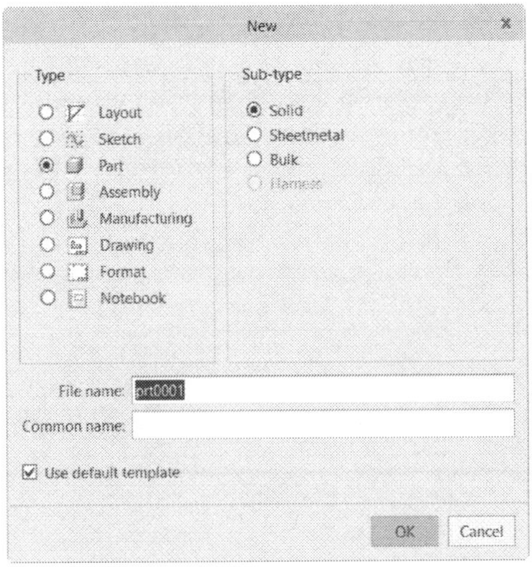

***Figure 1-15** The **New** dialog box*

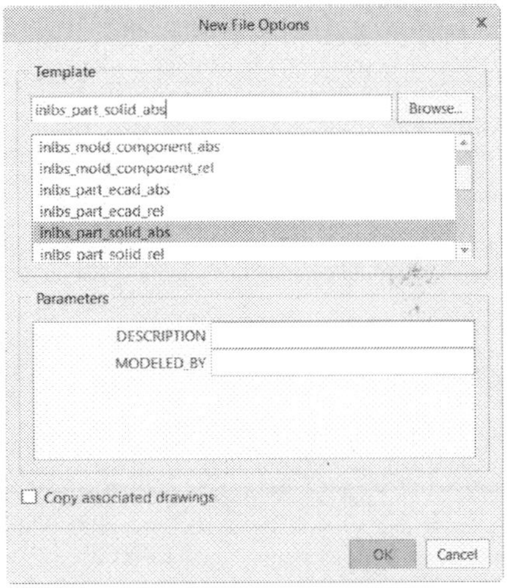

***Figure 1-16** The **New File Options** dialog box*

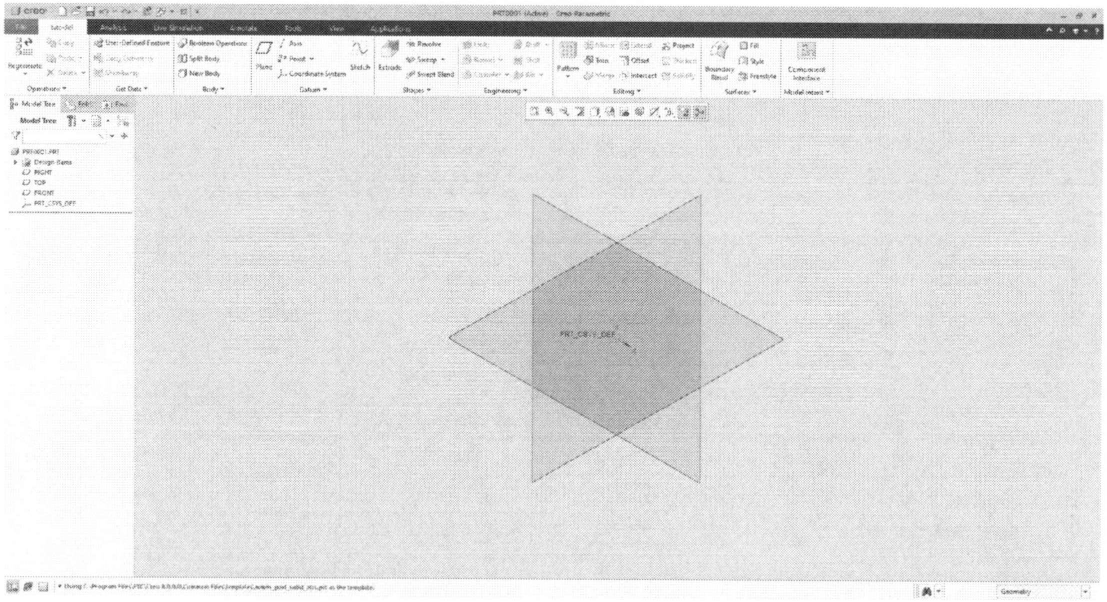

Figure 1-17 The initial screen appearance in the ***Part*** *mode*

Open

The **Open** button is used to open an existing object file. When you choose the **Open** option from the **File** menu or choose the **Open** button from the **Quick Access** toolbar, the **File Open** dialog box will be displayed, as shown in Figure 1-18. The selected working directory will be displayed in it. Note that the **Preview** area is not displayed by default. To view the **Preview** area, choose the **Preview** button. Most of the options in this dialog box are same as discussed in the **Select Working Directory** dialog box. The rest of the options in this dialog box are discussed next.

Note

In Creo Parametric, the name of the tools in some of the workbenches will be displayed with them only when the system screen is set to high resolution, such as 1900x1200 or 2880x1800 and so on. Otherwise only the tool icon will be displayed.

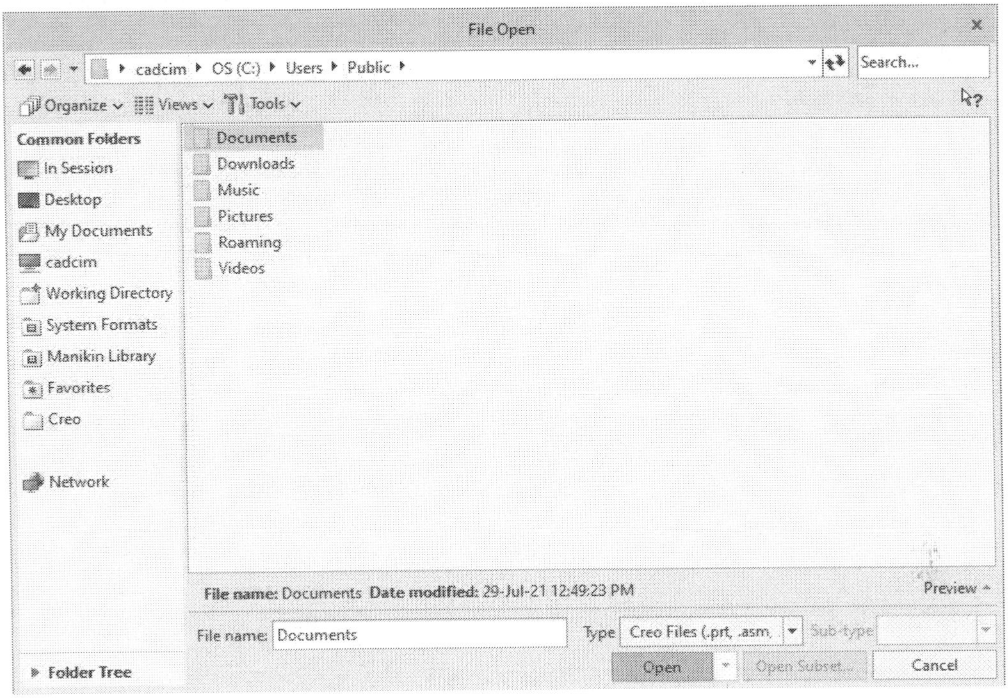

*Figure 1-18 The **File Open** dialog box*

Tools

On choosing this button, a flyout will be displayed. The options available in this flyout are same as the options discussed in the **Tools** button of the **Select Working Directory** dialog box, except the **All Versions** and **Show Instances** options. These two options are discussed next.

All Versions

This option, when chosen, displays all versions of an object file. In Creo Parametric, once a file is saved, its new version is generated with an extension that is incremented by 1. An object file is not copied on another object file but a new version of it is created. Therefore, every time you save an object using the **Save** option, a new version of it is created on the disk in the current working directory.

Show Instances

The **Show Instances** option, when chosen, displays all instances of the object file. Select the required file and then choose the **Show Instances** option from the **Tools** flyout; all the instances of the selected file will be displayed.

File name

In the **File name** edit box, you can enter the name of the existing object file that you want to open.

Type

The **Type** drop-down list contains the file formats of various modes available in Creo Parametric. It also contains many other file formats that can be imported in Creo Parametric.

These file formats include IGES, SET, STEP, DWG/DWF, Medusa, Inventor, Parasolid, Rhino, and so on.

By default, the **Creo Files** option is selected in this drop-down list. As a result, you can open the files created in any mode of Creo Parametric. However, if you select a specific mode from this drop-down list, only the files of the corresponding mode will be displayed. For example, if you select **Part** from the drop-down list, then only the *.prt* files will be displayed. This makes the selection and opening of the files easy.

Preview

The **Preview** button is used to preview the model before opening it. On choosing this button, you can preview the selected model in the **File Open** dialog box. You can zoom, pan, and rotate the model in the preview. Also, you can change the appearance (shaded, wireframe, no hidden, and hidden line) of the model, switch the model between the orthographic and perspective views, change the orientation type (dynamic, anchored, delayed, velocity, fly through, and standard) of the model, set the number of frames per second, and refit the preview in the preview screen by right-clicking in the preview area.

Note

Assembly files with the file extension .asm can also be previewed by using the ***Preview*** *button. If you are not able to see preview of the assembly files in the* ***Preview*** *area, then choose the* ***Refresh*** *button on the upper right of the* ***File Open*** *dialog box to resize the assembly according to the* ***Preview*** *area.*

There is no Command prompt in Creo Parametric. However, you are provided with prompts in the message area. Whenever you have to enter a numerical value or text, a message input window will be displayed in the message area.

In Session

The **In Session** folder is available in the **Common Folders** on the upper left of the **File Open** dialog box. When you choose the **In Session** folder, all the object files that are in the current session will be displayed in the display area. The object files that you open in Creo Parametric in the current session are stored in its temporary memory. This temporary memory is stored in a folder named **In Session**. Once you exit Creo Parametric, the contents of this folder are deleted automatically. However, the original files are not removed from their actual location.

Erase

As discussed earlier, all files opened in a session of Creo Parametric are saved in the temporary memory. There are three options to erase objects in the temporary memory: **Erase Current**, **Erase Not Displayed**, and **Erase Unused Model Reps**. These options are available in the **Manage Session** flyout in the **File** menu. The options that are displayed in this flyout are discussed next.

> **Tip**
> *Suppose you open an assembly that has a component named Nut. Close the assembly and now open another assembly that also has a component named Nut. Now, there are chances that the second assembly you choose to open may open with the Nut that was present in the previous assembly. This is because the component with the file named Nut was already present in the memory of Creo Parametric (in session).*
>
> *To avoid such type of errors, you should erase the files in the current session of Creo Parametric before opening the next assembly.*

Erase Current

The **Erase Current** option is used to erase the file opened and displayed in the drawing area. On choosing this option, the **Erase Confirm** message box will be displayed, prompting you to confirm the erasing of the file, as shown in Figure 1-19.

Erase Not Displayed

The **Erase** option is used to delete the files stored in the temporary memory. To do so, choose **File > Manage Session > Erase Not Displayed** from the menu bar; the **Erase Not Displayed** dialog box will be displayed, as shown in Figure 1-20. The files that are not open in the current session will be displayed in this dialog box. Choose the **OK** button from this dialog box to remove these files.

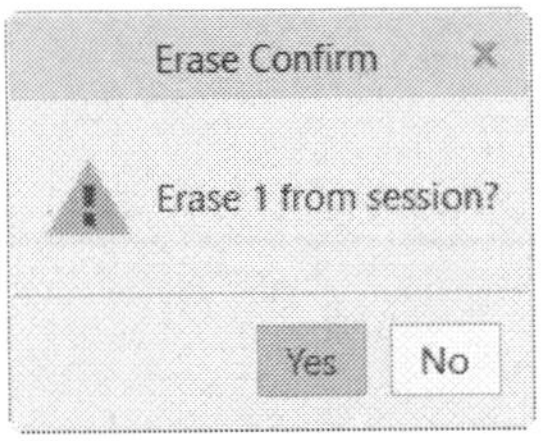

*Figure 1-19 The **Erase Confirm** message box*

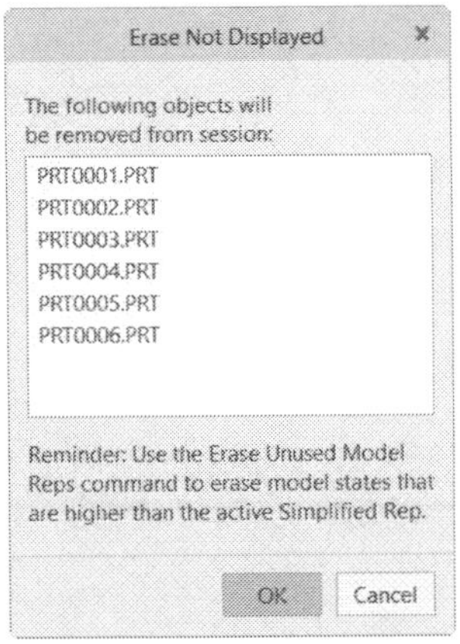

*Figure 1-20 The **Erase Not Displayed** dialog box*

Erase Unused Model Reps

This option is used to remove the unused simplified representations from the **In Session** folder. When you choose this option from the **File** menu, a message box will be displayed with the message that all the objects which were not displayed have been erased.

Delete

There are two options to delete files permanently from the hard disk. These options are available in the **Manage File** flyout in the **File** menu. The options in this flyout are discussed next.

Delete Old Versions

This option is used to delete all old versions of the current file. When you choose the **Delete Old Versions** option, the **Delete Old Versions** message box is displayed confirming to delete all old versions files. Choose **Yes** from the message box. All versions of that file will be deleted from the hard disk except the latest version.

Note

To delete the older versions of the model, set the directory containing the model as the working directory.

Delete All Versions

This option is used to delete all versions including the current file from the hard disk. When you choose the **Delete All Versions** option, a warning is displayed stating that performing this function can result in the loss of data. This option is chosen when the file is opened and is displayed in the drawing area.

Play Trail File

The **Play Trail File** option is used to recover data in case of a program crash or if you exit the session without saving the data. While working in Creo Parametric, a file called *trail.txt* is created in its default working directory (for example, *C:\Users\Public\Documents*). The file *trail.txt* contains all the working steps used in the last session.

Trail files are editable text (.txt) files. To retrieve the data, open the file *trail.txt* in Notepad, refer to Figure 1-21. If there is more than one such file in the directory, open the latest one and save this file by choosing **File > Save As** with another name like *crash.txt* and exit.

Figure 1-21 *The **trail.txt** file modified in Notepad*

Now, restart Creo Parametric and choose the **Play Trail File** option from **Manage Session**. Select the file *crash.txt* from the **Open** dialog box and then choose the **Open** button; Creo Parametric will repeat every step you made in the last session and restore the data.

Save

The **Save** option is used to save the objects present in the **In Session** folder or an object in the drawing area. When you choose the **Save** option from the **File** menu or the **Save** button from the **Quick Access** toolbar, the **Save Object** dialog box will be displayed. Also, the name of the current object will be displayed in the **File name** edit box. Choose the **OK** button from the **Save Object** dialog box to save the object.

Save a Copy

The **Save a Copy** option is used to save a copy of the current object in the same working directory or in some other directory. When you choose this option from **File > Save As**, the **Save a Copy** dialog box will be displayed. Now, you need to specify the new name of the object file to be saved as a copy and the name of the target directory in the **Save a Copy** dialog box. You can browse through the directories and select the target directory. The file will be saved in the selected directory.

Using this option, you can also export a file in other file formats such as Inventor file, pdf, ACIS, Wavefront file, and so on. After specifying the name of the new file and the target directory, choose the file format in which you want to export the file from the **Type** drop-down in the **Save a Copy** dialog box.

Save a Backup

The **Save a Backup** option in the **Save As** flyout is used to create a backup copy of an object file in the memory. When you choose this option, the **Backup** dialog box will be displayed, as shown in Figure 1-22.

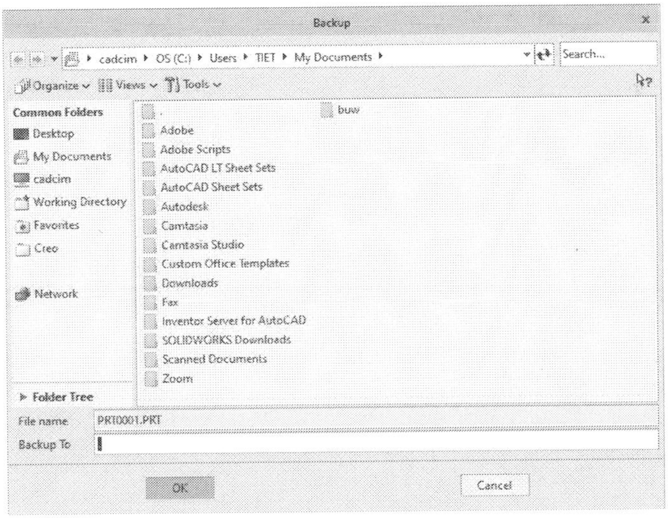

*Figure 1-22 The **Backup** dialog box*

In the **File name** edit box of the **Backup** dialog box, the name of the file for which you want to create a backup is displayed. In the **Backup To** edit box, the name of the directory is specified where the object will be saved as a backup. If you create the backup of an assembly or a drawing object, Creo Parametric will save all its dependent files in the specified directory. If you create the backup of a drawing file in a different directory, then the part file of the drawing will also be created automatically in the same directory where backup of the drawing file is created.

Rename

The **Rename** option in the **Manage File** flyout of the **File** menu is used to rename the currently active object on the screen. To rename an object, choose this option; the **Rename** dialog box will be displayed, as shown in Figure 1-23. Specify the new name of the object in the **New file name** edit box and choose the **OK** button.

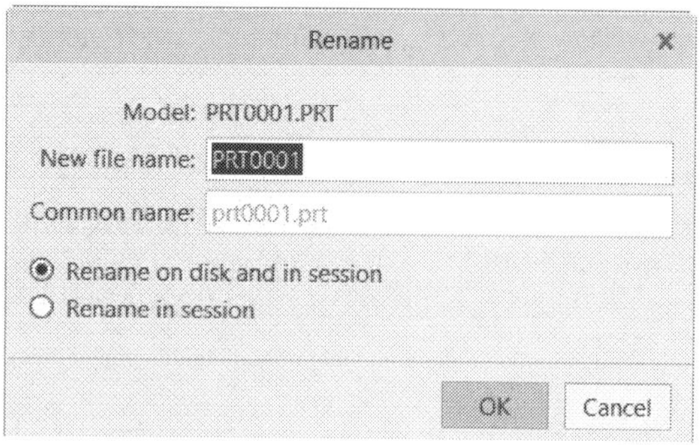

***Figure 1-23** The **Rename** dialog box*

MANAGING FILES

As discussed earlier, a new file is generated whenever you save an object. The number of files generated are directly proportional to the number of times you save that object. So, these files occupy a lot of disk space. The latest version of the file which is currently being used should be stored. Latest version refers to the highest number that is suffixed with the file name of that object. The rest of the files are old versions and should be deleted from the hard disk, if they are not required.

Note

*To save disk space, you should keep deleting the old versions of a file. This can be done by using the **File > Manage File > Delete Old Versions** option from the menu bar.*

MENU MANAGER

The **Menu Manager** consist of a cascading menu in which a set of menus and submenus are embedded. The display of the menus depends on the task chosen.

While using the **Menu Manager**, you need to choose the **Done** or **Done Sel** option to execute and complete the selected option. This is important when you are in the **Drawing** mode of Creo Parametric. If you are directly selecting one option after another, then it is easy to lose track of commands or options in the **Menu Manager**.

MODEL TREE

The **Model Tree** stores and displays all features in a chronicle. You can select any desired feature of a model or an assembly from the **Model Tree** and apply different operations on the selected feature. You can also select the feature by right-clicking on it; a shortcut menu will be displayed. Move the cursor on the shortcut menu and choose the required option from it by using the left mouse button.

When you create a new object file, the **Model Tree** appears and is attached to the drawing area by default, as shown in Figure 1-24. Therefore, the drawing area becomes small. You can hide the **Model Tree** by choosing the **Show Navigator** button available below the **Model Tree**. The **Model Tree** can slide in or slide out, thus increasing or decreasing the drawing area. It can also be stretched horizontally to cover the drawing area.

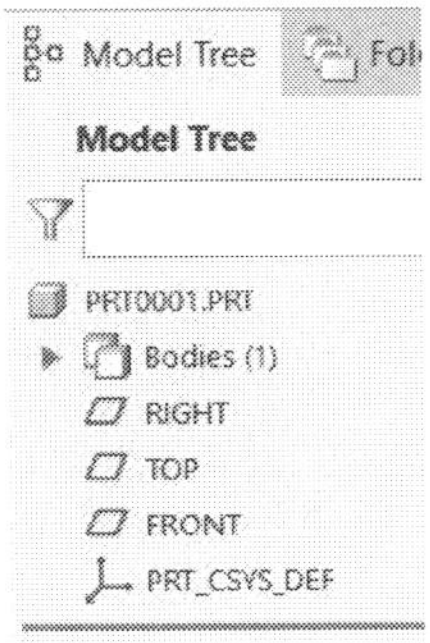

Figure 1-24 Partial view of the ***Model Tree***

Note

In Creo Parametric, most of the features are created using the ***Dashboard****. The* ***Dashboard*** *is displayed below the* ***Quick Access*** *toolbar and it contains all the options to complete the related operations on the model.*

Tip

Looking at the ***Model Tree****, you can understand the method and approach used to create the model. Using the* ***Model Tree****, you can modify the features of a model. Generally, when you import a model in a different file format in Creo Parametric, the features of the model are not displayed in the* ***Model Tree*** *and therefore, you will not be able to modify it.*

UNDERSTANDING THE FUNCTIONS OF THE MOUSE BUTTONS

While working with Creo Parametric, it is important to understand the function of the three buttons of the mouse to make an efficient use of this device. The various combinations of the keys and three buttons of the mouse are listed below:

1. Figure 1-25 shows the functions of the left mouse button. The left mouse button is used to make a selection. Using CTRL+left mouse button, you can add or remove items from the selection set.

2. Figure 1-26 shows the functions of the right mouse button. The right mouse button is used to invoke the shortcut menus and to query select the items. When you click the cursor on an item, it is highlighted in green color. Now, if you hold the right mouse button, a shortcut menu is displayed. Choose the **Pick From List** option from the shortcut menu; the **Pick From List** dialog box will be displayed. You can make selections from this dialog box.

3. Figure 1-27 shows the functions of the scroll wheel of the mouse in the 3D mode. The scroll wheel is used to spin the model in the drawing area and view it from different directions. The CTRL+scroll wheel is used to dynamically zoom in and out the view. When you press and hold the CTRL+scroll wheel and move the cursor up, the view is zoom out. When the mouse is moved down, the view is zoom in.

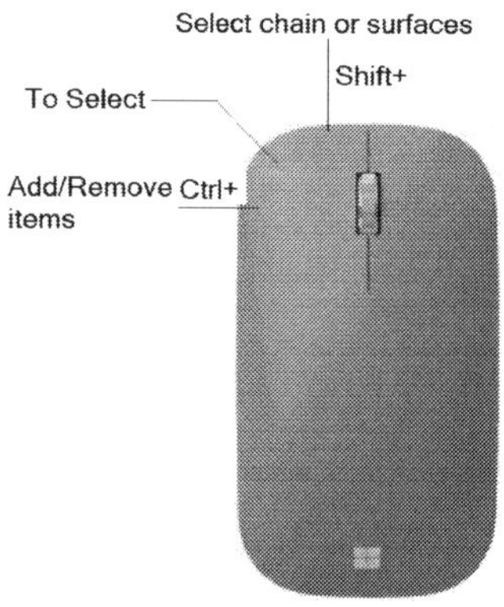

Figure 1-25 *Functions of the left mouse button*

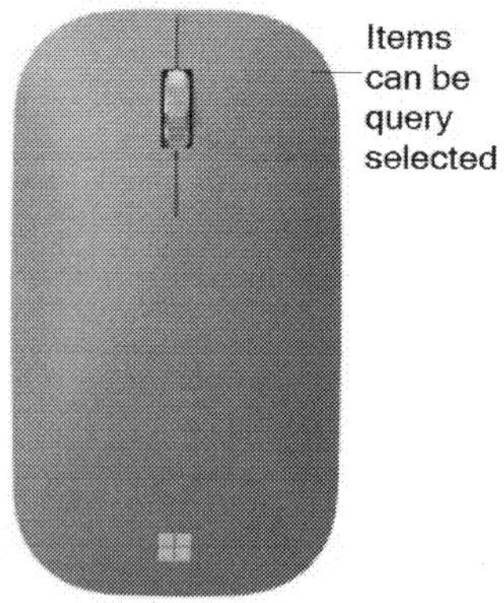

Figure 1-26 *Functions of the right mouse button*

When you press and hold the CTRL+scroll wheel and move the mouse horizontally, the model is rotated about a point that is specified as center.

The SHIFT+scroll wheel is used to pan the object on the screen.

4. Figure 1-28 shows the functions of the scroll wheel in the 2D mode (sketcher environment and **Drawing** mode). It is used to place dimensions in the drawing area. It is also used to confirm an option or to abort the creation of an entity.

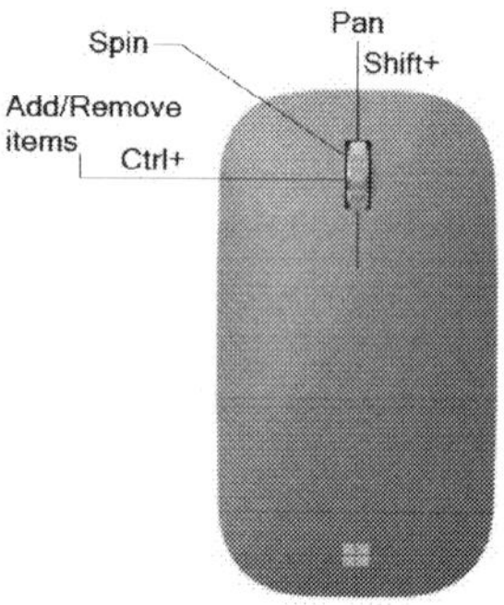

Figure 1-27 *Functions of the scroll wheel in the 3D mode*

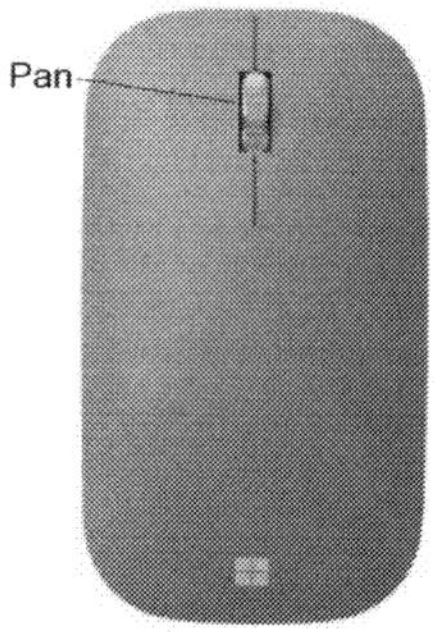

Figure 1-28 *Functions of the scroll wheel in the* ***Sketch*** *mode*

The scroll wheel is used to pan view in the **Sketch** mode and the **Drawing** mode.

Note

When you spin the model with the ***Spin Center*** *button, which is located in the Graphics bar, turned on, the model rotates about the spin center origin. If this button is turned off, then the model rotates about the specified point.*

RIBBON

Before you start working on Creo Parametric, it is very important for you to understand the default **Ribbon** and tools in the main window. Figure 1-29 shows various default interface components in Creo Parametric. The **Ribbon** is composed of a series of groups, which are organized into tabs depending on their functionality. The groups in the **Ribbon** that initially appear on the screen are shown in Figure 1-29. You will notice that all the tools in the groups are not enabled. These tools will be enabled only after you create a part or open an existing file. However, the tools that are required for the current session are already enabled. As you proceed to enter one of the modes provided by Creo Parametric, you will notice that the tools required by that mode are enabled. Additionally, to make the designing easy and user-friendly, this software package provides you with a number of groups. Different modes of Creo Parametric display different groups. Some of the frequently used groups are shown in Figure 1-29.

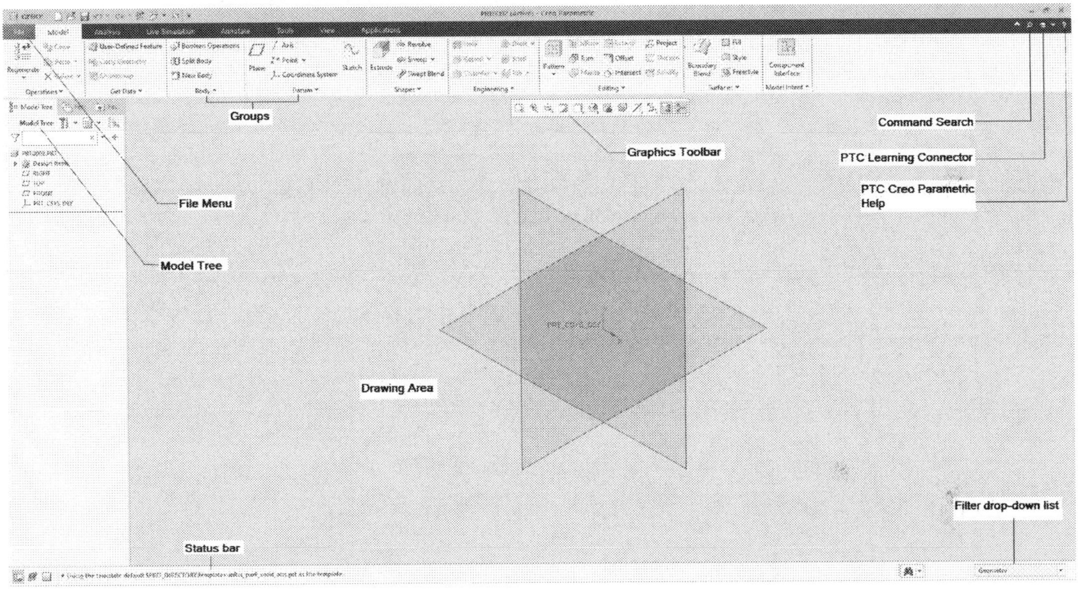

***Figure 1-29** Default interface components*

TOOLBARS

In Creo Parametric, there are two toolbars, **Quick Access** and **Graphics**. The toolbar on the top of the window is called the **Quick Access** toolbar and the toolbar on the top of the drawing area is called the **Graphics** toolbar.

Graphics Toolbar

The **Graphics** toolbar, shown in Figure 1-30, has twelve buttons by default. The first button is **Refit**, which is used to fit a model on the screen. The other buttons are **Zoom In** and **Zoom Out**, which are used to enlarge or diminish the model view, respectively. The **Repaint** button is used for repainting the screen, which helps in removing any temporary information from the drawing area. You can change the display style by using the **Display Style** drop-down.The **Perspective View** button toggles the perspective view. Using the **Saved Orientations** drop-down, you can change the view.

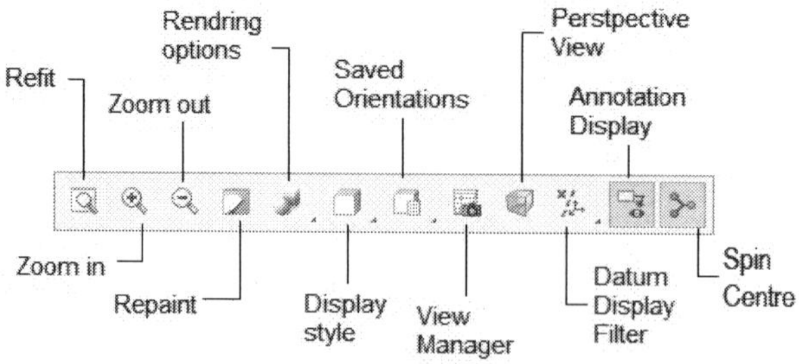

*Figure 1-30 The **Graphics** toolbar*

The next drop-down is **Datum Display Filters**. Using the options in this drop-down, you can toggle the display of datums to on/off. You can toggle the display of annotations to on/off using the **Annotation Display** button. The last button in this toolbar is **Spin Center**, which is used to toggle the visibility of the spin center to on/off.

File Menu

The **File** menu contains options shown in Figure 1-31. These options are used to create a new file, save a file, print the current file, or open an existing file. Various other options are also available in the menu to manage the current session and files. The **Options** tool is also available in the **File** menu. When you choose this tool, the **Creo Parametric Options** dialog box will be displayed. Using this dialog box, you can configure various parameters of Creo Parametric. You can also customize the **Ribbon** interface.

*Figure 1-31 The **File** menu*

Customizing the Ribbon

To customize the **Ribbon**, choose **Options** from the **File** menu; the **Creo Parametric Options** dialog box will be displayed. Select the **Customize > Ribbon** option from the left area of the dialog box; the dialog box will be modified. Choose the **New Group** or **New Tab** option, depending on your requirement, from the **New** drop-down list; the new group or tab will be added in the list box on the right side of the dialog box. Now, choose the button to be added in the new group from the list box on the left side of dialog box and then choose the **Adds selected item to the Ribbon** option from the dialog box; the selected button will be added to the group, refer to Figure 1-32.

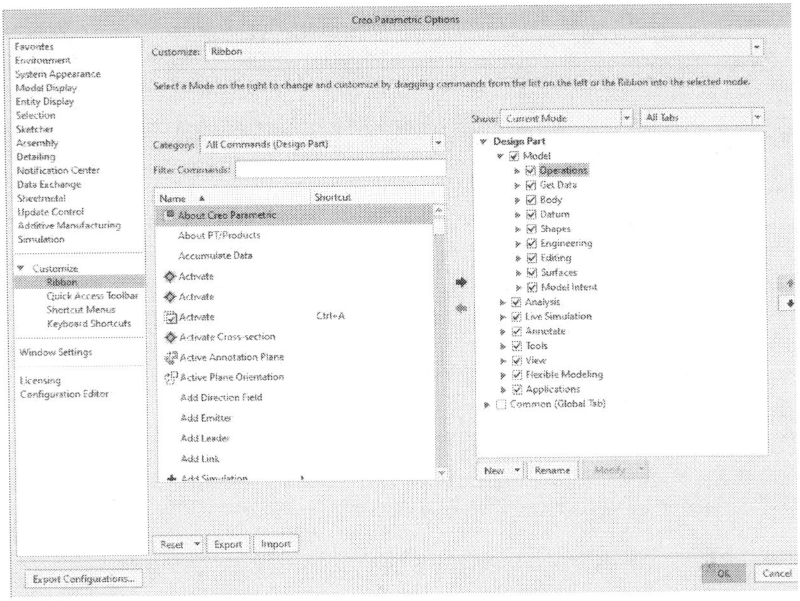

*Figure 1-32 Customizing the **Ribbon***

Creo Parametric Help

When you choose the **Creo Parametric Help** button, you are directed to **support.ptc.com** and the online Help page is displayed in your default browser. In this page, you can search for any help topic on Creo. The help database is accessible only to the users who have customer account with PTC.

Command Search

The **Command Search** button is available at the top right corner of the program window. When you choose this button, the search box is displayed, refer to Figure 1-33. You need to enter the first few letters of the command to be searched in the box. When you enter the first letter in the search box, a list of commands starting from that letter will be displayed. You can invoke any command by selecting it from the list, or you can enter the next letter to refine your search. On moving the cursor on any of the commands in the list, the command will get highlighted in the application window, as shown in Figure 1-34. Also, the path of the command will be displayed below the cursor.

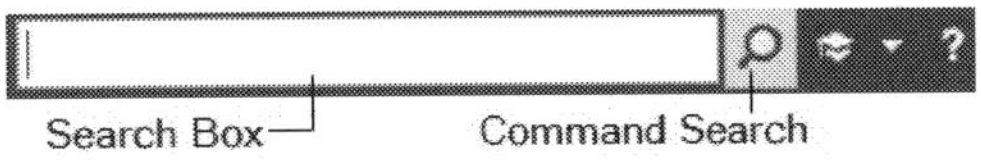

Figure 1-33 The command search box

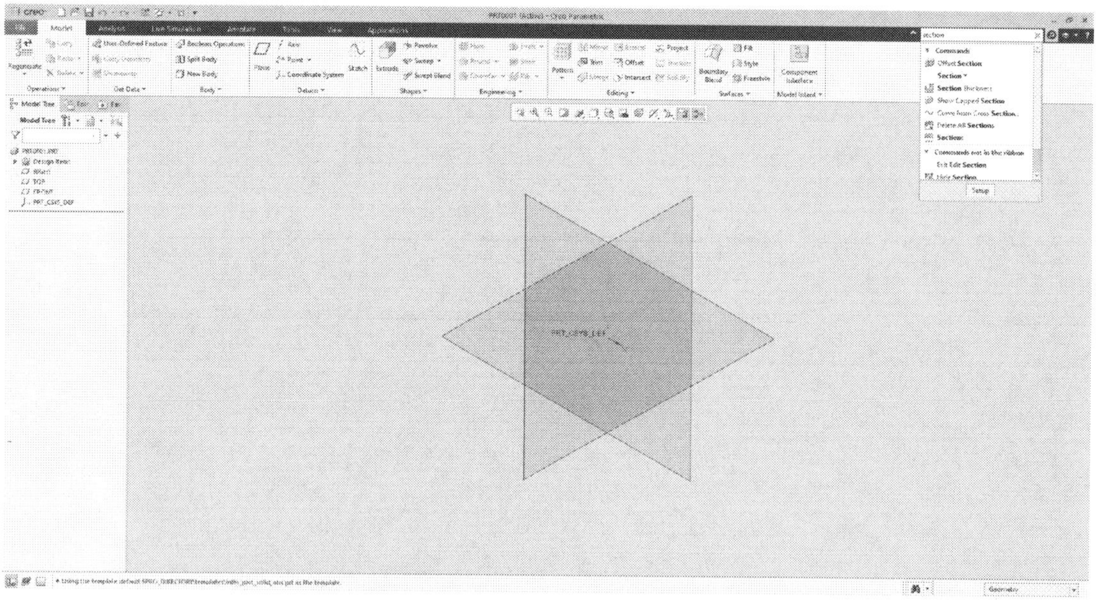

*Figure 1-34 The **Command Search** button*

PTC Learning Connector Button

The **PTC Learning Connector** button is used to access video tutorials provided by PTC. When you choose this button, the **PTC Learning Connector** separate window will be displayed, as shown in Figure 1-35. When you invoke any tool from the **Ribbon**, the corresponding tutorial will be displayed in the **PTC Learning Connector** window.

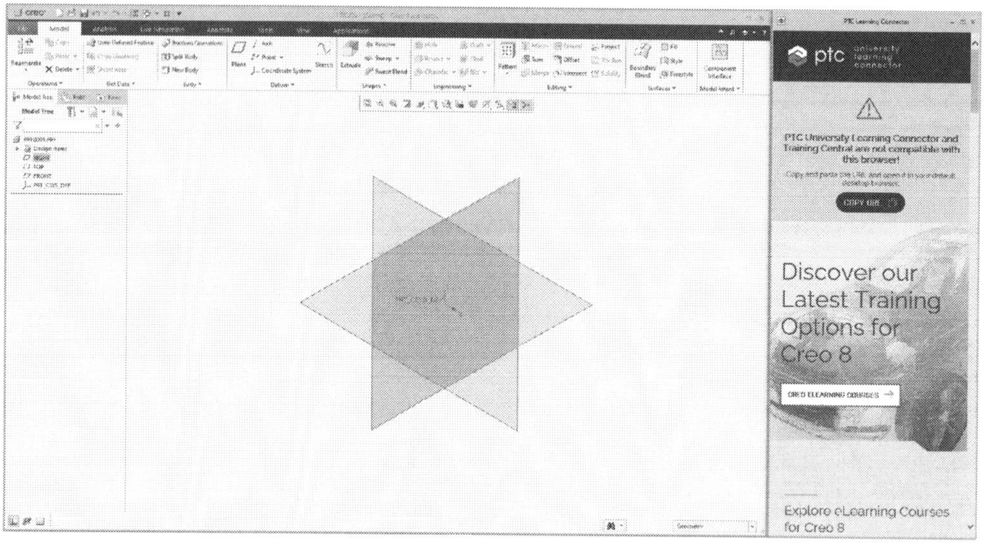

*Figure 1-35 Creo Parametric 8.0 screen with the **PTC Learning Connector** window*

Note

If you want to change the predefined unit system, choose ***File > Prepare > Model Properties*** *from the menu bar to display the* ***Model Properties*** *dialog box. In this dialog box, click on the* ***change*** *on the right of the* ***Units*** *option in the* ***Materials*** *head; the* ***Units Manager*** *dialog box will be displayed with the* ***System of Units*** *tab chosen. Next, select the desired unit system from the list box and then choose the* ***Set*** *button; the* ***Changing Model Units*** *message box will be displayed. Choose the* ***OK*** *button from the message box; the new unit system will be set and a blue arrow will appear on the left of the unit system selected.*

NAVIGATOR

The navigator is present on the left of the drawing area and can slide in or out of the drawing area. To make the navigator slide in or out, you need to select the **Show Navigator** button on the bottom left corner of the program window. A partial view of the navigator is shown in Figure 1-36. It has the following functions:

Figure 1-36 *Partial view of the navigator*

1. When you browse files using the navigator, the browser expands and the files in the selected folder are displayed in the browser.

2. When you open a model, the **Model Tree** is displayed in the navigation area.

3. The buttons on the top of the navigator are used to display different items in the navigation area. The **Model Tree** button is used to display the **Model Tree** in the navigation area. This button is available only when a model is opened. The **Folder Browser** button is used to display the folders that are in the local system. The **Favorites** button is used to display the contents of the **Personal Favorites** folder.

4. Any other location, if available on your system, can also be accessed by using the navigator.

Creo Parametric BROWSER

The Creo Parametric browser is present on the right of the navigator. You can also stretch the browser window horizontally. When the Creo Parametric session starts, the browser is displayed on the screen. The browser has four tabs. By default, the homepage is linked to *https://www.ptc.com/en/creo/generic?version=8_0&product=parametric*, the second tab is linked to *ptc.partcommunity.com*, the third tab is linked to *www.3dmodelspace.com/ptc,* and the last one is *home.punditas.com.* You can add more tabs in it, switch between tabs, and view the browsing history similar to any other internet explorer. The main advantage of using this browser is that you can work on the Creo Parametric files and navigate through the browser, simultaneously. Figure 1-37 shows the browser with some part files displayed in it. You can also browse to the required folder, change the display of files to thumbnails for previewing the file, dynamically view the models in a pop-up window, or drag and drop the selected file into the Creo Parametric window.

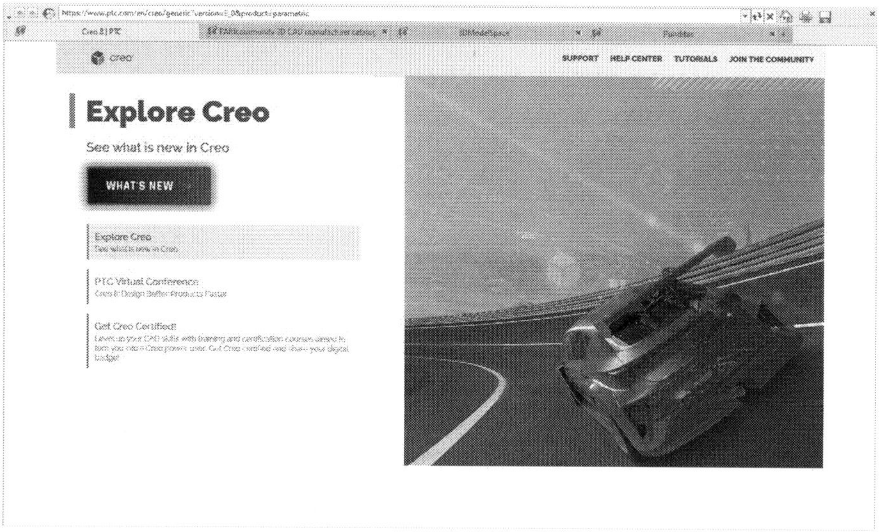

Figure 1-37 *Creo Parametric browser*

The Creo Parametric browser can be used for various purposes.

1. It is used to preview the online Creo Parametric files and browse the file system.

2. Using the browser, you can open a Creo Parametric file by double-clicking on it or by dragging it to the drawing area. When you open a file using the browser, the model is displayed on the screen and the browser is closed automatically.

3. You can access the Creo Parametric user community site using the browser.

4. You can connect to your client's computer and jointly work with them using the browser.

APPEARANCE GALLERY

A new library of appearances and real world materials has been added in Creo Parametric. You can also add shadows and reflections to the background of the model by using the appearance gallery. This is useful while rendering with a background image. This gives a realistic view to the model.

The **Appearance Gallery** button is available in the**Appearance** group of the **View** tab in the **Ribbon**. It is used to assign different colors to the model and change its appearances. To change the appearance of the model, click on the down arrow beside the **Appearance Gallery** button; a palette will be displayed, as shown in Figure 1-38. Select the required appearance from the palette; the **Select** box will be displayed. Select the entities from the drawing area whose appearance you want to change; the appearance of the entities will change. You can assign a desired appearance to the entire model or to the individual face of the model, or entire assembly or individual component of an assembly. Various areas and options in the palette are discussed next.

Search

This option is used for searching a specific appearance you need. To search an appearance, type a keyword related to the appearance to be searched in the **Search** text box and press ENTER; all the appearances matching with the keyword will be searched from all the system libraries and they will be displayed in the palette. For example, if you type **glass** in the text box, then **ptc-glass** will be displayed in the **My Appearances** area and the **Glass** library is displayed in the **Library** area.

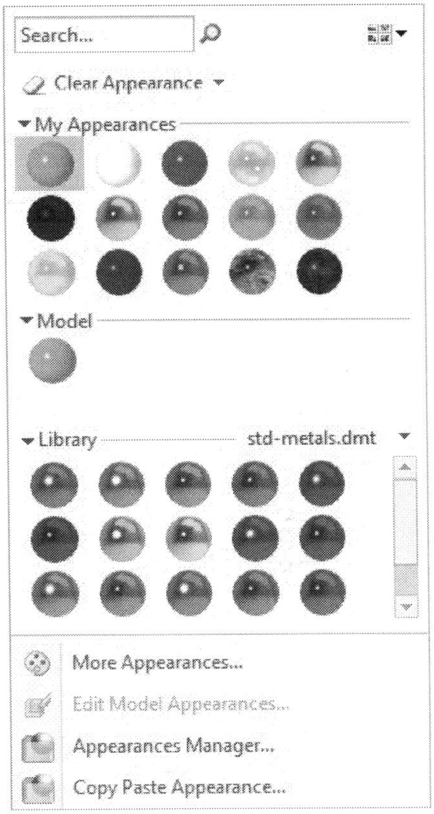

Figure 1-38 *Appearance Gallery*

View Options

You can set the display of the appearance icons in the palette by using the **View Options** flyout. These options will be available when you click on the button at the top-right corner of the palette. The icons can be displayed as small thumbnails, large thumbnails, names and thumbnails, or only names. To do so, invoke the **View Options** flyout and then choose the required option from it. In this flyout, the **Show Tooltips** check box is selected by default. As a result, tooltips are always displayed for appearances.

Clear Appearance

This option is used to remove the selected appearance from the model. To do so, select the appearance and then choose this option. Alternatively, first choose this option and then select the appearance to be removed from the model.

To remove all appearances assigned to a model, click on the black arrow displayed besides the **Clear Appearance** button; a flyout will be displayed. Next, choose the **Clear All Appearances** option from this flyout; the **Confirm** message box will be displayed. Choose the **Yes** button from the message box to clear appearances.

My Appearance

This area displays all appearances stored in the startup directory. You can also add new appearances to the palette, edit the existing appearance, or delete the selected appearance in the **My Appearance** area. To do so, right-click on the appearance icon and then choose the required option from the shortcut menu displayed. Note that you cannot delete the active appearance.

Model

This area consists of the appearances stored and used in an active model. You can edit the selected appearance, add new appearances, or select the objects to apply the selected appearance by using the shortcut menu as discussed in the **My Appearance** area.

Library

This area displays the predefined appearances in the current library. The library contains appearances from the System library and the Photolux library. Appearances in the library are grouped in various classes such as metals, glass, wood, and so on, based on their similar characteristics. The name of the current library is displayed on the right of the **Library** head. To assign appearances from other libraries, click on the library name; a flyout will be displayed with a tree list of classes and libraries. Browse through the tree and select the required library; all appearances in the selected library will be listed. From the list displayed, you can assign an appearance to the model as well as create a new appearance, as discussed earlier.

More Appearances

Choose the **More Appearances** button from the **Appearance Gallery**; the **Appearance Editor** will be displayed, as shown in Figure 1-39. The **Appearance Editor** provides options for editing the appearance as well as for advanced rendering, thereby providing much more realistic images.

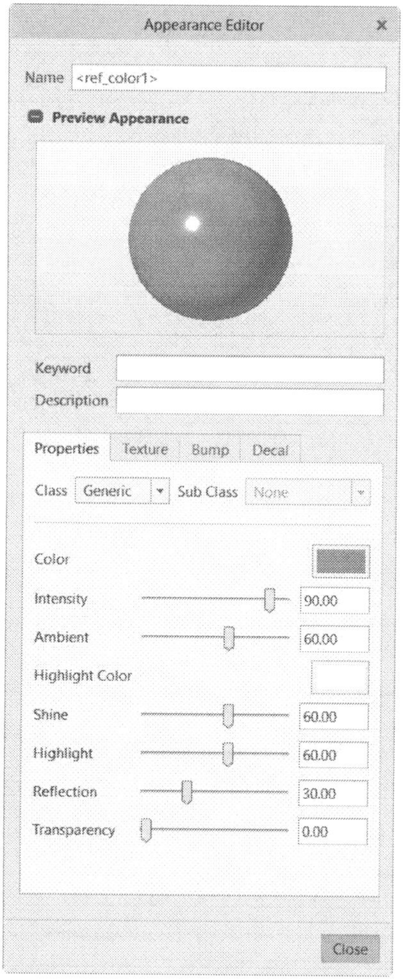

Figure 1-39** The **Appearance Editor

Using the **Appearance Editor**, you can edit the name, description, and keywords of the selected appearance, except the default one. The advanced options are given under the tabs of this dialog box. The **Properties** tab provides various classes and sub-classes of appearances. By default, **Generic** is selected in the **Class** drop-down list. However, you can select the required class and their sub-classes. Also, the properties related to selected class such as color, illumination, reflection, transparency, glossiness, and so on are displayed in the **Properties** area. These properties can be modified by adjusting the sliders given next to them.

You can use the **Texture** tab to define and apply texture such as wood, fabrics, metals, and so on to the model. You can preview the changes made to the model on changing the settings in the **Appearance Editor**.

Edit Model Appearances

On choosing the **Edit Model Appearances** button; the **Model Appearance Editor** will be displayed. In this editor, all appearances used in the model will be displayed along with their properties. The **Model Appearance Editor** is same as the **Appearances Editor**. Using this editor, you can modify the properties such as color, texture, shine, and gloss of the model.

Appearances Manager

Choose this button to display the **Appearances Manager** window. The **Appearances Manager** window displays two panels. The left panel displays the palette with appearances and libraries and the right panel displays the editor with the properties of the active appearance. You can set the required values in the right panel. Also, you can open an existing library file and add and save your own textured file.

RENDERING IN Creo Parametric

Rendering is a process of generating two-dimensional image of a three-dimensional scene or an object to make it more realistic. A rendered image makes it easier to visualize the shape and size of 3D object as compared to a wireframe or a shaded image. Rendering also helps you express your design intent to other people. You need to alter environments, lights, textures, and so on to get a high quality rendered image. Generally, the rendering process includes the following stages:

Loading a File

The first step is to load the solid model or the assembly to be rendered.

Applying Appearances

Next step is to apply the appearances to the model. You can do so by using the **Appearance Gallery**, which has already been discussed.

Applying Scenes

After applying the appearances, you need to apply scenes to the model. Scenes are predefined render settings that include lights, rooms, and environmental effects. To set scenes, choose the **Render Studio** tool from the **Rendering** group of the **Applications** tab; the **Render Studio** tab will be invoked. Next, choose the **Scenes** tool from the **Project** group of the **Render Studio** tab; the **Scenes** slide-down panel will be displayed. In the panel, choose the **Edit Scene** option to invoke the **Scene Editor** dialog box, as shown in Figure 1-40. Using this window, you can apply the predefined settings, edit the current settings, create new scenes, copy a scene, and save the scenes for future use. To apply a scene to the model, double-click on the required scene from the **Gallery** area of the **Scene Editor** dialog box.

Modifying the Scene Environment

The environment of a scene can be modified using the options available in the **Environment** tab of the **Scene Editor** dialog box. You can also apply shadows and reflections to the model by selecting the **Shadow** and **Reflection** check boxes in the **Floor Effects** area of the **Scene Editor** dialog box.

Applying Lights

The next step is to apply the lighting effects on the model. By adding lights to the model, you can illuminate the scene from various angles. To add lights, choose the **Lights** tab from the

Scenes Editor dialog box. You can create a light source as well as position it by using the options in the **Lights** tab.

Shadows are enabled as soon as you add lights to the model. The direction of shadows cast by a light bulb, spot light, skylight, or environment type of lights is towards the center of the room.

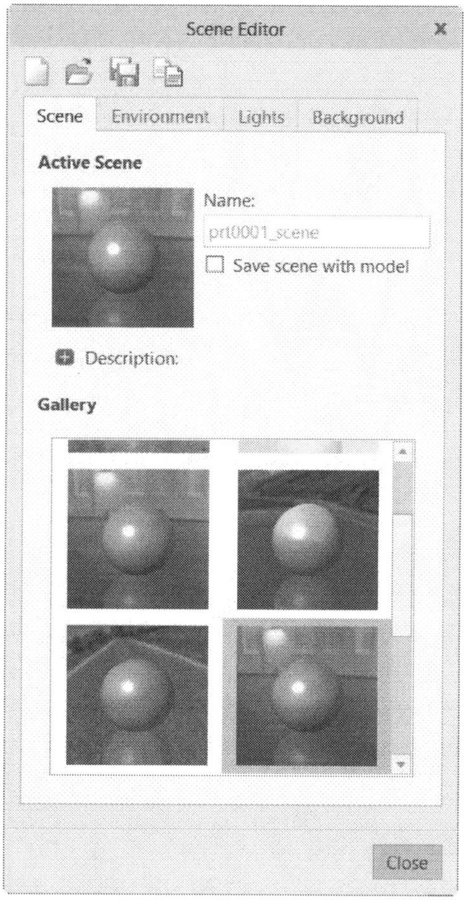

Figure 1-40 *The* ***Scene Editor*** *dialog box*

Adding Effects

Finally, you can add the surrounding effects such as ray bounces, indirect bounces, shadow quality, caustics, illumination, and self shadows. To do so, choose the **Real-Time Settings** option from the **Real-Time** drop-down of the **Render Studio** tab; the **Real-Time Rendering Settings** dialog box will be displayed. Using the options available in this dialog box, you can add desired effects to the model.

Setting the Perspective View

You can set the perspective view of the model that can be used for rendering the model. To do so, choose the **Perspective view** button from the **Project** group in the **Render Studio** tab.

To modify the settings of the perspective view, invoke the **View** dialog box by choosing the **Reorient** tool from the **Saved Orientations** flyout of the **Graphics** toolbar. Next, choose the **Perspective** tab in the dialog box and specify the settings for the perspective view.

Rendering the Model

The last step is to render the model. To render the model, choose the **Render** button from the **Render Output** group in the **Render Studio** tab; the **Render** dialog box will be displayed. Set the required rendering options in this window and choose the **Render** button to start the rendering process. If required, you can modify the render settings and render the model again.

Note

You can toggle between the display of the rendering effects on the model by choosing the ***Real-Time Rendering*** *button from the* ***Real-Time*** *group of the* ***Render Studio*** *tab.*

COLOR SCHEME USED IN THIS BOOK

Creo Parametric allows you to use various color schemes for the background screen color and for displaying the entities on the screen. This textbook will follow the white background of Creo Parametric environment at some places for the purpose of printing and image clarity. However, for better understanding and clear visualization at various places, you can follow other color schemes too. To change the color scheme of the background screen, choose **File > Options** from the menu bar; the **Creo Parametric Options** dialog box will be displayed. Select the **System Appearance** category from the left pane of this window. Next, choose the **Light (Previous Creo Default)** option from the **System Colors** drop-down list; different object names will be displayed with color drop-downs adjacent to them, refer to Figure 1-41. Next, choose the **OK** button to apply the color scheme.

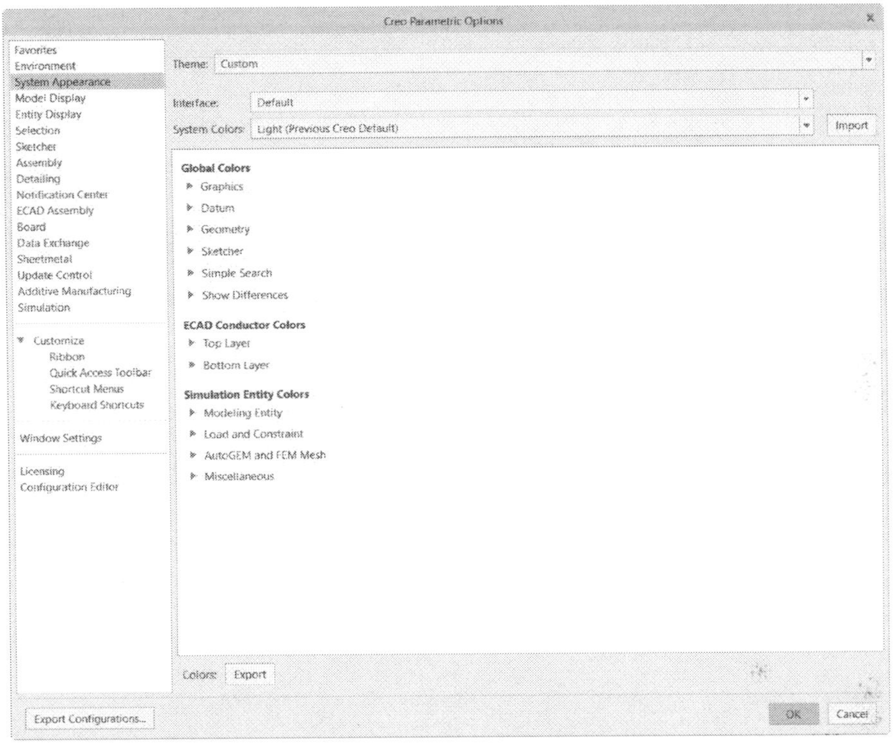

***Figure 1-41** The **System Colors** category in the **Creo Parametric Options** dialog box*

Self-Evaluation Test

Answer the following questions and then compare them to those given at the end of this chapter:

1. Which one is the default theme of Creo Parametric?

   a) Light (Previous Creo Default) b) Default
   c) Black on White d) White on Black

2. For which of the following purposes is the **Play Trail File** option used?

   a) Recover Data b) Play Tutorial Video
   c) Browse Files c) None of the above

3. You can change the system appearance from the__________ dialog box.

4. The __________ stores and display all features of a current model.

5. There is a bidirectional associativity between all modes of Creo Parametric. (T/F)

6. In Creo Parametric, you can use the SHIFT+scroll wheel to pan the object on the screen. (T/F)

Answers to Self-Evaluation Test

1. b, **2.** a, **3.** **Creo Parametric Options**, **4.** **Model Tree**, **5.** T, **6.** T

Chapter 2

Creating Sketches in the Sketch Mode-I

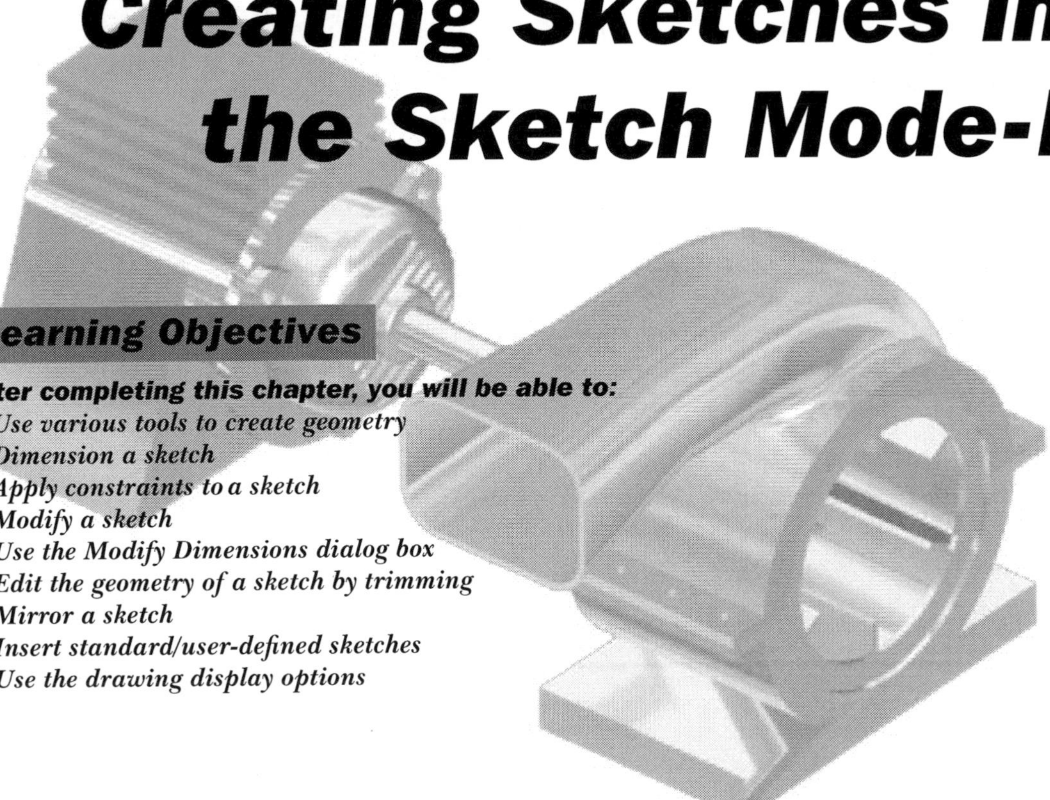

Learning Objectives

After completing this chapter, you will be able to:

- *Use various tools to create geometry*
- *Dimension a sketch*
- *Apply constraints to a sketch*
- *Modify a sketch*
- *Use the Modify Dimensions dialog box*
- *Edit the geometry of a sketch by trimming*
- *Mirror a sketch*
- *Insert standard/user-defined sketches*
- *Use the drawing display options*

THE SKETCH MODE

A sketch is a 2D entity that graphically captures an idea using lines, curves, constraints, and dimensions. To create a three-dimensional (3D) feature, it is necessary to draw or import a two-dimensional (2D) sketch. Almost all models designed in Creo Parametric consist of datums, sketched features, and placed features. When you enter the **Part** mode and select options to create any sketched feature, the system automatically takes you to the sketcher environment. In the sketcher environment, the sketch of the feature is created, dimensioned, and constrained. The sketches created in the **Sketch** mode are stored in the *.sec* format. After creating the sketch, you need to return to the **Part** mode to create the required feature.

Note

You will learn about datums and placed features in later chapters.

In Creo Parametric, a sketch can be drawn using the **Sketch** mode in the sketcher environment or can be imported from other softwares. You can draw a 2D sketch of the product and assign the required dimensions and constraints to capture the design intent. By assigning dimensions and constrains to sketches, you can ensure predictable results when a model is modified.

Working with the Sketch Mode

To create any sketch in the **Sketch** mode of Creo Parametric, certain basic steps have to be followed. The following points outline the steps to draw a sketch in the **Sketch** mode:

1. Sketch the Required Section

The different tools available in this mode can be used to sketch the required geometry.

2. Add Constraints and Dimensions to the Sketched Section

While sketching the section geometry, weak dimensions are automatically added to the section. The sketch can also be dimensioned and constrained manually. The dimensions that are applied manually are called strong dimensions. After adding the dimensions, you can modify them as required.

Invoking the Sketch Mode

To invoke the **Sketch** mode, choose **New** from the **File** menu or choose the **New** button from the **Data** group in the **Home** tab of the **Ribbon**; the **New** dialog box will be displayed with different Creo Parametric modes in the **Type** area. Select the **Sketch** radio button to start a new file in the **Sketch** mode, refer to Figure 2-1; a default name of the sketch file appears in the **File name** edit box. You can change the sketch name as required and then choose the **OK** button to enter the **Sketch** mode.

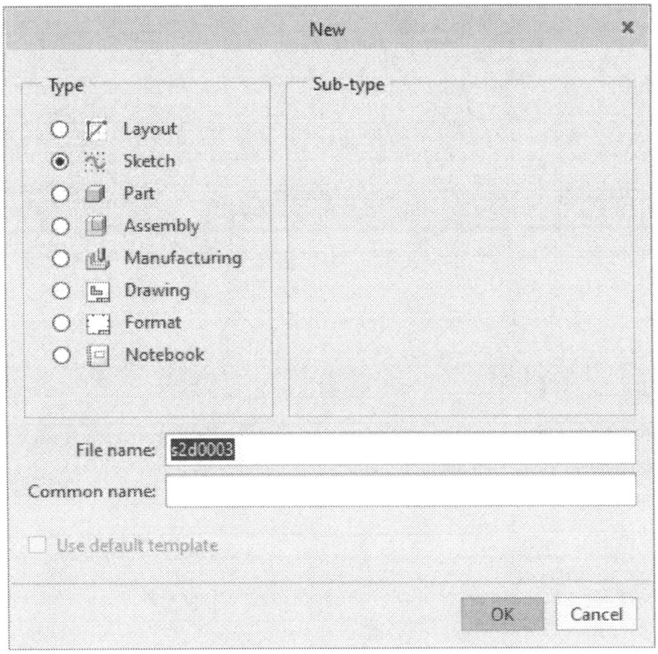

*Figure 2-1 The **New** dialog box*

THE SKETCHER ENVIRONMENT

When you invoke the **Sketch** mode, the initial screen displayed is similar to the one shown in Figure 2-2. The **Sketch** tab is chosen by default in the **Ribbon**. The drawing tools are available in the **Sketching** group of the **Sketch** tab.

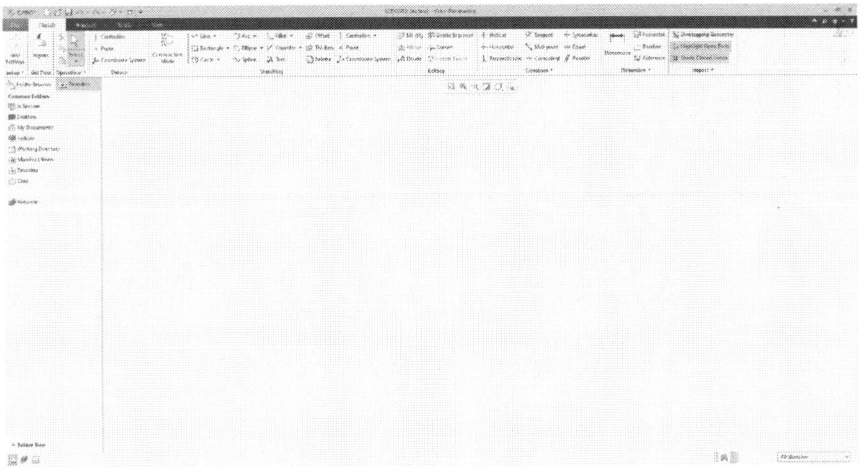

*Figure 2-2 Initial screen appearance in the **Sketch** mode*

The navigator is displayed on the left of the drawing area. In navigator, the **Folder Browser** tab is activated by default. It covers a part of the drawing area and therefore the drawing area is decreased. You can increase the drawing area by choosing the **Show Navigator** button, which is available at the bottom left corner of the window.

Note

*1. Datum planes are not displayed in the **Sketch** mode.*

*2. The **Folder Browser** tab is divided into two areas, **Common Folders** and **Folder Tree**. The functions of the **Common Folders** and **Folder Tree** areas have already been discussed in Chapter 1.*

WORKING WITH A SKETCH IN THE SKETCH MODE

When you invoke the sketcher environment, the **One-by-One** selection filter is selected by default in the **Select** drop-down list of the **Operations** group. You can select other selection filters from the **Select** drop-down list. Other selection filters available in this drop-down list are **Chain**, **All Geometry**, and **All**. By using the **One-by-One** selection filter, you can select each individual entity from the drawing area. By using the **Chain** selection filter, you can select a complete chain of entities linked with the selected entity. If the **All Geometry** selection filter is selected then all the geometric entities available in the drawing area are selected automatically. If you selected the **All** selection filter is selected then all entities available in the drawing area are selected automatically.

Note

*1. The **One-by-One** selection filter is activated by default. If you activate another selection filter, then that selection filter will be deactivated automatically after using once, and again the **One-by-One** selection filter will be activated automatically.*

2. A sketch in Creo is saved with .sec file extension.

3. You can create a simple sketch by using the options available in the shortcut menu which is displayed by pressing and holding the right mouse button in the drawing area. Once the shortcut menu is displayed, the right mouse button can be released.

DRAWING A SKETCH USING THE TOOLS AVAILABLE IN THE SKETCH TAB

In the sketcher environment, the **Sketch** tab is active in the **Ribbon** by default. There are various groups in this tab which contain tools to draw and modify a sketch and its dimensions. In this section, you will draw sketched entities using the tools available in the **Sketch** tab.

Creating a Point

Points are used to specify locations. You can use a point as reference for creating other geometric elements. In the sketcher environment of Creo, you can create construction points and geometry points. The procedures of creating these points are discussed next.

Creating a Construction Point

Ribbon: Sketch > Sketching > Point

The construction points can be used only in the sketcher environment. To create a construction point, choose the **Point** tool from the **Sketching** group and click in the drawing area to place the point at the specified location.

Creating a Geometric Point

Ribbon: Sketch > Datum > Point

The **Point** tool available in the **Datum** group is used to create a geometric point. The points created by using this tool can also be used as a reference point in the Part modeling environment . The procedure to create a geometric point by using the **Point** tool from the **Datum** group is similar to the procedure of creating point from the **Sketching** group.

Note

1. To increase the number of visible command prompt lines in the message area, select the upper boundary line of the message area using the left mouse button and drag it upward, toward the screen.

2. When you place a single point, no dimension appears. But when you place two points, they are dimensioned with respect to each other.

Drawing Lines

You can draw a line by using the tools available in the **Line** drop-down. The **Line Chain** tool is used to create a line or chain of lines by selecting two points in the drawing area and the **Line Tangent** tool is used to create a tangent line between two entities.

The procedure to create lines by using these tools is discussed next.

Drawing a Line Using the Line Chain Tool

Ribbon: Sketch > Sketching > Line drop-down > Line Chain

The **Line Chain** tool is used to create a line or chain of lines. The following steps explain the procedure to create a line using the **Line Chain** tool available in the **Line** drop-down:

1. Choose the **Line Chain** tool or press the L key from the keyboard to invoke the tool. Click in the drawing area to start the line; a rubber-band line appears starting from the selected point with the other end attached to the cursor.

2. After specifying the start point of the line, move the cursor in the drawing area to the desired location. You will notice that the sketcher begins to dynamically apply constraints in your sketch. For example, if you sketch the line approximately vertical or horizontal, sketcher dynamically applies a vertical or horizontal constraint to that line, helping you to lock its position. Next, click to specify the endpoint of the line. The rubber-band line continues and you can draw the second line.

3. Repeat step 2 until all lines are drawn. You can end the line creation by pressing the middle mouse button. To abort the line creation, press the middle mouse button again. The start point and end point of the line or line chain are displayed in red colored dots.

Note

1. After drawing a line, when you press the middle mouse button twice to end the line creation, the line drawn is highlighted in green color. In the sketcher environment, the green color of an entity indicates that it is selected. If you press the DELETE key, the line will be erased from the drawing area.

2. After drawing a line, weak dimensions are applied to the sketch and they appear in light blue color. These weak dimensions are applied automatically to the sketched entities as you draw them. The concept of weak dimensions is discussed later in this chapter.

Drawing a Line Using the Line Tangent Tool

Ribbon: Sketch > Sketching > Line drop-down > Line Tangent

The **Line Tangent** tool is used to draw a tangent line between two entities such as arcs, circles, or ellipses. The following steps explain the procedure to draw a tangent using this tool:

1. Choose the **Line Tangent** tool from the **Line** drop-down in the **Sketching** group; you will be prompted to select the start point on an arc, a circle, or an ellipse.

2. Select the first entity from where the tangent line will be drawn; a rubber-band line appears with the cursor. Also, you will be prompted to select the end point on an arc, a circle, or an ellipse. On selecting the second entity, a line tangent to both the selected entities will be drawn.

Note

You can draw a tangent line only if there are two or more than two arcs, circles, or ellipses already drawn in drawing area.

Drawing a Centerline

You can draw a centerline by using the tools available in the **Centerline** drop-down. Click on the down arrow on the right of the **Centerline** tool; a menu will appear with two tools. The first tool is the **Centerline** tool which is used to create a centerline by selecting two points in the drawing area. A centerline is used for creating revolved features, mirroring, and so on. The second tool is the **Centerline Tangent** tool that enables you to create a tangent centerline which can be referenced in the sketcher environment. The procedures to draw centerlines using these two tools are discussed next.

Drawing a Centerline Using the Centerline Tool

Ribbon: Sketch > Sketching > Centerline drop-down > Centerline

You can draw horizontal, vertical, or inclined centerlines using the **Centerline** tool. The centerline in a sketch is used as an axis of rotation for mirroring, aligning, and dimensioning entities.

The steps to draw a centerline are discussed next.

1. In the **Sketching** group, choose the **Centerline** tool from the **Centerline** drop-down; you will be prompted to select the start point.

2. Click in the drawing area to specify the start point; you will be prompted to select the end point.

3. Click in the drawing area to specify the endpoint; a centerline of infinite length is drawn.

Drawing a Centerline Using the Centerline Tangent Tool

Ribbon: Sketch > Sketching > Centerline drop-down > Centerline Tangent

The **Centerline Tangent** tool is used to draw a centerline tangent to two entities such as arcs, circles, ellipses or a combination of them. To draw a tangent centerline using this tool, you need to follow the steps given below:

1. Choose the **Centerline Tangent** tool from the **Centerline** drop-down in the **Sketching** group; you will be prompted to select the start location on arc or circle.

2. Select the first entity from where the tangent line has to be drawn; a rubber-band line with the cursor will appear. Also, you will be prompted to select the end point on an arc or circle. As soon as you select the second entity, a centerline with infinite length tangent to both the selected entities is drawn.

Drawing a Geometry Centerline

Ribbon: Sketch > Datum > Centerline

The **Centerline** tool available in the **Datum** group is used to create centerlines as a part of the geometry. The centerline created by using this tool can be referenced outside the sketcher environment. The procedure to create a centerline by using the **Centerline** tool from the **Datum** group or from the **Sketching** group is same. The only difference between them is that the centerline created by using the **Centerline** tool is a geometric entity, not a sketch. To draw a geometric centerline, choose the **Centerline** tool from the **Datum** group. Next, specify the start and end points of the geometry centerline; a centerline of infinite length will be created in the drawing area.

Drawing a Rectangle

Ribbon: Sketch > Sketching > Rectangle drop-down

In Creo Parametric, there are four tools available in the **Rectangle** drop-down that can be used to draw different types of rectangles. These tools are **Corner Rectangle**, **Slanted Rectangle**, **Center Rectangle**, and **Parallelogram** which are explained next.

Note

*When a non-overlapping closed sketch is drawn then it appears to be filled with light orange color. It indicates that the sketch is closed. This happens because the **Shade Closed Loops** button in the **Inspect** group of the **Sketch** tab is chosen by default. You can deactivate this button by clicking on it.*

Creating a Corner Rectangle

Ribbon: Sketch > Sketching > Rectangle drop-down > Corner Rectangle

You can create a rectangle by using the two corner points. To create a rectangle by using the **Corner Rectangle** tool, you need to follow the steps given below:

1. Invoke the **Corner Rectangle** tool from the **Rectangle** drop-down; you will be prompted to select two points as corners of the rectangle. Click to specify the first point; a rubber-band box appears with the cursor attached to the opposite corner of the box.

2. Move the cursor diagonally in the drawing area and then click to specify the second point for the diagonal of the rectangle.

Creating a Slanted Rectangle

Ribbon: Sketch > Sketching > Rectangle drop-down > Slanted Rectangle

You can create an inclined rectangle by using the **Slanted Rectangle** tool. To create an inclined rectangle, you need to follow the steps given next.

1. Invoke the **Slanted Rectangle** tool from the **Rectangle** drop-down; you will be prompted to click LMB to define the start point of the first side of the parallelogram. Click to specify the first point; an orange rubber-band line will be displayed with the cursor attached to the end point of the line.

2. Next, click at any point to create the end point of first side of the box; you will be prompted to specify the end point of the second side.

3. Move the cursor perpendicular to the first line in the drawing area and then click to specify the second side for the rectangle.

Creating a Center Rectangle

Ribbon: Sketch > Sketching > Rectangle drop-down > Center Rectangle

You can create a rectangle with the help of a center and end points by using the **Center Rectangle** tool. To create a rectangle by using the **Center Rectangle** tool, you need to follow the steps given next:

1. Invoke the **Center Rectangle** tool from the **Rectangle** drop-down available in the **Sketching** group; you will be prompted to specify the center point of the rectangle. Click to specify the center point; an orange rubber-band box appears with the cursor attached to the end point of the diagonal of the box.

2. Click at any desired point in the drawing area to create the rectangle.

Creating a Parallelogram

Ribbon: Sketch > Sketching > Rectangle drop-down > Parallelogram

You can create a parallelogram by using the **Parallelogram** tool. To create a parallelogram, you need to follow the steps given next:

1. Invoke the **Parallelogram** tool from the **Rectangle** drop-down available in the **Sketching** group; you will be prompted to define the start point of the first side of parallelogram. Click to specify the start point; an orange rubber-band line will be displayed with the cursor attached to the end point of the line.

2. Click at any desired point to draw the first side of parallelogram; an orange rubber-band box will be displayed with the cursor attached to the end point of the second side of the parallelogram. Also, you will be prompted to specify the end point of the second line.

3. Click at any desired point to create the parallelogram.

Drawing a Circle

In Creo Parametric, you can draw circles by using the **Center and Point**, **Concentric**, **3 Point**, and **3 Tangent** tools. These tools are available in the **Circle** drop-down of the **Sketching** group. The procedure to create a circle using these tools is discussed next.

Drawing a Circle Using the Center and Point Tool

Ribbon: Sketch > Sketching > Circle drop-down > Center and Point

The **Center and Point** tool is used to draw a circle by specifying the center of the circle and a point on its circumference. The following steps explain the procedure to draw a circle using this tool.

1. Choose the **Center and Point** tool; you will be prompted to select the center of the circle.

2. Click in the drawing area to specify the center point of the circle; you will be prompted to select a point on the circumference of the circle. Also, an orange rubber-band circle will be displayed with the center at the specified point and the cursor attached to its circumference.

3. Move the cursor to specify the size of circle. Click at a desired point to complete the creation of the circle; you will be prompted again to select the center of the circle.

4. Repeat steps 2 and 3 if you want to draw more circles, else press the middle mouse button to abort the process.

Drawing a Circle Using the Concentric Tool

Ribbon: Sketch > Sketching > Circle drop-down > Concentric

The following steps explain the procedure to draw a concentric circle using the **Concentric** tool:

1. Choose the **Concentric** tool from the **Circle** drop-down in the **Sketching** group; you will be prompted to select an arc to determine the center. You can select any arc or a circle to specify the center point.

2. Click on an arc or a circle to determine the concentricity of the circle to be drawn. Move the cursor and click at the required location to specify the size of circle.

3. To finish the creation of the circle, press the middle mouse button.

Drawing a Circle Using the 3 Point Tool

Ribbon: Sketch > Sketching > Circle drop-down > 3 Point

The following steps explain the procedure to draw a circle using the **3 Point** tool:

1. Choose the **3 Point** tool from the **Circle** drop-down; you will be prompted to specify the first point on the circle.

2. Click to specify the first point at the desired location in the drawing area; you will be prompted to select the second point on the circle. Move the cursor and click to specify the second point in the drawing area.

3. As you select the second point, an orange rubber-band circle appears with the cursor attached to it and you are prompted to select the third point. Move the mouse to size the circle and click to specify the third point; a circle is drawn and you are prompted again to select the first point on the circle to draw the next circle.

4. To abort the process of circle creation, you can press the middle mouse button at any stage.

Drawing a Circle Using the 3 Tangent Tool

Ribbon: Sketch > Sketching > Circle drop-down > 3 Tangent

The **3 Tangent** tool is used to draw a circle tangent to three existing entities. This tool references other entities to draw a circle. The circle created using this tool is drawn irrespective of the points selected on the entities. The following steps explain the procedure to draw a circle using the **3 Tangent** tool:

1. Choose the **3 Tangent** tool from the **Circle** drop-down; you will be prompted to select the start location on an arc, circle, or line.

2. Select the first entity; the color of the entity changes to green and you will be prompted to select the end location on an arc, circle, or line. Select the second tangent entity; you will be prompted to select the third location on an arc, circle, or line. Select the third tangent entity; a circle tangent to these three entities is drawn.

3. To end the process of circle creation, press the middle mouse button.

Drawing an Ellipse

You can draw an ellipse by using the tools available in the **Ellipse** drop-down. The tools available in this drop-down are **Axis Ends Ellipse** and **Center and Axis Ellipse**. The steps to draw ellipse using these tools are discussed next.

Drawing an Ellipse Using the Axis Ends Ellipse Tool

Ribbon: Sketch > Sketching > Ellipse drop-down > Axis Ends Ellipse

The following steps explain the procedure to draw an ellipse by using the **Axis Ends Ellipse** tool:

1. Choose the **Axis Ends Ellipse** tool from the **Ellipse** drop-down; you will be prompted to select the start point of the major axis of the ellipse.

2. Click in the drawing area to specify the start point of the major axis; you will be prompted to select the endpoint of the major axis.

3. Click in the drawing area to specify the endpoint; an orange rubber-band ellipse will appear with the cursor attached to it. Also, you will be prompted to select a point on the minor axis to define the ellipse.

4. Click to specify the point; an ellipse will be created. You can press the middle mouse button to end the creation of the ellipse.

Drawing an Ellipse Using the Center and Axis Ellipse Tool

Ribbon: Sketch > Sketching > Ellipse drop-down > Center and Axis Ellipse

The following steps explain the procedure to draw an ellipse by using the **Center and Axis Ellipse** tool:

1. Choose the **Center and Axis Ellipse** tool from the **Ellipse** drop-down; you will be prompted to specify the center of the ellipse.

2. Click at the desired location in the drawing area to specify the center point; you will be prompted to select the endpoint of the major axis of the ellipse. Click in the drawing area to specify the point; an orange rubber-band ellipse will appear with the cursor attached to the ellipse. Move the cursor in the drawing area to size the ellipse.

3. Specify the endpoint on the minor axis of the ellipse; the ellipse is drawn. After exiting the command, the dimensions for the major radius and the minor radius will be displayed in light blue color. The light blue color indicates that the dimensions are weak.

Drawing an Arc

Creo Parametric provides five tools to draw an arc. These tools can be invoked from the **Arc** drop-down in the **Sketching** group. The procedures to draw arcs using these tools are discussed next.

Drawing an Arc Using the 3-Point / Tangent End Tool

Ribbon: Sketch > Sketching > Arc drop-down > 3-Point / Tangent End

The **3-Point / Tangent End** tool is used to draw arcs that are tangent from the endpoint of an existing entity, or by defining three points in the drawing area.

When you choose this tool to draw an arc from an endpoint, the **Target** symbol is displayed as you select an endpoint. The **Target** symbol is a green colored circle that is divided into four quadrants. The following steps explain the procedure to draw an arc from the endpoint of an existing entity by using this tool:

1. Choose the **3-Point / Tangent End** tool from the **Arc** drop-down; you will be prompted to select the start point of the arc.

2. Specify three points in the drawing area to draw an arc. If you want to draw an arc from the endpoint of an existing entity, select the endpoint of that entity. As you select the endpoint, the **Target** symbol appears at the endpoint of the entity. Move the cursor along the tangent direction through a small distance, a rubber-band arc appears with one end attached to the endpoint of the entity and the other end attached to the cursor. Note that when you move the cursor out of the **Target** symbol perpendicular to the endpoint, an arc is drawn by specifying three points. In this case, the rubber-band arc does not appear, refer to Figure 2-3.

   On the other hand, if you move the cursor out horizontally from one of the quadrants of the **Target** symbol, an arc is drawn tangent to the endpoint, as shown in Figure 2-4.

3. Move the cursor to the desired position in the drawing area to size the arc. Use the left mouse button to complete the arc.

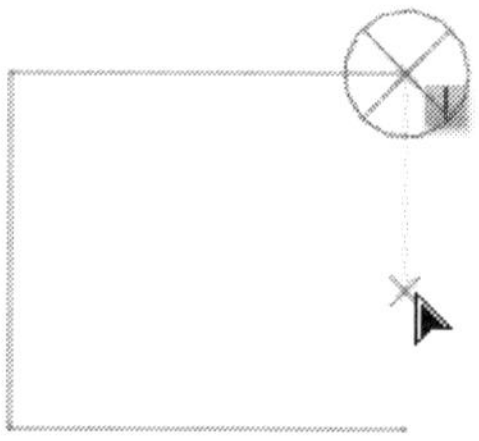

Figure 2-3 *Cursor moved out of the Target symbol perpendicular to the endpoint*

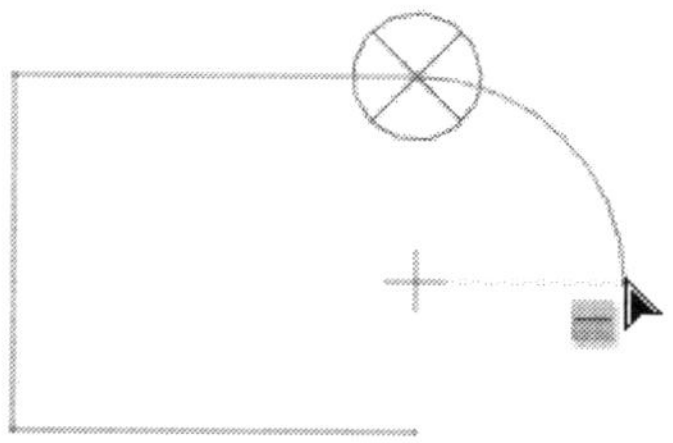

Figure 2-4 *Cursor moved out of the Target symbol along the tangent direction*

Tip

If you do not want to draw a tangent arc, move the cursor out of the ***Target*** *symbol perpendicular to the endpoint.*

Drawing an Arc Using the Center and Ends Tool

Ribbon: Sketch > Sketching > Arc drop-down > Center and Ends

The following steps explain the procedure to draw an arc using the **Center and Ends** tool:

1. Choose the **Center and Ends** tool from the **Arc** drop-down; you will be prompted to select the center of the arc.

2. Click to specify a center point for the arc in the drawing area; a violet colored center mark will appear at that point and you will be prompted to select the start point of the arc. As you move the cursor, a dotted circle appears attached to the cursor.

3. Specify the start point of the arc on the circumference of the dotted circle; an orange rubber-band arc will appear from the start point. The length of the arc will change dynamically as you move the cursor and you will be prompted to select the endpoint of the arc.

4. Move the cursor to specify the arc length and then click to select the endpoint of the arc; an arc will be drawn between the two specified points.

Note

You can draw only one arc with one center. If you want to draw another arc, you will have to select the center again.

Drawing an Arc Using the 3 Tangent Tool

Ribbon: Sketch > Sketching > Arc drop-down > 3 Tangent

The **3 Tangent** tool is used to draw an arc that is tangent to three selected entities. The following steps explain the procedure to draw an arc using this tool:

1. Choose the **3 Tangent** tool from the **Arc** drop-down; you will be prompted to select the start location on an arc, circle, or line.

2. As you move the cursor on the first entity or select the first entity, the color of the entity gets changed to green and you are prompted to select the end location on an arc, circle, or line.

3. On selecting the end location, you will be prompted to select a third location on an arc, circle, or line. Select the third entity; an arc is drawn tangent to the three selected entities.

You can continue drawing arcs or press the middle mouse button to abort arc creation.

Drawing an Arc Using the Concentric Tool

Ribbon: Sketch > Sketching > Arc drop-down > Concentric

The **Concentric** tool is used to draw an arc concentric to an existing arc. The entity selected must be an arc or a circle. The following steps explain the procedure to draw an arc using this tool:

1. Choose the **Concentric** tool from the **Arc** drop-down; you will be prompted to select an arc to determine the center of the arc to be created.

2. On moving the cursor on the first entity or selecting the entity a dotted circle will appear on the screen and you will be prompted to select the start point of the arc. Click to specify the start point; an orange rubber-band arc will appear with one end attached to the start point. As you move the cursor, the length of the arc will change and you will be prompted to select the endpoint of the arc.

3. Click to specify the endpoint; the arc will be created.

You can continue drawing another arc or end the arc creation by pressing the middle mouse button.

Drawing an Arc Using the Conic Tool

Ribbon: Sketch > Sketching > Arc drop-down > Conic

The **Conic** tool is used to draw a conic arc. The following steps explain the procedure to draw a conic arc using this tool:

1. Choose the **Conic** tool from the **Arc** drop-down; you will be prompted to specify the start point of the conic entity.
2. Click to specify the start point in the drawing area; you will be prompted to specify the endpoint of the conic entity.
3. Click to specify the endpoint; a centerline will be drawn between the two points and you will be prompted to specify the shoulder point of the conic. Specify a point on the screen; the conic arc will be drawn.

Note

1. If the conic arc is the only entity in the drawing area, then you cannot delete its centerline.

2. If you delete the centerline of the conic arc, the arc will not be deleted.

Drawing Construction Geometries

Ribbon: Sketch > Sketching > Construction Mode

Construction geometries are used to reference the solid geometries in sketches. Construction geometries can be dimensioned and constrained in the same manner as solid geometries. In Creo Parametric, the **Construction Mode** tool is used to toggle between the mode of creating solid geometry and construction geometry. When the **Construction Mode** is activated, any new geometry will be created as construction geometry. Note that you can also convert any solid sketched geometry entity into construction geometry and vice versa. To do so, select any sketched geometry and choose the **Construction** button from the mini toolbar.

DIMENSIONING THE SKETCH

The basic purpose of dimensioning is to locate and control the size of geometric entities with some reference. Dimensions are created to capture the design intent because these dimensions are displayed when you edit the model and when you create drawings of the model.

In Creo Parametric, sketched entities are dimensioned and constrained automatically while sketching. However, you need to add some additional dimensions and constraints to the sketch to make it fully constrained. The dimensioning tools are available in the **Dimension** group of the **Sketch** tab, refer to Figure 2-5. The **Dimension** tool in this group is used to manually dimension the entities.

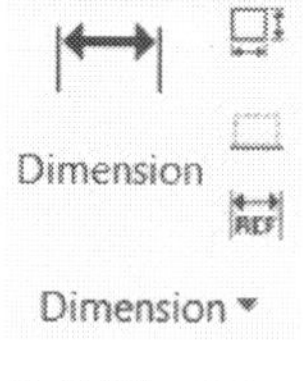

***Figure 2-5** The **Dimension** group*

Note

If you do not want the weak dimensions to be applied automatically, then clear the ***Show weak dimensions*** *check box from* ***File > Options > Sketcher > Object display settings****.*

Converting a Weak Dimension into a Strong Dimension

As discussed earlier when you draw a sketch, some weak dimensions are automatically applied to the sketch. These dimensions are displayed in light blue color. As you proceed to manually dimension the sketch, these weak dimensions are automatically deleted without any confirmation.

When you select a weak dimension from the drawing area the selected dimension will be highlighted in green color and a mini popup toolbar will be displayed near the selected dimension. From the mini popup toolbar, choose the **Strong** option, as shown in Figure 2-6, and press the middle mouse button; the select dimension will be converted into strong dimension. Alternatively, press CTRL+T to convert the selected dimension into a strong dimension. Note that a strong dimension is displayed in violet color.

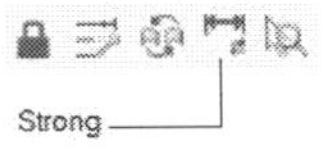

Figure 2-6 *The* ***Strong*** *option in the dimension mini popup toolbar*

Dimensioning a Sketch Using the Dimension Tool

Ribbon: Sketch > Dimension > Dimension

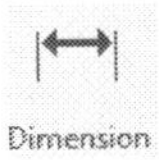

The **Dimension** tool in the **Dimension** group of the **Sketch** tab is used for dimensioning the sketched entities, refer to Figure 2-5. The following steps explain the procedure to dimension a sketch using this option:

1. Choose the **Dimension** tool from the **Dimension** group. Click on the entity you want to dimension; the color of the entity changes from orange to green.

2. Move the cursor and place the dimension at a desired place by double-clicking the middle mouse button. Before placing the dimension, you can change its value by clicking MMB once and then specifying the value.

Note that you can also change the dimension value using the **Modify** option that will be discussed later in this chapter.

Tip

You can also manually dimension the entities by using the dimensioning options available in the mini popup toolbar which appears when you select one or more sketched geometries.

Dimensioning the Sketched Entities

To dimension the sketched entities, you need to choose the **Dimension** tool from the **Dimension** group of the **Sketch** tab and follow the procedures given below.

Linear Dimensioning of a Line

You can dimension a line by selecting its endpoints or by selecting the line itself. After selecting the two endpoints or the line, press the middle mouse button twice to place the dimension at desired location. If the line is inclined and you select the two endpoints to dimension, then the location where you press the middle mouse button defines the orientation of the dimension that will be displayed on the screen.

Figure 2-7 shows the three possible orientations of dimension that can be displayed when you dimension a line.

Note

It is not possible to dimension a line in three orientations simultaneously in the sketcher environment. The dimensions in Figure 2-7 are shown for explanation purpose only.

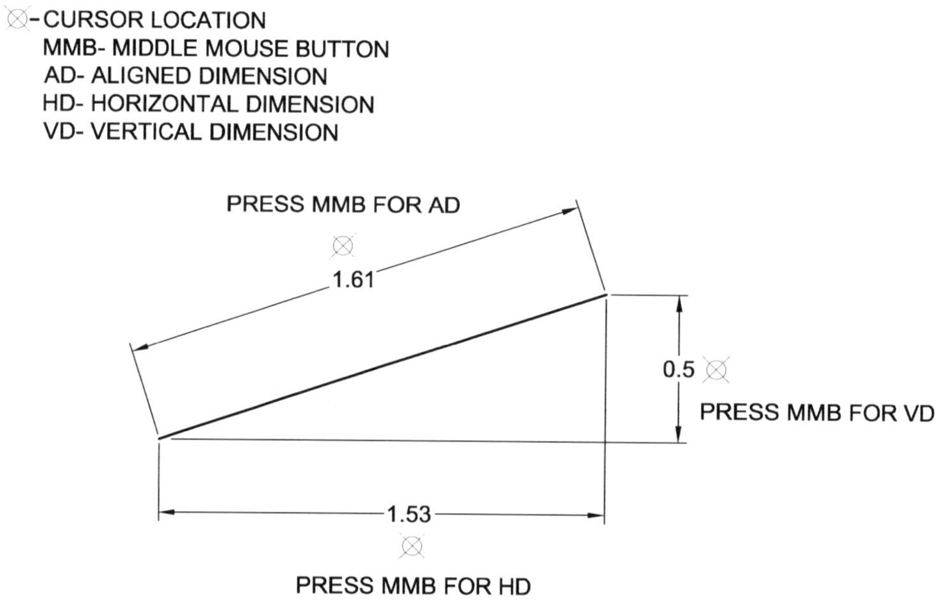

***Figure 2-7** Approximate locations of the cursor to achieve different dimensions between the end points of a line*

Dimensioning of an Arc

An arc is dimensioned by specifying the arc length or arc angle. To add angular dimension to an arc, select one end, centre point, and then select the other end of the arc. Next, place the dimension at the desired point by pressing the middle mouse button twice, as shown in Figure 2-8. If you select both ends of an arc and then select the arc itself, the dimension placed will be length of the arc, as shown in Figure 2-9. If you want to convert the angular dimension into arc length and vice versa, then select the dimension and choose the **Angle** or **Length** option from the mini popup toolbar.

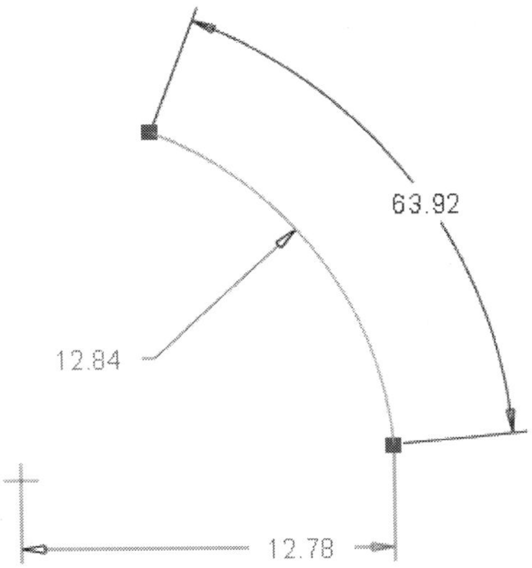

Figure 2-8 Angular dimension of an arc

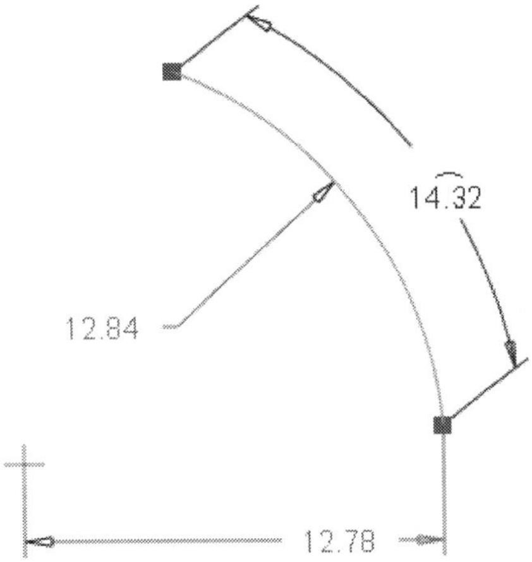

Figure 2-9 Dimensioning of arc length

Dimensioning of a Conical Arc

A conic is dimensioned by specifying its rho value that represents its conical sections. The rho value varies from 0 to 1. If the value is less than 0.5, the section will be elliptical; if rho value is 0.5, the section will be parabola; and if rho is greater than 0.5, the section will be hyperbola. To add rho dimension to a conical arc, select the conical arc. Next place the dimension at desired point by pressing the middle mouse button twice, as shown in Figure 2-10.

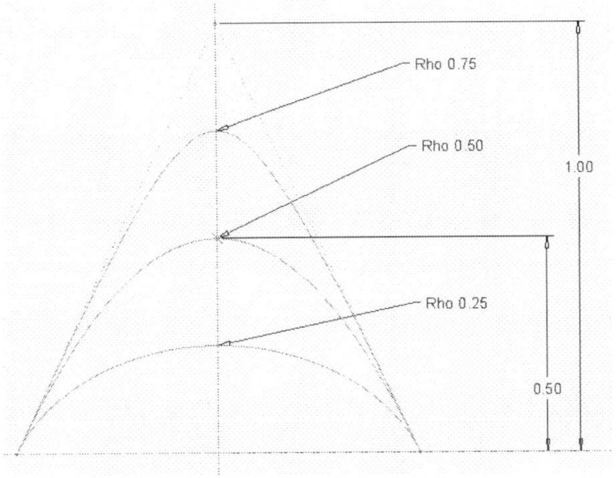

Figure 2-10 Conical dimensioning

Note that in Creo Parametric, you can enter the dimension value for a conical section varies from 0.05 to 0.95.

Diameter Dimensioning

For diameter dimensioning, click on a circle. Then place the dimension at the desired location by pressing the middle mouse button twice. The diameter dimension will be displayed, as shown in Figure 2-11. Alternatively, you can also click on the circle and choose the **Diameter** option from the mini toolbar, as shown in Figure 2-12. This method can also be used for dimensioning arcs.

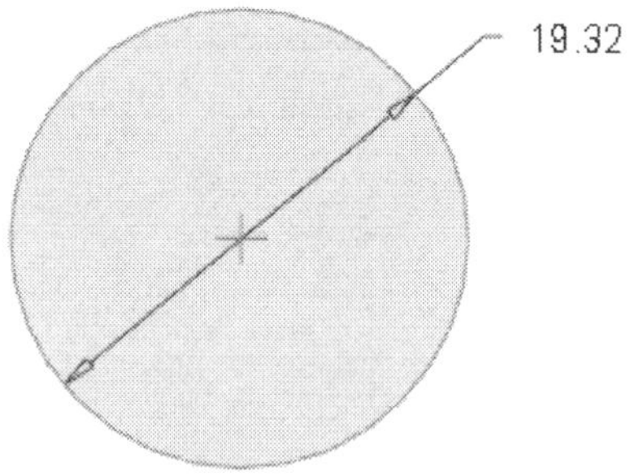

Figure 2-11 *Diameter dimensioning*

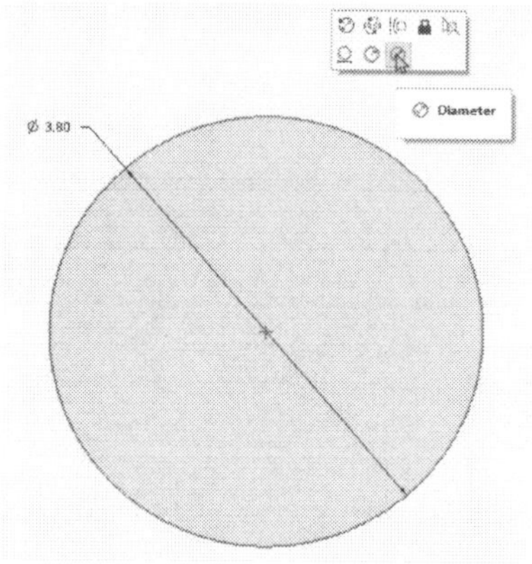

Figure 2-12 *Dimensioning using the **Diameter** option of the mini toolbar*

Radial Dimensioning

For radial dimensioning, click on the arc once. Then, place the dimension at the desired location by pressing the middle mouse button twice. The radial dimension will be displayed, as shown in Figure 2-13.

Dimensioning of Revolved Sections

Revolved sections are used to create revolved features such as flanges, couplings, and so on. To dimension a revolved section, click on the entity to be dimensioned. Next, select the centerline about which you want the section to be revolved and again, select the original entity that you want to dimension. Now, place the dimension at the desired location by pressing the middle mouse button twice. Figure 2-14 shows the dimension placed in a revolved section. This dimension represents the diameter of a revolved section.

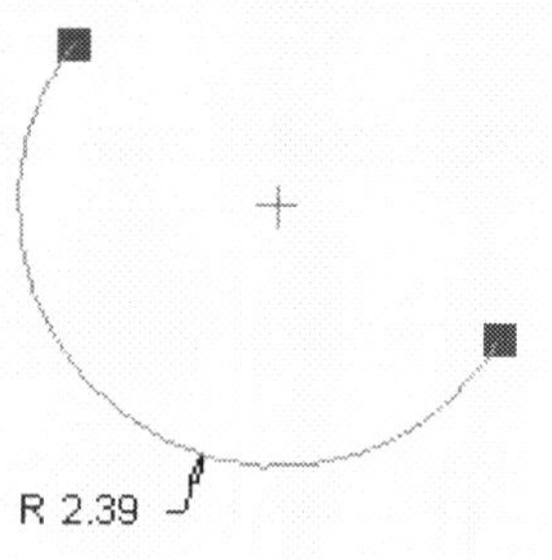

Figure 2-13 *Radial dimensioning*

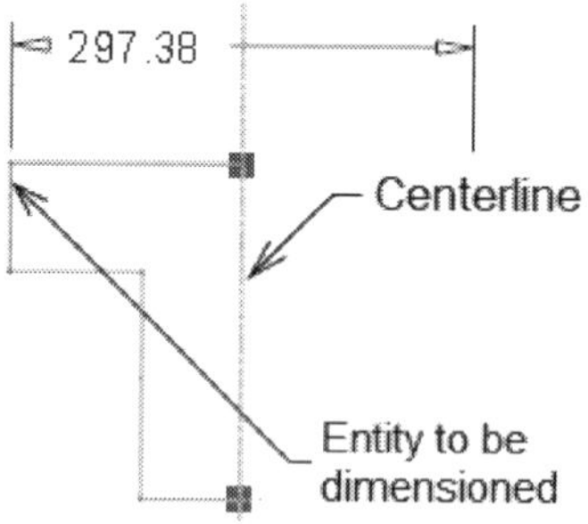

Figure 2-14 *Dimensioning the revolved sections*

Angular Dimensioning

To specify angular dimensions, you need to select two lines and then place the dimension at the desired location in the drawing area by pressing the middle mouse button twice. Note that the location where you press the middle mouse button specifies the outer or inner angle between the lines.

WORKING WITH CONSTRAINTS

Constraints are rules that are enforced on the sketched geometries. In other words, constraints are the logical operations that are performed on the selected geometries to fully define their size, shape, orientation, and location with respect to other geometries. One of the key benefits of using constraints in a sketch is that it reduces the number of dimensions that are required to fully constraint that sketch.

Creo Parametric automatically applies some constraints to geometries while sketching. For example, if you are creating a line which is nearly parallel to another line, Creo Parametric will automatically place and highlight the parallel constraint symbol on the line being created. If you confirm the line creation, a line will be drawn parallel to the other line. Alternatively, you can also apply constraints manually. The procedure to do so is discussed next.

Types of Constraints

There are two types of constraints in Creo Parametric: **Geometry** and **Assembly**. In this chapter, you will learn about the Geometry constraints only and the **Assembly** constraints will be discussed in later chapters.

The geometric constraints are available in the **Constrain** group of the **Sketch** tab. The tools available in this group are shown in Figure 2-15.

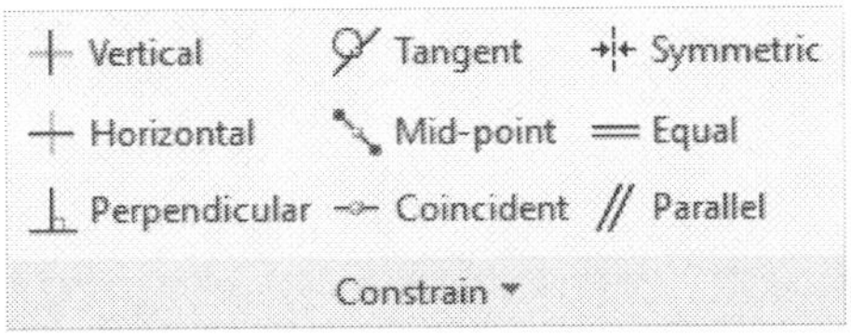

***Figure 2-15** The constraints in the **Constrain** group*

Note

*In Creo Parametric, the name of tools in several workbenches is displayed only in high screen resolution systems, such as 1900*1200 and 2880x1800. If the resolution is low, then only symbols are displayed.*

The constraints in this group are used to apply constraints manually. Although, some constraints are applied automatically as you draw the sketch. You can use the tools in this group to manually apply additional constraints to the sketch. The constraints in the group are discussed next.

Vertical

This constraint forces the selected line segment to become a vertical line. This constraint also forces the two vertices to be placed along a vertical line.

Horizontal

This constraint forces the selected line segment or two vertices that are apart by some distance to become horizontal or to lie in a horizontal line.

Perpendicular

This constraint forces the selected entity to become normal to another selected entity.

Tangent

This constraint forces the two selected entities to become tangent to each other.

Mid-point

This constraint forces a selected point or vertex to lie on the middle of a line.

Coincident

This constraint can be used to force the two selected points to become coincident. You can also apply this constraint to make two selected entities collinear, so that they lie on the same line.

Symmetric

This constraint makes a section symmetrical about the centerline. When you select this constraint, you will be prompted to select a centerline and two vertices to make them symmetrical.

Equal

This constraint forces any two selected entities to become equal in dimension. When you select this constraint, you will be prompted to select two lines to make their lengths equal, or you will be prompted to select two arcs, circles, or ellipses to make their radii equal.

Parallel

This constraint is used to force two lines to become parallel. When selected, this constraint prompts you to select two entities that you want to make parallel.

Disabling the Constraints

The need to disable a constraint arises while drawing an entity. For example, consider a case where you want to draw a circle at some distance apart from another circle. While drawing it, the system tends to apply the equal radius constraint when the sizes of the two circles become equal. At this moment, if you do not want to apply the equal radius constraint, right-click twice to disable the equal radius constraint; a cross will appear with the constraint notifying that the constraint is disabled. However, if you right-click once, the constraint will get locked and a lock symbol will appear with the constraint.

Tip

You can also press and hold the SHIFT key to disable the snapping of new constraint during sketching. If there are more than one snapping constraints while sketching, you can use the TAB key to toggle between the constraints.

MODIFYING THE DIMENSIONS OF A SKETCH

There are three ways to modify the dimensions of a sketch. These methods are discussed next.

Using the Modify Tool

Ribbon: Sketch > Editing > Modify

You can select one or more dimensions from the sketch to modify them. When you select dimension(s) from a sketch, they are highlighted in green. If you want to select more than one dimension, hold the CTRL key and select the dimensions by clicking on them. You can also use the CTRL+ALT+A keys or define a window to select all dimensions in the sketch. After selecting the dimension(s), choose the **Modify** tool from the **Editing** group; the **Modify Dimensions** dialog box will be displayed, as shown in Figure 2-16.

To modify dimensions by using the **Modify Dimensions** dialog box, you can either enter a value in the edit box or use the thumbwheel available on the right of the edit box. The **Sensitivity** slider is used to set the sensitivity of the thumbwheel.

By default, the **Regenerate** check box is selected. As a result, any modifications in the dimensions are automatically updated in the sketch. If you want to modify the dimensions of the sketch without regenerating the sketch, you need to clear this check box. If this check box is cleared, the dimensions will not be modified until you exit this dialog box. This means that Creo Parametric allows you to make multiple modifications before updating the sketch.

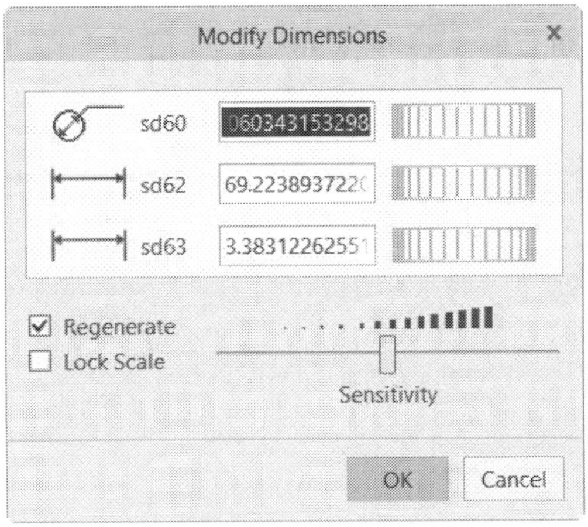

Figure 2-16 *The **Modify Dimensions** dialog box*

The **Lock Scale** check box is used to lock the scale of the selected dimensions. After locking the scale, if you modify any dimension, all other dimensions will also be modified by the same scale.

> **Tip**
> *If you need to modify more than one dimension or when the dimensions you specify vary drastically with the original ones, it is recommended that you clear the **Regenerate** check box and then modify the dimensions.*

Modifying a Dimension by Double-Clicking

You can also modify a dimension by double-clicking on it. When you double-click on a dimension, the pop-up text field appears. Enter a new dimension value in this field and press ENTER or use the middle mouse button. Remember that you can select a dimension only when you choose the **One-by-One** selection filter.

Modifying Dimensions Dynamically

In the sketcher environment, Creo Parametric is always in the selection mode, unless you have invoked some other tool. When you bring the cursor to an entity, the color of the entity changes to green. Now, if you hold down the left mouse button, you can modify the entity by dragging the mouse. You will notice that as the entity is modified, the dimensions referenced to the selected entity are also modified.

> **Tip**
> *In Creo Parametric, you can modify the dimensions by using the mini popup toolbar which appears when you select any dimension(s) or entity(s).*

LOCKING DIMENSIONS AND ENTITIES

By locking of dimensions and entities, you can avoid modifications made to the sketches by accidentally dragging a vertex or an entity. In the locked state, the dimension(s) or entity(-ies) remains unchanged when you quit or re-enter the **Sketch** mode. To lock or unlock dimension(s) or entity(-ies), select dimension(s) or entity(-ies) and choose the **Toggle Lock** option from the mini popup toolbar; the dimension(s) or entity(-ies) will be locked. The locked dimension is highlighted in red color. You can also set the **Sketch** mode to lock all user-defined dimensions automatically by selecting the **Lock user defined dimensions** check box in the **Sketcher** tab of the **Creo Parametric Options** dialog box. After setting this option, all dimensions that you subsequently create or modify will be locked automatically.

RESOLVE SKETCH DIALOG BOX

While applying constraints or dimensions, sometimes the system may prompt you to delete one or more highlighted dimensions or constraints. This is because while adding dimensions or constraints, some strong dimensions or constraints conflict with the existing dimensions or constraints. As the conflict occurs, the **Resolve Sketch** dialog box is displayed, as shown in Figure 2-17. When you select a dimension or a constraint from the **Resolve Sketch** dialog box, the corresponding dimension or constraint in the drawing area is enclosed in a blue box.

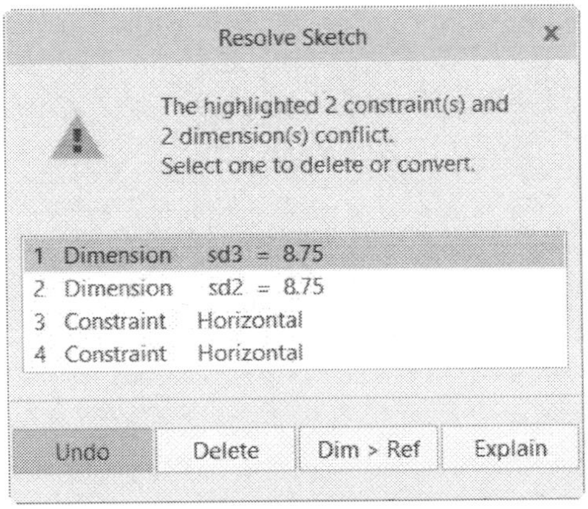

*Figure 2-17 The **Resolve Sketch** dialog box*

The buttons in the **Resolve Sketch** dialog box are discussed next.

Undo

When you choose the **Undo** button, the section is brought back to the state that it was in just before the conflict occurred.

Delete

The **Delete** button is used to delete a selected dimension or constraint that is enclosed within the blue box. To delete a dimension or a constraint, select it from the blue box and choose the **Delete** button from the **Resolve Sketch** dialog box.

Dim > Ref

On choosing the **Dim > Ref** button, the selected dimension is converted to a reference dimension.

Note

The reference dimensions are used only for reference and are not considered in feature creation.

Explain

When you choose the **Explain** button, the system provides you with the information about the selected constraint or dimension. The information will be displayed in the message area.

DELETING THE SKETCHED ENTITIES

To delete one or more sketched entities, select entities and hold down the right mouse button in the drawing area to invoke a shortcut menu. Next, choose the **Delete** option from this menu to delete the selected item. You can also delete multiple items by specifying a window. To specify a window, press and hold the left mouse button and drag the cursor to the opposite corner in the drawing area. The entities that are enclosed inside the window will get selected and the color of the selected entities changes to green. After selecting the entities, hold down the right mouse button in the drawing area to invoke a shortcut menu and choose the **Delete** option to delete the selected item.

You can also delete one or more than one items from the drawing area by using the DELETE key. To do so, press and hold the CTRL key and click on the entities to be deleted. Next, press the DELETE key to delete the selected entities. Alternatively, you can specify a window around the selected entities and then press the DELETE key to delete entities.

Note

It is necessary to be in the selection mode while selecting the items. The term "items" used in this chapter refers to dimensions and entities. The ***Sketcher Geometry****,* ***Dimension****, and* ***Constraint*** *filters are available in the drop-down list located in the* ***Status Bar****. These filters narrow down your search and help you to select the exact item. This means if you want to select all constraints in the sketch, choose the* ***Constraint*** *filter and specify a window to select. You will notice that only the constraints are selected in the sketch.*

TRIMMING THE SKETCHED ENTITIES

While creating a design, there are a number of places where you need to remove the unwanted and extended entities. You can do this by using the trimming tools that are available in the **Editing** group. These tools are discussed next.

Delete Segment

Ribbon: Sketch > Editing > Delete Segment

This tool is used to trim the entities that extend beyond the point of intersection. You can also use this tool to delete the selected entities. To trim or delete any entity, choose the **Delete Segment** tool from the **Editing** group and click on the entity to be trimmed or deleted. With the help of this tool, you can also trim and delete the entities dynamically. To do so, press and hold the left mouse button and move the cursor over the entities to be trimmed or deleted, refer to Figure 2-18 and Figure 2-19.

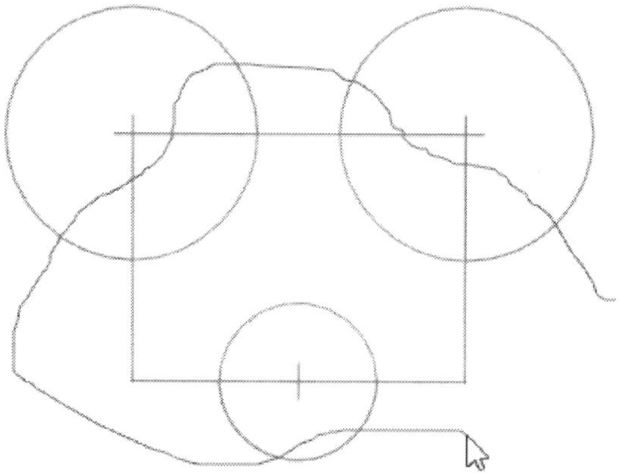

Figure 2-18 *Trimming the entities dynamically*

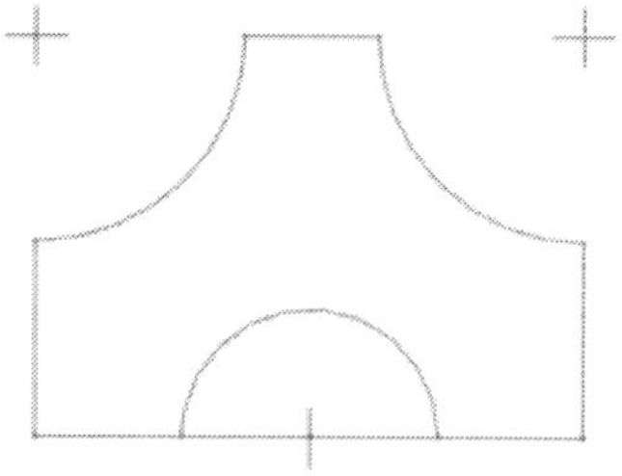

Figure 2-19 *The resulting sketch*

Corner

Ribbon: Sketch > Editing > Corner

The **Corner** tool is used to trim two entities at the corners. Note that when you trim entities using this option, the portion from where you select the entities is retained and the other portion is trimmed. The following steps explain the procedure to trim entities using this button:

1. Choose the **Corner** tool from the **Editing** group; you will be prompted to select two entities to be trimmed.

2. Click to select the two entities on the sides that you want to keep after trimming, refer to Figure 2-20. These two entities must be intersecting entities. The entities are trimmed from the point of intersection.

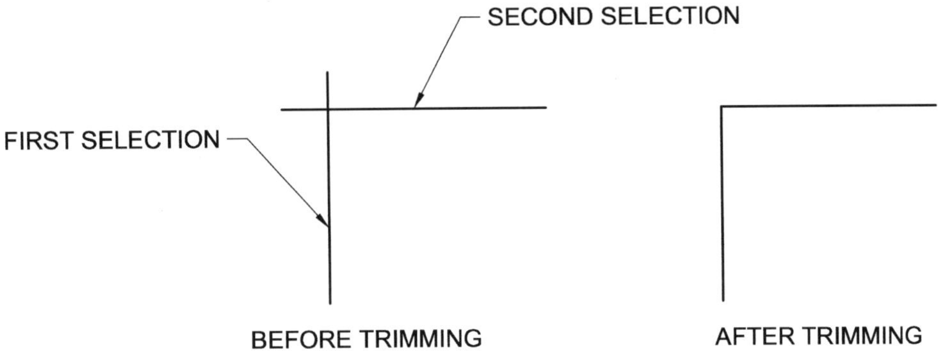

*Figure 2-20 Trimming the lines by using the **Corner** tool*

Divide

Ribbon: Sketch > Editing > Divide

The **Divide** tool is used to divide an entity into a number of parts by specifying points on the selected entity.

The following steps explain the procedure to divide an entity:

1. Choose the **Divide** tool from the **Editing** group; you will be prompted to specify a point on an entity.

2. Select a point on an entity to divide it. The divided entities can now be treated as two separate entities. Similarly, you can break other entities like circles or arcs into several small entities.

MIRRORING THE SKETCHED ENTITIES

Ribbon: Sketch > Editing > Mirror

The **Mirror** tool is used to mirror the sketched geometries about a centerline. This tool helps you to reduce the time utilized in creating symmetrical geometries and dimensions.

The procedure to mirror the sketched geometry is explained next:

1. Sketch a geometry and then sketch a centerline about which you need to mirror the geometry.
2. Select the entities that you need to mirror; the selected entities turn green in color.
3. Choose the **Mirror** tool from the **Editing** group; you will be prompted to select the centerline about which you need to mirror. Select the centerline; the selected entities will be mirrored about the centerline.

> **Tip**
> *In case of symmetrical parts, you can save time involved in dimensioning and constraining a sketch by dimensioning half of the section and then mirroring it. Creo Parametric will assume that the mirrored half has the same dimensions and constraints as the sketched half.*

INSERTING STANDARD/USER-DEFINED SKETCHES

Ribbon: Sketch > Sketching > Palette

The **Palette** enables you to quickly insert commonly used basic shapes, such as polygons, profiles, shapes, stars, and other previously created sketches in the **Sketch** mode, thus minimizing the time for repetitive sketching.

The steps that explain the procedure to insert a foreign entity in the **Sketch** mode are given next.

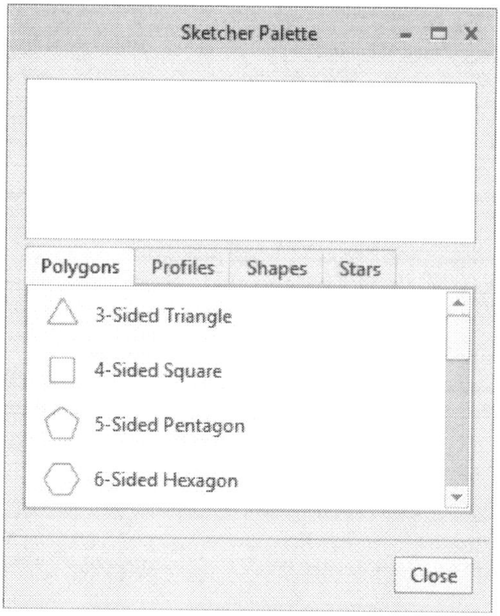

*Figure 2-21 The **Sketcher Palette** dialog box*

1. Choose the **Palette** tool from the **Sketching** group; the **Sketcher Palette** dialog box will be displayed, as shown in Figure 2-21. The options in this dialog box are used to insert a previously created sketch.
2. To insert a sketch from the sketcher palette in the current sketch, drag and drop it, or double-click on the required sketch from the list in the dialog box. Next, click anywhere in the drawing area; the **Import Section** tab will be displayed in the **Ribbon**. Also, the move, rotate, and scale handles will be displayed on the imported sketch automatically.
3. In the **Import Section** tab, enter the scale value in the **Scale** edit box and the rotational angle value in the **Rotate** edit box. Alternatively, left-click on the required handle on the sketch and drag; the corresponding action will take place dynamically. The move handle also acts as the pivot point for rotation and scale. However, you can relocate the pivot point. To move the pivot point, right-click on the pivot point and drag it to the required location and then release the right-click; the pivot point will be relocated.

4. Next, choose the **OK** button to exit.

5. Choose the **Close** button from the **Sketcher Palette** dialog box to accept the sketch inserted. Else, repeat the steps 2 - 4 to continue inserting more sketches in the drawing area.

   Similarly, you can add the sketches from the **Polygons**, **Profiles**, **Shapes**, and **Stars** tabs.

Note

If you have selected a working directory which contains only .sec files, a tab with the name of the working directory is displayed in the ***Sketcher Palette*** *dialog box and also will be chosen by default. On the other hand, if you have selected a working directory which contains other types of files, the same tab will be displayed in the end and the* ***Polygons*** *tab will be chosen by default, refer to Figure 2-21.*

DRAWING DISPLAY OPTIONS

While working with complex sketches, sometimes you need to increase the display of a particular portion of a sketch so that you can work on the minute details of the sketch. For example, you are drawing sketch of a piston and you have to work on the minute details of the grooves for the piston rings. To work on these minute details, you have to enlarge the display of these grooves. You can enlarge or reduce the drawing display using various drawing display tools provided in Creo Parametric. These tools are available in the **Graphics** toolbar as well as in the **View** tab. Some of these drawing display options are discussed next, and the remaining drawing display options will be discussed in later chapters.

Zoom In

Ribbon: View > Orientation > Zoom In

This tool is used to enlarge the view of the drawing on the screen. After choosing the **Zoom In** tool from the **Orientation** group, you will be prompted to define a box. The area that you will enclose inside the box will be enlarged and displayed in the drawing area. Note that when you enlarge the view of the drawing, the original size of the entities will not be changed. To exit the **Zoom In** tool, right-click in the drawing area.

Zoom Out

Ribbon: View > Orientation > Zoom Out

This tool is used to reduce the view of the drawing on the screen, thus increasing the drawing display area. Each time you choose this button, the display of the sketch in the drawing area is reduced. This button is available in the **Orientation** group.

Tip

You can use the mouse scroll wheel to zoom in and out the view. One more method to zoom in and out is to use the middle mouse button and the CTRL key. To zoom in and zoom out, press and hold CTRL+middle mouse button and drag the mouse downward and upward.

Refit

Ribbon: View > Orientation > Refit

This tool is available in the **Orientation** group. This tool is used to reduce or enlarge the display such that all entities that comprise the sketch are fitted inside the current display. Note that the dimensions may not be necessarily included in the current display.

Repaint

Ribbon: View > Display > Repaint

While working with complex sketches, some unwanted temporary information is retained on the screen. The unwanted information may include the shadows of the deleted sketched entities, dimensions, and so on. This unwanted information can be removed from the drawing area by using the **Repaint** button available in the **Display** group or from the **Graphics** toolbar in model display.

Sketcher Display Filters

While working with the sketches, sometimes you need to disable the display of some of the sketcher components such as dimensions, constraints, vertices, and so on. To disable or enable their display, you can use the toggle buttons available in the **Graphics** toolbar. These buttons are discussed next.

Dimensions Display

Ribbon: View > Display > Dimensions Display

You can use this button to enable or disable the display of dimension on the screen.

Constraints Display

Ribbon: View > Display > Constraint Display

You can use this button to enable or disable the display of geometric constraints.

Vertices Display

Ribbon: View > Display > Vertex Display

You can use this button to enable or disable the display of vertices on a sketch or a model.

Grid Display

Ribbon: View > Display > Grid Display

You can use this button to enable or disable the display of sketching grid. Sometimes you may not be able to see the grid as it becomes very dense. To see the grid in such cases, you need to zoom in the screen.

Locks Display

Ribbon: View > Display > Locks Display

You can use this button to enable or disable the display of locks on a sketch.

Note

*1. To remove the temporary information, you can repaint the screen by choosing the **Repaint** tool from the **Graphics** toolbar or pressing the CTRL+R keys.*

*2. In the **Sketch** mode, you can pan the sketch using the middle mouse button but in the **Part** mode, use SHIFT + middle mouse button to pan the model.*

TUTORIALS

Tutorial 1

In this tutorial, you will draw the sketch for model shown in Figure 2-22. The sketch of the model is shown in Figure 2-23. **(Expected time: 30 min)**

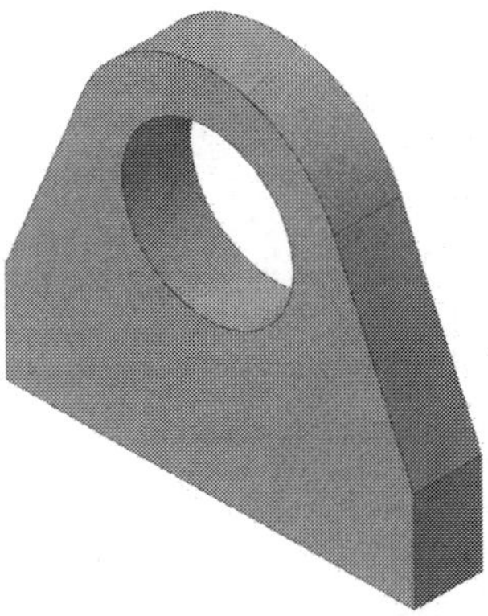

***Figure 2-22** Model for Tutorial 1*

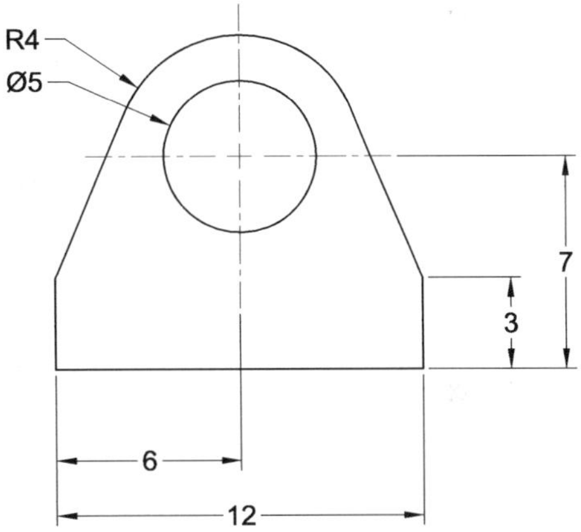

***Figure 2-23** Sketch of the model*

The following steps are required to complete this tutorial:

a. Start Creo Parametric session.
b. Set the working directory and create a new sketch file.
c. Draw lines by using the **Line Chain** tool.
d. Draw an arc and a circle.
e. Dimension the sketch and then modify the dimensions of the sketch.
f. Save the sketch and close the file.

Starting Creo Parametric

1. Start Creo Parametric by double-clicking on the **Creo Parametric** icon on the desktop of your computer.

Setting the Working Directory

After the Creo Parametric session is started, the first task is to set the working directory. A working directory is a directory on your system where you can save the work done in the current session of Creo Parametric. You can set any existing directory on your system as the working directory. Since it is the first tutorial of this chapter, you need to create a folder with the name *c02*.

1. Choose the **Select Working Directory** option from the **Manage Session** flyout of the **File** menu; the **Select Working Directory** dialog box is displayed.

2. In this dialog box, browse to *C:\Creo_8.0* folder. If this folder does not exist, create this folder before setting the working directory.

   For selecting *C:\Creo_8.0,* click on the arrow at the right of the **C:** option in the top address box of the **Select Working Directory** dialog box; the folders in the **C** drive are displayed in a flyout. Now, choose **Creo_8.0** from the flyout, refer to Figure 2-24. Alternatively, you can use the **Folder Tree** node available at the bottom left corner of the screen to set the working directory. To do so, click on the **Folder Tree** node; the **Folder Tree** expands. In the **Folder Tree**, browse to the desired location using the nodes corresponding to the folders. Select the required folder and choose the **OK** button; the selected folder will become the current working directory.

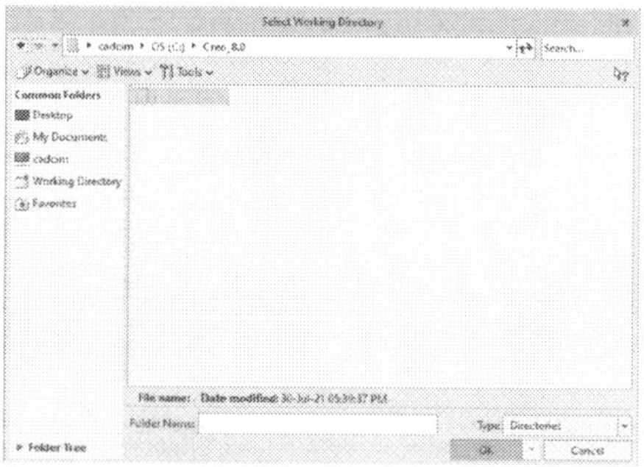

Figure 2-24 The ***Creo_8.0*** *folder chosen from the flyout*

3. Choose the **Organize** button from the top pane of the **Select Working Directory** dialog box to display the flyout. From the flyout, choose the **New Folder** option; the **New Folder** dialog box is displayed.

4. Enter **c02** in the **New directory** edit box and choose the **OK** button from the **New Folder** dialog box; a new folder named *c02* is created at ***C:\Creo_8.0*** location.

5. Choose the **OK** button from the **Select Working Directory** dialog box to set the working directory to *C:\Creo_8.0\c02*; the message **Successfully changed to C:\Creo_8.0\c02 directory** is displayed in the message area.

Starting a New Object File

Any sketch drawn in the **Sketch** mode is saved with the *.sec* file extension. This file format is one of the file formats available in Creo Parametric.

1. Choose the **New** button from the **Data** group in the **Ribbon** or **Quick Access** toolbar or press CTRL+N; the **New** dialog box is displayed. In this dialog box, select the **Sketch** radio button from the **Type** area; the default name of the sketch appears in the **File name** edit box.

2. Enter **c02tut01** in the **File name** edit box and choose the **OK** button.

   You are in the sketcher environment of the **Sketch** mode. When the **Sketch** mode is invoked, the **Show Navigator** is displayed on the left in the drawing area.

3. Choose the **Show Navigator** button available at the bottom left corner of the screen to close the **Show Navigator**. On closing the tree, the drawing area is increased.

Drawing the Lines of the Sketch

You need to start drawing the sketch with the right vertical line.

1. Choose the **Line Chain** tool from the **Line** drop-down available in the **Sketching** group.

2. Specify the start point by clicking on the right in the drawing area. One end of the line is attached to the cursor. Move the cursor down to an approximate length.

   Notice that when the cursor moves vertically downward, a blue colored symbol appears in the drawing area, next to the line. Now, if you draw a line, the vertical constraint will be applied to it.

3. Click to specify the endpoint of the line. The vertical constraint is applied to the line, but it is not visible in the drawing area until the line creation is active.

   Also, another rubber-band line is attached to the cursor with its start point at the endpoint of the last line.

4. Move the cursor horizontally toward the left; a horizontal rubber-band line extends to the left as you move the mouse.

   Notice that when the cursor moves horizontally toward the left, a blue colored symbol appears in the drawing area next to the line. Now, if you draw a line, a horizontal constraint will be applied to it.

5. After getting the desired size of the line created, click to end the line. The horizontal constraint is applied to the line, but it is not visible in the drawing area until the line creation is active.

6. Move the cursor upward in the drawing area; a vertical rubber-band line extends as you move the mouse. As you move the cursor upward, you will notice that at a particular point where the length of the left vertical line is equal to the length of the right vertical line, a symbol is displayed on the vertical line being created and the right line is highlighted in blue color. This symbol suggests that the equal length constraint is applied to the two vertical lines.

7. When equal constraint appears on the vertical line, click to specify the endpoint of the vertical line. The rubber-band line is still attached to the cursor.

You can also apply constraints later. However, to save an extra step of adding the constraints, you will use the constraints that are applied automatically while drawing.

8. Move the cursor to size the line and specify the endpoint of the left inclined line, as shown in Figure 2-25.

9. Press the middle mouse button to end the line creation.

10. The line option is still active. Move the cursor close to the top end of the right vertical line; the cursor snaps to the point that is at equal length of the left vertical line. Select the point by clicking.

11. Size the inclined line and specify the endpoint of the right inclined line. Press the middle mouse button twice; lines are created and all the constraints that you have applied become visible, refer to Figure 2-26. Now, you need to draw the arc and the circle.

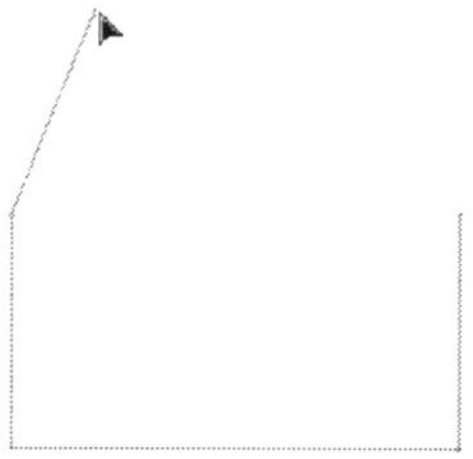

***Figure 2-25** Sketch with left inclined line*

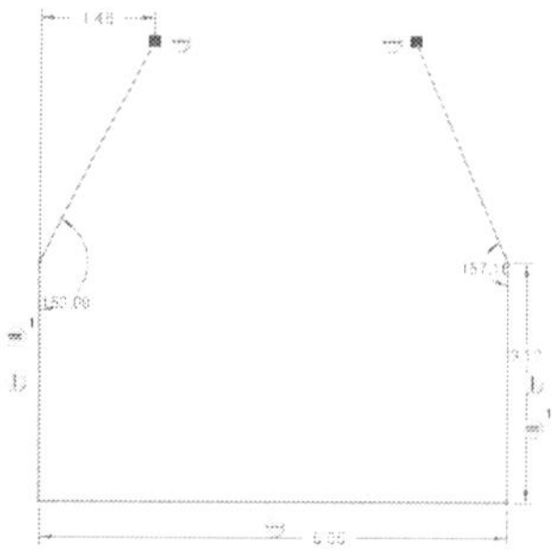

***Figure 2-26** Sketch of the line drawn with weak dimensions*

Note

1. The numbers in blue color with the constraint symbols refer to the label assigned to same type of constrained geometries in the sketch.

2. The horizontal and equal constraint appear in blue color when selected. The label with the constraint appears in blue color which indicates that the constraint is strong. This means you cannot change the orientation of the line until you delete the constraint applied to the line.

Drawing the Arc

1. Choose the **3-Point/Tangent End** tool from the **Arc** drop-down in the **Sketching** group; you are prompted to select the start point of the arc.

2. Select the endpoint of the left inclined line; the Target symbol appears in green color.

3. Move the cursor along the tangent direction through a small distance; a rubber-band arc that is tangent to the inclined line appears. Move the cursor to the endpoint of the right inclined line and click on the end point of the right inclined line to create the arc. As you exit the **Arc** tool, the tangent constraint is applied to the end points of the arc which is indicated by the symbol .

Note

1. The weak dimensions in Figure 2-26 are not displayed in Figure 2-27 because after an arc is drawn, some of the weak dimensions automatically get deleted.

*2. If the tangent constraint symbol is not displayed on any of the inclined lines, apply the constraint manually by choosing the **Tangent** button from the **Constrain** group.*

Drawing the Circle

1. Choose the **Concentric** tool from the **Circle** drop-down; you are prompted to select an arc.

2. Select the arc by clicking on it. Move the mouse; a circle appears.

3. Click to select a point inside the sketch to draw the circle.

4. Press the middle mouse button to end the circle creation. The sketch is completed.

Dimensioning the Sketch

The right vertical line, the bottom horizontal line, the arc, and the circle are dimensioned automatically and the weak dimensions are applied to them. You will use these dimensions. Hence, there is no need to dimension these entities again.

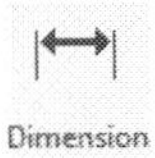

1. Choose the **Dimension** tool from the **Dimension** group.

2. Select the center of the arc and then the bottom horizontal line; the center turns red and the line turns green in color.

3. Place the dimension on the right of the sketch by pressing the middle mouse button twice.

4. Select the center of the arc and then the left vertical line; the center turns red and the vertical line turns green in color.

5. Press the middle mouse button to place the dimension below the sketch, refer to Figure 2-28.

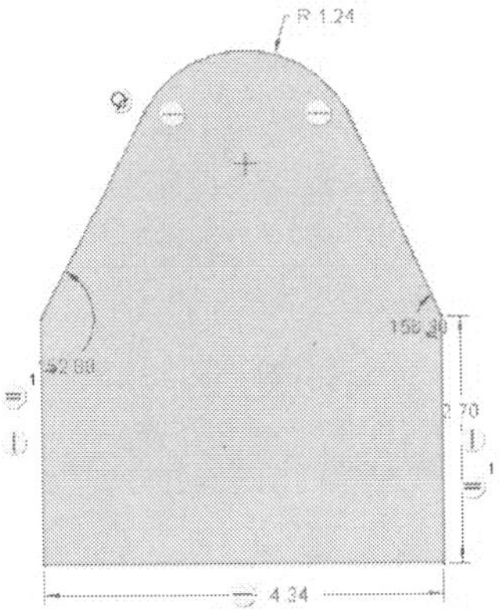

Figure 2-27 *Sketch with arc*

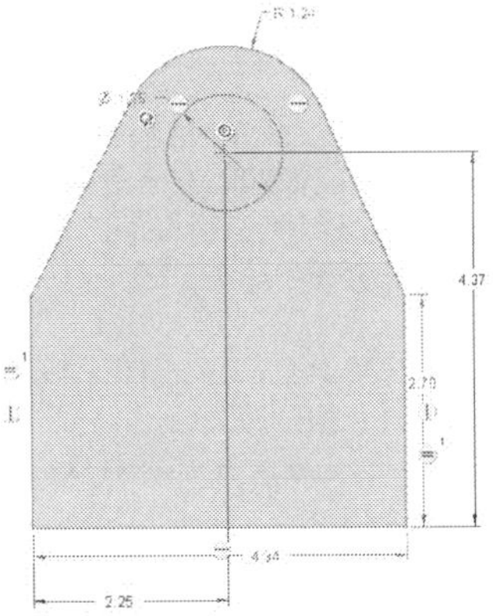

Figure 2-28 *Sketch with all the entities, weak dimensions, and weak constraints*

Modifying the Dimensions

The sketch is dimensioned with default values. You need to modify these values to the given values.

1. Select all dimensions by specifying a window around them.

Note
You can also use CTRL+ALT+A to select the entire sketch with dimensions.

2. When all dimensions turn green in color, choose the **Modify** tool from the **Editing** group; the **Modify Dimensions** dialog box is displayed.

   All dimensions in the sketch are displayed in this dialog box. Each dimension has a separate thumbwheel and an edit box. You can use the thumbwheel or the edit box to modify the dimensions. It is recommended that you use the edit boxes to modify the dimensions, if the change in the dimension value is large.

3. Clear the **Regenerate** check box and then modify the values of the dimensions.

   Once you clear this check box, any modification in a dimension value is not updated in the sketch. It is recommended that you clear the **Regenerate** check box when more than one dimension has to be modified.

   Notice that the dimensions that you select in the **Modify Dimensions** dialog box get enclosed in a violet box in the drawing area.

4. Modify all dimensions according to the dimensions shown in Figure 2-23. After modifying the dimensions, choose the **OK** button from the **Modify Dimensions** dialog box; the message **Dimension modifications successfully completed** is displayed in the message area.

Tip
You can modify the location of the dimensions as they appear on the screen by selecting and dragging them to a new location.

The completed sketch is shown in Figure 2-29.

Saving the Sketch

Now, the sketch needs to be saved because you may need the sketch later in the **Part** mode to create a 3D model.

1. Choose the **Save** button from the **Quick Access** toolbar; the **Save Object** dialog box is displayed with the name of the sketch that you had entered earlier.

2. Choose the **OK** button; the sketch is saved.

3. After saving the sketch, choose the **Close** button from the **Quick Access** toolbar.

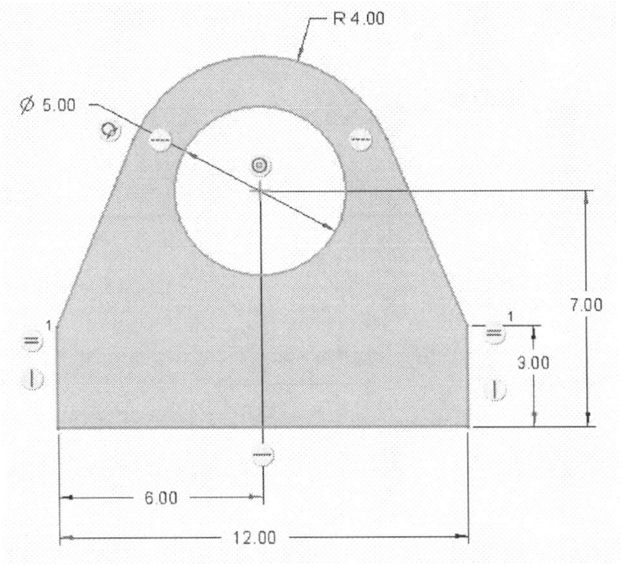

Figure 2-29 The complete sketch with dimensions and constraints

Tutorial 2

In this tutorial, you will draw the sketch for the model shown in Figure 2-30. The sketch of the model is shown in Figure 2-31. For your reference, all entities in the sketch are labeled alphabetically. **(Expected time: 30 min)**

Figure 2-30 Model for Tutorial 2

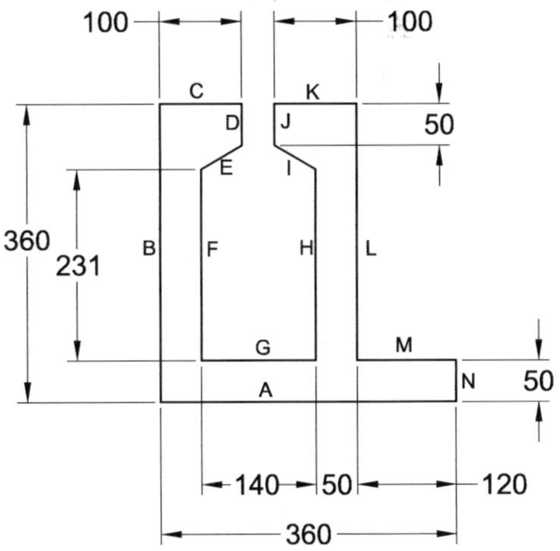

Figure 2-31 Sketch of the model

The following steps are required to complete this tutorial:

a. Set the working directory and create a new sketch file.
b. Draw the sketch by using the **Line Chain** tool.

c. Dimension the required entities and then modify the dimensions of the sketch.
d. Save the sketch and close the file.

Setting the Working Directory

The working directory was selected in Tutorial 1, therefore you need not to select the directory again. But, if a new session of Creo Parametric is started, you need to set the working directory again by following the steps given next.

1. Open the Navigator (if it is in collapsed state) by clicking on the **Show Navigator** button on the bottom left corner of the Creo Parametric main window; the Navigator slides out. At the bottom of the Navigator, the **Folder Tree** is displayed in the **Folder Browser** tab. Click on the **Folder Tree**; the **Folder Tree** expands.

2. Click on the arrow adjacent to the *Creo_8.0* folder in the Navigator; the contents of the *Creo_8.0* folder are displayed.

3. Now, right-click on the *c02* folder to display a shortcut menu. From the shortcut menu, choose the **Set Working Directory** option; the working directory is set to *c02*.

4. Close the Navigator by clicking on the **Show Navigator** button located at the bottom left corner of the main window; the Navigator slides in.

Starting a New Object File

1. Choose the **New** button from the **Data** group; the **New** dialog box is displayed. Select the **Sketch** radio button from the **Type** area of the **New** dialog box; the default name of the sketch appears in the **File name** edit box.

2. Enter **c02tut02** in the **File name** edit box and choose the **OK** button; you are in the sketcher environment of the **Sketch** mode.

Drawing the Sketch

The sketch in Figure 2-31 consists of only lines. For ease of understanding, all lines in the sketch are labeled alphabetically.

1. Choose the **Line Chain** tool from the **Line** drop-down of the **Sketching** group. Select a point close to the lower right corner of the drawing area by clicking and start drawing the horizontal line A. You will notice that as you draw line A, the symbol is displayed on the line. This indicates that the line is horizontally constrained. Move the cursor toward left and specify the endpoint of the line.

2. Move the cursor vertically upward so that the constraint appears on the line. When you get the appropriate size of the line, click to specify the endpoint of line B; line B is completed.

3. Move the cursor to the right in the drawing area and click to specify the endpoint of line C.

4. Now, to draw line D, move the cursor down and click to specify the endpoint of line D.

5. Move the cursor to size the line and click to specify the endpoint of the inclined line E.

6. The next line you need to draw is line F. Move the cursor vertically downward and click to specify the endpoint of line F.

7. Now, to draw line G, move the cursor horizontally toward the right and click to specify the endpoint of line G.

8. Move the cursor vertically upward and click to specify the endpoint of line H.

9. Now, continue drawing the remaining lines that are shown in Figure 2-31. When the sketch is complete, end the line creation by pressing the middle mouse button twice. Notice that the sketched entities are dimensioned automatically as you draw them. These dimensions are weak dimensions and appear in light blue color.

Applying Constraints to the Sketch

Constraints are applied to the sketch to maintain the design intent of the feature and this might sometimes result in less dimensions in the sketch.

1. Choose the **Equal** tool from the **Constrain** group and select lines F and H. The equal length constraint **1** is applied to both the lines. The constraint labels such as 1 or 2 vary from sketch to sketch.

2. Now, select lines J and N; the equal length constraint is applied to both the lines. Press the middle button to make other selections.

3. Select lines C and K; the equal length constraint is applied to both the lines. Press the middle mouse button to make other selections.

4. Select lines A and B; the equal length constraint is applied to both the lines. Press the middle mouse button twice to exit.

5. Choose the **Horizontal** tool from the **Constrain** group; you are prompted to select a line or two points.

6. Select the vertex that is joining the lines L and M. Now, select the vertex that is joining the lines G and H. For the placement of lines, refer to Figure 2-31. Both the vertices are aligned horizontally, as shown in Figure 2-32.

7. Select the vertex that is joining lines C and D and the vertex that is joining lines J and K, refer to Figure 2-30. Both the vertices are aligned horizontally, refer to Figure 2-32.

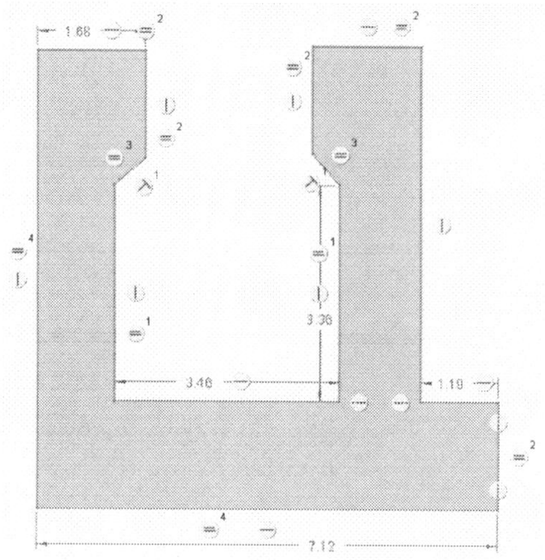

Figure 2-32 *Vertices aligned horizontally*

Dimensioning the Sketch

Weak dimensions have already been applied to the sketch while drawing. In this sketch, you need to dimension only the angle between lines D and E and lines J and I.

1. Choose the **Dimension** tool from the **Dimension** group.

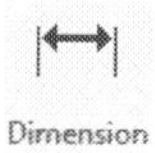

2. Select lines D and E by using the left mouse button; the selected lines turn green in color. Now, press the middle mouse button twice to place the dimension close to the vertex where lines D and E join.

3. Similarly, dimension the angle between lines J and I.

   Figure 2-33 shows the sketch after applying dimensions. If your sketch does not have all dimensions shown in this figure, apply them by using the **Dimension** tool.

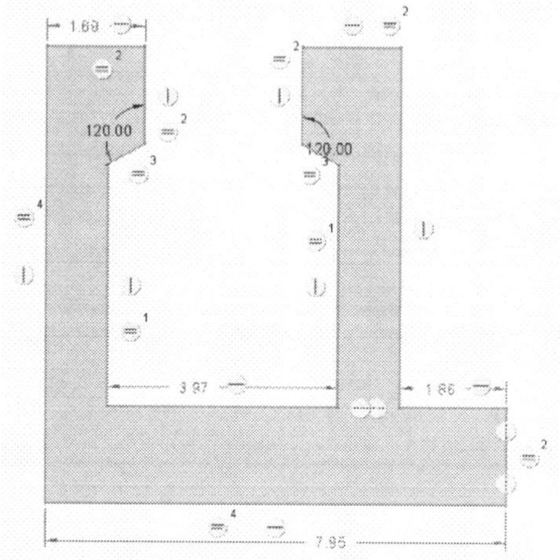

Figure 2-33 *Sketch after applying dimensions*

Modifying the Dimensions

The dimensions that are applied to the sketch need modification in dimension values.

1. Select all dimensions by specifying a window around them.

2. When dimensions turn green in color, choose the **Modify** tool; the **Modify Dimensions** dialog box is displayed.

3. Clear the **Regenerate** check box and then modify the values of the dimensions. On clearing this check box, the sketch is not regenerated while you modify the dimensions.

   Notice that the dimension that you select in the **Modify Dimensions** dialog box is enclosed in a violet box in the drawing area.

4. When all dimensions are modified, choose the **OK** button from the **Modify Dimensions** dialog box; the message **Dimension modifications successfully completed** is displayed in the message area. The completed sketch is shown in Figure 2-34.

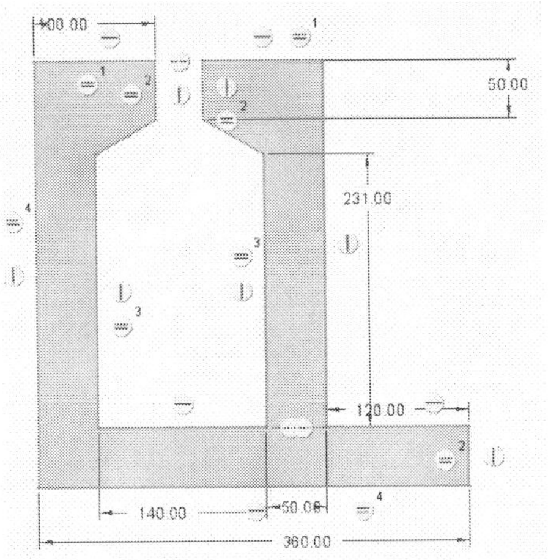

Figure 2-34 Complete sketch with dimensions and constraints

5. Save the sketch as discussed earlier. Next, choose the **Close** button from the **File** menu to exit the **Sketch** mode.

Note

You can also modify dimensions individually. However, individual modification of dimensions is recommended only when there is a minor change in the dimension value or when only one dimension is required to be modified.

Tutorial 3

In this tutorial, you will draw the sketch of the model shown in Figure 2-35. The sketch of the model is shown in Figure 2-36. For your reference, all entities in the sketch are labeled alphabetically. Also, you have to print the sketch. **(Expected time: 30 min)**

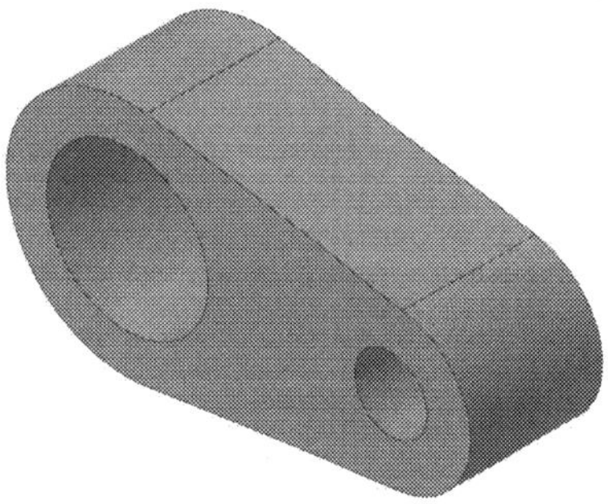

Figure 2-35 Model for Tutorial 3

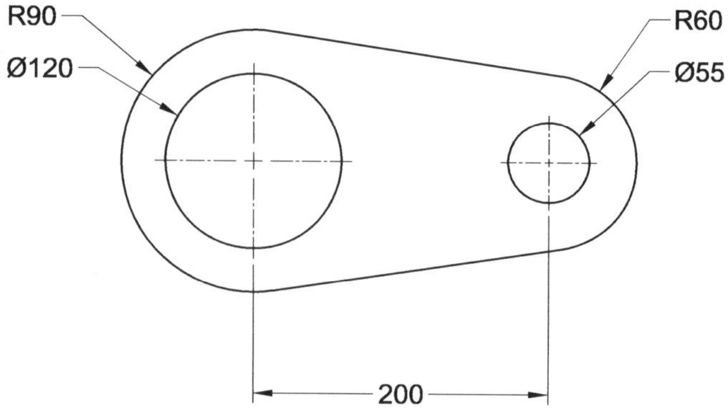

Figure 2-36 Sketch of the model

The following steps are required to complete this tutorial:

a. Set the working directory and create a new sketch file.
b. Draw the sketch by using the sketcher tools.
c. Dimension the sketch and then modify the dimensions of the sketch.
d. Save the sketch and print it.

Setting the Working Directory

The working directory was selected in Tutorial 1, therefore, you do not need to select the working directory again. But if a new session of Creo Parametric is started, then you have to set the working directory again by following the steps given next.

1. Open the Navigator by sliding it out. In the Navigator, the **Folder Tree** is displayed at the bottom. Click on the black arrow available on the right of the **Folder Tree**; the **Folder Tree** expands. Click on the arrow adjacent to the *Creo_8.0* folder in the Navigator; the contents of the *Creo_8.0* folder are displayed.

2. Now, right-click on the *c02* folder to display a shortcut menu. From the shortcut menu, choose the **Set Working Directory** option; the working directory is set to *c02*. Close the Navigator.

Starting a New Object File

1. Choose the **New** button from the **Data** group; the **New** dialog box is displayed. Select the **Sketch** radio button from the **Type** area of the **New** dialog box; the default name of the sketch appears in the **File name** edit box.

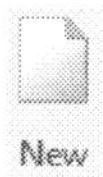

2. Enter **c02tut03** in the **File name** edit box. Choose the **OK** button to enter the sketcher environment of the **Sketch** mode.

Drawing the Circles

1. Choose the **Center and Point** tool from the **Circle** drop-down in the **Sketching** group and specify the center of the circle.

2. Move the cursor to size the circle and then click to complete the circle.

3. Draw another circle whose center is collinear with the center of the previous circle.

   Figure 2-37 shows the two collinear circles drawn by using the **Center and Point** tool.

Drawing the Tangent Lines

1. Choose the **Line Tangent** tool from the **Line** drop-down in the **Sketching** group; you are prompted to select the start location on the arc or the circle.

2. Select the left circle at the top; a rubber-band line appears whose one end is attached to the circle and the other end is attached to the cursor.

3. Click on the top of the right circle; a tangent connecting the two circles is drawn.

4. Similarly, draw a tangent by selecting the two circles at the bottom.

   Figure 2-38 shows the sketch after drawing the tangent lines.

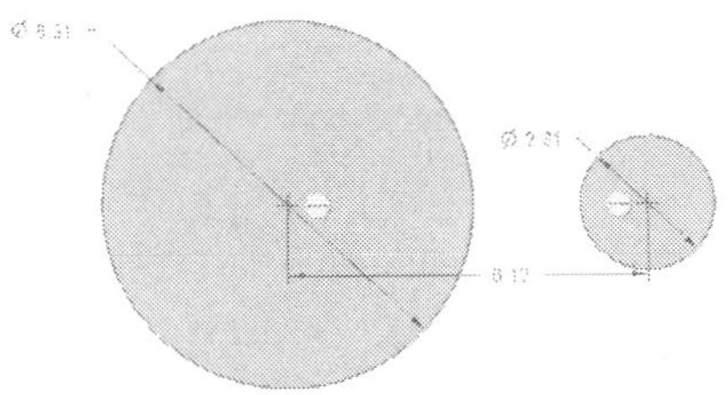

Figure 2-37 *Two collinear circles drawn using the* ***Centre and Point*** *tool*

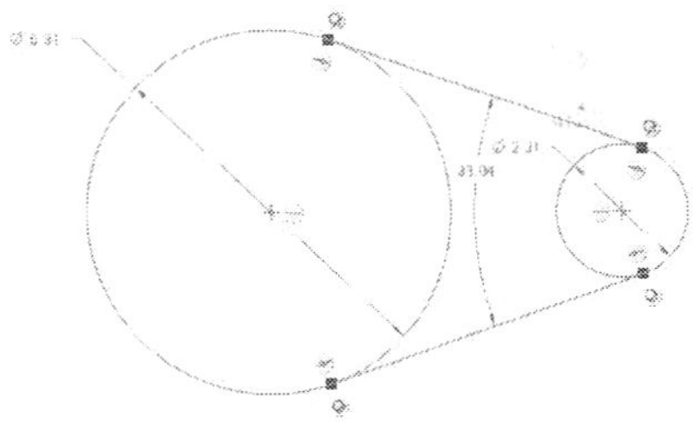

Figure 2-38 *Circles joined by tangent lines*

Trimming the Circles

As evident from Figure 2-37, the tangents that are drawn intersect the circles at the point where they meet the circle. Therefore, the part of the circle that is not required can be dynamically trimmed.

1. Choose the **Delete Segment** tool from the **Editing** group.

2. Select the two circles individually to trim them at the locations shown in Figure 2-39.

   Figure 2-40 shows the two circles after deleting the unwanted portions of the circle.

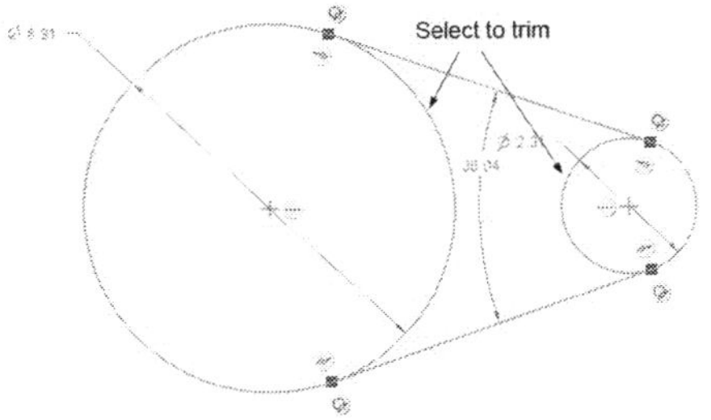

Figure 2-39 *Locations to trim*

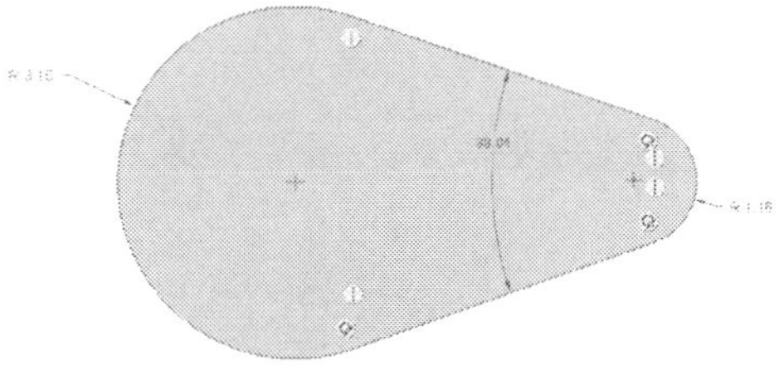

Figure 2-40 *Sketch after trimming*

Drawing the Circles

1. Choose the down arrow on the right of the **Center and Point** tool to display the flyout. Choose the **Concentric** tool from the flyout; you are prompted to select an arc.

2. Select the left arc and create a circle concentric to the arc. Similarly, select the right arc to create a concentric circle (refer to Figure 2-36).

   Notice that the radius dimension is applied to the two arcs, whereas the diameter dimension is applied to the circles. It is so because the arcs are applied with radius dimension and circles are applied with diameter dimension by default.

Dimensioning the Sketch

In order to fully define a sketch, you need to dimension it.

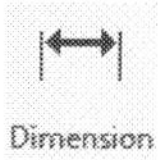

1. Choose the **Dimension** tool.

2. Select the centers of the two circles and place the dimension at the bottom of the sketch.

Modifying the Dimensions

1. Select all dimensions by defining a window.

> **Note**
> *You can also use CTRL+ALT+A from the keyboard to select all entities and items in the sketch.*

2. When all dimensions turn green in color, choose the **Modify** tool; the **Modify Dimensions** dialog box is displayed.

3. Clear the **Regenerate** check box and then modify the values of the dimensions. You will notice that the dimension that you edit in the **Modify Dimensions** dialog box is enclosed by a violet box in the drawing area.

4. When all dimensions are modified, choose the **OK** button from the **Modify Dimensions** dialog box; the message **Dimension modifications successfully completed** is displayed in the message area.

   The completed sketch is shown in Figure 2-41.

5. Save the sketch as discussed earlier. Next, you need to print the sketch.

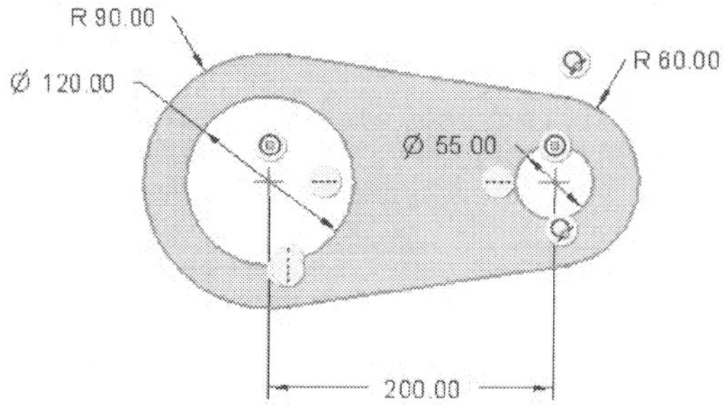

Figure 2-41 *Complete sketch with dimensions and constraints*

Printing the Sketch Using the Plot Option

1. Choose the **Print** option from the **File** menu or press CTRL+P; the **Printer Configuration** dialog box is displayed, as shown in Figure 2-42. 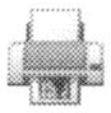

2. Choose the **Commands and Settings** button from this dialog box; a shortcut menu is displayed, as shown in Figure 2-43. 

*Figure 2-42 The **Printer Configuration** dialog box*

***Figure 2-43 Commands and Settings** shortcut menu*

3. Choose the **Add Printer Type** option from the shortcut menu; the **Add Printer Type** dialog box is displayed.

4. From the printers listed in the **Add Printer Type** dialog box, select the printer that is installed on your system and choose the **OK** button.

5. From the **Printer Configuration** dialog box, choose the **Page** tab and select the **A** option from the **Size** drop-down list, if not already selected; the dimensions of the sheet are set, by default.

6. Also, select the desired image resolution and the image depth from the dialog box under the **Resolution** area of the **Printer Configuration** dialog box.

7. Next, choose the **OK** button from the **Printer Configuration** dialog box to complete the printing.

Self-Evaluation Test

Answer the following questions and then compare them to those given at the end of this chapter:

1. The **Modify** tool is located in the__________ group of the **Ribbon**.
2. A sketch can be modified by changing its __________.
3. There are two types of constraints in Creo Parametric: __________ and __________.
4. In the **Sketch** mode, a weak dimension is highlighted by __________ color.
5. The file created in the **Sketch** mode is saved with a __________ file extension.
6. When you draw a sketch, the dimensions and constraints are automatically applied to it. (T/F)
7. In Creo Parametric, you can create lines that are tangent to two circles. (T/F)
8. You can convert a solid geometry into a construction geometry. (T/F)
9. You can convert a weak dimension into strong dimension by using the mini popup toolbar that is displayed when you click on the dimension. (T/F)
10. For drawing a circle, first you need to specify its diameter. (T/F)

Review Questions

Answer the following questions:

1. The tools available in the __________ group are used to apply constraints manually.
2. You can dynamically modify the geometry of a sketch. (T/F)
3. You can use the **Rectangle** tool from the **Sketching** group to draw a square. (T/F)
4. You cannot undo a previous operation in the sketcher environment. (T/F)
5. You can use the options in the shortcut menu to draw a sketch, when no other tool is chosen. (T/F)
6. You cannot disable constraints in the **Sketch** mode. (T/F)
7. Construction points are used to specify locations in sketch. (T/F)

8. Two types of lines can be sketched by using the tools available in the **Sketching** group. (T/F)

9. The **Dimension** tool is used for normal dimensioning of the sketch. (T/F)

10. You can select one or more dimensions from the sketch to modify them. (T/F)

EXERCISES

Exercise 1

In this exercise, you will draw the sketch of the model shown in Figure 2-44. The dimensions of the model are shown in Figure 2-45. **(Expected time: 30 min)**

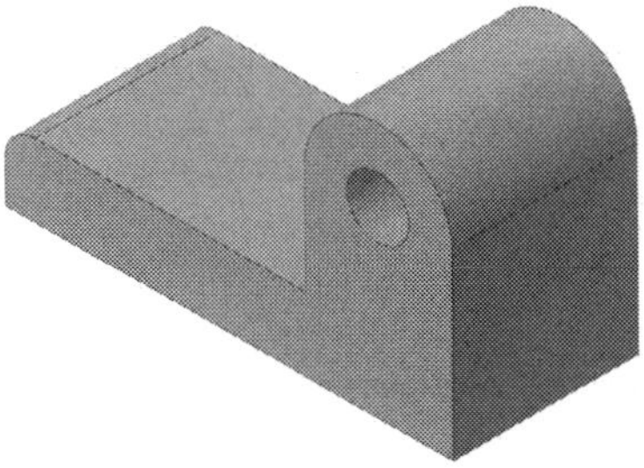

Figure 2-44 *Solid model for Exercise 1*

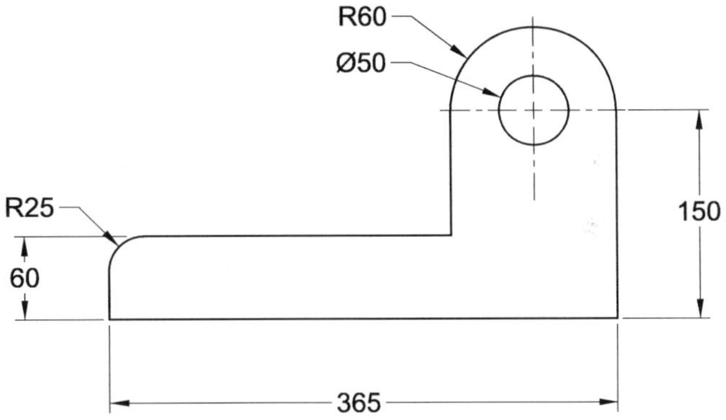

Figure 2-45 *Dimensions of the model*

Exercise 2

In this exercise, you will draw the sketch of the model shown in Figure 2-46. The dimensions of the model are shown in Figure 2-47. **(Expected time: 30 min)**

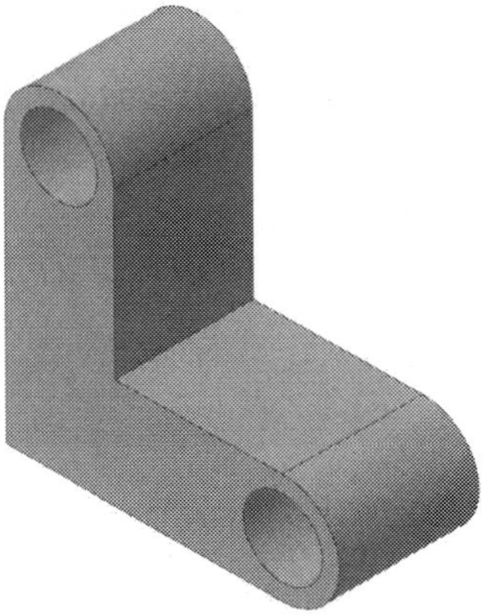

Figure 2-46 *Solid model for Exercise 2*

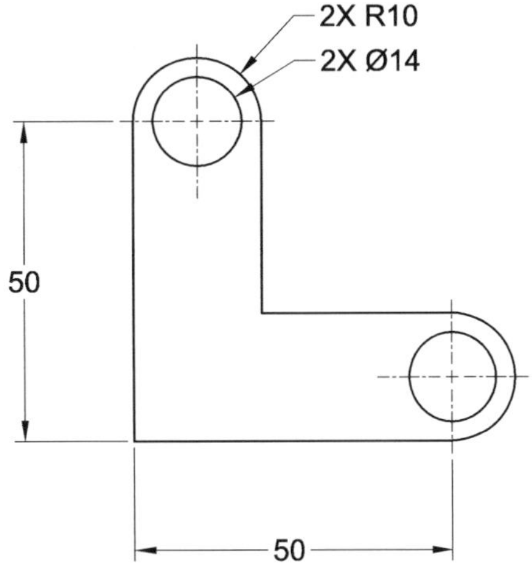

Figure 2-47 *Dimensions of the model*

Exercise 3

In this exercise, you will draw the sketch of the model shown in Figure 2-48. The dimensions of the model are shown in Figure 2-49. **(Expected time: 30 min)**

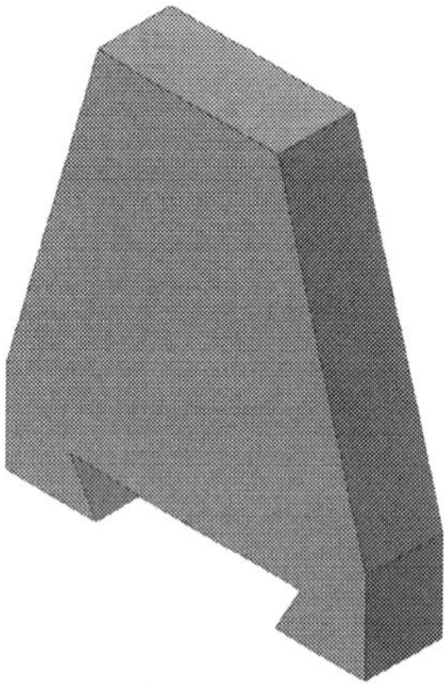

Figure 2-48 Solid model for Exercise 3

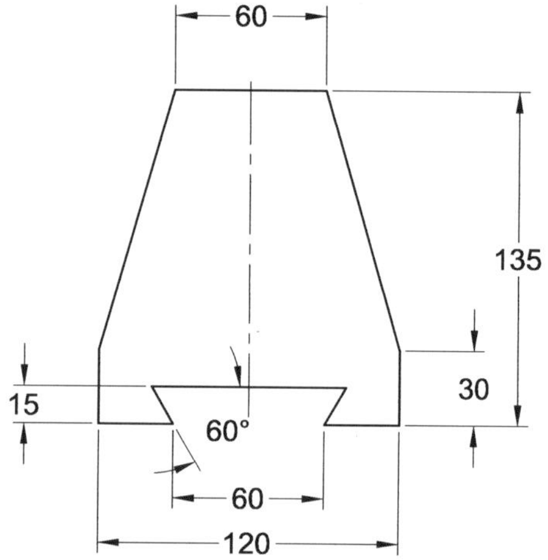

Figure 2-49 Dimensions of the model

Exercise 4

In this exercise, you will draw the sketch of the model shown in Figure 2-50. The dimensions of the model are shown in Figure 2-51. **(Expected time: 30 min)**

Figure 2-50 Solid model for Exercise 4

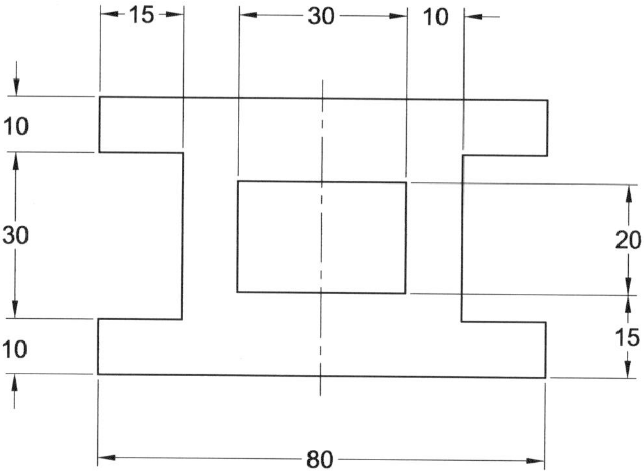

Figure 2-51 Dimensions of the model

Exercise 5

In this exercise, you will draw the sketch of the model shown in Figure 2-52. The dimensions of the model are shown in Figure 2-53. **(Expected time: 30 min)**

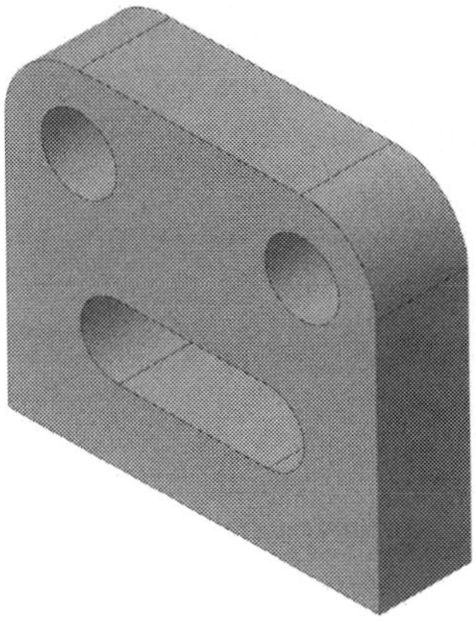

Figure 2-52 *Solid model for Exercise 5*

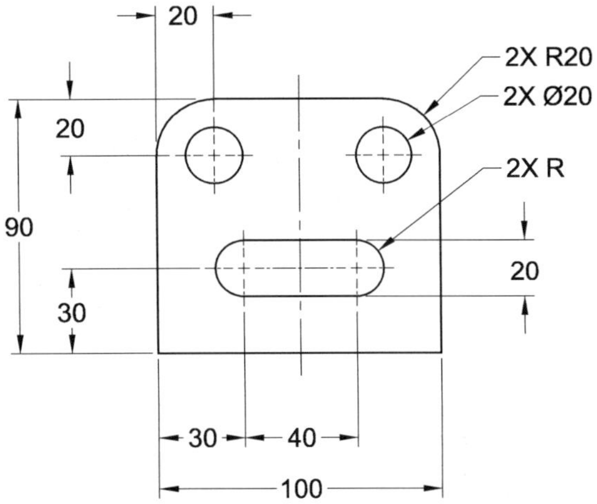

Figure 2-53 *Dimensions of the model*

Exercise 6

In this exercise, you will draw the sketch of the model shown in Figure 2-54. The dimensions of the model are shown in Figure 2-55. **(Expected time: 30 min)**

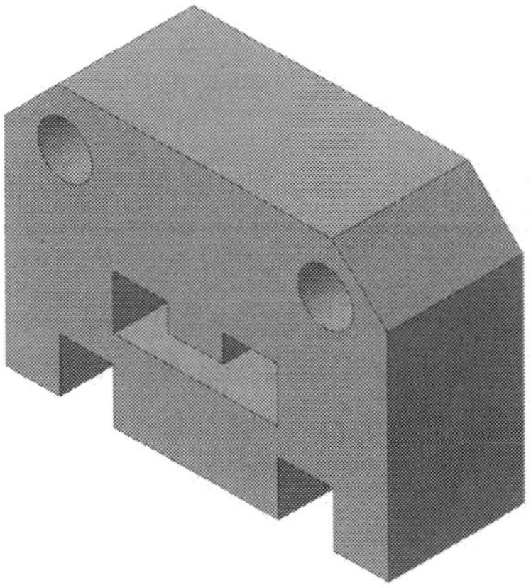

Figure 2-54 *Solid model for Exercise 6*

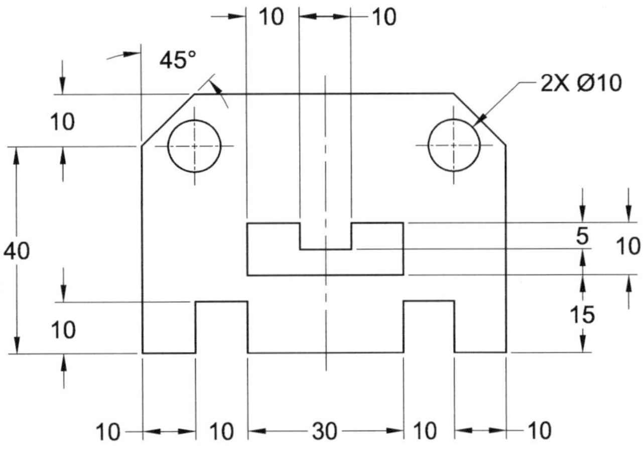

Figure 2-55 *Dimensions of the model*

Answers to Self-Evaluation Test

1. Editing, **2.** dimensions, **3. Geometry**, **Assembly**, **4.** light blue, **5.** *.sec,* **6.** T, **7.** T, **8.** T, **9.** T, **10.** F

Chapter 3

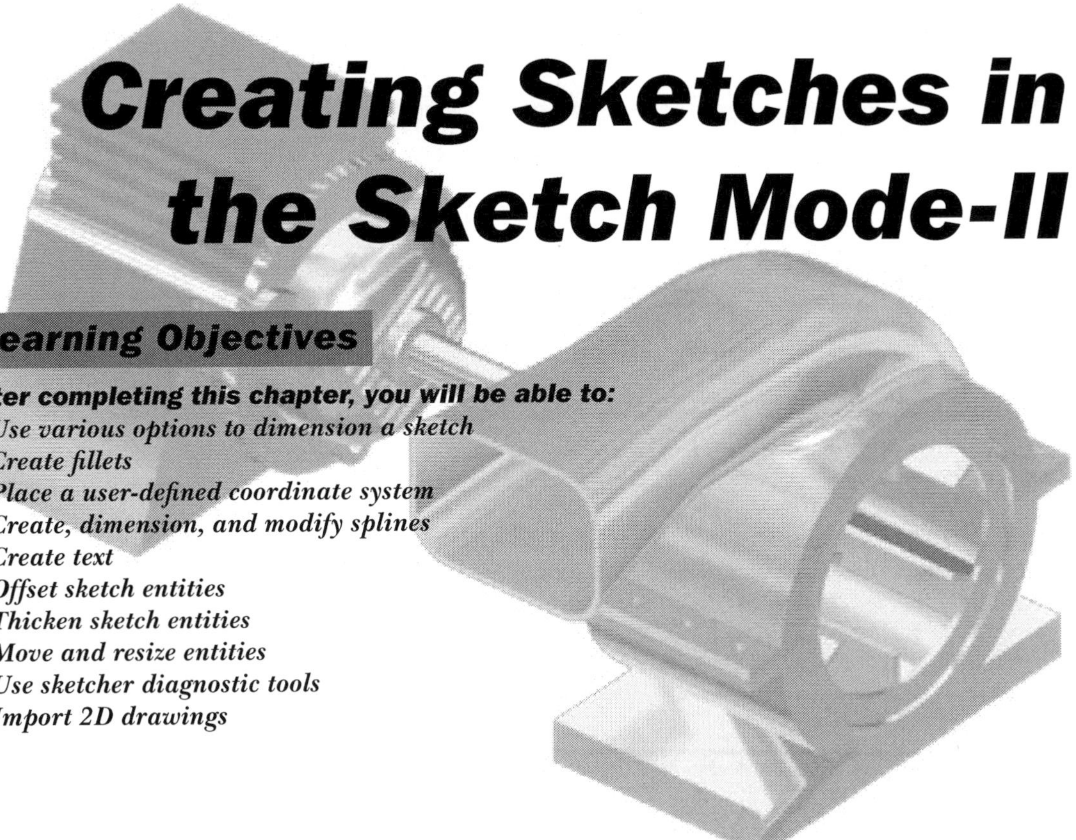

Creating Sketches in the Sketch Mode-II

Learning Objectives

After completing this chapter, you will be able to:

- *Use various options to dimension a sketch*
- *Create fillets*
- *Place a user-defined coordinate system*
- *Create, dimension, and modify splines*
- *Create text*
- *Offset sketch entities*
- *Thicken sketch entities*
- *Move and resize entities*
- *Use sketcher diagnostic tools*
- *Import 2D drawings*

DIMENSIONING THE SKETCH

In Chapter 2, you learned dimensioning a sketch using the **Dimension** tool from the **Dimension** group. In this chapter, you will learn the use of the **Baseline** tool for dimensioning a sketch.

Dimensioning a Sketch Using the Baseline Tool

Ribbon: Sketch > Dimension > Baseline

In Creo Parametric, the **Baseline** tool is used to create dimensions in terms of horizontal and vertical distance values of an entity with respect to a specified baseline. This type of dimensioning in a drawing makes writing a CNC program for manufacturing a component easy.

The **Baseline** tool is available in the **Dimension** group. Using this tool, you can dimension a line, arc, conic, and so on. The following steps explain the procedure to create dimensions using the **Baseline** tool:

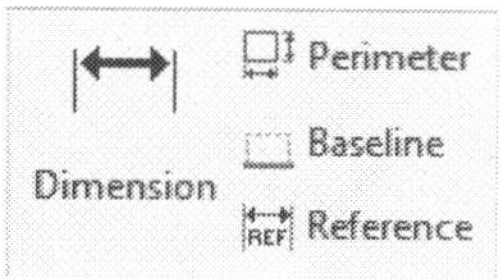

Figure 3-1 *The* ***Dimension*** *group*

1. Choose the **Baseline** tool from the **Dimension** group, refer to Figure 3-1.

2. Select the entity that will act as the baseline (origin or reference). Press the middle mouse button to place the dimension; the dimension **0.00** will be displayed where you place the dimension. Note that since the location value of the baseline is taken as the origin, the dimension value of the baseline entity will become 0.00. The dimension values of all other entities dimensioned with reference to the baseline will be measured from this origin.

   Depending upon the entity selected as the baseline, the horizontal or vertical location values of the entity will be placed. For example, if you select a vertical line, the value of its location will be placed vertically. Similarly, if you select a horizontal line, the value of its location will be placed horizontally.

   You can dimension arcs, circles, and splines by using the two options that are displayed on invoking the **Baseline** tool. If you select the center of a circle or an arc for baseline dimensioning and press the middle mouse button, the **Dim Orientation** dialog box will be displayed, as shown in Figure 3-2. Also, you will be prompted to select the orientation. Select the required radio button from this dialog box and choose the **Accept** button; the dimension will be placed based on the orientation selected.

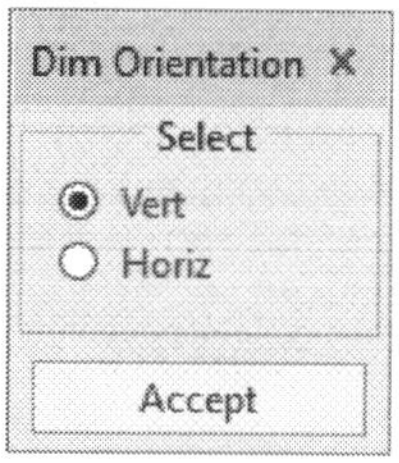

Figure 3-2 *The* ***Dim Orientation*** *dialog box*

3. Next, choose the **Dimension** tool from the **Dimension** group. Select the baseline dimension that was placed earlier and then select the entity to dimension. Now, press the middle mouse button to place the dimension.

   The orientation of the dimension will depend upon the baseline dimension and the entity selected. Figure 3-3 shows a sketch dimensioned using the above-mentioned method. In this figure, the two baselines are dimensioned using the **Baseline** tool. Therefore, the dimensions of these lines are displayed as 0.00. The remaining lines are dimensioned by selecting the baseline dimension and then the required entity by using the **Dimension** tool.

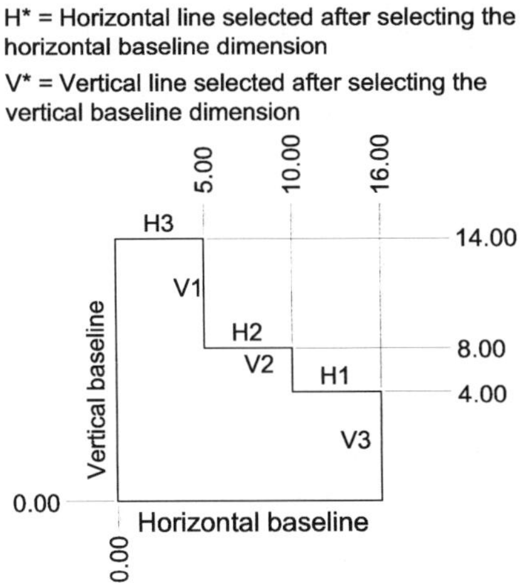

***Figure 3-3** Baseline dimensioning of a sketch*

Replacing the Dimensions of a Sketch Using the Replace Tool

Ribbon: Sketch > Operations > Replace

The **Replace** tool is used to replace a dimension with a new dimension in a sketch. To use this option, you must have a dimensioned sketch. The following steps explain the procedure to dimension a sketch using the **Replace** tool:

1. Select the dimension; a mini popup toolbar appears. Choose the **Replace** tool from the mini popup toolbar; the selected dimension will be deleted and you will be prompted to create a replacement dimension.

2. Select the entities between which you want to create a new dimension; the previous dimension will be replaced by a new dimension.

CREATING FILLETS

In the sketcher environment, you can create the following two types of fillets:

1. Circular fillets
2. Elliptical fillets

Creating Circular Fillets

Ribbon: Sketch > Sketching > Fillet drop-down > Circular

A circular fillet is the arc formed at the intersection of two lines, a line and an arc, or two arcs. This type of fillet is controlled by the radius or diameter dimension of the fillet. The resulting fillet will depend on the location where the elements are selected. Figure 3-4 shows two non-parallel lines and Figure 3-5 shows the circular fillet created between them and construction lines are added that extend to the intersection of both lines. The circular fillet thus created is an arc with its endpoints tangent to the two lines.

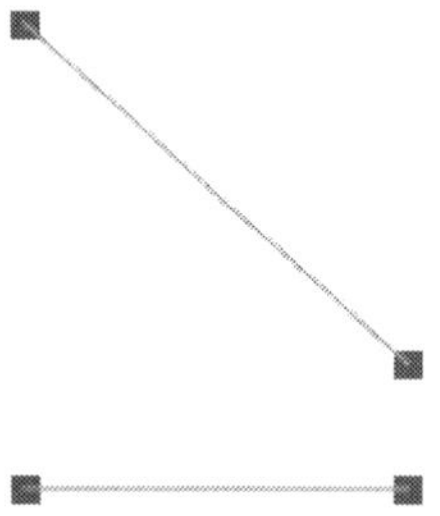

Figure 3-4 *Two lines that do not join*

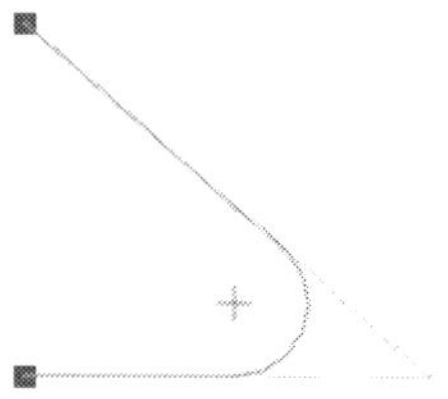

Figure 3-5 *Fillet created between the two lines*

Figure 3-6 shows two lines that join at a point and Figure 3-7 shows the circular fillet created at the joint.

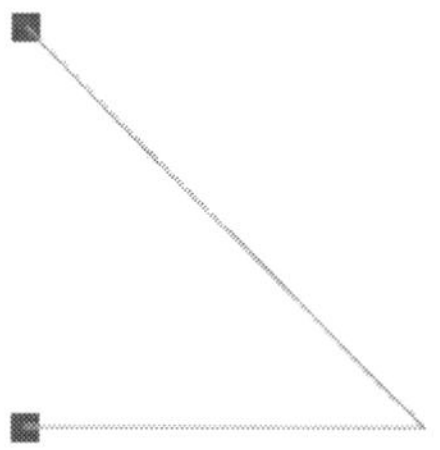

Figure 3-6 *Two lines joining at a point*

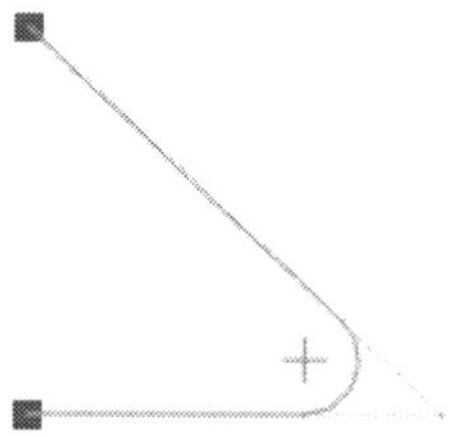

Figure 3-7 *Filleted corner*

Figure 3-8 shows two sets of arcs and Figure 3-9 shows the circular fillet created between the two arcs. The location where you select the arcs to create the fillet is important because the fillet is created tangent to the selection points on the arcs.

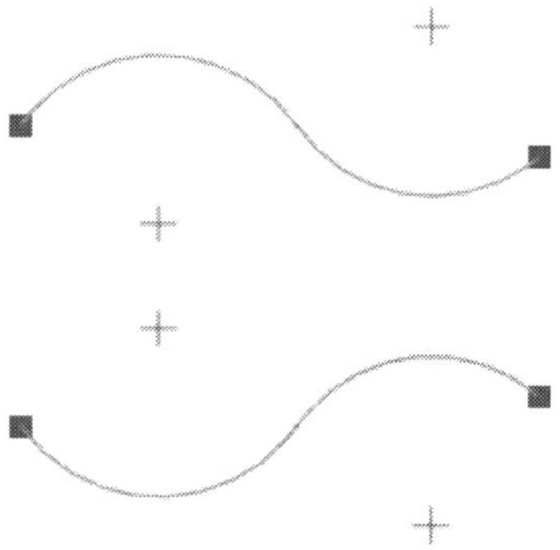

Figure 3-8 *Two sets of arcs*

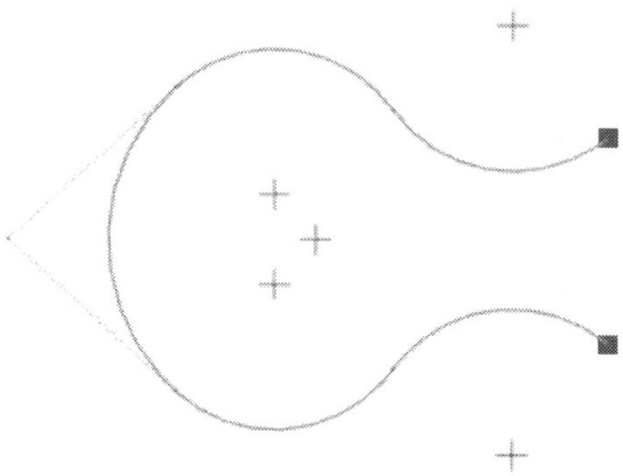

Figure 3-9 *Fillet created by selecting the endpoints*

To create circular fillets, there are two options, **Circular** and **Circular Trim**, available in the **Fillet** drop-down. When you create a fillet using the **Circular** option, the filleted entities are converted into construction geometries. When you create a fillet using the **Circular Trim** option, the portion extending beyond the entities gets trimmed automatically.

The procedure to create circular fillets using both the options is same and is discussed next.

1. Choose the **Circular** or **Circular Trim** option from the **Fillet** drop-down in the **Sketching** group; you will be prompted to select two entities.

2. Select the first entity for filleting by using the left mouse button; the red color of the first entity changes to green. Now, select the second entity. It will be drawn between the two selected entities as soon as you select the second entity.

3. Repeat step 2 until you have created all fillets.

Creating Elliptical Fillets

Ribbon: Sketch > Sketching > Fillet drop-down > Elliptical

An elliptical fillet is the arc in the form of an ellipse that joins two lines, two arcs, or a line and an arc. The geometry of the elliptical fillet depends on the location where you select the entities to create a fillet.

The advantage of elliptical fillets over circular fillets is that the geometry of elliptical fillets can be controlled by dimensions in two directions. Therefore, when an elliptical fillet is dynamically modified, its geometry can be controlled in either the x-direction or the y-direction resulting in more curved geometric shape than a circular fillet.

Figures 3-10 and 3-11 illustrate the elliptical fillet. Notice that a tangent constraint is automatically applied when you create a fillet.

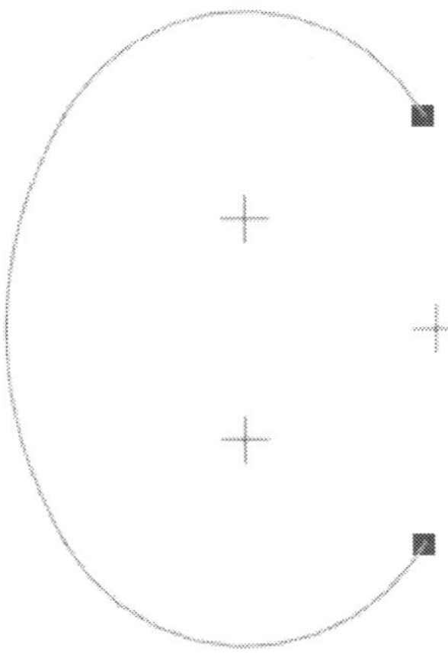

Figure 3-10 *Arcs to be filleted*

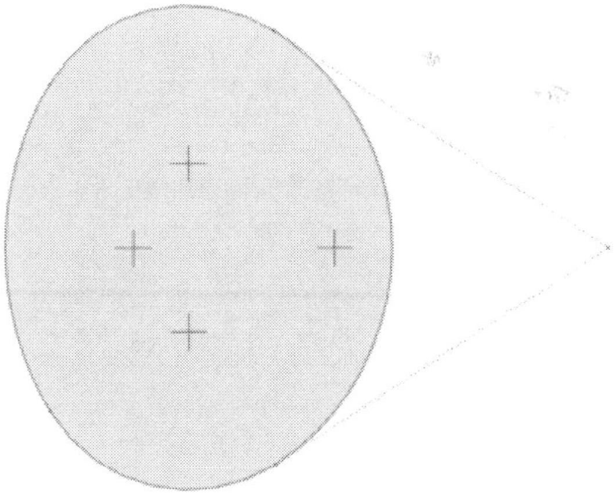

Figure 3-11 *Elliptical fillet created*

Similar to circular fillet, there are two options to create elliptical fillets, **Elliptical** and **Elliptical Trim**. The procedure to create elliptical fillets using any of these options is same. The procedure to create elliptical fillets is discussed next.

1. Choose the **Elliptical** or **Elliptical Trim** tool from the **Fillet** drop-down; you will be prompted to select two entities.

2. Select the first entity by clicking; the color of the entity changes to green.

3. Select the second entity. As soon as you select the second entity, the elliptical fillet is created. The shape of the elliptical fillet depends upon the specified points. After the fillet is created, you will be again prompted to select two entities for elliptical fillet.

4. Repeat steps 2 and 3 until you have created all fillets.

When you select an elliptical fillet to dimension, the **Ellipse Rad** dialog box will be displayed, as shown in Figure 3-12. There are two radio buttons in this dialog box. When the **Major Axis** radio button is selected, the elliptical fillet will be dimensioned radially along the X-direction. The **Minor Axis** radio button, when selected, dimensions the elliptical fillet radially in the Y-direction.

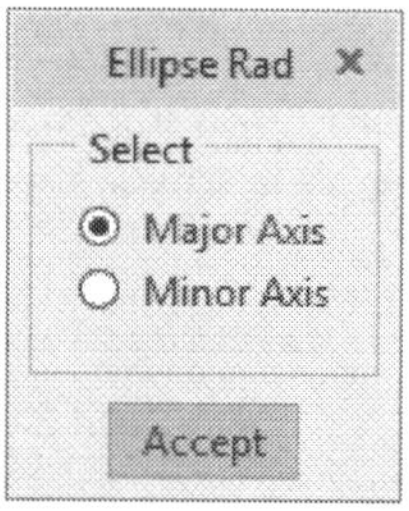

***Figure 3-12** The **Ellipse Rad** dialog box*

CREATING A REFERENCE COORDINATE SYSTEM

There are two types of coordinate systems, construction and geometric. As discussed earlier, the construction entities cannot be referenced outside the Sketcher environment whereas, the geometric entities can be. The **Coordinate System** tool in the **Datum** group is used to create a geometric coordinate system that will act as a reference for dimensioning. You can dimension the splines using the coordinate system. Thus, it provides you the flexibility to modify the spline points by specifying different coordinates with respect to the coordinate system.

The user-defined coordinate system is used in blend features to align different sections in a blend. It is also used in the **Assembly** and **Manufacturing** modes of Creo Parametric.

The following steps explain the procedure to create a coordinate system:

1. Choose the **Coordinate System** tool from the **Datum** group; you will be prompted to select the location for the coordinate system. The coordinate system symbol is attached to the cursor.

2. Place the coordinate system at the desired points on the screen by clicking the left mouse button. The coordinate system will be placed at as many places as you click in the graphics window. You can end the coordinate system creation by using the middle mouse button.

Note

If you add a coordinate system to a sketch, it must be dimensioned. But if the coordinate system is placed at the endpoints of a line, an arc, a spline, or at the center of an arc or a circle, it need not be dimensioned. In other words, a coordinate system must be referenced to an entity in a sketch.

WORKING WITH SPLINES

Splines are curved entities that pass through a number of intermediate points. Generally, splines are used to define the outer surface of a model. This is because the splines can provide different shape to curves and the flexibility to modify the surfaces that result from the splines. The application of splines is widely found in vehicle body designing.

Creating a Spline

Ribbon: Sketch > Sketching > Spline

To draw a spline, choose the **Spline** tool from the **Sketching** group. The steps to create a spline are discussed next.

1. Choose the **Spline** tool from the **Sketching** group; you will be prompted to select the location for spline.

2. Use the left mouse button to select the start point for the spline. Similarly, select additional points in the graphics window; a spline will be drawn passing through all specified points. Press the middle mouse button to end the creation of spline. All points through which the spline passes are called interpolation points.

Dimensioning of Splines

When a spline is drawn, the weak dimensions are automatically applied to the spline. A spline can be dimensioned manually by:

1. Dimensioning the endpoints
2. Radius of curvature dimensioning
3. Tangency dimensioning
4. Coordinate dimensioning
5. Dimensioning the interpolation points

Dimensioning the Endpoints

To dimension a spline by selecting the endpoints, you need to follow the steps given below:

1. Choose the **Dimension** tool from the **Dimension** group.

2. Select the two endpoints of the spline and place the horizontal or vertical dimension by pressing the middle mouse button twice. Figure 3-13 shows a spline that is dimensioned by selecting the endpoints.

Radius of Curvature Dimensioning

The radius of curvature of a spline can be dimensioned only if its tangency is defined. In other words, radius of curvature of a spline can be dimensioned only if the spline is tangent to an entity. For dimensioning the radius of curvature of a spline, you need to follow the steps given below:

1. Choose the **Dimension** tool from the **Dimension** group.

2. Select the endpoint of the spline where the tangency is defined.

3. Press the middle mouse button to place the dimension. Figure 3-14 shows the radius of curvature dimensioning of a spline.

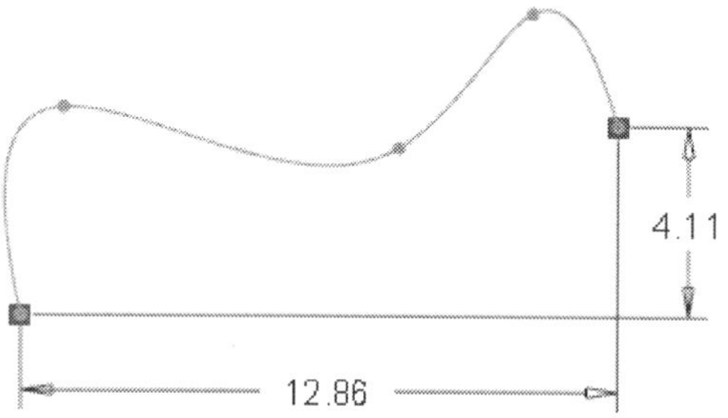

Figure 3-13 Endpoint dimensioning

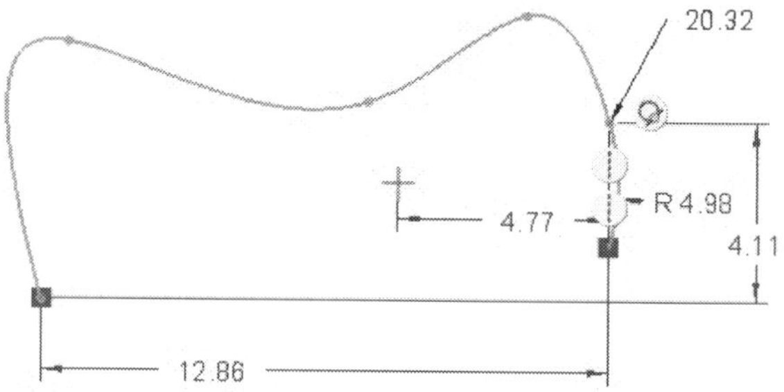

Figure 3-14 Radius of curvature dimensioning

Tangency or Angular Dimensioning

A spline can be dimensioned angularly with respect to a line tangent to it. This type of dimensioning is also called angular dimensioning. To angular dimension a spline and a line tangent to it, you need to follow the steps given below:

1. Choose the **Dimension** tool from the **Dimension** group.
2. Select the spline by clicking the left mouse button.
3. Select the entity tangent to the spline by clicking the left mouse button.
4. Select the interpolation point of the spline that is to be dimensioned tangentially.
5. Press the middle mouse button to place the dimension.

Coordinate Dimensioning

The spline can be dimensioned with respect to a user-defined coordinate system. Choose the **Coordinate System** tool from the **Sketching** group. On doing so, the coordinate system is attached to the cursor. Place the coordinate system in the graphics window. Now, the spline can be dimensioned with respect to the coordinate system.

Dimensioning the Interpolation Points

A spline can be dimensioned by dimensioning its interpolation points or vertices. This type of dimensioning is used when the designer wants the spline to be standard for all designs. This is because the exact curve can be duplicated if the interpolation points or the vertices of a spline are dimensioned.

Tip

A dimension can be moved by pressing and holding the left mouse button on the dimension and moving it. The dimension text is replaced by a green colored text. You can drag the dimension to the desired location in the graphics window and release the left mouse button to place the dimension at that point.

Modifying a Spline

In Creo Parametric, a spline can be modified by:

1. Moving the interpolation points of the spline.
2. Adding points to a spline.
3. Deleting points of a spline.
4. Creating a control polygon and moving its control points.
5. Modifying the dimensions of the spline.

Moving the Points of a Spline

The position of the interpolation points can be dynamically modified. To modify a spline, select an interpolation point on the spline and then drag it to modify the shape of the spline.

Alternatively, select the spline and choose the **Modify** option from the mini popup toolbar or double-click on the spline; the **Spline** tab will be displayed in the **Ribbon** with the options and buttons to modify a spline.

The interpolation points of the spline appear in black crossmarks in the graphics window. Drag the interpolation points to modify the shape of the spline.

Adding Interpolation Points to a Spline

To add interpolation points to a spline, double-click on the spline to invoke the **Spline** tab. Next, right-click on the spline and choose the **Add Point** option from the shortcut menu; a point is added to the spline where the spline was selected. The new point appears in black crossmark. Remember that, you cannot increase the length of the spline by adding points before the start point and after the endpoint of the spline.

Deleting Interpolation Points of a Spline

To delete a point of a spline, double-click on it. Next, right-click on the point to be removed to invoke the shortcut menu. Choose the **Delete Point** option from the shortcut menu; the selected point is deleted. You can continue deleting vertices or points from a spline until only two end points are left in the spline.

Creating a Control Polygon and Moving its Control Points

When you draw a spline, it is associated with a control frame. The vertices of this frame are called control points. To create a control polygon, choose the **Modify spline using control points** button from the **Settings** area of the **Spline** tab. The control polygon will be displayed in the graphics window. If you have already added a strong dimension(s) to the spline, then on choosing the **Modify spline using control points** button in the **Spline** tab, the **Modify Spline** confirmation box will be displayed, as shown in Figure 3-15. Next, choose **Yes** to display the control polygon. The spline shape can be modified by dragging the control points on the polygon.

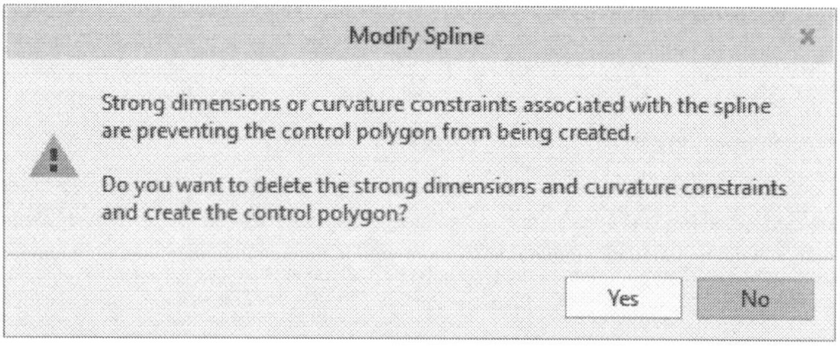

*Figure 3-15 The **Modify Spline** confirmation box*

Modifying the Dimensions of the Spline

The shape of the spline is controlled by the position of its interpolation points. Hence by modifying the dimensions, the position of the interpolation points are changed, which results in modification of the shape of the spline.

WRITING TEXT IN THE SKETCHER ENVIRONMENT

Ribbon: Sketch > Sketching > Text

In the sketcher environment, the text is written using the **Text** tool from the **Sketching** group. The following steps explain the procedure to write text in the sketcher environment:

1. Choose the **Text** tool from the **Sketching** group; you will be prompted to select the start point of line to determine the text height and orientation.

2. Specify the start point on the screen by left-clicking; you will be prompted to select the second point of line to determine the text height and orientation. Note that to write the text upright, the second point should be above the start point and in a straight line. If the second point is below the start point, the text will be written downward from right to left.

3. Specify the second point on the screen by left-clicking; the **Text** dialog box will be displayed, as shown in Figure 3-16 and a construction line will be drawn having height equal to the distance between the two points. The height and orientation of the text depends on the height and angle of the construction line. If the construction line is drawn at an angle, then the text will be written at that angle. By default, the dimension of the construction line is a strong dimension.

4. Enter the text in the **Text** edit box. The length of the text can be up to 79 characters. As you enter the text, the text will be displayed dynamically in the graphics window. To insert symbols with the text, choose the button next to the **Text** edit box; the **Text Symbol** palette will be displayed, as shown in Figure 3-17. Next, choose the desired symbol from the palette to insert it with the text. You can choose the desired font of the text from the **Font** drop-down list. The aspect ratio, slant angle, and spacing of the text can be controlled by using the slider bars available in the **Options** area. You can also modify the position of the text using the options available in the **Alignment** area of the dialog box.

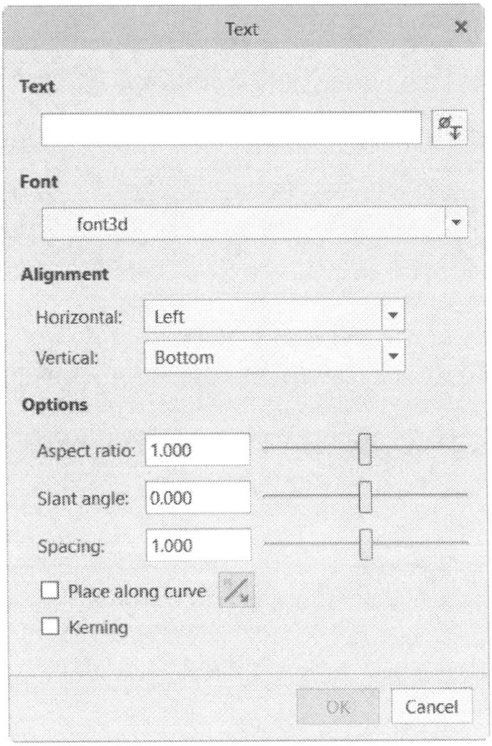

***Figure 3-16** The **Text** dialog box*

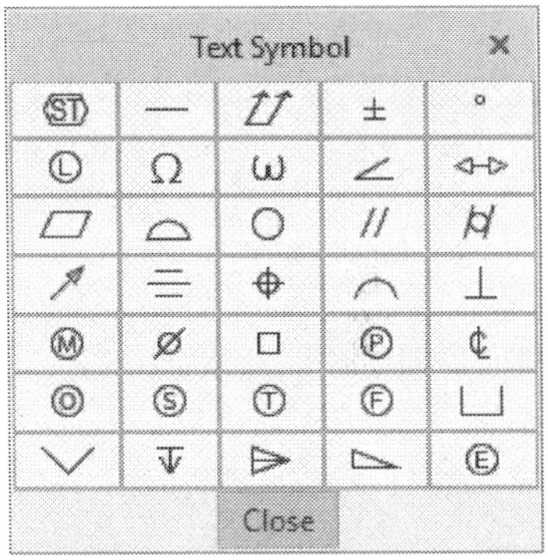

***Figure 3-17** The **Text Symbol** palette*

5. If you want to place the text along a curve, then first select the **Place along curve** check box and then select the curve from the drawing area; the text will be placed along the selected curve. You can flip the text to the other side of the selected curve by choosing the **Flip text to other side of curve** button. This button will get activated on selecting the **Place along curve** check box.

6. Choose the **OK** button in the **Text** dialog box to exit it.

Tip
You can also change the height and width of the text by choosing the Width A and Height AI options available in the mini popup toolbar.

OFFSETTING SKETCHED ENTITIES

Ribbon: Sketch > Sketching > Offset

In the **Sketch** mode, the entities in a sketch can be offset by a specified distance. You can create offset entities of lines, arcs, or splines by using the **Offset** tool available in the **Sketching** group. The offset can be a positive or a negative value. The options available in this tool can be used to create offset of a single entity, a chain of two or more entities, or a loop of two or more entities. The following steps explain the procedure to create offset sketched entities:

1. Choose the **Offset** tool; the **Type** dialog box will be displayed with the **Single** radio button selected by default, as shown in Figure 3-18. To create an offset of a single sketched entity, choose the **Single** radio button and then select the entity to be offset; the **Enter offset in the direction of the arrow [Quit]** edit box appears on the top of the drawing area. Also, an arrow is displayed on the selected entity. This arrow shows the direction in which the resulting entity will be created.

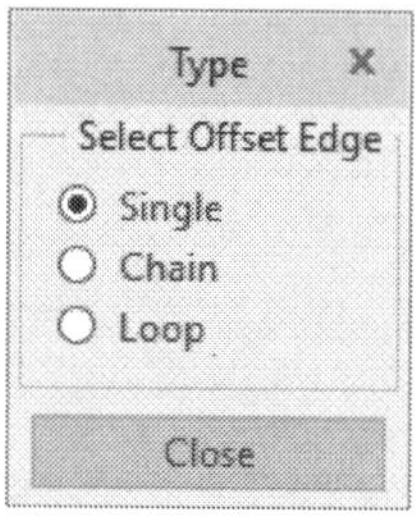

Figure 3-18 *The* ***Type*** *dialog box*

2. Enter the offset value in the **Enter offset in the direction of the arrow [Quit]** edit box and choose the **OK** button. Depending upon the value specified, a copy of the selected geometry will be created in the drawing area. Note that, if you specify a positive value, the entity will get offset in the direction of arrow, refer to Figure 3-19. If you specify a negative value in the edit box, the offset geometry will be created in a direction opposite to the direction of arrow, refer to Figure 3-20.

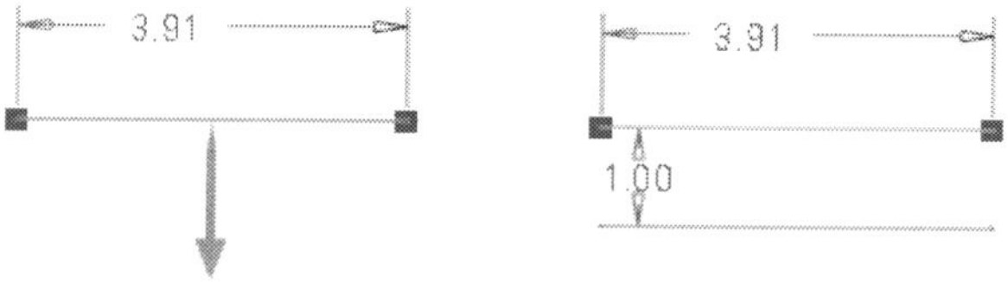

Figure 3-19 *Offset created with a positive value*

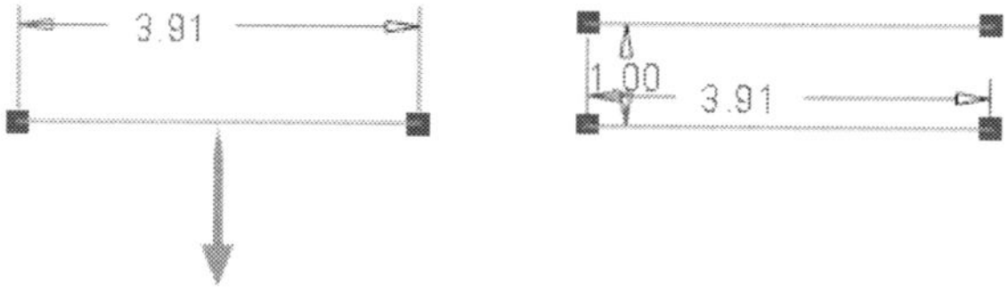

Figure 3-20 *Offset created with a negative value*

3. To create offset of a chain of two or more entities, select the **Chain** radio button from the **Type** dialog box and then select the entities in chain. Specify the value of offset in the **Enter offset in the direction of the arrow [Quit]** edit box and choose the **OK** button to create the offset of the selected entities, refer to Figure 3-21.

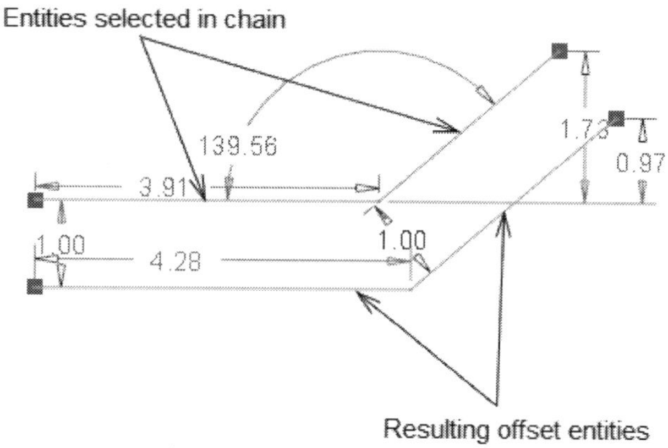

Figure 3-21 *Offset created in chain mode*

4. You can create offset of a closed loop generated by two or more entities. To do so, select the **Loop** radio button from the **Type** dialog box and then select an entity of the loop; the **Enter offset in the direction of the arrow [Quit]** edit box will be displayed above the drawing area. Specify the offset value in this edit box and choose the **OK** button; the offset of the selected loop will be created, refer to Figure 3-22.

5. If you select two chained entities of a loop with the **Chain** radio button selected; the **Menu Manager** will be displayed, refer to Figure 3-23. Choose the **Next** or **Previous** option to highlight the entities creating the loop. Next, choose the **Accept** option from the **Menu Manager** to select the highlighted entities and specify the offset value in the **Enter offset in the direction of the arrow [Quit]** edit box to create the offset of selected entities.

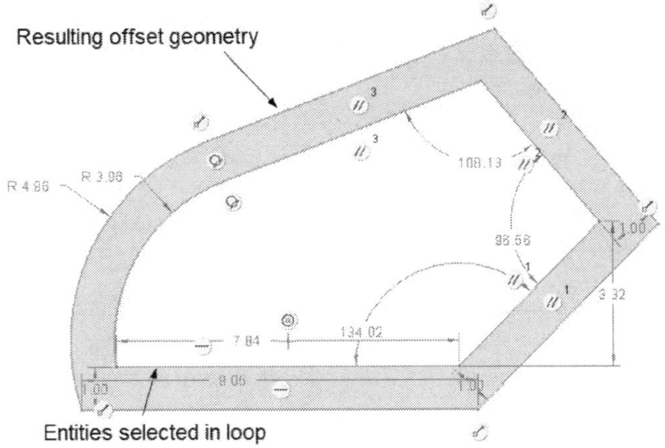

Figure 3-22 *Offset created in loop mode*

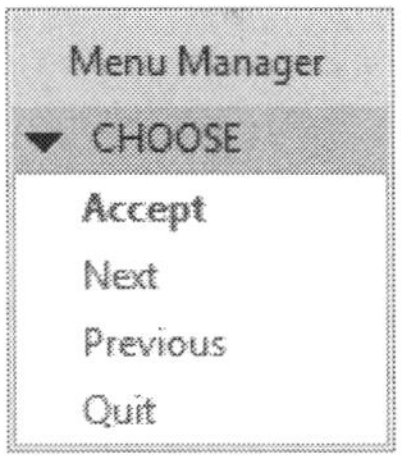

Figure 3-23 *The **Menu Manager***

6. After you have created offset sketched entities, press the middle mouse button or choose the **Close** button in the **Type** dialog box to complete the procedure.

Note

If you offset entities by a large dimension, it is possible that Creo Parametric creates an offset chain that might have a different number of entities. For example, offsetting a spline by a large value causes the resulting spline to be broken into several pieces. If the offset value is changed, the system can piece together the broken spline so it becomes a single entity again.

THICKENING SKETCHED ENTITIES

Ribbon: Sketch > Sketching > Thicken

A thickened entity is a bidirectional offset entity whose components are separated by a user-defined distance. You can also add flat or circular end caps to connect the two offset entities. Depending on the offset and thickness values you enter, the resulting thickened entities can be on the either side, or both sides of the parent geometry. The following steps explain the procedure to create thickened sketched entities:

1. Choose the **Thicken** tool from the **Sketching** group of the **Sketch** tab; the **Type** dialog box is displayed. By default, the **Single** radio button is selected in the **Select Thicken Edge** area and the **Open** radio button is selected in the **End caps** area of the **Type** dialog box, as shown in Figure 3-24.

2. To thicken a single sketched entity, choose the **Single** radio button; to thicken a chain of two or more entities, choose the **Chain** radio button; and to thicken a closed loop generated by two or more entities, choose the **Loop** radio button in the **Select Thicken Edge** area of the **Type** dialog box.

3. You can create entities with flat or circular end caps by selecting the options in the **End caps** area of the **Type** dialog box, refer to Figure 3-24. To do so, choose the desired radio button from the **End caps** area of the **Type** dialog box before clicking on the entities. Next, follow the step 2 and step 3 to create geometries with flat or circular end caps, as shown in Figure 3-25.

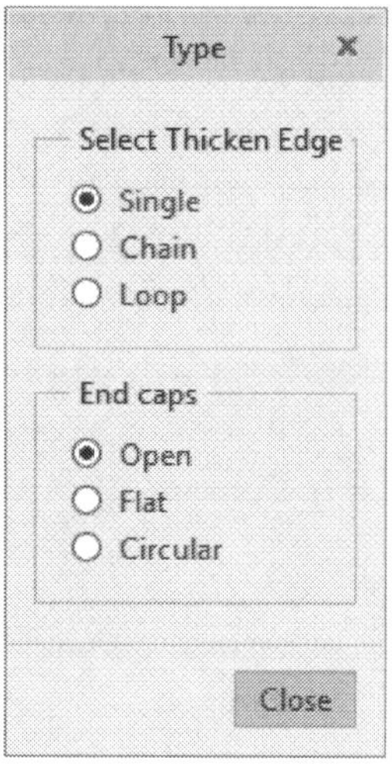

Figure 3-24 *The **Type** dialog box of the **Thicken** tool*

4. Select the entity(-ies) to thicken; the **Enter thickness [-Quit-]:**edit box is displayed on the top of the drawing area. Specify the value of thickness in the edit box and choose the **OK** button; an arrow is displayed on the entity and the **Enter offset in the direction of the arrow [Quit]** edit box is displayed. Specify the offset distance in the edit box and choose the **OK** button; the selected entity(ies) will get thicken. Next, press the middle mouse button or choose the **Close** button from the **Type** dialog box to close the **Type** dialog box.

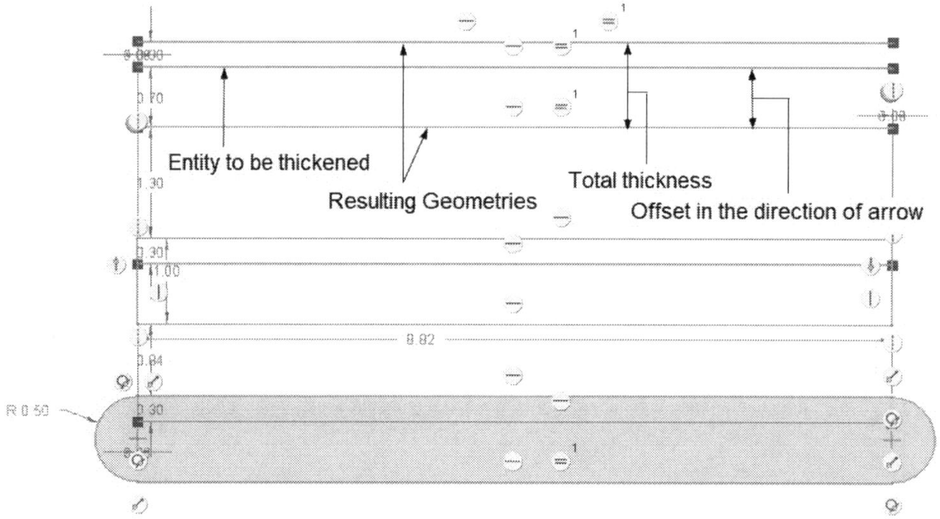

Figure 3-25 *Entities created using the **Open**, **Flat**, and **Circular** options of the **Thicken** tool*

ROTATING AND RESIZING ENTITIES

Ribbon: Sketch > Editing > Rotate Resize

The sketches can be translated, scaled, or rotated by using the **Rotate Resize** tool available in the **Editing** group. Select a sketch and then choose the **Rotate Resize** tool from the **Editing** group. On choosing this tool, the sketch, which consists of various entities, will act as a single entity. Also, the sketch appears green in color and is enclosed within a boundary box, as shown in Figure 3-26.

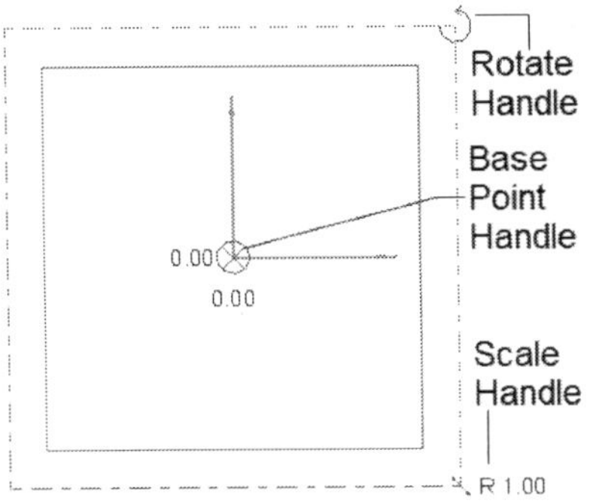

Figure 3-26 *Selected entities with move, rotate, and resize handles*

There are three handles that facilitate in scaling, rotating, and moving the selected sketch. The rotate handle is used to dynamically rotate the selected entities. The scale handle is used to dynamically scale the selected entities. The base point handle is used to pick the sketch and place it at any other location in the graphics window. To change the location of any of the three handles, click on the handle and drag it by pressing and holding the left mouse button; the selected handle will move along with the cursor. Place the symbol at the desired location. The following steps explain the procedure to move and resize a sketch:

1. Select the sketch to be rotated and scaled, and then choose the **Rotate Resize** tool; the **Rotate Resize** tab will be activated in the **Ribbon**. The options in these areas are used to move, scale, and rotate the sketch dynamically or by entering a value in the respective edit boxes. You can also select a reference about which you want to translate, rotate, or scale the sketch.

2. To move the sketch dynamically, select the move handle and then drag the handle to the required location; the sketch will be repositioned at the new location. You can also select a reference from the drawing area about which you want to translate, rotate, or scale the sketch and the reference will be displayed in move collector of the dashboard.

3. To dynamically rotate the sketch, select the rotate handle and then move the cursor; the sketch will rotate as you move the cursor. You can also enter the rotation angle in the **Rotate** edit box.

4. To scale the sketch, select the scale handle and then move the cursor. As you move the cursor, the sketch is scaled dynamically in the graphics window. You can also enter the scale value in the **Scale** edit box.

5. After the sketch has been moved and resized, choose the **OK** button from the **Rotate Resize** tab.

SKETCH DIAGNOSTIC TOOLS

As discussed earlier that a sketch-based feature uses a sketch to create desired geometry. Hence it is very necessary that the sketch should fulfill the requirements of a sketch-based feature. Creo Parametric provides you tools to diagnose and analyze the requirements of the sketch in the **Sketch** mode. The tools available in the **Inspect** group of the **Sketch** mode are used to diagnose the sketches. These tools are discussed next.

Feature Requirements

Ribbon: Sketch > Inspect > Feature Requirements

The **Feature Requirements** tool will be available only when you invoke the sketcher environment in the **Part** mode. It helps you to analyze whether the sketch is appropriate for the feature or not. This tool is very useful in creating internal sketches. In the sketcher environment of the **Part** mode, choose the **Feature Requirements** tool from the **Inspect** group of the **Sketch** mode to invoke the **Feature Requirements** dialog box, refer to Figure 3-27. The **Feature Requirement** dialog box lists the requirements of the current feature in the **Requirements** column and status of the feature in the **Status** column. The information displayed in this dialog box varies from feature to feature.

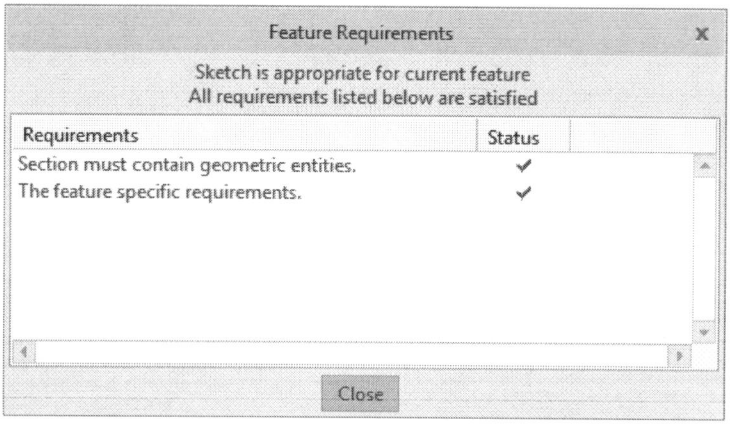

*Figure 3-27 The **Feature Requirements** dialog box*

Overlapping Geometry

Ribbon: Sketch > Inspect > Overlapping Geometry

This tool helps you to detect and highlight overlapping geometries in an active sketch. When you choose the **Overlapping Geometry** tool, the entities that are overlapping with each other will be highlighted in merlot color.

Highlight Open Ends

Ribbon: Sketch > Inspect > Highlight Open Ends

This tool is activated by default in the **Sketch** mode. The **Highlight Open Ends** tool highlights the end points of entities that do not coincide with end points of other entities. In other words, only the open ends of entities are highlighted when this tool is turned on.

Shade Closed Loops

Ribbon: Sketch > Inspect > Shade Closed Loops

This tool is used to detect closed loops formed by sketched entities. A closed loop is a chain of entities that forms a section and can be used for creating a solid feature. In the **Sketch** mode, the **Shade closed Loops** is activated by default. As a result, only the closed loops created with valid entities appear shaded in the drawing area. If you are creating a sketch with several closed loops one inside the other, the outermost loop will be shaded, and the shading of the internal loops will be alternated.

Tip

Sketch diagnostic tools assist you in creating sketches. It is recommended that you do not turn off the ***Shade Closed Loops*** *and* ***Highlight Open Ends*** *buttons while you are creating sketches for solid features.*

IMPORTING 2D DRAWINGS IN THE SKETCH MODE

Ribbon: Sketch > Get Data > File System

The two-dimensional (2D) drawings when opened in the sketcher environment can be saved in the *.sec* format. The *.sec* file can be used to create a solid model. You can use a prestored sketch by importing it in the modeling environment. The **File System** button in the **Get Data** group is used to import 2D sketches from other type of files stored in your computer. Using this button, you can save time in drawing the same or similar section again. The file formats from which the data can be imported are shown in Figure 3-28.

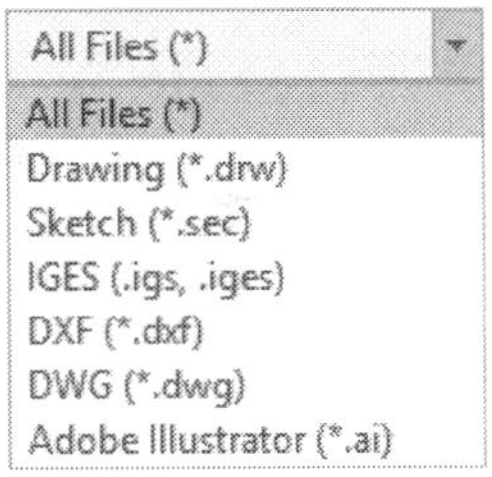

Figure 3-28 *File formats*

When you choose the **File System** button from the **Get Data** group; the **Open** dialog box will be displayed, as shown in Figure 3-29. You can use this dialog box to select and open the file.

When you select a drawing file created in the **Drawing** mode of Creo Parametric, the draft entities of that file will be imported and selected drawing will be opened in a window. Also, you will be prompted to select the entities to copy from the window. Select the draft entities and then press the middle mouse button; the window disappears and a plus sign gets attached to the cursor indicating that you need to select a point on the Creo Parametric screen to insert the file. Select a point on the screen; the selected entities get inserted and are displayed within an enclosed boundary. Also, the **Import/Drawing** tab will be invoked. You can use this tab to set the position, scale, and orientation of the imported sketch. Note that if the *.drw* file does not consist of draft entities, no data will be imported.

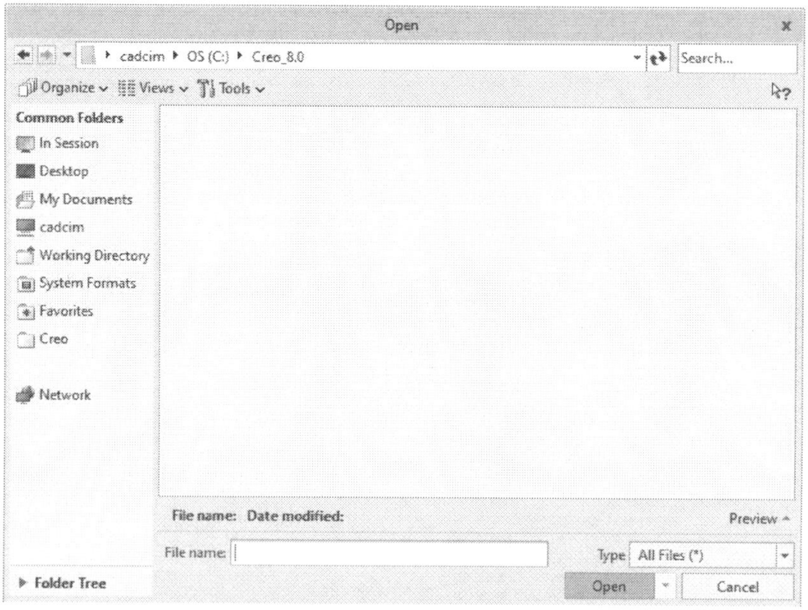

Figure 3-29 *The **Open** dialog box*

The section imported using the **File System** button in the current sketch is an independent copy. The imported section will be no longer associated with the source section. The units, dimensions, grid parameters, and accuracy are acquired from the current sketch.

TUTORIALS

Tutorial 1

In this tutorial, you will import an existing sketch that you had drawn in Tutorial 3 of Chapter 2. After placing the sketch, draw the keyway, as shown in Figure 3-30.

(Expected time: 15 min)

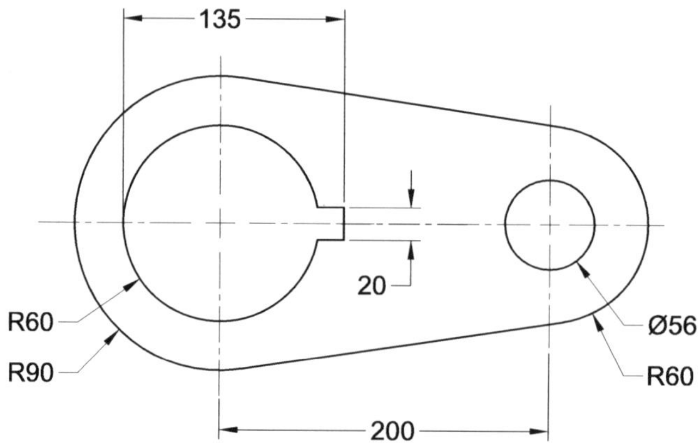

Figure 3-30 *Sketch for Tutorial 1*

The following steps are required to complete this tutorial:

a. Start Creo Parametric.
b. Set the working directory and create a new object file.
c. Import the section by using the **File System** button.
d. Draw the keyway and dimension it.
e. Modify the dimensions.
f. Save the sketch and exit the sketcher environment.

Starting Creo Parametric

1. Start Creo Parametric by double-clicking on the Creo Parametric icon on the desktop of your computer.

Setting the Working Directory

When the Creo Parametric session starts, the first task is to set the working directory. As mentioned earlier, working directory is a directory on your system where you can save the work done in the current session of Creo Parametric. You can set any existing directory on your system as the working directory. Since this is the first tutorial of this chapter, you need to create a folder named *c03* in the *C:\Creo_8.0* folder.

1. Choose **Manage Session > Select Working Directory** option from the **File** menu; the **Select Working Directory** dialog box is displayed.

2. Select *C:>Creo_8.0*. If this folder does not exist, then first create it and then set the working directory.

   Alternatively, you can use the **Folder Tree** available on the bottom left corner of the screen to set the working directory. To do so, click on the **Folder Tree** node; the Folder Tree will expand. In the Folder Tree, browse to the desired location using the nodes corresponding to the folders and select the required folder. After selecting the folder, the **Folder Content** window will be displayed. Close this window and the selected folder will become your current working directory.

3. Choose the **Organize** button from the **Select Working Directory** dialog box or right-click in this dialog box to display a shortcut menu. From the shortcut menu, choose the **New Folder** option; the **New Folder** dialog box is displayed.

4. Enter **c03** in the **New directory** edit box of the **New Folder** dialog box and then choose the **OK** button; a folder with the name *c03* is created at *C:\Creo_8.0*.

5. Choose the **OK** button from the **Select Working Directory** dialog box; *C:\Creo_8.0\c03* is set as the working directory.

Starting a New Object File

1. Choose the **New** button from the **Data** group; the **New** dialog box is displayed. Select the **Sketch** radio button from the **Type** area of the **New** dialog box; the default name of the sketch appears in the **File name** edit box.

2. Enter *c03tut1* in the **File name** edit box and choose the **OK** button.

You are in the sketcher environment of the **Sketch** mode. When you enter the sketcher environment, the Navigator is displayed on the left in the graphics window. Slide-in the Navigator by clicking on the **Show Navigator** button present on the bottom left corner of the drawing area. Now, the drawing area is increased.

Importing the Section

1. Choose **Get Data > File System** from the **Ribbon**; the **Open** dialog box is displayed with the working directory as the current directory.

2. Click on the black arrow beside the **Creo_8.0** option in the address bar and choose **c02** from the flyout displayed. Make sure the **Sketch (*.sec)** option is selected in the **Type** drop-down list. Select *c02tut3.sec* and choose the **Open** button from the **Open** dialog box.

3. Move the cursor in the drawing area. Notice that the cursor is attached with a plus sign. Now, click anywhere in the drawing area to place the sketch. The sketch is displayed in the drawing area and the **Import Section** tab is displayed in the **Ribbon**.

4. Enter **1** as a scale factor in the scale edit box and choose the **OK** button to complete importing the sketch.

5. Choose the **Refit** button from the **Graphics** toolbar. The sketch, similar to the one shown in Figure 3-31, is displayed in the drawing area.

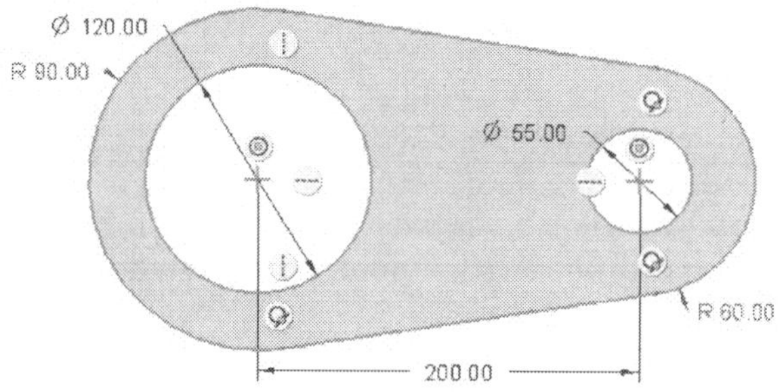

Figure 3-31 *Sketch imported and placed in the current file*

Drawing the Keyway

To create the keyway, you need to sketch a small rectangle and then remove the portion of the circle that lies between the rectangle.

1. Choose the **Line Chain** tool from the **Line** drop-down in the **Sketching** group.

2. Draw the keyway, as shown in Figure 3-32; the weak dimensions and constraints are automatically applied to the sketch of the keyway.

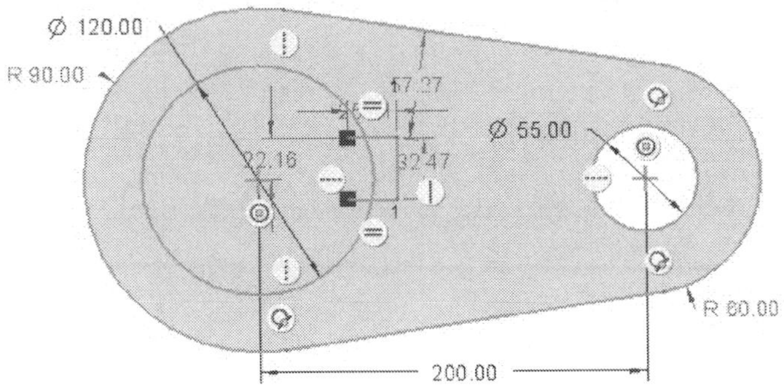

Figure 3-32 *Sketch of the keyway with weak dimensions and constraints*

The horizontal lines of the keyway and the circle intersect at the points where the lines meet the circle. The portion of the circle that lies between the two horizontal lines of the keyway needs to be deleted from the circle.

3. Choose the **Zoom In** button from the **Orientation** group of the **View** tab; the cursor is converted into a magnifying glass symbol. 

4. Draw a window around the keyway to zoom in it. Now, the display of the keyway is enlarged.

5. Choose the **Delete Segment** tool from the **Editing** group. 

6. Click to select the part of the circle that lies between the two horizontal lines; the selected part is deleted.

7. Choose the **Refit** button from the **Graphics** toolbar to view the full sketch. 

Dimensioning the Keyway

Now, you need to apply dimensions to the keyway.

1. Choose the **Dimension** tool from the **Dimension** group. 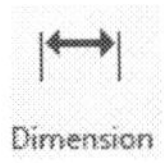

2. Dimension the keyway, as shown in Figure 3-33.

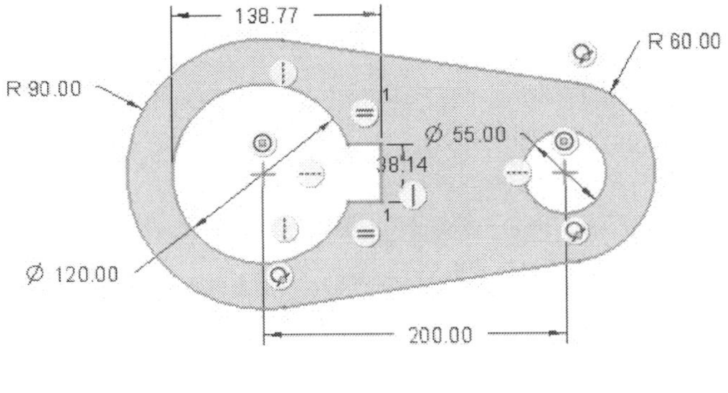

Figure 3-33 *Sketch after dimensioning the keyway*

Modifying the Dimensions

The dimensions of the keyway need to be modified as per the given dimension values.

1. Select the two dimensions of the keyway by pressing CTRL+left mouse button.

2. Choose the **Modify** tool from the **Editing** group; the **Modify Dimensions** dialog box is displayed.

3. Clear the **Regenerate** check box and then modify the dimensions of the keyway. When you clear the check box, the sketch does not regenerate as you modify the dimensions.

   The dimension that you edit in the **Modify Dimensions** dialog box gets enclosed in a violet box in the sketch.

4. Modify all dimensions. Refer to Figure 3-30 for dimension values.

5. After the dimensions are modified, choose the **OK** button from the **Modify Dimensions** dialog box; the message **Dimension modifications successfully completed** is displayed in the message area.

   The sketch after modifying the dimension values of the sketch is shown in Figure 3-34.

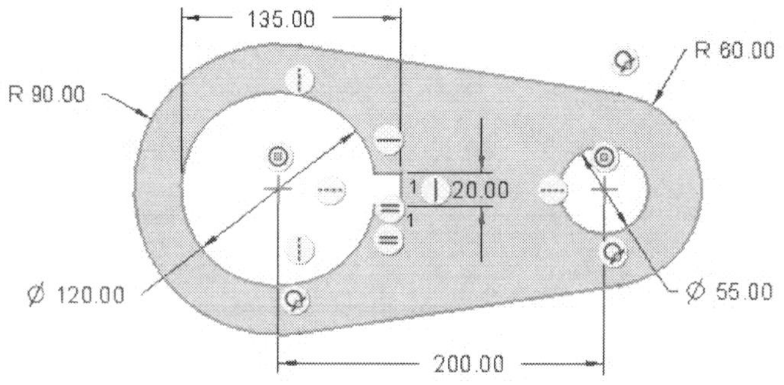

Figure 3-34 *Sketch after modifying the dimensions*

Saving the Sketch

As you may need the sketch later, you must save it.

1. Choose the **Save** button from the **File** menu; the **Save Object** dialog box is displayed with the name of the sketch entered earlier.

2. Choose the **OK** button; the sketch is saved.

3. After saving the sketch, choose the **Close** button from the **Quick Access** toolbar to exit the **Sketch** mode.

Tutorial 2

In this tutorial, you will draw the sketch for the model shown in Figure 3-35. The sketch is shown in Figure 3-36. **(Expected time: 30 min)**

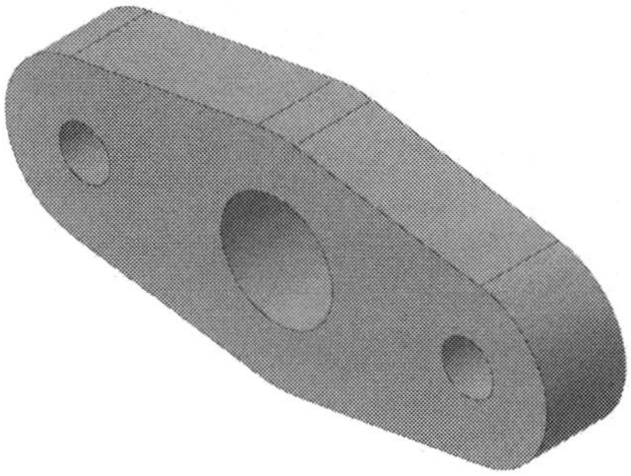

Figure 3-35 Model for Tutorial 2

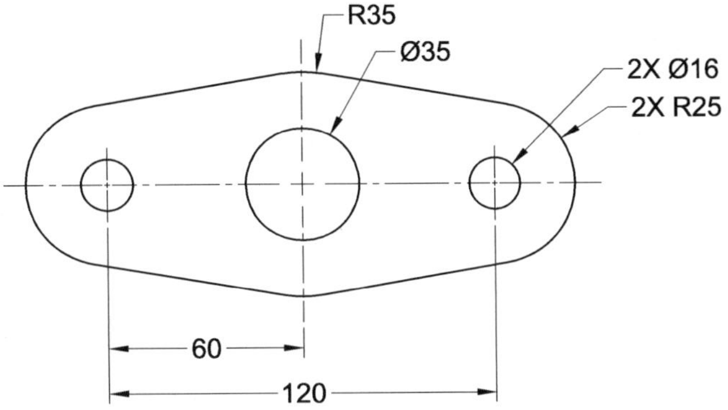

Figure 3-36 Sketch of the model

The following steps are required to complete this tutorial:

a. Set the working directory and create a new object file.
b. Draw the sketch using sketcher tools.
c. Apply the required constraints and dimensions to the sketched entities.
d. Modify the dimensions of the sketch.
e. Save the sketch and exit the **Sketch** mode.

Setting the Working Directory

The working directory was selected in Tutorial 1, and therefore there is no need to select the working directory again. But if a new session of Creo Parametric is started, then you need to set the working directory again by following the steps given next.

1. Open the Navigator by choosing the **Show Navigator** button on the bottom left corner of the Creo Parametric screen; the Navigator slides out. In the Navigator, the **Folder Tree** is displayed at the bottom. Click on the black arrow available on the left of the **Folder Tree**; the **Folder Tree** expands.

2. Click on the node adjacent to the **Creo_8.0** folder in the Navigator; the contents of the **Creo_8.0** folder are displayed.

3. Now, right-click on the *c03* folder to display a shortcut menu. From this shortcut menu, choose the **Set Working Directory** option; *c03* is set as the working directory.

4. Hide the Navigator by clicking on the **Show Navigator** button on the bottom-left of the Status Bar.

Starting a New Object File

1. Start a new object file in the **Sketch** mode. Name the file as *c03tut2*.

Drawing the Sketch

To draw the outer loop, you need to draw three circles and then draw lines tangent to them.

1. Choose the **Center and Point** tool from the **Circle** drop-down. Draw the circles in a horizontal line, as shown in Figure 3-37.

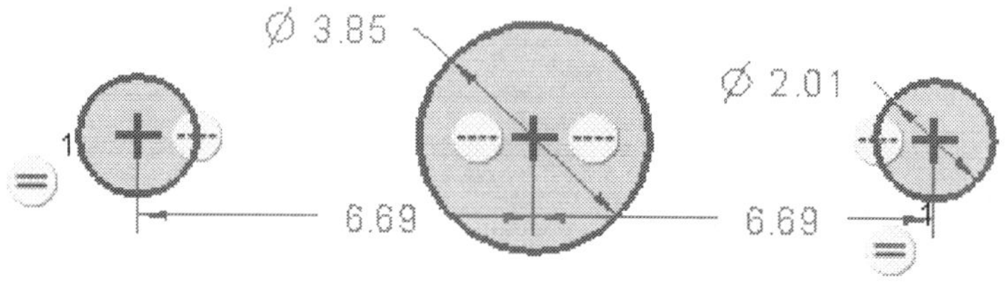

Figure 3-37 Three circles with weak dimensions and constraints

2. Choose the **Line Tangent** tool from the **Line** drop-down in the **Sketching** group.

3. Select the left and middle circles on their respective top points; a tangent is drawn from the top of the left circle to the top of the middle circle.

4. Next, select the right and middle circles on their respective top points; a tangent is drawn from the top of the right circle to the top of the middle circle.

5. Similarly, using the **Line Tangent** tool, draw the other tangents through the bottom-most points of the left, middle, and right circles, as shown in Figure 3-38.

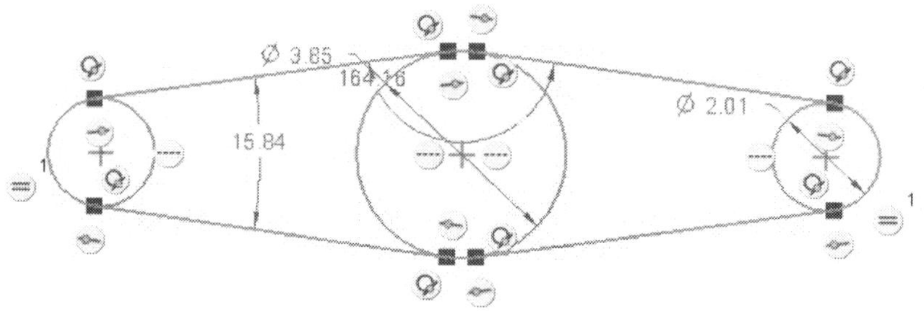

Figure 3-38 Tangent lines drawn on the circles

Trimming the Circles

The inner portions of the circles are not required. Therefore, you need to trim them.

1. Choose the **Delete Segment** tool from the **Editing** group.

2. Move the cursor close to the right portion of the left circle; the right part of the circle turns green in color. Next, click on it to delete it.

3. Similarly, trim the parts of the middle and right circles that are not required. The sketch after trimming the circles is shown in Figure 3-39.

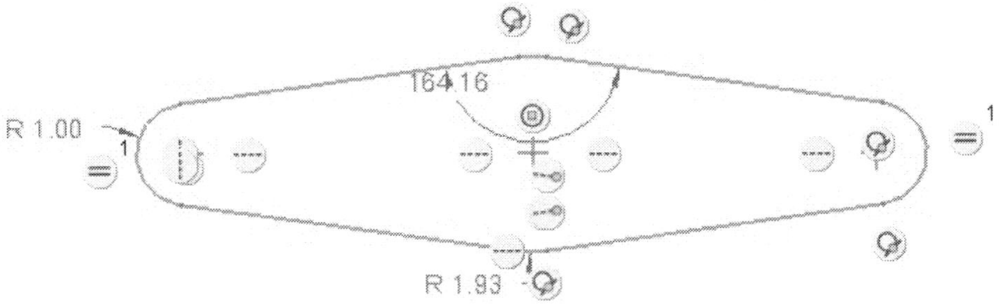

Figure 3-39 Sketch after trimming the circles

Drawing the Circles

1. Choose the **Concentric** tool from the **Circle** drop-down in the **Sketching** group; you are prompted to select an arc.

2. Click on the left arc and move the mouse; a circle appears. Select a point inside the sketch to complete the circle. Press the middle mouse button.

> **Tip**
> *While drawing a concentric circle, sometimes the circle snaps to the other circle or arc and it becomes difficult to draw a circle of the size you need. In such a case, you can disable the snapping of the circle to the other circle or arc. To do so, hold the SHIFT key to disable the snapping or to disable the equal radii constraint that the system tends to apply while drawing the circle.*

> **Note**
> *You may need to zoom in the drawing to select the top arc in the next step.*

3. Click on the top arc and move the mouse; a circle appears. Select a point inside the sketch to complete the circle. Press the middle mouse button to end the creation of circle.

4. Click on the right arc and move the mouse; a circle appears. Select a point inside the sketch to complete the circle. Press the middle mouse button to end the creation of circle.

The sketch after drawing all three circles is shown in Figure 3-40.

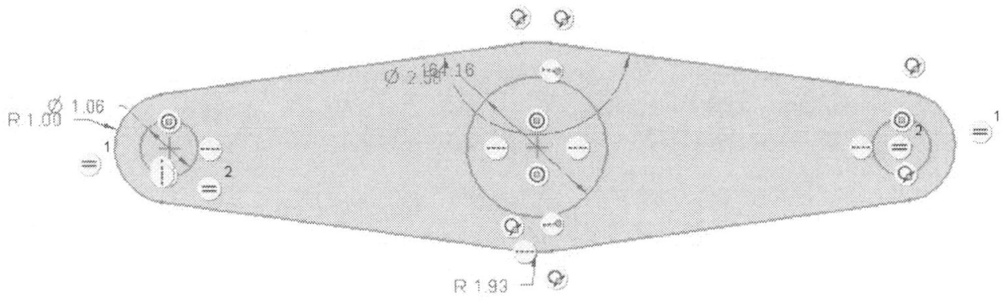

Figure 3-40 Sketch after drawing all three circles

Applying the Constraints

1. Choose the **Equal** tool from the **Constrain** group.

2. Select the left arc and then select the right arc to apply the equal radius constraint.

3. Select the left circle and the right circle to apply the equal radius constraint.

4. Choose the **Parallel** tool from the **Constrain** group.

5. Click to select the tangent line that connects the left arc and the middle arc at the top and then click to select the tangent line that connects the right arc and the middle arc at the bottom. It is evident from the parallel constraint symbol that the parallel constraint has been applied to the two tangent lines in the sketch.

Dimensioning the Sketch

Creo Parametric applies weak dimensions to the sketch automatically. These dimensions are not the needed dimensions because these dimensions do not help in machining the model. Therefore, you need to dimension the sketch with the dimensions that will be used to machine the model. To do so, follow the steps given next.

1. Choose the **Dimension** tool from the **Dimension** group.

2. Select the center of the right and left circles; the centers of the circles turn black in color. Now, using the middle mouse button, place the dimension below the sketch. The sketch after applying constraints is shown in Figure 3-41.

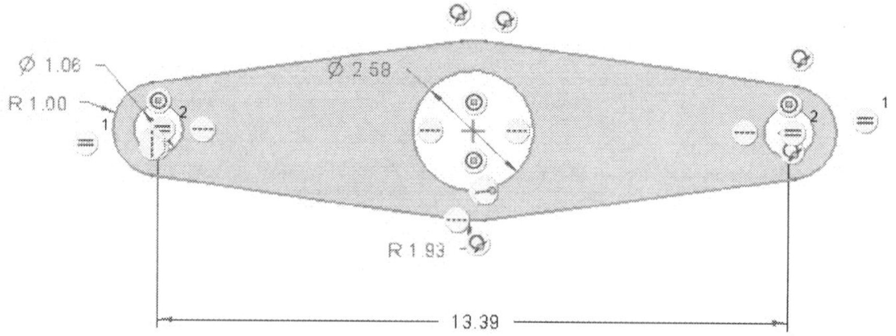

***Figure 3-41** Sketch with constraints*

The rest of the weak dimensions are the needed dimensions.

Modifying the Dimensions

All constraints and dimensions have been applied to the sketch. Now, you need to modify the dimensions.

1. Select all dimensions using the CTRL+ALT+A keys.

2. Choose the **Modify** tool from the **Editing** group; the **Modify Dimensions** dialog box is displayed.

3. Clear the **Regenerate** check box and then modify the values of the dimensions, refer to Figure 3-36 for dimension values.

When you clear the check box, the sketch does not regenerate while modifying the dimensions. The dimension that you edit in the **Modify Dimensions** dialog box is enclosed in a blue box in the sketch.

4. After the dimensions are modified, choose the **OK (Regenerate the section and close the dialog)** button from the **Modify Dimensions** dialog box; the message **Dimension modifications successfully completed** is displayed in the message area.

The sketch after modifying the dimension values is shown in Figure 3-42.

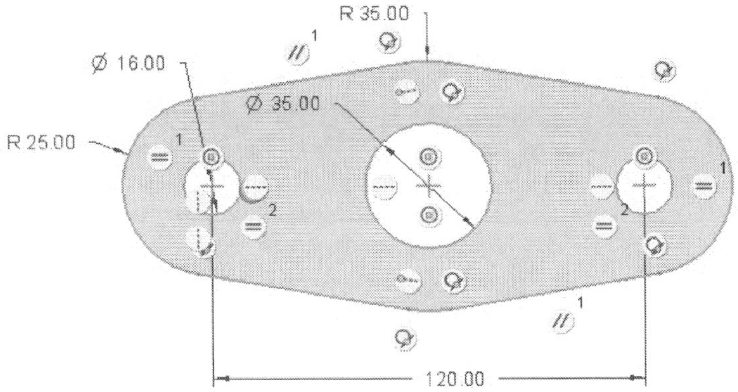

Figure 3-42 *Sketch after modifying the dimensions*

Saving the Sketch

1. Choose the **Save** button from the **File** menu and save the sketch.

Exiting the Sketch Mode

1. Choose the **Close** button from the **Quick Access** toolbar to exit the sketch.

Tutorial 3

In this tutorial, you will draw the sketch of the model shown in Figure 3-43. The sketch is shown in Figure 3-44. **(Expected time: 30 min)**

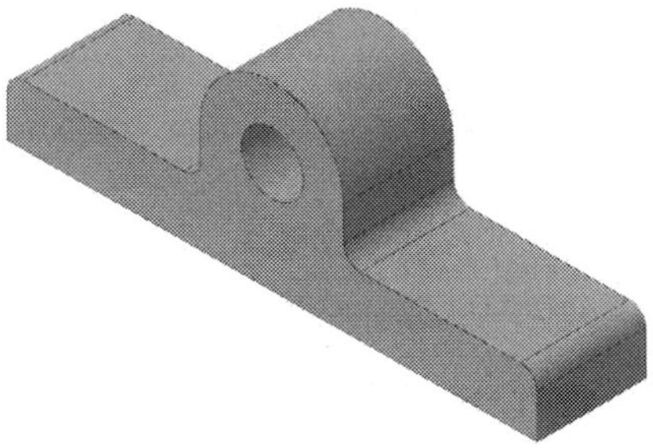

Figure 3-43 *Model for Tutorial 3*

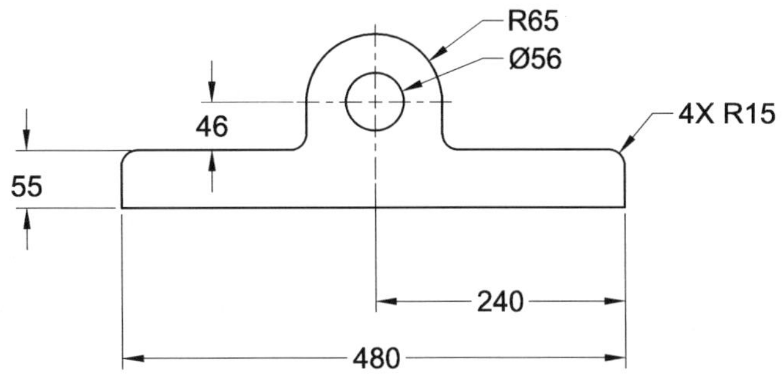

Figure 3-44 *Sketch of the model*

The following steps are required to complete this tutorial:

a. Set the working directory and create a new object file.
b. Draw the sketch using sketcher tools.
c. Apply fillets at two corners of the sketch.
d. Dimension the sketch.
e. Modify dimensions of the sketch.
f. Save the sketch and exit the **Sketch** mode.

Setting the Working Directory

The working directory was selected in Tutorial 1, and therefore there is no need to select the working directory again. But if a new session of Creo Parametric start, then you have to set the working directory again by following the steps given next.

1. Open the Navigator by clicking on the **Show Navigator** button in the left edge of the Creo Parametric window; the Navigator slides out. In the Navigator, the Folder Tree is displayed at the bottom. Click on the black arrow that is available at the left-side of the Folder Tree to expand it.

2. Click on the node adjacent to the *Creo_8.0* folder in the Navigator to display the content of this folder.

3. Now, right-click on the *c03* folder to display a shortcut menu. From this shortcut menu, choose the **Set Working Directory** option; *c03* is set as the working directory.

4. Close the Navigator by clicking on the **Show Navigator** on the right edge of the Navigator; the Navigator slides in.

Starting a New Object File and Drawing the Sketch

1. Start a new object file in the **Sketch** mode. Name the file as *c03tut3*.

2. Choose the **Line Chain** tool from the **Line** drop-down of the **Sketching** group.

3. Draw the lines with constraints, as shown in Figure 3-45.

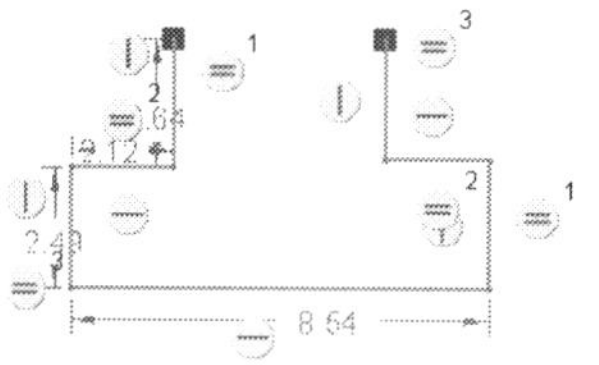

Figure 3-45 *Lines in the sketch with the dimensions turned off for clarity*

4. Choose the **3-Point / Tangent End** tool from the **Arc** drop-down available in the **Sketching** group. 

5. Select the endpoint of the left vertical line as the start point of the arc. Complete the arc at the endpoint of the right vertical line.

6. Choose the **Concentric** tool from the **Circle** drop-down available in the **Sketching** group; you are prompted to select an arc.

7. Click on the arc; a rubber-band circle appears. Size the circle by moving the cursor and click to complete it. The sketch after drawing the circle is shown in Figure 3-46.

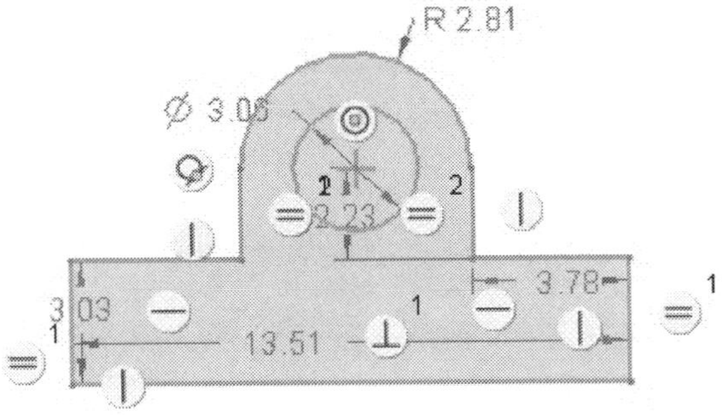

Figure 3-46 *Sketch after drawing the arc and the circle*

Note

*1. Choose the **Dimensions Display** button from the **Display** group in the **View** tab to turn the dimensions on or off.*

2. Creo Parametric does not have the options like midpoint, endpoint, or center of an arc or a circle. However, while drawing a sketch, these options are applied in the form of weak constraints. For example, while drawing an entity, the endpoint of the entity snaps to the cursor. The middle point constraint appears when you bring the cursor near to the middle point of the line to draw another line.

Filleting the Corners

1. Choose the **Circular Trim** option from the **Fillet** drop-down in the **Sketching** group; you are prompted to select the two entities to be filleted. The corners that you need to fillet are shown in Figure 3-47.

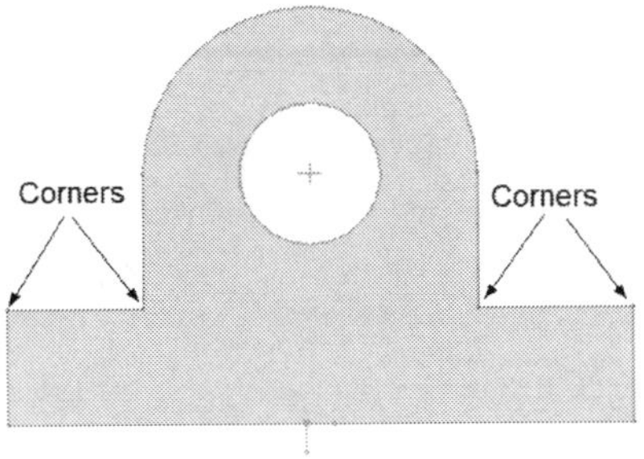

Figure 3-47 *Corners to be filleted*

2. Select the two entities one by one using the left mouse button to fillet the corners of these entities.

The sketch after creating the fillets is shown in Figure 3-48.

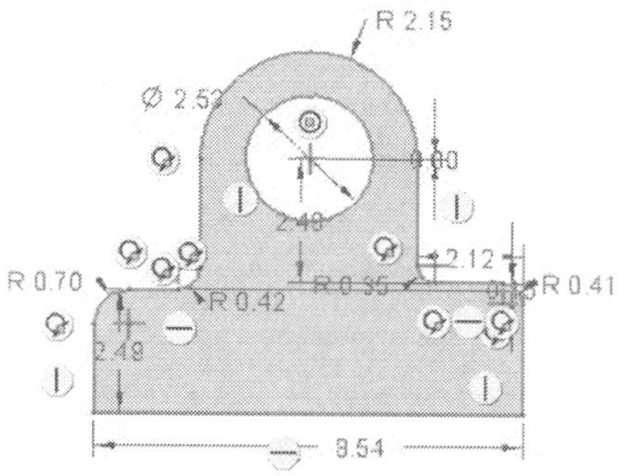

Figure 3-48 Sketch after creating fillets

Applying the Constraints

1. Choose the **Equal** tool from the **Constrain** group.

2. Click to select the fillets and apply the equal constraint to all fillets.

Dimensioning the Sketch

The weak dimensions are applied to the sketch automatically. These are not the required dimensions and therefore, you need to dimension the sketch manually.

1. Choose the **Dimens**ion tool from the **Dimension** group.

2. Dimension the sketch, as shown in Figure 3-49.

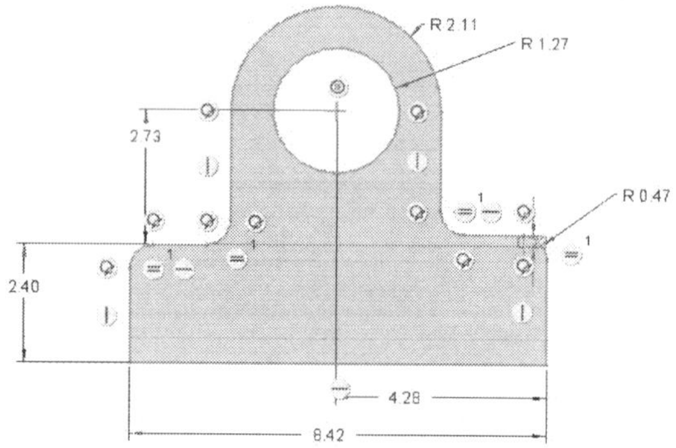

Figure 3-49 Sketch after dimensioning

Modifying the Dimensions

You need to modify the dimension values that are assigned to the sketch.

1. Select all dimensions using CTRL+ALT+A.

2. Choose the **Modify** tool from the **Editing** group; the **Modify Dimensions** dialog box is displayed.

3. Clear the **Regenerate** check box and then modify the values of the dimensions. If this check box is cleared, the sketch does not regenerate while modifying the dimensions.

   The dimension that you edit in the **Modify Dimensions** dialog box is enclosed in a violet box in the sketch.

4. Modify all dimensions. Refer to Figure 3-43 for dimension values.

5. Choose the **OK (Regenerate the section and close the dialog)** button from the **Modify Dimensions** dialog box.

   The sketch after modifying the dimension values is shown in Figure 3-50.

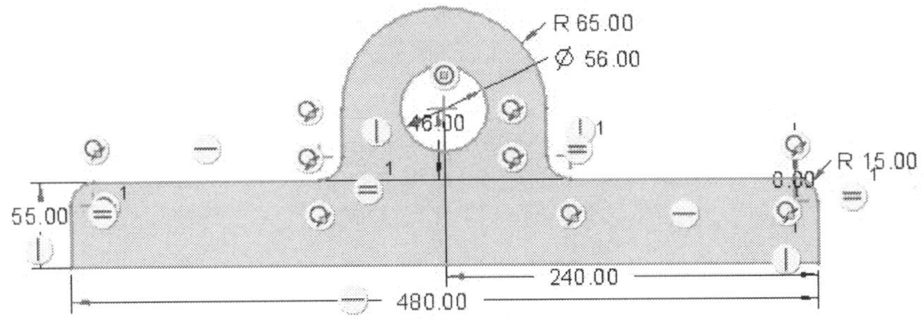

***Figure 3-50** Sketch after modifying the dimensions*

Saving the Sketch and Exiting the Sketch Mode

1. Choose the **Save** button from the **File** menu and save the sketch.

2. Choose the **Close** button from the **Quick Access** toolbar to exit the **Sketch** mode.

Self-Evaluation Test

Answer the following questions and then compare them to those given at the end of this chapter:

1. The __________ dialog box is used to modify dimensions.

2. The display of dimensions and constraints can be turned on/off by choosing the __________ and __________ buttons respectively from the **Display** group.

3. The __________ tool is used to rotate selected entities.

4. You can delete entities by selecting them and then using the __________ key on the keyboard.

5. The shape of a spline can be modified dynamically. (T/F)

6. You can increase the length of the spline by adding points before the start point and after the endpoint of the spline. (T/F)

7. You can modify the dimensions dynamically. (T/F)

8. When you modify a weak dimension, it becomes strong. (T/F)

9. You can dimension the length of a centerline. (T/F)

10. The font of the text written in the sketcher environment cannot be modified. (T/F)

Review Questions

Answer the following questions:

1. How many handles are displayed in the graphics window while rotating and resizing entities?

   (a) one (b) two
   (c) three (d) four

2. Which of the following mouse buttons is used to place the dimension?

   (a) left (b) middle
   (c) right (d) mouse is not used for dimensioning

3. Which of the following is the default font for the text in the sketcher environment?

   (a) font (b) filled
   (c) font3d (d) isofont

4. Which of the following groups is used to toggle the display of dimensions and constraints in the sketcher environment?

   (a) **Sketching** (b) **Constrain**
   (c) **Display** (d) **Dimension**

5. In which type of dimensioning, the **Dim Orientation** dialog box is displayed while dimensioning the arcs and circles?

   (a) **Normal** (b) **Perimeter**
   (c) **Baseline** (d) None of the above

6. For placing a section in a new sketch, you can use the right mouse button. (T/F)

7. You can create elliptical fillets in Creo Parametric. (T/F)

8. While creating text in the sketcher environment, you need to draw a line that will define the height of the text. (T/F)

9. While copying the sketched entities, the **Paste** tab is also displayed. (T/F)

10. You can modify a spline by moving its interpolation points. (T/F)

EXERCISES

Exercise 1

In this exercise, you will draw the sketch of the model shown in Figure 3-51. The sketch to be drawn is shown in Figure 3-52. **(Expected time: 30 min)**

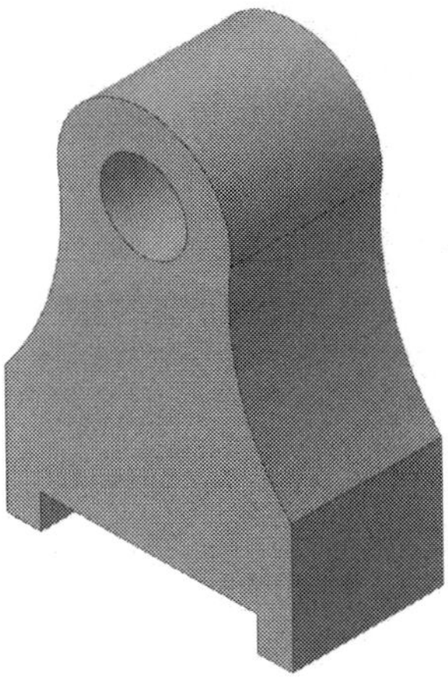

Figure 3-51 Solid model for Exercise 1

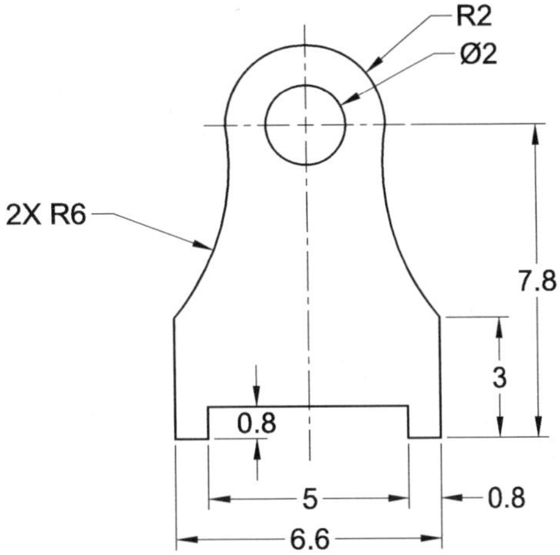

Figure 3-52 Sketch of the model

Exercise 2

In this exercise, you will draw the sketch of the model shown in Figure 3-53. The sketch to be drawn is shown in Figure 3-54. **(Expected time: 15 min)**

Figure 3-53 Solid model for Exercise 2

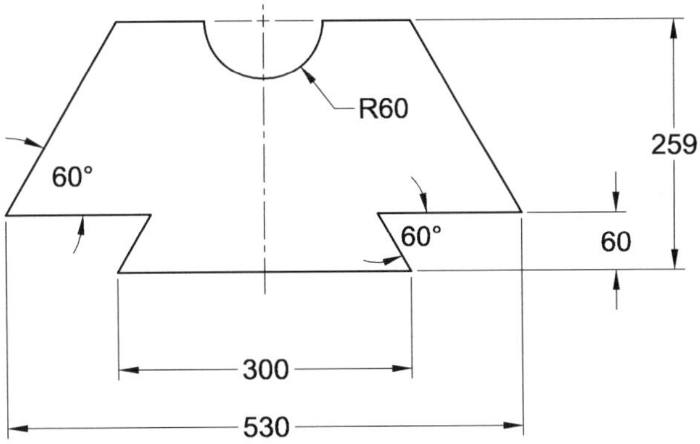

Figure 3-54 Sketch of the model

Exercise 3

In this exercise, you will draw the sketch of the model shown in Figure 3-55. The sketch to be drawn is shown in Figure 3-56. **(Expected time: 30 min)**

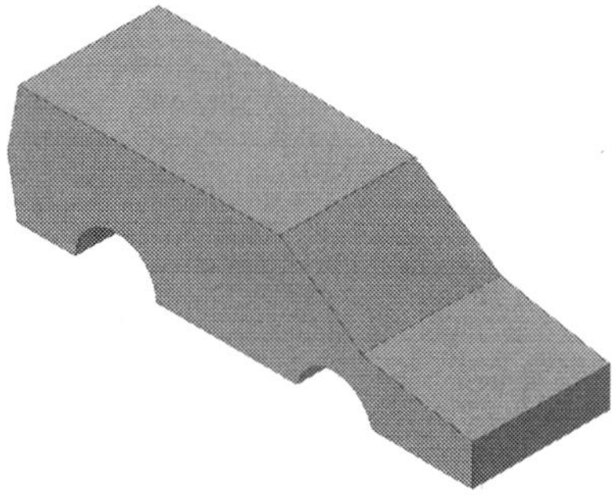

Figure 3-55 Solid model for Exercise 3

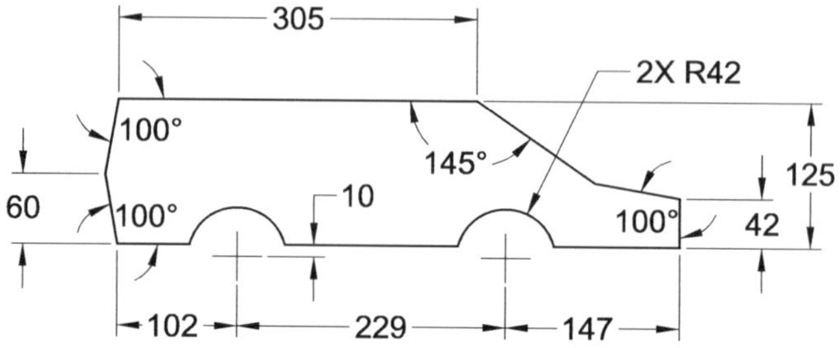

Figure 3-56 Sketch of the model

Answers to Self-Evaluation Test

1. **Modify Dimensions**, **2.** **Dimensions Display**, **Constraints Display**, **3.** **Rotate Resize**, **4.** DELETE, **5.** T, **6.** F, **7.** T, **8.** T, **9.** F, **10.** F

Chapter 4

Creating Base Features

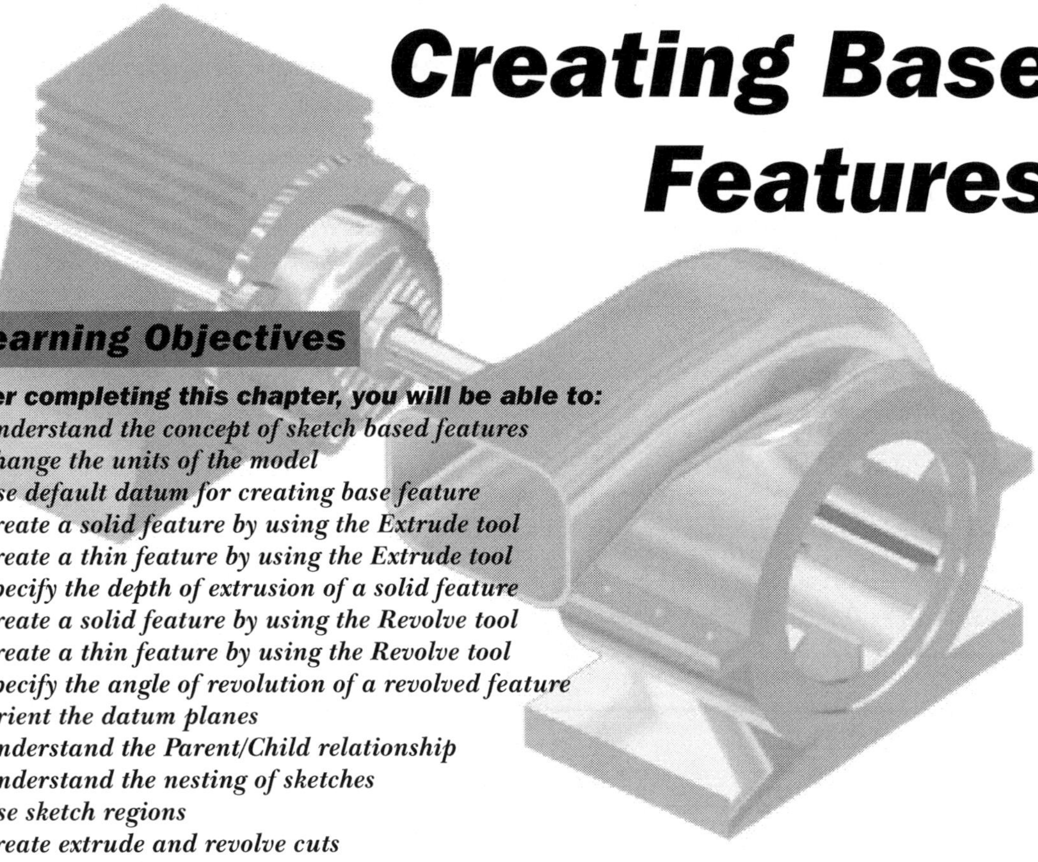

Learning Objectives

After completing this chapter, you will be able to:

- *Understand the concept of sketch based features*
- *Change the units of the model*
- *Use default datum for creating base feature*
- *Create a solid feature by using the Extrude tool*
- *Create a thin feature by using the Extrude tool*
- *Specify the depth of extrusion of a solid feature*
- *Create a solid feature by using the Revolve tool*
- *Create a thin feature by using the Revolve tool*
- *Specify the angle of revolution of a revolved feature*
- *Orient the datum planes*
- *Understand the Parent/Child relationship*
- *Understand the nesting of sketches*
- *Use sketch regions*
- *Create extrude and revolve cuts*

CONCEPT OF SKETCH BASED FEATURES

Sketch-based feature uses a sketch to define its shape, dimensions, and general placements. All the features that are used to create solid models are known as sketch based features. A sketch based feature uses a sketch in the following two ways:

1. As an internal section, the sketch is created within a feature. This sketch is contained in the feature it defines. So, the changes that you make to an internal section affect only the sketch-based feature. An internal section cannot be unlinked from the feature hence it cannot be used for defining other features. Using internal sections to define features is advantageous because it reduces the feature count in a part model.

2. As an external sketch, the existing sketch is used to define the feature. You can unlink an external sketch from the feature so that it can be used to define one or more features in a solid model.

Note

1. In this book, the sketch based features are defined using the internal sections.

2. Sketch-based features may have requirements such as a closed loop section (sketch), reference of a horizontal or a vertical axis, or a coordinate system. If the sketch is invalid, Creo Parametric displays a warning dialog box. If regeneration of the feature is failed, a notification is displayed in the notification area.

CREATING BASE FEATURES

The base feature is the first feature created while creating a model in the **Part** mode. The base feature is created using the datum planes. Although you can create a base feature without using the datum planes, but in that case, you will not have proper control over the orientation of feature and direction of feature creation.

While creating the base feature of a model, the designer should be extra careful in specifying its attributes. This is because if the base feature is not created correctly, then the features created on it will also be inappropriate. This results in waste of time and effort. Although Creo Parametric provides you the options to redefine a feature, doing that also consumes additional time and effort. In next chapters, you will learn to create these features in the most efficient manner.

Tip

It is recommended that you set the working directory before starting a new file.

Invoking the Part Mode

The **Part** mode in Creo Parametric helps you to create a part from the conceptual sketch through solid feature-based modeling. In the **Part** mode, you can build and modify parts through direct and intuitive graphical manipulation. To invoke the **Part** mode, choose the **New** option from the **File** menu or choose the **New** button from the **Data** group; the **New** dialog box will be displayed with various modes of Creo Parametric. The **Part** radio button in the **Type** area and the **Solid** radio button in the **Sub-type** area of the **New** dialog box are selected by default. The default name of the part file also appears in the **File name** edit box. You can change the part name as desired and then choose the **OK** button to enter the **Part** mode. Note that you cannot use spaces in the name of a file.

In the **New** dialog box, the **Use default template** check box is selected by default. As a result, the default template provided in Creo Parametric will be used for the part file. This template has certain parameters that will be assigned to the part file that you will create. For example, the units of this model will be Inch lbm Second; the length will be in inch; mass in lb; time in seconds, and temperature in Fahrenheit. If you clear the **Use default template** check box and choose the **OK** button from the **New** dialog box, the **New File Options** dialog box will be displayed, as shown in Figure 4-1.

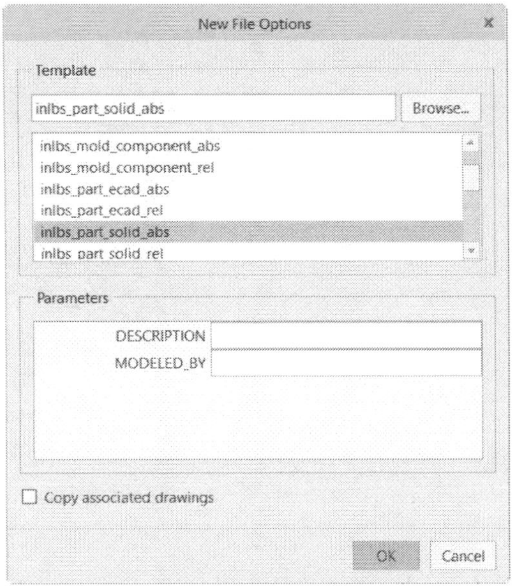

***Figure 4-1** The **New File Options** dialog box*

From the **New File Options** dialog box, you can select the required template file. If you want the default system of units to be mmns (millimeter Newton sec), then select the **mmns_part_solid_abs** template and choose the **OK** button from this dialog box. Note that, the **mmns_part_solid_abs** template has been used throughout this book.

The **Part** mode is the most commonly used mode of Creo Parametric. This is because solid modeling is done in this mode. It should be noted that a solid model is the start of a product development cycle. Product development cycle refers to the development of a product from scratch to its prototype. If you have created a solid model, then it can further be used to generate its drawing views, numerically controlled (NC) machining codes, core and cavity, analysis of the solid model, and so on.

The two-dimensional (2D) sketch drawn in the **Sketch** mode can be converted into a three-dimensional (3D) model in the **Part** mode. The **Part** mode contains the same sketcher environment with similar options to sketch as those available in the **Sketch** mode. There are some sketcher options in the **Part** mode that are not available in the **Sketch** mode because they do not have any use in the **Sketch** mode.

Figure 4-2 shows the initial screen appearance in the **Part** mode. It displays **Model Tree**, three default datum planes, **Ribbon**, and toolbars.

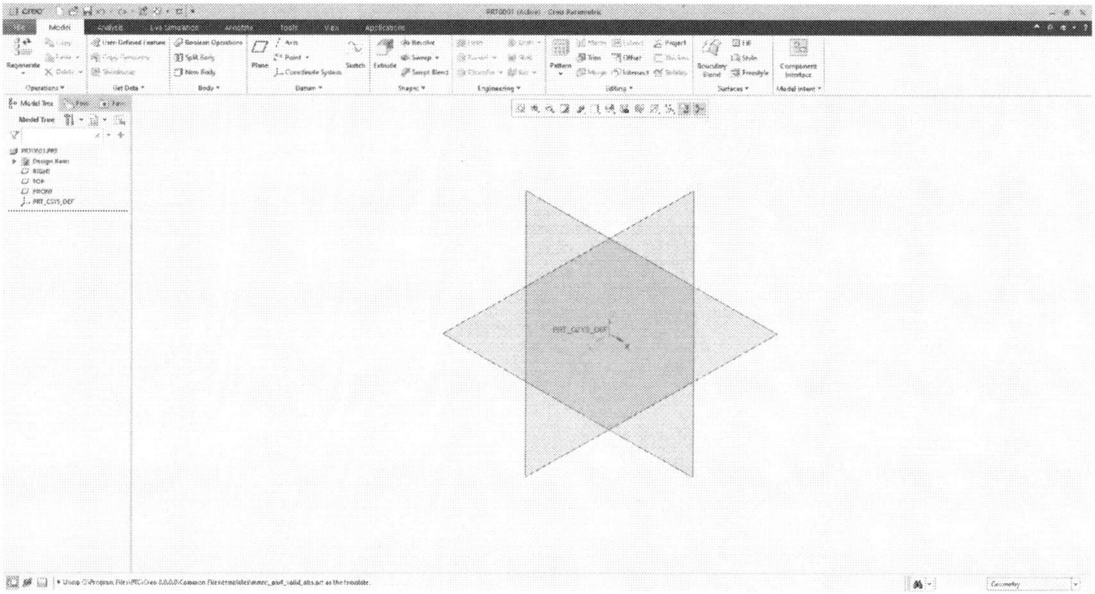

***Figure 4-2** The initial screen appearance in the **Part** mode*

Changing the Units of the Part Mode

Each model that you create has a basic system of metric and non-metric units to ensure that all material properties of that model are consistently measured and defined. In Creo Parametric, you can select a predefined system of units or you can create your own system of units. As discussed earlier, the units assigned to the models are specified according to the template selected for the mode (Part, Assembly, Drawing) in which you are working. The system of units that you select is called the principal system of units for that mode.

You can change the principal unit system of Creo Parametric at any stage of designing process. To do so, choose **File > Prepare > Model Properties**; the **Model Properties** dialog box will be displayed. Next, select the **change** option adjacent to **Units**; the **Units Manager** dialog box will be displayed with the **Systems of Units** tab chosen by default, as shown in Figure 4-3. The unit system which is currently active will be highlighted in light blue color. Select the system of units that you want to set as the principal system of units and choose the **Set** button; the **Changing Model Units** dialog box will be displayed with the **Model** tab chosen by default, as shown in Figure 4-4. By default, the **Convert dimensions (for example, 1" becomes 25.4mm)** radio button is chosen in this dialog box. Next, choose the **OK** button; the existing data will be converted to the new system of units without changing the physical size of the model. If your model has correct values but an undesired system of units, select the **Interpret dimensions (for example, 1" becomes 1mm)** radio button; the existing data will not be converted to the new system of units and the physical size of the model will change but dimensional values will remain same. Note that the angular units are not affected by either of these options. Now, choose the **OK** button from the **Changing Model Units** dialog box and then, choose the **Close** button from the

Units Manager dialog box; the **Model Properties** dialog box will be displayed again. Choose the **Close** button to complete the process of changing the units.

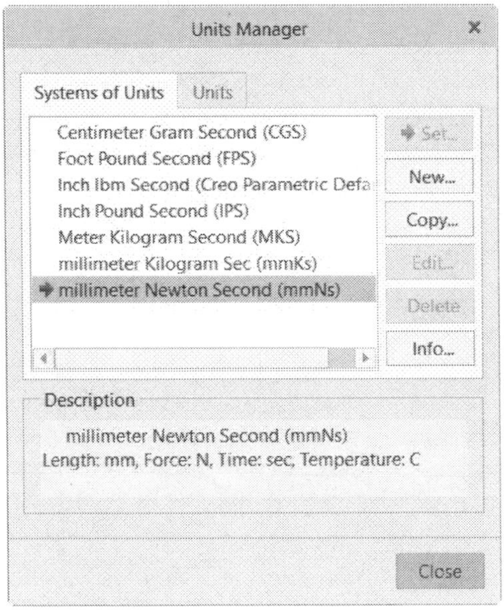

*Figure 4-3 The **Units Manager** dialog box*

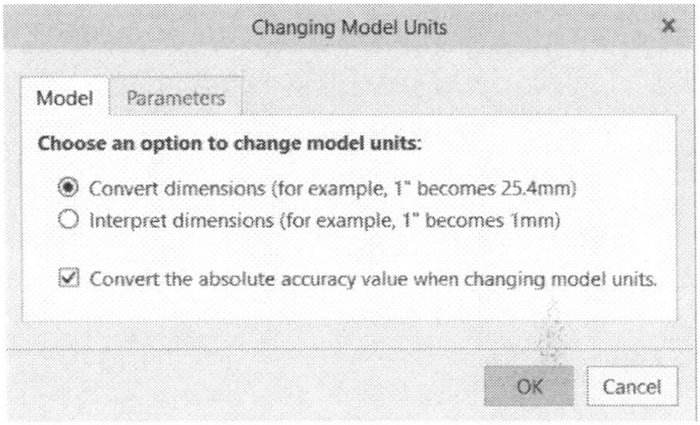

*Figure 4-4 The **Changing Model Units** dialog box*

Creating a Custom System of Units

In Creo Parametric, you can create a custom system of units by combining individual units. You can also combine predefined units and custom units to create a custom system of units. The system of units depends upon the basic physical quantities and quantities derived from the basic dimensions. For example, if you define force as one of the basic physical quantity, mass becomes a derived unit for your model because the connection between force and mass is derived as Force = Mass X Acceleration ($F = ML/T^2$).

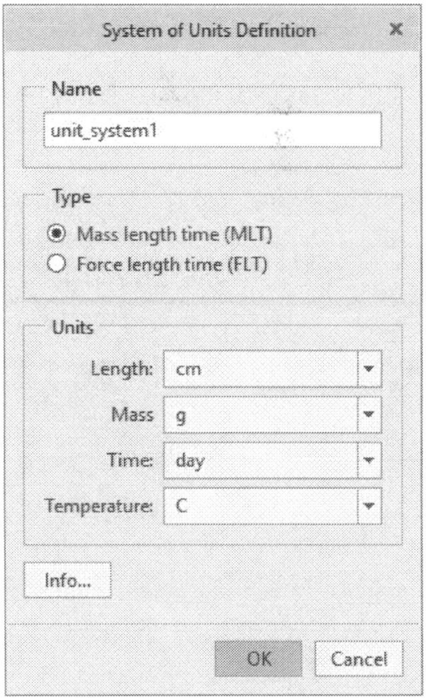

*Figure 4-5 The **System of Units Definition** dialog box*

To create a custom system of units, choose the **New** button from the **Units Manager** dialog box, refer to Figure 4-3; the **System of Units Definition** dialog box appears, as shown in Figure 4-5. Specify the name of unit system in the **Name** edit box. The radio buttons available in the **Type** area are used to specify the basic physical quantity. To specify the system of units based on mass, select the **Mass length time(MLT)** radio button and to specify the system of units based on force, select the **Force length time(FLT)** radio button. Next, specify the units for length, mass, time, and temperature in the **Units** area of the **System of Units Definition** dialog box and choose the **OK** button to close the dialog box. As you close this dialog box, the **Units Manager** dialog box will be display again and you will notice that the unit system you created is displayed with the other unit systems in the **Systems of Units** tab.

The Default Datum Planes

Generally, the three default datum planes act as the first feature in the **Part** mode. These datum planes are used to create the base feature. Also, they are used to draw a 2D sketch and then convert it into a 3D model by using the feature creation tools. The base feature that you create is referenced with default datum planes.

Tip
It is recommended that you always use the datum planes to create a base feature. This is because the model created using the datum planes can be easily oriented. The uses of datum planes are discussed in Chapter 5.

Note
Although, the three default datum planes, ***RIGHT****,* ***TOP****, and* ***FRONT****, are the primary features in the* ***Part*** *mode. In the* ***Model Tree****, datum planes appear as three separate features. If you delete any one of them, only that datum plane will be deleted.*

These three default datum planes are mutually perpendicular to each other. They are not referenced to each other and are individual features. When a solid model is created, these datum planes adjust their size according to the size of the model. You can create any number of datum planes as per your requirement. The creation of additional datum planes is discussed in Chapter 5.

In Creo Parametric, a default template is provided with three default datum planes. However, if you do not need the default datum planes, then at the time of creating a new file, you need to clear the **Use default template** check box in the **New** dialog box and select the **Empty** template from the **New File Options** dialog box.

Note
It is important to create a model in the correct orientation. The correct orientation is the one in which the model will be used later either in assembly or other modes. For example, the Casting component of the Plummer block is modeled in such a way that it always stands vertical.

CREATING A PROTRUSION

Ribbon: Model > Shapes

Protrusion is defined as a process of adding material defined by a sketched section. In Creo Parametric, there are various options available for adding material to a sketch such as **Extrude**, **Revolve**, **Sweep**, and so on. These options are also known as sketch based features and can be chosen from the **Shapes** group in the **Model** tab. Figure 4-6 shows the options in the **Shapes** group.

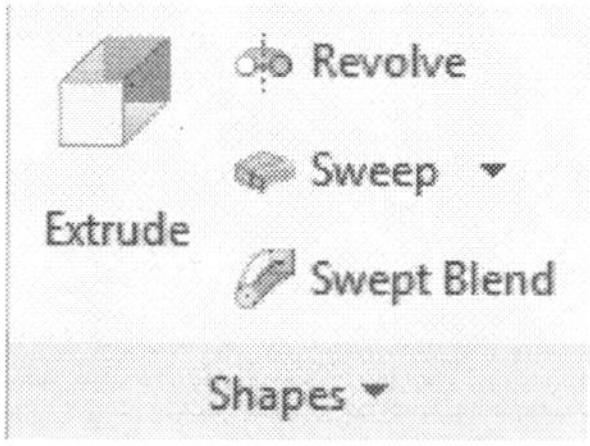

Figure 4-6 *The* ***Shapes*** *group*

Tip

1. If you are creating a protrusion (a material addition process), then irrespective of the direction of view chosen, the protrusion takes place in the direction toward the user, that is, out of the screen.

*2. In the **Part** mode, you can invoke the sketch based features from the mini popup toolbar which will appear when you click on any sketched entity.*

Note

1. In Creo Parametric, there are three methods to create a sketch based feature. In the first method, you can select a sketch or a sketch plane and then invoke the required tool. In the second method, you can invoke the tool and then select the sketch or sketch plane. In the third method, you can invoke the tool and then create the sketch.

*2. You can also invoke the **Sketch** mode directly by pressing the **S** key from the keyboard provided by the sketch plane.*

Extruding a Sketch

Ribbon: Model > Shapes > Extrude

Extrusion is a method of defining three-dimensional geometry by translating a two-dimensional sketch normal to the sketch plane for a pre-defined distance or up to a specified reference. The **Extrude** tool in the **Shapes** group adds material to the area defined by a closed sketch drawn. The procedure of creating a base feature by using the **Extrude** tool is explained next.

1. Choose the **Extrude** tool from the **Shapes** group or press the **X** key from the keyboard; a dashboard named **Extrude** will be displayed on the top of the drawing area. This dashboard has all options to define the extrude feature, refer to Figure 4-7.

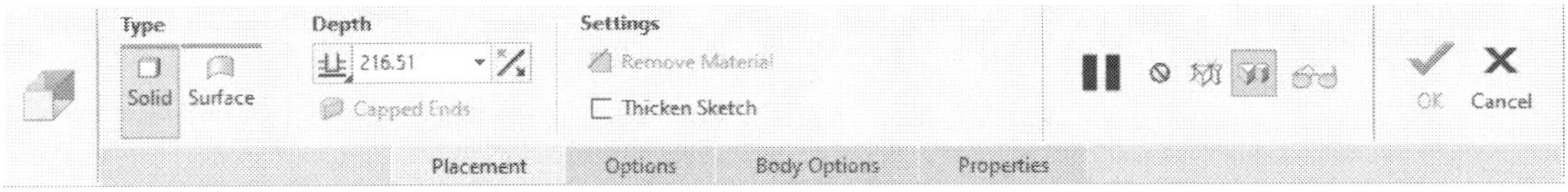

***Figure 4-7** Partial view of the **Extrude** dashboard*

Tip

*You can invoke the **Extrude** tool from the mini popup toolbar. The mini popup toolbar appears on selecting any sketch from the drawing area or from the model tree.*

2. Choose the **Placement** tab from the dashboard to display the slide-down panel. Choose the **Define** button from the slide-down panel to display the **Sketch** dialog box. Alternatively, press the right mouse button in the drawing area to display the shortcut menu. Choose the **Define Internal Sketch** option from the shortcut menu; the **Sketch** dialog box will be displayed. Also, you will be prompted to select a sketching plane. Select the datum plane named **FRONT** from the drawing area as the sketching plane, as shown in Figure 4-8.

***Figure 4-8** The **Sketch** dialog box*

In the **Sketch** dialog box, the name of the plane that you have selected appears in the **Plane** collector. At the same time, the reference plane is selected by default. The name of the reference datum plane appears in the **Reference** collector of the **Sketch** dialog box. Creo Parametric sets the orientation of the reference plane by default. You can also set the orientation of the reference plane as required by selecting an appropriate option from the **Orientation** drop-down list.

3. After selecting the sketching plane and the reference plane, choose the **Sketch** button in the **Sketch** dialog box.

   Now, you have entered in the sketcher environment, refer to Figure 4-9. The selected datum plane is the sketching plane and is not oriented parallel to the screen. To make it parallel to the screen, choose the **Sketch View** button from the **Graphics** toolbar. Alternatively, choose the **Sketch View** button from the **Setup** group of the **Sketch** tab. In Creo Parametric, the references for the sketch are applied by default. So, you need not to select references for the sketches.

   If you want to increase the drawing space in the drawing area, close the **Model Tree** by choosing the **Show Navigator** button at the bottom left corner of the screen, refer to Figure 4-9.

Tip

*1. You can directly enter the sketching environment in the **Part** mode by selecting the sketching plane first and then invoking the **Extrude** tool.*

*2. You can orient the sketching plane parallel to screen automatically. To do so, choose the **Model Display** button from the **Home** tab of the start screen. Next, choose **Sketcher** from the left pane of the dialog box and select the **Make the sketching plane parallel to the screen** check box in the **Sketcher startup** area and save the settings. Next time when you invoke the sketcher environment in part mode, the sketching plane will be oriented parallel to the screen automatically.*

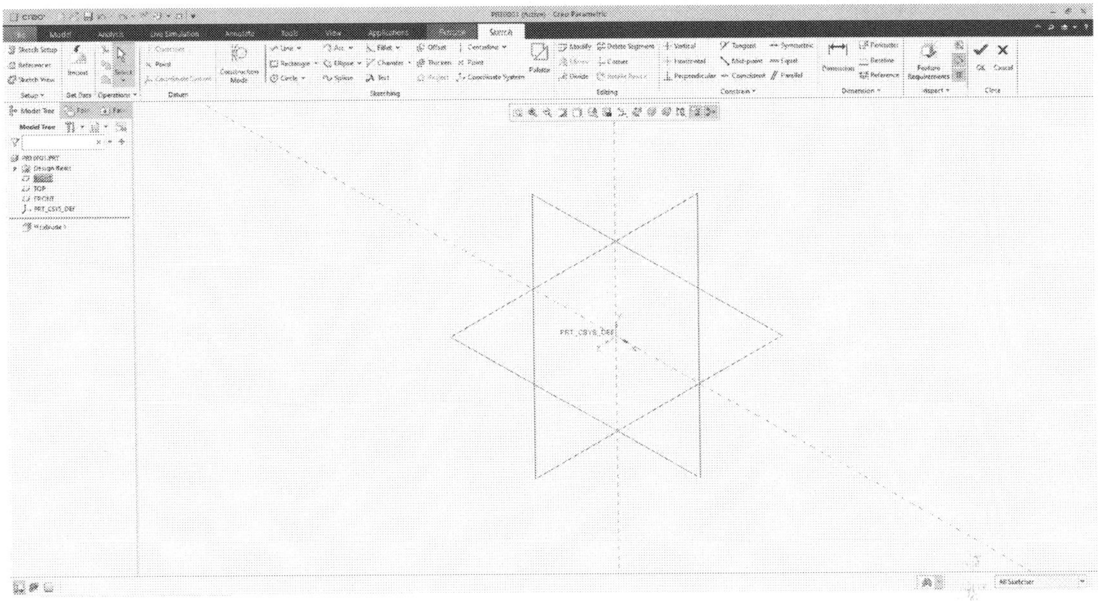

Figure 4-9 The drawing area after entering the sketcher environment

4. Now, choose the **Line Chain** tool from the **Line** drop-down in the **Sketching** group to draw the right-half of the I-section and then draw a vertical centerline aligned with the **RIGHT** datum plane. Select all lines on the right-half of the I-section and choose the **Mirror** button from the **Editing** group; you will be prompted to select a centerline. Select the centerline to mirror the right-half and to create the I-section, as shown in Figure 4-10. After the I-section is sketched, dimension it and choose the **OK** button from the **Sketch** dashboard.

After doing so, the **Extrude** dashboard will be enabled and displayed above the drawing area. The model is created by assuming some default attributes and its preview is displayed in orange color. The attributes of a model are available on the dashboard and are discussed later in this chapter.

On the dashboard, the **Depth** edit box contains a default value. This value is the depth of extrusion of the sketch that you have created. Enter the desired value in the edit box and press ENTER. You can also hold and drag the handle displayed on the model to dynamically specify the depth of extrusion. This drag handle will be visible if the model is rotated slightly using the middle mouse button.

Three buttons are also available in the dashboard, **No Preview**, **Un-attached**, and **Attached**. When you invoke the **No Preview** tool, preview of the model will not be displayed while modeling. You can invoke the **Un-attached** tool when you want to see only the outlines of the model while working with the **Extrude** tool. The preview displayed after choosing this tool will be a transparent geometry. You can invoke the **Attached** tool when you want to see the final visualization of the model to be created.

5. Choose the **OK** button; the required 3D model will be displayed in the drawing area. However, it will appear as a 2D entity. You need to change the display of the model such that the depth of the model can also be viewed. To do so, choose the **Saved Orientations** button from the **Graphics** toolbar and choose the **Default Orientation** option from the flyout displayed. Alternatively, press CTRL+D to change the view to default orientation. Figures 4-10 and 4-11 show the sections after extruding up to a certain depth.

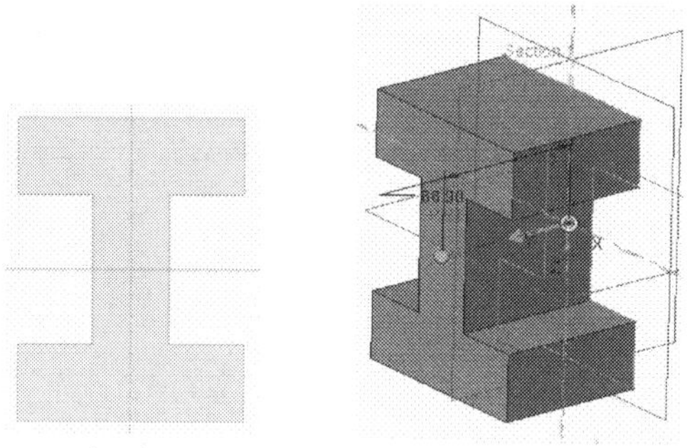

Figure 4-10 *The I-Section extruded to a certain depth*

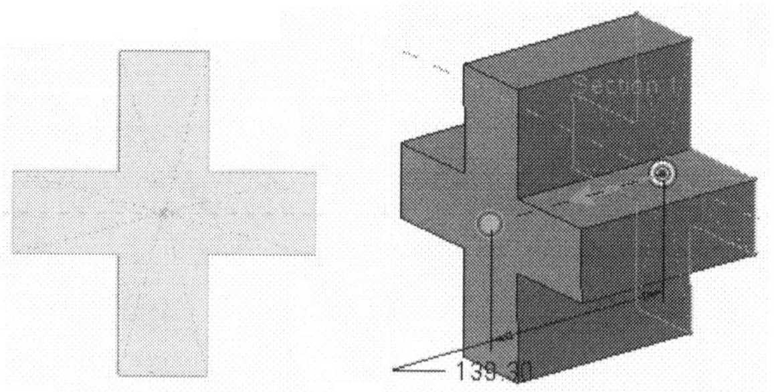

Figure 4-11 *Model created using the **Extrude** tool*

The previous steps explain how to construct a 3D model using the **Extrude** tool. The options available in the **Extrude** dashboard and the **Sketch** dialog box that were used while creating the 3D model from the I-section are discussed next.

The Extrude Dashboard

The options and buttons in the **Extrude** dashboard of the **Ribbon**, shown in Figure 4-12, are used to extrude a sketch and to specify certain attributes related to the model. These attributes can be assigned to the model before or after drawing the sketch of the model. The tabs and the buttons available in the dashboard are discussed next.

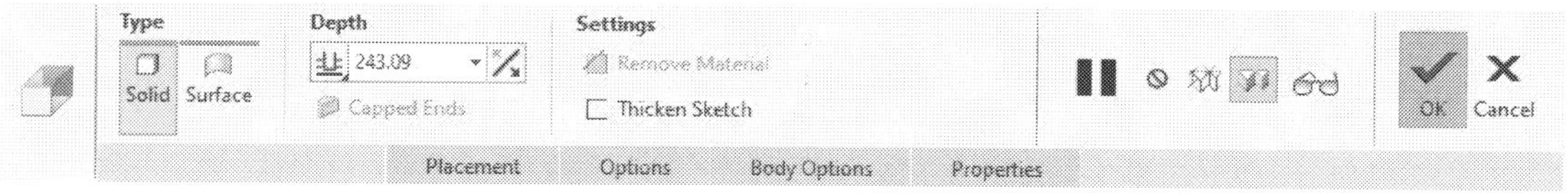

Figure 4-12 *Partial view of the **Extrude** dashboard*

Placement Tab

Choose the **Placement** tab; a slide-down panel will be displayed, as shown in Figure 4-13. The collector in this panel displays **Select 1 item** because the section has not been defined yet. You can select a sketch that is drawn using a datum curve or you can choose the **Define** button to draw a sketch. Choose the **Define** button from this slide-down panel; the **Sketch** dialog box will be displayed. If you have selected an external sketch then the **Unlink** button will replace the **Define** button. Choose the **Unlink** button; the **Unlink** message box will be displayed, as shown in Figure 4-14. Choose the **OK** button to unlink the external sketch. Unlinking the external sketch will create a copy of the selected sketch as an internal sketch, which will be linked with the feature. You can also edit the copied sketch by choosing the **Edit** button next to the sketch collector. Note that editing the internal sketch will not affect the dimensions of the external sketch.

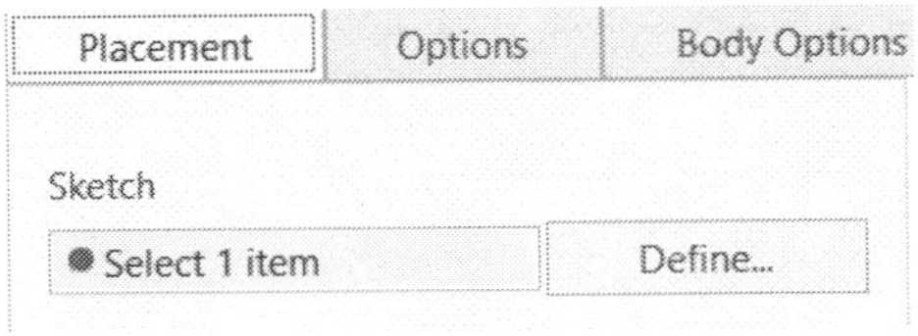

Figure 4-13 *The slide-down panel of the* ***Placement*** *tab*

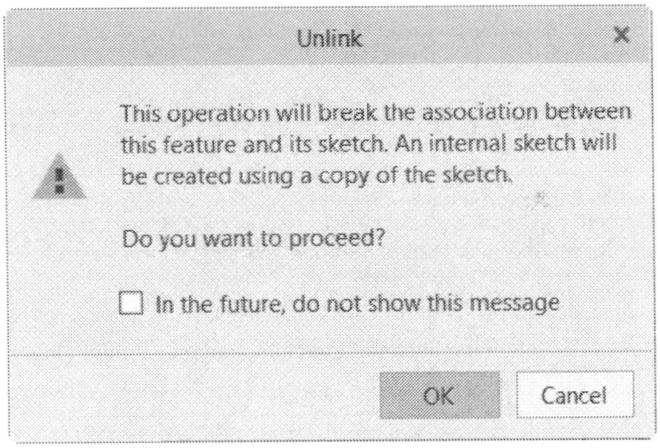

Figure 4-14 *The* ***Unlink*** *message box*

Options Tab

Choose the **Options** tab; a slide-down panel will be displayed, as shown in Figure 4-15. The slide-down panel is used to specify whether you want the sketch to extrude to one side or to both sides of the sketching plane. It has **Side 1** and **Side 2** drop-down lists. The options in the **Side 1** drop-down list are discussed next.

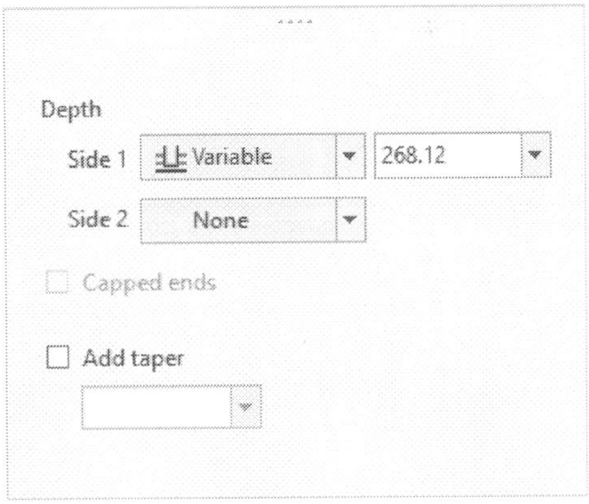

Figure 4-15 *The slide-down panel of the* ***Options*** *tab*

Variable: The **Variable** option is the most commonly used option to define the extrusion depth of the sketch by specifying a depth value. When you select this option from the **Side 1** drop-down list, a default value appears in the edit box that is adjacent to this drop-down list. Enter a value in this edit box and press ENTER to modify the depth of extrusion.

Symmetric: If you select the **Symmetric** option from the **Side 1** drop-down list, the material will be equally added in both the directions of the sketching plane. When you select this option, the **Side 2** drop-down list becomes inactive that means you cannot select any option from this drop-down list.

To Next: The **To Next** option is used to set the extrusion depth of the sketch up to the next surface of the model in the direction of extrusion.

Through All: The **Through All** option is used to set the extrusion depth of the sketch through all the surfaces of the model in the direction of extrusion.

Through Until: The **Through Until** option is used to select the surface up to which the sketch is to be extruded. While using this option, you cannot select the point or plane for reference.

To Reference: The **To Reference** option is used to select a point, surface, plane, or a curve up to which the section is to be extruded. When you select the **To Reference** option from **Side 1** drop-down list, a drop-down list is activated below this option. These options are:

This option is disabled by default and it extrudes the sketch to a selected point, curve, plane, or surface.

Choose this option to extrude the selected sketch to the offset of a selected point, curve, plane, or surface. You can specify an offset or translation value in the Offset edit box. You can also use the drag handle to change the value of offset. To flip the direction of offset reference, choose the **Change offset or translation direction to the other side** button.

Choose this option to extrude the selected sketch upto the selected point, curve, plane, or surface. To specify the translation value, enter a value in the edit box next to this option. Remember that an offset is always done in parallel direction and a translation is done in linear direction.

Note

*The **To Next**, **Through All**, and **Through Until** options are available in the drop-down list only after creating the base feature.*

The options in the **Side 2** drop-down list are similar to the options in the **Side 1** drop-down list. The **None** option in this drop-down list allows you to extrude the sketch in the first direction only. The **Variable** and **To Reference** options provide the depth of extrusion to the sketch in the second direction.

Tip

*1. The features created by using the **To Reference** option do not have a dimension associated with them. Therefore, they cannot be modified by changing the dimension value. However, changing the terminating surface changes the depth of the feature. You will understand it better when you learn the modification of an existing feature.*

2. On specifying a negative value in any edit box, the direction of feature creation will be flipped.

Capped ends

The **Capped ends** check box is used only in case of surfaces. When a surface is created using the **Extrude** tool, then its ends are open. If you select this check box, then the open ends of the surface created will be closed by an auto generated surface. You will learn more about this option in later chapters.

Add Taper

If you select the **Add taper** check box, an edit box below this check box will get activated. This edit box is used to add draft in the solid or surface created by using the **Extrude** tool. The value entered in this edit box is referenced to the datum on which the sketch was created for extrusion. The taper value ranges from -89.9 to 89.9.

Body Options

Choose the **Body Options** tab; a slide-down panel will be displayed. If you select the **Create new body** check box in the **Add geometry to body** area, the feature will be added to the Body2(new body). If you select the **Remove Material** button in the **Settings** area, then the **All** and **Selected** radio buttons will become available under the **Cut geometry from body** area. Select the **All** radio button to remove the material from all the bodies through which the extruded feature passes. Select the **Selected** radio button to remove material from the selected bodies through which the extruded feature passes.

Properties Tab

Choose the **Properties** tab; a slide-down panel will be displayed. This slide-down panel displays the feature name in the **Name** collector. When the **i** button is chosen from the slide-down panel, the browser will be opened and all the information about the protrusion feature gets displayed. The browser and its properties have been discussed in Chapter 1.

> **Tip**
> *In Creo Parametric, you can also access the extrusion depth options from the shortcut menu which are displayed when you right-click on the extrude handle.*

Solid

The **Solid** button is chosen by default and is used to create a solid by adding material to the section. When you select a sketch for extrusion using this button, the sketch should be a single closed loop.

Surface

The **Surface** button is used to create a surface by adding material to the section. Using this button, you can create surface models. The sketch that is drawn for the surface model need not be a closed loop. The surface modeling is discussed later in Chapter 15.

> **Note**
> *The tabs and the buttons available in the* ***Extrude*** *dashboard can either be used before drawing the sketch or after drawing it.*

Change depth direction of extrude to other side of sketch

The **Change depth direction of extrude to other side of sketch** button is used once the sketch is completed. When you choose this button, it toggles the direction of extrusion with reference to the sketch plane. The direction of extrusion is defined as the direction in which the feature is created with respect to the sketching plane. This direction is displayed by an arrow in the drawing area. When you choose this button, the arrow points in the reverse direction suggesting that the direction of feature creation has been reversed. You can also click on the arrow to flip the direction of extrusion.

Note

The direction of the arrow depends on the type of extrude feature being created. If material has to be added to a feature, then the arrow, by default, points in the direction toward the user, that is out of the screen. But, if a cut feature is being created in which material has to be removed from a feature, then by default, the arrow points into the screen.

If you spin the model using the middle mouse button, then the direction of the arrow can be easily recognized.

Remove Material

The **Remove Material** button is available in the **Extrude** dashboard only after the base feature is created. This is because this button is used to remove the material from an existing feature. Therefore, when you create a base feature, this button is not available. The use of this button is explained in later chapters.

Thicken Sketch

The **Thicken Sketch** button is used to create thin parts with specified thickness. When you choose this button, the **Change direction of extrude between one side, other side, or both sides of sketch** button and a dimension edit box appear on the right of this button.

The **Change direction of extrude between one side, other side, or both sides of sketch** button is used to set the direction with respect to the sketch where the thickness should be added. This button appears on the dashboard only when the sketch is completed and you exit the sketcher environment. This button serves three functions. It is used to set the thickness direction to either side of the sketch and even to both sides of the sketch symmetrically. When the sketch is completed, the preview of the model will be displayed in the drawing area. By default, the thickness to the sketch is applied on one side. Now, when you choose the **Change direction of extrude between one side, other side, or both sides of sketch** button, the thickness will be applied to the other side of the sketch. When you choose this button for the second time, the thickness is applied symmetrically to both sides of the sketch.

Choose the **Thicken Sketch** button to enter the thickness value of the feature to be created. This edit box is available only when you exit the sketcher environment.

By default, the thickness is added to the outer side of the sketch boundary, as shown in Figure 4-16. To change the direction of the material, choose the **Change direction of extrude between one side, other side, or both sides of sketch** option, refer to Figure 4-17.

If you choose the **Change direction of extrude between one side, other side, or both sides of sketch** button again, the thickness of the material added gets symmetric to both sides of the section wall, as shown in Figure 4-18. Figure 4-19 shows the model created by using the **Thicken Sketch** button.

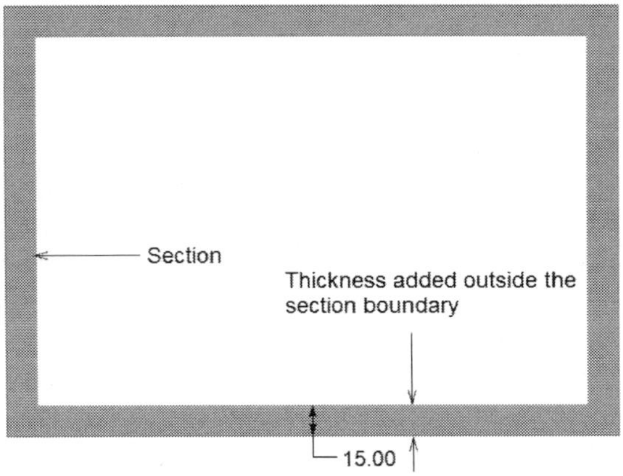

Figure 4-16 *Thickness of the material added to outer side of the sketch*

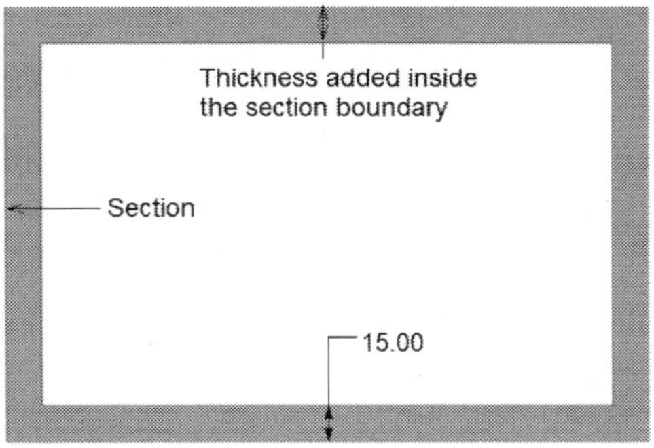

Figure 4-17 *Thickness of the material added to inner side of the sketch*

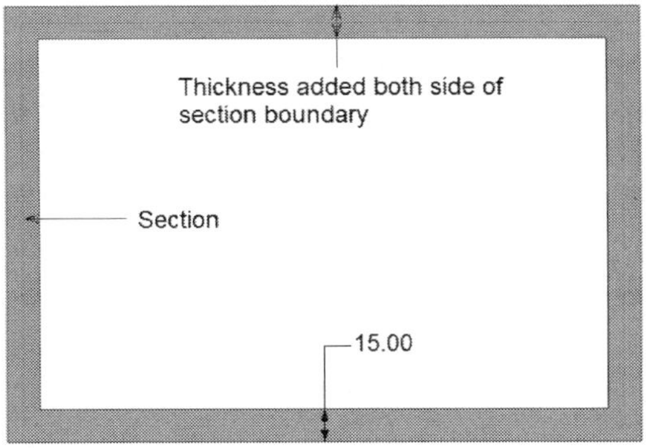

Figure 4-18 *Thickness of the material added to both sides of the sketch*

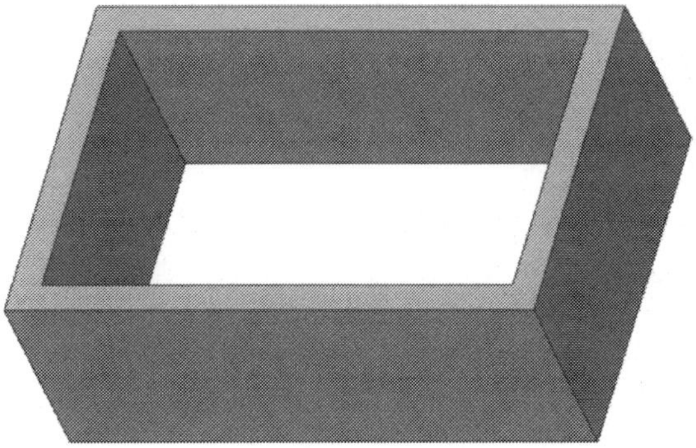

Figure 4-19 *Resulting model after applying thickness*

Pause

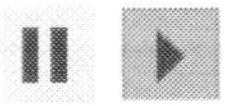

This button is used to pause the current feature creation. To resume the current feature creation, you can choose the **Resume** button that appears in place of the **Pause** button.

Geometry preview/Feature preview

There are four buttons available in the **Extrude** dashboard to control the preview of a model while using the protrusion commands which are discussed next.

No Preview

Choose this button to turn off the preview of the current geometry creation. To preview the geometry again, you can choose the **Un-attached**, **Attached**, or **Verify** button.

Un-attached

Choose this button to preview the protrusion in transparent mode.

Attached

Choose this button to preview the protrusion in attached mode.

Verify mode

Choose this button to preview the final attached geometry. By choosing this button, system allows you to preview the feature and its relationship with other features after finalizing the protrusion. Note that the preview of the feature appears to be a solid model. To exit the verify mode, press this button again.

OK

This button is used to confirm the feature creation and exit the current feature creation tool.

Cancel

This button is used to abort the creation of feature. When you choose this button, the dashboard is closed.

Note that you can modify the depth of extrusion of the model dynamically when the **Extrude** tool is active. To do so, you need to drag the small, yellow coloured circle symbol displayed on the model. You can also move the sketch by dragging the green coloured dot. This method is called dynamic editing and is discussed later. You can also modify the direction of extrusion dynamically. To do so, click on the pink arrow displayed on the model.

The Sketch Dialog Box

The **Sketch** dialog box will be displayed when you choose the **Define** button from the **Placement** slide-down panel. Alternatively, right-click in the drawing area to display the shortcut menu. Choose the **Define Internal Sketch** option from it; the **Sketch** dialog box will be displayed, as shown in Figure 4-20. You can also invoke the **Sketch** dialog box by choosing the **Sketch** option from the **Datum** flyout of the **Extrude** dashboard. Thus, the sketch created will be an external sketch.

This dialog box is used to select the sketching plane and to set its orientation. As soon as you select the sketching plane, the reference plane and its orientation are selected by default. If you want to change the sketching plane that is already selected, then hold the right mouse button in the drawing area to display a shortcut menu, as shown in Figure 4-21. There are three options in this shortcut menu. By default, the **View Orientation** option is selected. If you choose the **Clear** option, the reference plane and its orientation which were selected by default will be cleared from the **Plane** and **Reference** collector in the **Sketch** dialog box. Next, you can select the planes manually. Similarly, to change the sketching plane, choose the **Placement** option in the shortcut menu and select the desired sketching plane.

However, after you select the sketching plane, no matter which reference plane is selected by default, you can select the reference plane manually. The reference you select manually replaces the old reference. If the background of the **Plane** collector or any other collector is displayed in green color, then it indicates that you can select its reference from the model. You can also clear the selection by right-clicking on the **Plane** collector and selecting the **Remove** option from the shortcut menu. These points are also applicable to the collectors of other dialog boxes that are available in Creo Parametric. You will learn more about these collectors later in the book.

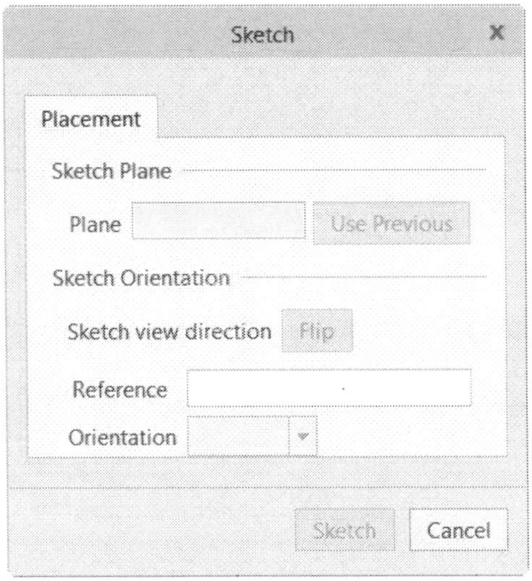

***Figure 4-20** The **Sketch** dialog box*

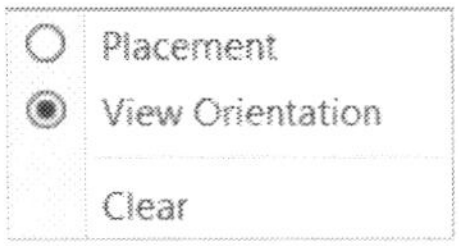

***Figure 4-21** Shortcut menu displayed when the **Sketch** dialog box is invoked*

The **Flip** button in the **Sketch** dialog box can be used after you select the sketching plane and its orientation. This button is used to set the view direction of the sketching plane. When you choose this button, the direction of the arrow which is displayed on the sketching plane will be flipped in either direction.

In the **Sketch** dialog box, the options in the **Orientation** drop-down list are used to specify the orientation of the horizontal or vertical reference for the sketching plane. Generally, the view is normal to the sketching plane you have selected. In order to orient the sketching plane normal to the viewing direction, you have to specify a plane or a planar surface that is perpendicular to the sketching plane. For example, if you select the **RIGHT** datum plane as the reference plane and then select the **Top** option from the drop-down list, then the **RIGHT** datum plane will be on the top while sketching. The options in this drop-down list are common to other feature creation tools and are available whenever you need to draw a sketch. Before you proceed further, you need to understand the three default datum planes that are displayed when you open a new part file. You can view the feature that you have created from different directions. To do so, choose the **Saved Orientations** button from the **Orientation** group in the **View** tab or from the **Graphics** toolbar; a flyout will be displayed, as shown in Figure 4-22. In this flyout, there are some standard preset views provided by Creo Parametric. If you see the feature from the right, your viewing direction will be normal to the **RIGHT** datum plane. If you see the feature from the top, your viewing direction will be normal to the **TOP**

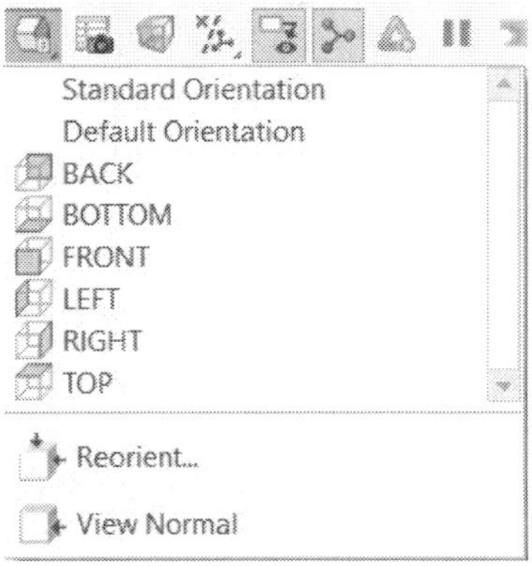

***Figure 4-22** The **Saved Orientations** flyout*

datum plane. Similarly, if you see the feature from the front, your viewing direction will be normal to the **FRONT** datum plane.

The options in this flyout are used to set the orientation of the model in the drawing area.

The options in the **Orientation** drop-down list in the **Sketch** dialog box are very important for the orientation of the base feature. So, you need to select the reference plane very carefully, especially for the base feature, refer to Figure 4-23. This figure shows two cases in which all the factors are same except the reference plane selected for sketch. Note the difference in orientation of the resulting models in their default trimetric orientations. The model can be oriented in its default orientation using CTRL+D or by selecting the **Default Orientation** option from the **Saved Orientations** flyout in the **Graphics** toolbar.

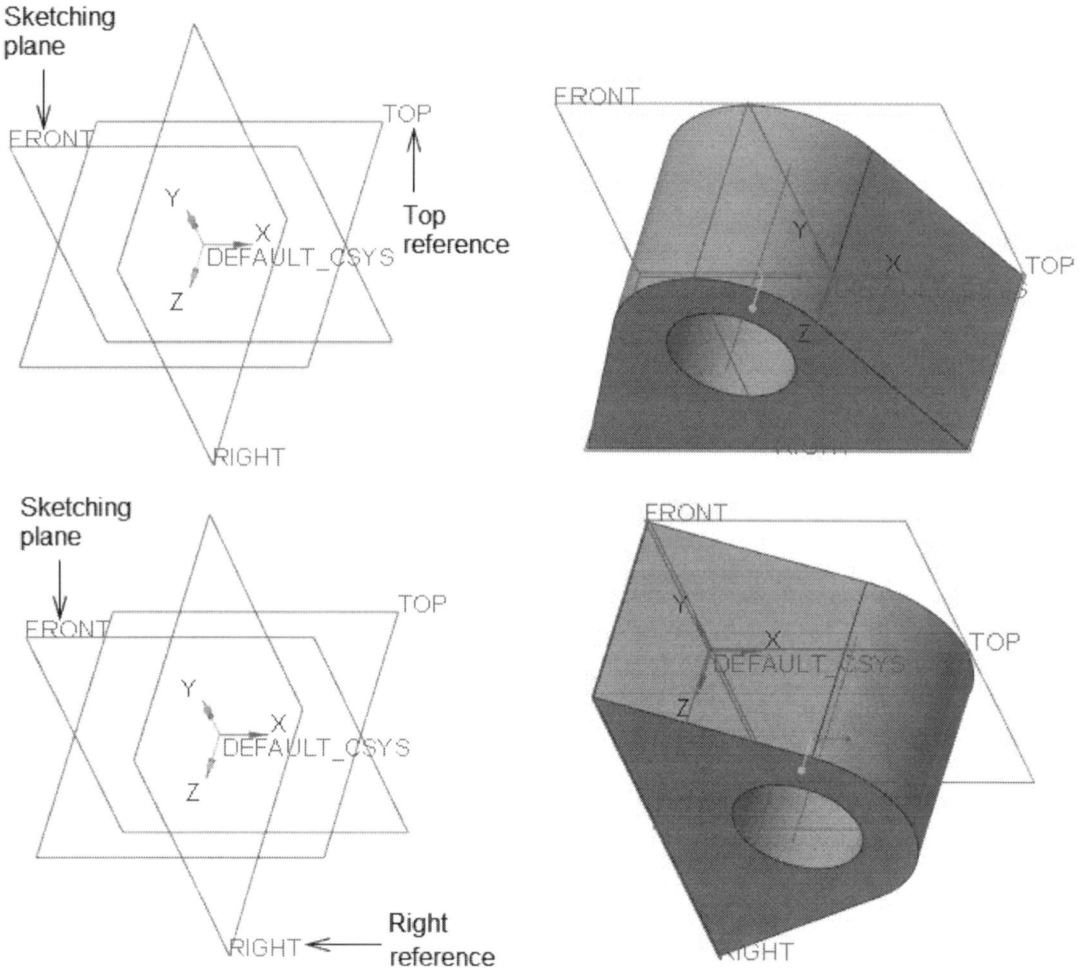

Figure 4-23 *Figure displaying importance of selecting reference planes*

Revolving a Sketch

Ribbon: Model > Shapes > Revolve

The **Revolve** tool allows you to create a protrusion by revolving a sketched section around a centerline. The axis of revolution for the feature can be defined as an external reference or as an internal centerline. If you are selecting any existing linear geometry (datum axis, straight edge, straight curve, or axis of a coordinate system) as the axis of revolution then it will be defined as an external reference. If you are using a centerline that was created in the sketch, it will be defined as an internal centerline. You can use the **Revolve** tool to create a solid or surface feature and to add or remove material. The sketch can be revolved on one side of the sketching plane or on both sides of the sketching plane. Besides this, other attributes can be specified on the sketch by using the **Revolve** dashboard. The **Revolve** dashboard is shown in Figure 4-24 and it will be displayed when you choose the **Revolve** tool from the **Shapes** group in the **Model** tab. The tools available in this dashboard are discussed next.

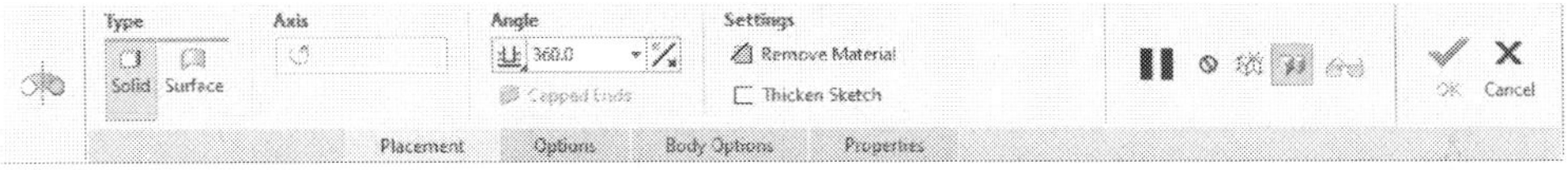

***Figure 4-24** Partial view of the **Revolve** dashboard*

Some of the points to be kept in mind while creating a revolve feature are given next.

1. If the revolve feature is a base feature, then the section drawn should be a closed section for revolving the sketch as a solid. If the revolve feature is the second feature and it is intersecting with a solid geometry, you can use an open section to revolve around the axis.

2. If a sketch contains more than one centerline, the first geometry centerline created will be used as the axis of revolution. If the sketch does not contain a geometry centerline, the first construction centerline created will be used as the axis of revolution.

3. If there are more than one geometric centerlines in a sketch then you can select the desired centerline from the drawing area, right-click and choose the **Axis of Revolution Collector** option from the shortcut menu displayed and then select the required axis.

4. The section should be drawn on one side of the centerline.

5. You can change the axis of revolution while defining a revolved feature. For example, you can select an external axis instead of the centerline.

Solid

The **Solid** button in the **Revolve** tab is used to revolve a sketch to create a solid feature. To revolve a sketch using this button, the sketch should be a closed loop. Figures 4-25 and 4-26 show two examples of solid features created by using the **Solid** button.

Surface

The **Surface** button in the **Revolve** tab is used to revolve a sketch to create a surface feature. The sketch that is drawn for the surface model need not be a closed loop. Surface modeling is discussed later in Chapter 15.

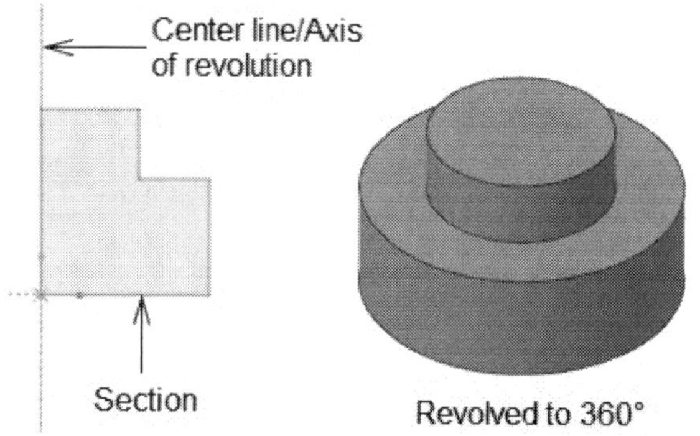

Figure 4-25 Sketch revolved to 360 degree

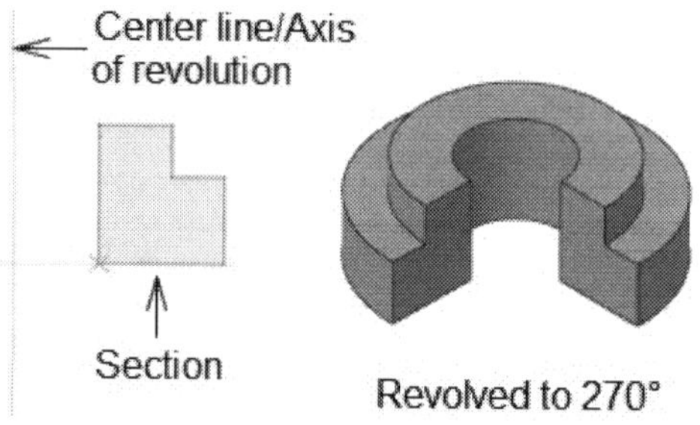

Figure 4-26 Sketch revolved to 270 degree

Revolving a Sketch with Thickness

The **Thicken Sketch** button is used in combination with the **Solid** button to create revolved features having certain thickness. Unlike revolving the sketch to create a solid feature, when you revolve a sketch with thickness, the section need not be a closed loop. When you exit the sketcher environment after the section is created, you can specify the side of the section where the material will be added. After specifying the side for the material addition, you can enter the thickness value in the dimension edit box that appears on the dashboard.

Figures 4-27 and 4-28 show the examples of the revolved sketch having some thickness.

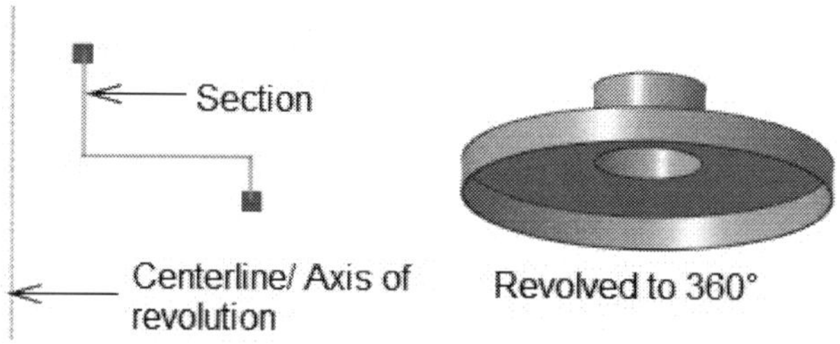

Figure 4-27 Sketch revolved to 360 degree

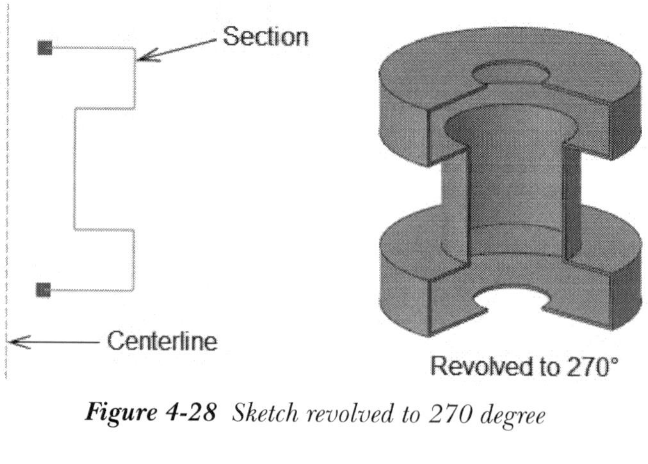

Figure 4-28 Sketch revolved to 270 degree

UNDERSTANDING THE ORIENTATION OF DATUM PLANES

Consider a case where you need to revolve a sketch through an angle of 270-degree. The trimetric view of the solid model is shown in Figure 4-29. Now, when you start to create the solid model, the first step is to decide the datum plane on which you need to create the sketch.

Figure 4-29 *The trimetric view of the solid model*

As evident from the model, the axis of revolution of the model is perpendicular to the **TOP** datum plane. Therefore, the **TOP** datum plane will be selected to be at the top while drawing the sketch. Now, only two options are left for selecting the sketching plane, the **FRONT** and the **RIGHT** datum planes. You can draw the section of the revolved model on any of the two planes. This is because when you view the model, you will notice that the cross-section of the model is parallel to the **RIGHT** datum plane as well as the **FRONT** datum plane.

The procedure to draw the cross-section sketch on the two datum planes to achieve the desired orientation of the model is discussed next.

Case-1

Drawing a Sketch on the RIGHT Datum Plane

1. Choose the **Revolve** tool from the **Shapes** group in the **Model** tab and choose the **Define** button from the **Placement** slide-down panel. When the **Sketch** dialog box is displayed, select the **RIGHT** datum plane as the sketching plane.

   On doing so, the selected datum plane will get highlighted and the **TOP** datum plane will be automatically selected as the reference plane. Also, an arrow will appear on the sketching plane. This arrow indicates the direction of the feature creation. You can change the direction of arrow, but at this stage, you have to accept its default direction.

2. From the **Orientation** drop-down list, select the **Top** option, if it is not already selected.

   Now, when you draw the sketch, the **RIGHT** datum plane will be parallel to the screen and the **TOP** datum plane will be at the top.

3. Choose the **Sketch** button from the **Sketch** dialog box to enter the sketcher environment.

4. Next, draw the sketch of the cross-section, dimension it, and then modify the dimensions, as shown in Figure 4-30.

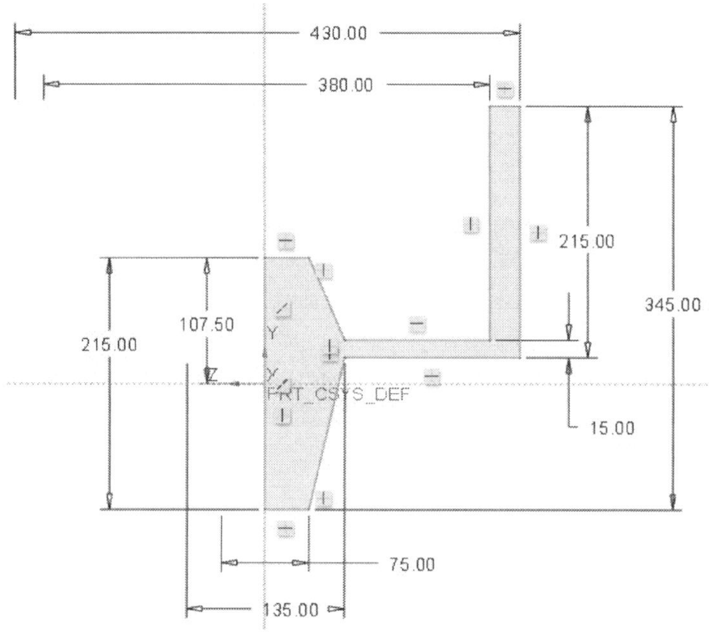

***Figure 4-30** Sketch with dimensions and constraints*

5. Exit the sketcher environment by choosing the **OK** button from the **Sketch** dashboard.

6. If you have drawn the geometric centerline while sketching, then the preview will be displayed. However, if you have not drawn the centerline, then you will be prompted to select a straight edge or an axis of the coordinate system. In this figure, the geometric centerline is drawn along the Y axis.

7. Select the angle value **270** in the drop-down list available on the dashboard.

8. From the **Saved Orientations** flyout, choose the **Default Orientation** option or press CTRL+D. The model will orient to its default orientation, that is, trimetric view, as shown in Figure 4-31.

   However, the view that is shown in Figure 4-31 is not the required one. The model is not oriented correctly. This means the direction of feature creation that was selected is not correct in order to get the desired orientation.

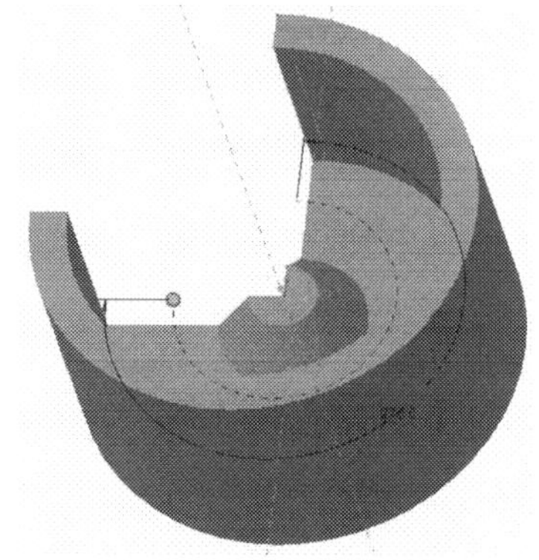

***Figure 4-31** Default trimetric view of the model*

Figure 4-32 shows the sketch drawn on the **RIGHT** datum plane. In such a case, the sketch always rotates in the clockwise direction. Note that the arrow in Figure 4-32 shows the direction in which you will be viewing the sketching plane while in the sketcher environment. Considering the mentioned facts, to orient the sketching plane correctly, the arrow direction should have been flipped by using the **Flip** button in the **Sketch** dialog box. Since you have not achieved the desired orientation of the model, exit this feature creation tool. Then, again invoke the **Revolve** tool, and this time after you select the sketching plane, choose the **Flip** button to reverse the direction of viewing the sketch. Remember that the case discussed here is only applicable when you select the **RIGHT** datum plane as the sketching plane.

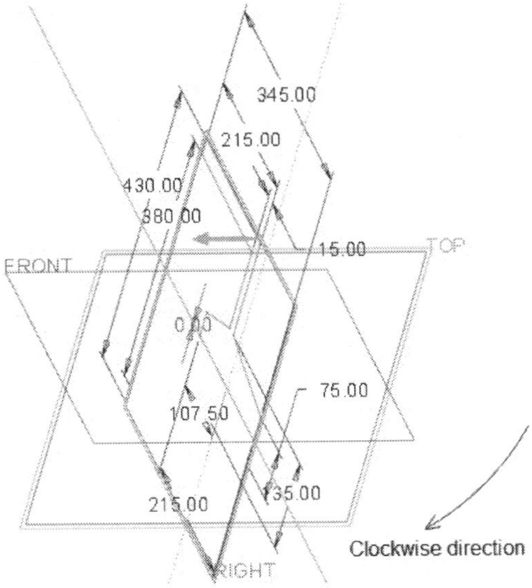

***Figure 4-32** Sketch drawn on the **RIGHT** datum plane*

Tip

1. You might have difficulty in visualizing the cross-section of a revolved model that has an angle of revolution of 360-degree. This is because the cross-section of the model is not visible. Therefore, you cannot decide the sketch to be drawn for the cross-section of the model.

2. Whenever you come across a revolved model, just imagine that if you cut the model along its axis of revolution and remove one quarter of the revolved feature, then the section obtained in one quarter of the revolved feature is the section of the revolved model. Therefore, you need to draw this section in the sketcher environment to create the desired model.

Figure 4-33 shows the sketch on the sketching plane with the direction of the arrow reversed. In this case, the sketch when rotated in the clockwise direction results in the desired orientation of the model, as shown in Figure 4-34.

Tip

*In Creo Parametric, when a revolved model is created, the material is added in the clockwise direction. You can revolve the model in the counterclockwise direction by using the **Change angle direction of revolve to other side of sketch** button from the **Revolve** tab. Also, to revolve the model in the counterclockwise direction, you can enter a negative value for the angle of revolution.*

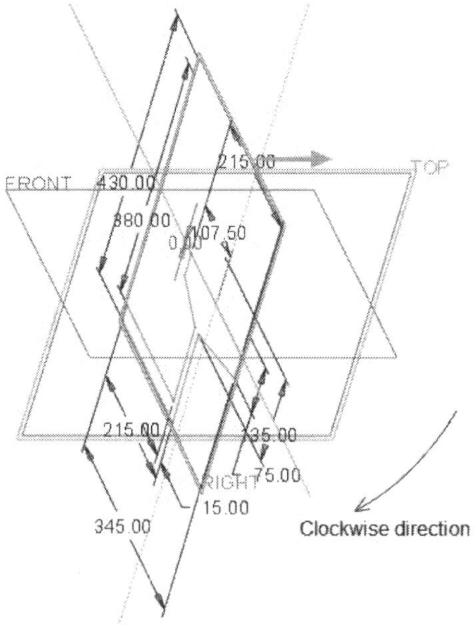

Figure 4-33 *Sketch drawn on the **RIGHT** datum plane with reverse direction of arrow*

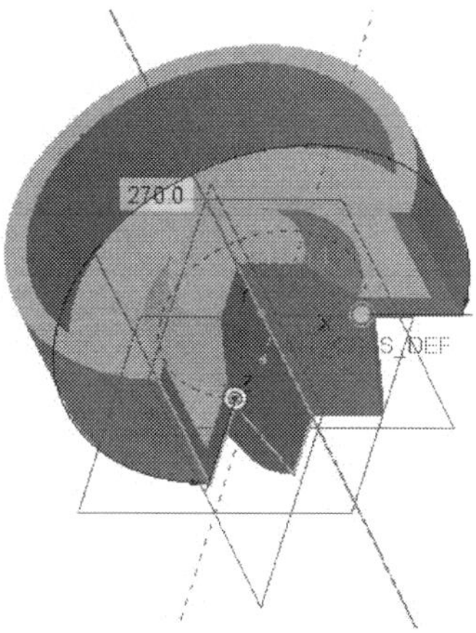

Figure 4-34 *The desired trimetric view of the model*

Case-2

Drawing a Sketch on the FRONT Datum Plane

In this case, you will select the **FRONT** datum plane as the sketching plane and the **RIGHT** datum plane as the reference plane. It is recommended that you decide the direction of the arrow at this stage so that you can create the model in the required orientation. Figure 4-35 shows the sketch drawn on the **FRONT** datum plane and the default direction of viewing the sketching plane. Note that the arrow points in the direction which is normal to the sketching plane. Figure 4-36 shows the default trimetric view of the solid model created in this case.

This is not the desired default orientation of the model. You can still get the desired orientation of the model by choosing the **Change angle direction of revolve to other side of the sketch** button from the **Revolve** dashboard, but you should understand the default direction of the arrow on the sketching plane. The desired orientation of the model is shown in Figure 4-37.

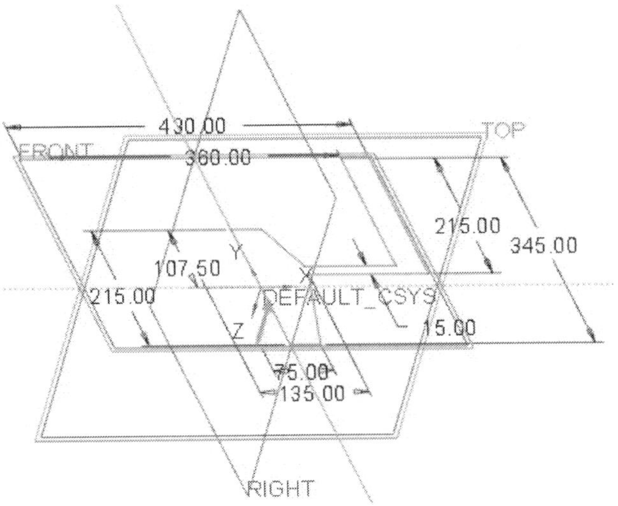

*Figure 4-35 Sketch drawn on the **FRONT** datum plane*

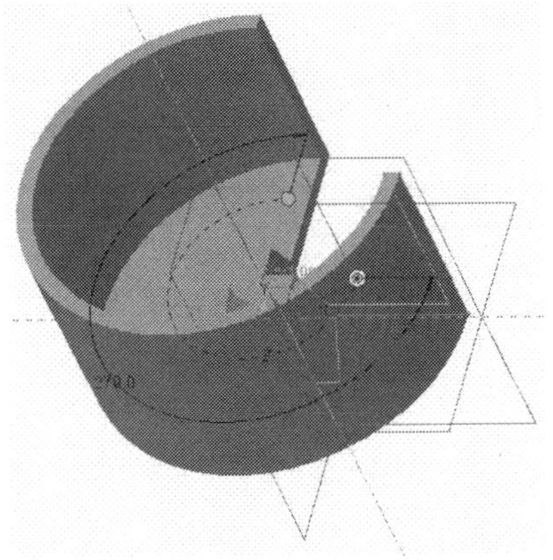

Figure 4-36 Default trimetric view of the model

Figure 4-38 shows the sketch drawn on the **FRONT** datum plane and the arrow is showing the direction of viewing the sketching plane. The direction of the arrow is reversed by choosing the **Flip** button in the **Sketch** dialog box. Figure 4-39 shows the default trimetric view of the model in this case.

In Figure 4-38, the sketch is revolved in the clockwise direction to create the model shown in Figure 4-39.

From the two cases discussed, the conclusion is that the arrow that will be displayed when you select the sketching plane is of great importance in orienting the model. Also, the direction of feature creation can be changed once it is created using the **Change angle direction of revolve to other side of the sketch** button from the **Revolve** dashboard.

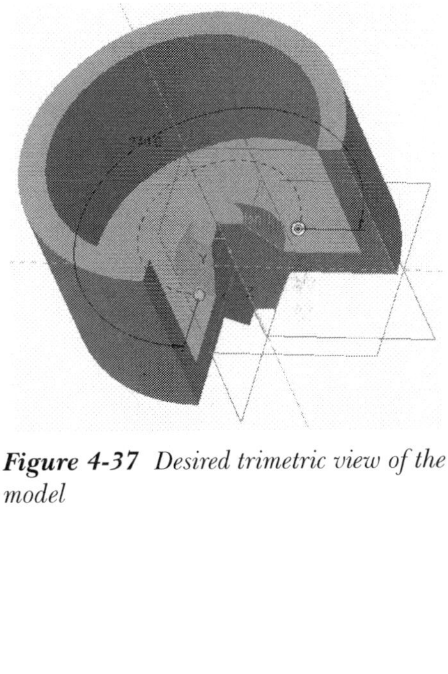

Figure 4-37 Desired trimetric view of the model

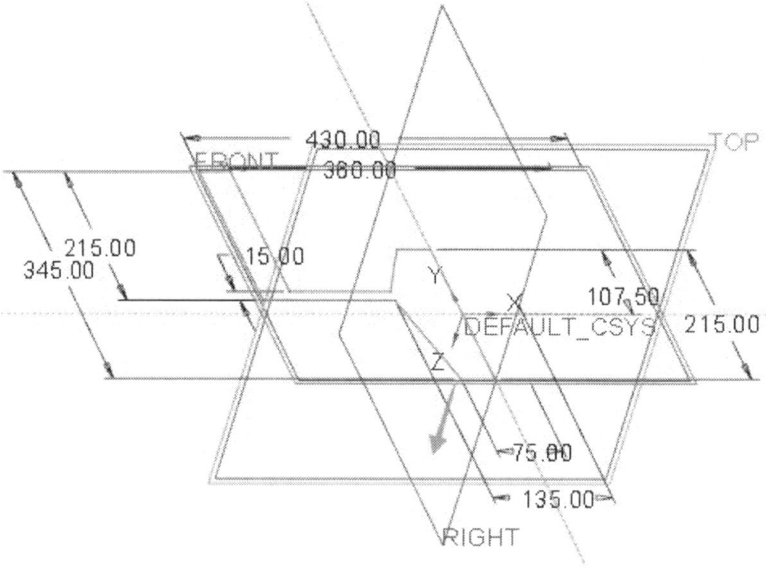

Figure 4-38 *Sketch drawn on the* ***FRONT*** *datum plane*

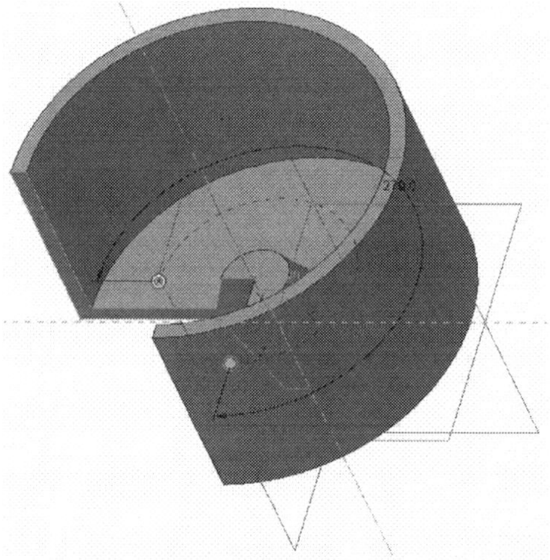

Figure 4-39 *Default trimetric view of the model*

CHANGING THE ORIENTATION OF THE MODEL

As discussed earlier, the orientation of a model depends upon the orientation of the base feature created. However, you can change the orientation of your model or create new orientations by using the **View** dialog box. To invoke the **View** dialog box, you can choose the **Reorient** option either from the **Saved Orientations** drop-down list in the **Orientation** group of the **View** tab or from the **Saved Orientations** drop-down list in the graphics toolbar, refer to Figure 4-40. The different options available in this dialog box are discussed next.

View Name

The **View Name** edit box displays the name of the view which you select from the **Saved Orientations** list box. This edit box also allows you to change the name of the saved orientation. The **Save** button next to the **View Name** edit box is used to save the settings that you change in the **View** dialog box.

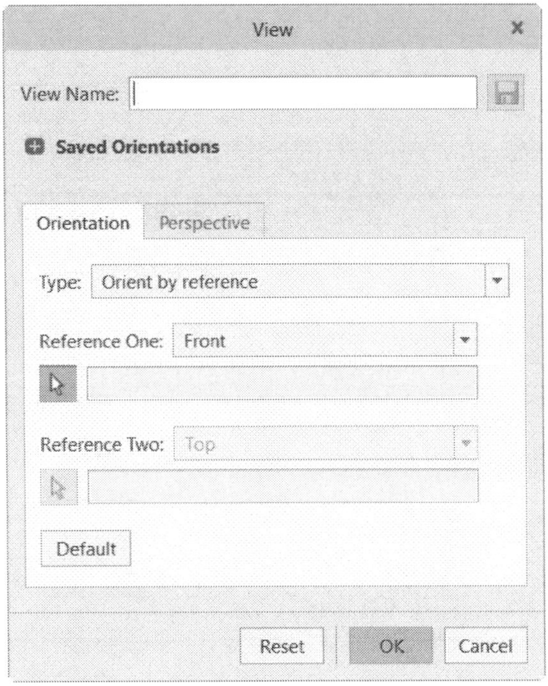

*Figure 4-40 The **View** dialog box*

Saved Orientations

This area displays the list of saved orientations in the current model. To expand the saved orientations list box, choose the ⊞ button if it is not expanded by default. To activate an orientation view, double-click on the desired view; the model will be oriented according to the selected view in the drawing area.

Name

The **Delete** button in this area is used to delete any saved orientation in the model. This button will be activated only when you select any saved orientation from the **Saved Orientations** list box.

Orientation Tab

The options available under this tab are used to orient the model in the drawing area. By choosing the options available in the **Type** drop-down list of this tab, you can orient your model in different ways. Choose the **Dynamic orient** option to reorient the model using the sliders available in the **Pan** and **Spin** areas. If you choose the **Orient by reference** option from the **Type** drop-down list, you can specify references such as axis and planes for orienting the model. If you choose the **Preferences** option, you can specify the spin center of the model by selecting the respective radio buttons available in the **Spin Center** area. You can orient the model in **Isometric**, **Trimetric**, or **User Defined** orientation by selecting the options available in the drop-down list of the **Default Orientation** area.

Perspective Tab

The options in this tab are used to customise the settings of perspective view. You can change the view mode by selecting the **Perspective** or **Orthographic** radio button in the **View** area of this tab. You can customize various settings of the perspective view mode by using the options such as **Focal Length (mm)**, **Maintain Field of View**, **Eye Distance (Dolly)**, and **Image Zoom**. In the perspective view, the object that are far away appear smaller than those nearby. In the orthographic view, all objects appear at the same scale.

PARENT-CHILD RELATIONSHIP

Every model created in Creo Parametric is composed of features that in some way or the other are related to other features in the model. The feature that occurs first in the **Model Tree** is called the parent feature and any feature(s) that is related to this feature is called the child feature(s).

> **Tip**
> *If you want to check the parent-child relationship of features in a model, choose the* ***Feature Information*** *tool from the* ***Investigate*** *group of the* ***Tools*** *tab in the* ***Ribbon;*** *you will be prompted to select a feature. After you select the feature, the* ***Feature info*** *window will be displayed. The child features of the selected feature will be displayed in the* ***Children*** *area and all parent features will be displayed in the* ***Parents*** *area.*

There are two types of relationships that can exist between the features, Implicit relationship and Explicit relationship.

Implicit Relationship

This type of relationship exists when the two features are related through equations. These equations are formed using relations.

Explicit Relationship

The explicit relationship is developed when a feature is used as a reference to create another feature. For example, one of the planar surfaces of a feature is used to create another feature, or an edge of a feature is used to dimension the other feature. In this case, the first feature is called the parent feature and the second feature is called the child feature of the first feature. Another example of this type of relationship is, when a hole is referenced to the edges of the surface it is placed on, or to the edges of some other surfaces. In this case, the hole is the child feature of the feature it is referenced to.

Remember that if the parent feature is modified, then the child feature will also be modified. Similarly, if the parent feature is deleted, then the child feature will also be deleted. For example, if you have used the three default datum planes to create the base feature and you delete any one of the datum planes, the base feature will also be deleted. In Chapter 6, you will learn how to break the parent-child relationship.

SKETCH REGION

You can use the **Sketch Region** option to make quick selection to create geometry with the selected sketch-based features. A sketch region is a closed contour defined by sketched entities and their intersection with coplanar 3D edges in the part geometry. Sketch-based feature geometry creation is faster and easier. The **Sketch Region** option reduces the need to perform project and trim operations within the sketch. It also offers a flexible way to use portions of a single sketch as the basis for several sketch-based features.

You can select the **Sketch Region** option from the selection filter area then select the sketch region, refer to Figure 4-41. Sketched entities that are involved can belong to different sketches of the same model. After you make your selection, a contextual mini toolbar provides access to the features such as Extrude, Revolve, Fill, and Sketch, refer to Figure 4-42.

> **Tip**
> • *Press SHIFT+S to switch to the Sketch Region selection filter.*
> • *Press SHIFT+G to switch back to the Geometry selection filter.*

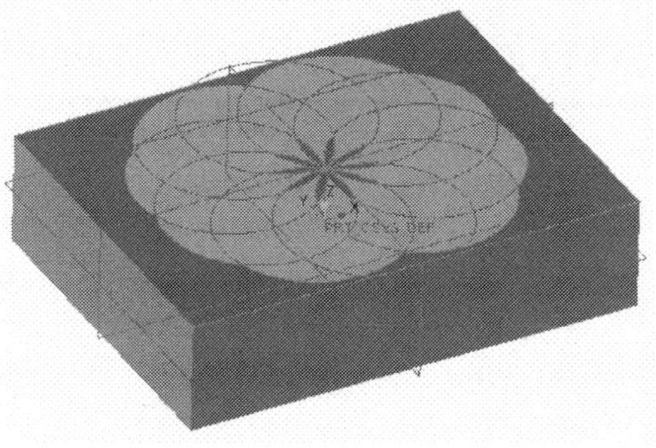

Figure 4-41 *Selected sketch region*

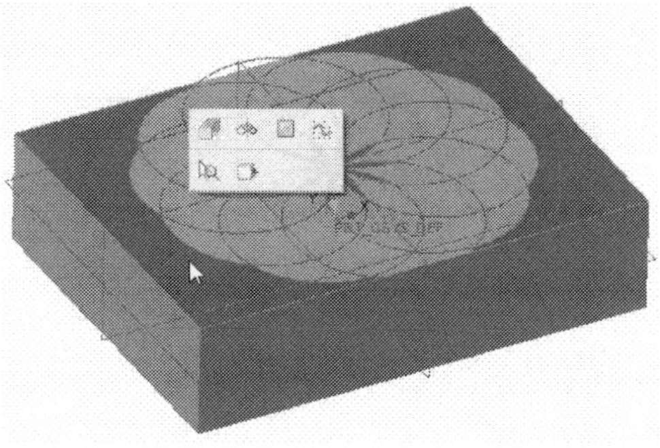

Figure 4-42 *The **mini** toolbar*

NESTING OF SKETCHES

The process of creating more than one closed loops in a single sketch for a single feature is known as nesting of sketches. These sketches are drawn in the sketcher environment. Figure 4-43 shows a sketch in which three circles are created on the base profile.

Advantages of Nesting the Sketches

1. One of the advantages of nesting the sketches is that the number of features used to create a model are reduced. In Figure 4-44, the model has two features. The base feature is the base plate and the second feature is the holes in the model. But, when you nest the two sketches to create the model, you are using only one feature to create the model.

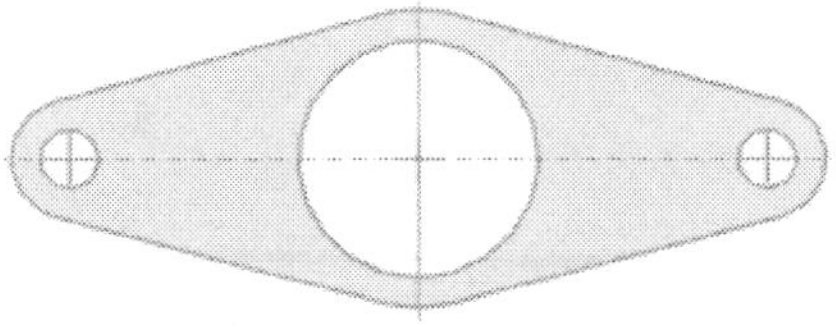

Figure 4-43 *Nested sketch*

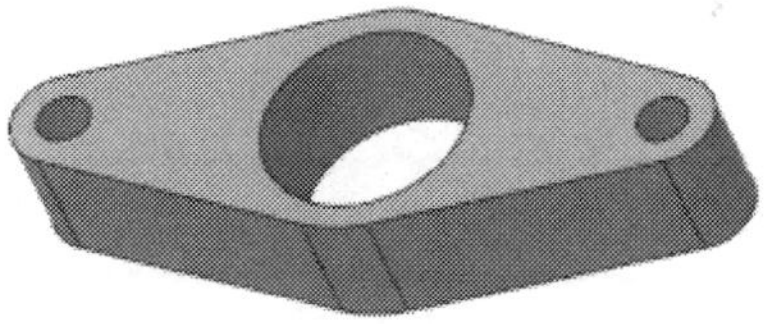

Figure 4-44 *Solid model of the sketch*

2. There is no parent-child relationship that exists in a nested sketch.

3. The depth of the hole is equal to the depth of extrusion of the base feature. This is because the two circles and the base profile are part of the same sketch. This depends on the designer whether he wants to use the nested sketch to create a part.

Disadvantages of Nesting the Sketches

1. In nesting, since the two features on the model are combined into one feature, therefore there is no flexibility in editing the features of a model.

2. If at a later stage, the designer needs to convert the circular holes into elliptical holes with depth as half the depth of extrusion of the base feature, it consumes a lot of time to edit the model.

After understanding the advantages and disadvantages of nesting the sketches, it is recommended to divide a model into separate features. Draw all features as individual features so that the created model is flexible. However, it depends on the need of the designer and the need of the model how the model is approached for creating.

CREATING CUTS

In Creo Parametric, you can create cuts in a model. Cutting is a material removal process and the option is available only when at least one base feature exists in the drawing area. You can create cuts in a feature by using the **Remove Material** button that is available on the dashboard of the feature creation tools. In this section, you will learn to create extrude and revolve cuts. The procedure to create a cut on a feature is similar to that of adding material or protrusion.

Removing Material by Using the Extrude Tool

You can create a cut by removing material from an existing feature using the **Extrude** tool. The material removed from the feature is defined by a sketch.

To create an extrude cut in a model, choose the **Extrude** tool from the **Shapes** group; a dashboard will be displayed. Choose the **Remove Material** button and enter the sketcher environment as discussed earlier in this chapter. After drawing the sketch for the cut feature, you can specify the direction of material removal with respect to the sketch using the arrow displayed in drawing area. For example, the arrow in Figure 4-45 shows the direction of material removal. After the required settings have been specified, choose the **OK** button to create the cut feature, as shown in Figure 4-46.

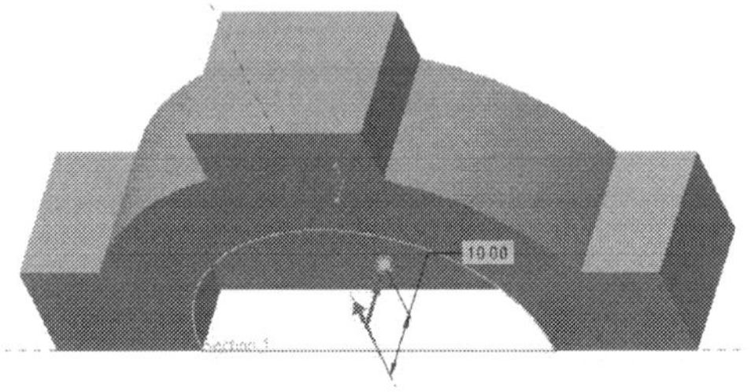

Figure 4-45 *Sketch for the extrude cut and arrow showing the direction of material removal*

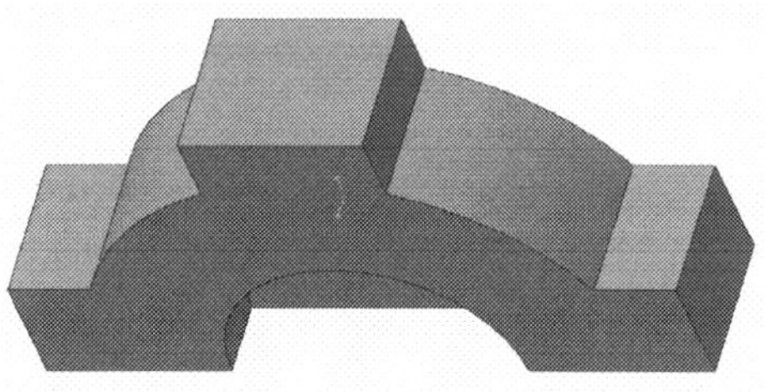

Figure 4-46 *Cut feature created on the model*

Tip

In the model shown in Figure 4-45, the sketching plane selected for the creation of extruded cut is the front face of the base feature and not a datum plane. In Chapter 5, you will learn to create a datum plane on the surface of an existing feature and select it as the sketching plane. But it is not recommended to create a datum plane if the planar surface of the feature can be used as the sketching plane.

You can also change the direction of material removal by choosing the **Change material direction of extrude to other side of sketch** button from the **Extrude** dashboard. On doing so, the arrow will point in the direction shown in Figure 4-47. In this case, the material on the plane selected for sketching will be removed, leaving the extruded cut feature, as shown in Figure 4-48.

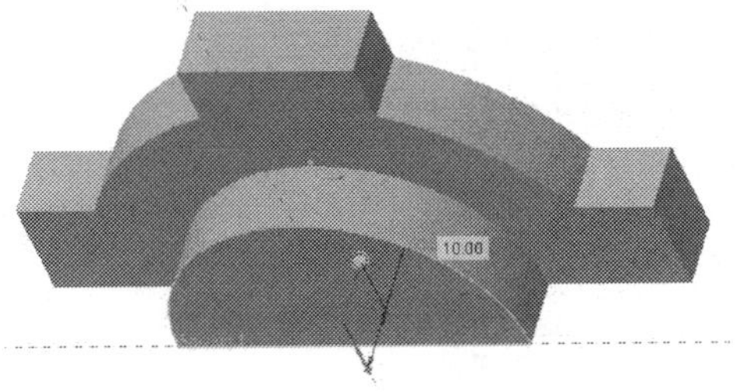

Figure 4-47 *Arrow showing the direction of material removal*

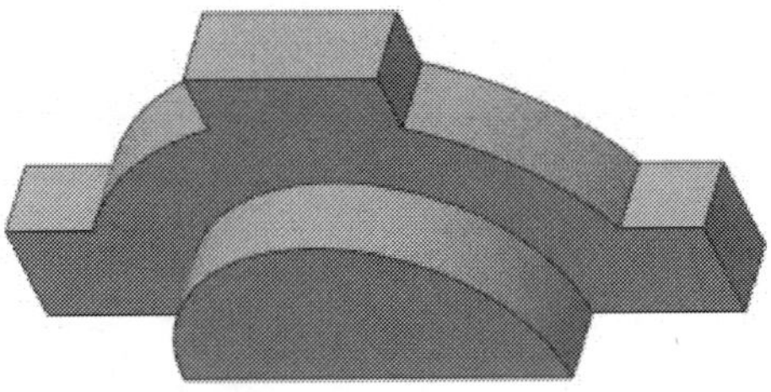

Figure 4-48 *Cut feature created in the direction of arrow*

Note

A simple hole can also be created by drawing its cross-section, which is a circle, and then creating an extrude cut. But Creo Parametric provides predefined placement for a hole feature, which is more desirable than dimensioning the cross-section of a cut feature. Simple holes do not require a sketch if you use the ***Hole*** *dashboard. The* ***Hole*** *dashboard is discussed in Chapter 6.*

Removing Material by Using the Revolve Tool

You can create a cut by removing material from an existing feature using the **Revolve** tool. The material to be removed will be defined by the sketch you draw.

To remove the material from an existing feature, choose the **Revolve** tool from the **Shapes** group, a dashboard will be displayed. Choose the **Remove Material** button to enter the sketcher environment, as discussed earlier in this chapter. Remember that the geometry centerline is necessary in a revolve feature. Figure 4-49 shows the section drawn to be revolved. The front face of the second extruded feature is selected as the sketching plane. Figure 4-50 shows the revolve cut created on the selected surface.

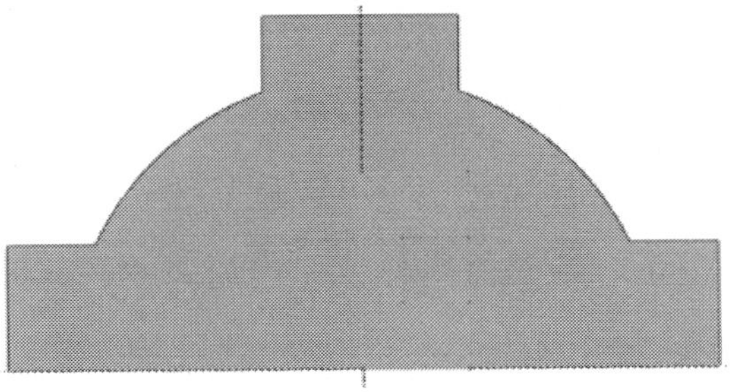

Figure 4-49 *The section for revolve cut*

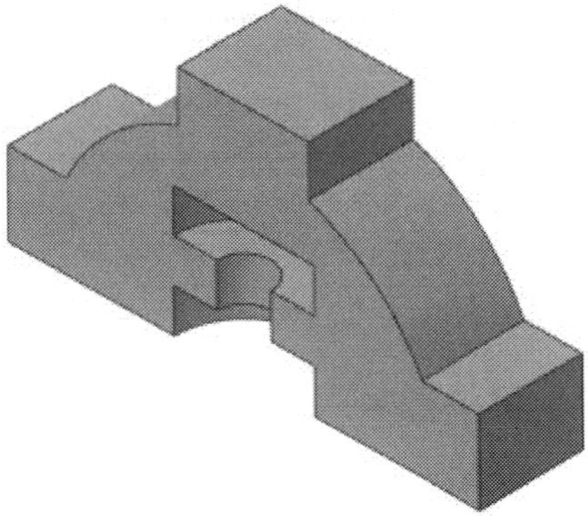

Figure 4-50 *Revolve cut created*

TUTORIALS

Tutorial 1

In this tutorial, you will create the model shown in Figure 4-51. The dimensions of the model are shown in Figure 4-52. **(Expected time: 30 min)**

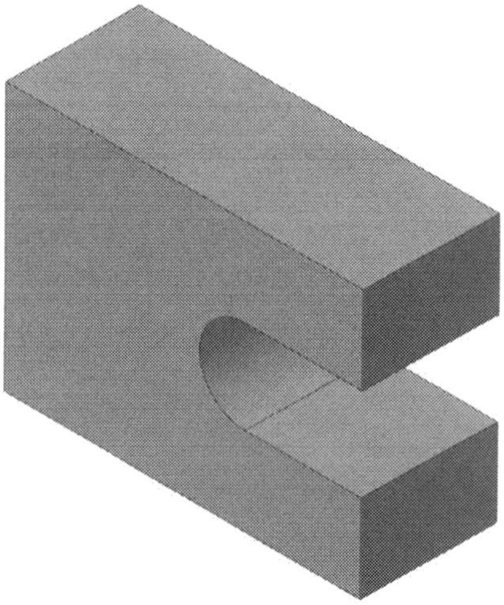

Figure 4-51 *The isometric view of the model*

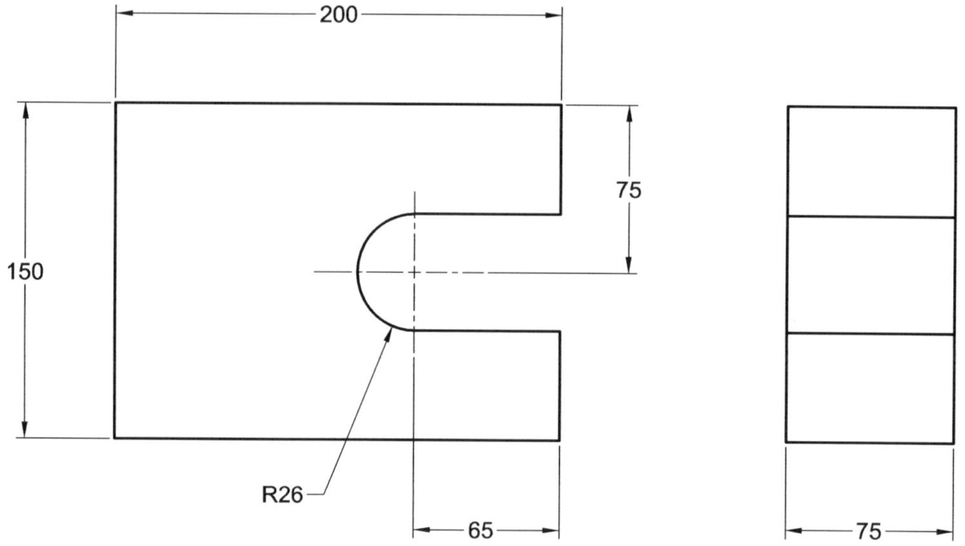

Figure 4-52 *The front and right views of the model with dimensions*

The following steps are required to complete this tutorial:

a. Set the working directory and create a new object file in the **Part** mode.
b. First examine the model and then determine the type of protrusion for the model. Select the sketching plane for the model and orient it parallel to the screen.
c. Draw the sketch by using the sketching tools and apply constraints and dimensions.
d. Exit the sketcher environment and define the model attributes.

Setting the Working Directory

After starting the Creo Parametric session, the first task is to set the working directory. A working directory is a directory on your system where you can save the work done in the current session of Creo Parametric. You can set any existing directory on your system as the working directory.

1. Choose the **Manage Session > Select Working Directory** from the **File** menu; the **Select Working Directory** dialog box is displayed. Browse to the *C:\Creo_8.0* folder.

2. Choose the **Organize** button from the **Select Working Directory** dialog box to display the flyout. Next, choose the **New Folder** option from the flyout; the **New Folder** dialog box is displayed.

3. Enter **c04** in the **New Directory** edit box and choose the **OK** button from the **New Folder** dialog box. Now, you have created a folder named *c04* in *C:\Creo_8.0.*

4. Next, choose the **OK** button from the **Select Working Directory** dialog box; the working directory is set to *C:\Creo_8.0\c04* and a message **Successfully changed to C:\Creo_8.0\c04 directory** is displayed in the message area.

Starting a New Object File

Solid models are created in the **Part** mode of Creo Parametric. The file extension for the files created in this mode is *.prt*.

1. Choose the **New** button from the **File** menu; the **New** dialog box is displayed. The **Part** radio button is selected by default in the **Type** area and the **Solid** radio button is selected by default in the **Sub-type** area of the **New** dialog box.

2. Enter the file name as *c04tut1* in the **File name** edit box and choose the **OK** button. The three default datum planes are displayed in the drawing area. Also, the **Model Tree** appears on the left of the drawing area in the Navigator.

> **Tip**
> *By default, a Creo file opens in* ***inch lbm second*** *unit. However, you can change it by choosing* ***Prepare > Model Properties > Units*** *option from the* ***File*** *menu.*

Selecting the Extrude Option

The given solid model is created by extruding the sketch to a distance of 75 units. Therefore, the sketch will be extruded as a solid to create the model.

1. Choose **Extrude** from the **Shapes** group in the **Model** tab; the **Extrude** dashboard is displayed on the top of the drawing area. All the attributes needed to create the model will be defined after the sketch is drawn.

Selecting the Sketching Plane

To create a sketch for the extruded feature, first you need to select the sketching plane for the model. The **FRONT** datum plane will be selected as the sketching plane. The sketching plane is selected such that the direction of extrusion of the solid model is perpendicular to it. From the isometric view of the model shown in Figure 4-51, it is evident that the direction of extrusion of the model is perpendicular to the **FRONT** datum plane.

1. Choose the **Placement** tab from the **Extrude** dashboard to display the slide-down panel. Next, choose the **Define** button from the slide-down panel; the **Sketch** dialog box is displayed.

2. Select the **FRONT** datum plane from the drawing area as the sketching plane. As you select the sketching plane, the reference plane and its orientation are set automatically. The reference plane is selected to orient the sketching plane.

   The arrow appearing on the sketching plane indicates the direction of viewing the sketch.

   In the **Sketch** dialog box, the **Reference** collector displays **RIGHT:F1(DATUM PLANE)**. This indicates that the **RIGHT** datum plane is selected as the reference plane. In the **Orientation** drop-down list, the **Right** option is selected by default. This means while drawing the sketch, the **RIGHT** datum plane will be on the right. The **RIGHT** datum plane will be perpendicular to the sketching plane and the sketching plane will be parallel to the screen.

3. Choose the **Sketch** button in the **Sketch** dialog box to enter the Sketcher environment.

Drawing the Sketch

You need to draw the sketch of the solid model that will be extruded later to create the 3D model.

1. Choose the **Corner Rectangle** tool from the **Sketching** group.

2. Draw a rectangle by defining its lower left corner and upper right corner. The rectangle is created and weak dimensions are applied to it.

   You will notice that some of the constraints are applied to the lines composing the rectangle. This is because while drawing the rectangle some constraints are applied automatically to the lines composing the rectangle.

3. Select the right vertical line; the line turns green. Press the DELETE key to delete it. The sketch after deleting the vertical line is shown in Figure 4-53.

4. Now, draw the lines and arc. Some weak dimensions and constraints are applied to the sketch, as shown in Figure 4-54.

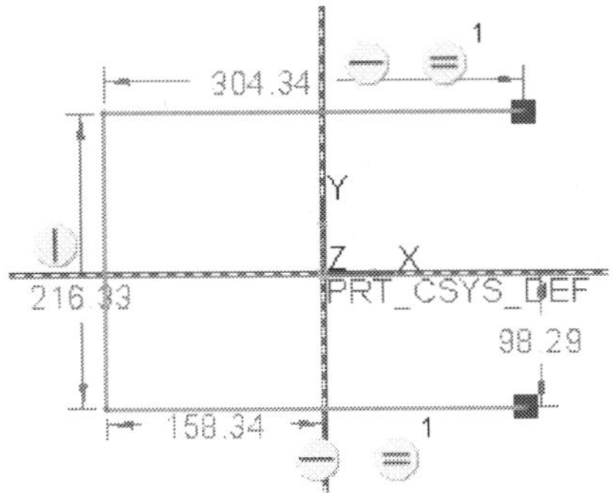

Figure 4-53 The sketch after deleting the vertical line

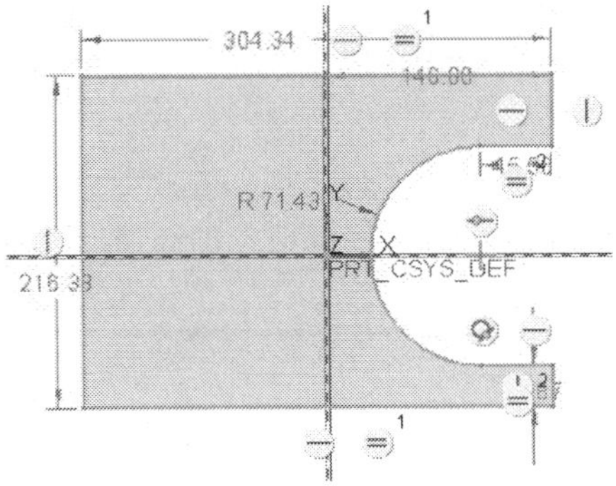

Figure 4-54 The sketch with weak dimensions

Note

The center of the arc and the ***TOP*** *datum plane are aligned by default. In case, they are not aligned, you need to align them. To do so, choose the* ***Coincident*** *button from the* ***Constrain*** *group. Select the center of the arc and then select the* ***TOP*** *datum plane. Now, the center and the datum plane are aligned.*

Tip

Since you are creating a solid sketch based feature, it is suggested that you do not turn off the ***Shade Closed Loops*** *and* ***Highlight Open Ends*** *buttons available in the* ***Inspect*** *group of the* ***Sketch*** *tab.*

Applying Constraints to the Sketch

You need to apply equal length constraints to the sketch in order to maintain the design intent of the model.

1. Choose the **Equal** tool from the **Constrain** group in the **Sketch** tab and select the two vertical lines on the right of the sketch. Now, the equal length constraint = is applied to both the lines. Next, double-click on the middle mouse button to exit the tool.

2. Select the two horizontal lines that are connected to the arc to apply the equal length constraint. As you apply the constraint to the lines, some of the weak dimensions will disappear from the drawing area and a label will be assigned to the equal constraint. The sketch after applying the equal length constraints is shown in Figure 4-55.

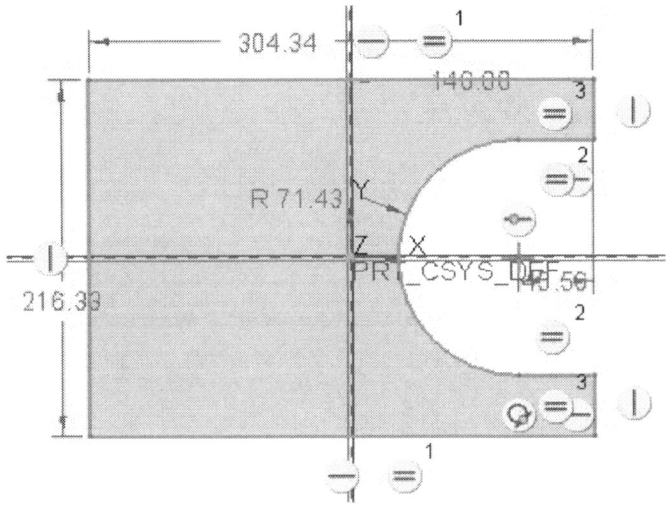

***Figure 4-55** Equal length constraints applied to the sketch*

Dimensioning the Sketch

Although some weak dimensions are applied to the sketch, you need to add a dimension to the sketch.

1. Choose the **Dimension** tool from the **Dimension** group.

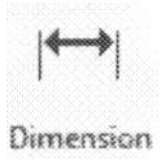

2. Select the center of the arc and the upper right vertical line and then place the dimension by using the middle mouse button, as shown in Figure 4-56.

   You need to dimension only these entities because the rest of the weak dimensions are useful dimensions and can be modified directly.

Note

If the dimensions in your sketch are different from those shown in Figure 4-50, add the missing dimensions and delete the dimensions that are not required.

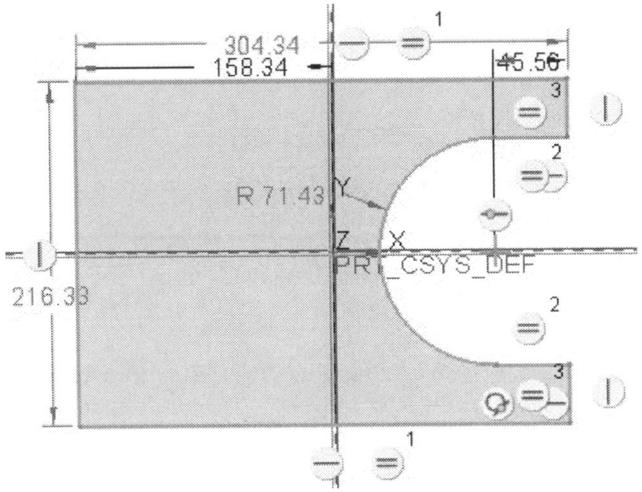

Figure 4-56 Dimension added to the sketch

Modifying the Dimensions

You need to modify the dimension values of the sketch. The default dimensions shown in Figure 4-56 also include the length and width of the rectangle and the distance of the sketched section from the selected references. It is recommended that you draw the base feature symmetrical with the other two datum planes. As a result, the distance from the **RIGHT** datum plane is 100 (200 divided by 2 is equal to 100).

1. Select the sketch and dimensions using the CTRL+ALT+A keys.

2. Choose the **Modify** tool from the **Editing** group; the **Modify Dimensions** dialog box is displayed. All dimensions in the sketch are displayed in this dialog box, and each dimension has a separate thumbwheel and an edit box. You can use the thumbwheel or the edit box to modify dimensions. It is recommended to use the edit boxes to modify dimensions, if the desired value is not obtained using the thumbwheel. 

3. Clear the **Regenerate** check box; any modification done in a dimension value does not update the sketch during modification. The dimensions get modified after you exit the **Modify Dimensions** dialog box. It is recommended to clear the **Regenerate** check box when more than one dimension has to be modified.

4. Modify all dimensions one by one, as shown in Figure 4-57. You will notice that the dimension you select in the **Modify Dimensions** dialog box is enclosed in a violet box in the drawing area.

5. After modifying the dimensions, choose the **OK** button from the **Modify Dimensions** dialog box; the message **Dimension modifications successfully completed** is displayed in the message area.

6. Choose the **OK** button from the **Close** group to exit the sketcher environment.

Specifying the Model Attributes

Next, you need to specify the attributes to create the model.

1. Choose the **Standard Orientation** option from the **Saved Orientations** flyout in the **Graphics** toolbar or press CTRL+D; the default trimetric view of the model is displayed in the drawing area. 

   This display gives you a better view of the sketch in the 3D space. The model is displayed in orange. Also, the pink colored arrow is displayed on the model indicating the direction of extrusion. The model may not fit fully in the drawing area.

2. Press CTRL+middle mouse button and drag the mouse downward in the drawing area. Notice that a red rubber band line is attached to the cursor. The length of the rubber band line gives an idea about the extent you need to zoom the model. After the model is visible in the drawing area, release the middle mouse button and the CTRL key.

Note

You can also fit the model into the drawing area by choosing the ***Refit*** *button from the* ***Graphics*** *toolbar.*

The model appears, as shown in Figure 4-58. All attributes selected by default in the **Extrude** tab will be used to create the model. You need to change only the depth of extrusion.

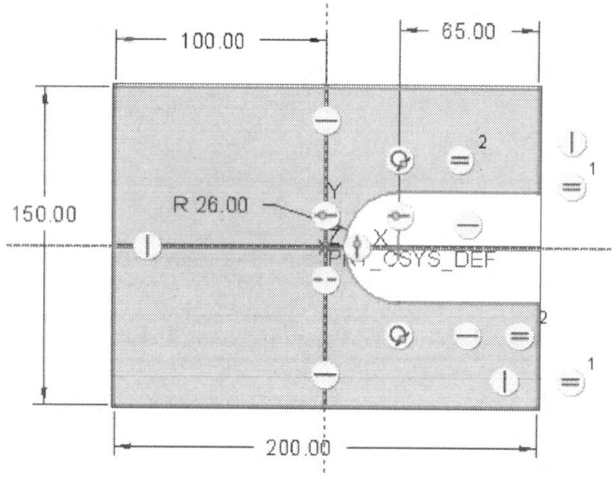

Figure 4-57 *Sketch after modifying the dimensions*

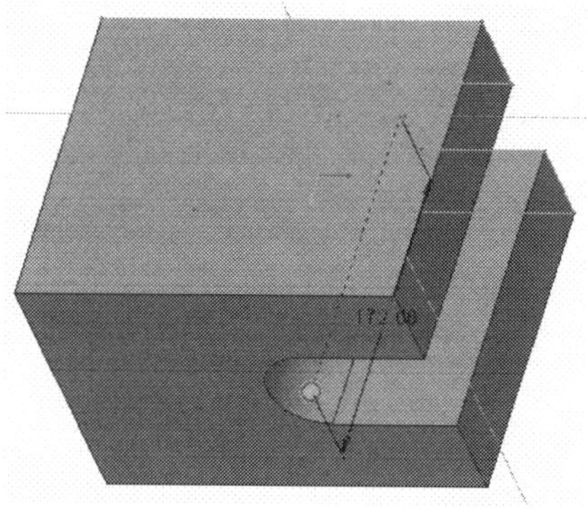

Figure 4-58 *Arrow showing the direction of feature creation*

3. Enter **75** in the edit box present on the **Extrude** tab, and press ENTER; the model in the drawing area is displayed with the specified depth of extrusion.

4. Choose the **OK** button from the **Extrude** dashboard to confirm the feature creation and exit the current feature creation tool. The trimetric view of the model is shown in Figure 4-59.

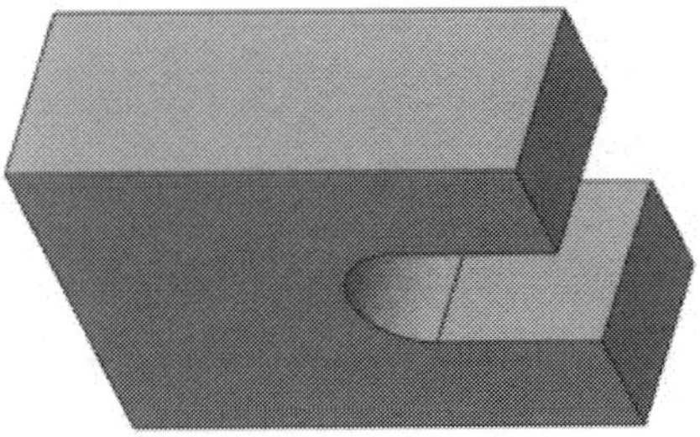

Figure 4-59 Default trimetric view of the model

Note

*In Figure 4-59, the display of datum planes and the coordinate system are turned off by clearing the **Plane Display** and **Csys Display** check boxes, respectively from the **Datum Display Filters** drop-down list in the **Graphics** toolbar.*

Saving the Model and Closing the File

1. Choose the **Save** option from the **File** menu or choose the **Save** button from the **Quick Access** toolbar; the **Save Object** dialog box is displayed with the name of the object file that you have specified earlier.

2. Choose the **OK** button from the **Save Object** dialog box to save the file.

Note

*You can also change the current name of the file before saving it by choosing the **Rename** option from the **Manage File** flyout in the **File** menu. If you do so, the **Rename** dialog box is displayed. Enter the required name of the file in the **New file name** edit box and then choose the **OK** button to accept the new name.*

3. Choose **File > Close** from the menu bar or choose the **Close** button from the **Window** group in the **View** tab to close the current window.

Tutorial 2

In this tutorial, you will create the model shown in Figure 4-60. The dimensions of the model are shown in Figure 4-61. **(Expected time: 30 min)**

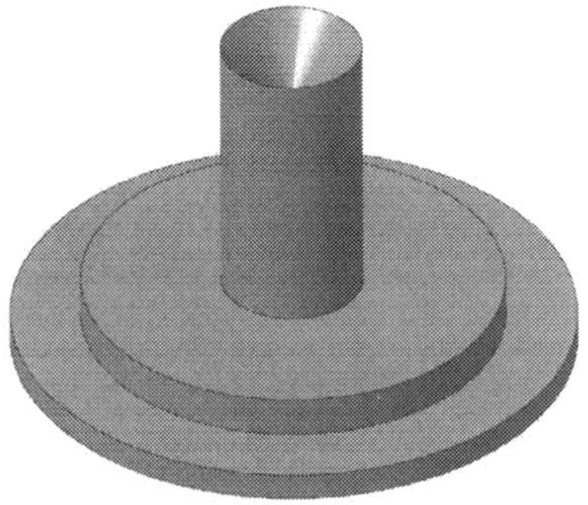

Figure 4-60 The isometric view of the solid model

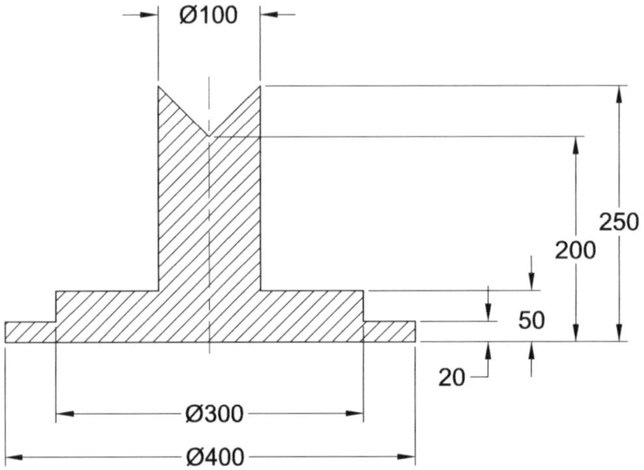

Figure 4-61 The front section view of the solid model

The following steps are required to complete this tutorial:

a. Create a new object file in the **Part** mode.
b. First examine the model and then determine the type of protrusion required for the model. Next, select the sketching plane for the model.
c. Draw the sketch for the revolved feature and a centerline to revolve it using the sketching tools. Next, apply dimensions to the model.
d. Exit the sketcher environment and define the model attributes.

Setting the Working Directory

The working directory was selected in Tutorial 1; therefore, there is no need to select it again. But if you are starting a new session of Creo Parametric, you need to set the working directory again by following the steps discussed earlier.

Starting a New Object File

1. Open a new object file in the **Part** mode and then name the file as *c04tut2*.

The three default datum planes are displayed in the drawing area. However, if the default datum planes were turned off in the previous tutorial, then they will not appear in the drawing area.

2. Turn on the display of datum planes by selecting the **Plane Display** check box available in the **Datum Display Filter** drop-down list in the **Graphics** toolbar.

Selecting the Revolve Tool

The given solid model is a revolved feature that will be created by revolving the sketch through an angle of 360-degree about an axis. Therefore, the **Revolve** tool will be used to create the model. The procedure to create the solid model using the revolve tool is given next.

1. Choose the **Revolve** tool from the **Shapes** group in the **Model** tab; the **Revolve** dashboard is displayed in the **Ribbon**.

   Some of the attributes like rotation angle and direction of rotation of the sketch will be defined after the sketch has been created.

Selecting the Sketching Plane

To create the sketch of the model, first you need to select a sketching plane for the model. Note that the axis of revolution of the revolved feature is normal to the **TOP** datum plane in the model. Therefore, any of the other two datum planes, other than the **TOP** datum plane, can be selected as the sketching plane. Here, you will select the **FRONT** datum plane as the sketching plane. To do so, you need to follow the steps given next.

1. Choose the **Placement** tab from the dashboard to display the slide-down panel. Choose the **Define** button from the side-down group; the **Sketch** dialog box is displayed.

2. Select the **FRONT** datum plane as the sketching plane. As you select the sketching plane, the reference plane and its orientation are set automatically. The reference plane is selected in order to orient the sketching plane.

   In the **Sketch** dialog box, the **Reference** collector displays **RIGHT:F1(DATUM PLANE)**. This indicates that the **RIGHT** datum plane is selected as the reference plane. In the **Orientation** drop-down list, the **Right** option is selected by default. As a result, while drawing the sketch, the **RIGHT** datum plane will be on the right.

   The **RIGHT** datum plane will be perpendicular to the sketching plane and the sketching plane will be parallel to the screen.

3. Choose the **Sketch** button from the **Sketch** dialog box.

Drawing the Sketch

Now, you need to draw the sketch of the revolved feature. The sketch to be drawn is the cross-section of the revolved feature, which will be revolved about the centerline.

1. Choose the **Line Chain** tool from the **Sketching** group and then draw the sketch, refer to Figure 4-60. The sketch should be a closed loop and the bottom horizontal line should be aligned to the **TOP** datum plane.

   As you draw the sketch, weak dimensions and strong constraints are applied to the sketch.

2. Choose the **Centerline** tool from the **Datum** group or from the **Sketching** group. Draw a centerline for the axis of revolution. The centerline should be drawn such that it is aligned with the **RIGHT** datum plane, refer to Figure 4-62.

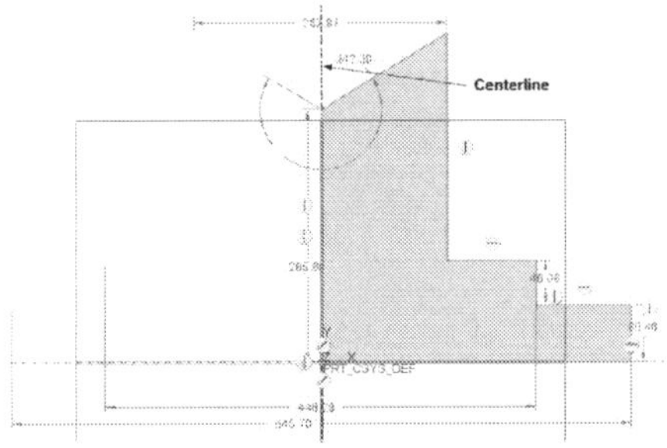

***Figure 4-62** Sketch with weak dimensions*

> **Tip**
> *If you have drawn a centerline on both axes, then press and hold right mouse button on the desired centerline to display a shortcut menu. Choose the **Designate Axis of Revolution** option from the shortcut menu to revolve the sketch around that particular centerline.*

Dimensioning the Sketch

Weak dimensions are automatically applied to the sketch. Since the model is a revolved feature, you need to manually apply the linear diameter dimensions to the sketch. The linear diameter dimensions are applied by using the centerline that was drawn in the sketch.

> **Tip**
> *The linear diameter dimensioning is necessary for all revolved features because mostly all revolved models are machined on a lathe. Therefore, while machining a revolved model, it is necessary that the operator of the machine has a drawing of the model that is diametrically dimensioned.*

1. Choose the **Dimension** tool from the **Dimension** group.

2. Select the centerline and the first right vertical line.

3. Now, use the middle mouse button to place the dimension on the top of the sketch; the linear dimension is placed. You can also convert a linear dimension into a diameter dimension. To do so, exit the **Dimension** tool, select the linear dimension and choose the **Diameter** option from the mini popup toolbar; the linear dimension is converted into the diameter dimension.

4. Select the centerline, the second right vertical line, and then again the centerline.

5. Now, press the middle mouse button to place the dimension below the sketch.

6. Select the centerline, the third right vertical line, and then again the centerline. Now, press the middle mouse button to place the dimension below the previous dimension.

   Dimension the remaining entities in the sketch, as shown in Figure 4-63.

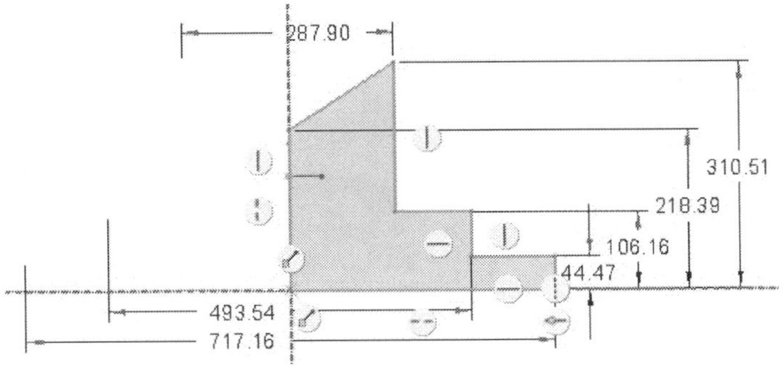

Figure 4-63 Sketch after dimensioning

Modifying the Dimensions

When you dimension a sketch, default dimension values are applied to the sketch. You need to modify the dimension values of the sketch.

1. Select the sketch and dimensions by using the CTRL+ALT+A keys.

2. Choose the **Modify** tool from the **Editing** group; the **Modify Dimensions** dialog box is displayed. 

3. In this dialog box, clear the **Regenerate** check box and then modify the values in the dimensions, as shown in Figure 4-64. If you clear this check box, any modification of the dimension value will not update the sketch. It is recommended that you clear the **Regenerate** check box if more than one dimension has to be modified.

   You will notice that the dimension that you have selected in the **Modify Dimensions** dialog box is enclosed in a violet box in the drawing area.

4. After modifying all dimensions, choose the **OK** button from the **Modify Dimensions** dialog box; the message **Dimension modifications successfully completed** is displayed in the message area.

5. Choose the **OK** button to exit the sketcher environment.

Specifying the Model Attributes

When you exit the sketcher environment, the **Revolve** dashboard above the drawing area is enabled again. Using this dashboard, you can specify the angle of revolution for the revolved feature.

1. Choose the **Saved Orientations** button from the **Orientation** group in the **View** tab to display a flyout. Choose the **Default Orientation** option from the flyout or press CTRL+D; the model is oriented in its default orientation, that is, trimetric view, as shown in Figure 4-65.

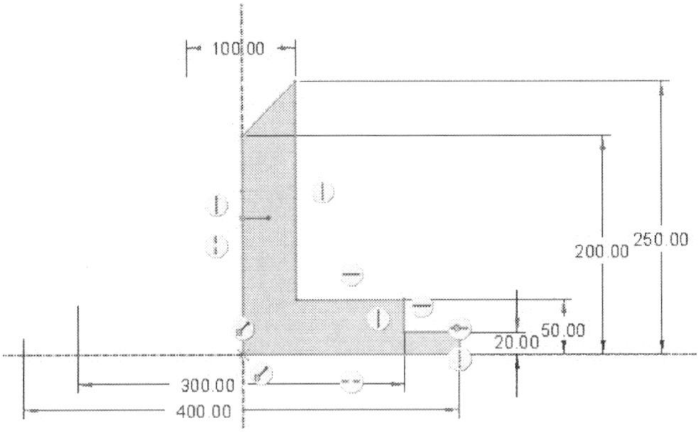

Figure 4-64 Sketch after modifying the dimensions

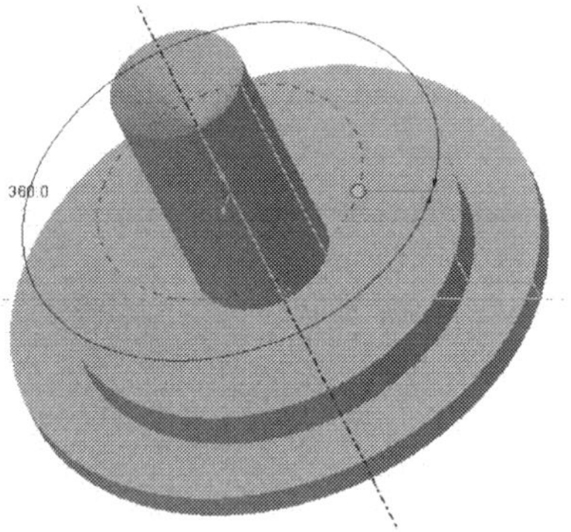

Figure 4-65 Preview of the model in the default trimetric view

This display gives you a better view of the sketch in the 3D space and the model appears in orange color. The drag handle is also available on the model that can be used to modify the angle of revolution dynamically.

All attributes needed to create a solid model are selected by default in the **Revolve** dashboard.

Note

*If the model is not fully displayed in the drawing area, you can choose the **Refit** option from the graphics toolbar to fit the model in the drawing area.*

2. Choose the **OK** button in the **Revolve** dashboard. The model appears as the one shown in Figure 4-66.

Figure 4-66 The default trimetric view of the model

Saving the Model

1. Choose the **Save** option from the **File** menu or choose the **Save** button from the **Quick Access** toolbar; the **Save Object** dialog box is displayed with the name of the object file specified earlier.

2. Choose the **OK** button to save the file.

Closing the Current Window

The given model is completed and is also saved. Now, you can close the current window.

1. Choose **File > Close** from the menu bar.

Tutorial 3

In this tutorial, you will create the model shown in Figure 4-67. The dimensions of the sketch are shown in Figure 4-68 The thickness of the model is 1 unit and depth of extrusion is 25 units.

(Expected time: 30 min)

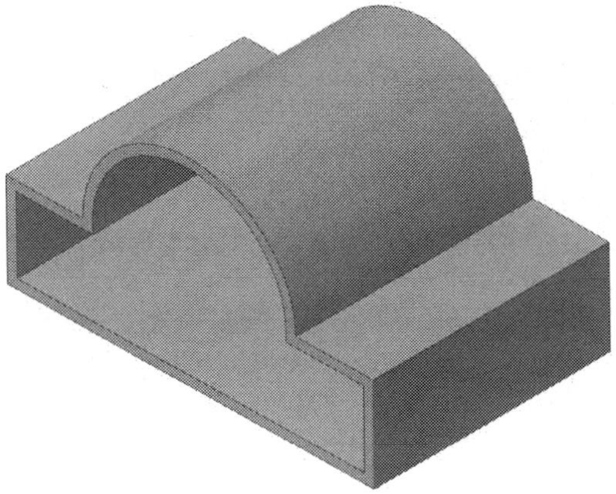

Figure 4-67 *The isometric view of the model*

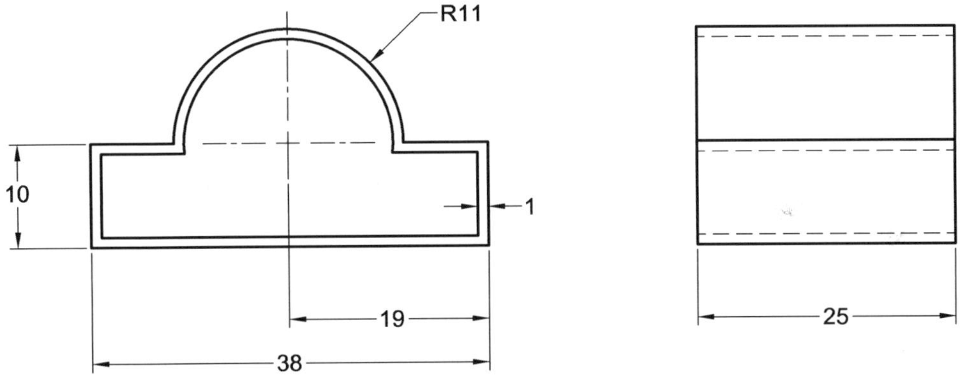

Figure 4-68 *The front and the right views*

The following steps are required to complete this tutorial:

a. Create a new object file in the **Part** mode.
b. First examine the model and then determine the type of protrusion for the model. Next, select the sketching plane for the model.
c. Draw the sketch by using the sketching tools and then apply dimensions to it.
d. Specify the attributes to create the model and then save it.

The working directory was selected in Tutorial 1, and therefore, you do not need to select the working directory again. But if a new session of Creo Parametric is started, then set the working directory using the Navigator.

Starting a New Object File

1. Open a new object file in the **Part** mode and then name it as *c04tut3*.

If the default datum planes were not turned off in the previous tutorial, they will appear in the drawing area. If the datum planes are not displayed, turn them on by using the **Plane Display** button.

Selecting the Extrude Option

The given thin solid model is created by extruding the sketch to a distance of 25 units. So, you need to use the **Thicken Sketch** button on the **Extrude** dashboard to create the model.

1. Choose the **Extrude** tool from the **Shapes** group; the **Extrude** tab is displayed in the **Ribbon**.

Most of the attributes needed to create the model will be defined after the sketch is drawn.

Selecting the Sketching Plane

From the isometric view of the model shown in Figure 4-67, it is clear that the direction of extrusion of the solid model is perpendicular to the **FRONT** datum plane. Therefore, you need to select the **FRONT** datum plane as the sketching plane.

1. Choose the **Placement** tab from the **Extrude** dashboard to display the slide-down panel and choose the **Define** button from the slide-down panel; the **Sketch** dialog box is displayed.

2. Select the **FRONT** datum plane as the sketching plane; the reference plane and its orientation are set automatically.

In the **Sketch** dialog box, the **Reference** collector displays **RIGHT:F1(DATUM PLANE)**. This indicates that the **RIGHT** datum plane is selected as the reference plane. In the **Orientation** drop-down list, the **Right** option is selected by default. As a result, while drawing the sketch, the **RIGHT** datum plane will be on the right.

The **RIGHT** datum plane will be perpendicular to the sketching plane and the sketching plane will be parallel to the screen.

3. Choose the **Sketch** button from the **Sketch** dialog box to enter the sketching environment.

Drawing the Sketch

Using the sketcher tools, you need to draw the sketch for the thin extruded model.

1. Draw the sketch of the model, as shown in Figure 4-69, by using the **Line Chain** and **3-Point / Tangent End** tools from the **Sketching** group.

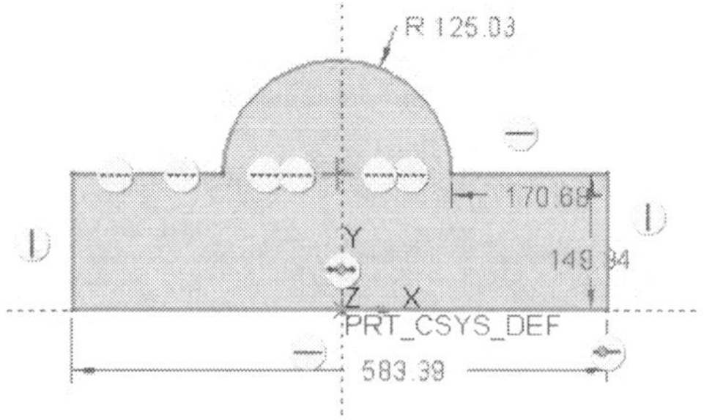

Figure 4-69 *Sketch of the thin extruded model*

Applying Constraints to the Sketch

While drawing the sketch, some weak dimensions are applied to the sketch. These dimensions appear in gray color. Although the weak dimensions are applied, you need to apply other constraints to the sketch.

1. Choose the **Coincident** tool from the **Constrain** group.

2. Select the center of the arc and then the **RIGHT** datum plane; the center of the arc is aligned with the **RIGHT** datum plane.

3. Select the bottom horizontal line and then the **TOP** datum plane. The bottom horizontal line is aligned with the **TOP** datum plane. Skip this point, if the two horizontal lines on the top are aligned with the **TOP** datum plane.

4. Choose the **Equal** tool from the **Constrain** group and then select the two horizontal lines on the top to apply the equal length constraint.

5. Choose the **Perpendicular** tool and select the left horizontal line and the arc; the perpendicular constraint symbol is applied. Similarly, select the right horizontal line and the arc; the perpendicular constraint symbol is applied. If these constraints have already been applied, the **Resolve Sketch** dialog box is displayed. Choose **Undo** from this dialog box.

   The sketch after applying constraints is shown in Figure 4-70.

Note

When you apply constraints to the sketch, some of the weak dimensions are automatically deleted.

Modifying Dimensions

You need to modify the dimension values of the weak dimensions.

1. Select the sketch and dimensions by using the CTRL+ALT+A keys.

2. Choose the **Modify** tool from the **Editing** group; the **Modify Dimensions** dialog box is displayed.

3. In this dialog box, clear the **Regenerate** check box and then modify the values of the dimensions, as shown in Figure 4-71. As mentioned earlier, it is recommended that you clear the **Regenerate** check box when more than one dimension has to be modified.

   You will notice that the dimension selected in the **Modify Dimensions** dialog box is enclosed in a violet box in the drawing area.

4. After modifying all dimensions, choose the **OK** button from the **Modify Dimensions** dialog box; the message **Dimension modifications successfully completed** is displayed in the message area.

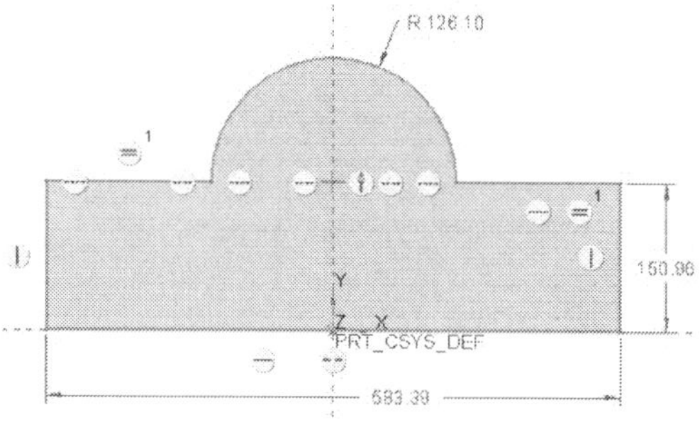

Figure 4-70 Sketch after applying constraints

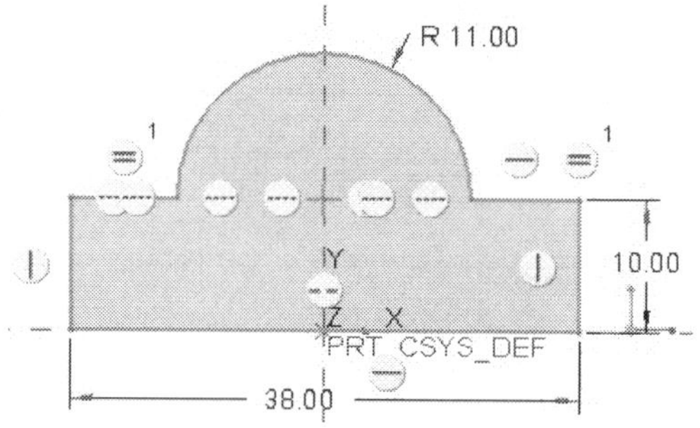

Figure 4-71 Sketch after modifying the dimensions

5. Choose the **OK** button to exit the sketcher environment.

Specifying the Model Attributes

The attributes required to create the model need to be selected from the **Extrude** dashboard.

1. Choose the **Thicken Sketch** button from the **Extrude** dashboard; the model changes to a thin model of default thickness, as shown in Figure 4-72.

2. Choose the **Saved Orientations** button from the **View** tab; a flyout is displayed. Choose the **Default Orientation** option from the flyout or you can press CTRL+D.

   The default trimetric view of the model is displayed. The model appears orange in color. If the model does not fit completely on the screen, then you need to zoom out. This display gives you a better view of the model in the 3D space.

   There are two drop-down lists in the **Extrude** dashboard. The drop-down list on the left is used to enter the depth of extrusion and the drop-down list on the right appears only when the **Thicken Sketch** button is chosen. This drop-down list is used to specify the thickness value of the thin model.

3. Enter **25** in the left edit box and press ENTER.

4. Enter **1** in the right edit box and press ENTER. The model appears, as shown in Figure 4-73. The model is completed and now you need to exit the **Extrude** dashboard.

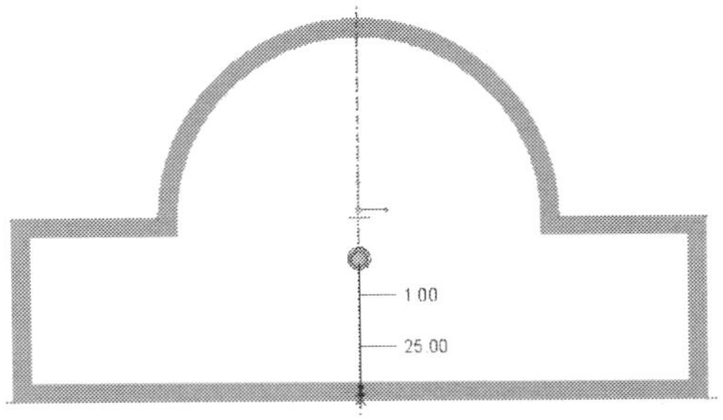

*Figure 4-72 Thickness added outside the sketch after choosing the **Thicken Sketch** button*

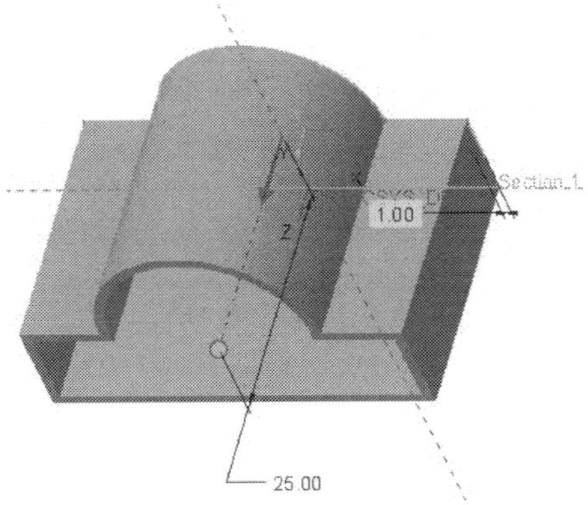

Figure 4-73 The default trimetric view of the model

5. Choose the **OK** button from the **Extrude** dashboard; the default trimetric view of the model is displayed, as shown in Figure 4-74.

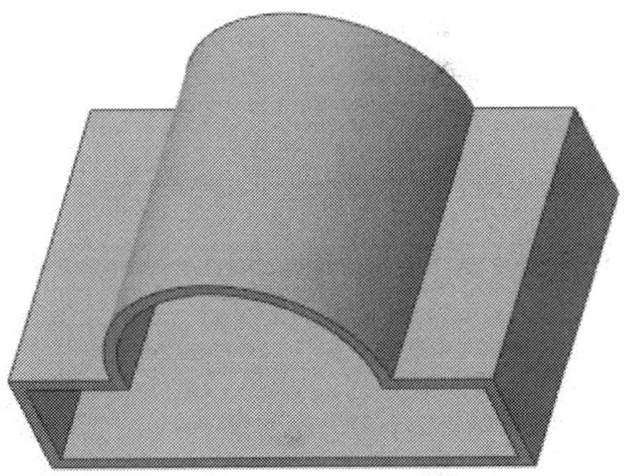

Figure 4-74 The default trimetric view of the solid model

Saving the Model

1. Choose the **Save** option from the **File** menu or choose the **Save** button from the **Quick Access** toolbar; the **Save Object** dialog box is displayed with the name of the object file specified earlier.

2. Choose the **OK** button to save the file.

Closing the Current Window

The given model is completed and is also saved. Now, you need to close the current window.

1. Choose **File > Close** from the menu bar.

Tutorial 4

In this tutorial, you will create the model shown in Figure 4-75. The dimensions of the sketch are shown in Figure 4-76. **(Expected time: 30 min)**

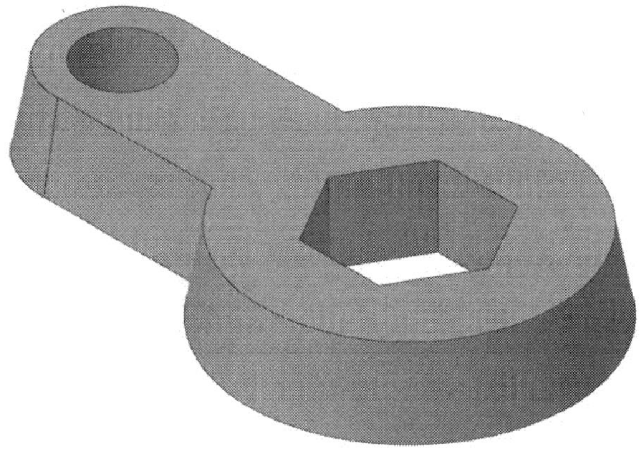

Figure 4-75 The isometric view of the model

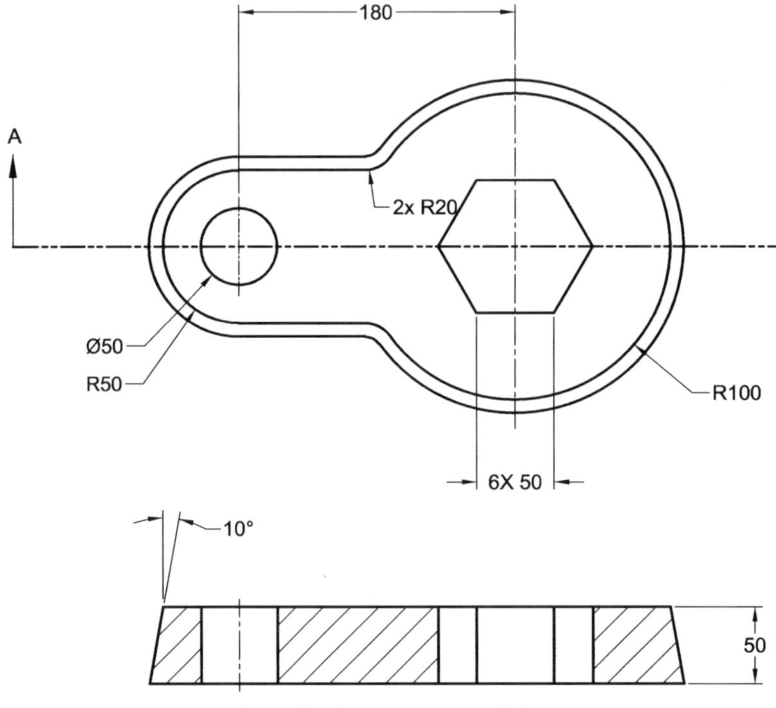

Figure 4-76 The front section and the top views

The following steps are required to complete this tutorial:

a. Create a new object file in the **Part** mode.
b. Determine the type of protrusion for the model.
c. Create the sketch for base feature.
d. Specify the attributes to create the base feature.
e. Create the cut feature by selecting the top face as the sketching plane.
f. Specify the attributes to create the cut feature and then save it.

The working directory was selected in Tutorial 1, and therefore, you do not need to select the working directory again. But if a new session of Creo Parametric is started, then set the working directory using the Navigator.

Starting a New Object File

1. Open a new object file in the **Part** mode and then name it as *c04tut4*.

   If the default datum planes were not turned off in the previous tutorial, they will appear in the drawing area. If the datum planes are not displayed, turn them on by using the **Plane Display** check box.

Selecting the Extrude Tool

The desired solid model will be created by extruding the sketch to a distance of 50 units. Therefore, the sketch will be extruded as a solid to create the model.

1. Choose the **Extrude** tool from the **Shapes** group; the **Extrude** tab is displayed in the **Ribbon**.

   Most of the attributes needed to create the model will be defined after the sketch is drawn.

Selecting the Sketching Plane

From the isometric view of the model shown in Figure 4-76, it is clear that the direction of extrusion of the solid model is perpendicular to the **TOP** datum plane. Therefore, you need to select the **TOP** datum plane as the sketching plane.

1. Choose the **Placement** tab from the **Extrude** dashboard to display the slide-down panel. Choose the **Define** button from the slide-down panel; the **Sketch** dialog box is displayed.

2. Select the **TOP** datum plane as the sketching plane; the reference plane and its orientation are set automatically.

   In the **Sketch** dialog box, the **Reference** collector displays **RIGHT:F1(DATUM PLANE)**. This indicates that the **RIGHT** datum plane is selected as the reference plane. In the **Orientation** drop-down list, the **Right** option is selected by default. As a result, while drawing the sketch, the **RIGHT** datum plane will be on the right.

3. Choose the **Sketch** button from the **Sketch** dialog box to enter the sketching environment.

Drawing the Sketch

Using the sketcher tools, you need to draw the sketch for the extruded model.

1. Draw the sketch of the model by using the **Line Chain**, **3-Point / Tangent End**, and **Circle** tools from the **Sketching** group, as shown in Figure 4-77.

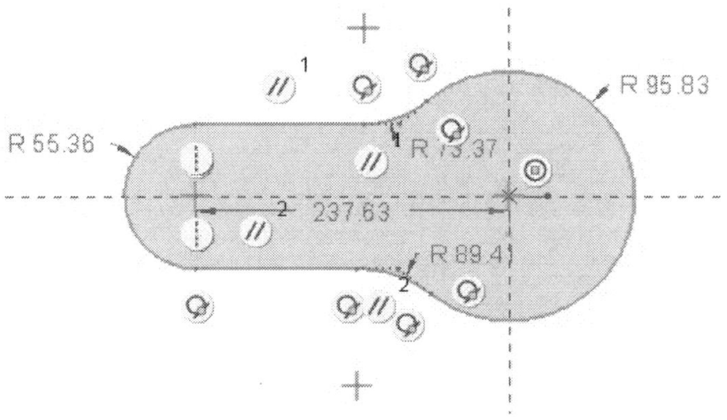

Figure 4-77 First sketch of the model

Applying Constraints to the Sketch

While drawing the sketch, some weak dimensions are applied to the sketch. These dimensions appear in blue color. Although the weak dimensions are applied, you need to apply other constraints to the sketch.

1. Choose the **Equal** tool from the **Constrain** group and select the two lines and circular fillet to apply the equal length constraint.

2. Choose the **Tangent** tool and select the lines and arcs one by one; the tangent constraint symbol is applied. If these constraints have already been applied, the **Resolve Sketch** dialog box is displayed. Choose **Undo** from this dialog box, the given constraint will be deleted.

   The sketch after applying constraints is shown in Figure 4-78.

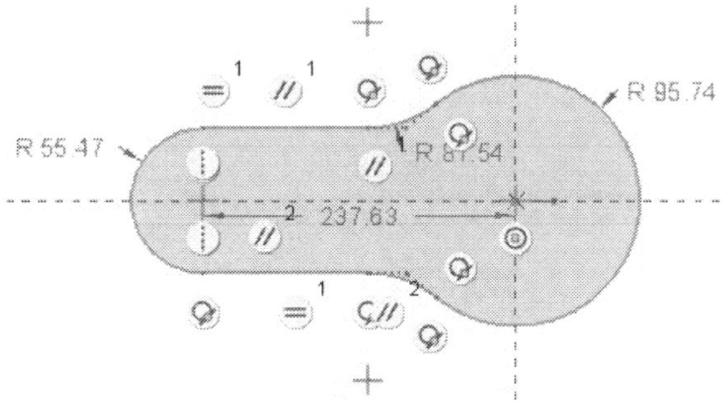

Figure 4-78 Sketch after applying constraints

Modifying Dimensions

You need to modify the dimension values of the weak dimensions.

1. Select the sketch and dimensions by using the CTRL+ALT+A keys.

2. Choose the **Modify** tool from the **Editing** group; the **Modify Dimensions** dialog box is displayed.

3. In this dialog box, clear the **Regenerate** check box and then modify the values of the dimensions, refer to Figure 4-79. As mentioned earlier, it is recommended that you clear the **Regenerate** check box when multiple dimension have to be modified.

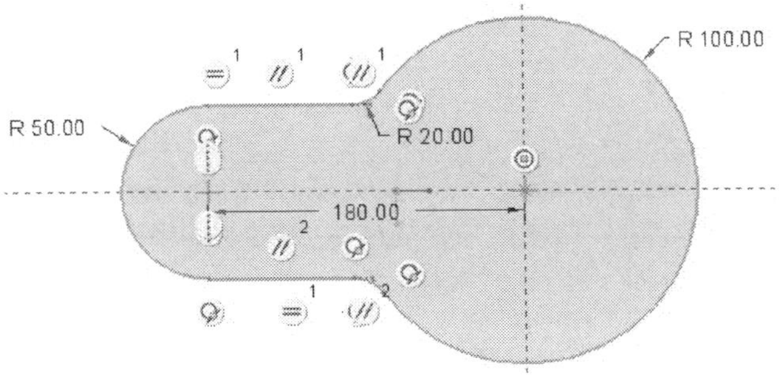

Figure 4-79 *Sketch after modifying dimensions*

You will notice that the dimension selected in the **Modify Dimensions** dialog box is enclosed in a blue box in the drawing area.

4. After modifying all dimensions, choose the **OK** button from the **Modify Dimensions** dialog box; the message **Dimension modifications successfully completed** is displayed in the message area.

5. Choose the **OK** button to exit the sketcher environment.

Specifying the Model Attributes

The attributes required to create the model need to be selected from the **Extrude** dashboard.

1. Choose the **Saved Orientations** button from the **Orientation** group in the **View** tab; a flyout is displayed. Choose the **Default Orientation** option from the flyout or press CTRL+D; the default trimetric view of the model is displayed in the drawing area, as shown in Figure 4-80.

   This display gives you a better view of the sketch in the 3D space. The model is displayed in orange. Also, the pink colored arrow is displayed on the model indicating the direction of extrusion. You can choose the **Change depth direction of extrude to other side of sketch** button to change the direction of extrusion.

2. Enter **50** in the **Depth** value edit box and press ENTER.

3. Choose the **Options** tab; a flyout is displayed. Select the **Add taper** check box from the flyout. Enter the value **10** in the **Add taper** edit box. If required, you can enter a negative value in the **Add taper** edit box to reverse the direction of taper. The base feature is completed as shown in Figure 4-81, and now you need to exit the **Extrude** dashboard.

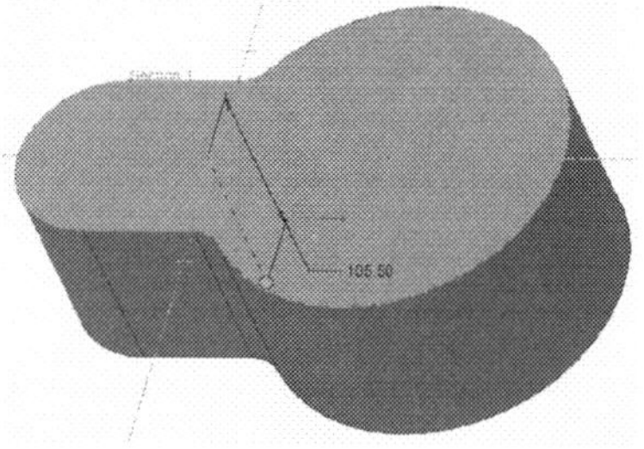

Figure 4-80 *The default trimetric view of the model*

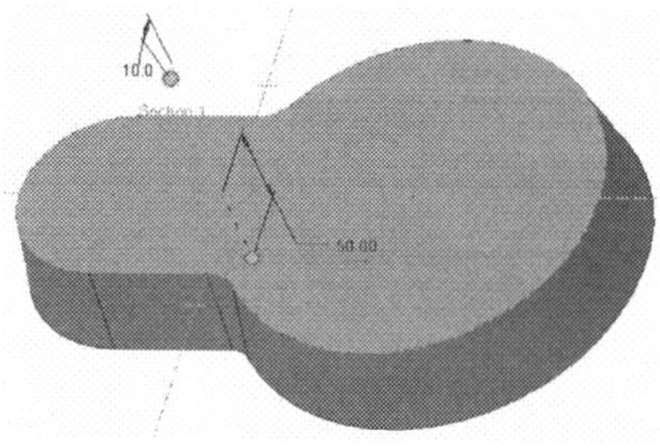

Figure 4-81 *Model after applying taper and depth*

Selecting the Sketching Plane for the Cut Feature

Next feature to be created is an extruded cut. This feature will be created on top of the model.

1. Choose the **Extrude** tool from the **Shapes** group; the **Extrude** dashboard is displayed in the drawing area.
2. Choose the **Remove Material** button from the **Extrude** dashboard.
3. Choose the **Placement** tab from the dashboard; a slide-down panel is displayed. Next, choose the **Define** button from it; the **Sketch** dialog box is displayed.
4. Select the top face of the model as the sketching plane, as shown in Figure 4-82.

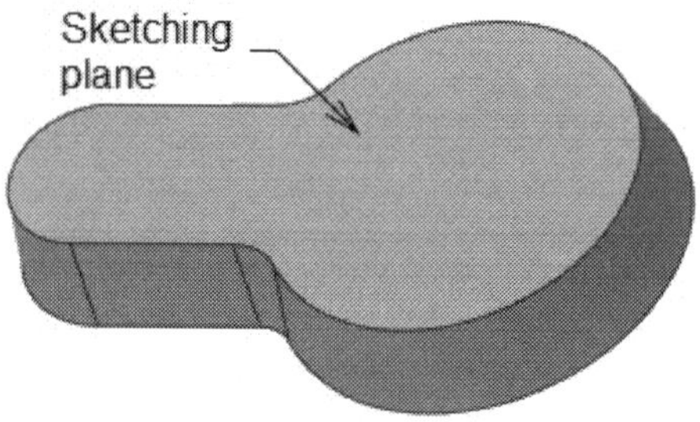

Figure 4-82 *Top face selected as the sketching plane*

5. Using the left mouse button, select the **RIGHT** datum plane as the reference plane and then select the **Right** option from the **Orientation** drop-down list.
6. Choose the **Sketch** button; the sketcher environment is displayed.

Creating the Sketch for the Cut Feature

1. Turn the model display to **No hidden** from the **Display Style** drop-down list in the **Graphics** toolbar. Choose the **Sketch View** button from the **Graphics** toolbar to orient the sketching plane parallel to screen. Draw the sketch of the cut feature and add dimensions to it, as shown in Figure 4-83.

2. Choose the **OK** button and turn the model display to **Shading With Edges**. Choose the **Default Orientation** option from the **Saved Orientations** flyout in the **Graphics** toolbar or press CTRL+D; the default trimetric view of the model is displayed in the drawing area, as shown in Figure 4-84.

3. Two arrows also appear in the model, refer to Figure 4-84. One arrow indicates the direction of feature creation and the other arrow indicates the direction along which the material will be removed. Click on **Change depth direction of extrude to other side of sketch** to change the direction of material removal.

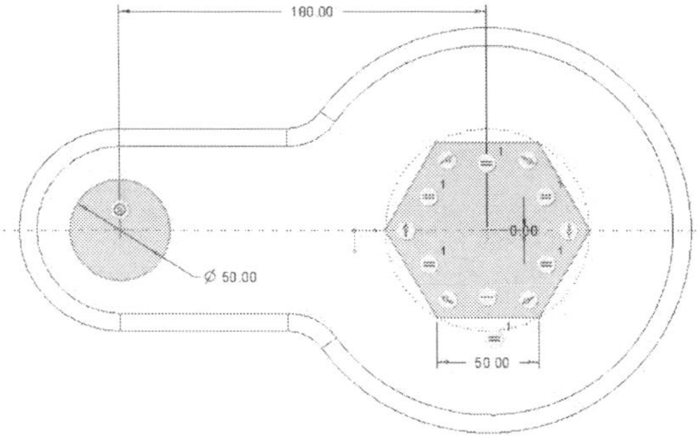

Figure 4-83 Sketch of the cut feature of the model

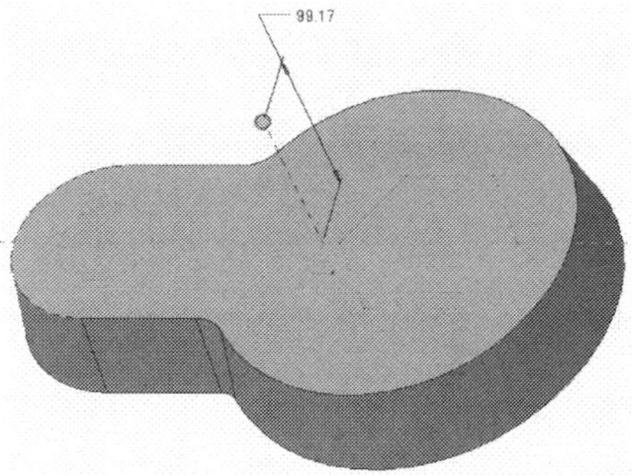

Figure 4-84 The default trimetric view of the model

4. Choose the **OK** button from the **Extrude** dashboard; the default trimetric view of the model is displayed, as shown in Figure 4-85.

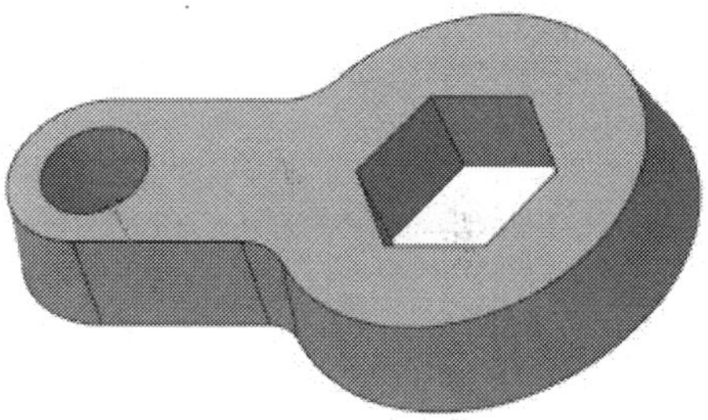

Figure 4-85 The default trimetric view of the solid model

Saving the Model

1. Choose the **Save** option from the **File** menu or choose the **Save** button from the **Quick Access** toolbar; the **Save Object** dialog box is displayed with the name of the object file specified earlier.

2. Choose the **OK** button to save the file.

Closing the Current Window

The given model is completed and is also saved. Now, you need to close the current window.

1. Choose **File > Close** from the menu bar.

Tutorial 5

In this tutorial, you will create the model shown in Figure 4-86. Figure 4-87 shows the front view and dimensions of the model. The thickness of the model is 1 unit.

(Expected time: 30 min)

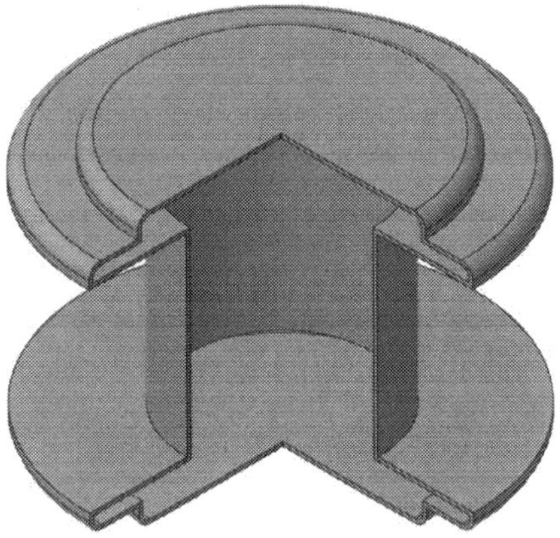

Figure 4-86 The isometric view of the solid model

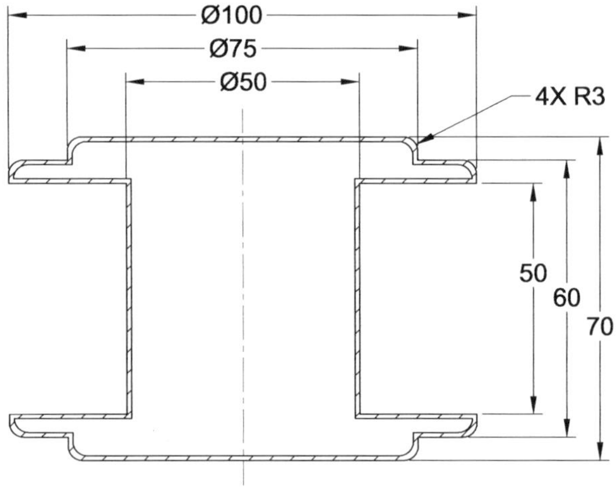

Figure 4-87 The front section view of the solid model

The following steps are required to complete this tutorial:

a. Create a new object file in the **Part** mode.
b. First examine the model and then determine the type of protrusion for the model. Next, select the sketching plane for the model.
c. Draw the sketch by using sketching tools, apply dimensions, and modify dimension values.
d. Specify the attributes for the model and then save it.

The working directory was selected in Tutorial 1, therefore, you do not need to select the directory again. But if a new session of Creo Parametric is started, then set the working directory using the Navigator.

Starting a New Object File

1. Start a new object file in the **Part** mode and then name the file as *c04tut5*.

The default datum planes are displayed, if they are not turned off. If the default datum planes were not turned off in the previous tutorial, they will appear in the drawing area. If the datum planes are not displayed, turn them on by using the **Plane Display** button from the **View** tab.

Selecting the Revolve Tool

Now, you need to create a revolved thin model by revolving the sketch through a given angle. Therefore, you need to use the **Thicken Sketch** button from the **Revolve** dashboard to create the model. The procedure is given next.

1. Choose the **Revolve** tool from the **Shapes** group; the **Revolve** dashboard is displayed on the top of the drawing area.

2. Choose the **Thicken Sketch** button from the dashboard. Make sure that the **Solid** button is chosen in the dashboard.

In this tutorial, the **Thicken Sketch** button is chosen before drawing the sketch because the sketch to be drawn is an open section. It is not possible to draw an open sketch with the default attributes selected in the **Revolve** dashboard.

3. Choose the **Placement** tab; a slide-down panel is displayed. Choose the **Define** button from the slide-down panel; the **Sketch** dialog box is displayed.

Selecting the Sketch Plane

Note that in Figure 4-86, the imaginary axis of revolution of the revolved feature is normal to the **TOP** datum plane. Therefore, the **TOP** datum plane will be selected to be at the top while drawing the sketch. Now, you need to decide the sketching plane from the **RIGHT** and **FRONT** datum planes. Any of the two datum planes can be selected as the sketching plane. In this case, the **RIGHT** datum plane will be selected as the sketching plane.

1. Select the **RIGHT** datum plane as the sketching plane. As you select the sketching plane, the reference plane and its orientation are set automatically by default.

   In the **Sketch** dialog box, the **Reference** box displays **TOP:F2(DATUM PLANE)**. This indicates that the **TOP** datum plane is selected as the reference plane. But you need to change the orientation of the plane.

2. From the **Orientation** drop-down list, select the **Top** option.

   The **TOP** datum plane will be perpendicular to the sketching plane and the sketching plane will be parallel to the screen.

3. Choose the **Flip** button to change the direction of the pink arrow that appears on the sketching plane; the direction of viewing the sketching plane is reversed.

4. Choose the **Sketch** button in the **Sketch** dialog box.

   Choose the **Sketch View** button from the **Graphics** toolbar to make the sketching plane parallel to the screen if it is not oriented by default.

Drawing the Sketch

You need to draw the sketch of the thin extruded model. The sketch can be a closed loop or an open loop. Here, you need to draw an open sketch.

1. Draw geometric centerlines and then draw the sketch, as shown in Figure 4-88. One of these geometric centerlines will be used to mirror the sketched entities. The second geometric centerline will be designated as the axis of revolution. To fillet the corners, choose the **Circular** tool from the **Fillet** drop-down in the **Sketching** group.

2. While drawing the sketch, weak dimensions will be applied to the sketch. If required, you can turn off the display of dimensions and constraints for better visibility. To do so, clear the **Dimensions Display** and **Constraints Display** check boxes in the **Sketcher Display Filters** drop-down.

3. Select all the sketched entities and then choose the **Mirror** tool from the **Editing** group of the **Sketch** tab. The selected entities will be mirrored about the selected axis, as shown in Figure 4-89.

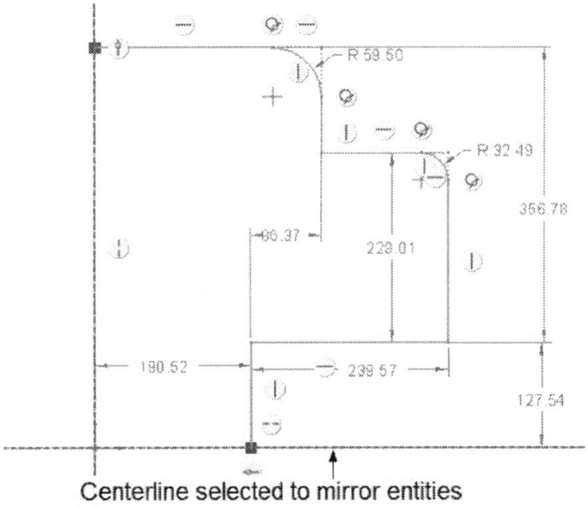

Figure 4-88 *Sketch drawn with geometric centerlines*

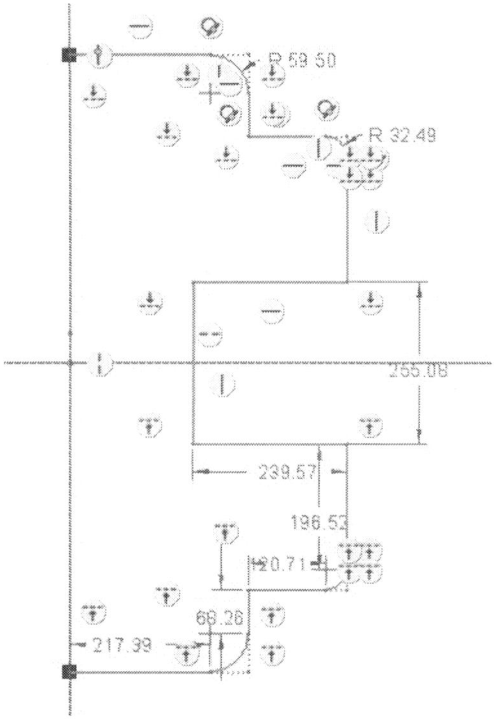

Figure 4-89 *Sketch after mirroring the entities*

4. Select the vertical centerline then press and hold right mouse button on it to display a shortcut menu. Choose the **Designate Axis of Revolution** option from the shortcut menu to revolve the sketch around that particular centerline.

Applying Constraints to the Sketch

Although some of the constraints are applied to the sketch while drawing but you need to apply some of the constraints manually. Note that after applying constraints, some of the weak dimensions may disappear from the drawing area.

1. Choose the **Equal** tool from the **Constrain** group.

2. Select the two horizontal lines to apply the equal length constraint, refer to Figure 4-90.

3. Select the two vertical lines to apply the equal length constraint, refer to Figure 4-90.

4. Select all fillets to apply equal radii constraint, refer to Figure 4-90.

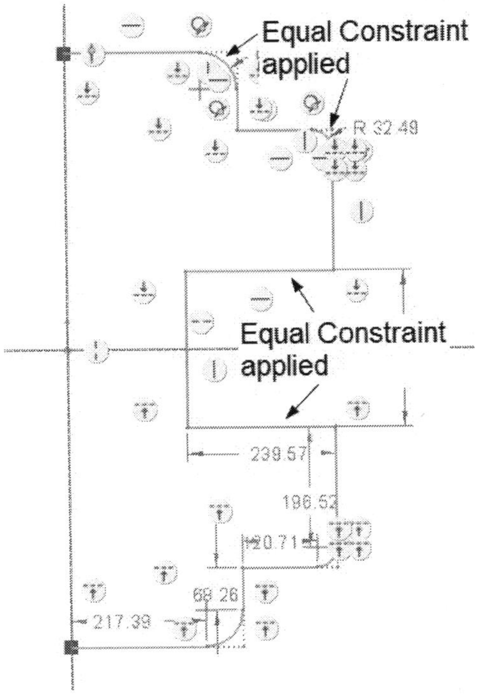

Figure 4-90 Sketch after applying the equal constraint

Applying Dimensions to the Sketch

The weak dimensions are automatically applied to the sketch. Remember that since the model is a revolved feature, you need to apply the linear diameter dimensions to the sketch manually. You need to apply the linear diameter dimensions by using the centerline drawn in the sketch.

1. Choose the **Dimension** tool from the **Dimension** group.

2. Dimension the sketch, as shown in Figure 4-91.

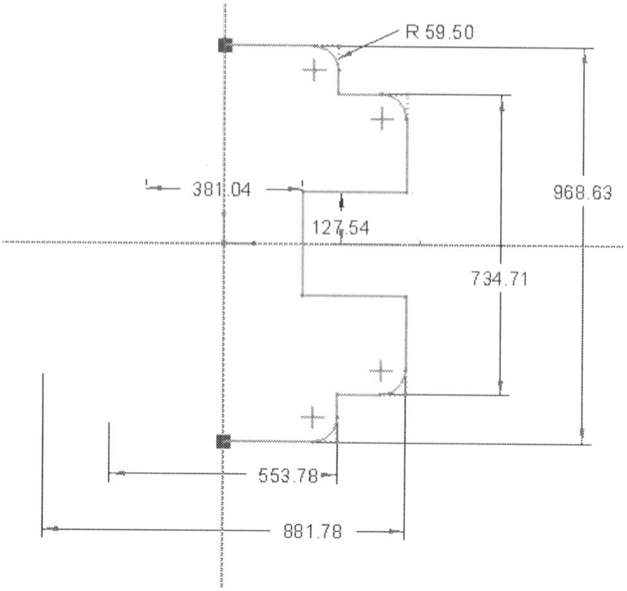

Figure 4-91 Sketch after dimensioning with the constraints turned off for clarity

Modifying the Dimensions

You need to modify the dimension values of the weak dimensions.

1. Select the sketch and dimensions by using the CTRL+ALT+A keys.

2. Choose the **Modify** tool from the **Editing** group; the **Modify Dimensions** dialog box is displayed.

3. In this dialog box, clear the **Regenerate** check box and then modify the values of the dimensions, as shown in Figure 4-92.

   You will notice that the dimension you select in the **Modify Dimensions** dialog box is enclosed in a blue box in the drawing area.

4. After modifying all dimensions, choose the **OK** button from the **Modify Dimensions** dialog box; the message **Dimension modifications successfully completed** is displayed in the message area.

5. Choose the **OK** button to exit the sketcher environment.

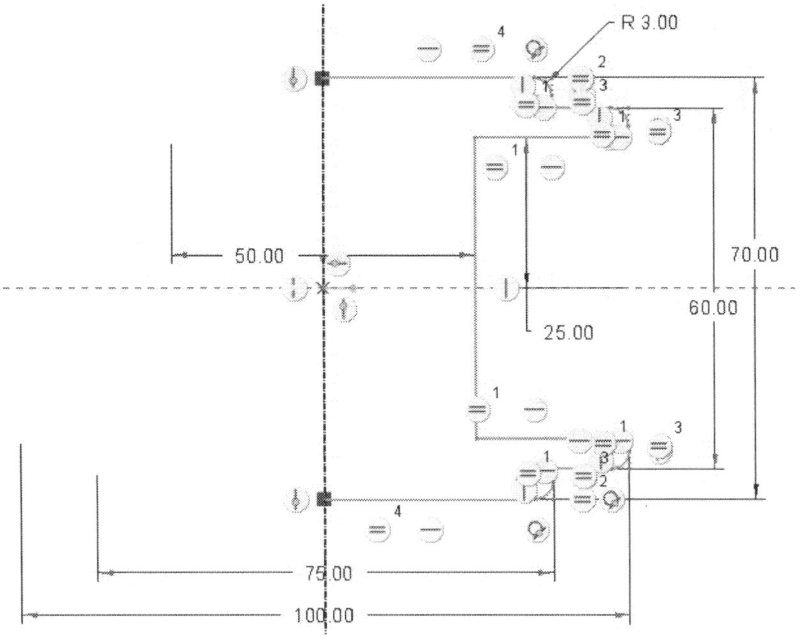

Figure 4-92 Sketch after modifying the dimensions with the constraints turned off for clarity

Specifying the Model Attributes

After completing the sketch, you need to specify the side to which thickness of material will be applied. The thickness can be applied outside the section, inside the section, or symmetrically to both sides of the section boundary. In this model, you need to apply the thickness inside the section.

When you exit the sketcher environment, the model assumes some default attributes and therefore it is not displayed in the drawing area.

1. Enter the thickness value as **1** in the drop-down list available on the **Revolve** dashboard, and press ENTER.

2. Enter **270** in the edit box present on the left of the **Revolve** dashboard and press ENTER. The default value of angle of revolution is displayed as 360.

3. Choose the **Saved Orientations** button from the **View** tab to display the flyout. From the flyout, choose the **Default Orientation** option or press CTRL+D; the trimetric view of the model after specifying all attributes is displayed, as shown in Figure 4-93.

4. Choose the **OK** button from the **Revolve** dashboard. The default trimetric view of the model is shown in Figure 4-94.

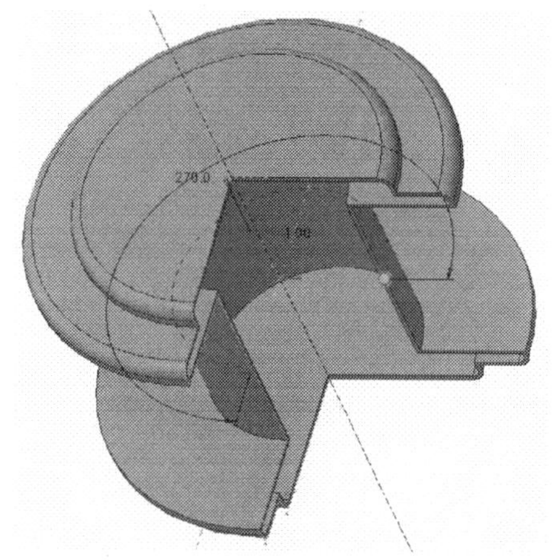

Figure 4-93 Model after specifying attributes

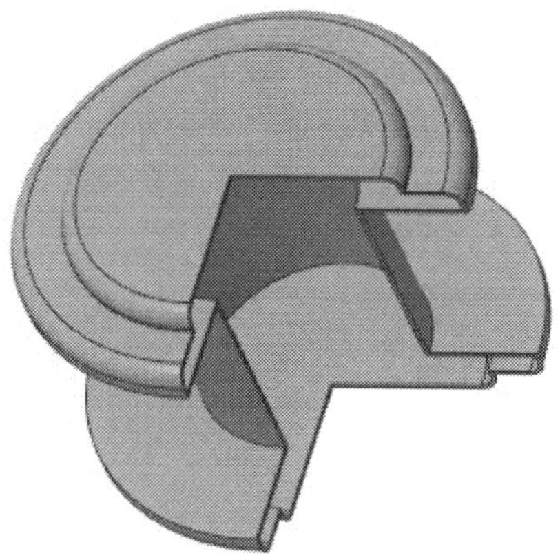

Figure 4-94 Default trimetric view of the model

Saving the Model

1. Choose the **Save** option from the **File** menu or choose the **Save** button from the **Quick Access** toolbar; the **Save Object** dialog box is displayed with the name of the object file specified earlier.

2. Choose the **OK** button to save the file.

Closing the Current Window

The given model is completed and saved. Now, you can close the current window.

1. Choose **File > Close** from the menu bar; the window is closed.

Self-Evaluation Test

Answer the following questions and then compare them to those given at the end of this chapter:

1. The __________ button is chosen by default in the **Extrude** dashboard.

2. The __________ button in the **Revolve** dashboard is used to create a thin revolved model.

3. The **View** dialog box is displayed when you choose the __________ button.

4. After you exit the sketcher environment, the __________ appear on the model to dynamically modify the extrusion depth or the angle of revolution.

5. The **Revolve** tool revolves the sketched section about the __________ through the specified angle.

6. All features created in the **Part** mode are called the base features. (T/F)

7. You can extrude a sketch to both sides of the sketching plane symmetrically. (T/F)

8. When the sketcher environment is invoked, the references required for creating a sketch are automatically selected. (T/F)

9. The **Sketch** dialog box is used to select the sketching plane. (T/F)

10. After selecting the sketching plane, the pink arrow displayed on it shows the direction in which you will view the sketching plane. (T/F)

Review Questions

Answer the following questions:

1. By default, in which direction does a sketch revolve about a geometric centerline?

(a) Clockwise (b) Counterclockwise
(c) Both (a) and (b) (d) None of these

2. Which of the following groups in the **Model** tab contains the **Extrude** tool?

(a) **Editing** (b) **Get Data**
(c) **Shapes** (d) **Operations**

3. How many datum planes are available when you enter the **Part** mode?

(a) 4 (b) 3
(c) 2 (d) None of these

4. Which of the following groups is used to turn off the display of datum planes?

(a) **Orientation** (b) **Window**
(c) **Model Display** (d) **Show**

5. Which of the following combinations is used to change the orientation of the model to the default orientation?

(a) CTRL+D (b) CTRL+left mouse button
(c) CTRL+right mouse button (d) None of these

6. Features created by using the **Variable** option do not have a dimension associated with them. Hence, they cannot be modified by changing the dimension value. (T/F)

7. It is recommended not to use the default datum planes to create the base feature. (T/F)

8. The section drawn for revolving as a solid should be a closed loop. (T/F)

9. The revolved section should have a centerline. (T/F)

10. A revolved section can be drawn on both sides of a centerline. (T/F)

EXERCISES

Exercise 1

Create the model shown in Figure 4-95. The dimensions of the model are shown in Figure 4-96. **(Expected time: 20 min)**

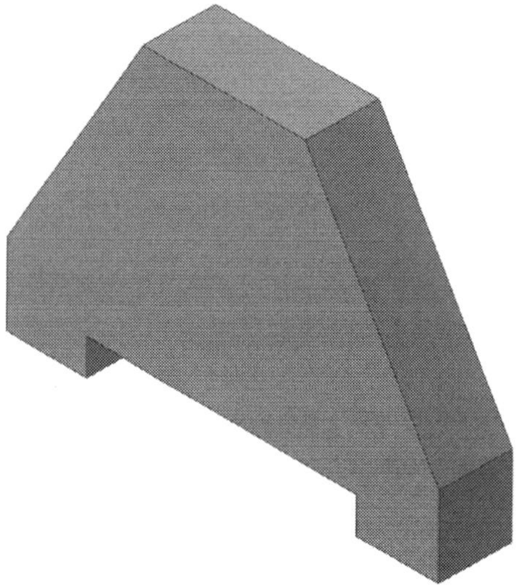

Figure 4-95 *The isometric view of the model*

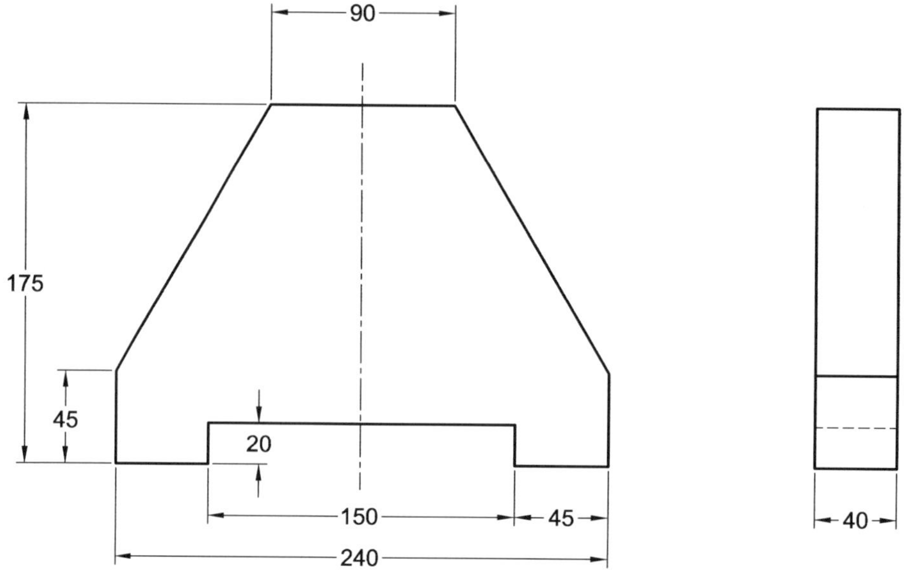

Figure 4-96 *The front and right views of the model*

Exercise 2

Create the model shown in Figure 4-97 The dimensions of the model are shown in Figure 4-98. **(Expected time: 30 min)**

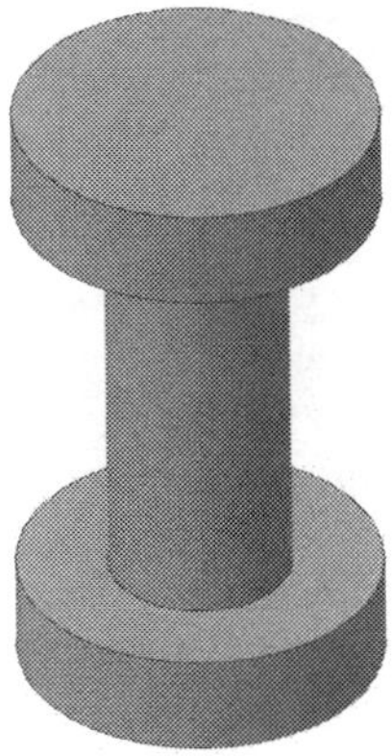

Figure 4-97 *The isometric view of the model*

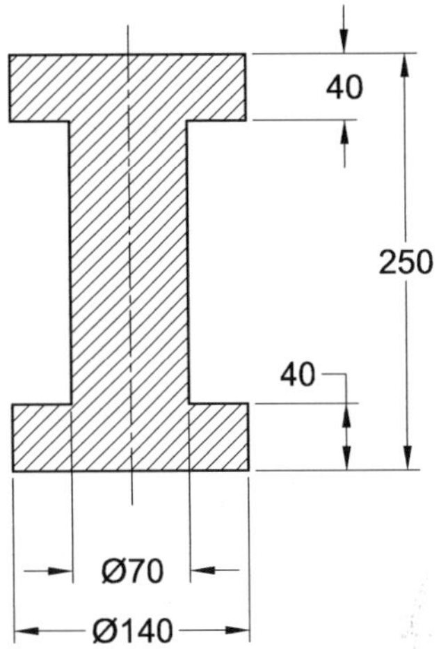

Figure 4-98 *The front section view of the solid model*

Exercise 3

Create the model shown in Figure 4-99. The dimensions of the model are shown in Figure 4-100. **(Expected time: 30 min)**

Figure 4-99 *The isometric view of the model*

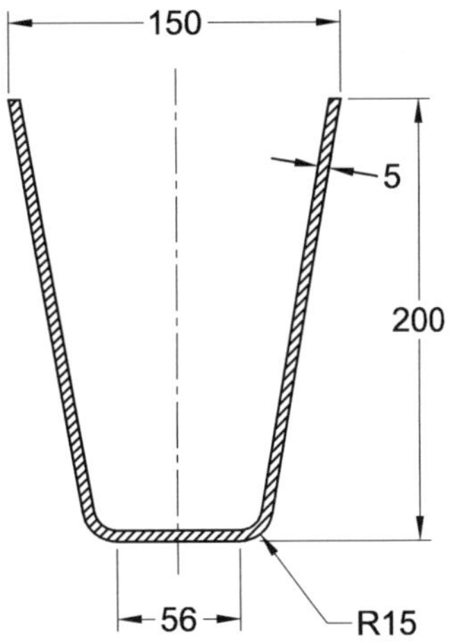

Figure 4-100 *The front section view of the model*

Exercise 4

Create the model shown in Figure 4-101. The dimensions of the model are shown in Figure 4-102. **(Expected time: 30 min)**

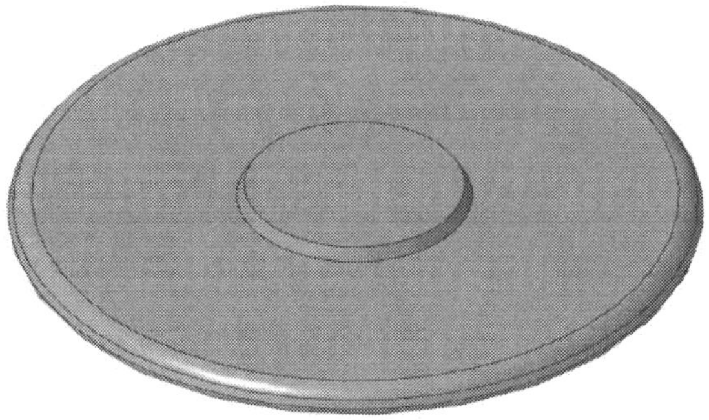

Figure 4-101 The isometric view of the model

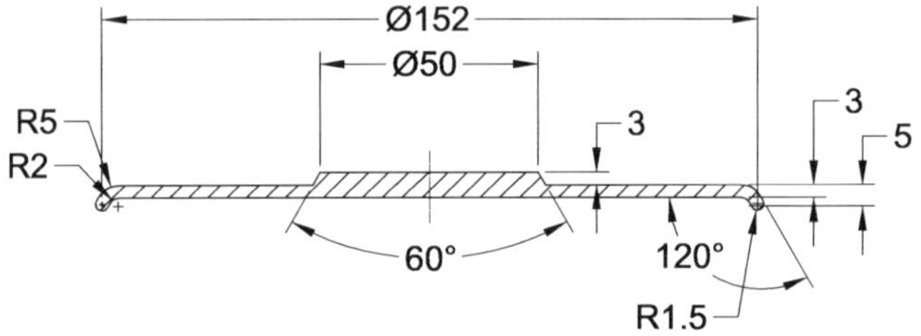

Figure 4-102 The front section view of the model

Exercise 5

Create the model shown in Figure 4-103. The dimensions of the model are shown in Figure 4-104. **(Expected time: 30 min)**

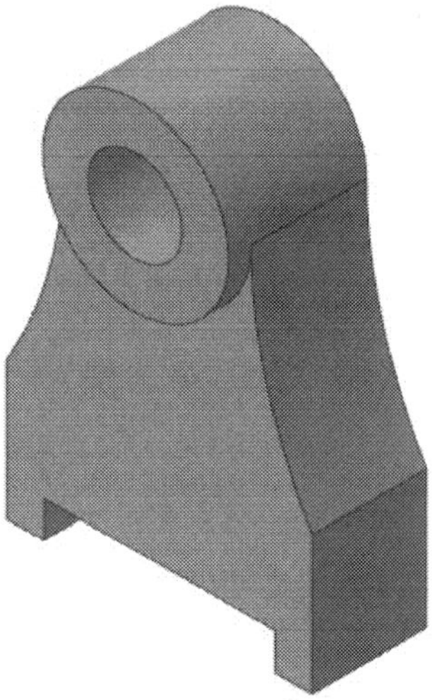

Figure 4-103 The isometric view of the model

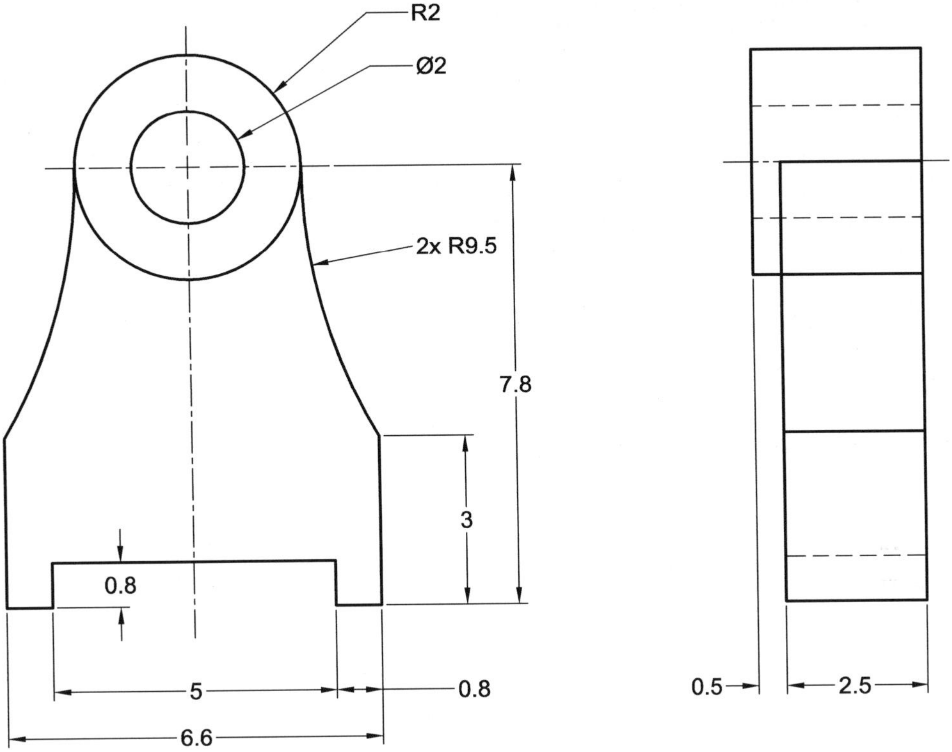

Figure 4-104 Orthographic views of the model

Answers to Self-Evaluation Test

1. **Solid**, **2**. **Thicken Sketch**, **3**. **Reorient**, **4**. handles, **5**. Geometric Centerline, **6**. F, **7**. T, **8**. T, **9**. T, **10**. T

Chapter 5

Datums

Learning Objectives

After completing this chapter, you will be able to:

- *Understand the need of datums in modelling*
- *Work with three default datum planes*
- *Understand selection methods in Creo Parametric*
- *Create datum planes using different constraints*
- *Create datum axes using different constraints*
- *Create datum points*
- *Create datum coordinate system*
- *Create datum curves*
- *Understand asynchronous datum features*

DATUMS

Datums are references used to describe the position of a feature(s). They act as reference for sketching a feature, orienting a model, assembling components, and so on. Datums are imaginary features with no mass or volume. Remember that datums play a very important role in creating complex models in Creo Parametric; therefore, you must have a good understanding of datums. Datums are considered to be features but not model geometry. In Creo Parametric, datums exist as datum plane, datum curve, datum point, datum coordinate system, datum graph, and so on.

NEED FOR DATUMS IN MODELING

Generally, most of the engineering components or designs consist of more than one feature. First the base feature of the model is created and then other features of the model are added. Since all features of a model cannot be drawn on a single plane, therefore, to draw rest of the features, sometimes, additional planes have to be created or selected. Also, most of the times, the three default datum planes are not enough to create a complex model having many features. For example, Figure 5-1 shows a simple model that consists of two features that require two different planes.

Tip

Whenever you come across any solid model, first try to visualize the number of features in that model and then decide which feature is to be considered as the base feature.

In Figure 5-1, any one of the two features defined on two different planes can be considered as the base feature. However, in this discussion, the feature that is selected as the base feature is shown in Figure 5-2. After creating the base feature, the next feature will be created. For the next feature, a sketching plane has to be defined. Therefore, an additional plane is required on which you can draw the sketch for the second feature.

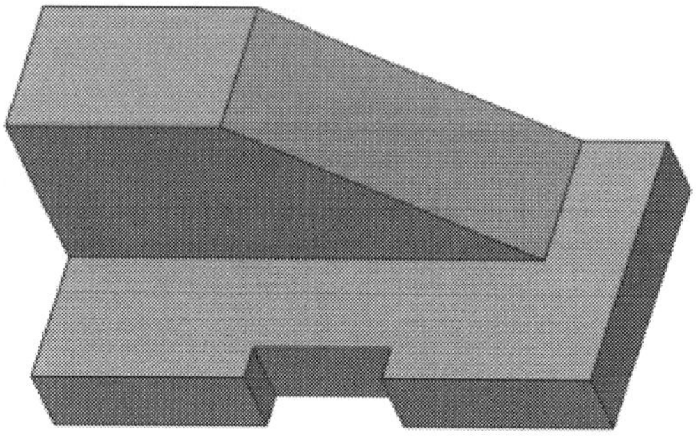

Figure 5-1 *Model having two extruded features*

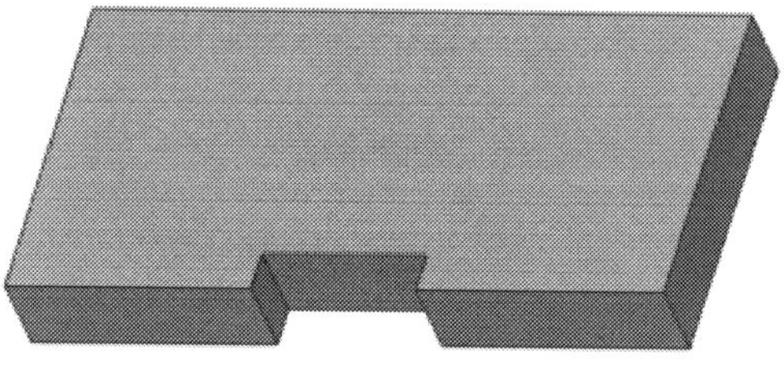

Figure 5-2 *Base feature of the model*

As shown in Figure 5-3, the top datum plane is used to create the base feature. To create the second feature, front datum plane is used, as shown in Figure 5-4. The sketch of the second feature will be drawn on this plane.

To create different types of datums, you need to select the references on the model. Based on the selection you make on the model, Creo Parametric applies constraints to create the datum.

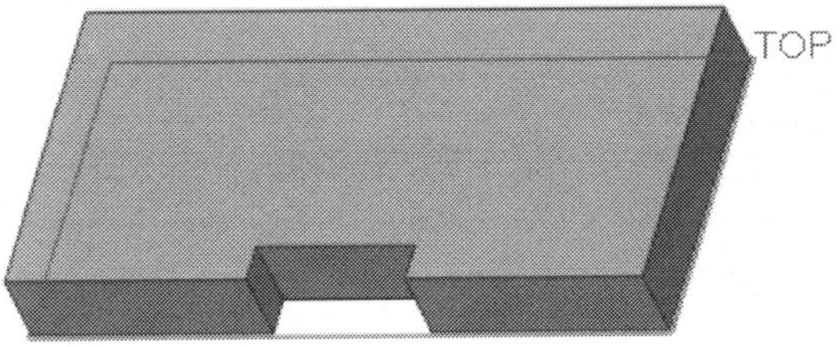

Figure 5-3 Base feature created on Top plane

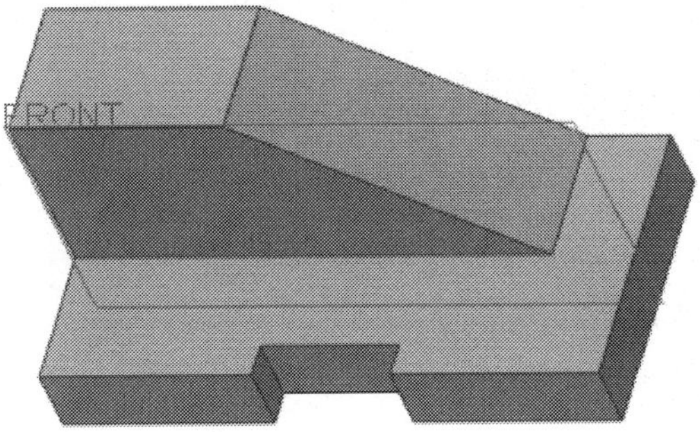

Figure 5-4 Second feature created on Front plane

SELECTION METHOD IN Creo Parametric

Before creating any geometry or datum, it is very important to understand the selection methods used in Creo Parametric. There are two types of selection methods in Creo Parametric:

1. Selection
2. Collection

Selection

In Creo Parametric, selection refers to an action in which you first select entities such as edge, plane, face, axis, coordinate system, and so on, and then invoke a feature creation tool. This property of a design package is also known as Object Action. To make your selection process easier, filters are available in Creo Parametric. Filters are available in the Filter drop-down list located at the right corner of the Status Bar at the bottom. The options in the Filter drop-down list change depending upon the mode that is active. The options in this drop-down list, when the **Part** mode is active, are shown in Figure 5-5.

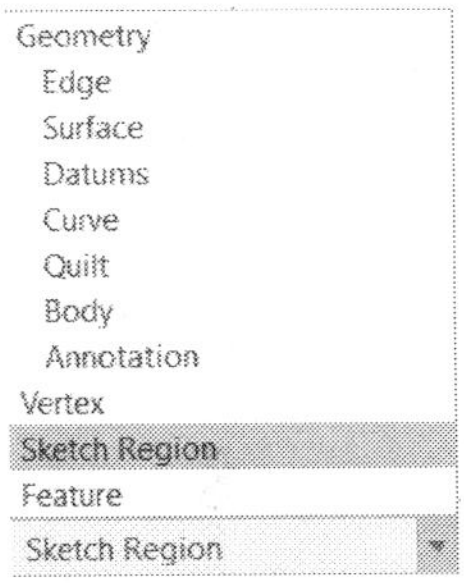

Figure 5-5 The Filter drop-down list

As you select entities on the model, the numbers of selected entities appear on the left of the Filter drop-down list in the Status Bar. When you double-click on the number, the **Selected Items** dialog box will be displayed, as shown in Figure 5-6. All selections that you make on the model are available in this dialog box. You can use the dialog box to remove selected entities. These selections appear highlighted on the model. To remove the selection from the model, press the CTRL key and then select the highlighted entity. The list of filters available in the Filter drop-down list is determined by the mode and tab chosen.

There are two type of selection filters available in Creo parametric: compound selection filters and individual selection filters. Compound selection filters enable the selection of more than one type of object. For example, Geometry selection filter is a pre-defined compound selection filter which includes selection filters for edge, surface, datums, curves, quilts, and annotation. Individual selection filters enable the selection of only one type of object such as vertices and features. All the filters are context-sensitive, hence Creo Parametric automatically activates the best filter according to the active mode or tab.

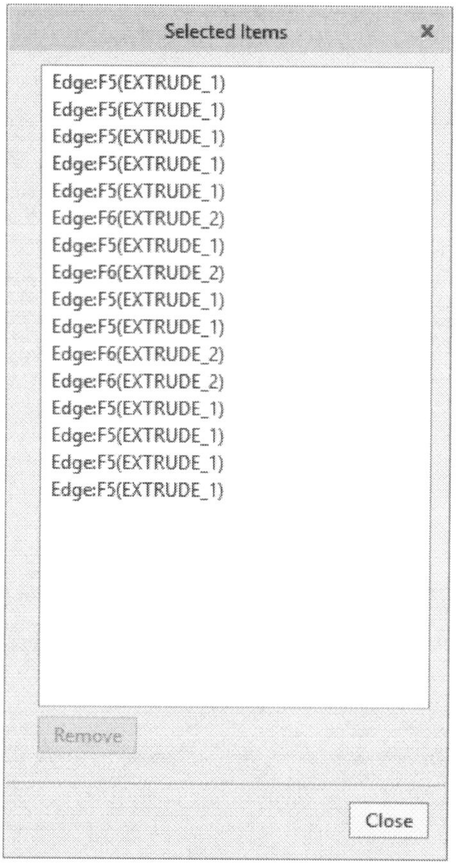

Figure 5-6 *The* ***Selected Items*** *dialog box*

Creo Parametric allows you to create a user defined selection filter. The filter you define will be named **My Filter** and will appear at the top in the Filter drop-down list. More than one filter can be added to **My Filter** and for each mode, a filter can be defined. To create a user defined filter, choose the **Options** option in the **File** menu, the **Creo Parametric Options** dialog box will be displayed. Next, choose the **Selection** option from the **Creo Parametric Options** dialog box, the **Set the selection options** page will be displayed, as shown in Figure 5-7. Select a filter from the **Choose filter from** list and click on the **Add** button to add the selected filter in the **My Filter** list. Next, choose the **OK** button to finish the process. You can also import or export the filter configuration by using the options in the **My filter** drop-down list available at the bottom right corner of the **Creo Parametric Options** dialog box.

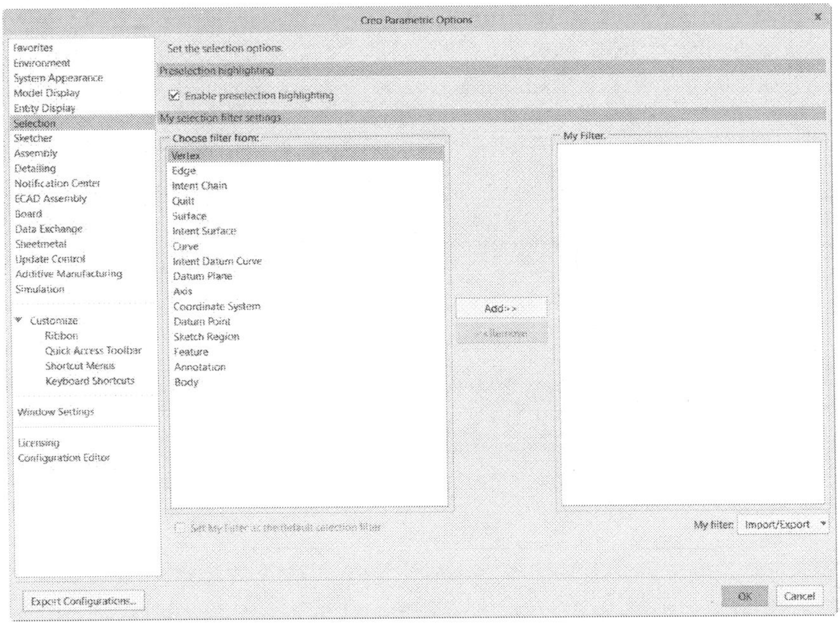

*Figure 5-7 The **Creo Parametric Options** dialog box*

Collection

In Creo Parametric, collection refers to an action in which you select the entities to collect the references for creating a feature. Collectors also gather all the entities that you select before invoking a geometry creation tool. These entities can be used as references for creating features. To make your selection process easy, filters can be used. The filters available in the Filter drop-down list depend on the feature creation tool that you have invoked.

DEFAULT DATUM PLANES

When you enter the **Part** mode or the **Assembly** mode, the three datum planes are displayed by default in the drawing area. These datum planes are known as the default datum planes and they are mutually perpendicular to each other. The only difference between the default datum planes of the **Part** mode and those of the **Assembly** mode lies in the names of the datum planes.

The default datum planes in the **Part** mode are named as **FRONT**, **TOP**, and **RIGHT**. In case of the **Assembly** mode, the default datum planes are named as **ASM_FRONT**, **ASM_TOP**, and **ASM_RIGHT**. However, the names of the default datum planes can be changed if required.

To change the name of a default plane, choose **File > Prepare > Model Properties** from the menu bar; the **Model Properties** window will be displayed. In this window, click on the **change** link in front of the **Names** line under the **Features and Geometry** head; the **Rename** dialog box will be displayed, as shown in Figure 5-8. Next, click on the datum plane you want to rename and enter the desired name to change the default name of the plane. Next, choose the **OK** button from the **Rename** dialog box and then close the **Model Properties** window.

Alternate method to rename a datum plane is to select it and right-click. On doing so, a shortcut menu will be displayed. Choose **Properties** or **Rename** from it and then enter the desired name. You can also use the **Edit Definition** option from the mini popup toolbar to change the properties of the selected datum plane.

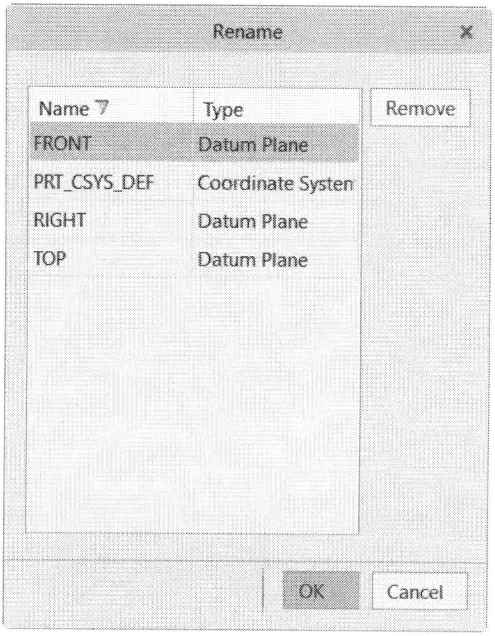

*Figure 5-8 The **Rename** dialog box*

Note

*The **Model Properties** window can also be used to set the material, units, and other parameters related to the model.*

DATUM OPTIONS

After discussing the default datum planes, which are the first features in the **Part** mode, you must know various other features created using the datum options. Datums are also considered as features having no geometry. Figure 5-9 shows the **Datum** group and various datum types options from the group.

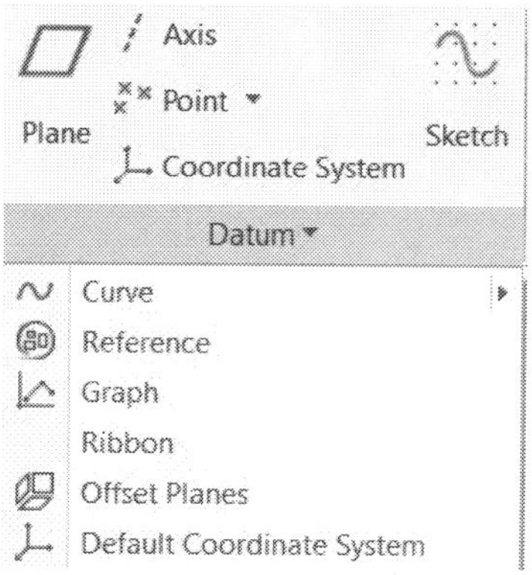

*Figure 5-9 The **Datum** group*

Datum Planes

Ribbon: Model > Datum > Plane

You can create datum planes other than the three default datum planes by using the **Plane** tool available in the **Datum** group. You can create a datum plane even when any other tool is invoked. You can turn on or off the display of the datum planes by using the **Plane Display** button from the **Show** group of the **View** tab or from the **Graphics** toolbar. Before discussing the procedure to create datum planes, it is important to understand the use of datum planes. Some of the uses of datum planes in Creo Parametric are listed next.

1. Datum planes are used as sketching planes to create sketches for the features of a model.
2. Datum planes are used as reference planes for sketching.
3. Datum planes are used as references for placing holes and for assembling components.
4. Datum planes are used as a reference for mirroring features, copying features, creating a cross-section, and for orientation of references.
5. Datum planes are used as a reference to place datum tag annotations. These datum tag annotations are used in geometric dimensioning and tolerancing of features of the part.

Tip
Generally, the three default datum planes are used for creating the base feature. As the part becomes complex or in other words, as the number of features increases, the need for additional datum planes arises.

Creo Parametric provides you various options to create additional datum planes. Additional datum planes are created with the help of constraints and the filters available in the Filter drop-down list. Datums can also be created while you are in the sketcher environment. To create a datum plane, choose the **Plane** tool from the **Datum** group of the **Model** tab; the **Datum Plane** dialog box will be displayed, as shown in Figure 5-10. Separate buttons for creating datum axis, datum curve, datum point, and datum coordinate system are available in the **Datum** group.

Figure 5-10 shows the **Datum Plane** dialog box with various options to create datum planes. The options in different tabs of the **Datum Plane** dialog box are discussed next.

***Figure 5-10** The **Datum Plane** dialog box with the **Placement** tab chosen*

Placement Tab

The **Placement** tab in the **Datum Plane** dialog box is chosen by default. In this tab, the **References** area displays the references selected to create the datum plane. All valid constraint types for these references are displayed in a drop-down list on the right of the selected reference. These constraints are applied automatically based on the references you select. To change a constraint, select the constraint in the dialog box; you will notice that a drop-down list appears in its place. From this drop-down list, select a different constraint. This drop-down list is available only for those constraints in the dialog box that can be substituted by another constraint. The constraints that are used to create a datum plane are discussed later in the chapter.

The **Offset** area will be available only when you use the **Offset** constraint to create a datum plane. The **Translation** edit box in this area is used to specify the offset distance of the new datum plane. This distance can also be set dynamically on the model by using the drag handle. The **Rotation** edit box will be available only when you create a datum plane at an angle with the selected edge or axis. This edit box is used to specify the rotation angle of the new datum plane. You can also set the rotation angle dynamically by using the drag handle displayed on the model.

Display Tab

The **Display** tab of the **Datum Plane** dialog box is shown in Figure 5-11. The **Flip** button in this dialog box is used to change the normal direction of the datum plane being created. The arrow direction and hence the direction of the plane can be changed by using the **Flip** button. By default, the **Use display reference** check box is selected. This option allows you to specify the size of the datum plane with respect to a reference such as part, feature, edge, or surface. If you clear this check box, you can set the size of the datum plane without any reference. You can select the **Lock aspect ratio** check box to maintain the ratio between the height and width of the datum plane. You can adjust the size of the plane by specifying the values for width and height in their respective edit boxes.

Properties Tab

The **Datum Plane** dialog box with the **Properties** tab chosen is shown in Figure 5-12. This tab, when chosen allows you to name the datum plane you are creating. By default, the first datum plane you create is named DTM1 and then other datum planes created are successively numbered.

Tip
*When the **Datum Plane** dialog box is active, right-click in the drawing area to access the options such as **Flip Normal direction, Clear, Switch to Offset, Placement References,** and **Fit Outline**.*

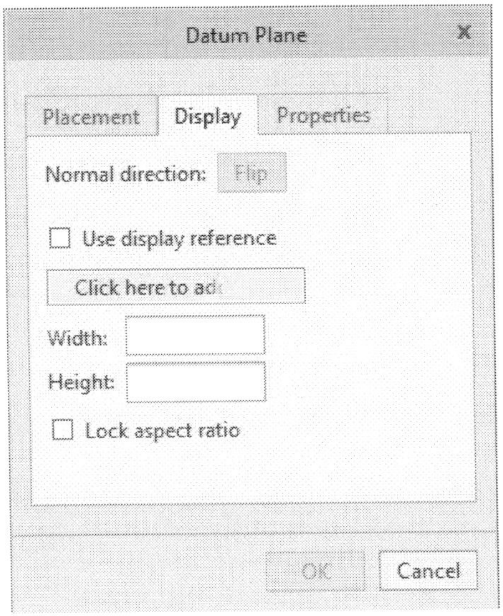

*Figure 5-11 The **Datum Plane** dialog box with the **Display** tab chosen*

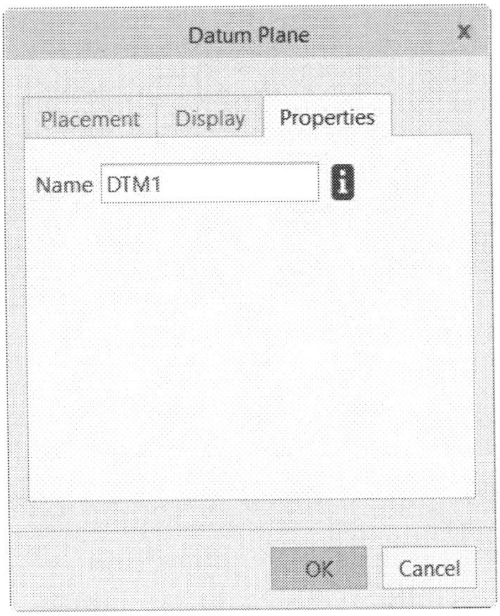

*Figure 5-12 The **Datum Plane** dialog box with the **Properties** tab chosen*

Creating Datum Planes

When you create a datum plane, the constraints are applied automatically based on the selection you make on the model. These constraints appear in the **Datum Plane** dialog box. Sometimes, while applying constraints to define a datum plane, a single constraint is enough to define the datum plane and sometimes, you may need more than one constraint to do the same. When only one constraint is used to create a datum plane, it is called stand-alone constraint. The stand-alone constraints are sufficient by themselves to constrain a datum plane definition. The constraints that are used to create a datum plane are discussed next.

Through Constraint

The Through constraint is used to create a datum plane through any specified axis, edge, curve, point/vertex, plane, cylinder, non cylinder, datum coordinate system, sketch-based features, or coordinate system. If you select a coordinate system as the placement reference, the **Plane** drop-down list gets activated. Next, select the **XY**, **YZ**, or **ZX** from the drop-down list to place the datum through the selected reference. By default, **YZ** is selected in the **Plane** drop-down list. This constraint can be used as a stand-alone constraint or in combination with other constraints.

The options in the Filter drop-down list shown in Figure 5-13 are used to make a selection on a model. This drop-down list is located at the right corner of the main window. The options in this drop-down list change depending on the operation being performed. For example, when you are creating a datum plane, the options shown in the Filter drop-down list will become available. Once you exit the datum plane creation process, the options in this drop-down list will change.

Figure 5-13 *The Filter drop-down list when the* ***Datum Plane*** *dialog box is displayed*

The cylindrical surface and the default datum plane is used to create a datum plane at an angle to the selected default datum plane and passing through the center of the cylindrical surface, as shown in Figure 5-14 and Figure 5-15. When you select the cylindrical surface, it is highlighted in green indicating that it is selected. To make the second selection, select the **Datum Plane** filter from the Filter drop-down list. Now, using the CTRL button and then select the **RIGHT** datum plane. This method of selecting by using the CTRL button is known as collection. Click on the constraint displayed against the selected plane in the **Datum Plane** dialog box and then select the **Offset** option from the drop-down list. Enter the angle value in the **Rotation** dimension box. Choose the **OK** button to create the datum plane.

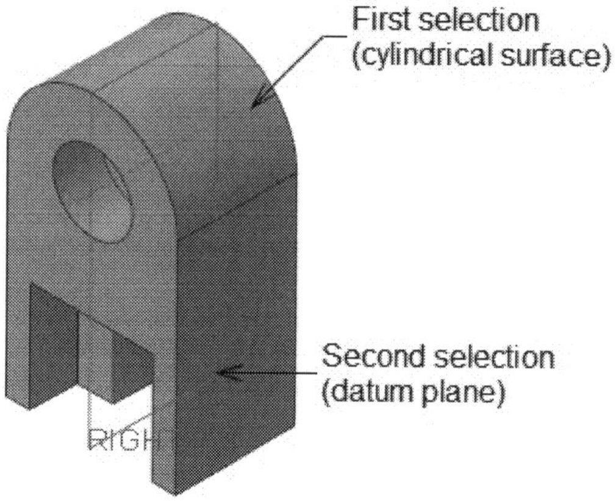

Figure 5-14 *References selected to create a datum plane*

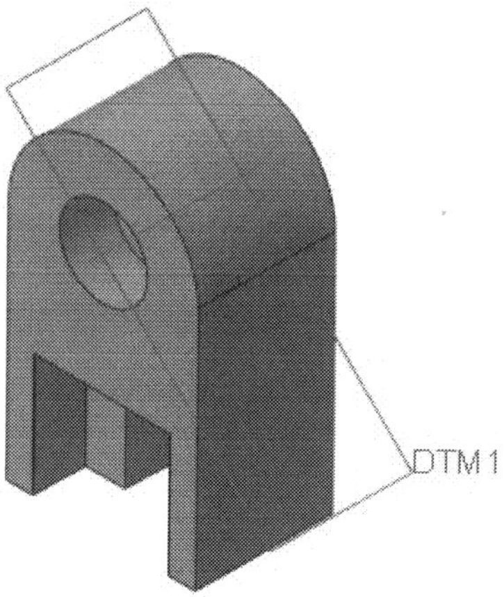

Figure 5-15 *Resulting datum plane passing through the center of the cylinder*

Note

1. The first reference to constrain a datum plane is selected by using the left mouse button. The second reference is selected by using the CTRL+left mouse button.

2. Make sure you change the option in the Filter drop-down list when you invoke any other tool.

Normal Constraint

The Normal constraint is used to create a datum plane normal to any specified axis, edge, curve, or plane. The Normal constraint cannot be used as a stand-alone constraint but in combination with other constraints. Figure 5-16 shows an example where the Right datum plane is selected as the normal surface and the cylindrical face is selected to be tangent to the datum plane. The following steps explain the procedure to create this datum plane:

1. Invoke the **Datum Plane** dialog box by choosing the **Plane** tool from the **Datum** group in the **Model** tab. Select the first reference shown in Figure 5-16. On doing so, the Through constraint is automatically applied and will be displayed in the **References** collector.

2. Click on the **Through** constraint in the collector; a drop-down list appears. Select the **Tangent** option from this drop-down list.

3. Now, use the CTRL+left mouse button to select the second reference shown in Figure 5-16. On doing so, the **Normal** constraint will be displayed in the **References** collector.

   The resulting datum plane that is created is shown in Figure 5-17.

4. Choose the **OK** button from the **Datum Plane** dialog box.

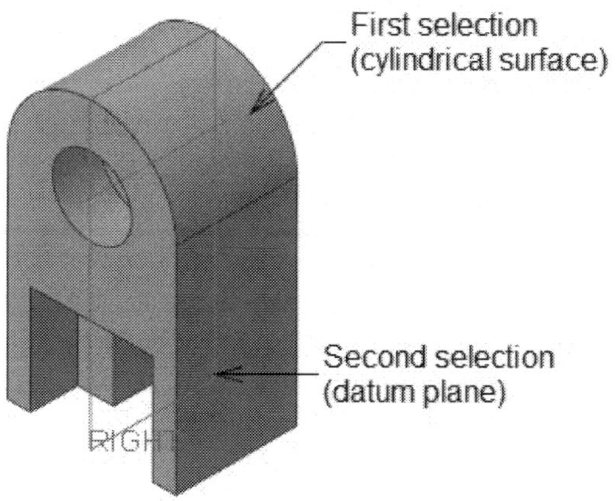

***Figure 5-16** Selection of references to create a datum plane*

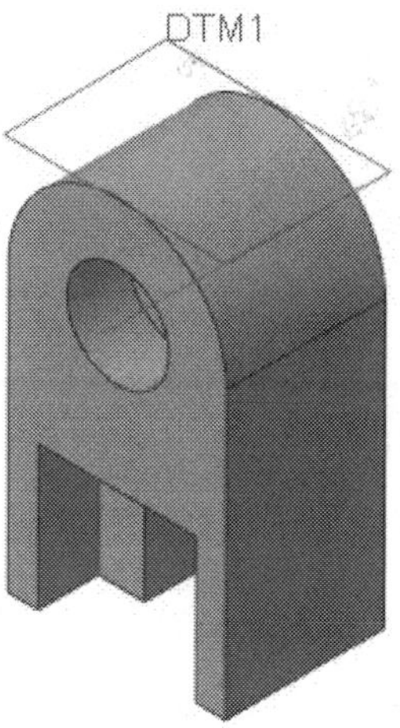

***Figure 5-17** Resulting datum plane*

Parallel Constraint

The Parallel constraint is used to create a datum plane parallel to any specified datum plane or planar face. The Parallel constraint also cannot be used as a stand-alone constraint but in combination with other constraints. Figure 5-18 shows the selection of a default datum plane and an axis to create a datum plane. The resulting datum plane is parallel to the selected datum plane and passes through the axis, as shown in Figure 5-19. The steps explaining the procedure to create this datum plane are given next.

1. Invoke the **Datum Plane** dialog box. Select the first reference, refer to Figure 5-18; the Offset constraint is automatically applied and will be displayed in the **References** collector.

2. Click on the **Offset** option in the collector; a drop-down list appears. Select the **Parallel** option from this drop-down list; the plane is oriented parallel to the selected reference.

3. Now, using CTRL+left mouse button and select the second reference, refer to Figure 5-18; the **Through** constraint will be displayed in the **References** collector.

4. Choose the **OK** button from the **Datum Plane** dialog box. The resulting datum plane is shown in Figure 5-19.

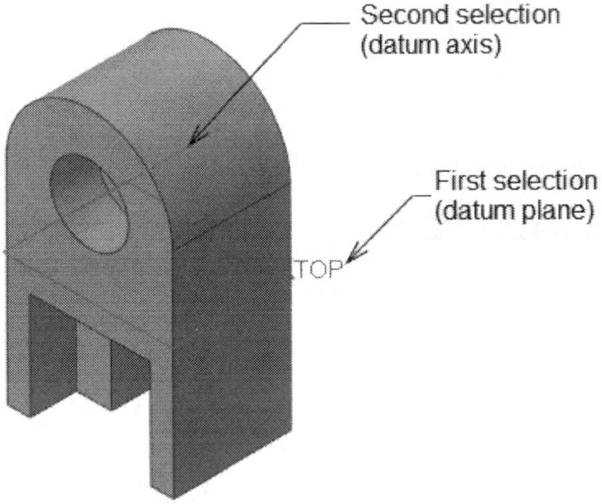

Figure 5-18 Selecting a datum plane and an axis as reference

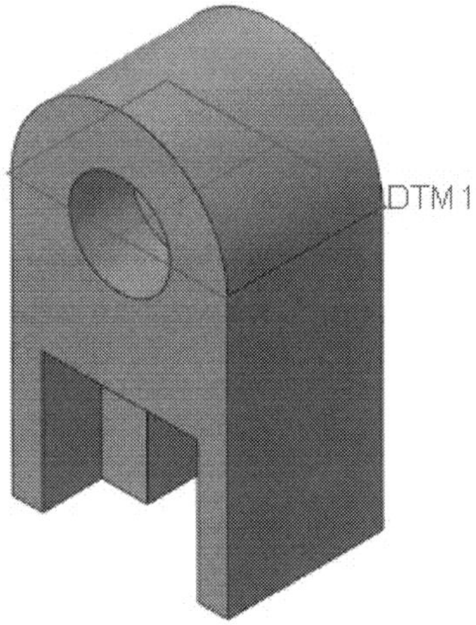

Figure 5-19 Resulting datum plane

Offset Constraint

The Offset constraint creates a datum plane that is at an offset distance to any specified plane or coordinate system. This option is used in combination with other constraints. However, the Offset constraint can be used as a stand-alone constraint when you select a plane to offset from.

When you select a plane or a planar surface as the reference, the Offset constraint is applied to the plane being created and the **Translation** dimension edit box is activated in the **Datum Plane** dialog box. This edit box is used to specify the offset distance. If you are selecting any linear geometry (axis, edge) as the second reference then the **Translation** dimension edit box converts into the **Rotation** dimension edit box in which you need to specify the angle. If you select a datum coordinate system as the placement reference, you can specify the translation of the plane with respect to X, Y, or Z direction. The arrow which appears on the plane shows the positive direction of the offset distance or angle.

The planar surface is selected as reference in Figure 5-20. Figure 5-21 shows the resulting datum plane created with an offset value.

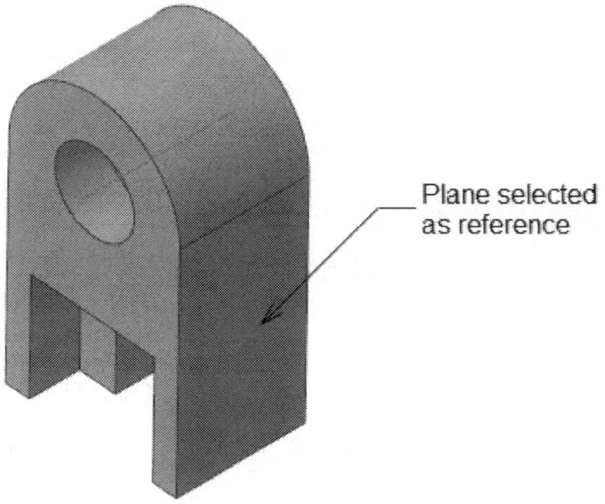

Figure 5-20 *Selecting a planar surface as reference*

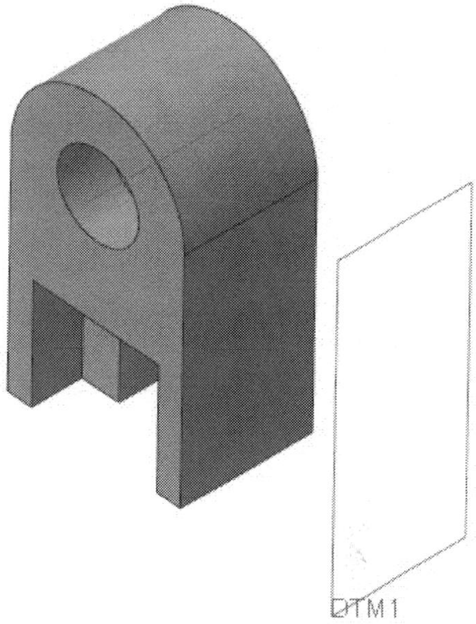

Figure 5-21 *Resulting datum plane*

The following steps explain the procedure to create this datum plane:

1. Invoke the **Datum Plane** dialog box. Select the first reference refer to Figure 5-20. The Offset constraint is automatically applied and will be displayed in the **References** collector. Also, the **Translation** dimension box appears.

2. Specify an offset value in the **Translation** box and choose the **OK** button from the **Datum Plane** dialog box; the datum plane shown in Figure 5-21 gets created.

Note
*The **OK** button in the **Datum Plane** dialog box will be enabled only when the datum plane you are creating is fully constrained.*

Tangent Constraint

The Tangent constraint creates datum planes tangent to cylindrical surfaces or conical surfaces. The Tangent constraint cannot be used as a stand-alone constraint but is always used in combination with other constraints. While creating a datum plane with Tangent constraint, you will require two references. One should be a cylindrical or conical surface and the other should be a datum point, a vertex, or an endpoint of an edge.

Midplane Constraint

The Midplane constraint creates a datum plane between two parallel planes or at the bisector of two non parallel references. While creating a datum plane with Midplane constraint you will require two references. You can select datum planes, planar surfaces, part edges, or vertexes as reference to create a datum plane with Midplane constraint. This constraint is used in combination with other constraints and cannot be used as a stand-alone constraint.

The following steps explain the procedure to create a mid plane between two parallel references:

1. Invoke the **Datum Plane** dialog box and select the first reference, refer to Figure 5-22; the Through constraint is automatically applied and gets displayed in the **References** collector.

2. Click on the **Through** option in the collector and select the **Midplane** option from the drop-down list.

3. Now, using CTRL+left mouse button, select the second reference, refer to Figure 5-22; the **Parallel** option is displayed in the **References** collector.

4. Choose the **OK** button from the **Datum Plane** dialog box. The resulting datum plane is shown in Figure 5-23.

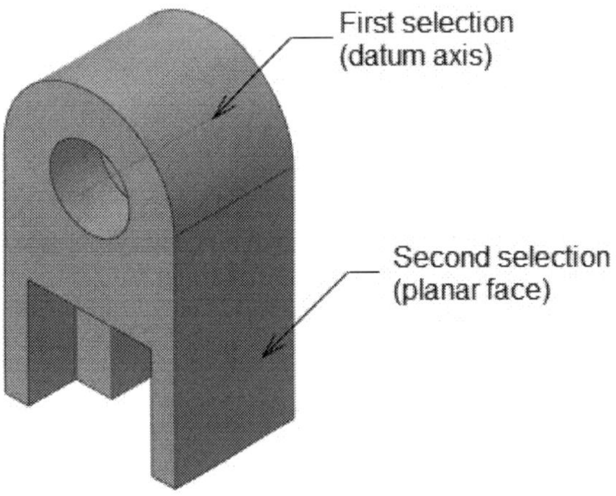

Figure 5-22 *Selecting an axis and planar face as reference*

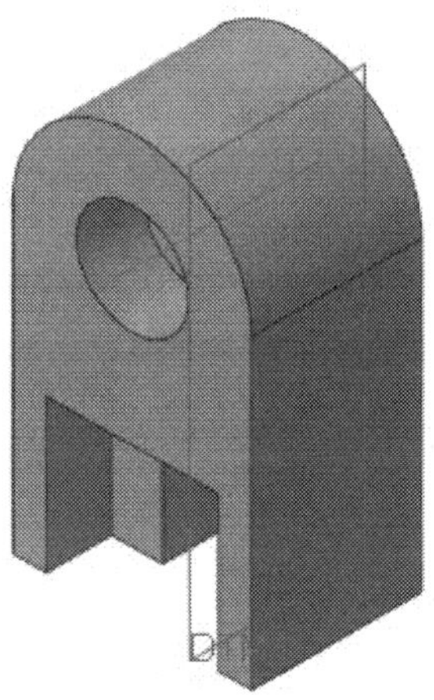

Figure 5-23 *Resulting datum plane created between the axis and planar face*

The following steps explain the procedure to create a mid plane between two nonparallel planes:

1. Invoke the **Datum Plane** dialog box and select any one of the two planes as the first reference, refer to Figure 5-24; the Offset constraint is automatically applied and gets displayed in the **References** collector.

2. Click on the **Offset** option in the collector and select the **Midplane** option from the drop-down list and select the second reference plane using the CTRL key; the **Midplane** option is automatically displayed in the **References** collector of the first reference. Also, as you select the second reference, **Bisector1** is displayed in the **References** collector of the second reference. Select **Bisector1** in the drop-down list of the second reference; a datum plane will be created bisecting the angle formed between two reference planes, refer to Figure 5-25. If you select **Bisector2** in the drop-down list of the second reference, the resulting plane

will be placed in the perpendicular direction of the plane placed by **Bisector1** between two reference planes, refer to Figure 5-26.

3. Choose the **OK** button from the **Datum Plane** dialog box to complete the procedure.

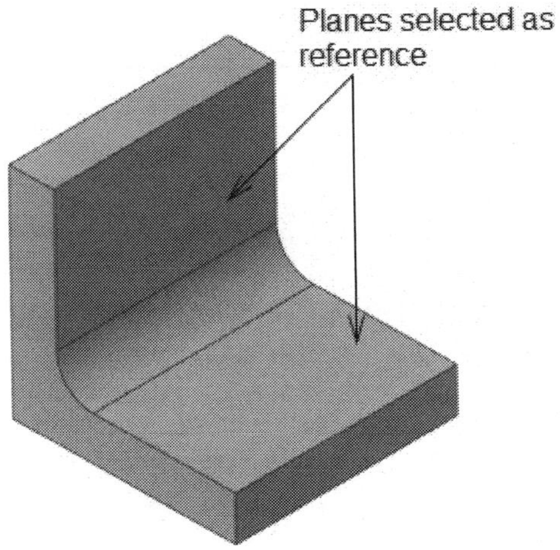

Figure 5-24 Selecting two non parallel planar face as reference

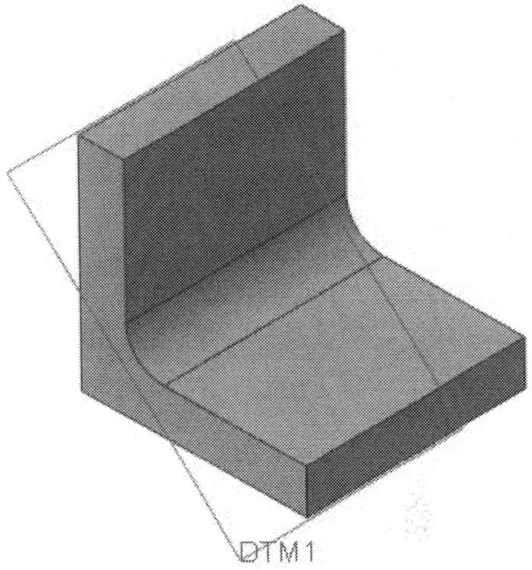

Figure 5-25 Datum plane created by choosing the ***Bisector1*** *option*

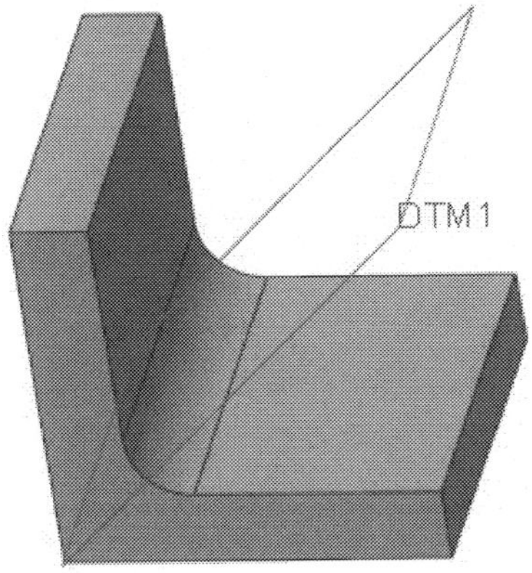

Figure 5-26 Datum plane created by choosing ***Bisector2*** *option*

Note

If you want to remove the reference that was selected while creating the datum plane, select the reference in the ***Datum Plane*** *dialog box and press the DELETE key. Alternatively, right-click on the reference in the dialog box and then choose* ***Remove*** *from the shortcut menu displayed.*

Offset Planes

Ribbon: Model > Datum > Offset Planes

You can use the **Offset Planes** tool to create datum planes at an offset from the model origin. When you invoke this tool from the **Datum** group of the **Model** tab; you are prompted to specify the offset distance in x-direction, y-direction, and z-direction. Specify the offset values in the respective edit boxes and choose the **Accept value** button; the datum planes will be created in the drawing area. In case, your model does not have any default coordinate system, this tool automatically places a default coordinate system at a predetermined location to create offset planes.

Datum Axes

Ribbon: Model > Datum > Axis

Similar to datum planes, datum axes can also be used as references for feature creation. Datum axes are created automatically when any cylindrical feature is created but you can also create Datum axes manually by using the **Axis** tool available in the **Datum** group. The display of the datum axis can be turned on or off by selecting the **Axis Display** check box from the **Datum Display Filters** drop-down in the **Graphics** toolbar. The uses of datum axes are given next.

1. Datum axes act as reference for feature creation.
2. They are used in creating a datum plane along with different constraint combinations.
3. They are used in placing features co-axially.
4. They are also used to create rotational patterns. You will learn to create patterns in later chapters.

In Creo Parametric, datum axes are named by default. The default name of a datum axis is **A_#**, where **#** represents the number of datum axis. However, the default name of the datum axes can be changed in the same way as that of the datum planes.

When you choose the **Axis** tool from the **Datum** group, the **Datum Axis** dialog box is displayed with the **Placement** tab chosen by default, as shown in Figure 5-27. The options in this dialog box are discussed next.

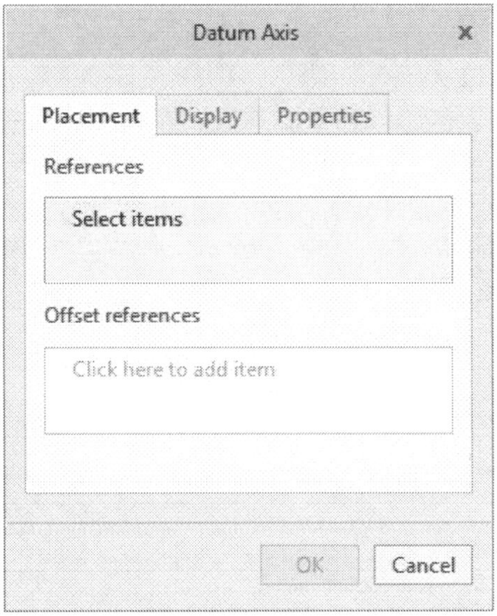

***Figure 5-27** The **Datum Axis** dialog box with the **Placement** tab chosen*

Placement Tab

When you invoke the **Datum Axis** dialog box, the **Placement** tab is chosen by default. Under this tab, the **References** collector allows you to select the references that will be used to create the datum axis. The constraints are displayed on the right of the references. These constraints are applied automatically based on the reference you select. The constraints that are used to create a datum axis are discussed later in this chapter.

The **Offset references** collector will be available only when the Normal constraint is used in the **References** collector to define a datum axis. To define the datum axis, you need to select two references. These references can be an edge, datum plane, face, or axis. The references can be specified dynamically on the model by dragging the handles displayed on the datum axis. The handles on the datum axis are displayed only when you select a reference. You can also select the references by using the **Datum Axis** dialog box.

There are three drag handles available on the model, as shown in Figure 5-28. The middle handle is used to move the position of the axis you are creating and it appears like a yellow circle. The other two handles are used to specify references for dimensioning and appear like a red brackets.

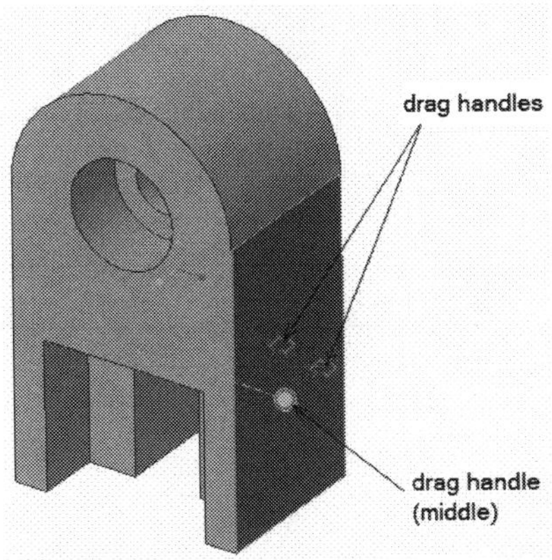

Figure 5-28 *Solid model with the datum axis and three drag handles*

Display Tab

The options in this tab are used to control the length of an axis. The **Datum Axis** dialog box with the **Display** tab chosen is shown in Figure 5-29. Select the **Adjust outline** check box to activate the **Length** edit box. By default, the **Length** edit box is in the inactive mode. You need to select the **Adjust outline** check box to activate it. You can enter a value in this edit box or drag the handles displayed at both ends of the datum axis to modify its length.

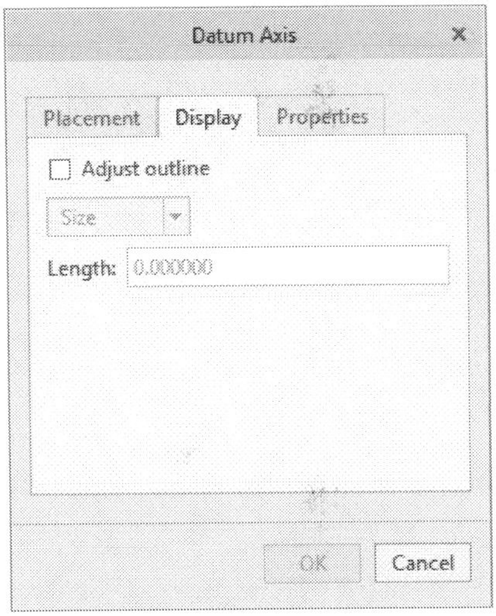

Figure 5-29 *The **Datum Axis** dialog box with the **Display** tab chosen*

Properties Tab

The **Properties** tab of the **Datum Axis** dialog box is shown in Figure 5-30. This tab, when chosen, allows you to name the datum axis you are creating. By default, the first datum axis you create is named A_1 and then additional datum axes created are named successively. As mentioned earlier, a datum axis is created using constraints that are applied automatically. The options available in the Filter drop-down list in the Status Bar while creating a datum axis, are shown in Figure 5-31. The constraints used to create a datum axis are discussed next.

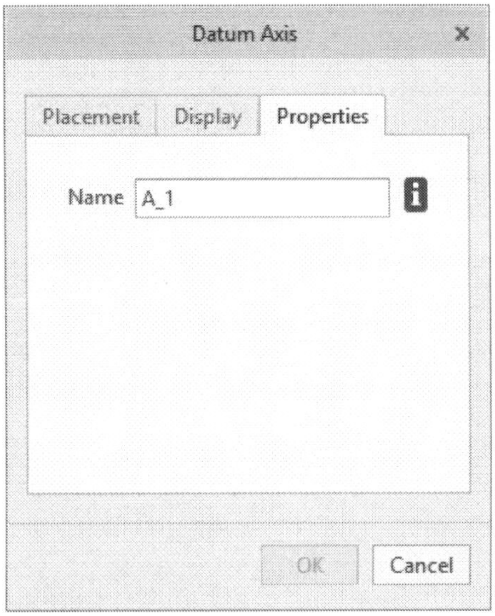

*Figure 5-30 The **Datum Axis** dialog box with the **Properties** tab chosen*

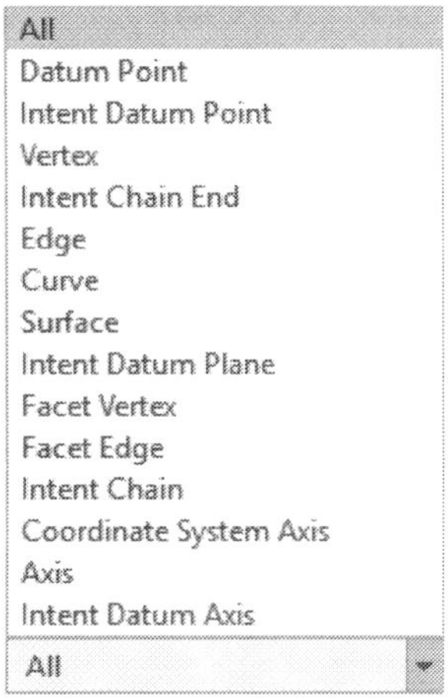

Figure 5-31 Options available in the Filter drop-down list

Datum Axis Passing along an Edge

The Through constraint is used to create a datum axis passing through any selected edge. This constraint will be displayed in the dialog box when you select an edge, as shown in Figure 5-32. In Figure 5-32, the datum axis **A_1** is created using the Through constraint.

Note

While creating a datum axis, you may need to create datum points passing through the axis. Therefore, in various cases that are discussed next, you may need to create datum points. Datum points are discussed in detail later in this chapter.

Datum Axis Normal to a Plane

The Normal constraint is used to create a datum axis normal to a selected face or datum plane. When you select a face or a datum plane, the **Normal** option will be displayed in the dialog box and you are prompted to select two references to place the axis. Notice that three drag handles appear on the model. Use the middle handle to change the location of the axis on the selected face. After you specify its placement location, you need to select two edges, axes, datums, or faces to specify the linear dimension for the placement of the datum axis. Click in the **Offset references** collector; the collector turns green. When you select the first edge for the placement dimensions of the axis, the offset value will be displayed in the **Offset references** collector. You can accept the default dimension or click on it to change its value. Similarly, select the second edge for dimensioning by using CTRL+left mouse button and enter the dimension value. Figure 5-33 shows preview of the datum axis with drag handles. It should be noted that the references for dimensioning can also be selected dynamically using these drag handles.

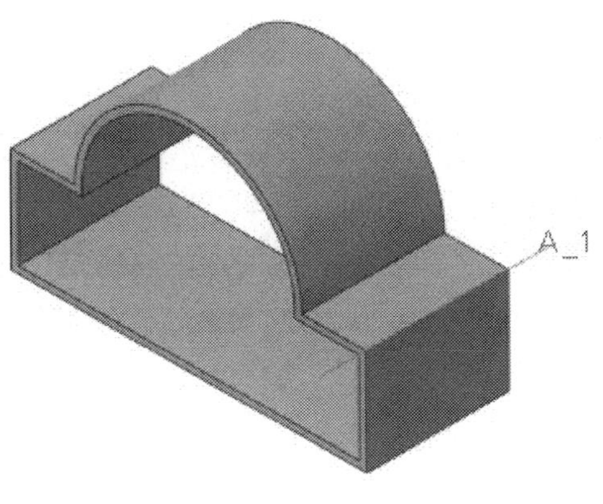

Figure 5-32 Datum axis created along the edge

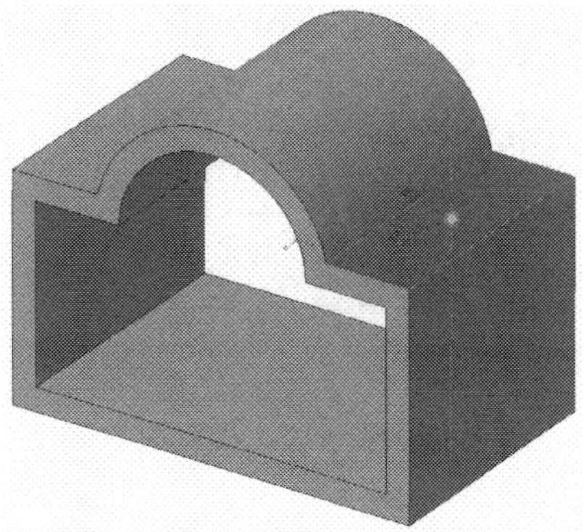

Figure 5-33 Datum axis created normal to the plane

Note

When you select a reference for creating the datum axis, you are prompted to select the offset references. For selecting the offset references, you need to activate the ***Offset references*** *collector by clicking on it. On doing so, the area in the* ***Datum Axis*** *dialog box will be highlighted, which indicates that you can select references to place the axis. Until the* ***Offset references*** *collector is activated, you will not be able to select placement references.*

Datum Axis Passing through a Datum Point and Normal to a Plane

The Normal constraint creates a datum axis passing through a datum point or vertex and normal to any face or datum plane. When you select a face or a datum plane to which the datum axis will be normal, you will be prompted to select two references to place the axis. Use CTRL+left mouse button to select the datum point or vertex to create an axis passing through it. Choose the **OK** button to exit the **Datum Axis** dialog box. Figure 5-34 shows preview of the datum axis and the datum point.

Datum Axis Passing through the Center of a Cylindrical Surface

The Through constraint is used to create a datum axis passing through the center of a cylindrical or round surface. To create this datum axis, select the round surface through which you need to pass the datum axis; an axis will be created automatically and will pass through the center. Figure 5-35 shows preview of the datum axis. In this figure, the selected cylindrical surface is highlighted.

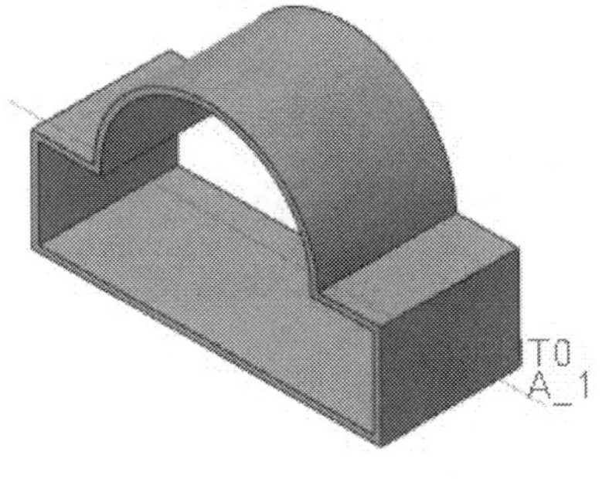

Figure 5-34 Datum axis passing through the datum point and normal to the plane

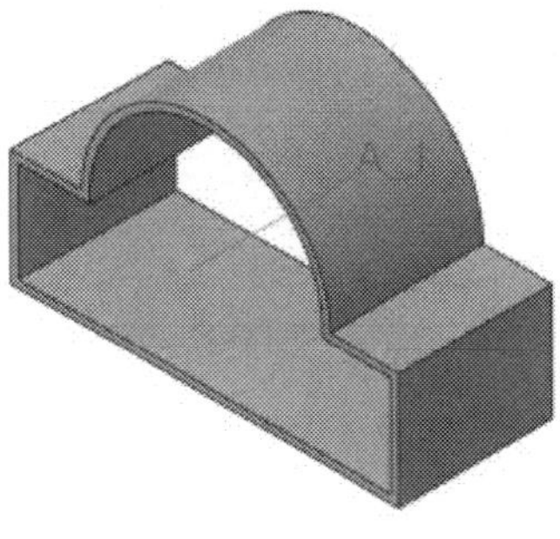

Figure 5-35 Datum axis passing through the center of a cylinder

Datum Axis Passing through the Edge Formed by Two Planes

The Through constraint is used to create a datum axis passing through the intersection edge of two planar faces or planes. When you select the face, the Normal constraint is applied automatically. Now, select the second face using the CTRL+left mouse button; the Through constraint is applied to both faces automatically and a datum axis is created passing through the edge formed by the two faces or planes. Figure 5-36 shows preview of the datum axis. The two faces that were selected are also highlighted in this figure.

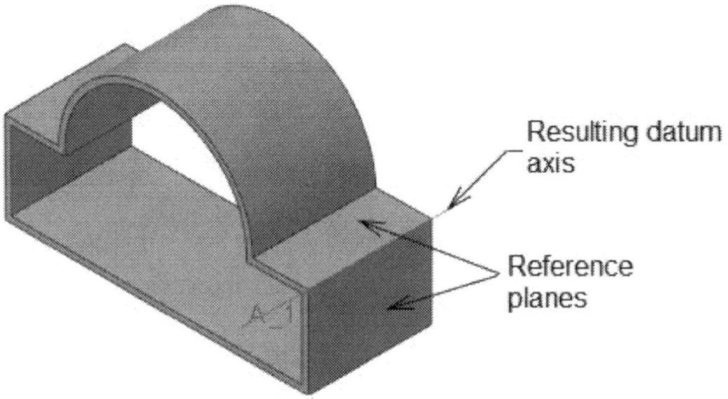

Figure 5-36 *Datum axis created at the edge where the two selected planes meet*

Datum Axis Passing through Two Datum Points or Vertices

The Through constraint is used to create a datum axis between two datum points or vertices. To create this datum axis, select the first vertex and then select the second vertex using CTRL+left mouse button; the datum axis will be created along the two selected datum points or vertices. Figure 5-37 shows preview of the datum axis with the two vertices highlighted.

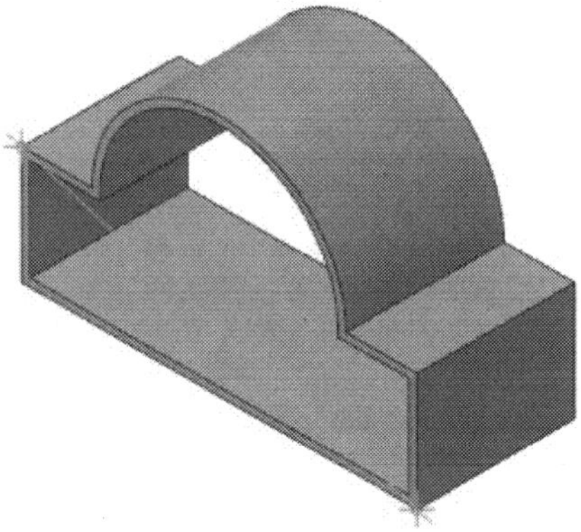

Figure 5-37 *Datum axis created between the two selected vertices*

Datum Axis Tangent to a Curve and Passing through its Vertex

The Tangent and Through constraints create a datum axis tangent to a curve and passing through one of its vertex. Select the edge of the cylindrical surface as the first selection. Make sure you do not select the cylindrical surface. After you select a curve or an edge, use CTRL+left mouse button to select one vertex of the edge. The datum axis is created tangent to the curve and passing through its selected vertex. Figure 5-38 shows the preview of the datum axis created by using the Tangent and Through constraints. The curved edge is also highlighted in this figure.

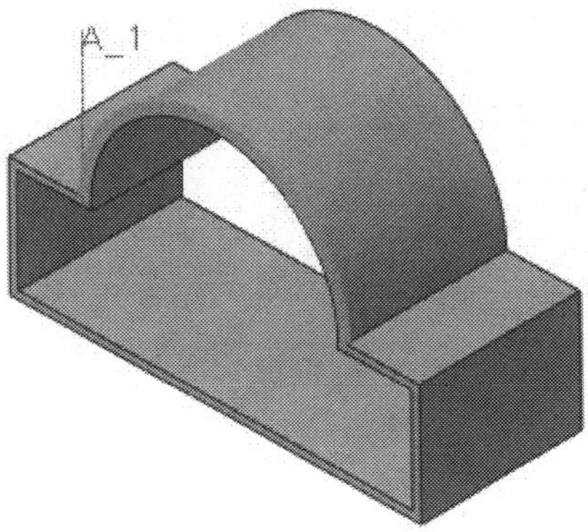

Figure 5-38 Datum axis created tangent to the highlighted curve

Datum Points

Ribbon: Model > Datum > Point > Point

Datum points are imaginary points created to help creating models and drawings, analyzing models, and so on. Creo Parametric creates datum point as a feature. A datum point feature can contain multiple datum points that are created during the same operation. Datum points created in one operation appear under one feature in the **Model Tree**. To edit or delete a point of a datum point feature you need to edit the definition of that feature. You can add points using different placement methods within one datum point feature.

The datum points are used:

1. To create datum planes and axes
2. To associate note in the drawings and attach datum targets
3. To create coordinate system
4. To specify point loads for mesh generation
5. To create pipe features

The default name associated with a datum point in Creo Parametric is **PNT(#)** where **#** indicates the number of datum points created in a particular model. However, you can change the default name associated with the datum points.

When you choose the **Point** tool from the **Point** drop-down in the **Datum** group of the **Ribbon**, the **Datum Point** dialog box will be displayed with various options to create datum points, as shown in Figure 5-39. The options in the **Datum Point** dialog box are discussed next.

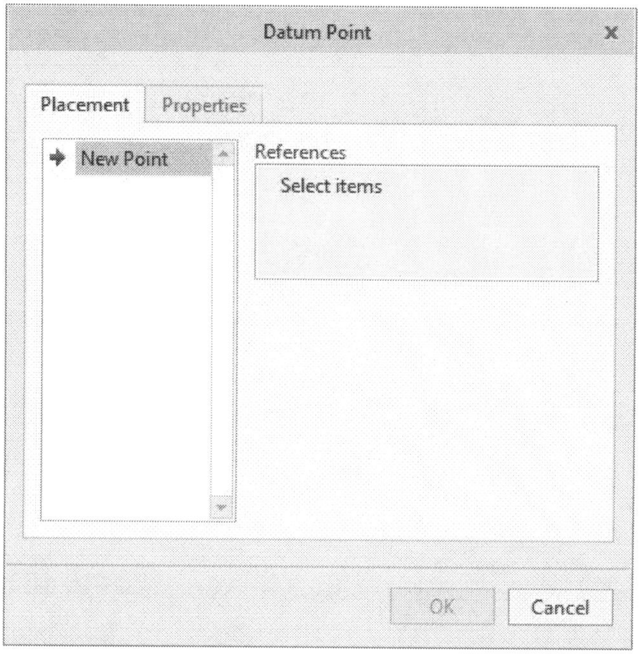

Figure 5-39 *The* ***Datum Point*** *dialog box with the* ***Placement*** *tab chosen*

Tip
On selecting references in the model while creating the datum feature, some constraints appear in the ***References*** *area automatically. You need to choose the required constraint from these constraints to constrain the datum feature. In this process, you will notice that the range of available constraints reduces, thereby enabling you to make a faster decision.*

Placement Tab

The **Placement** tab in the **Datum Point** dialog box has the options that change with the reference selected. However, the **References** collector shown in Figure 5-39 is always present. This area is used to select the references which aid in placing the datum point. The constraints available in the drop-down list in the **References** collector depend on the references you select from the model. This means that the references narrow down the type of constraints you can apply on the datum point that you are creating.

Various methods of creating datum points are discussed next.

Datum Point on a Face or a Datum Plane

The On constraint is used to create datum points on a face or a datum plane. When you select a face or a datum plane to place the datum point, a yellow colored point will be displayed at the selected location on the surface with the three drag handles, as shown in Figure 5-40.

Next, click in the **Offset references** collector to make it green. You are prompted to select two references to specify the linear dimensions for the placement of the datum point. Note that the second selection to select the reference should be made by using CTRL+left mouse button.

After you select the two planes or edges for the placement of dimension of the point, a default value will be displayed in the **Offset references** collector. You can accept the default value or change it to the required value. After the datum point is located at the desired position on the face or datum plane, choose the **OK** button to exit the dialog box.

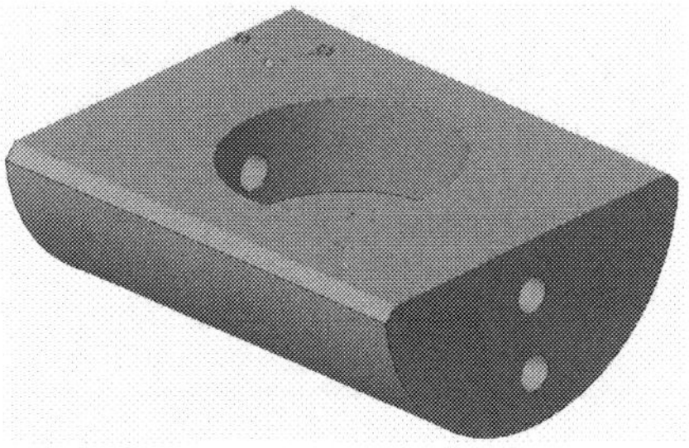

__Figure 5-40__ Datum point on a face and the three drag handles

Datum Point Offset to a Face or a Datum Plane

The Offset constraint creates datum points at an offset distance from a specified face or a datum plane in a specified direction. Select a face or a datum plane from where the offset distance for the placement of the datum point will be measured. The **On** option is displayed automatically in the **Datum Point** dialog box. From the drop-down list in the **References** collector, select the **Offset** option and make required settings, as shown in Figure 5-41. Figure 5-42 shows the datum point created at an offset from the selected face.

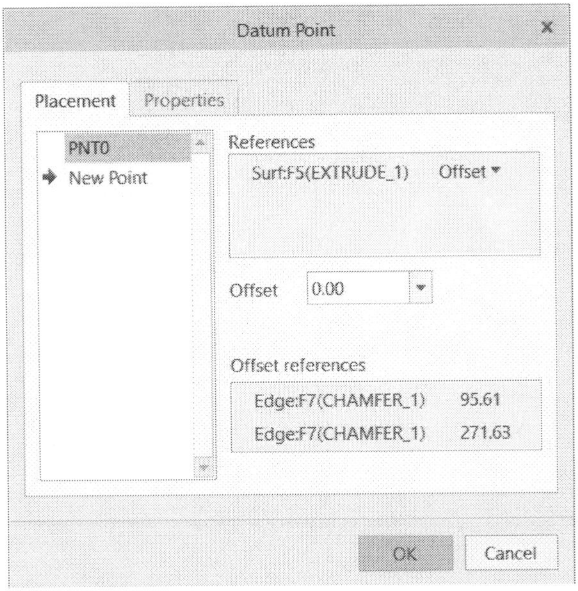

*__Figure 5-41__ The **Datum Point** dialog box with the **Offset** option selected*

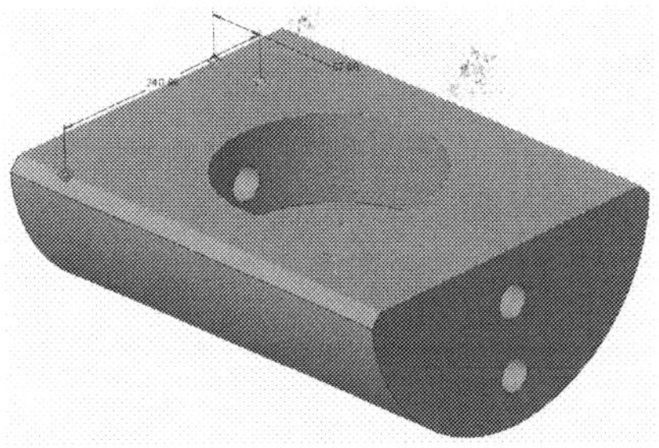

__Figure 5-42__ Datum point on the top face

In the **Offset** edit box under the **References** collector, enter the offset value; you will be prompted to select the planes or the edges for dimensioning the point. Select two planes or edges for dimensioning and enter the distances from the highlighted references.

You can also specify the offset distance dynamically by using the middle drag handle. The two references for dimensioning the datum point can be specified by using the drag handles or by directly pointing them on the model.

Note

*When you select the **Offset** option, the middle drag handle is used to specify the offset distance and when you select the **On** option, the same drag handle is used to specify the location of the datum point on the selected face.*

Datum Point Projected on a Face or a Datum Plane

You can project a datum point onto a face or datum plane by using the Project constraint. When you choose the **Point** tool from the **Point** drop-down in the **Datum** group of the **Ribbon**, the **Datum Point** dialog box will be displayed. Select a point which is to be projected; the selection will appear in the **References** collector. By default, the **On** option is selected in the drop-down list in the **References** collector in the **Datum Point** dialog box. In this drop-down list, select the **Project** option, refer to Figure 5-43. Next, press the CTRL key and select the surface on which the point is to be projected; the surface selected will appear in the **References** collector with the constraint set to **On**. After the datum point is located at the desired face or datum plane, choose the **OK** button to exit the dialog box. Figure 5-44 shows the source point selected and projected point with the reference surface selected.

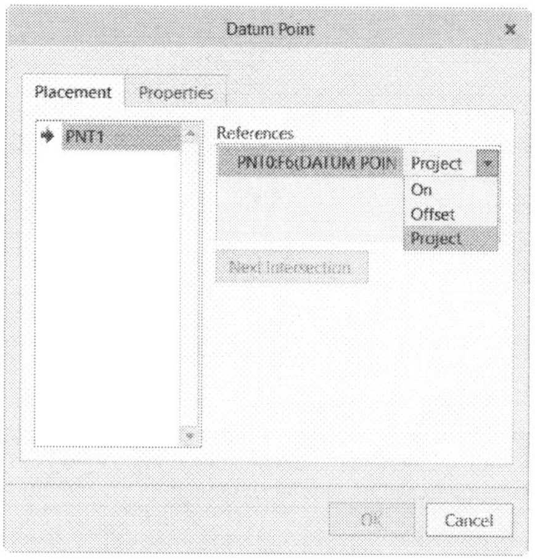

Figure 5-43 *The **Datum Point** dialog box with **Project** option selected*

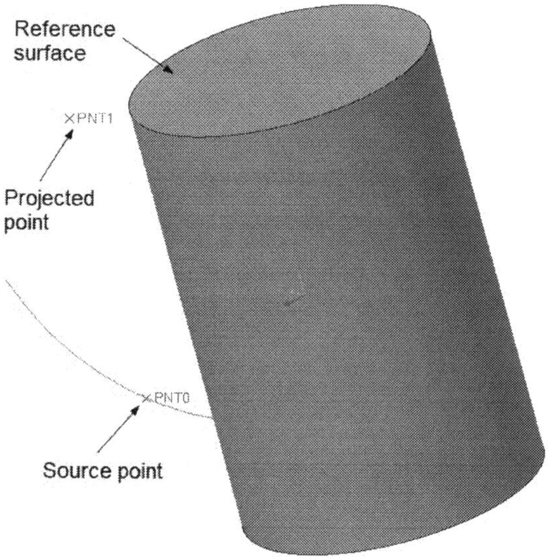

Figure 5-44 *The three highlighted inputs for projection datum point*

Datum Point at the Intersection of Three Surfaces

The **On** constraint is used to create a datum point at the intersection of three surfaces. To create this datum point, select the three surfaces; a datum point will be created, as shown in Figure 5-45.

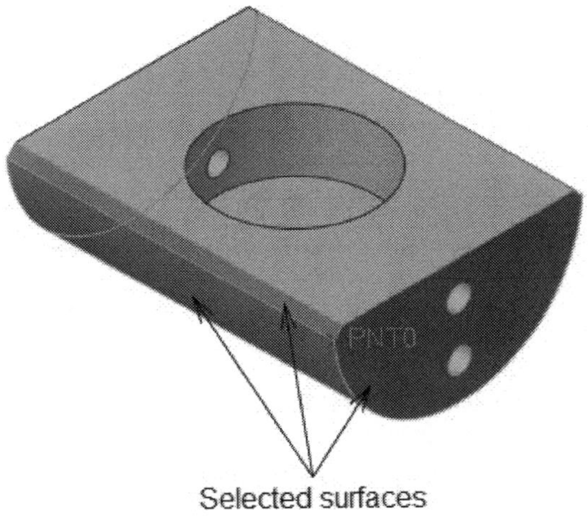

__Figure 5-45__ The three highlighted surfaces and the datum point at the intersection of the three surfaces

The datum point is created on the vertex that is common to the three surfaces. Remember that the first reference is selected using the left mouse button. The second and third references are selected using CTRL+left mouse button. The **Datum Point** dialog box that will be displayed after selecting the three surfaces is shown in Figure 5-46.

Sometimes, more than one intersection point exists with the current selection of references. In such a case, use the **Next Intersection** button in the **Datum Point** dialog box to select the other intersection points to create the datum point on them.

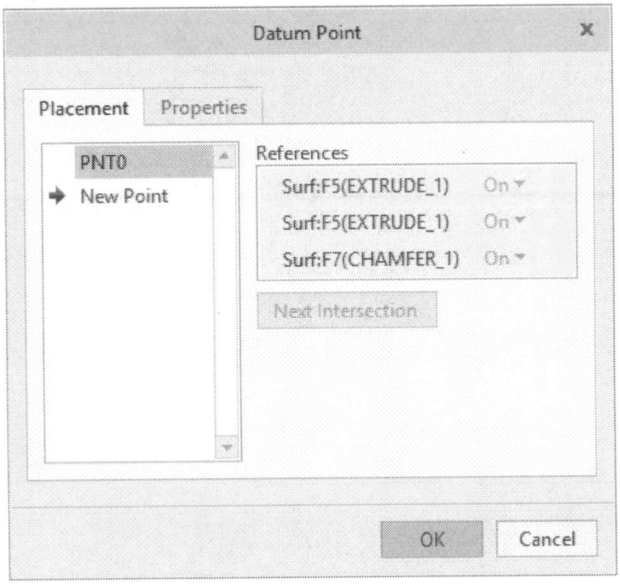

__Figure 5-46__ The __Datum Point__ dialog box

Datum Point on a Vertex

The **On** option is used to create a datum point on the vertex of a face, an edge, or a datum curve. Invoke the **Datum Point** dialog box and then select a vertex to create the datum point. Choose the **OK** button to exit the dialog box.

Note
You can create more than one datum point by simply selecting the desired vertex.

Datum Point at the Center of a Curved Edge

The **Center** option is used to create a datum point at the center of an arc or a curved edge. Invoke the **Datum Point** dialog box and select the curve edge; the **On** constraint is applied by default. To modify the constraint, select the constraint in the **References** collector; a drop-down list appears to the right of the constraint. Select the **Center** constraint from it to create the datum point at the center of the curved edge.

Datum Point on an Edge or a Curve

The **On** option is used to create a datum point on an edge or a curve. Invoke the **Datum Point** dialog box, and select an edge or a curve from the drawing area. On doing so, the datum point will be placed on the selected edge, as shown in Figure 5-47, and the **Datum Point** dialog box will be displayed.

Figure 5-47 *The highlighted edge and the datum point at some offset distance from one end of edge*

In the **Datum Point** dialog box, after the point is placed, the dimension type can be defined. The **Offset** edit box displays the offset distance of the datum point from one end of the edge. The offset distance is measured as a ratio or a real value from the end of the edge which can be selected from the drop-down list, refer to Figure 5-48.

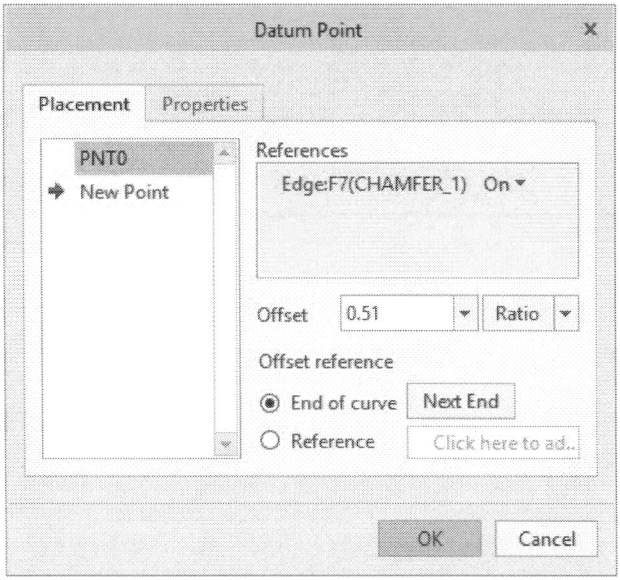

***Figure 5-48** The **Datum Point** dialog box with the **Ratio** option selected*

In the **Offset reference** area, the **Next End** button is used to flip the end of the edge from where the offset distance is measured. This button is available only when the **End of curve** radio button is selected. If you select the **Reference** radio button, you need to select a reference from which the datum point will be dimensioned.

Note

The geometry points created in the sketcher environment can also be used as datum points.

Datum Point On or Offset from a Coordinate System

To place a datum point on a coordinate system, select the coordinate system and choose the **OK** button from the **Datum Point** dialog box. To create a datum point that is offset from a coordinate system, click on the point; the **On** option will be displayed automatically in the **Datum Point** dialog box. From the drop-down list in the **References** collector, select the **Offset** option and select a coordinate system by using the CTRL+left mouse button. As you select the coordinate system a yellow colored circle will be displayed at the origin of the coordinate system, as shown in Figure 5-49. Also, the **Offset type** drop-down list gets activated in the **Datum Point** dialog box, as shown in Figure 5-50. By default, the **Cartesian** option is selected in this drop-down list. As a result, you can specify the offset values for the resulting datum point by specifying values in the **X**, **Y** and **Z** edit boxes. You can also choose the **Cylindrical** or **Spherical** option from this drop-down list as the offset type for creating datum point.

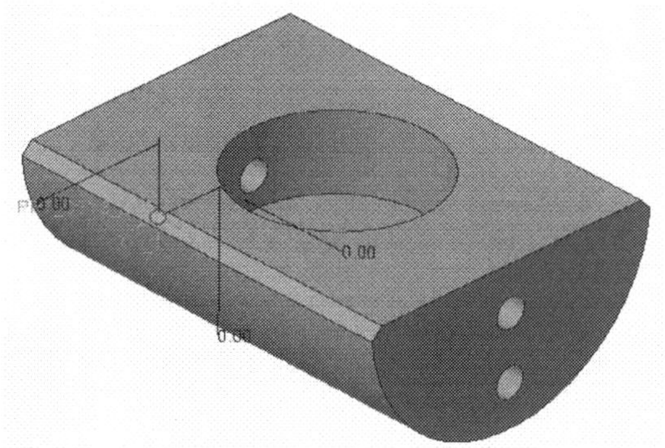

Figure 5-49 The datum point created at offset distance from a coordinate system

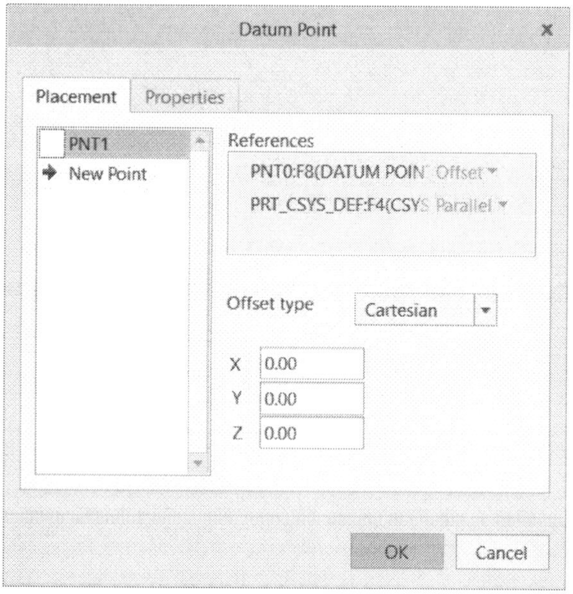

Figure 5-50 *The **Datum Point** dialog box with the **Offset** option selected in the **References** collector*

Creating an Array of Datum Points from a Coordinate System

Ribbon: Model > Datum > Point drop-down > Offset Coordinate System

The **Offset Coordinate System** tool available in the **Point** drop-down of the **Datum** group is used to create an array of datum points at an offset distance from a coordinate system. You can offset points using a Cartesian, Spherical, or Cylindrical coordinate system. You can change the location of the points by redefining the array. Note that the Datum Coordinate system must be defined before you create an array of datum points.

When you invoke the **Offset Coordinate System** tool, the **Datum Point** dialog box will be displayed and you will be prompted to select a coordinate system. After selecting a coordinate system, you can enter the values of coordinates for the datum points, refer to Figure 5-51.

To add a point, click under the **Name** column in the list box; the first row gets activated. Click under the **X Axis** column; the edit box appears in which you can enter an offset value. Similarly, enter the offset values for **Y Axis** and **Z Axis.** To add another point click in the next row. From the **Type** drop-down list, select the type of coordinate system: **Cartesian**, **Cylindrical**, or **Spherical**. The options in the list box change according to the coordinate system selected from the **Type** drop-down list.

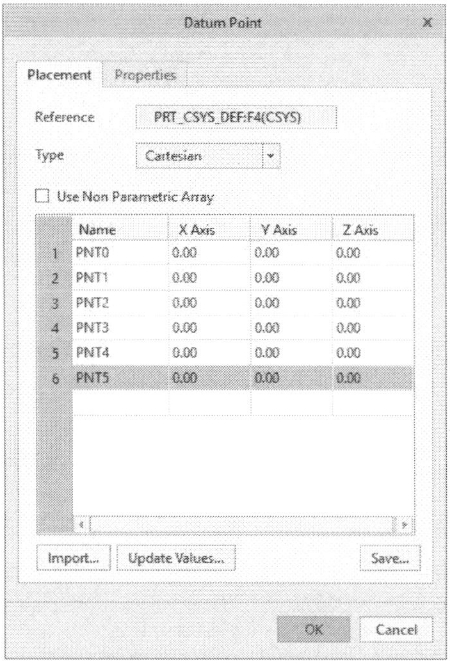

*Figure 5-51 The **Datum Point** dialog box displayed on choosing the **Offset Coordinate System** tool*

Creating a Field Point

Ribbon: Model > Datum > Point drop-down > Field

A field point is a type of datum point that is intended for use in conjunction with a user-defined analysis (UDA). The field point does not require dimensions for its placement because it belongs to the entire domain. The default name associated with a field datum point in Creo Parametric is **FPNT(#)** in part mode and **AFPNT(#)** in assembly mode. Here # represents number of the point. When you choose the **Field** tool from the **Point** drop-down of the **Datum** group; the **Datum Point** dialog box will be displayed, as shown in Figure 5-52. Using this dialog box, you can create a datum point anywhere on the surface or edge of the model. You just need to specify the location of the datum point by using the left mouse button to place a datum point.

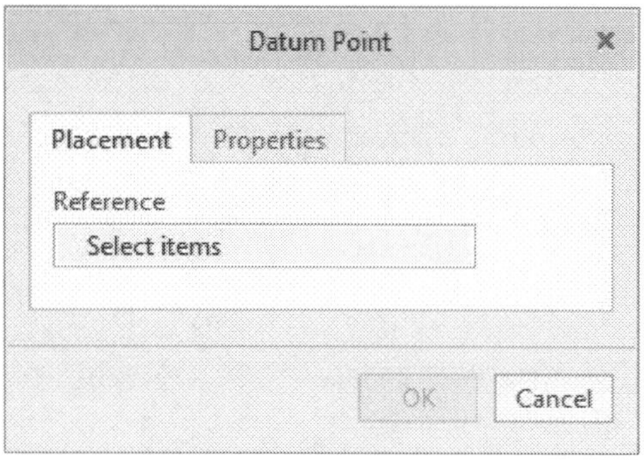

*Figure 5-52 The **Datum Point** dialog box displayed on choosing the **Field** tool*

Datum Coordinate System

Ribbon: Model > Datum > Coordinate System

Coordinate system is used to determine the position of a point, planes or other geometric element. A coordinate system helps you to create datum planes and points, calculate mass properties, set modeling and assembly references, create manufacturing operation references for toolpaths, and define specific location in space.

There are two types of coordinate systems that can be created in Creo Parametric: On-surface coordinate system and Offset coordinate system.

The on-surface coordinate system is defined by a point located on the surface of a geometry. This point is placed by means of a primary placement reference and two offset references or by two to three primary references. The primary placement reference can be a quilt, a surface, or a datum plane. Planar, non-planar, cylindrical, conical, or spherical surfaces can also be used as primary placement references. The offset values of a primary placement coordinate system can be defined in Linear, Radial, or Diameter measurements. If you select more than one primary reference, the **Offset references** collector gets disabled and the offset handles do not appear in the preview.

The offset coordinate system is defined by a point using an existing coordinate system as primary placement reference. The three offset values are referenced by using existing coordinate system. This type of coordinate system can be used to position sequential coordinate systems, for an instance, you can position a number of user-defined features (UDFs) on a model by using such coordinate systems. The offset values of an offset coordinate system can be defined in Cartesian, Cylindrical, Spherical, or from file measurements.

The default name associated with a coordinate system in Creo Parametric is **CS(#)**, where **#** indicates the number of coordinate systems created for a particular model. However, you can change the default name associated with the coordinate system.

For creating a coordinate system, choose the **Coordinate System** tool from the **Datum** group of the **Ribbon**, the **Coordinate System** dialog box will be displayed, as shown in Figure 5-53. The **Properties** tab similar to the one discussed in earlier sections of this chapter. Rest of the options and tabs in the **Coordinate System** dialog box are discussed next.

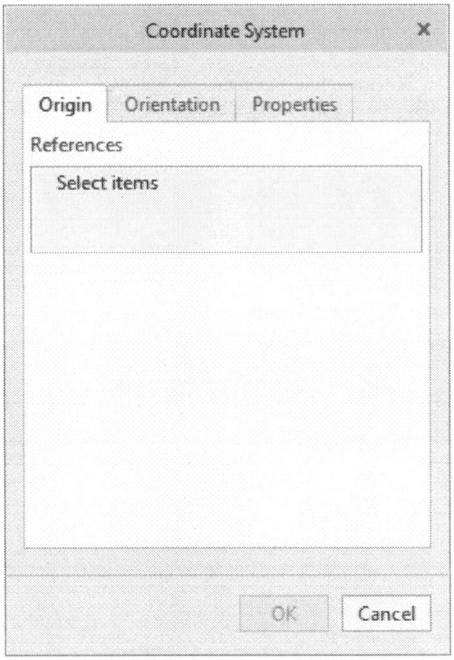

***Figure 5-53** The **Coordinate System** dialog box with the **Origin** tab chosen*

Origin Tab

The options in this tab change based on the reference selected. However, the **References** collector shown in Figure 5-53 will be always available in this tab. It is used to select the references which help in placing the coordinate system. As you select any geometry as the primary reference, the **Type** drop-down list and the **Offset references** collector list gets activated. But if you select Coordinate system, then the **Offset type** drop-down list gets activated. These options are discussed later in this chapter.

Orientation Tab

This tab is used to orient the coordinate system with respect to the selected reference. In this tab, the **References selection** and **Selected CSYS axes** radio buttons are used to orient the coordinate system with respect to the references selected in the **Origin** tab. The options for orienting a coordinate system are discussed later in this chapter.

Properties Tab

This tab is used to specify the display properties of the coordinate system. You can change the default name of the coordinate system by specifying a name in the **Name** text box. By default, the first coordinate system you create is named as **CS0** and then additional coordinate systems created are named successively. Select the **Display coordinate system name** check box to display the name of the coordinate system that you are creating. Selecting the **Display zoom-dependent** check box will increase or decrease the coordinate system on zooming in or out. You can also specify a length value for the axes of the coordinate system by specifying a value in the **Axis length** edit box.

Creating an On-surface Coordinate System

To create a coordinate system on a face, invoke the **Coordinate System** dialog box. Next, select the face on which you want to create the coordinate system; an orange colored point will be displayed at the selected location with three drag handles, refer to Figure 5-54. Also, the options in the **Coordinate System** dialog box will be modified. By default, the **On** constraint is selected in the **References** collector of the **Origin** tab. As a result, the coordinate system will be created on a face or a datum plane.

Next, click on the **Offset references** collector to highlight it; you will be prompted to select two references to specify the linear dimensions for the placement of the coordinate system. Select the first reference and then select the second reference by using CTRL+left mouse button. After you select the two planes or edges for the placement of dimension of the coordinate system, a default value will be displayed in the **Offset references** collector. You can accept the default value or change it to the required value.

When you select a flat face or a datum plane, the **Type** drop-down list will be displayed. The options in this drop-down list are **Linear**, **Radial**, and **Diameter**, refer to Figure 5-55. The options in this drop-down list are discussed next.

Linear

This option is used for linear reference selection. When you select any surface, plane, or edge, the **Linear** option will be selected by default in the **Type** drop-down list. The **Offset references** collector will display two options: **Offset** and **Align**. The **Offset** option is used to create a coordinate system at an offset distance from the selected reference. If the **Align** option is selected, the coordinate system created will be aligned with the selected reference.

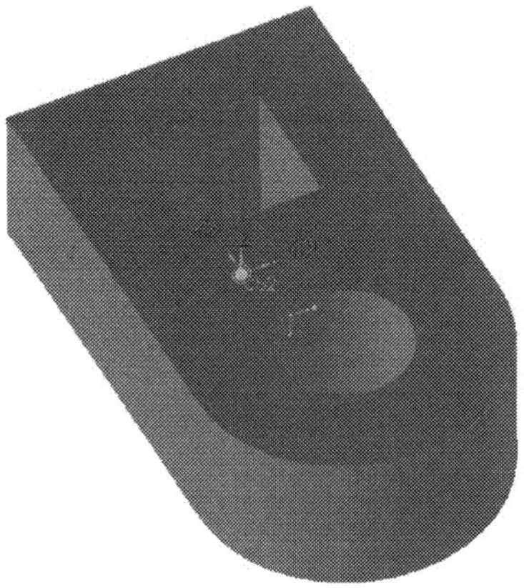

Figure 5-54 The Coordinate System on a face with three drag handles

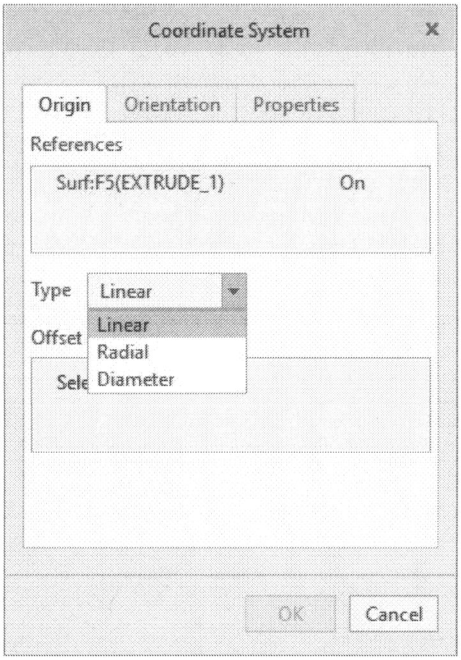

***Figure 5-55** The **Coordinate System** dialog box with the **Type** drop-down list*

Radial

This option is used for radial reference selection. In the radial referencing, one option must be surface or plane for angle reference and other must be edge or axis for radius reference.

Diameter

This option is used for diametrical reference selection. In the diameter referencing, one option must be surface or plane for angle reference and other must be edge or axis for diameter reference.

Note

*When you select any circular surface, only the **Linear** and **Radial** options will be displayed in the **Type** drop-down list and the **Radial** option will be selected by default.*

After specifying settings in the **Origin** tab, choose the **Orientation** tab. As you have selected a face as a reference, the **References selection** radio button will be automatically selected, as shown in Figure 5-56. The **Use** collectors displays the references to orient the first and second axis of the new coordinate system. The **to determine** drop-down list is used to select the axis of the new coordinate system in the first direction. The orientation of the axis will be in relation to the reference selected in the **Use** collector of the first direction. The **to project** drop-down list is used to select the axis of the new coordinate system in the second direction. The orientation of the axis will be in relation to the reference selected in the **Use** collector of the second direction. The **Add rotation about the first axis** check box in this tab is used to rotate the coordinate system with respect to the first offset reference.

After setting all the parameters in the **Coordinate System** dialog box, choose the **OK** button; the coordinate system will be created on the selected face.

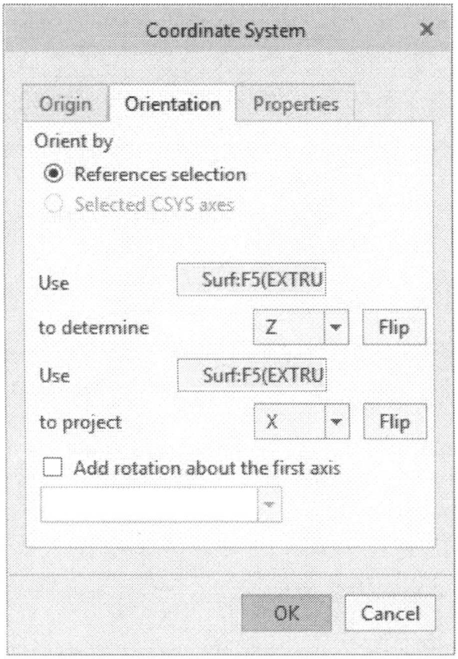

Figure 5-56 *The* ***Orientation*** *tab with* ***References selection*** *radio button selected*

Creating an Offset Coordinate System

To create an offset coordinate system using an existing coordinate system, invoke the **Coordinate System** dialog box. Next, select the existing coordinate system of the model; an orange colored point will be displayed at the selected coordinate system, refer to Figure 5-57. Also, the options in the **Coordinate System** dialog box will be modified, as shown in Figure 5-58. By default, the **Offset** constraint is selected in the **References** collector of the **Origin** tab. As a result, the coordinate system will be created at an offset distance.

There are four options in the **Offset type** drop-down list to create a coordinate system, refer to Figure 5-58. The options in this drop-down list are discussed next.

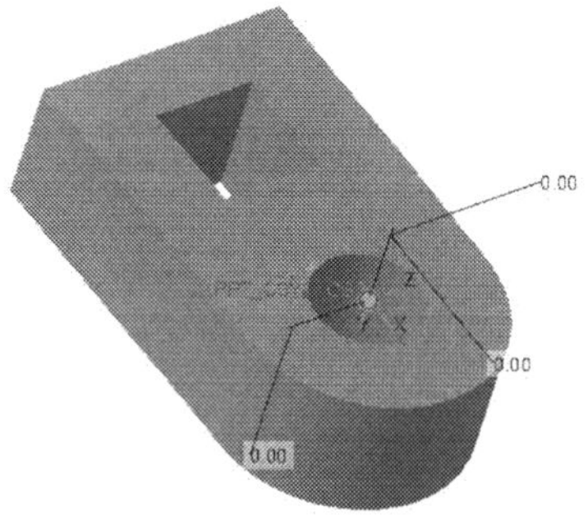

Figure 5-57 The coordinate system created on an existing coordinate system with offset 0 for all axes

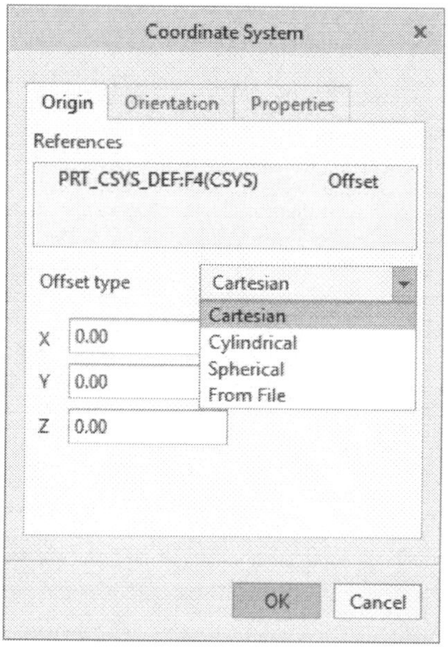

*Figure 5-58 The **Coordinate System** dialog box with the **Offset type** drop-down list*

Cartesian

When you select this option, the **X**, **Y**, and **Z** edit boxes will be displayed. You can specify the values in these edit boxes to change the location of the coordinate system.

Cylindrical

When you select this option, the **R**, Θ, and **Z** edit boxes will be displayed. You can specify the values in these edit boxes to change the location of the coordinate system.

Spherical

When you select this option, the **r**, Φ, and Θ edit boxes will be displayed. You can specify the values in these edit boxes to change the location of the coordinate system.

From File

When you select this option, a window will be displayed. In this window, you can browse to the ***.trf*** file (transformation file). This file is used to place the coordinate system.

After specifying settings in the **Origin** tab, choose the **Orientation** tab. As you have selected an existing coordinate system, the **Selected CSYS axes** radio button will be selected by default, refer to Figure 5-59. The orientation can be changed by specifying the values in the **About X**, **About Y**, and **About Z** edit boxes.

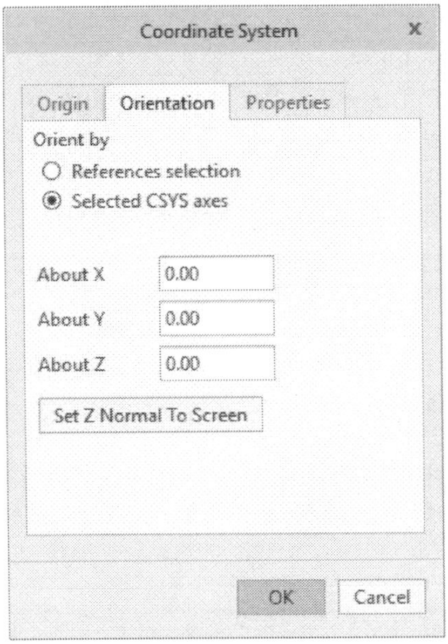

*Figure 5-59 The **Orientation** tab with **Selected CSYS axes** radio button selected*

After setting all the parameters in the **Coordinate System** dialog box, choose the **OK** button; the coordinate system will be created.

Creating a Coordinate System at the Center of an Arc

You can create a coordinate system at the center of an arc. To do so, choose the **Coordinate System** tool from the **Datum** group of the **Model** tab; the **Coordinate System** dialog box will be displayed. Next, select a circular arc or circular edge. As you select the arc, preview of the coordinate system will be displayed in the drawing area. This coordinate system coincides with the center of the arc and the z-axis of the coordinate system will be normal to the plane that contains the circular arc. To change the axis, select the **Orientation** tab, and select an axis in the **to determine** drop-down list of the first direction. Note that, you need to define a second reference for the axis of the coordinate system. Click on the **Use** collector of the second direction and select a reference from the drawing area. To change the orientation of the axis in the second direction, select the axis from the **to project** drop-down list. To flip the direction of the axes, you can click on the **Flip** button next to the corresponding reference collector. Next, choose the **OK** button to create the coordinate system.

Datum Curves

Ribbon: Model > Datum > Curve

Datum curves are 2D curves that can be used to create features such as extrusion or revolve feature. Datum curves can also be used to create trajectories for swept features. Swept features are very useful when you are working with complex surface design, piping design, or wiring and harness design. To access the tools for creating datum curves, click the down arrow in the **Datum** group of the **Model** tab and select the desired tool from the **Curve** flyout. The options and tools to create datum curve are discussed next.

Curve through Points

The **Curve through Points** tool allows you to create a datum curve passing through points. The resulting curve can be a spline or a sequence of alternating tangent lines and arcs. To create a datum curve through the number of points, select the **Curve through Points** tool from the **Curve** flyout; the **CURVE: Through Points** dashboard will be displayed. The options available in this dashboard are discussed next.

Use spline to connect the point to the previous point: Choose this button if you want to connect the selected points with a spline.

Use line to connect the point to the previous point: Choose this button if you want to connect the selected points with straight lines.

Rounds curve by specified value: This option allows you to add fillet to the corners of the datum curve. This tool is automatically activated in the dashboard when you choose the **Straight line** option in the slide-down panel of the **Placement** tab.

Placement Tab

Choose the **Placement** tab; a slide-down panel will be displayed, as shown in Figure 5-60. The **Point** collector in this tab is used to collect the points required for creating the datum curve. You can select a point, vertex, or any curve end by using this collector. To add more points, you can choose the **Add Point** option from the left pane of this tab or you can select points by clicking on them in the drawing area. To reorder the selected points upward or downward in the points list, you can choose the ⬇ or ⬆ button. When you add two or more points in the points list, the **Spline** and **Straight line** radio buttons in the **Connect to previous point by** area get activated. By default, the **Spline** radio button is selected in this area. As a result, a spline will connect all the selected points. Select the **Straight line** radio button if you want to use a straight line to connect a selected point to the previous point. Select the **Add fillet** check box, if you want to add a fillet to the curve. Select the **Group with equal radius points** check box to make the selected point a part of a logical group of points having same fillet radius. To add a point set to a group, select the point set from the model tree and then select the **Group with equal radius points** check box. You can select the **Place curve on surface** check box and select a surface in the **Surface** reference collector to force the entire curve to be placed on the selected surface.

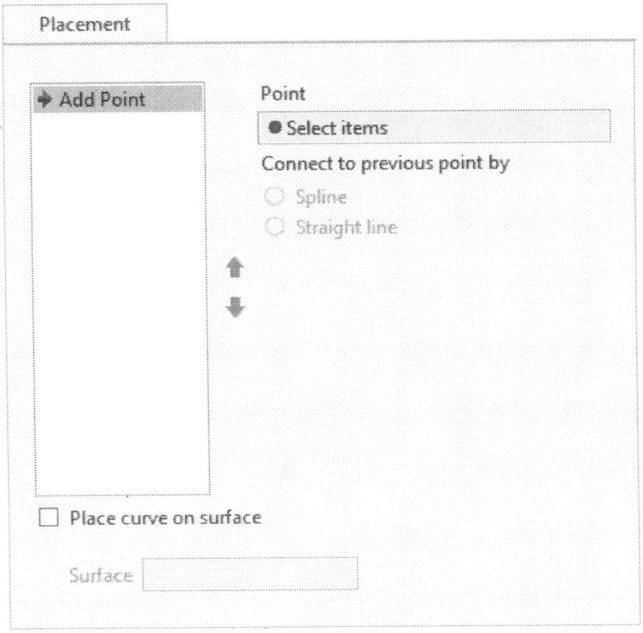

***Figure 5-60** The **Placement** tab of the **CURVE: Through Points** dashboard*

Ends Condition tab

The options in this tab are used to specify end conditions for the datum curve. Select either **Start Point** or **End Point** under the **Curve side** list and choose an end condition from the **End Condition** drop-down list. By default **Free** is selected in this drop-down list. As a result, the selected point of the curve is free from any tangency constraint. The other options in this drop-down are discussed next.

Tangent: As you choose the **Tangent** option from the **End Condition** drop-down list; the **Tangent to** collector gets activated, refer to Figure 5-61. You can select an axis, edge, curve, plane, or surface to make the end of the curve tangent to it. Choose the **Flip** button if you want to change the tangent direction to the other side of the reference. Select the **Make end curve to be perpendicular** check box to make the curve end perpendicular to a selected reference when the selected tangency reference is a surface or plane. Click in the **Perpendicular to** collector and select an edge to make the curve end perpendicular to it.

Curvature continuous: This option makes the end of the datum curve tangent to a selected reference and applies a continuous curvature condition to the selected point. Click in the **Curvature continuous to** collector, and select as axis, edge, curve, plane, or surface to which the curve end is tangent. Choose the **Flip** button if you want to change the tangent direction to the other side of the reference.

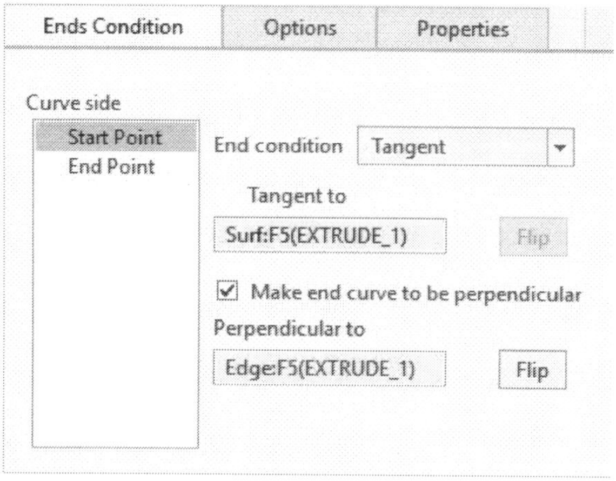

Figure 5-61 *The* ***Ends Condition*** *tab of the* ***CURVE: Through Points*** *dashboard*

Normal: This option makes the selected end of the curve normal to a selected reference. Click the **Normal to** collector, and select an axis, edge, curve, plane, or surface to which the curve end is normal.

Options tab

You can tweak and manipulate the shape of the curve in 3D space if you are creating a datum curve through two points.

Select the **Tweak curve** check box and click **Tweak Curve Settings**; the **Modify Curve** dialog box is displayed, as shown in the Figure 5-62. The options in this dialog box are discussed next.

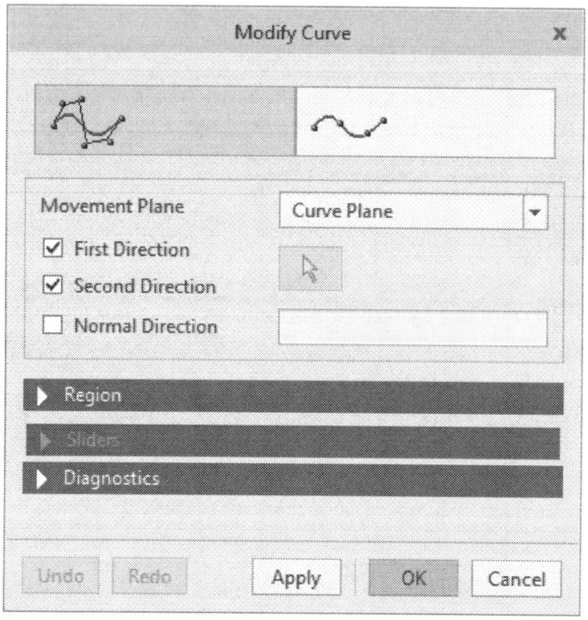

Figure 5-62 *The* ***Modify Curve*** *dialog box*

Control Polyhedron: This option allows you to move the datum curve by moving the points of the polyhedron surrounding the curve. To specify the movement plane for the datum curve tweaking, select required option from the **Movement Plane** drop-down list. The **First Direction** and **Second Direction** check boxes are selected by default. As a result, the interpolation points can be moved in two directions. If you select the **Normal Direction** check box; the **First Direction** and **Second Direction** check boxes get cleared, and the movement of the interpolation point is set normal to the selected plane. To limit the movement range of the curve, you can use the options available in the **Region** rollout. To tweak the shape of the datum curve, expand the **Sliders** rollout, drag the sliders or specify values in their respective edit boxes. You can analyze the curve by selecting analysis type from the **Diagnostics** rollout. To highlight the analysis result in drawing area, select an analysis type from the list and choose the Hide/Show button ( ). To change the display settings of the analysis, click the **Setting** button; the **Display Settings** dialog box is displayed. Adjust the settings and click **OK**. To change the computation settings, click the **Computation** button from the **Diagnostics** rollout; the **Computation** dialog box is displayed. To change the resolution of the analysis result, choose from the options available in the **Resolution** area of the **Computation Settings** rollout.

Control Points: Choose this option to move the control points on the datum curve. By default, the **Move** radio button is selected in the **Style Points** area. To add more points, select the **Add** radio button; you will be prompted to select a location to add an interpolation point. Click on the datum curve to add an interpolation point. You can add more than one point at a time by clicking on the datum curve. To delete a point, select the **Delete** radio button and click on the point; the point will be deleted from the drawing area. Select the **Redistribute** radio button to equally distribute the interpolation points along the entire length of the datum curve.

Curve from Equation

In Creo Parametric, you create a 1-D, 2-D, or 3-D datum curve defined by a mathematical equation. This equation is established according to the type of the selected coordinate system. To create a datum curve you can use either a parametric equation or an explicit equation.

In mathematics, parametric equations define a group of quantities as functions of one or more independent variables. Such variables are called parameters and are denoted in terms of *t*. For example, the parametric representation of a circle in the form of equations will be x=cos *t* and y= sin *t*, where *t* is a parameter.

Another method for defining curves is to use an explicit equation. An explicit equation defines a formula for a sequence in terms of variable *n*. Creo Parametric automatically finds independent variable within the specified range and internally parameterizes the equations to draw the curve.

To create a datum curve from equations, choose the **Curve from Equation** tool from the **Curve** flyout; the **CURVE: From Equation** dashboard will be displayed and you will be prompted to select a coordinate system to be referenced by the equation. Select a type of coordinate system from the Coordinate System drop-down in the dashboard and then select a coordinate system from the **Model Tree** or from the drawing area. You will notice that the selected coordinate system is highlighted in the **Coordinate system** collector of the **Reference** tab. To type the equation from the curve, click the **Equation** button in the dashboard; the **Equation** dialog box along with the

Equation information box will be displayed, refer to Figure 5-63. To close the information box, click on the **Close** button from the **Equation** information box. Next, type the equation for the curve in the text editor below the **Relations** rollout. You can insert a function in the equation by choosing the **Insert function from list** () button under the **Relations** rollout; the **Insert Function** dialog box is displayed. Double-click on a function in the functions list; the selected function is inserted in the equation text editor. After specifying the equations, you can choose the **Execute/Verify Relations and Create new Parameters by Relations** () button to verify the validity of the equations. Remember, to verify the validity of equations, each equation has to be written in a separate line in the text editor. To accept the equations, click the **OK** button; the curve will be displayed in the drawing area. If required, type a value for the lower and upper limit of the independent variable range in the **From** and **To** edit boxes, respectively. By default, the range of independent variable is set from 0 to 1, but you can specify any value from –1000000.0 to 1000000.0. Choose the **OK** button from the dashboard to complete the process.

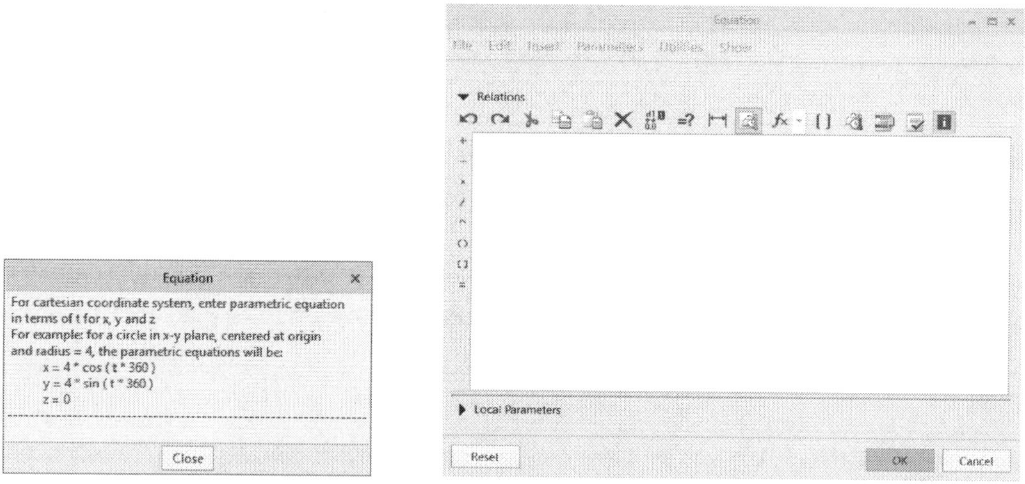

***Figure 5-63** The **Equation** information box displayed along with the **Equation** dialog box*

Tip

*The **Equation** information box displays the information according to the type of coordinate system selected.*

Figures 5-64 to 5-67 show examples of creating curves using parametric equations with the default range from 0 to 1. The equations for these curves are shown with respective figures.

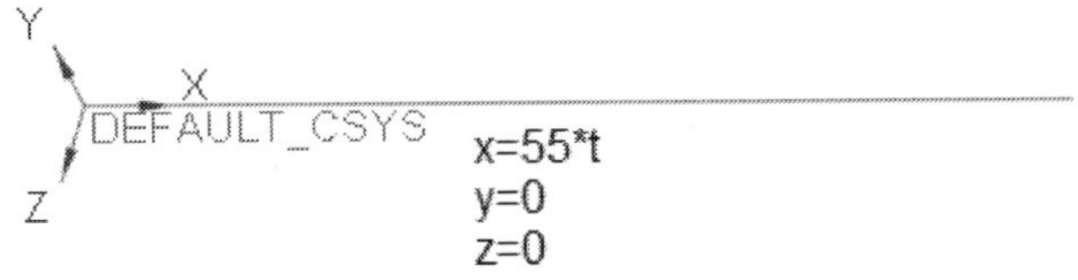

Figure 5-64 *A linear datum curve created in X direction*

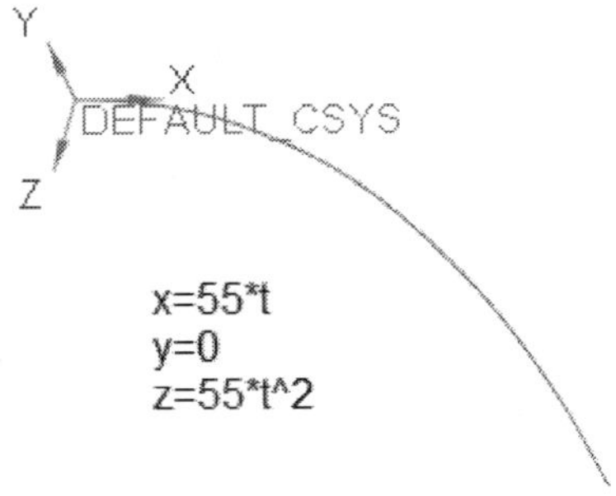

Figure 5-65 *A parabolic datum curve created in XZ plane*

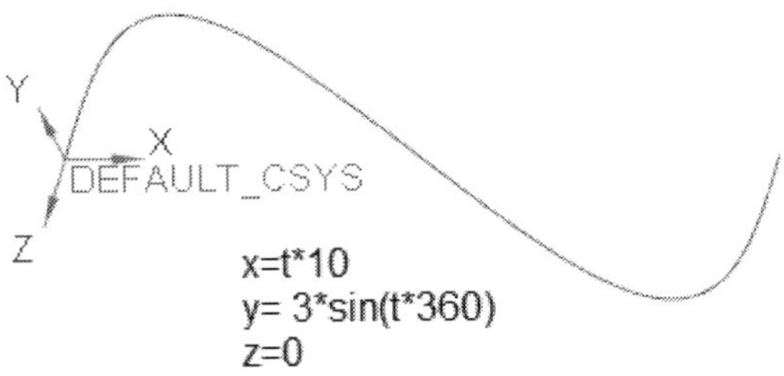

Figure 5-66 *A sine wave datum curve created in XY plane*

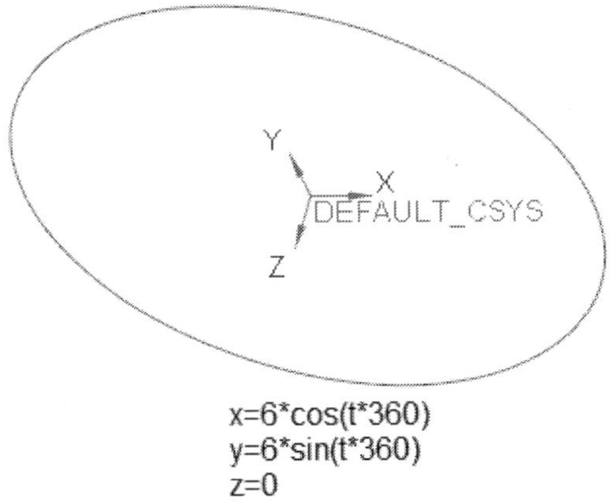

Figure 5-67 *A circular datum curve created in XY plane*

The curves shown in Figures 5-68 through 5-72 are created using explicit equations with varying ranges of upper limit and lower limit. The equation for each type curve is shown with its respective figure.

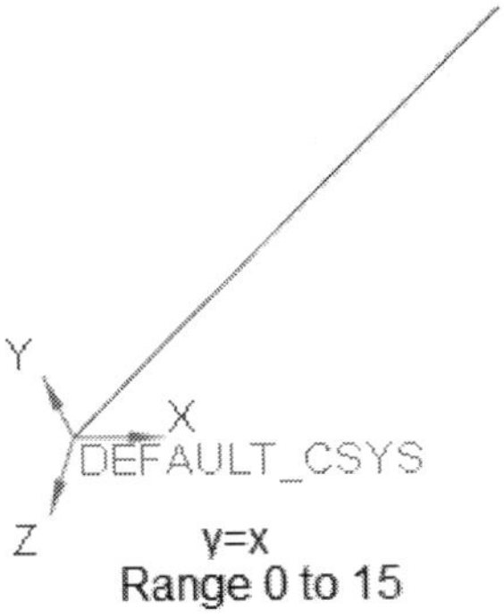

Figure 5-68 *A linear datum curve created from origin in XY plane*

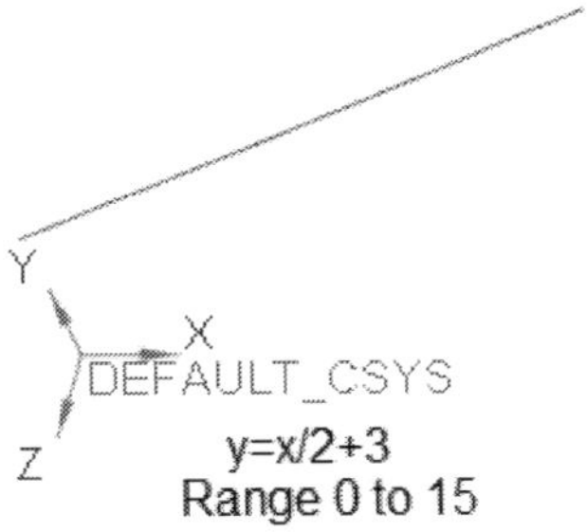

Figure 5-69 *A linear datum curve created in XY plane with an offset from origin*

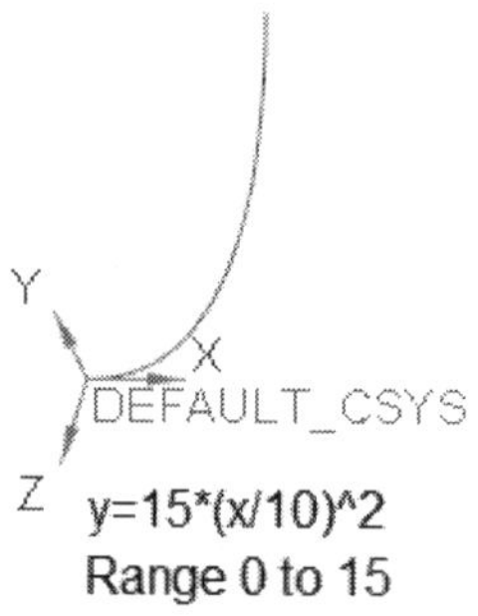

Figure 5-70 *A half parabolic curve created in XY plane*

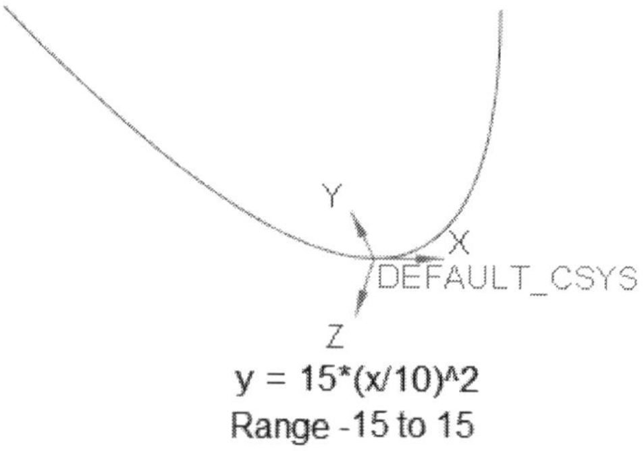

Figure 5-71 *A full parabolic curve created in XY plane*

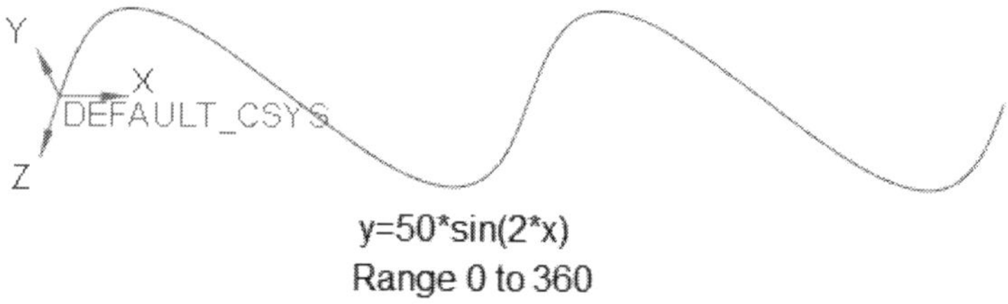

Figure 5-72 *A sine wave datum curve created in XY plane*

Curve from Cross Section

You can create a datum curve from the boundary of the planar cross section of a solid or surface model. If the cross-section has more than one chain, each chain is converted into a composite curve. To create a datum curve from cross section, choose the **Curve from Cross Section** tool; the **CURVE** dashboard will be displayed and you will be prompted to select a cross-section. Note that a cross-section should be present in the model before you invoke this tool. Select a cross-section from the **Cross-section** drop-down list; the name of the selected section will be displayed in the **Cross-Section** collector of the **References** tab. Next, choose the **OK** button from the dashboard to finish creating curve from the cross section.

Tip

*The **Sketch** tool can be used to sketch datum curves using the sketcher tools. This is the most commonly used tool to sketch a datum curve. Sketched curves consist of one or more sketched entities with one or more open or closed loops, whereas datum curves have single curve with single open or closed loop.*

Asynchronous Datum Features

Asynchronous datum features are those datum features which are created while feature-creation tools are active. To create an asynchronous feature, choose any tool from the **Datum** drop-down of the feature dashboard. The asynchronous feature thus created is embedded with the feature that you are creating. When you create an asynchronous datum feature, it is neither visible in the drawing area nor it is displayed under the default display of the **Model Tree**. But, it can be seen in the **Model Tree** when you click the arrow sign (▶) that appears on the left of the feature created. In case, you have converted an asynchronous feature into a stand-alone datum feature, it will automatically become visible in the drawing area and also in the model tree.

You can create an asynchronous datum plane by following these steps:

1. Choose the **Extrude** tool; the **Extrude** dashboard will be displayed.

2. In the **Extrude** dashboard, choose the **Placement** tab and then choose the **Define** button to invoke the **Sketch** dialog box.

   Now, you need to select a sketching plane. Instead of selecting an existing datum plane or a face as the sketch plane, you can create a datum plane on-the-fly.

3. Choose the **Plane** tool from the **Datum** flyout in the dashboard. Select the references and constraints to create the datum plane and choose the **Resume** button from the dashboard. This datum plane is created and automatically selected as the sketching plane. You can create more than one asynchronous datum planes to use them as reference for orienting the sketching plane. Remember that the datum plane thus created can not be used as reference for other features until it is embedded in any feature.

TUTORIALS

Tutorial 1

In this tutorial, you will create the model shown in Figure 5-73. The front and right views of the solid model are shown in Figure 5-74. **(Expected time: 30 min)**

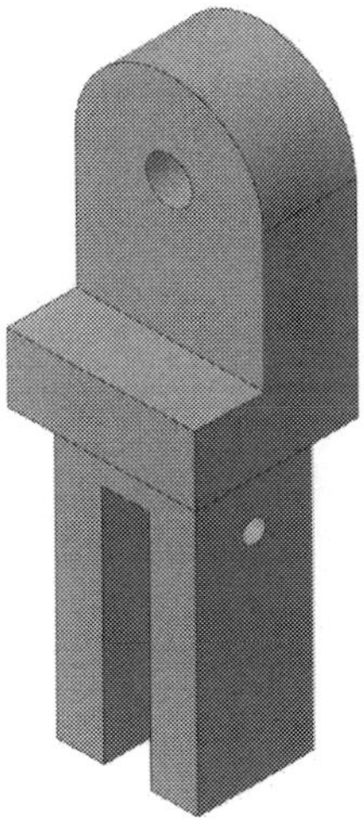

Figure 5-73 *Model for Tutorial 1*

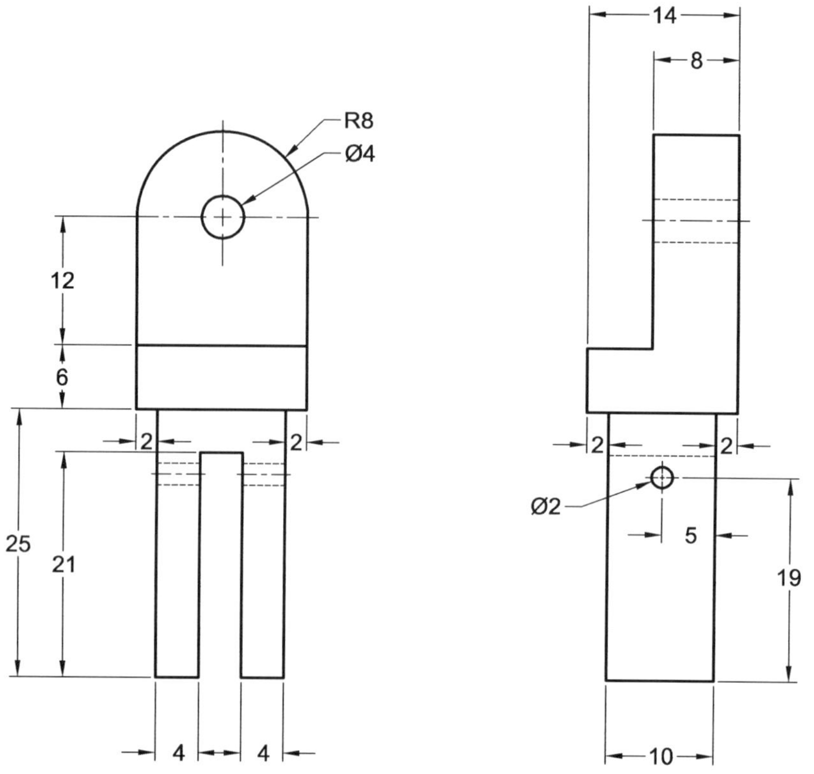

Figure 5-74 *Front and right views of the model*

The following steps are required to complete this tutorial:

Examine the model and determine the number of features in it.

a. Create the base feature.
b. Create the second extrude feature.
c. Create the third feature on an offset plane.
d. Create the circular cut feature.

Setting the Working Directory

After starting the Creo Parametric session, the first task is to set the working directory. A working directory is a directory on your system where you can save the work done in the current session of Creo Parametric. You can set any existing directory on your system as the working directory.

1. Choose **Manage Session > Select Working Directory** from the **File** menu; the **Select Working Directory** dialog box is displayed. Select the *C:\Creo_8.0* folder from this dialog box.

2. Choose the **Organize** tab from the **Select Working Directory** dialog box to display the flyout. Next, choose the **New Folder** option from the flyout; the **New Folder** dialog box is displayed.

3. Enter **c05** in the **New directory** edit box and choose the **OK** button from the dialog box; a folder with the name *c05* is created at *C:\Creo_8.0*.

4. Next, choose the **OK** button from the **Select Working Directory** dialog box. The working directory is set to *C:\Creo_8.0\c05*. Also, a message **Successfully changed to C:\Cre_8.0\c05 directory** is displayed in the message area.

Starting a New Object File

1. Choose the **New** button to invoke the **New** dialog box. Next, enter *c05tut1* in the **File name** edit box.

   The three default datum planes are displayed in the drawing area. If required, close the **Model Tree** by choosing the **Show Navigator** button on the bottom left corner of the program window so that the drawing area is increased.

Selecting the Sketching Plane for the Base Feature

To create the sketch for the base feature, you first need to select the sketching plane. In this model, you need to draw the base feature on the **FRONT** datum plane because it is evident from the isometric view of this model that the direction of extrusion for this feature is perpendicular to the **FRONT** datum plane.

Note

You can select any plane as the sketching plane for creating the base feature. The base feature thus created may not have proper orientation. So, you need to be careful while defining the sketching plane for creating the base feature. The desired orientation of the model is shown in Figure 5-73.

1. Choose the **Extrude** tool from the **Shapes** group; the **Extrude** dashboard is displayed above the graphics window.

2. Choose the **Placement** tab from the dashboard; a slide-down panel is displayed. Next, choose the **Define** button from the slide-down panel; the **Sketch** dialog box is displayed.

3. Select the **FRONT** datum plane as the sketching plane; an arrow pointing in the direction of the view is displayed on the **FRONT** datum plane.

4. Select the **TOP** datum plane from the drawing area and then select the **Top** option from the **Orientation** drop-down list.

   The **TOP** datum plane is selected in order to orient the sketching plane.

5. Choose the **Sketch** button from the **Sketch** dialog box to enter into the sketcher environment.

Note

If the sketching plane does not orient parallel to the screen then you need to choose the ***Sketch View*** *button from the* ***Graphics*** *toolbar. You can make the sketching plane parallel to the screen by default. To do so, choose* ***File > Options*** *from the menu bar. Then, choose* ***Sketcher*** *from the list shown in the left area of the* ***Creo Parametric Options*** *dialog box. Now, select the* ***Make the sketching plane parallel to the screen*** *check box from the* ***Sketcher startup*** *area and then* *choose the **OK** button from the window; the **Creo Parametric Options** message window will be displayed. Choose the **Yes** button from this window; the **Save As** dialog box will be displayed. Choose the **OK** button to save and exit.*

Creating and Dimensioning the Sketch for the Base Feature

The base feature can be created by drawing the sketch and then extruding it to the given distance.

1. Draw the sketch using various sketcher tools and then add required constraints and dimensions to the sketch, as shown in Figure 5-75. When you initially draw the sketch, it is dimensioned automatically and some weak dimensions are assigned to it.

> **Tip**
> *It is recommended that you use the **Modify** button to modify weak dimensions. In the **Modify Dimensions** dialog box that appears on choosing this button, clear the **Regenerate** check box and then modify dimensions by using the thumbwheel or the edit boxes. This way the sketch will not be regenerated while editing.*

2. Modify the dimension values, as shown in Figure 5-75.

3. After the sketch is completed, choose the **OK** button. Now, you are out of the sketcher environment.

4. Choose the **Saved Orientations** button from the **Graphics** toolbar; a flyout is displayed. Choose the **Default Orientation** option from the flyout; the model orients itself in default trimetric view, refer to Figure 5-76. The pink colored arrow is displayed on the model indicating the direction of extrusion.

5. In the dimension box of the **Extrude** dashboard, enter **8**; the model looks like the one shown in Figure 5-76.

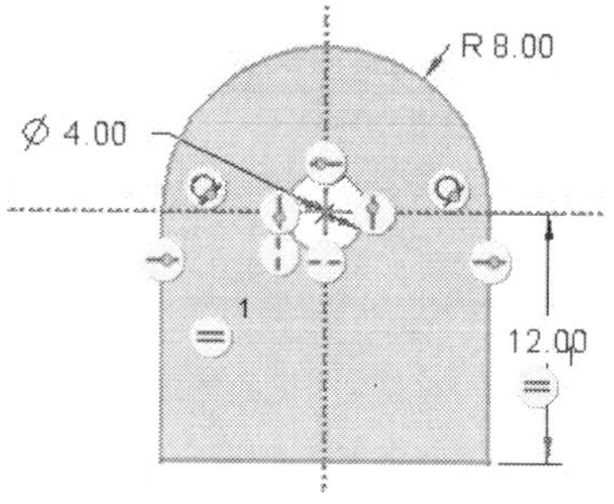

Figure 5-75 *Sketch for the base feature with dimensions and constraints*

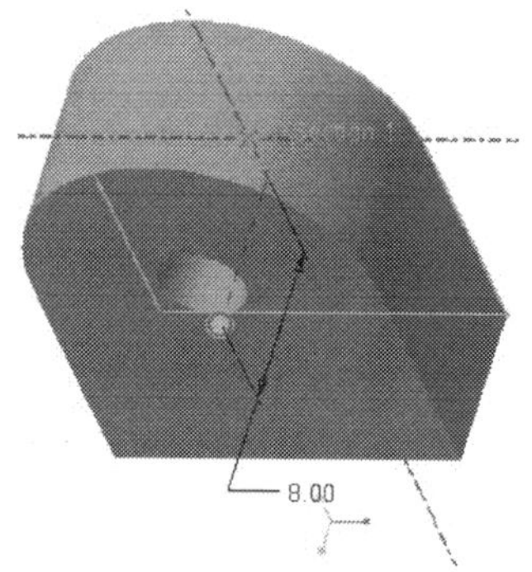

Figure 5-76 *Default orientation of the model and the arrow showing the direction of feature creation*

6. Choose the **OK** button from the **Extrude** dashboard; the base feature is created. You can use the middle mouse button to spin the model in order to view it from various directions.

Note

*When you choose the **Default Orientation** option from the **Saved Orientations** drop-down, the resulting orientation of the model is trimetric, not isometric. If you want the model to be displayed in the isometric view whenever you choose the **Default Orientation** option, use the **Creo Parametric Options** dialog box. To display this dialog box, choose **Files > Options** from the menu bar. In this dialog box, choose the **Model Display** from the left area and select the **Isometric** option from the **Default model orientation** drop-down list. Next, choose the **OK** button and save these settings. Now, the default orientation will be set to isometric.*

Tip

*It is recommended to check the orientation of the base feature of a model when it is completed. To check whether the plane you specified for sketching was correct, choose the **Saved Orientations** button from the **View** tab; a flyout is displayed. Next, choose the **FRONT** option from the flyout; the base feature will reorient in the drawing area such that you can view the front view of the base feature.*

Selecting the Sketching Plane for the Second Feature

The second feature is an extrude feature. It will be created on the plane that was used to create the base feature.

1. Choose the **Extrude** tool again from the **Shapes** group; the **Extrude** dashboard is displayed above the drawing area.

2. Choose the **Placement** tab from the dashboard. Next, choose the **Define** button from the slide-down panel; the **Sketch** dialog box is displayed.

3. Choose the **Use Previous** button from the **Sketch Plane** area in the **Sketch** dialog box; the **Top** datum plane with default orientation is selected automatically.

   When you choose the **Use Previous** button, the system selects the sketching plane that was used previously to create the base feature. You need to choose this button because the base feature and the second feature are on the same plane, but have different depths of extrusion. In case they had the same depth of extrusion, you could have drawn them on the same plane at the same time as a single feature. The **TOP** datum plane and its orientation are set automatically.

Drawing the Sketch for the Second Feature

The second feature has a rectangular section that will be extruded to a depth of 14 units.

1. Draw the sketch, as shown in Figure 5-77.

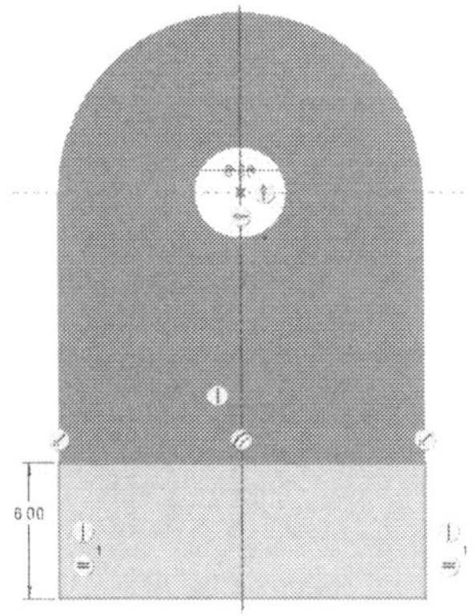

Figure 5-77 *Sketch for the second feature*

2. The sketch is automatically constrained and some weak dimensions are assigned to it. Add required constraints and modify weak dimensions, refer to Figure 5-77.

3. Choose the **OK** button to exit the sketcher environment; the **Extrude** dashboard is enabled above the drawing area.

> **Tip**
> *You can use the* ***Project*** *button from the* ***Sketching*** *group to use the bottom edge of the base feature for sketching. Else, you need to draw an aligned line on the edge.*

4. Use the middle mouse button to orient the model, as shown in Figure 5-78. This orientation of the model gives you a better view of the sketch in three-dimensional (3D) space. The colored arrow is also displayed on the model indicating the direction of extrusion.

5. Enter **14** in the dimension box available in the **Extrude** dashboard and using ENTER. The second extruded feature is completed and its preview is displayed in the drawing area.

6. Choose the **Saved Orientations** button from the **Graphics** toolbar; a flyout is displayed. Choose the **Default Orientation** option from the flyout; the model orients itself in default trimetric view.

7. Now, choose the **OK** button from the **Extrude** dashboard; the feature is created and oriented, as shown in Figure 5-79. You can also use the middle mouse button to spin the model.

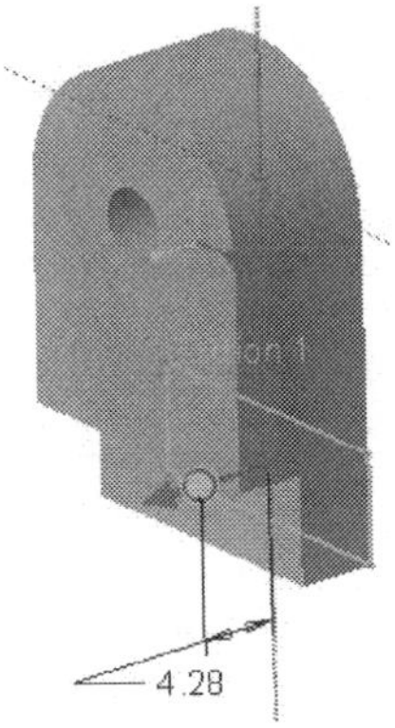

Figure 5-78 *Arrow showing the direction of feature depth*

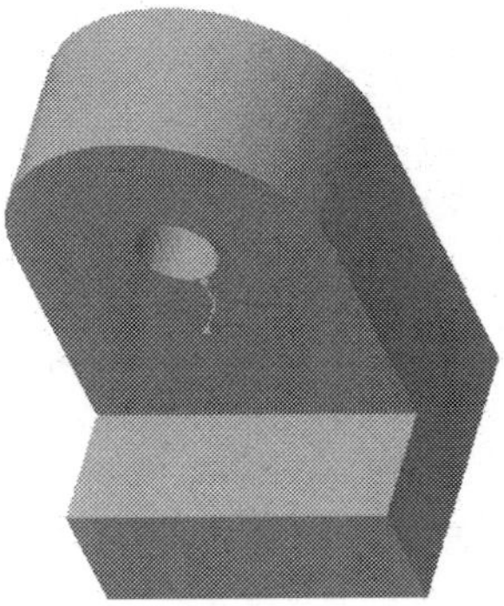

Figure 5-79 *Second extruded feature with the base feature*

Creating the Datum Plane for the Third Feature

A new datum plane is required to create the next feature. The datum plane will be created at an offset distance of 2 units from the front face of the second feature. You need to turn on the display of datum planes, if it is off.

1. Select the front face of the second feature; a mini popup toolbar appears in the drawing ares, as shown in Figure 5-80. Also, the selected face is highlighted in green.

2. Choose the **Plane** button from the mini popup toolbar; the **Datum Plane** dialog box appears. In the **Datum Plane** dialog box, the **Offset** option automatically becomes available under the **References** collector.

   Now, you need to specify the offset distance. If you enter a positive value, the datum plane will be created along the direction of arrow and if you enter a negative value then the datum plane will be created in the direction opposite to that shown by the arrow.

3. In the **Translation** edit box under the **Offset** area of the **Datum Plane** dialog box, enter **-2** and press ENTER.

   The negative value is entered because the datum plane has to be created in the direction opposite to that shown by the arrow.

4. Choose the **OK** button; the datum plane named **DTM1** is created, as shown in Figure 5-81.

Figure 5-80 The mini popup toolbar displayed on selecting the front face of the second feature

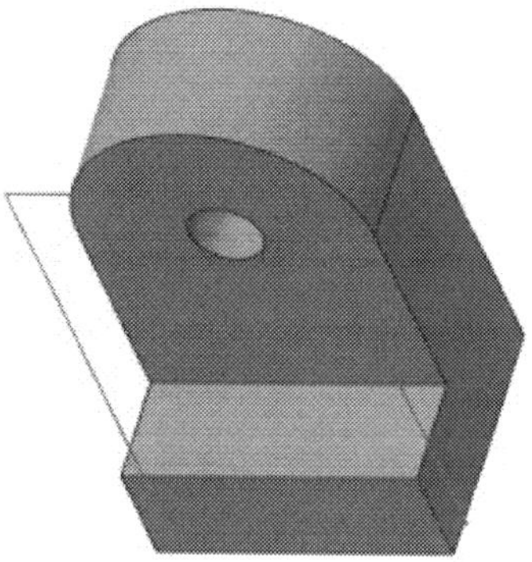

Figure 5-81 The datum plane created

Creating the Third Feature on DTM1

The datum plane **DTM1** is created and can be seen in the **Model Tree** as well as in the drawing area. The sketch of the next feature that will be extruded has to be created on the datum plane **DTM1**.

1. Choose the **Extrude** tool from the **Shapes** group of the **Model** tab; the **Extrude** dashboard is displayed above the drawing area.

2. Choose the **Placement** tab; a slide-down panel is displayed. Now, choose the **Define** button from the panel; the **Sketch** dialog box is displayed.

3. Select **DTM1** as the sketching plane for the third feature; a pink arrow is displayed on the selected datum plane, as shown in Figure 5-82. This arrow shows the direction of viewing the sketching plane.

   The **TOP** datum plane and its orientation are selected by default.

4. Choose the **Sketch** button from the **Sketch** dialog box to enter the sketcher environment. If required, choose the **No hidden** button from the **Display Style** drop-down in the **Graphics** toolbar to view the visible edges of the model.

5. Sketch the section of the third feature of the model and add constraints and dimensions to the sketch, as shown in Figure 5-83.

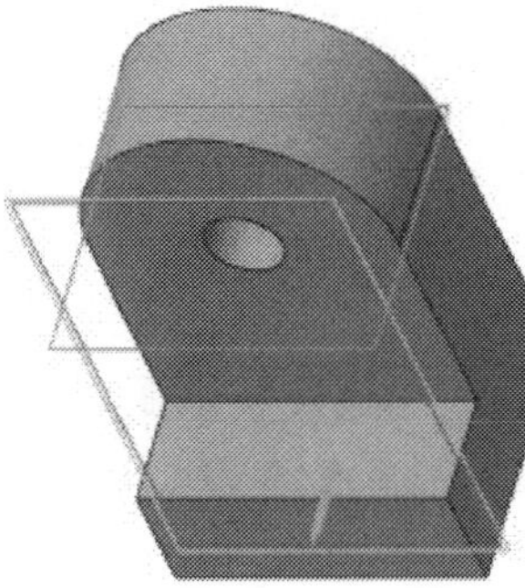

Figure 5-82 Arrow on DTM1 showing the direction of viewing the sketching plane

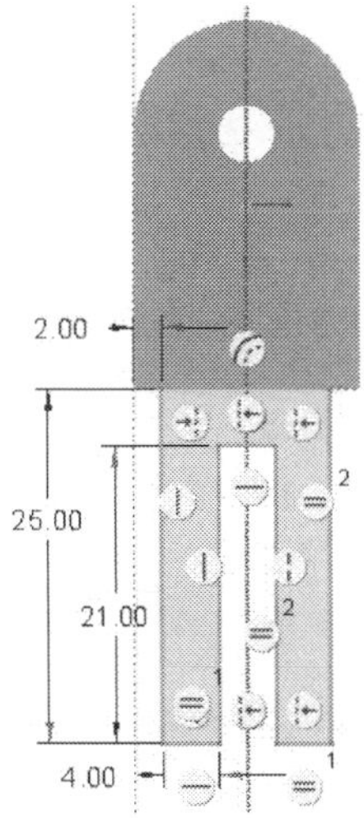

Figure 5-83 Sketch of the third feature with dimensions and constraints

6. Choose the **OK** button to exit the sketcher environment; the **Extrude** dashboard is enabled above the drawing area.

7. Turn the model display to **Shading With Edges**. Use the middle mouse button to orient the model, as shown in Figure 5-84. This orientation gives a better view of the model.

   Notice that the arrow is pointing toward the direction of feature creation. But, you need to extrude the sketch in the opposite direction.

8. Choose the **Change depth direction of extrude to other side of sketch** button from the **Extrude** dashboard to change the direction of arrow.

9. Enter **10** in the dimension box present on the **Extrude** dashboard; preview of the third feature is displayed in the drawing area.

10. Choose the **OK** button from the **Extrude** dashboard to confirm the feature creation.

    Choose the **Saved Orientations** button from the **Graphics** toolbar; a flyout is displayed. Choose the **Default Orientation** option from the flyout; the model gets oriented in the default trimetric view, as shown in Figure 5-85.

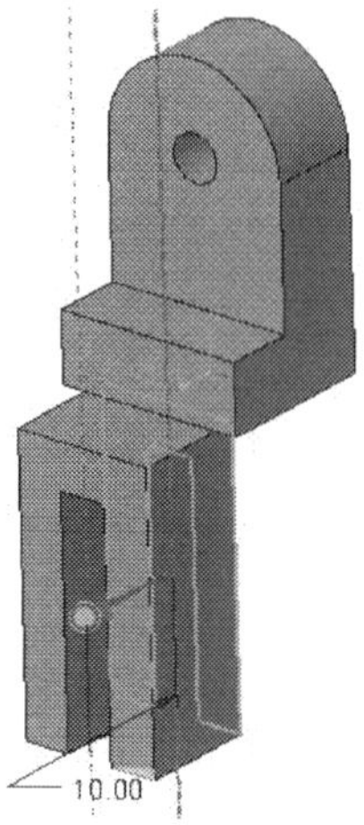

Figure 5-84 Arrow showing the direction of material addition

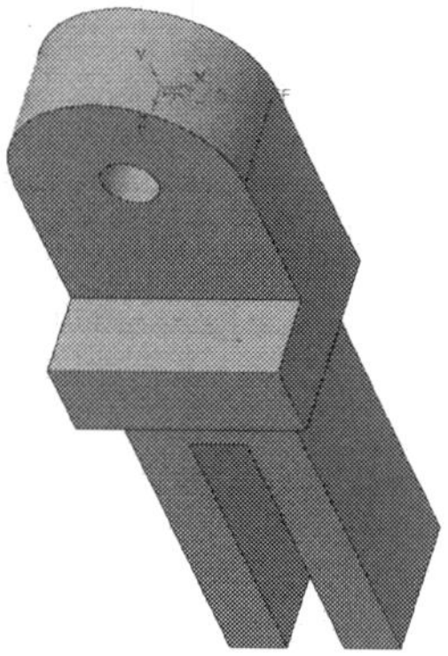

Figure 5-85 Model after creating the third feature

Selecting the Sketching Plane for the Cut Feature

You need to create a circular section for the circular cut feature. The sketching plane for the cut feature is shown in Figure 5-86.

Note

*The circular cut feature can also be created using the **Hole** tool which will be discussed in Chapter 6.*

1. Choose the **Extrude** tool from the **Shapes** group; the **Extrude** dashboard is displayed above the drawing area.

2. Choose the **Remove Material** button from the **Extrude** dashboard.

3. Choose the **Placement** tab from the dashboard; a slide-down panel is displayed. Next, choose the **Define** button from it; the **Sketch** dialog box is displayed.

4. Select the face shown in Figure 5-86 for sketching.

5. Using the left mouse button, select the **TOP** datum plane and then select the **Top** option from the **Orientation** drop-down list.

6. Choose the **Sketch** button; the system takes you to the sketcher environment.

Creating the Sketch for the Cut Feature

1. Draw the sketch of the cut feature and add dimensions to it, as shown in Figure 5-87.

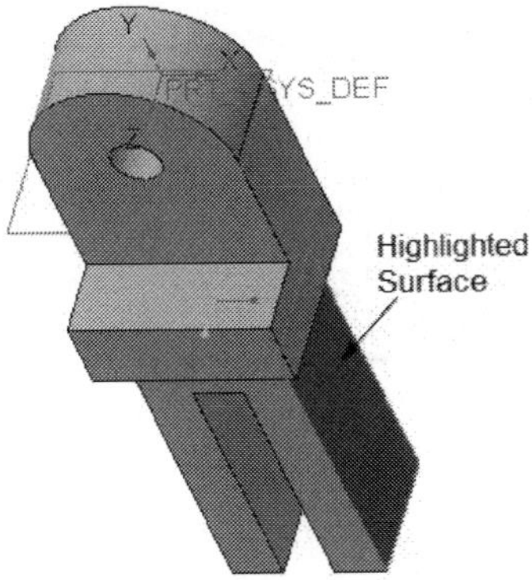

Figure 5-86 Sketching plane for the cut feature

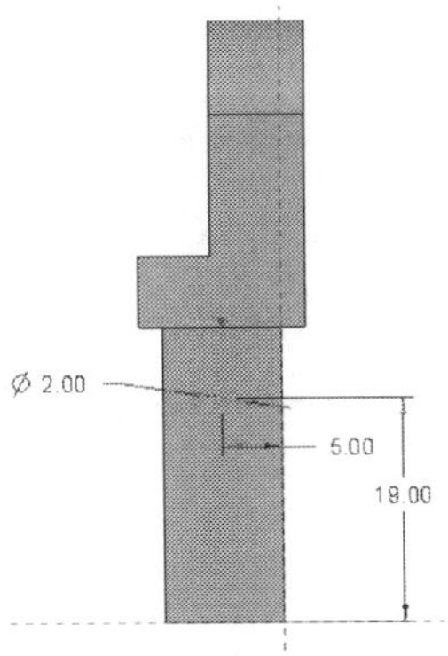

Figure 5-87 Sketch and dimensions for the cut feature

2. Choose the **OK** button.

3. Press the CTRL+D keys orient the model to its default orientation, as shown in Figure 5-88; two arrows appear.

   One arrow indicates the direction of feature creation and the other arrow indicates the direction along which the material will be removed.

4. Choose the **Options** tab in the **Extrude** dashboard; the **Depth** slide-down panel is displayed.

5. From the **Side 1** drop-down list, choose the **Through All** option; cut feature is created and can now be previewed in the drawing area.

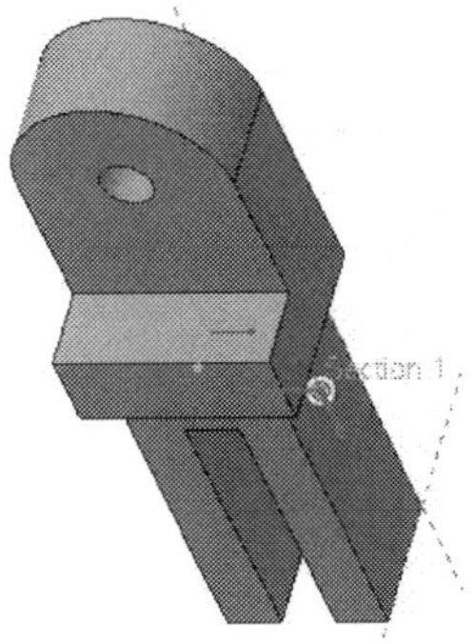

Figure 5-88 The two arrows on the cut feature

6. Choose the **OK** button from the **Extrude** dashboard to accept the feature creation. The trimetric view of the model is shown in Figure 5-89.

Saving the Model

1. Choose the **Save** button from the **File** menu and save the model.

The order of feature creation can be seen in the **Model Tree** shown in Figure 5-90.

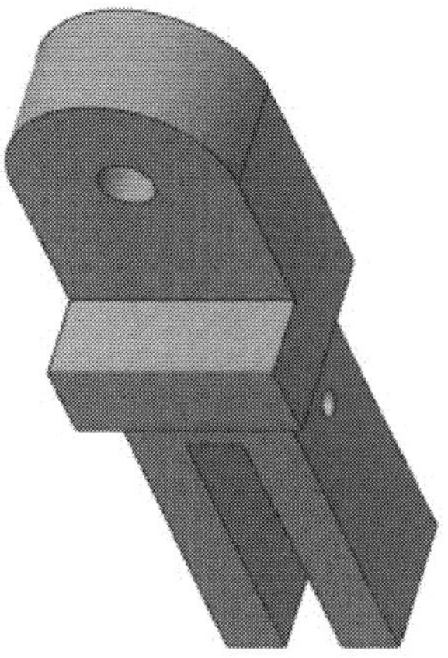

Figure 5-89 *Completed model for Tutorial 1*

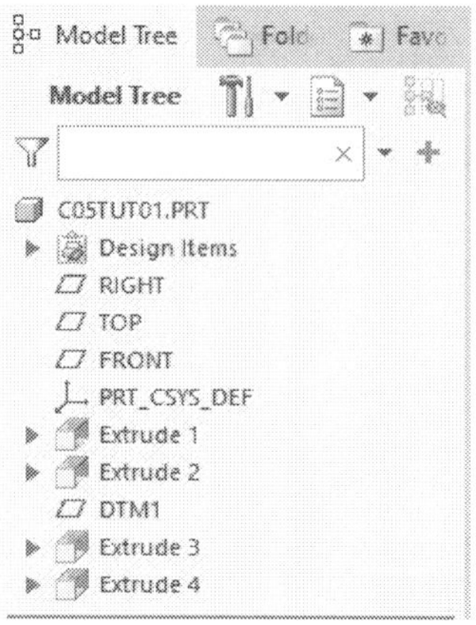

Figure 5-90 *The Model Tree for Tutorial 1*

Tutorial 2

In this tutorial, you will create the model shown in Figure 5-91. The front, top, and right views of the solid model are shown in Figure 5-92. **(Expected time: 30 min)**

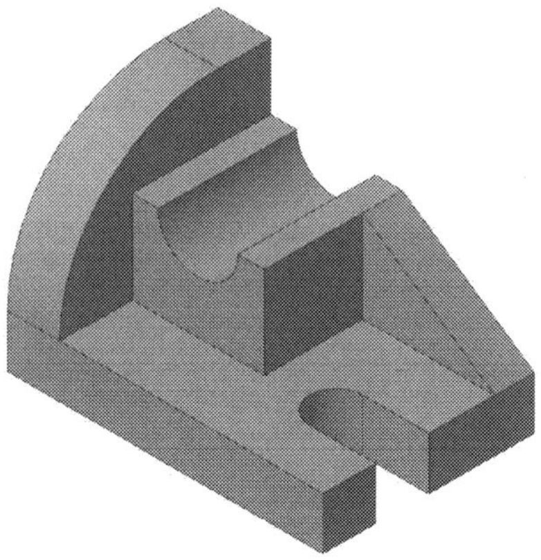

Figure 5-91 *Isometric view of the model*

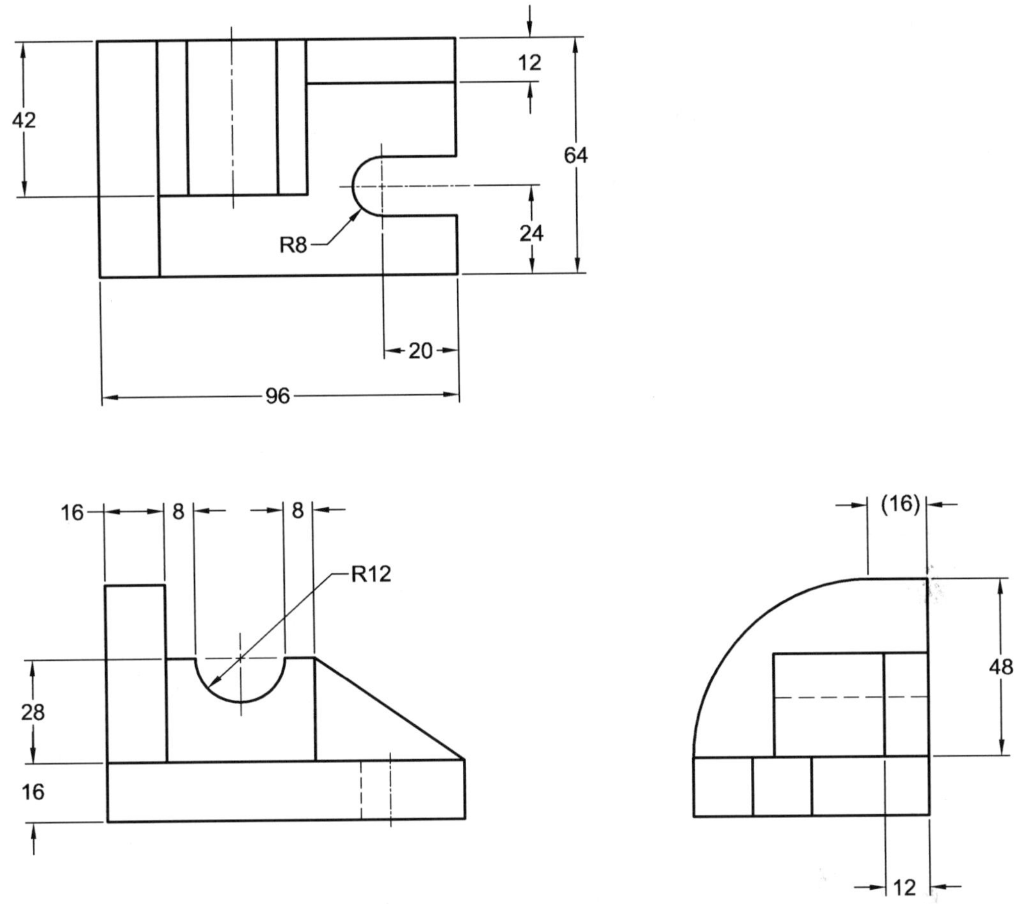

Figure 5-92 *Orthographic views of the model*

The following steps are required to complete this tutorial:

Examine the model and determine the number of features in it, refer to Figure 5-91.

a. Start a new file and create the base feature.
b. Create the second feature on the left face of the base feature.
c. Create the third and fourth features on the same plane but with different extrusion depths.
d. Save the model.

After understanding the procedure for creating the model, you are now ready to create it. Set the working directory, if required.

Starting a New Object File

1. Start a new part file and name it as *c05tut2*.

   Three default datum planes are displayed in the drawing area.

Selecting the Sketching Plane for the Base Feature

To create sketch for the base feature, you first need to select the sketching plane for the base feature. In this model, the direction of extrusion of the base feature is perpendicular to the **TOP** datum plane so you need to draw the base feature on the **TOP** datum plane.

1. Choose the **Extrude** tool from the **Shapes** group; the **Extrude** dashboard is displayed above the drawing area.

2. Choose the **Placement** tab from the dashboard to display a slide-down panel. Choose the **Define** button from it; the **Sketch** dialog box is displayed.

3. Select the **TOP** datum plane as the sketching plane; an arrow is displayed on the **TOP** datum plane pointing in the direction of viewing the sketch. The **RIGHT** datum plane and its orientation are set automatically.

4. Choose the **Sketch** button in the **Sketch** dialog box to enter into the sketcher environment.

Creating and Dimensioning the Sketch for the Base Feature

The sketch for the base feature has a rectangular shape with a slot, refer to Figure 5-93. When you extrude this sketch, the base feature will be created with the slot, as shown in Figure 5-94.

1. Draw the sketch using sketcher tools and add required constraints and dimensions to it. Modify these dimensions, as shown in Figure 5-93.

2. Choose the **OK** button from the dashboard to exit the sketcher environment.

3. Choose the **Saved Orientations** button from the **Graphics** toolbar; a flyout is displayed. Choose the **Default Orientation** option from the flyout; the model orients itself in default trimetric view.

   The default view of the model is displayed, but it does not fit on the screen. You may need to zoom in the model to view it properly in the drawing area. A pink arrow is also displayed on the model indicating the direction of extrusion.

4. Enter **16** in the dimension box present on the **Extrude** dashboard; preview of the base feature is displayed in the drawing area.

5. Choose the **OK** button from the **Extrude** dashboard; the base feature is created, as shown in Figure 5-94. You can use the middle mouse button to spin the object to view it from various directions.

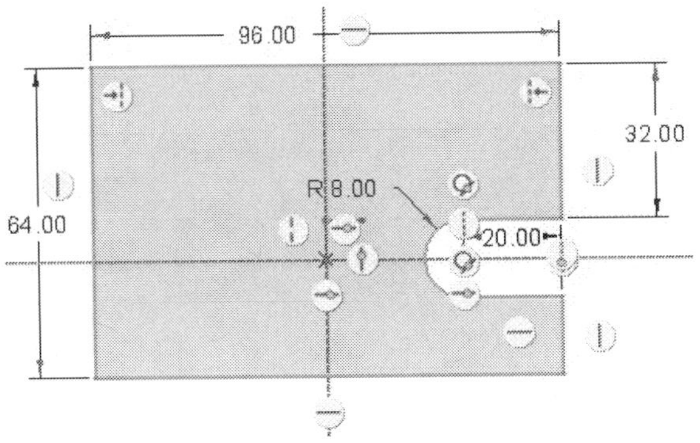

Figure 5-93 Sketch of the base feature with dimensions and constraints

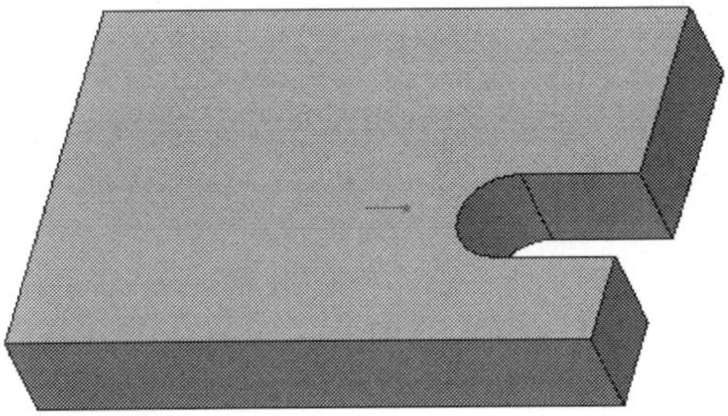

Figure 5-94 Base feature of the model

Selecting the Sketching Plane for the Second Feature

The next feature is an extruded feature and will be created on the left face of the base feature. Therefore, you need to select the left face of the base feature as the sketching plane and then draw the sketch.

1. Choose the **Extrude** tool from the **Shapes** group; the **Extrude** dashboard is displayed, above the drawing area.
2. Choose the **Placement** tab from the dashboard to display a slide-down panel and then choose the **Define** button from it; the **Sketch** dialog box is displayed.
3. Use the middle mouse button to spin the model to get the view shown in Figure 5-95.
4. Now, select the left face of the base feature as the sketching plane; an arrow pointing in the direction of view of the sketching plane appears on the left face of the base feature.
5. Choose the **Flip** button to flip the arrow and to change the direction of viewing the sketching plane, refer to Figure 5-95.

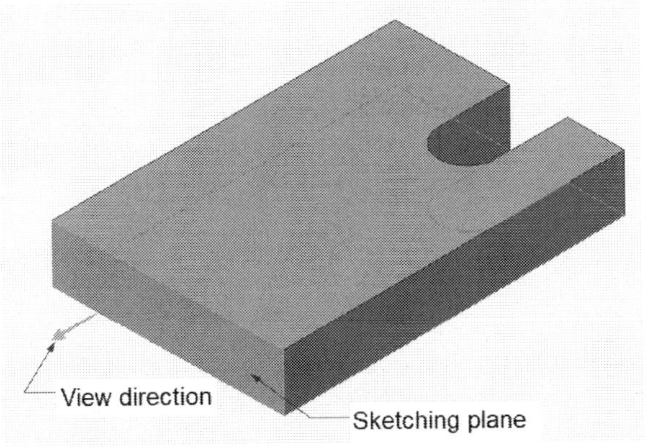

Figure 5-95 Arrow showing the direction of viewing the model

6. Click in the **Reference** collector and select the top face of the base feature, refer to Figure 5-96. Next, select the **Top** option from the **Orientation** drop-down list.

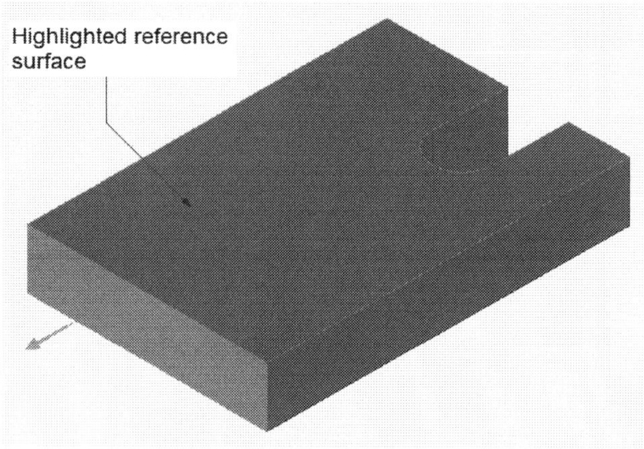

Figure 5-96 *Surface selected to be at top*

By selecting the top surface of the base feature, the model will be oriented in such a way that the highlighted planar surface will be at the top while sketching.

7. Choose the **Sketch** button from the **Sketch** dialog box to enter into the sketcher environment.

Creating and Dimensioning the Sketch for the Second Feature

The sketch for the second feature consists of two lines and an arc. The bottom edge of the sketch coincides with the top edge of the base feature. This sketch is extruded upto a depth of 16.

1. Draw the sketch using various sketcher tools, as shown in Figure 5-97. The sketch is dimensioned automatically and some weak dimensions are assigned to it.

2. Apply constraints and modify weak dimensions, as shown in Figure 5-97.

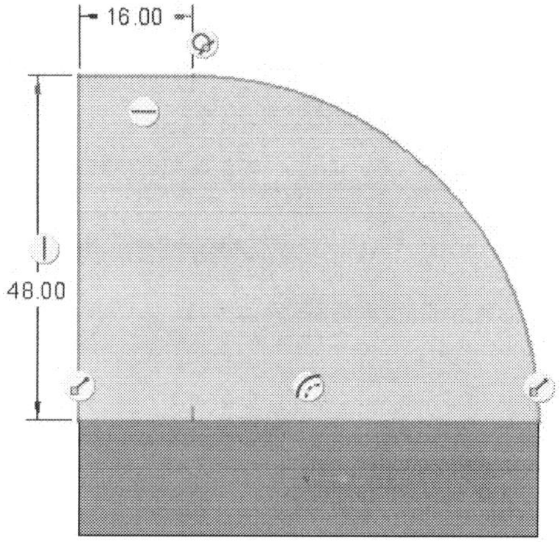

Figure 5-97 *Sketch of the second feature with dimensions and constraints*

3. After the sketch is complete, choose the **OK** button; the **Extrude** dashboard is displayed above the drawing area.

   Use the middle mouse button to orient the model, as shown in Figure 5-98. On doing so, a pink arrow is displayed on the model indicating the direction of extrusion.

4. Enter **16** in the dimension box present in the **Extrude** dashboard and using ENTER; its preview is displayed on the screen.

5. Choose the **OK** button from the **Extrude** dashboard; the second feature is completed and is shown in Figure 5-99. You can use the middle mouse button to spin the model to view it from various directions.

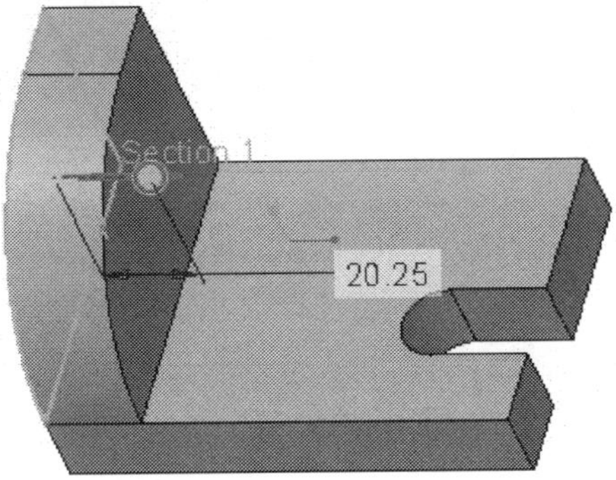

Figure 5-98 *Arrow showing the direction of material addition*

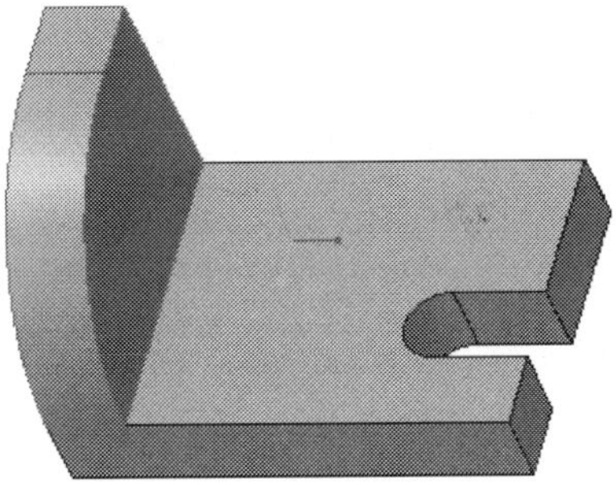

Figure 5-99 *Model with the second extruded feature*

Selecting the Sketching Plane for the Third Feature

The third feature is an extruded feature and will be created on the back planar surface of the base feature. Therefore, you need to define the back face of the base feature as the sketching plane and then draw the sketch.

1. Choose the **Extrude** tool from the **Shapes** group; the **Extrude** dashboard is displayed above the drawing area.

2. Choose the **Placement** tab from the dashboard to display a slide-down panel and then choose the **Define** button from it; the **Sketch** dialog box is displayed.

3. Use the middle mouse button to spin the model and then select the face of the base feature shown in Figure 5-100, as the sketching plane; an arrow is displayed on the selected face.

4. Choose the **Flip** button to flip the arrow if it is required; the arrow points in the direction of viewing the sketching plane, refer to Figure 5-100.

5. Select the top face shown in Figure 5-101 and then select the **Top** option from the **Orientation** drop-down list. Now, the top face of the base feature is selected to be at top while drawing the sketch.

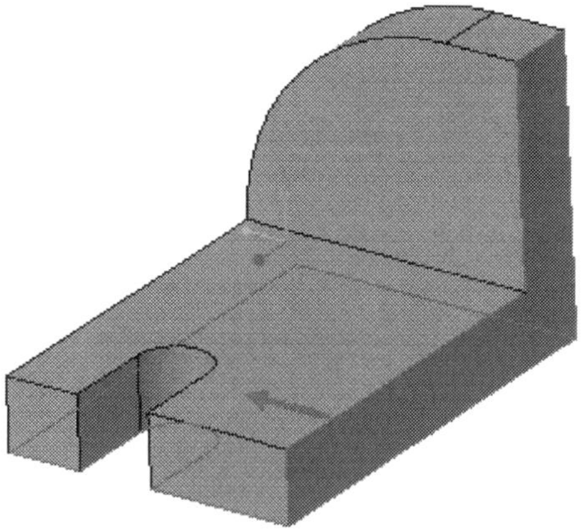

Figure 5-100 *Selected face and arrow indicating the direction of view of the model*

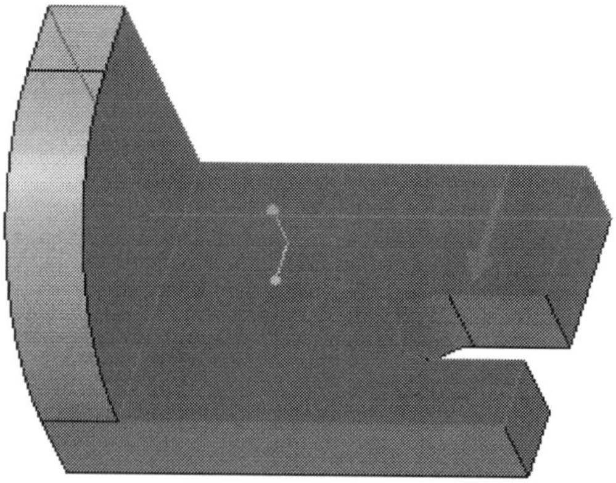

Figure 5-101 *Face selected to be at top*

6. Choose the **Sketch** button to enter the sketcher environment.

Creating and Dimensioning the Sketch for the Third Feature

The sketch for the base feature consists of a rectangular section with a semicircular cut at the top. When this section is extruded, a feature with a semicircular slot will be created.

1. Draw the sketch using the sketcher tools, as shown in Figure 5-102. The sketch is dimensioned automatically and some weak dimensions are assigned to it.

2. Add required constraints and modify weak dimensions to the dimensions shown in Figure 5-102.

3. Choose the **OK** button to exit the sketcher environment; the **Extrude** dashboard is enabled above the drawing area.

4. Use the middle mouse button to orient the model, as shown in Figure 5-103. This orientation of the model gives a better view of the sketch in 3D space. An arrow is displayed on the model indicating the direction of extrusion.

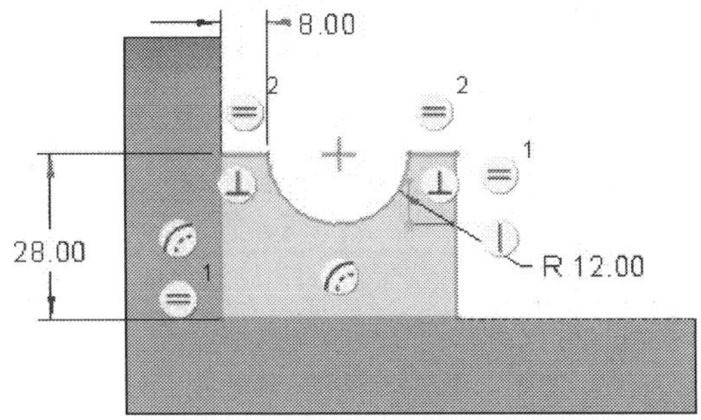

Figure 5-102 *Sketch with dimensions and constraints*

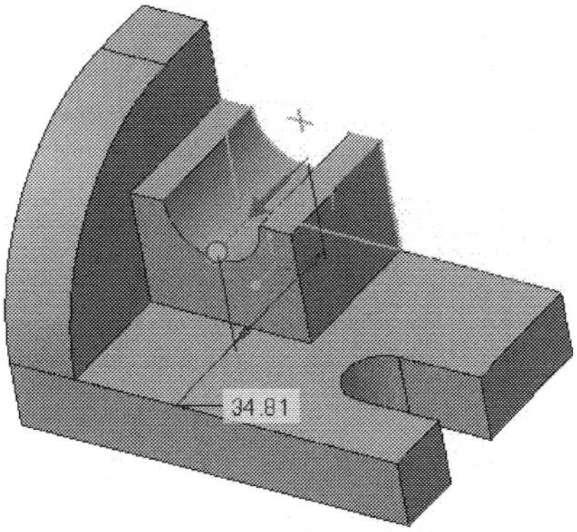

Figure 5-103 *Arrow showing the direction of material addition*

5. Enter **42** in the dimension box present in the **Extrude** dashboard and using ENTER. This feature is completed and its preview can be seen in the drawing area.

6. Choose the **OK** button from the **Extrude** dashboard to confirm the feature creation. Choose the **Saved Orientations** button from the **Graphics** toolbar; a flyout is displayed. Choose the **Default Orientation** option from the flyout; the default view of the model is displayed, as shown in Figure 5-104.

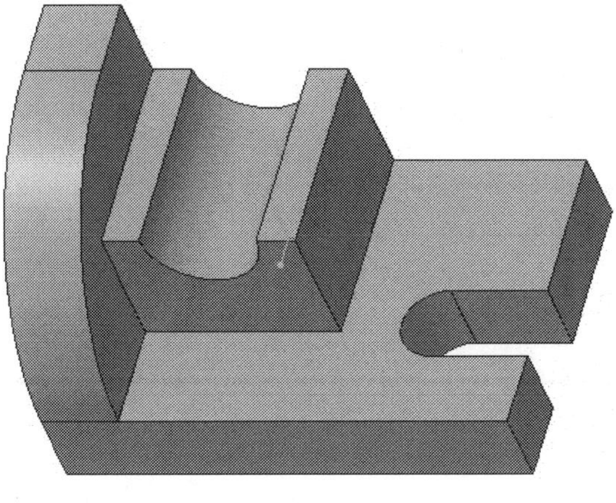

Figure 5-104 *Model with the third extruded feature*

Selecting the Sketching Plane for the Last Feature

The last feature and the third feature are created on the same plane. But the depth of extrusion is different for both of them. This is the reason they are considered as separate features. Therefore, for sketching this feature, you can use the sketching plane that was used for creating the third feature.

1. Choose the **Extrude** tool and then invoke the **Sketch** dialog box.

2. Choose the **Use Previous** button from the **Sketch Plane** area in the **Sketch** dialog box; the sketcher environment gets activated automatically.

When you choose the **Use Previous** button, the system selects the previous sketching plane that was used to create the base feature.

Creating and Dimensioning the Sketch for the Last Feature

The sketch of the last feature consists of three lines. The bottom edge of the sketch is aligned with the top edge of the base feature and the left edge is aligned with the right edge of the third feature. When this sketch is extruded, it creates a rib shape. However, you can also create this feature using the **Profile Rib** tool in the **Engineering** group. The use of this tool will be discussed in the later chapters of this book.

1. Draw the section sketch by using the sketcher tools. The sketch is dimensioned automatically and some weak dimensions are assigned to it. You can add required constraints (Coincident constraint) to align the lines and points in the sketch with other features, as shown in Figure 5-105. In this figure, the constraint symbol displayed on the line indicates that the edges of the adjacent features are used to close the section.

2. After the sketch is complete, choose the **OK** button; the **Extrude** dashboard is enabled. Use the middle mouse button to orient the model, as shown in Figure 5-106.

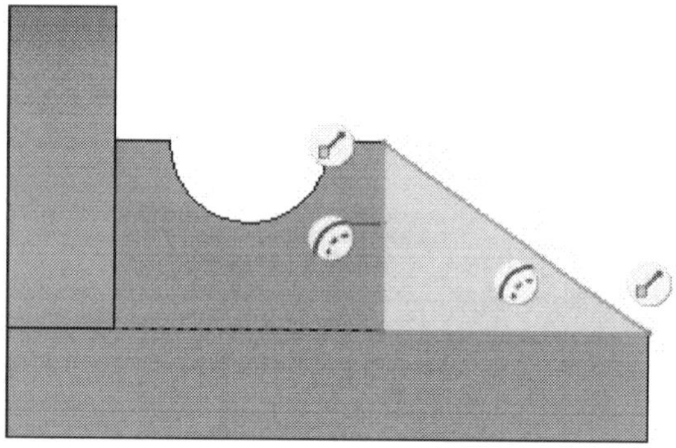

Figure 5-105 Sketch and constraints for the last feature

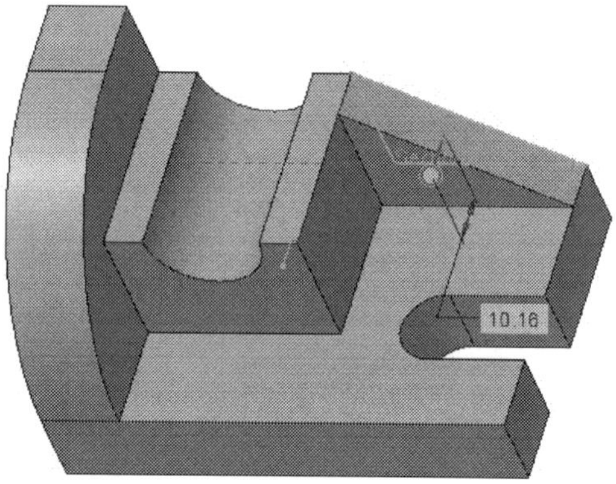

Figure 5-106 Arrow showing the direction of material addition

3. Enter the value **12** in the dimension box present in the **Extrude** dashboard; the feature is completed and its preview can be seen in the drawing area.

4. Choose the **OK** button in the **Extrude** dashboard.

The default view of the model is shown in Figure 5-107. To invoke the default view, choose the **Saved Orientations** button from the **View** tab; a flyout is displayed. Next, choose the **Default Orientation** option from this flyout.

Saving the Model

1. Choose the **Save** button from the **File** menu and save the model. The order of feature creation can be seen from the **Model Tree** shown in Figure 5-108.

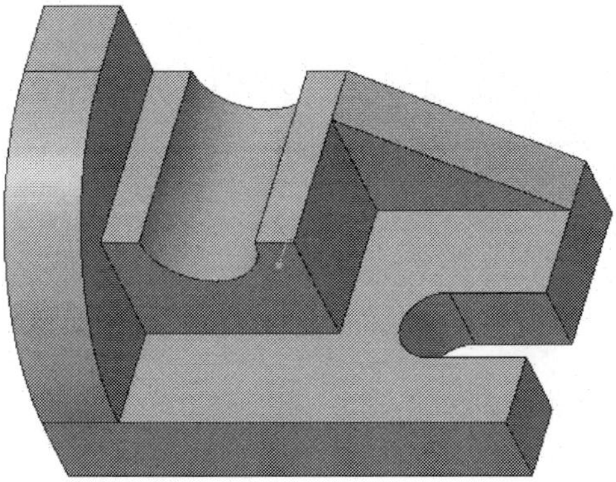

Figure 5-107 *Final model for Tutorial 2*

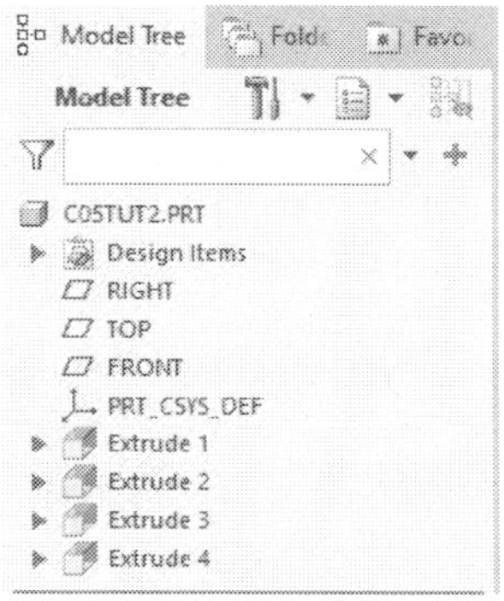

Figure 5-108 *The Model Tree for Tutorial 2*

Tutorial 3

In this tutorial, you will create the model shown in Figure 5-109. The front and top views of the solid model are shown in Figure 5-110. **(Expected time: 30 min)**

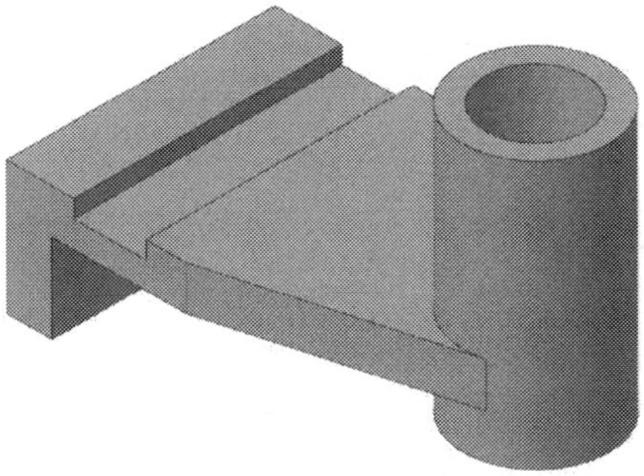

Figure 5-109 *Isometric view of the model*

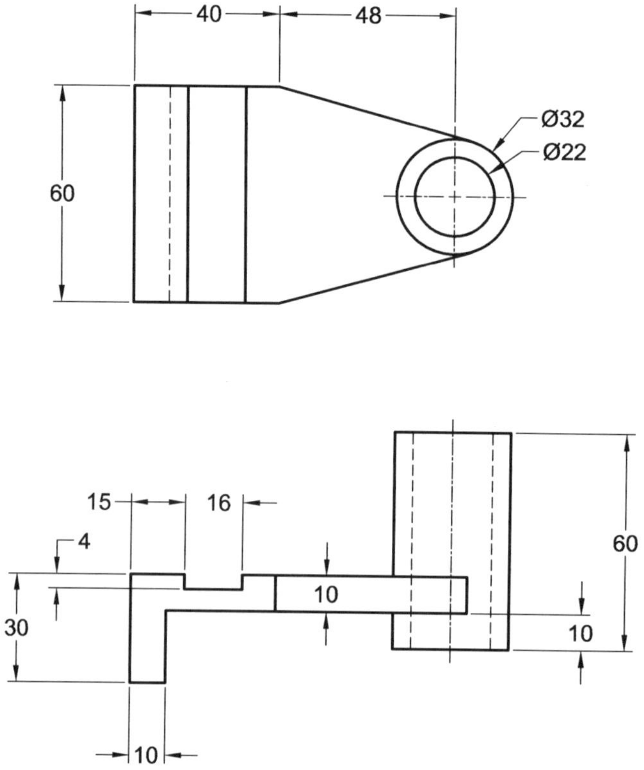

Figure 5-110 Orthographic views of the model

Examine the model and then determine the number of features in it, refer to Figure 5-109.

The following steps are required to complete this tutorial:

a. Start a new file and create the base feature.
b. Create the second feature.
c. Create the hollow cylindrical feature on an offset datum plane.

Starting a New Object File

1. Set the working directory, if required, and then start a new part file with the name *c05tut3*.

The three default datum planes are displayed in the drawing area.

Selecting the Sketching Plane for the Base Feature

To create the sketch for the base feature, you first need to select the sketching plane. In this model, you need to draw the base feature on the **FRONT** datum plane because the direction of extrusion is perpendicular to the **FRONT** datum plane.

1. Choose the **Extrude** tool from the **Shapes** group.

2. Choose the **Placement** tab from the dashboard to display a slide-down panel. Next, choose the **Define** button; the **Sketch** dialog box is displayed.

3. Select the **FRONT** datum plane as the sketching plane.

   A pink arrow is displayed on the **FRONT** datum plane and it points in the direction of viewing the sketch plane. The **RIGHT** datum plane and its orientation are selected automatically.

4. Choose the **Sketch** button to enter the sketcher environment.

Creating and Dimensioning the Sketch for the Base Feature

The section to be extruded for the base feature is evident from the model. The section sketch is shown in Figure 5-111. When this sketch is extruded, it will create the base feature.

1. Draw the sketch using various sketcher tools, as shown in Figure 5-111.

   The sketch is dimensioned automatically and some weak dimensions are assigned to it.

2. Add required constraints and modify weak dimensions, as shown in Figure 5-111.

3. Choose the **OK** button; the **Extrude** dashboard is displayed.

4. Choose the **Saved Orientations** button from the **Graphics** toolbar; a flyout is displayed. Choose the **Default Orientation** option from the flyout; the model orients itself in default trimetric view, and an arrow is also displayed on it indicating the direction of extrusion.

5. Enter **60** in the dimension box that is present on the **Extrude** dashboard and then using ENTER.

6. Choose the **OK** button from the **Extrude** dashboard.

   The base feature is completed, as shown in Figure 5-112. You can use the middle mouse button to spin the model to view it from various directions.

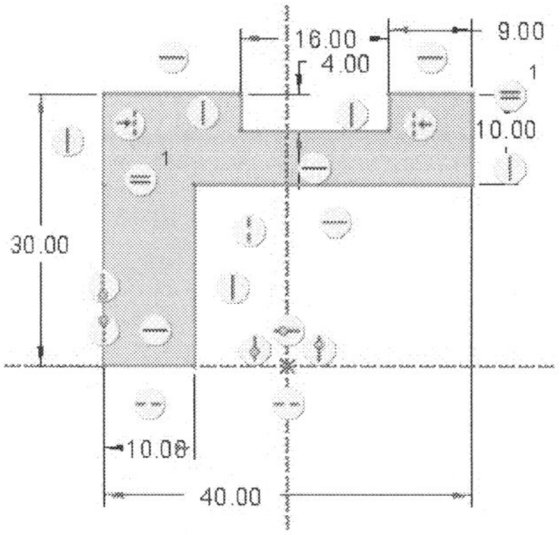

Figure 5-111 Sketch with dimensions and constraints for the base feature

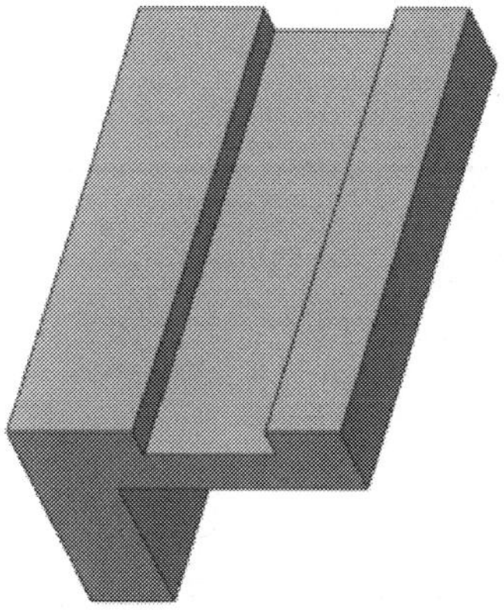

Figure 5-112 Base feature of the model

Selecting the Sketching Plane for the Second Feature

The next feature is an extruded feature. The sketching plane for this feature is the top face of the base feature.

1. Choose the **Extrude** tool from the **Shapes** group.
2. Choose the **Placement** tab from the dashboard; the slide-down panel is displayed. Choose the **Define** button from the slide-down panel; the **Sketch** dialog box is displayed.
3. Select the top face of the base feature as the sketching plane; an arrow pointing in the direction of viewing the sketch is displayed on the top face, refer to Figure 5-113.
4. Choose the **Flip** button to flip the arrow to point in the direction.
5. Select the **RIGHT** datum plane and then select the **Right** option from the **Orientation** drop-down list.
6. Choose the **Sketch** button to enter the sketcher environment.

Creating and Dimensioning the Sketch for the Second Feature

The next feature to be created is an extrude feature. The section for the extrude feature is shown in Figure 5-114.

1. Draw the sketch using the sketcher tools. In the sketch, draw a center line passing through the center of the arc, refer to Figure 5-114. This center line helps in constraining the sketch. Add required constraints and dimensions to the sketch, refer to Figure 5-114.

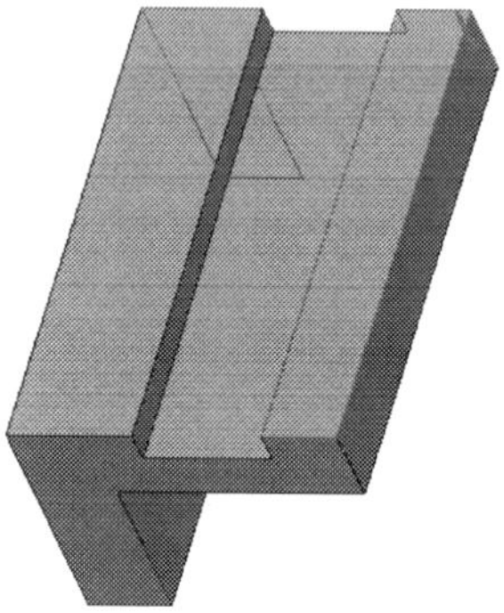

Figure 5-113 *Arrow pointing from the sketching plane in the direction of viewing the sketch*

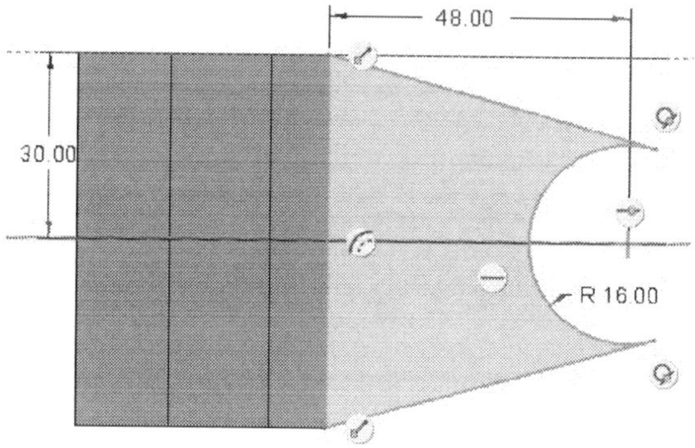

Figure 5-114 *Sketch with dimensions and constraints*

2. Choose the **OK** button; the **Extrude** dashboard is enabled. Use the middle mouse button to orient the model, refer to Figure 5-115.

3. Enter **10** in the dimension box present on the **Extrude** dashboard and using ENTER.

4. Now, choose the **OK** button to confirm the feature creation. The default trimetric view of the extruded feature is shown in Figure 5-116.

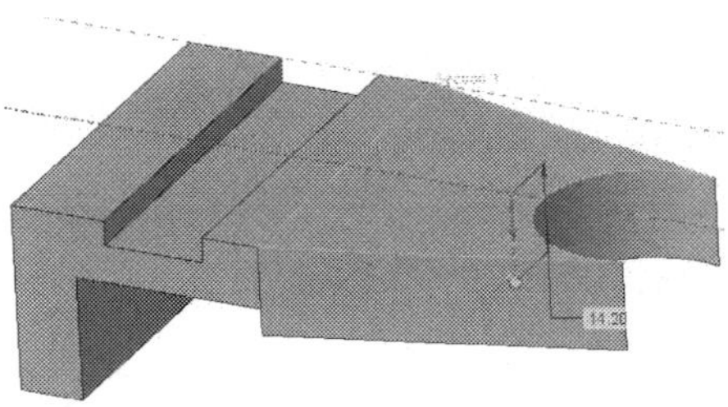

Figure 5-115 *Arrow showing the direction of material addition*

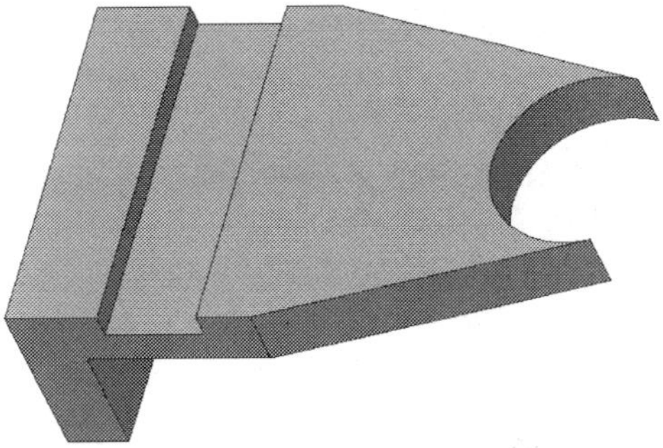

Figure 5-116 *Model with the second extruded feature*

Creating a Datum Plane for the Last Feature

To create the hollow cylindrical feature, you need a datum plane. This datum plane will be created at an offset distance of 10 units from the bottom face of the second feature shown in Figure 5-117.

Note

The other method to create this feature is to select the top planar surface of the second feature as the sketching plane and extrude the sketch on both sides of the sketching plane. The depth of extrusion will be different on both the sides. If you use this method to create this cylindrical feature, you do not need to create a datum plane.

1. Choose the **Plane** tool from the **Datum** group; the **Datum Plane** dialog box is displayed.

2. Spin the model using the middle mouse button and then select the bottom face of the second feature.

   As you select the face of the second feature, the **Offset** constraint is displayed in the **References** collector of the dialog box.

3. In the **Translation** dimension box, enter **10**.

4. Choose the **OK** button from the **Datum Plane** dialog box; the datum plane **DTM1** is created, as shown in Figure 5-118, and is selected as the sketching plane for creating the sketch.

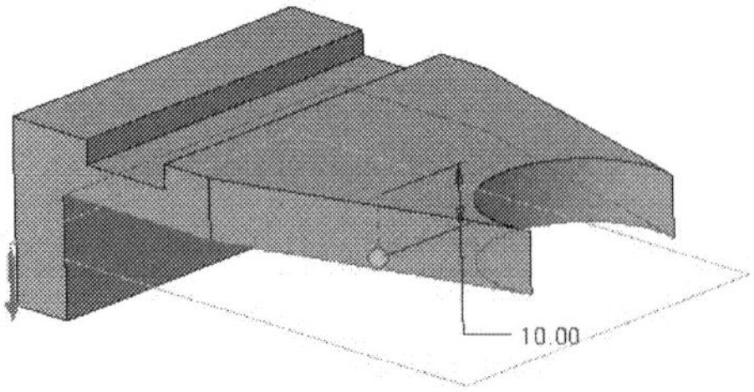

Figure 5-117 Creating the datum plane

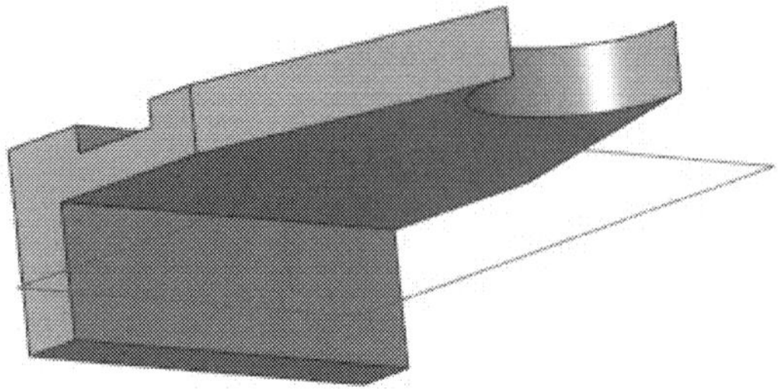

Figure 5-118 Model after creating the datum plane

Selecting the Sketching Plane for the Last Feature

The plane **DTM1** will be selected as the sketching plane and the depth of extrusion will be defined from this plane.

1. Choose the **Extrude** tool from the **Shapes** group.
2. Choose the **Placement** tab from the dashboard; the slide-down panel is displayed. Choose the **Define** button from the slide-down panel; the **Sketch** dialog box is displayed.
3. Select **DTM1** as the sketching plane; an arrow appears on the datum plane.
4. Choose the **Flip** button to reverse the direction of viewing the sketch. Figure 5-119 shows the plane with the direction of the arrow reversed.

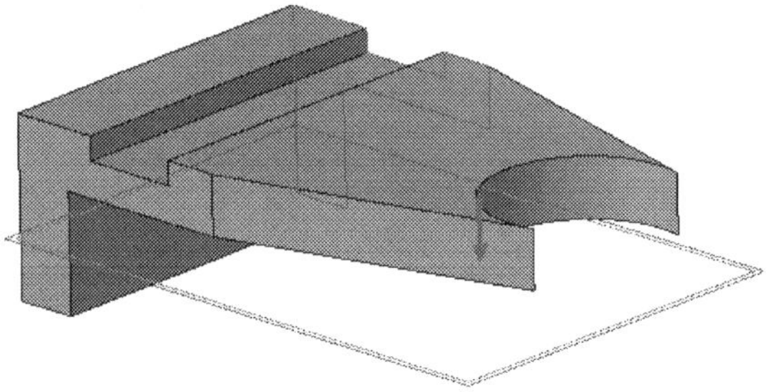

Figure 5-119 Direction of viewing the sketching plane

5. Select the **FRONT** datum plane and select the **Bottom** option from the **Orientation** drop-down list.
6. Choose the **Sketch** button to enter into the sketcher environment.

Creating and Dimensioning the Sketch for the Last Feature

The sketch for the hollow cylindrical feature will be drawn on the datum plane **DTM1**. The sketch for the hollow cylindrical feature consists of two concentric circles.

1. Draw the sketch using the sketcher tools and add required constraints and dimensions to it, as shown in Figure 5-120.

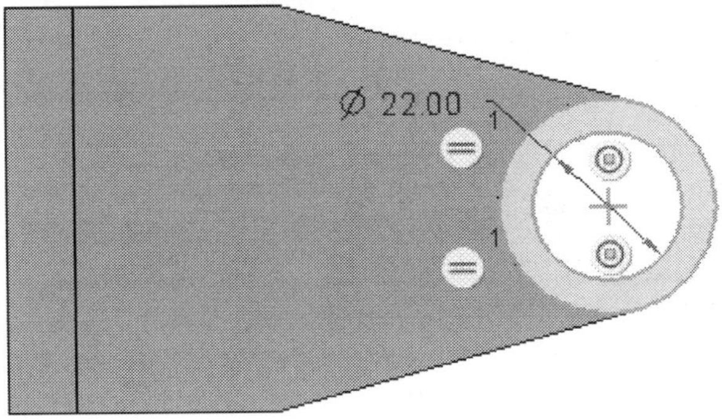

Figure 5-120 *Sketch with dimensions and constraints*

2. Choose the **OK** button; the **Extrude** dashboard is displayed. Now, turn the model display to **Shading With Edges**.

   Use the middle mouse button to orient the model, as shown in Figure 5-121. This orientation gives you a better view of the sketch in the 3D space.

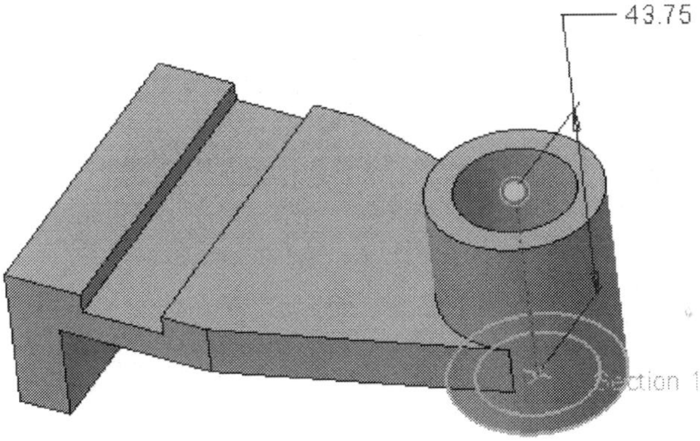

Figure 5-121 *Preview of the feature*

3. Enter **60** in the dimension box present on the **Extrude** dashboard and then press ENTER.

4. Choose the **OK** button to confirm the feature creation. The default trimetric view of the complete model is shown in Figure 5-122.

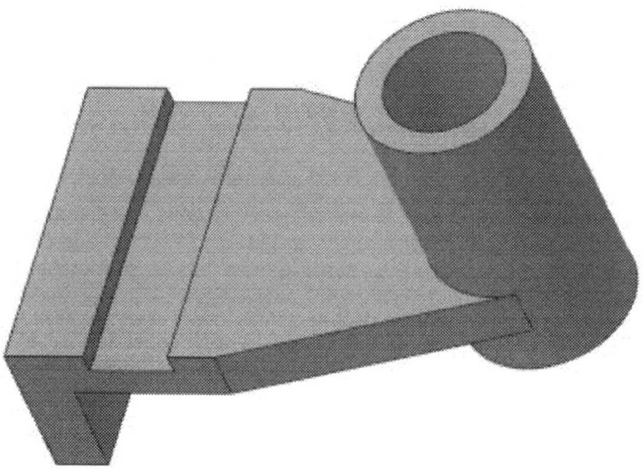

Figure 5-122 *Final model of Tutorial 3*

Saving the Model

1. Choose the **Save** button from the **File** menu and save the model.

Self-Evaluation Test

Answer the following questions and then compare them to those given at the end of this chapter:

1. While creating a datum feature, the __________ selected from model helps you choose a constraint from the available constraints to constrain the datum feature.

2. Datum axis is an __________ axis that is created in Creo Parametric.

3. The trimetric view of the model is displayed when you choose the __________ option from the **Saved Orientations** drop-down.

4. When you create a new object file, __________ default datum planes are displayed in the drawing area.

5. The default datum planes in the **Part** mode are named as __________, __________, and __________.

6. You can change the default names assigned to datum planes. (T/F)

7. Datum points are also used to add notes in drawings and attach datum targets to them. (T/F)

8. By using the **Plane** tool, you can create an offset plane through a cylinder. (T/F)

9. Generally, all features in a model are created on a single sketching plane. (T/F)

10. A sketching plane can be selected on a flat face of any feature. (T/F)

Review Questions

Answer the following questions:

1. In Creo Parametric, which one of the following is not a type of datum?

(a) Axis (b) Plane
(c) Circle (d) Curve

2. How many methods are available in Creo Parametric to create a datum plane?

(a) One (b) Two
(c) Three (d) Four

3. Which one of the following dialog boxes is displayed when you choose the **Point** tool from the **Datum** group while extruding a section sketch?

(a) **Datum Point** (b) **Datum Plane**
(c) **Datum Axis** (d) None of these

4. Which one of the following options is used to spin a model in the drawing area?

(a) CTRL+ALT (b) middle mouse button
(c) left mouse button (d) CTRL+right mouse button

5. Datum planes are considered as feature geometry and have mass and volume. (T/F)

6. To set the default orientation of a model to isometric, you need to use the **Environment** dialog box. (T/F)

7. Generally, the sketching plane for the base feature of any model is determined after viewing the isometric view or drawing views of the model. (T/F)

8. Datum planes are used as a reference for mirroring features, copying features, creating a cross-section as well as for changing the orientation of the references. (T/F)

9. Unlike the datum planes constraint option, all datum axes constraint options are stand-alone. (T/F)

10. Coordinate system is used to determine the position of a point or other geometric element. (T/F)

EXERCISES

Exercise 1

Create the model shown in Figure 5-123. The dimensions, front view, and right-side view of the model are shown in Figure 5-124. **(Expected time: 45 min)**

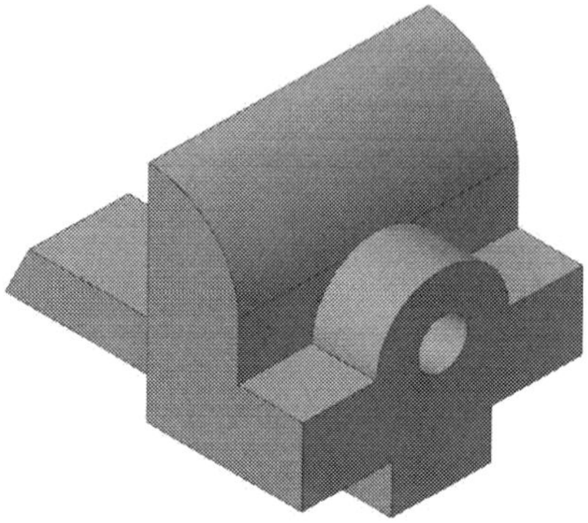

Figure 5-123 *Isometric view of the model*

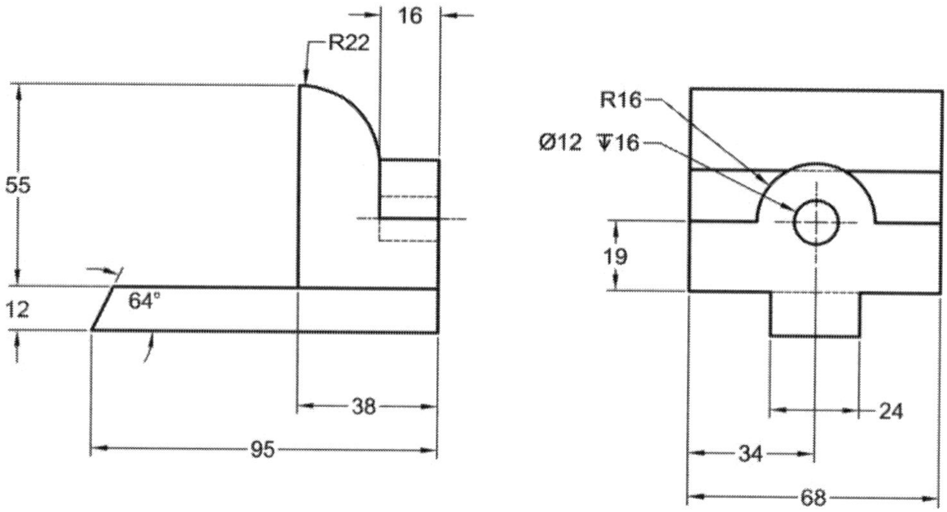

Figure 5-124 *The front and right view of the model*

Exercise 2

Create the model shown in Figure 5-125. The dimensions as well as the front, top, right-side, and isometric views of the model are also shown in Figure 5-126. **(Expected time: 45 min)**

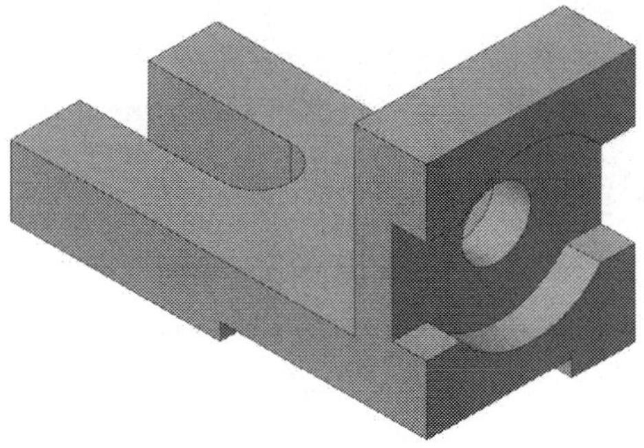

Figure 5-125 Isometric view of the model

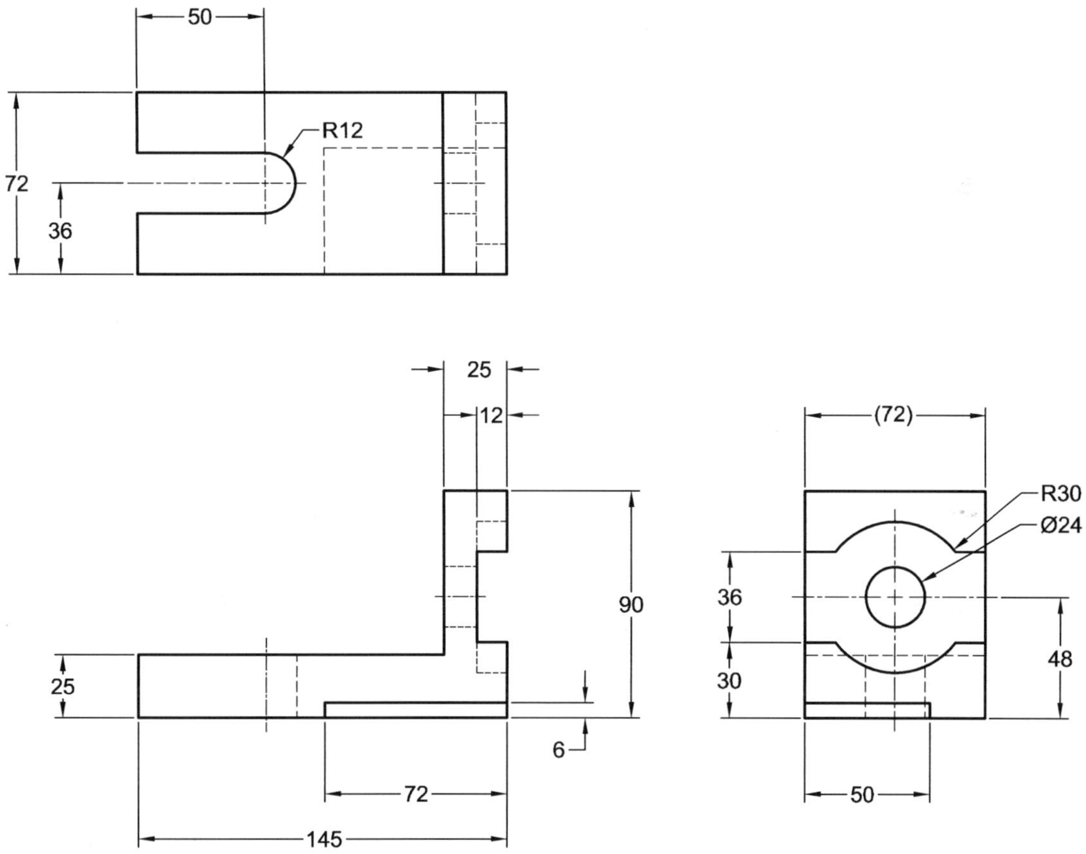

Figure 5-126 Orthographic views of the model

Exercise 3

Create the model shown in Figure 5-127. The orthographic and auxiliary views of the model are shown in Figure 5-128. **(Expected time: 1 hr)**

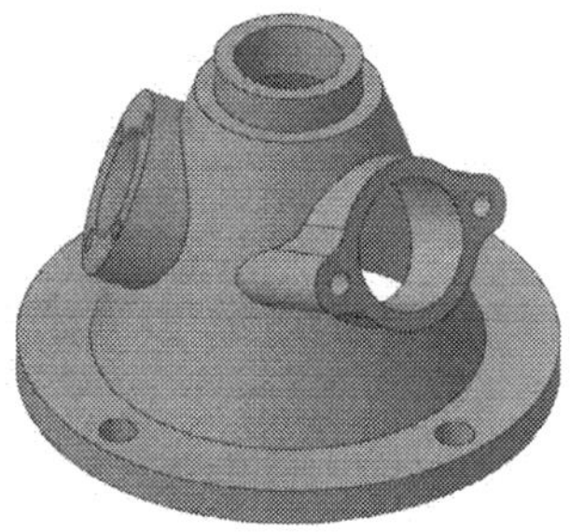

Figure 5-127 *Isometric view of the model*

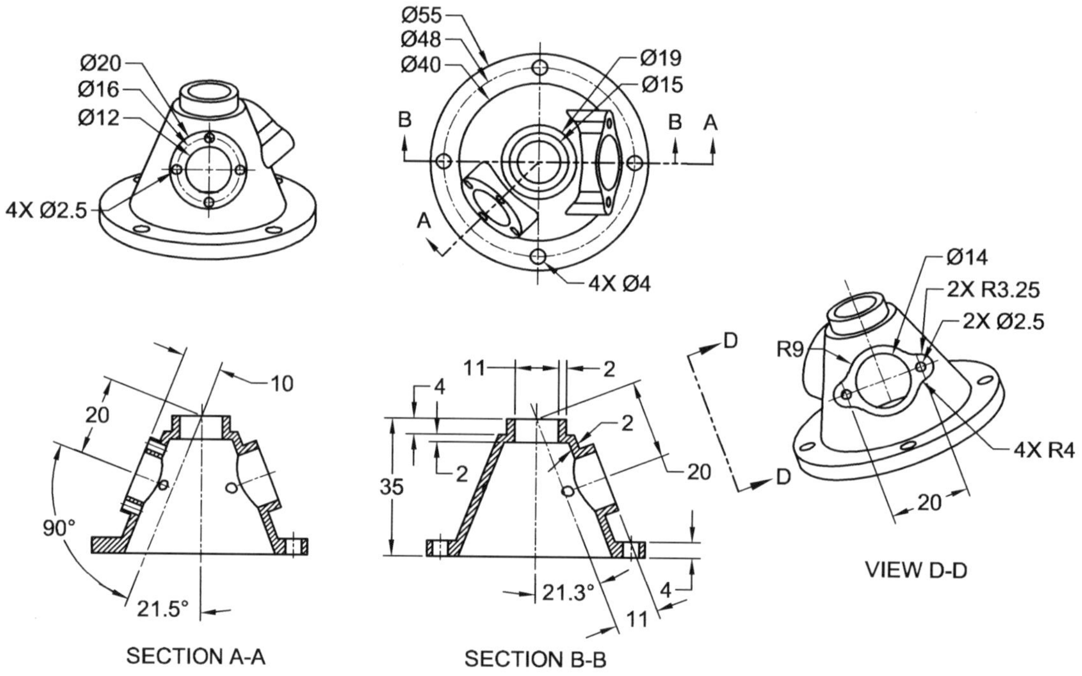

Figure 5-128 *Orthographic and auxiliary views of the model*

Hint to create the two side features in Exercise 3:

Feature on the right

1. Create an axis passing through the top face of the base feature and through the **RIGHT** datum plane.

2. Create a datum plane passing through the axis (created in the previous step) and at an angle of **21.3** degrees from the **RIGHT** datum plane.

3. Now, create an offset plane at a distance of 11 from the datum plane (created in the previous step). Select this datum plane as the sketching plane.

Feature on the left

1. Create a datum plane passing through the axis of revolution of the base feature and at an angle of 45 degrees from the **FRONT** datum plane

2. Create a datum axis passing through the top face of the base feature and through the datum plane (created in the previous step).

3. Create a datum plane passing through datum axis (created in the previous step) and at an angle of **21.5** degrees from the datum plane (created in step1).

4. Create a datum plane offset to a distance of 10 from the datum plane (created in the previous step).

Exercise 4

Create the model shown in Figure 5-129. The dimensions, the left-side view, auxiliary view, and front view of the model are also shown in Figure 5-130. **(Expected time: 45 min)**

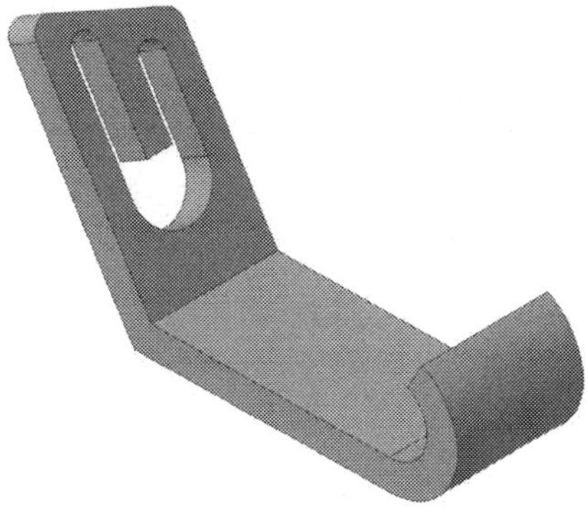

Figure 5-129 *Isometric view of the model*

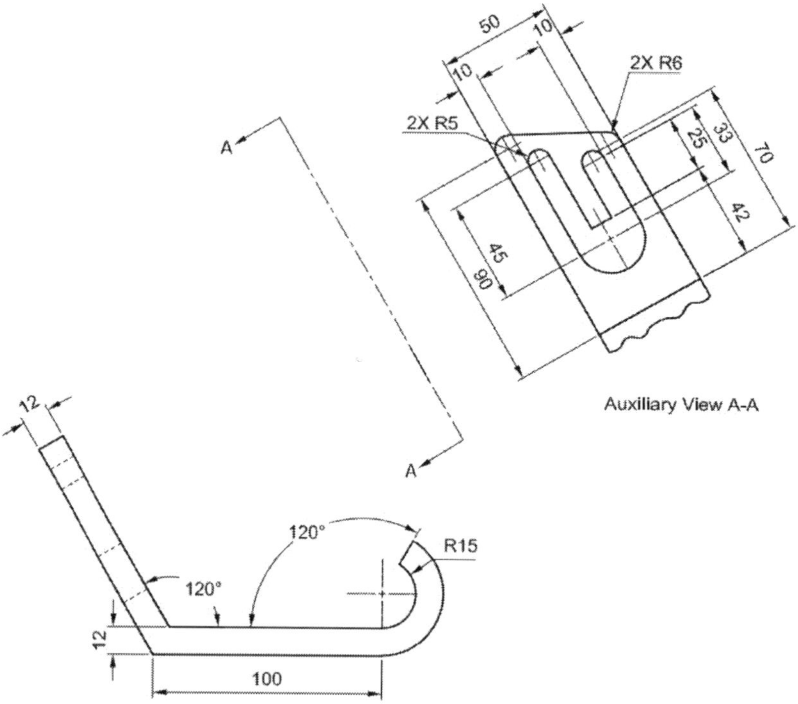

Figure 5-130 *Orthographic views of the model*

Answers to Self-Evaluation Test

1. geometry, **2.** imaginary, **3. Default Orientation**, **4.** three, **5. TOP**, **FRONT**, **RIGHT**, **6.** T, **7.** T, **8.** T, **9.** F, **10.** T

Chapter 6

Options Aiding Construction of Parts-I

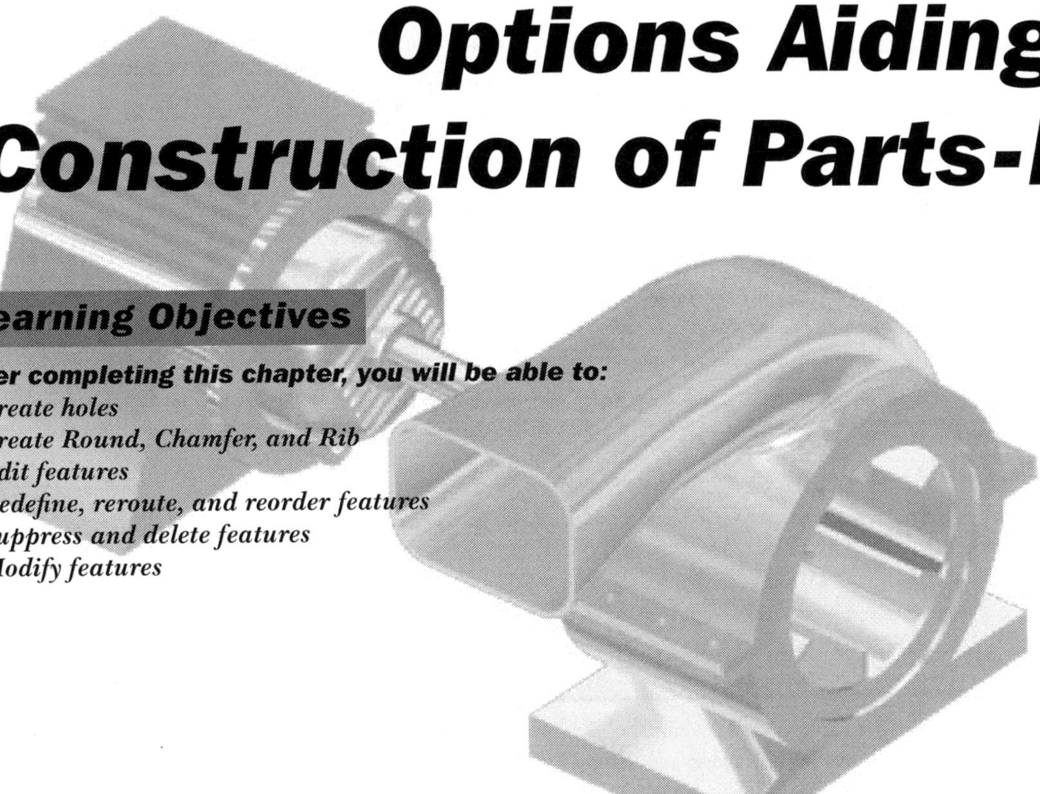

Learning Objectives

After completing this chapter, you will be able to:

- *Create holes*
- *Create Round, Chamfer, and Rib*
- *Edit features*
- *Redefine, reroute, and reorder features*
- *Suppress and delete features*
- *Modify features*

OPTIONS AIDING CONSTRUCTION OF PARTS

This chapter explains the feature creation tools provided in Creo Parametric that help in creating and editing a model. In this chapter, you will also learn to create various types of holes that are required in most of the engineering designs. In previous chapters, you learned to create holes using extrude cut, but in this chapter you will create holes using the **Hole** tool. The **Hole** tool enables you to add simple, custom, and industry-standard holes to your models. You can create and modify holes in an easier manner using the options in the **Hole** tool.

In this chapter, you will also learn to create rounds, chamfers, and ribs.

CREATING HOLES

Hole feature is similar to the cut features with only difference is that a hole feature does not require a sketch. A hole feature is preferred over other cut features because it requires less dimensioning than that is required in a cut feature. In engineering components, holes can be counterbore, countersink, tapered, or drilled. Creo Parametric allows you to create all such types. Creo Parametric also provides industry standard holes that have standard dimensions. In Creo Parametric, holes are created using the **Hole** tool.

The Hole Dashboard

Ribbon: Model > Engineering > Hole

The **Hole** dashboard is displayed when you choose the **Hole** tool from the **Engineering** group in the **Model** tab, as shown in Figure 6-1. You can create two types of holes using the **Hole** dashboard, the **Simple** hole and the **Standard** hole. The options to create simple and standard holes are discussed next.

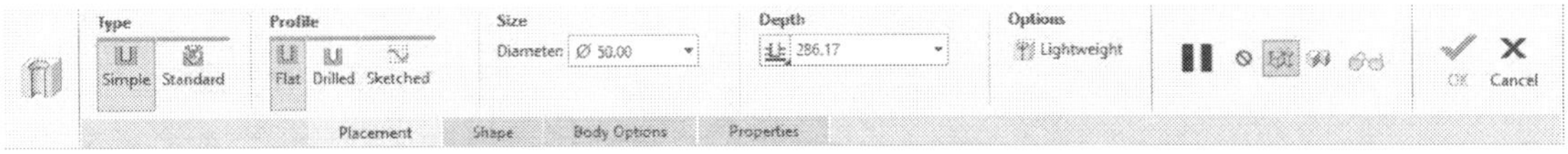

Figure 6-1 *Partial view of the* ***Hole*** *dashboard*

Creating Simple Holes Using Predefined Rectangle Profile

A simple hole consists of a revolved cut created using a rectangular cross-section. When you invoke the **Hole** dashboard, the **Simple** and **Flat** buttons are chosen by default, refer to Figure 6-1. The options and the buttons available in the dashboard are discussed next.

Settings Area

The **Diameter** edit box on the left in the **Hole** dashboard is used to specify the diameter of the hole.

The second dimension box from the left in the **Hole** dashboard is used to specify the depth of the hole. The options to specify the depth of the hole are the same as those available in the **Extrude** dashboard.

Placement Tab

When you choose the **Placement** tab, a slide-down panel will be displayed, as shown in Figure 6-2. This slide-down panel has the parameters that need to be specified to define a linear hole. The **Placement** collector in the slide-down panel is used to select the primary placement references. You can select the placement plane only when this collector is highlighted. Once you have selected the primary reference, its name appears in the box. The primary reference for the placement of holes can be a plane, axis, or point.

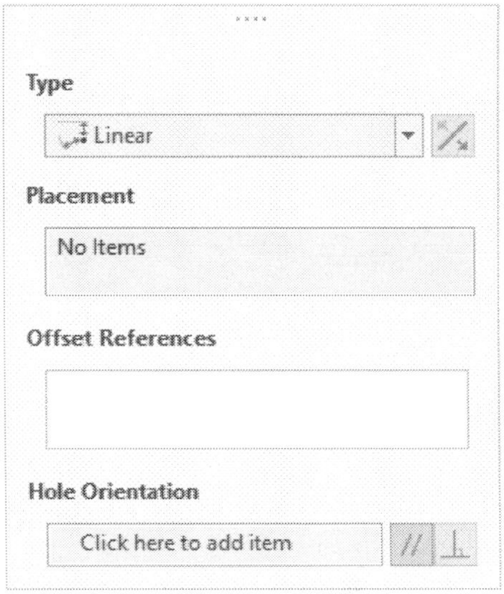

***Figure 6-2** The **Placement** slide-down panel*

The **Flip** button next to the **Placement** collector is used to change the direction of the hole creation with reference to the primary reference. When you choose this button, the direction of the hole creation is reversed and you can view its preview in the drawing area. Remember that the hole will be created only if it is possible to create the hole in the specified direction.

The **Type** drop-down list below the **Placement** collector contains the options that determine the type of placement for the hole. You can use the options in this drop-down list after selecting the placement plane or the primary reference. These options are discussed next.

Linear: This option is used to place a hole on a surface by using two linear dimensions. This option is selected by default when you select a surface or datum plane as the primary placement reference. When you select this option, you need to specify the distance of the hole axis from two linear references. Generally, these linear references are the edges of the planar surface on the model, any two planar surfaces or axes, or a combination of any of them. Figure 6-3 shows a hole placed on the top face of the block. The procedure to create this hole using the **Hole** dashboard is as follows:

1. Invoke the **Hole** dashboard.
2. Choose the **Placement** tab to open a slide-down panel.
3. Select the top face of the model to place the hole; the preview of the hole will be displayed.
4. Click in the **Offset References** collector to make the secondary selections.
5. Select the left face or the top left edge of the model.
6. Use CTRL+left mouse button to select the front face or the top front edge of the model. Now, you have selected the two references that are needed to dimension the position of hole.

7. Specify the exact location of hole in the edit boxes of the **Offset References** collector.

8. Specify the diameter of the hole in the **Diameter** edit box on the **Hole** dashboard.

9. From the depth drop-down, choose the **Drill to intersect with all surfaces** button.

10. Choose the **OK** button to create the hole and exit the **Hole** dashboard.

Note

*To select the secondary references, you need to click in the **Offset References** collector and then select the references on the model. Dimensions of the center of the hole from the selected references are displayed in the drawing area as well as in the **Offset References** collector. You can click on the dimensions to modify them.*

Radial: This option is used to create a hole that can be referenced to an axis. This option is selected by default when you select a circular surface as the primary placement reference. When you select this option, you need to select an axial reference and an angular reference to place the hole. The distance from the axis and the angle from the plane are entered in the boxes under the **Offset References** collector. You can select a datum plane, or a planar, cylindrical, or conical solid surface as the primary placement reference. Figure 6-4 shows a radial hole on a curved surface. The following steps explain the procedure to create this hole.

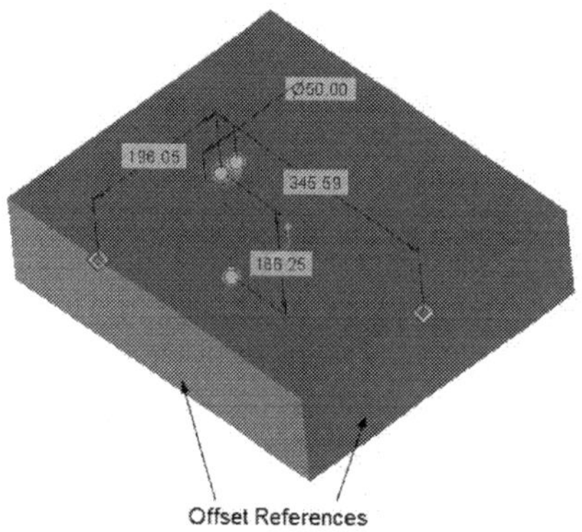

Figure 6-3 *Linear hole on the top face of the block*

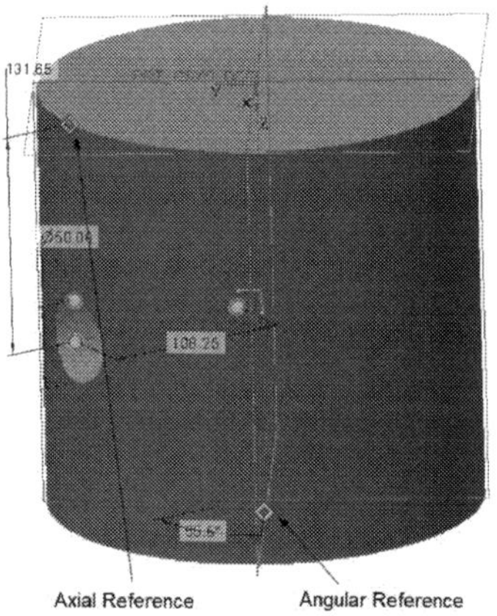

Figure 6-4 *Radial hole on the curved surface*

Tip

Note that since you have selected the curved surface as the placement plane for the hole, therefore, you need not to select the axis to specify the radial distance. However, if the placement surface is a plane surface, then you do need to select an axis to create a radial hole at some radius from the selected axis.

1. Invoke the **Hole** dashboard.

2. Choose the **Placement** tab to open the slide-down panel.

3. Select the curved surface of the model to place the hole; the preview of the hole will be displayed.

4. Click in the **Offset References** collector to make the secondary selections.

5. Select the datum plane to specify the angular value.

6. Use CTRL+left mouse button to select the top face or the top edge of the model to specify the axial offset for the hole placement. Now, you have selected the two references that were needed to dimension the hole.

7. Specify the diameter of the hole in the dimension box on the **Hole** dashboard.

8. From the depth drop-down, choose the **Drill to intersect with all surfaces** button.

9. Choose the **OK** button to create the hole and exit the **Hole** dashboard.

Diameter: This option is used to create a diametrically placed hole. When you select this option, you need to select an axial reference and an angular reference to place it. Figure 6-5 shows a diameter hole on a plane. The following steps explain the procedure to create this hole.

1. Invoke the **Hole** dashboard.

2. Choose the **Placement** tab to open the slide-down panel.

3. Select the top face of the model to place the hole; the preview of the hole will be displayed.

4. From the drop-down list, select the **Diameter** option.

5. Click in the **Offset References** collector to make the secondary selections.

6. Select the face or a datum plane to specify the angular value, refer to Figure 6-5.

7. Use CTRL+left mouse button to select a datum axis that is normal to the top face. You can also select the axis of any other hole as reference in the Offset collector. Now, you have selected the two references that are needed to dimension the hole.

8. Specify the distance for the hole measured from the axis of the reference hole and the angle in the **Offset References** collector.

9. From the depth drop-down, choose the **Drill to intersect with all surfaces** button.

10. Choose the **OK** button to exit the **Hole** dashboard.

Coaxial: This option is used to create a coaxial hole. When you select this option, you need to select an axis. No dimensions are required to place a coaxial hole. Figure 6-6 shows a coaxial hole on a plane. The procedure to create this hole is given next.

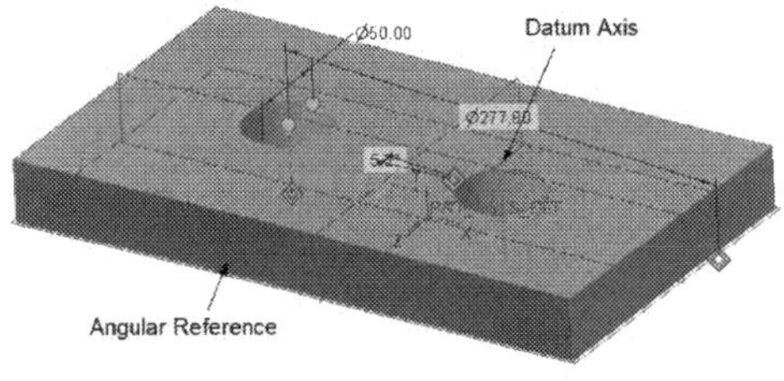

Figure 6-5 Diameter dimensioning of the hole

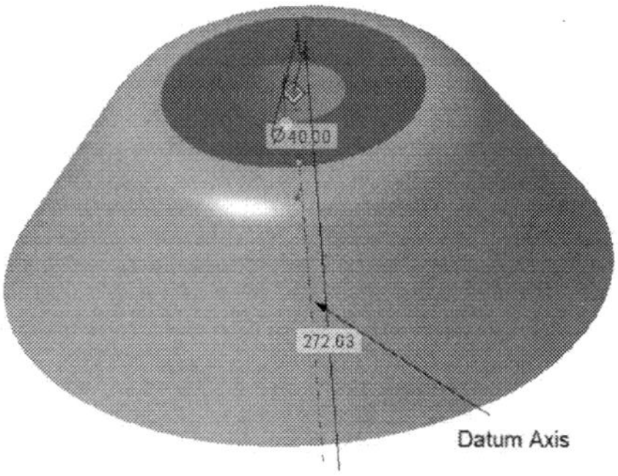

Figure 6-6 Coaxial hole

1. Invoke the **Hole** dashboard.

2. Choose the **Placement** tab to open the slide-down panel.

3. Select the axis of the model to place the hole; the preview of the hole will be displayed. In the **Type** drop-down list, the **Coaxial** option is selected by default.

4. Now, press the CTRL key to select the placement plane or the surface where you want to place the base of the hole.

5. Specify the diameter of the hole in the dimension edit box available on the **Hole** dashboard.

6. From the depth flyout, choose the **Drill to intersect with all surfaces** button.

7. Choose the **OK** button to exit the **Hole** dashboard.

On Point: This option is used to create holes on the datum point. Note that the datum point is located either on the surface or at an offset from it. The procedure of creating this hole is explained next.

1. Invoke the **Hole** dashboard.

2. Choose the **Placement** tab to open the slide-down panel.

3. Select the datum point to place the hole; the preview of the hole will be displayed.

   In the **Type** drop-down list, the **On Point** option is selected by default.

4. Specify the diameter of the hole in the first dimension edit box available on the **Hole** dashboard.

5. From the depth flyout, choose the **Drill to intersect with all surfaces** button.

6. Choose the **OK** button to exit the **Hole** dashboard.

> **Tip**
> *Remember that while placing the linear, diameter, and radial holes using the **Hole** dashboard, you need to define two types of references: primary and secondary. Primary reference is the placement plane or the surface on which the hole feature will be created and secondary references are dimensional references. In case of the **Coaxial**, **On Point**, and **Sketched** holes, only primary references are required for the placement of the hole feature.*

Sketched: This option is used to create a hole on sketched point, midpoint, or endpoint of sketched line.

Hole Orientation

The default orientation of the hole depends upon the selected placement references. The options available in the **Hole Orientation** area of the **Placement** tab can be used to change the default orientation of the hole geometry. You can orient the hole parallel to or perpendicular to the orientation reference. When the orientation reference is a plane, the hole can only be perpendicular to the hole orientation reference. When the orientation reference is an axis or an edge, the hole can be perpendicular or parallel to the selected reference. If the hole orientation is changed, the hole depth and other dimensions are measured with respect to the placement point and along the direction of either the default orientation or the selected orientation reference.

To change the default orientation of the hole geometry, click on the **Hole orientation** collector and select an edge, an axis, or a plane; the hole is oriented parallel or perpendicular to the selected reference, refer to Figure 6-7.

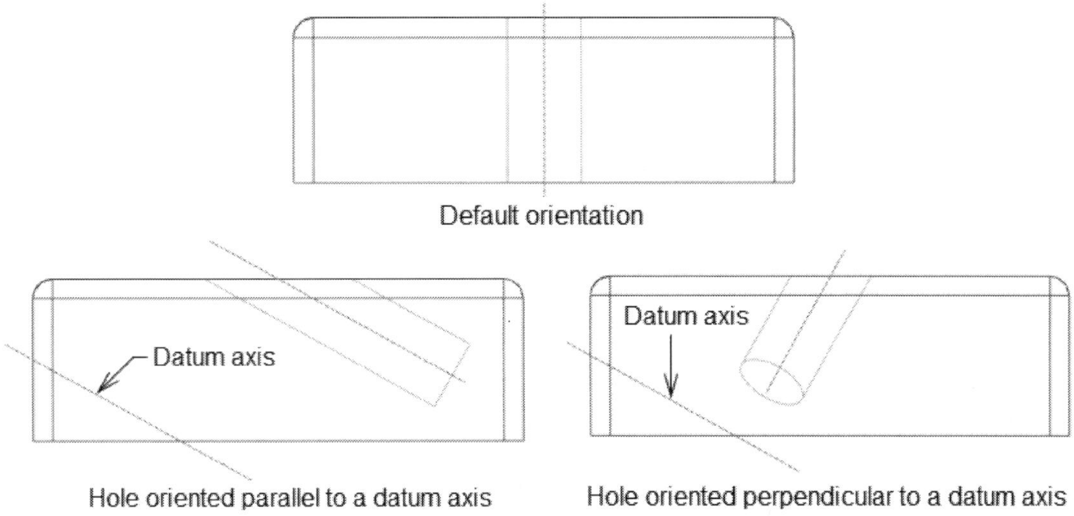

***Figure 6-7** Orientation of the hole geometry with respect to a datum axis*

Shape Tab

When you choose the **Shape** tab, a slide-down panel will be displayed, as shown in Figure 6-8. In this panel, some of the options are displayed based on the hole type selected. The options in this slide-down panel can be used to specify the depth and the diameter of the hole. The depth of the hole can be specified in two directions with respect to the placement plane. The options to specify the depth of the hole are similar to those that are available in the **Extrude** dashboard. In the slide-down panel, the sketch of the hole, gives an idea of the shape of the hole and its dimensions. The sketch that is shown in the slide-down panel shows the depth of the hole in the **Side 1** direction. The **Side 2** drop-down list is used to specify the depth of the hole in the direction opposite to **Side 1**. This means the hole can be created on both sides of the placement plane. By default, the **None** option is selected in this drop-down list. You can select the **Top Clearance** check box to provide some clearance between outer surface of the given object and starting point of the hole to be generated.

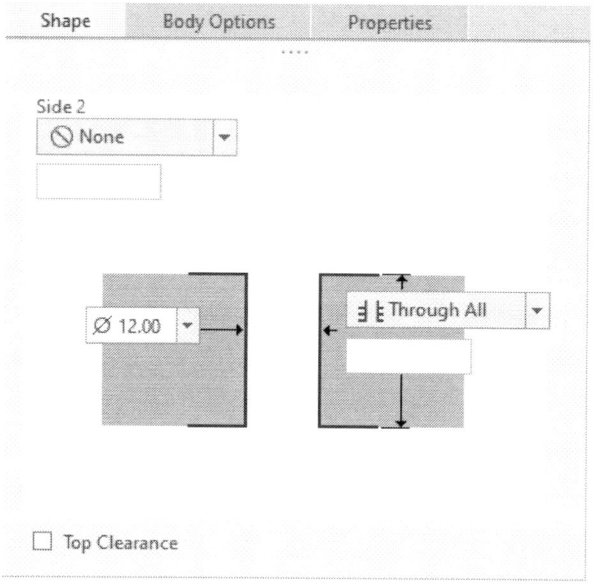

***Figure 6-8** The **Shape** slide-down panel*

Note

*1. The termination options in the **Shape** slide-down panel are similar to those discussed in the **Extrude** dashboard.*

2. When you specify the placement plane for a hole and then specify two references in case of a linear hole, you will notice that five drag handles are displayed on the hole. These five drag handles including three yellow circles and two red brackets, have four different functions to place the hole dynamically on the placement plane. Using these handle you can set the location, diameter, depth, and specify two references to place the hole. Remember that these drag handles appear only when you select the primary reference, or in other words, the placement plane.

Body Options Tab

Choose the **Body Options** tab; a slide-down panel will be displayed. The options in this slide-down panel are used to cut the geometry from the existing body. If you select the **All** radio button in the **Cut geometry from body** area, the geometry will be cut from all the bodies that the feature passes through. If you select the **Selected** radio button in the **Cut geometry from body** area, the geometry will be cut from the selected bodies.

Properties Tab

The **Properties** slide-down panel of the **Hole** dashboard is shown in Figure 6-9. This tab, when chosen, allows you to name the hole you are creating. By default, the first hole you create is named as HOLE_1 and then every hole created is successively numbered.

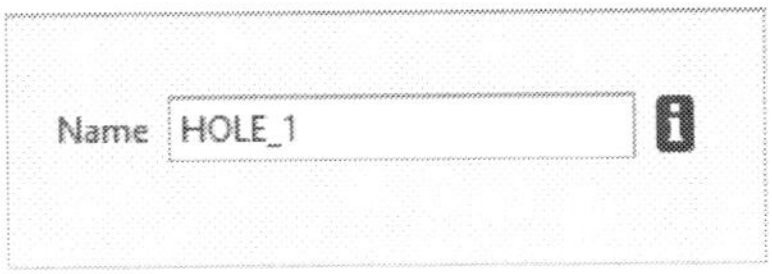

*Figure 6-9 The **Properties** slide-down panel*

Creating Lightweight Holes

A lightweight hole is a graphical representation of a standard hole. Lightweight holes are useful when you are creating a large number of holes in a model. Consider an example where you have created 5000 holes in a model. Such a large number of features can cause Creo Parametric to run out of memory during the regeneration of the model. To reduce memory usage while creating a large number of holes, you can use lightweight holes. A lightweight hole can be used exactly as a hole with solid geometry. You can convert the geometry of a simple hole from solid to lightweight, and vice versa, at any time in the **Part** mode. Lightweight holes are represented in the drawing area by a circular curve and an axis. The curve follows the hole circumference and lies on the hole placement plane.

To toggle between lightweight hole geometry and solid hole geometry, select the **Lightweight** button from the **Hole** dashboard. Alternatively, select a hole from the **Model Tree** and right-click on it. Next, select the **Hole Representation** > **Lightweight** or **Hole Representation** > **Solid** from the shortcut menu displayed to convert a lightweight hole into a standard hole, or a standard hole into a lightweight hole. In the **Model Tree**, lightweight holes are represented by the icon.

Creating Simple Holes Using Standard Hole Profile

To create a simple hole with standard profile, you need to choose the **Drilled** button from the **Hole** dashboard, as shown in Figure 6-10. On choosing this button, you can specify the countersink, counterbore, and depth of the hole. The options and tool buttons available in the dashboard are discussed next.

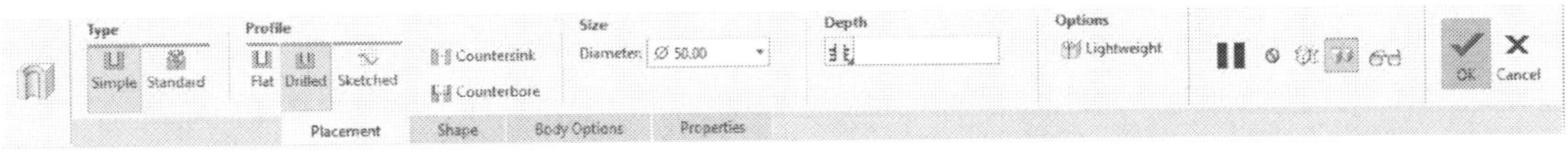

*Figure 6-10 Partial view of the **Hole** dashboard with the **Drilled** button selected*

Placement Tab

The options in this tab are same as discussed while creating the Simple hole using the **Drilled** button.

Shape Tab

The options in the slide-down panel will be displayed when you choose the **Shape** tab from the **Hole** dashboard. As discussed earlier that some of the options in the **Shape** tab are displayed based on the type of hole selected. These options are different from the options that were available when you created the Simple hole using the **Flat** button. The sketch for the hole in the slide-down panel gives an idea about the shape of the hole and its parameters, as shown in Figure 6-11. The **Shoulder** and **Tip** radio buttons are used to specify the measurement method for measuring the hole depth.

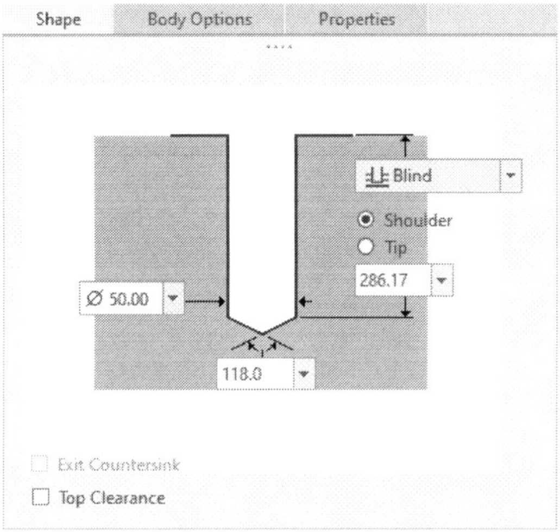

Figure 6-11 *The **Shape** slide-down panel*

Countersink

Choose this button to create a countersink hole. A countersink hole has two diameters and the transition between the larger diameter and the smaller diameter is in the form of a cone. In the slide-down panel of the **Shape** tab, a preview of the cross-section with parameters of the countersink hole will be displayed, as shown in Figure 6-12. In the preview of the section, the dimensions can be edited as per the requirement.

Counterbore

Choose this button to create a counterbore hole. A counterbore hole is a stepped hole, and it has two diameters, a larger diameter and a smaller diameter. The larger diameter is called counter diameter and the smaller diameter is called drill diameter. In the slide-down panel of the **Shape** tab, a preview of the cross-section with parameters of the counterbore hole will be displayed, as shown in Figure 6-13. In the preview of the section, the dimensions can be edited as required. Note that if both the buttons are chosen, then the hole created will be a combination of the counterbore and countersink holes.

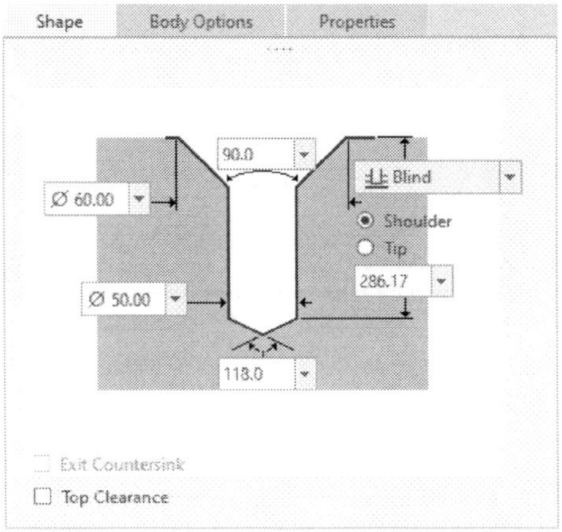

Figure 6-12 The preview of the cross-section with parameters of countersink hole

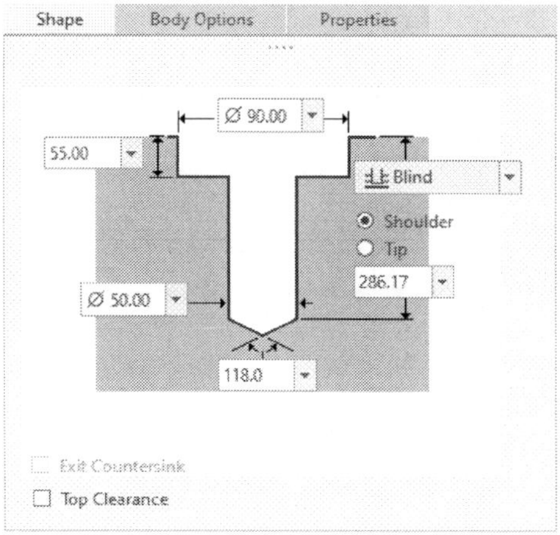

Figure 6-13 The preview of the cross-section with parameters of the counterbore hole

The **Exit Countersink** check box is used to add a countersink profile at the bottom of the hole. This check box becomes active only when you select the **Through All** option from the depth termination drop-down list in the **Shape** tab. As you select this check box, the cross-section of the hole is modified and the parameters of the countersink are added at the bottom of the hole geometry, as shown in Figures 6-14 and 6-15.

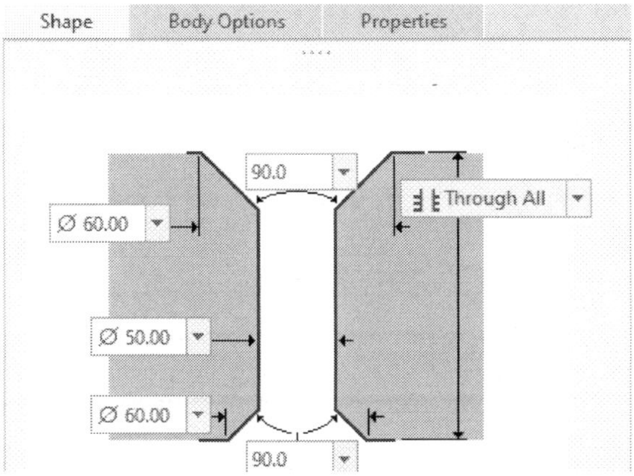

*Figure 6-14 The preview of the countersink cross-section with the **Exit Countersink** check box selected*

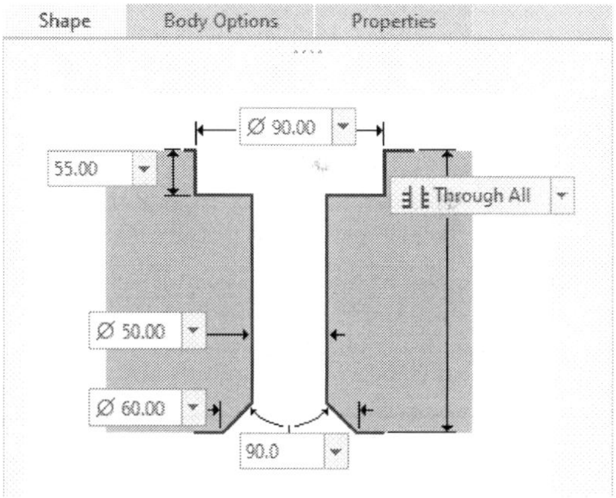

*Figure 6-15 The preview of the counterbore cross-section with the **Exit Countersink** check box selected*

Drilled hole shoulder depth

This button will only become available when the **Drilled** button is selected in the dashboard. Select this button to specify the depth of the hole. When you choose the **Drilled hole shoulder depth** button from the **Hole** dashboard, the depth of the hole will be measured upto the shoulder only. In the slide-down panel of the **Shape** tab, a preview of the cross-section with parameters of the hole, whose depth is measured upto the shoulder, will be displayed, as shown in Figure 6-16.

Drilled hole depth

When you choose the **Drilled hole depth** button from the **Hole** dashboard, the depth of the hole is measured upto the tip of the hole. To invoke this button, choose the black arrow on the right of the **Drilled hole shoulder depth** button; a flyout will be displayed. Choose the **Drilled hole depth** button from the flyout. In the slide-down panel of the **Shape** tab, a preview of the cross-section with parameters of the hole, whose depth is measured upto the tip, will be displayed, as shown in Figure 6-17.

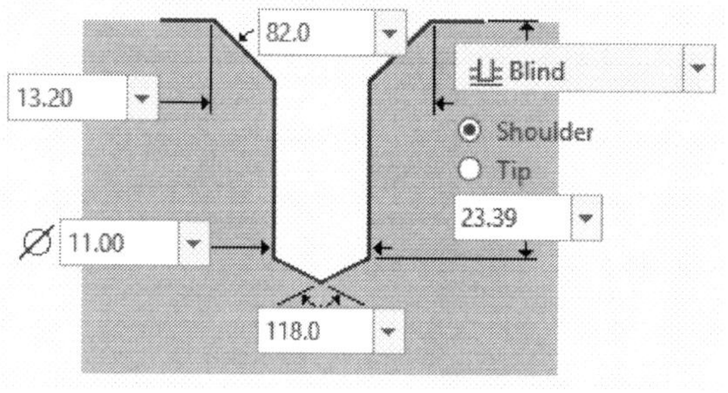

***Figure 6-16** Parameters of the hole with depth measured upto the shoulder*

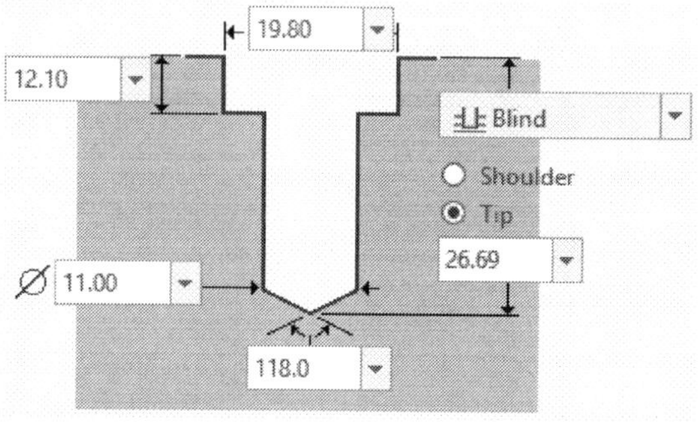

***Figure 6-17** Parameters of the hole with depth measured upto the tip*

Creating Simple Holes Using Sketched Profile

This option allows you to sketch the cross-section of the hole that is revolved about a center axis. This option is used to create custom shapes for the hole. When you choose the **Sketched** button from the **Hole** dashboard, the **Hole** dashboard appears, as shown in Figure 6-18. The tools and options in the dashboard are discussed next.

***Figure 6-18** Partial view of the **Hole** dashboard with the **Sketched** button selected*

Sketch

Choose the **Sketch** button to open a new window with the sketcher environment. The cross-section for the hole is sketched using the sketcher tools available. While drawing the sketch, a center line must be drawn that acts as the axis of revolution for the section of hole and the sketch must be a closed loop with no intersecting entities. The sketched holes can be blind or through all, depending on the dimensions of its sketch.

When you complete the sketch for the hole and choose the **OK** button to exit the sketcher environment, the **Hole** dashboard will be enabled above the drawing area. Now, you need to specify the placement options to place the hole. The placement options are the same as discussed earlier.

Note

The placement options to place a hole can be specified before or after drawing the sketch for the hole. However, when you specify the placement options after drawing the sketch, preview of the hole will not available in the drawing area because the primary reference won't be selected yet. Therefore, it is recommended to specify the placement options before drawing the sketch for the hole.

Open

On the **Hole** dashboard, the **Open** button is used to open an existing sketch that is saved in *.sec* file format. When you choose this button, the **OPEN SECTION** dialog box will be displayed. Select a sketch and choose the **Open** button to define the profile of the hole geometry.

Placement Tab

The options displayed in the slide-down panel on choosing the **Placement** tab from the **Hole** dashboard are the same as those discussed earlier. You can also use the same type of placement options as discussed earlier.

Shape Tab

The options displayed in the slide-down panel on choosing the **Shape** tab from the **Hole** dashboard are different than the options that were available in case of **Flat/Drilled** holes. When you choose the **Shape** tab, the slide-down panel displays the preview of the sketch that you have drawn in the sketcher environment. You can pan and zoom the sketch in the slide-down panel.

Note

*If you choose the **Shape** tab before drawing the sketch for the hole, then the slide-down panel becomes blank and does not show a sketch.*

Creating Standard Holes

A standard hole consists of an extruded cut based on industry-standard fastener table. Creo Parametric provides industry-standard hole charts and tapped or clearance diameters for the selected fastener. You can also create your own hole charts. When you create a standard hole, thread notes are automatically created with the feature. The standard holes can be created by choosing the **Standard** tool from the **Type** area of the **Hole** dashboard.

When you choose the **Standard** tool from the **Hole** dashboard, the dashboard appears, as shown in Figure 6-19. The tools and options in the dashboard are discussed next.

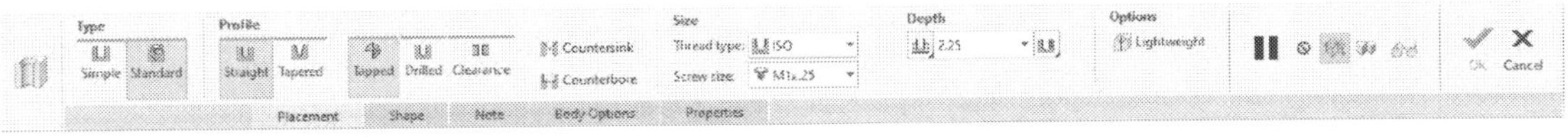

*Figure 6-19 Partial view of the **Hole** dashboard with the **Standard** button chosen*

Drop-down Lists

The **Thread type** drop-down list from the left in the dashboard is used to specify the type of thread on the hole. The options in this drop-down list are **ISO**, **UNC**, and **UNF**. The default option selected in this list depends upon the template selected in the **New File Options** dialog box. You can also insert custom hole charts using this drop-down list. The hole charts enable you to standardize the available thread, diameter, countersink, and counterbore options while creating standard hole feature. Hole chart files are designated with *.hol* extension. To load a hole chart in the current working environment, you can choose **File > Options > Configuration Editor**. Choose **Add** option, then enter **hole_parameter_file_path** in the **Option name** edit box. Next, browse for the *.hol* file, the parameters defined in the hole chart will be displayed in the **Properties** tab of the **Hole** dashboard and the name of the hole chart will be displayed in the thread type drop-down list.

 The **Screw size** drop-down list in the **Hole** dashboard is used to specify the size of the screw that corresponds to the hole. If you enter a screw size that is not listed, the system automatically selects the closest screw size. You can also drag the hole diameter handle to select a screw size.

The Depth drop-down list in the **Hole** dashboard is used to specify the depth of the hole. The options in this drop-down list are similar to those available in the **Extrude** dashboard.

Tapped

 Choose this button to tap a hole. Tapping is the process of cutting threads in the hole. In Creo Parametric, these threads are known as cosmetic threads. This means that the threads created will not be visible in the hole. In the slide-down panel of the **Shape** tab, a preview of the cross-section with parameters of the tapped hole will be displayed, as shown in Figure 6-20. In the preview of the section, the dimensions can be edited, as required. The dotted lines represent the threading outside the hole. By default, the **Include thread surface** check box is selected in the slide-down panel of the **Shape** tab. As a result, when you finish the hole creation process, the hole will be created with a surface feature indicating that the hole has internal threads. When the **Straight** button is selected the **Tapped**, **Drilled hole**, and **Clearance** tools will be displayed. Choose the required button and then set the parameters in the dashboard to create the desired hole.

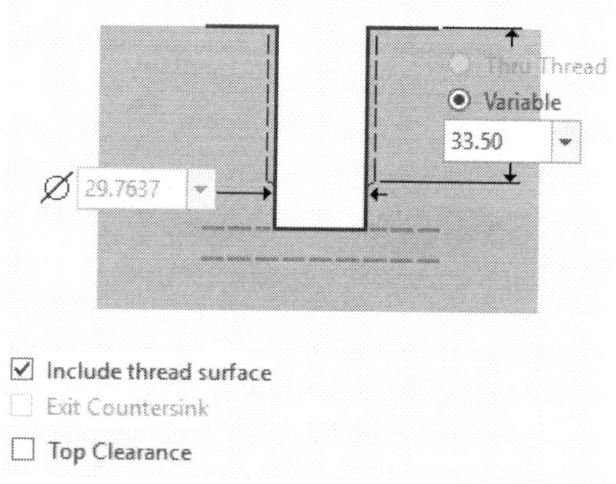

***Figure 6-20** Parameters of the tapped hole*

Clearance

 Choose this button to create clearance holes. This option is available only when the **Straight** button is chosen in the **Hole** dashboard. To add a clearance to a standard hole, select the type of clearance from the drop-down list available in the **Shape** tab. Remember that, when you create a clearance hole, the diameter of the hole is varied automatically after choosing a type of clearance from the drop-down list in the **Shape** tab.

Tapered

 Choose this button to create tapered threaded holes. The sizes of holes can conform to the ISO, JIS, and ANSI standards. There are three standards in the thread type drop-down list, **ISO_7/1**, **NPT**, and **NPTF**. Set the required standard, specify the screw size, and the drill depth value to create the tapered hole. You can also add countersink or counterbore to the tapered hole, if required.

Countersink

As discussed earlier, this button is used to create a countersink hole. A countersink hole has two diameters and the transition between the bigger diameter and the smaller diameter is in the form of a cone. In the slide-down panel of the **Shape** tab, a preview of the cross-section with threads and parameters of the countersink hole will be displayed, as shown in Figure 6-21. In the preview of the section, the dimensions can be edited as required.

Note that you need to deactivate the **Countersink** button and then choose the **Counterbore** button to create the counterbore hole. Else, the hole created will be a combination of the countersink and counterbore holes.

Counterbore

As discussed earlier, this button is used to create a counterbore hole. A counterbore hole is a stepped hole and has two diameters, a larger one and a smaller one. The larger diameter is called counter diameter and the smaller diameter is called the drill diameter. In the slide-down panel of the **Shape** tab, a preview of the cross-section with threads and parameters of the counterbore hole will be displayed, as shown in Figure 6-22. In the preview of the section, the dimensions can be edited as required.

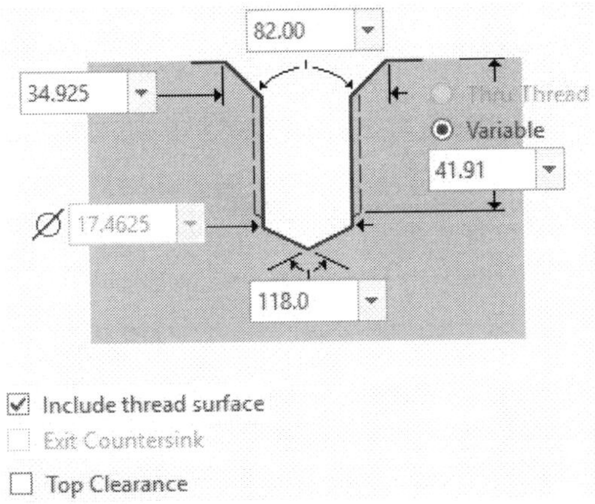

***Figure 6-21** The preview of the cross-section with threads and the parameters of the countersink hole*

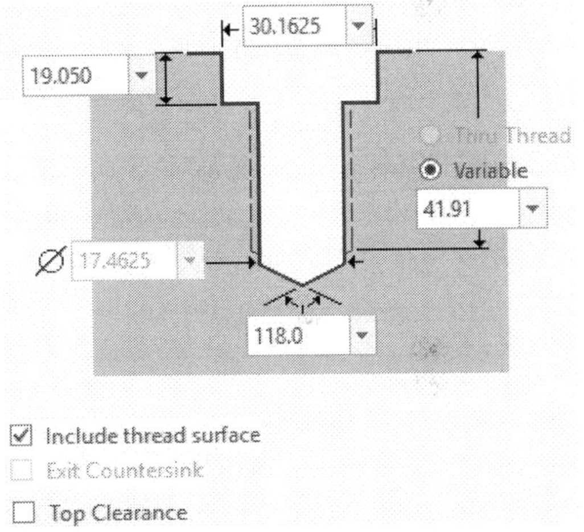

***Figure 6-22** The preview of the cross-section with threads and the parameters of the counterbore hole*

Placement Tab

The options in the slide-down panel that will be displayed when you choose the **Placement** option from the **Hole** dashboard are the same as those that were available in case of the **Simple** hole. You can use the same type of placement options that were available when you created the **Simple** hole.

Shape Tab

The options in the slide-down panel that will be displayed when you choose the **Shape** tab from the **Hole** dashboard are different from the ones that were available in case of **Simple** hole. When you choose the **Shape** tab, the slide-down panel displays the preview of the hole. You can specify the dimensions for the hole in the respective dimension boxes.

Note Tab
This tab displays the specifications and values of the hole in the text format. By default the Add a note check box is selected in the Note tab. As a result, the specifications and parameters of the standard hole will be displayed with a leader on the model. This option is available only when you create a Standard hole.

Body Options Tab
Choose the **Body Options** tab; a slide-down panel will be displayed. The options in this slide-down panel are used to cut the geometry from the existing body. If you select the **All** radio button in the **Cut geometry from body** area, the geometry cuts will be cut from all the bodies that the feature passes through. If you select the **Selected** radio button in the **Cut geometry from body** area, the geometry will be cut from the selected bodies only.

Properties Tab
This tab displays the parameters of the hole feature.

Previewing the Hole

The **Feature Preview** button is used to preview the hole created before confirming its creation. Changes and modifications in the hole parameters can be made once the hole is previewed.

Important Points to Remember While Creating a Hole
The points given next should be remembered while creating a hole.

1. While drawing the sketch of a hole, it should have an axis of revolution and at least one entity normal to it. Now, when you place this hole on the primary reference (**Placement** reference), then this normal entity is aligned with the plane (primary reference). If there are two entities normal to the axis, then the normal entity at the top of the sketch is aligned with the placement plane.

2. While creating a hole, the primary reference for the placement of hole can be selected without choosing any option from the **Hole** dashboard. However, if you need to specify the secondary references (**Offset references**), then choose the **Placement** tab from the **Hole** dashboard; the slide-down panel will be displayed. From the slide-down panel, click in the **Offset References** collector and when the **Placement** collector appears white in color, you can select the secondary references.

   Remember when the first reference is selected by using the left mouse button, then the second reference is selected by using CTRL+left mouse button.

3. It is recommended that if you are creating a Standard hole then the units of the model should be in mm. This can be achieved by using the appropriate template at the time of creating a new file or by setting the system of units.

4. An application of the Sketched hole is a tapered hole. Sketched holes are always blind and have depth in one direction only.

5. You can select a convex or concave surface as the placement plane. To create a hole on such a surface, you need to create a point or an axis at the desired location on the surface and then select this point or axis as the placement of hole.

Tip

1. The diameter of the hole and its depth in the ***Side 1*** *direction can be specified without using the slide-down panels. The options to specify the diameter and depth in the* ***Side 1*** *direction are available on the* ***Hole*** *dashboard. If you want to specify the depth in the* ***Side 2*** *direction, open the slide-down panel by choosing the* ***Shape*** *tab from the* ***Hole*** *dashboard.*

2. You can access the hole termination options from the shortcut menu displayed on right clicking on the depth handle.

3. You can use the ***Clear*** *option from the shortcut menu to clear all references from the active collector.*

CREATING ROUNDS

In mechanical engineering, round is a feature which creates a radius or chord on one or more edges, an edge chain, or at the space between surfaces. The radius added to the inside corner is called fillet and to the outside corner is called round. In Creo Parametric, the **Round** tool is used to create a fillet or a smooth rounded transition between two adjacent faces, with a circular or a conic profile. This feature creation tool can be invoked from the **Engineering** group in the **Model** tab. Using the **Round** tool, you can add or remove material, depending on the edge selected. You can select the edges or faces and create a round in a single set or define more than one set. A round set is an individual unit containing one or more pieces of round geometry.

Figures 6-23 and 6-24 show some examples of rounds. From the figures, it is evident that the geometry of the rounds is tangent to the references selected. These references can be edges or surfaces. When you select multiple references, the round propagates across tangent neighbors until it encounters a break in tangency. However, if you use a One-by-One chain selection method, the round does not propagate across tangent neighbors.

Figure 6-23 *Rounds created on the edges*

Figure 6-24 *Round created on surfaces*

Note

The round created after you select references on the model has a circular cross-section and rolling shape. In Creo Parametric, there is more than one shape that a round can have.

Creating Basic Rounds

Ribbon: Model > Engineering > Round

Choose the **Round** tool from the **Engineering** group; the **Round** dashboard will be displayed, as shown in Figure 6-25. Also, you will be prompted to select an edge or a surface set to create the round. Select an edge; a preview of the round with a default radius value and drag handles will be displayed on the selected edge. Similarly, you can go on selecting edges to add rounds to them. But if you select a surface, you are prompted to select another surface to include in the same set. Use CTRL+left mouse button to select another surface; the preview of the round with a default radius value and drag handles will appear on the model, as shown in Figure 6-26. Note that because the radius of all rounds in a set is equal, the drag handles are available on only one edge. You can drag the handles to dynamically specify the radius of the round or specify the radius in the **Round** dashboard. Once all parameters of the round are specified, choose the **OK** button to create it. The various options in the **Round** dashboard are discussed next.

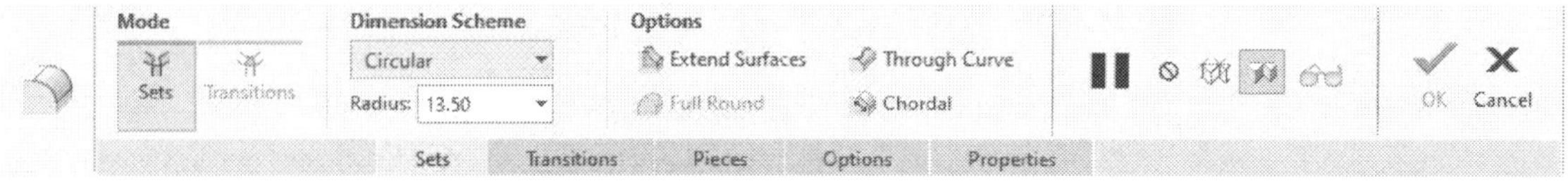

Figure 6-25 *Partial view of the* ***Round*** *dashboard*

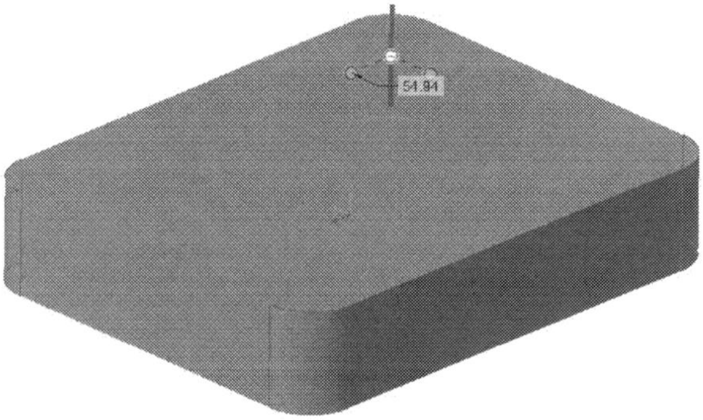

Figure 6-26 *Preview of the round feature along with the radius of round and the drag handles*

Sets

When you invoke the **Round** dashboard, the **Sets** button is chosen by default. This mode is used to create rounds by defining sets. When this button is chosen, the dimension box is displayed on its right and it is used to specify the radius of round.

Sets Tab

The options in the **Sets** tab are used to specify references and the geometry of the round. When you choose the **Sets** tab, a slide-down panel will be displayed, as shown in Figure 6-27. The options in this slide-down panel are discussed next.

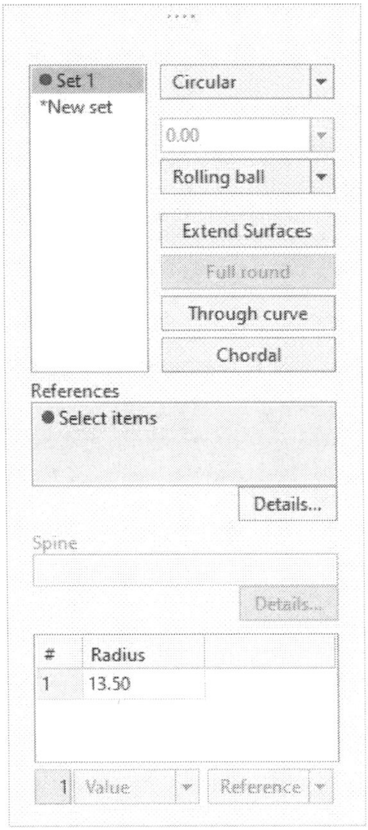

Figure 6-27 *The* ***Sets*** *slide-down panel*

First Drop-down List

The first drop-down, refer to Figure 6-28, that appears in the slide-down panel is used to specify the cross-section of the round you need to create. The options in this drop-down list are discussed next.

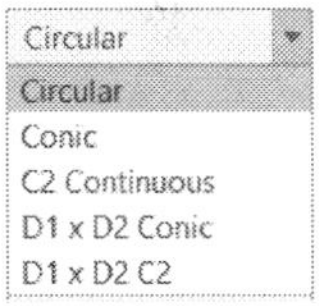

Figure 6-28 *First drop-down list*

Circular: When you choose this option from the drop-down list, the round that is created has a circular cross-section. This is the most widely used geometry type of rounds. The geometry is controlled by a radius value.

Conic: This type of round has geometry of a conic. The sharpness of the conic is controlled by the conical parameter. The value of the conic parameter lies between 0.05 and 0.95. The geometry of a conic round is controlled by two dimensions: the leg distance and the conic parameter. When you choose this option, the **Conic Parameter** edit box appears on the **Round** dashboard and also the edit box below the drop-down becomes active.

In Figure 6-29, the round on the model that is on the left is created using a circular cross-section and the round on the model that is on the right is created using a conic cross-section.

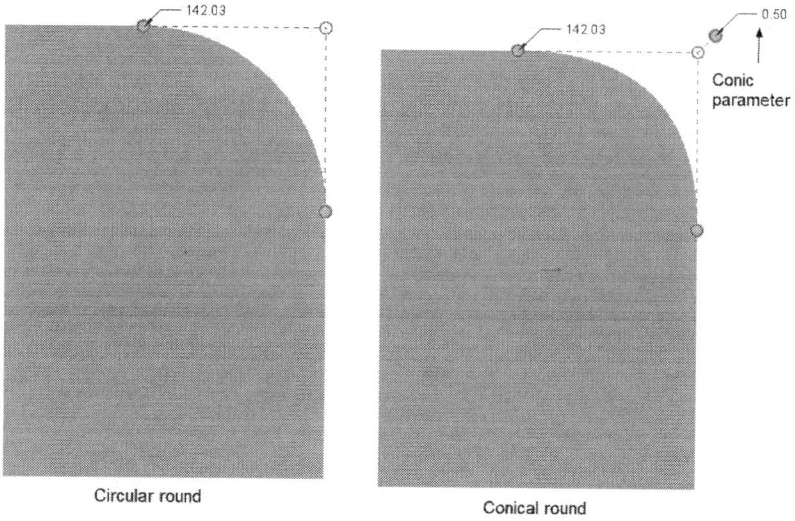

Figure 6-29 *Preview of the round feature along with the radius of round*

C2 Continuous: This type of round has a spline geometry. This type of round maintains the curvature continuity with adjacent surfaces. You can set the sharpness of the spline by specifying a value in the **C2 Shape Factor** edit box in the dashboard. The sharpness value varies between 0.05 - 0.95. The effect of the C2 shape factor on the shape of the splines is determined by the Conic Parameter of a conic cross-section. You can also set the value of the leg distance in the second edit box in the dashboard. Alternatively, you can drag the handles on the model to adjust the sharpness and leg distance.

D1 x D2 Conic: The round created using this option has a geometry that can be controlled by three dimensions. Figure 6-30 shows the round that is created using the **D1 x D2 Conic** option. This figure shows the two end legs distances and the middle dimension is the conic parameter. When you select this option to create a round, three edit boxes are displayed on the **Round** dashboard. The **Reverse side of conic distance** button is used to reverse the conic legs lengths. Figure 6-31 shows the geometry of the conic round and the dimensions of the two legs after choosing this button.

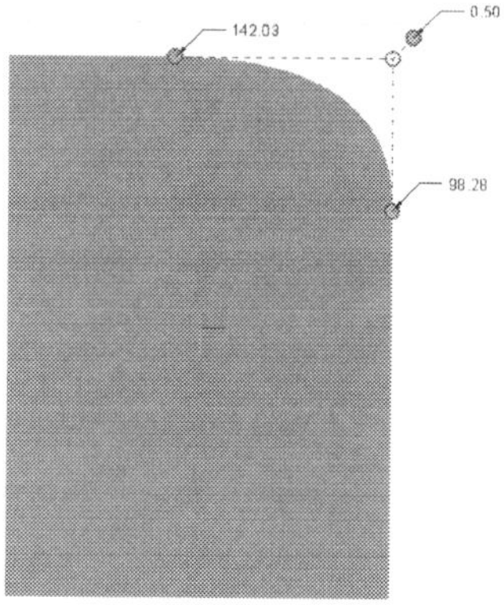

Figure 6-30 *The D1 x D2 Conic round*

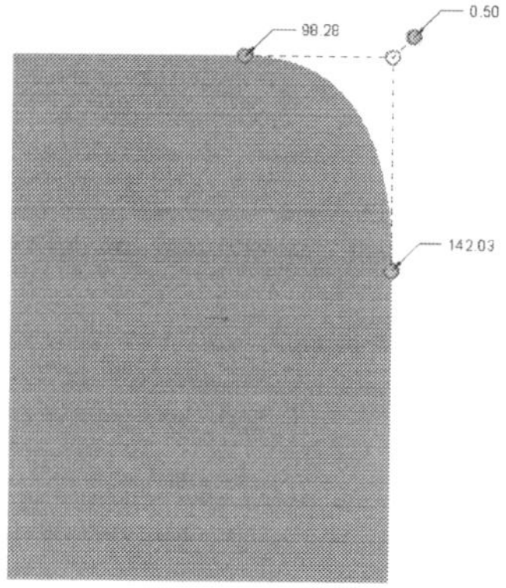

Figure 6-31 *The D1 x D2 Conic round after reversing the leg distances*

D1x D2 C2: This type of round has spline geometry and it maintains the curvature continuity with adjacent surfaces. In this case also, you can control the geometry by three dimensions as you did in D1 x D2 Conic.

Tip

*A conic uses equal leg distance on both sides. Length of each leg is automatically updated when any of the legs is modified. To toggle between the **D1 x D2 Conic** option and the **Conic** option, right-click on any of the drag handles and select the **Independent** check box from the shortcut menu.*

Note

The rounds and chamfers are defined in sets when there is a variation in their dimensions. For example, consider a situation, when you may want to create rounds on two sets of edges and the dimensions for both the rounds are different. In this case, you can define two sets of rounds; one set will define the rounds on one edge with some radius value and the second set will define the rounds on the other edge with a different radius value. The sets are also used in a similar way while creating chamfers.

Dimension Box

The edit box that is present below the first drop-down list is available only when you select the **Conic** or **D1 x D2 Conic** options. This is because this dimension box is used to enter the value for the sharpness of the conic round. The value lies between 0.05 and 0.95.

Second Drop-down List

The second drop-down list that is present in the slide-down panel is used to specify the method that Creo Parametric will use to create the round. The options in this drop-down list are **Rolling ball** and **Normal to spine**. These options are discussed next.

Rolling ball

When you select this option, the round is created by rolling a spherical ball along the surfaces to which it will possibly stay tangent. This is the most common type of round that you will create.

Normal to spine

This option creates a round by sweeping an arc or a conic cross-section normal to the selected spine curve where the spine curve is the edge you select for rounding. The Rolling ball method uses a standard round algorithm for creating rounds, and in some cases it fails to create round on the selected edge, refer to Figure 6-32. The **Normal to spine** method works well in situations where the direction of round changes quickly, refer to Figure 6-33. You can also use the Conic and D1 x D2 Conic profiles with the Normal to spine method.

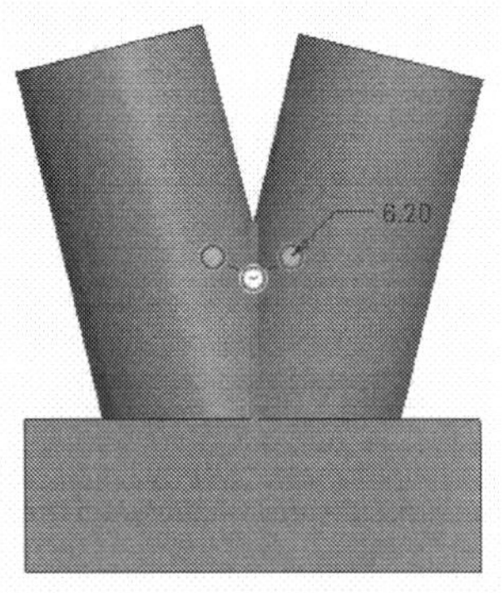

***Figure 6-32** Round feature failed when the **Rolling ball** option is selected*

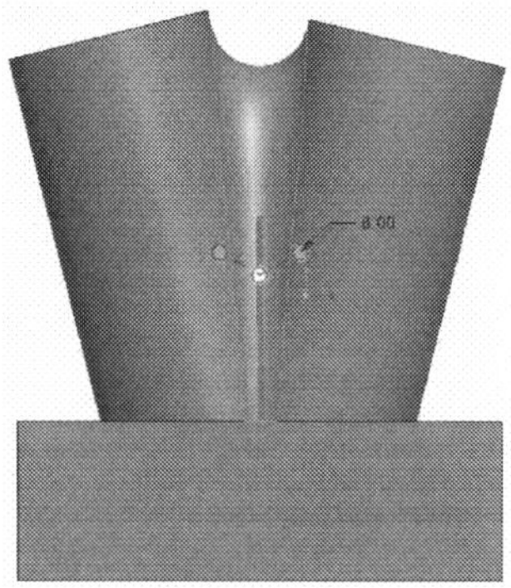

***Figure 6-33** Round feature created with Normal to spine method*

The area to the left of the drop-down lists in the slide-down panel lists the number of sets formed for creating rounds. This means you can define more than one set to create the rounds. Each set may have different values and references. To add a set, right-click in this area and choose the **Add** option from the shortcut menu or you can just click on the **New set**. The set selected in this area is the active set. It is recommended that you fully define one set of round and then define the other set.

Full Round

The **Full Round** button is used to create a complete round between two selected edges or two planar or non-planar surfaces. This button will be available only after selecting two references and will not be available when you choose the **Conic**, **C2 Continuous**, **D1 x D2 Conic**, and **D1 x D2 C2** options or the **Normal to spine** option.

The following steps explain the procedure to create a full round by selecting two edges.

1. Invoke the **Round** dashboard.

2. Select the first edge and then use CTRL+left mouse button to select the second edge, refer to Figure 6-34. Note that the edges selected should have a surface between them that can be converted into a round.

3. Make sure the **Circular** and **Rolling ball** options are selected from the first and third drop-down lists, respectively.

4. Choose the **Full Round** button in the **Sets** tab from the **Round** dashboard; a preview of the full round will be displayed, as shown in Figure 6-35. Choose the **OK** button from the dashboard to create the round.

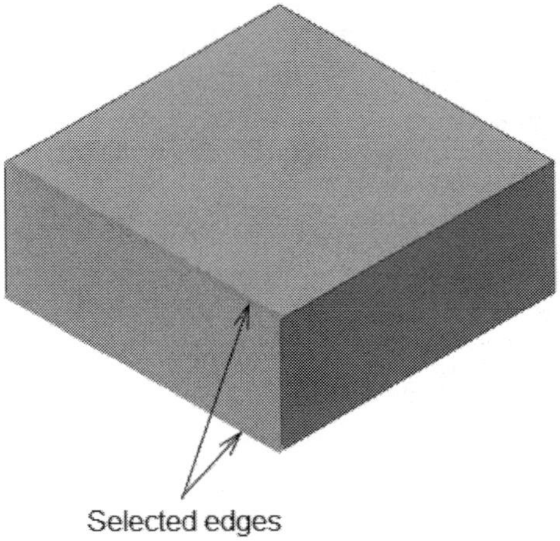

Figure 6-34 Edges selected in a set

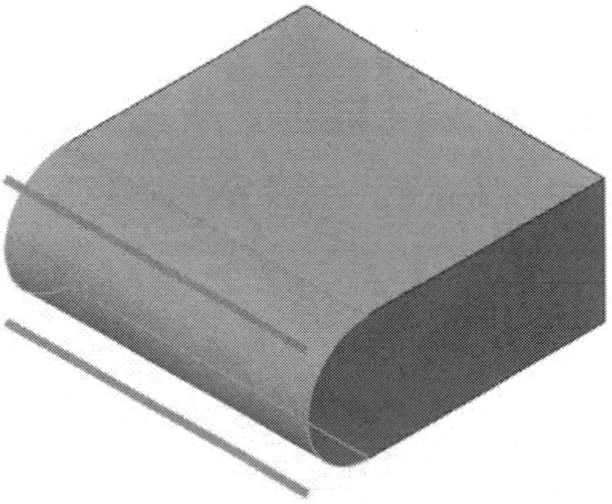

Figure 6-35 Preview of the full round feature

The following steps explain the procedure to create a full round by selecting the two surfaces.

1. Invoke the **Round** dashboard.

2. Select the first surface and then use CTRL+left mouse button to select the second surface, refer to Figure 6-36. Note that the surfaces selected should have a surface between them that can be converted into a round, for example, the top and bottom faces of a rectangular block.

3. If the **Full round** button is not chosen automatically then choose it from the **Sets** tab of the **Round** dashboard, or from the shortcut menu. On doing so, you will be prompted to select a surface to replace it with a surface-to-surface full round. Select the front face of the model using the **Driving surface** collector of the **Set** tab; a preview of the full round will be displayed, as shown in Figure 6-37. Choose the **OK** button from the dashboard to create to complete the procedure.

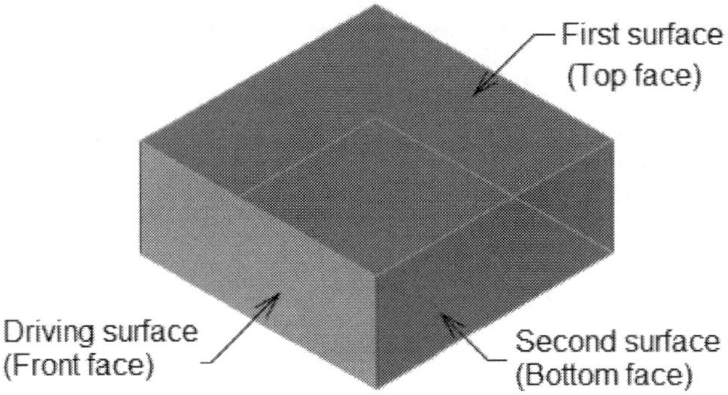

Figure 6-36 Surfaces selected for creating a full round

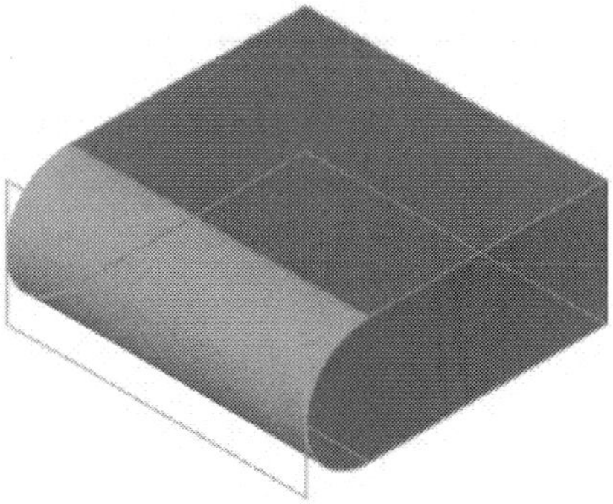

Figure 6-37 Preview of the full round

Extend Surfaces

The **Extend Surfaces** option allows you to create round feature by specifying a radius larger than the width of the adjacent surface of the selected edge. You can create an extended surface round only on edges. You can select closed loop edges or open loop edges to create an extended round feature. When you create an extended round, the system determines the width of the neighboring surfaces to be extended or trimmed to close any gaps and attaches the round to the model. Figures 6-38 and 6-39 show the round created with and without the **Extend Surfaces** option used. You will notice that the neighboring surfaces of the round are extended to fill the gaps formed between surfaces.

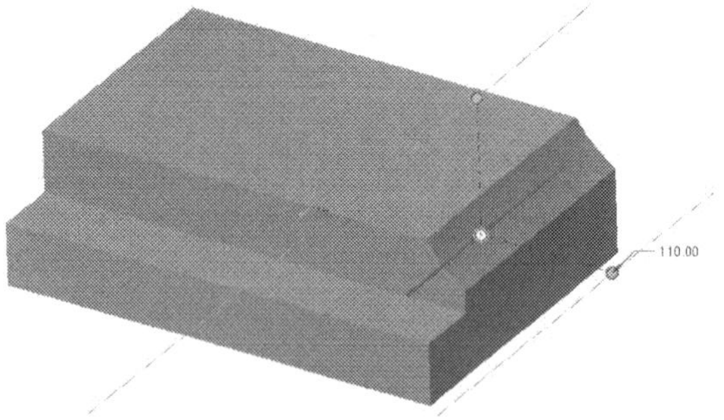

Figure 6-38 *Preview of the round with the* ***Extend Surfaces*** *option selected*

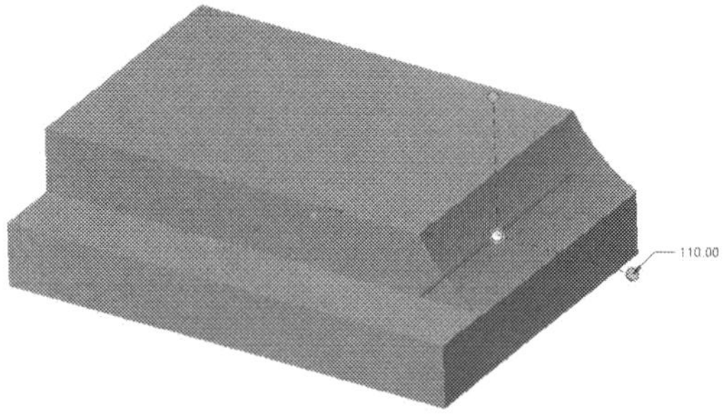

Figure 6-39 *Preview of the round without the* ***Extend Surfaces*** *option selected*

Through Curve

The **Through curve** button is used to create a round whose radius is defined by an existing curve. This button is available with the **Circular**, **Conic**, or **C2 Continuous** options only. To create a round using the **Through curve** option, select an edge or, a set of two surfaces and then choose the **Through curve** option in the **Sets** tab of the **Round** dashboard. On choosing this button, the **Driving curve** collector is activated and you are prompted to select the chain of tangent curves to create the round through curve. Select an existing curve in the **Driving curve** collector; preview of the round will be displayed in the drawing area, as shown in Figure 6-40.

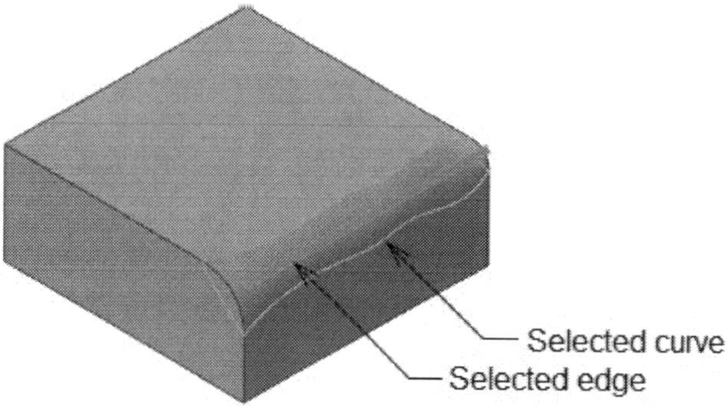

Figure 6-40 *Preview of the round created using the* ***Through curve*** *option*

Chordal

The **Chordal** button is used to create a round with constant chordal length. The advantage to use chordal round over a constant radius round is that the tangent edges of the resulting chordal round feature appear parallel which is not always a case in rounds created using a radius, refer to Figure 6-41. Notice that the edges of the chordal round appear parallel at all points on the round whereas the edges of the round created using a radius do not appear parallel.

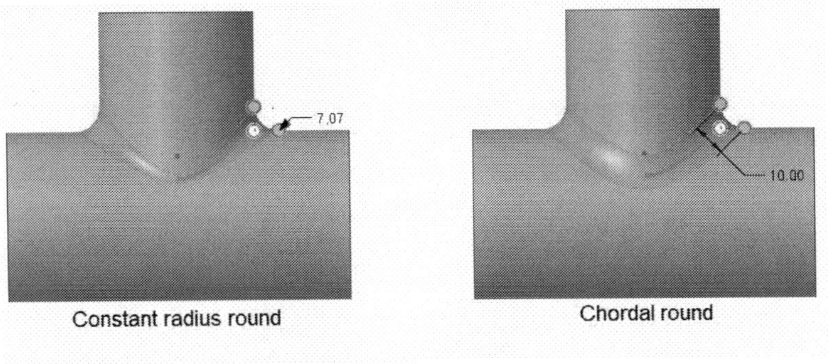

Figure 6-41 *Example of constant radius round and chordal round*

References Collector

The **References** collector displays the references selected on the model. These references can be deleted by selecting them in the collector and then holding down the right mouse button to display the shortcut menu. Choose the **Remove** or **Remove All** option from this shortcut menu as per the requirement.

Spine Collector

The **Spine** collector displays the spine selected on the model. This collector will get active only when you select the **Normal to spine** option from the second drop-down list. The **Spine** collector gets modified to the **Driving surface** collector when you select two surfaces and then choose the **Full Round** button. The **Spine** collector is modified to the **Driving Curve** collector when you choose the **Through curve** button.

The area under the **Spine** collector is used to define the dimension values and references for driving dimensions of the rounds.

Tip

1. You can create a round on an edge without invoking the ***Round*** *tool. To do so, select an edge and press the* ***R*** *key from the keyboard to invoke the* ***Round*** *dashboard*

2. The need for creating rounds by defining more than one set arises when you want to create rounds that have different radii. The rounds created by using this method will appear as a single feature in the ***Model Tree****.*

Transitions Tab

Transitions connect overlapping or discontinuous round pieces. When you choose the **Transitions** tab, a slide-down panel will be displayed, as shown in Figure 6-42. This slide-down panel displays list of transitions if present in the given round. The **Transitions** list will display the transitions only when the **Transitions** button is chosen in the **Round** dashboard. The transition types available in the Transition type drop-down list will be according to the geometry selected.

Note that some transitions created are context-sensitive and they will not appear under the **Transitions list**. The use of transitions with round feature will be discussed later in this chapter.

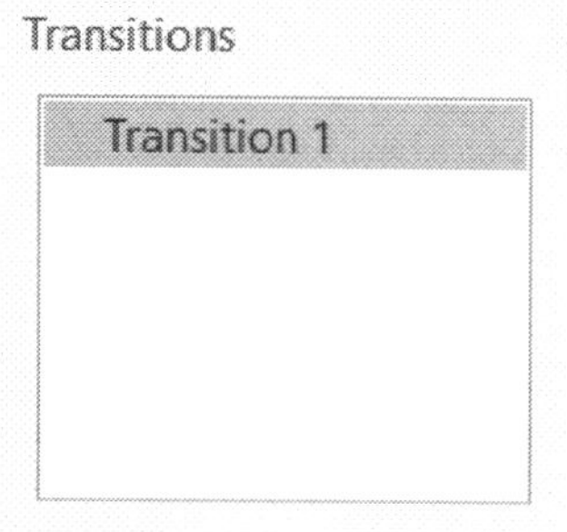

***Figure 6-42** The **Transitions** slide-down panel*

Pieces Tab

When you choose the **Pieces** tab, a slide-down panel will be displayed. It provides possible locations of the round in the form of pieces. Management of round pieces becomes essential when you work with advanced part modelling. By using the options available in this tab, you can manage the pieces of round features in a better way. For example, consider a case where you select two faces to fillet. One of the faces has a cut feature such that it has two disjointed surfaces, as shown in Figure 6-43. When you move the cursor over any piece under the **Pieces** area, that piece is highlighted on the model. The options available in this tab are useful in resolving ambiguity which arises if multiple placement locations for rounds with the current selection of the references exist. This happens if two surfaces intersect each other at more than one place. From the pieces listed in the **Pieces** area, you can select the pieces to exclude by selecting the **Excluded** option from the drop-down list that will be displayed when you click on the **Included** field, as shown in Figure 6-44.

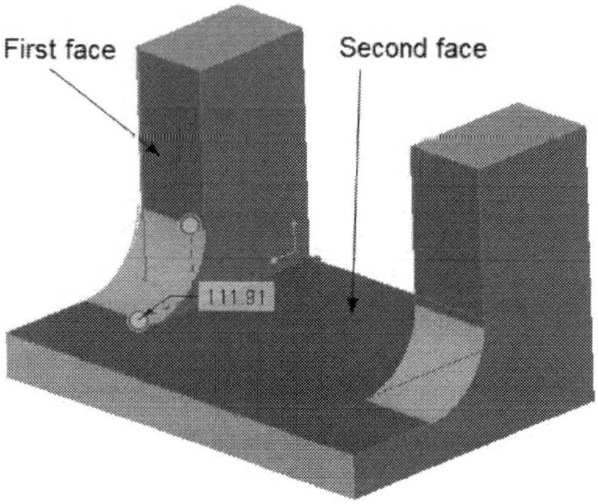

***Figure 6-43** Faces selected for creating a round*

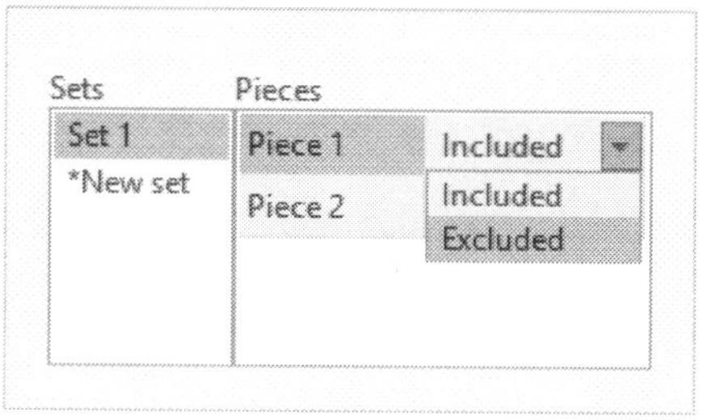

***Figure 6-44** The **Pieces** slide-down panel*

As you create a round, the round geometry automatically propagates across tangent points and then stops at a non-tangent point. With the help of this tab, you can trim or extend the round pieces at tangent and non-tangent points.

Refer to Figure 6-45, the round feature on the selected edge propagates along the edge because of the existing round geometry. The round feature on the selected edge consists of three patches. If you need to terminate the round geometry before the non-tangent point, you can trim the round piece up to a patch that you select. To do so, drag the Trim/Extend handle up to a desired point. As you drag the Trim/Extend handle, preview of the round geometry is displayed in the drawing area, as shown in Figure 6-46. Notice that the piece of round is displayed as **Edited** in the **Pieces** area, as shown in Figure 6-47. To quickly restore the round geometry, choose the **Included** option from the drop-down list of the respective round piece.

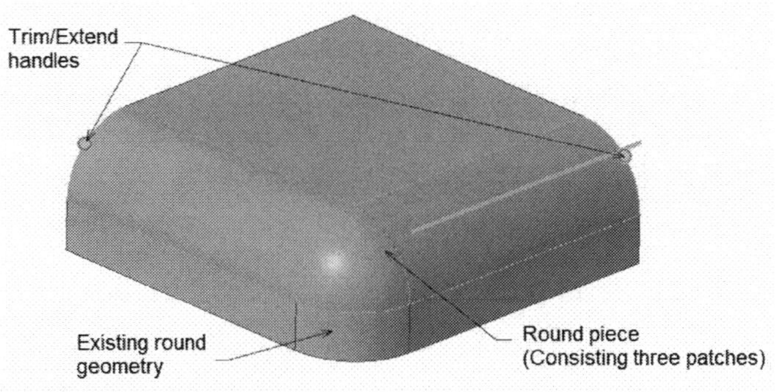

***Figure 6-45** Round feature with Trim/Extend drag handles*

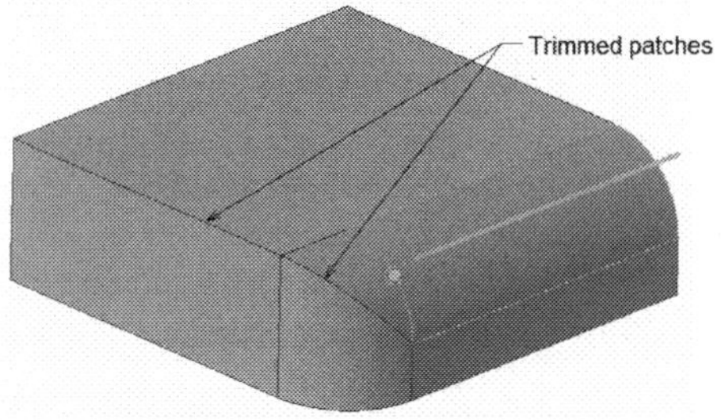

***Figure 6-46** Preview of the round feature with trimmed patches*

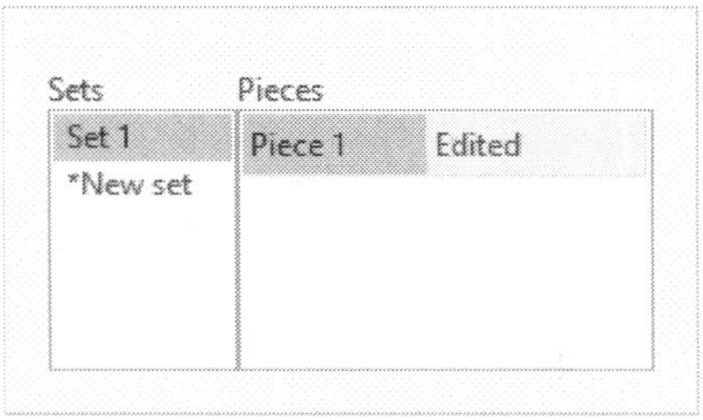

***Figure 6-47** Round piece displayed as edited in the **Pieces** area*

Options Tab

When you choose the **Options** tab, a slide-down panel will be displayed that has two radio buttons. These radio buttons allow you to make the round a surface or a solid. When you select the **Surface** radio button, the **Create end surfaces** check box will be enabled, refer to Figure 6-48. This check box is used to cap all round ends of the model by surfaces.

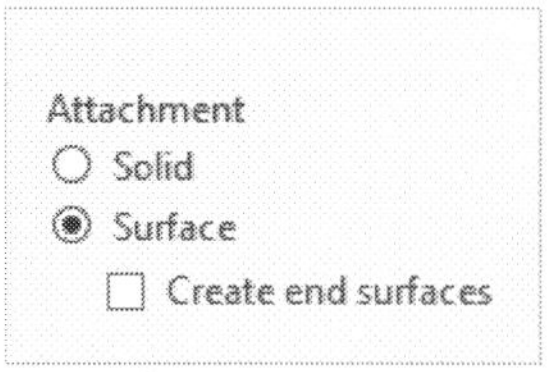

***Figure 6-48** The **Options** tab*

Properties Tab

When you choose this option, the slide-down panel will be displayed. This slide-down panel displays the feature identity. Click on the **i** button to read the information about the feature. Click on the cross on the top-right of the **Feature info** window to close it. You can also rename the feature.

Creating a Variable Radius Round

A variable radius round is the one in which the radius of the round varies through the length of the edge. For example, refer to Figure 6-49, which shows a constant radius round and Figure 6-50, which shows a variable radius round. To create a variable radius round, choose the **Round** tool and select the edge. When the preview of the round appears on the edge, hold down the right mouse button anywhere in the drawing window to display a shortcut menu. Choose the **Make variable** option from this shortcut menu. Now, in the slide-down panel that will be displayed when you choose the **Sets** tab, the two default values of the radius are available, which can be modified as per requirement. You can also drag the handles displayed at both ends of the selected edge to dynamically modify the radius of round.

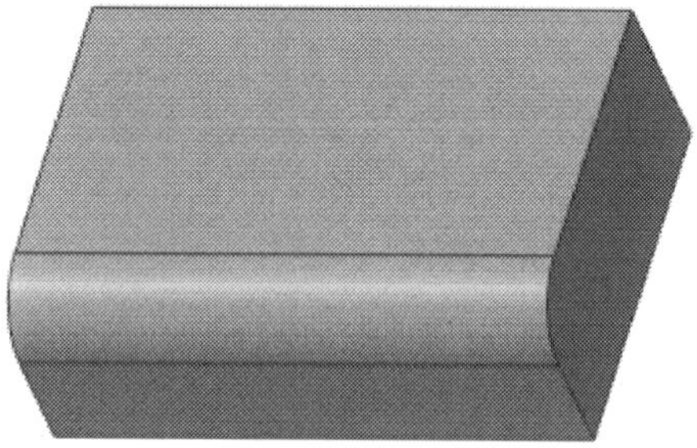

Figure 6-49 *Constant radius round*

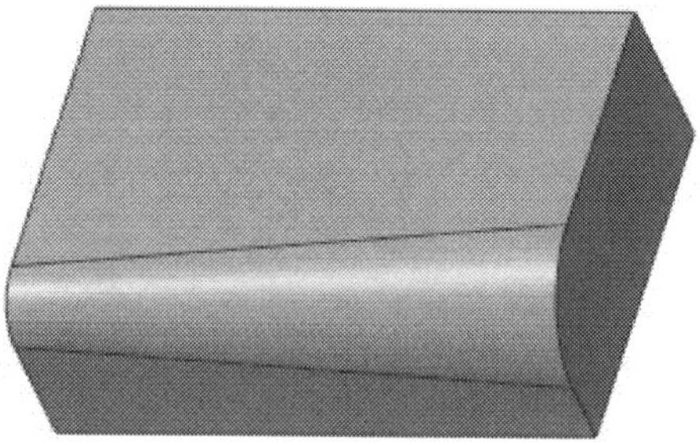

Figure 6-50 *Variable radius round*

Note

You can also create a variable radius round by selecting two surfaces. In this case, the edge that is common to the two surfaces is converted into a round.

You can also add additional points between the two end points and define a variable radius in these additional points. To define additional points, choose the **Sets** tab to invoke the slide-down panel. Bring the cursor to the area in the slide-down panel where the radius value and the location of the points are displayed and right-click to invoke the shortcut menu. Choose the **Add radius** option; one additional point appears in the same area of the slide-down panel and also on the selected edge. This point by default is placed on the selected edge by a ratio value with respect to the end point. You can select the **Reference** option from the drop-down list in this area and select a location on the edge to place the point. The location of the point can also be modified by dragging the circular handle displayed on the edge.

Note

Rounds can be created on one edge, chain of edges, between two surfaces, an edge and a surface, and between two edges.

Transitions

The **Transitions** button will be available only if at least one reference for the round is defined. This button has primarily two applications. The first one is to connect overlapping or discontinuous round pieces using various transition methods and the other is to define a limit up to which the round will be created.

When you choose the **Transitions** button, the transitions of the round feature are highlighted on the model. Select the piece to activate the Transition type drop-down list options in the

Transition Settings area. The Transition type drop-down is displayed in the **Round** dashboard next to the **Transitions** button. When you select a transition on the model, the default transition for the selected round feature is displayed in the Transition type drop-down list, as shown in Figure 6-51. You can select default transition and any other valid transition available from the drop-down list.

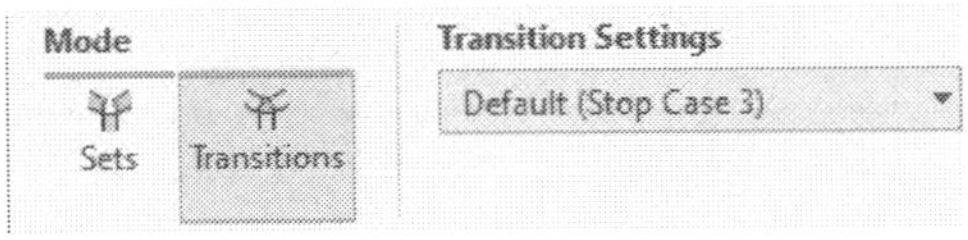

***Figure 6-51** The Transition type drop-down list*

As discussed earlier, different types of transitions are created according to the geometry selection. By default, Creo Parametric determines the transition type that is best fit for the geometrical context. Note that not all transitions types listed are available for a given context. These transitions types are discussed next.

Stop Cases

This type of transition terminates the round using one of the three different stop cases. In each stop case, Creo Parametric configures the geometry based on the geometrical context.

For example, consider a case where you want to create a round feature on the selected edge, as shown in Figure 6-52. When you choose the **Transitions** button in the **Round** dashboard, the transitions of the round feature will be highlighted on the model. Select the transition you want to modify, refer to Figure 6-53; the **Default (Stop Case 3)** will be displayed in the Transition type drop-down list. Select **Stop Case 1** or **Stop Case 2** as required to change the transition of the round feature, as shown in Figure 6-54.

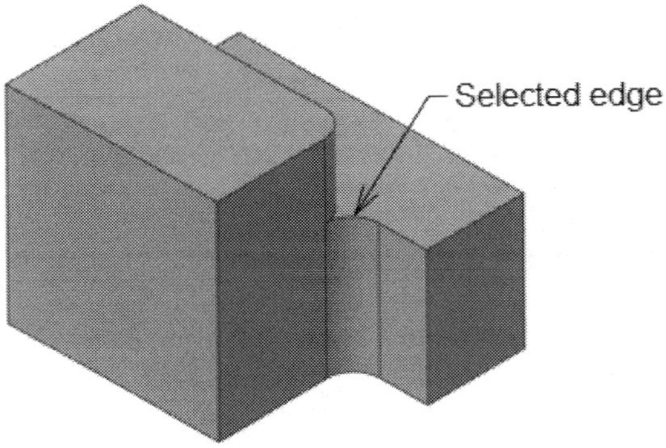

***Figure 6-52** Edge selected to create the round feature*

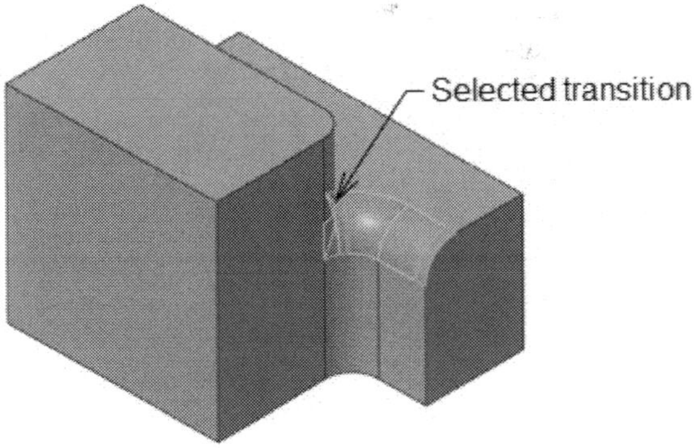

***Figure 6-53** Transition of the round feature selected*

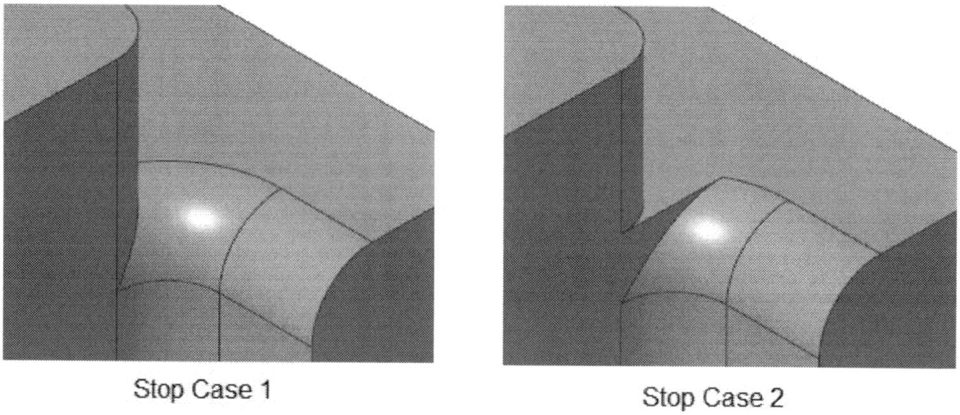

Figure 6-54 *Round created using Stop Case transition*

Stop at Reference

Select this transition type to terminate the round geometry at the specified datum point or datum plane. After selecting the edge for creating the round, choose the **Transitions** button to display the two transitions at both the ends. Select and activate any one of them, refer to Figure 6-55. Now, select the **Stop at Reference** option from the Transition type drop-down list available in the **Round** dashboard. Select a plane or a datum point that will act as the stopping reference; the selected reference will be displayed in the **Stop reference** collector in the **Round** dashboard. Note that the edge on the side in which the transition is active will be retained and round will be created on the other side of the stopping reference, refer to Figure 6-56.

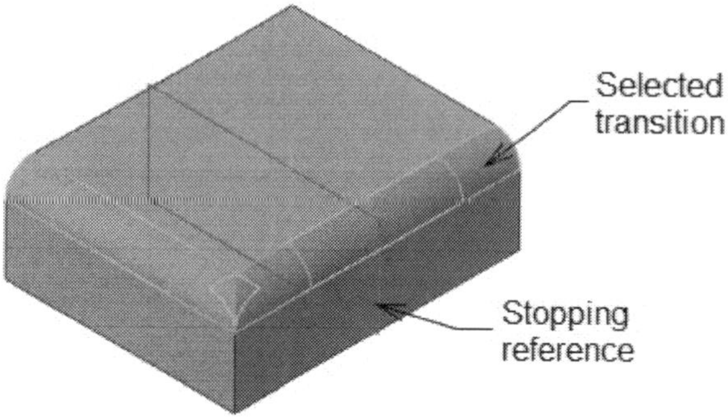

Figure 6-55 *Transition selected to stop at reference*

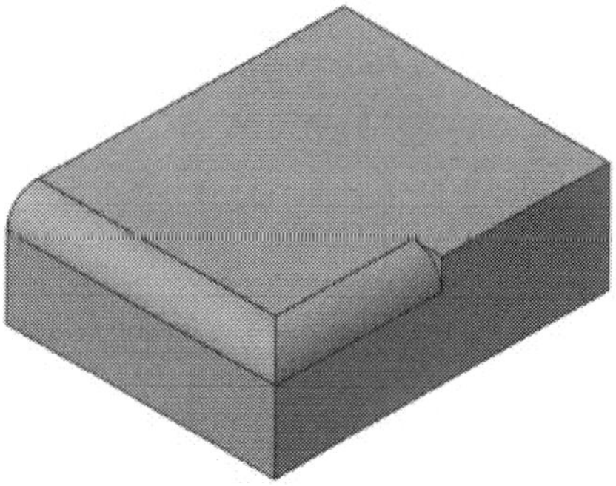

Figure 6-56 *Resulting round feature*

Blend

Use this transition type to blend a round set with its adjacent round set. Consider an example shown in Figure 6-57. By default the system applies the Intersect transition on the corner and it is displayed as **Default (Intersect)** in the Transition type drop-down list of the **Round** tab. The tangent round geometry will terminate at sharp corners while using this type of transition method. To change the transition type, choose the **Blend** option from the Transition type drop-down list; the resulting round will be created as shown in Figure 6-58.

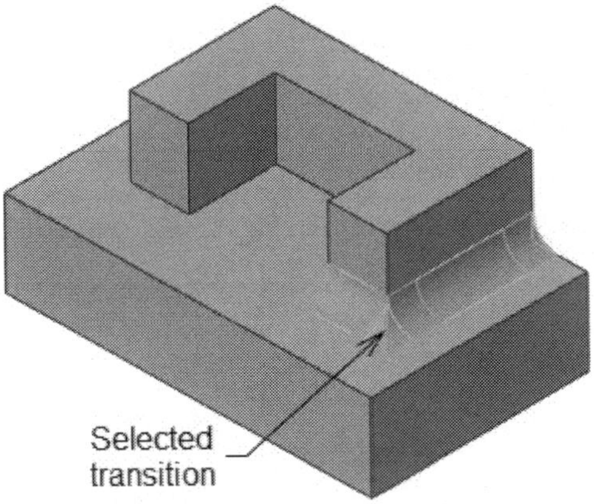

Figure 6-57 *Default transition between round sets*

Figure 6-58 *Resulting round feature with Blend transition*

You can also use the blend transition in case of two collinear round pieces. As shown in Figure 6-59, the system by default applies the Continue transition on the edge. When you select the **Blend** option from the drop-down list, both pieces of the round are blended together in order to maintain the shape of the round geometry, as shown in Figure 6-60.

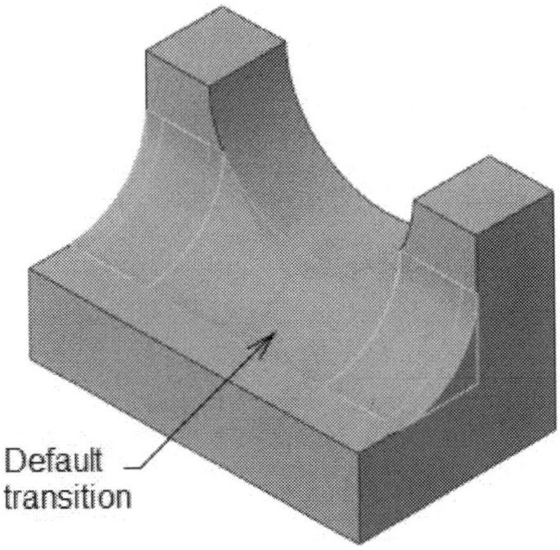

Figure 6-59 *Preview of the round with default transition*

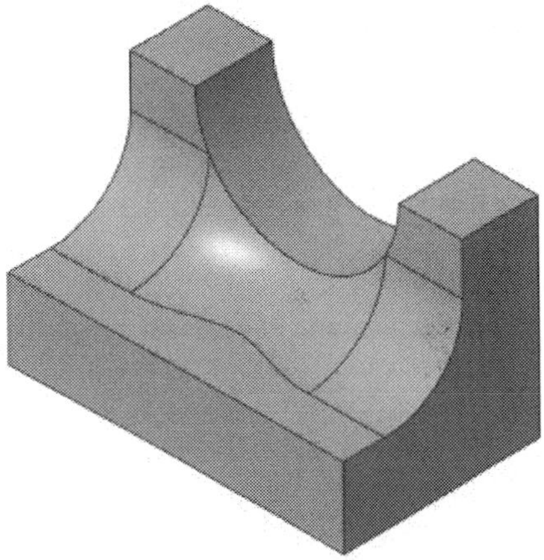

Figure 6-60 *Resulting round after selecting the Blend transition*

Continue

This type of transition extends the round geometry into two round pieces. The tangent round geometry does not stop at sharp edges, unlike the Blend transition. The resulting geometry looks as if the round was placed first, and then the geometry was cut away after creating the round feature, refer to Figure 6-59. The neighboring surfaces are extended to meet round geometry where applicable.

Corner Sphere

Corner Sphere transition applies only where three round pieces overlap at a corner. This type of transition rounds the corner transition formed by three overlapping round pieces with a spherical corner. By default, the sphere has the same radius as the largest overlapping round piece. However, you can modify the radius of the sphere as well as the transition distance along each edge. Refer to Figure 6-61, when you create a round feature on the selected edges, the Round Only type of transition is automatically applied to the corner. As you select the **Corner Sphere** option from the Transition type drop-down, the radius of the sphere is displayed in the model and the **R** edit box is displayed in the **Round** dashboard, refer to Figure 6-62. Note that the new radius value must be greater than the radius of the largest overlapping round piece. As you modify the value of radius; the transition length edit boxes are displayed in the **Round** dashboard. To modify the transition length along each edge, you can specify a value in the edit box or drag the length handle displayed on the model, refer to Figure 6-63.

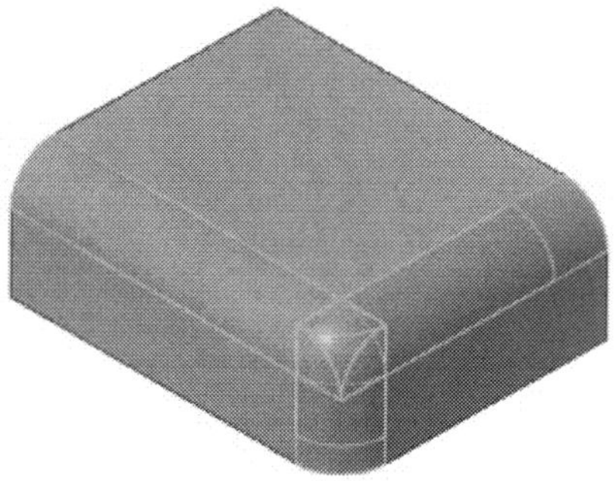

Figure 6-61 *Round feature created using default transition*

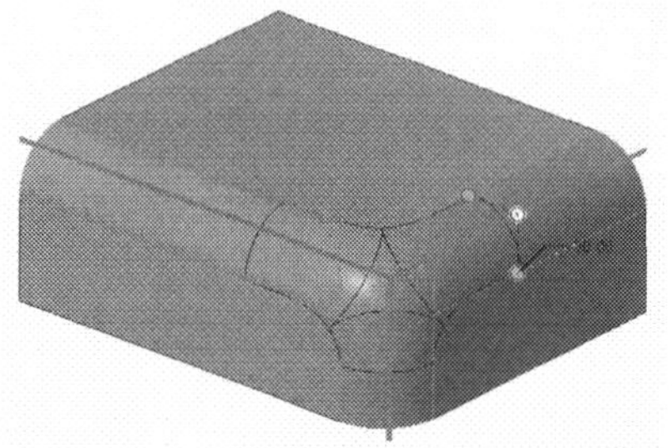

Figure 6-62 *Resulting round feature created using Corner Sphere transition*

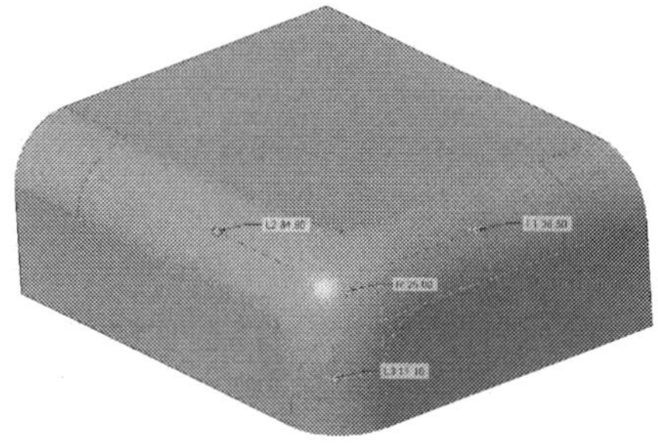

Figure 6-63 *Preview of the round feature with transition length drag handles*

Note

You need to select a piece to activate the options of the Transition type drop-down list in the ***Settings*** *area.*

Intersect

Intersect transitions are only applied on two or more overlapping round pieces. Here the overlapping round pieces get extended toward each other until they merge and form a sharp boundary at their intersection, refer to Figure 6-64. While creating a round feature using Intersect transition, you can control the length of the patches that are created to fill the gaps between surfaces. To do so, select the **Intersect at Surface - side 1** or **Intersect at Surface - side 2** option from the Transition type drop-down. Notice that the **L** edit box is displayed in the **Round** dashboard, and a drag handle is displayed on the model when you select the surface of intersection from the transitions tab, refer to Figure 6-65. You can change the length of the surface patch by dragging the handle or by specifying the length in the **L** edit box.

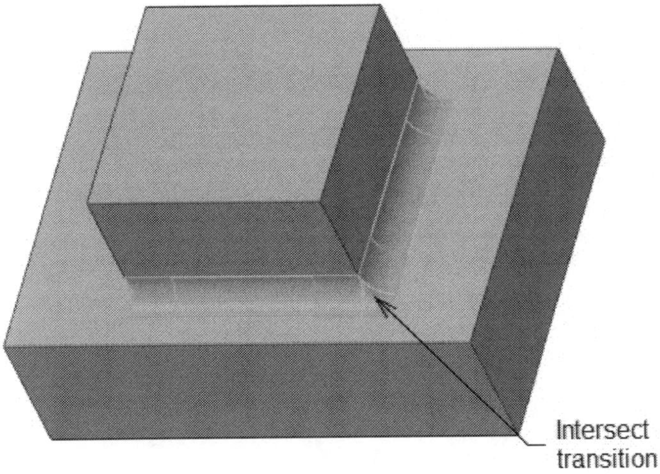

Figure 6-64 *Preview of the round feature using Intersect transition*

When you select the **Intersect at Surface - side 1** or **Intersect at Surface - side 2** option from the Transition type drop-down list, the **One side** radio button gets selected in the **Transition Span** area of the **Transitions** tab. As a result, you can control the length of the patch only in one side of the transition. If you select the **Two sides** radio button from the **Transition Span** area; the **Conic Parameter**, Leg distance, and Patch length edit boxes will be displayed in the **Round** dashboard. Also, respective drag handles will be displayed on the model, as shown in Figure 6-66.

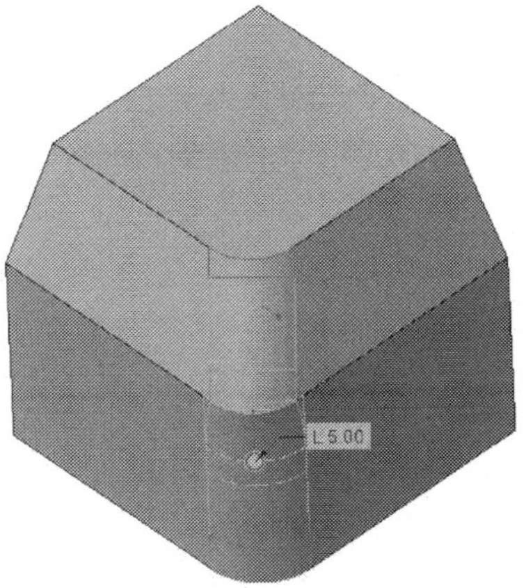

Figure 6-65 *Preview of the round feature with the* ***One side*** *radio button selected*

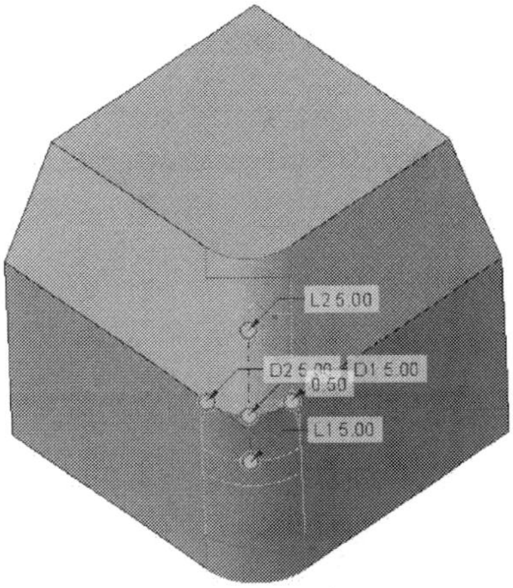

Figure 6-66 *Preview of the round feature with the* ***Two side*** *radio button selected*

Patch

This type of transition can be applied to a geometry where three or four round pieces overlap at a corner. The geometry between such converging round sets is created as a patch. When you select the **Patch** option from the Transition type drop-down list, the **Optional surface** collector is displayed in the **Round** dashboard. Click in the collector to activate it and then select a surface on which you want to place a fillet for patch transition, refer to Figure 6-67.

As you select a surface in the **Optional surface** collector; the **Radius** edit box is displayed in the **Round** dashboard. You can specify the value of radius in the **Radius** edit box or drag the handle on the model to modify the value of transition radius.

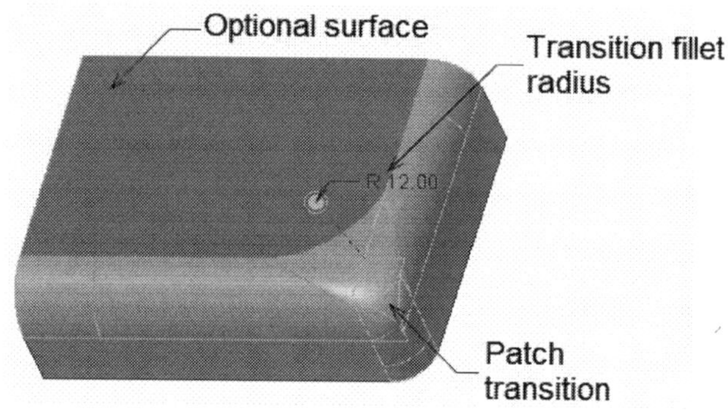

***Figure 6-67** Preview of the round feature with the **Patch** option selected*

Round Only

Round only transition is applied when each round piece has a different radius value. This type of transition is created by sweeping the smallest radius round set along the edge of the largest radius round set. The transition thus created will be a compound round geometry.

In geometrical context, the Round Only transitions are of two types: Round Only 1 and Round Only 2, refer to Figure 6-68. These transitions can be used to achieve round shape according to the convexity of the round pieces.

Tip

*1. You can switch to Transition mode by choosing the **Transitions** button from the **Round** dashboard.*

2. In Transition mode, you switch between transition types by clicking on a transition and selecting another transition from the shortcut menu.

*3. You can delete a transition by clicking on a transition and selecting the **Delete transition** option from the shortcut menu. To add a transition, hold down the CTRL key and select all the round piece ends freed by the deleted transitions. Next, select the **Make transition** option from the shortcut menu; the newly created transition will be listed under the **Transitions** area of the **Transitions** tab.*

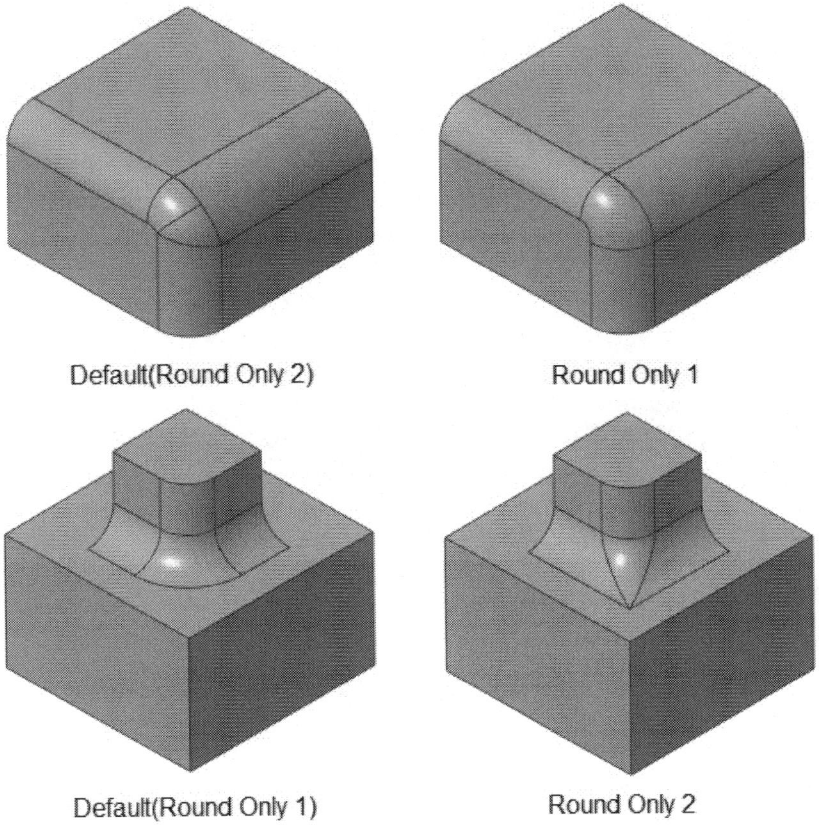

Figure 6-68 Round feature created using the Round Only transition

Points to Remember While Creating Rounds

The following points should be remembered while creating rounds:

1. To avoid conflict due to the parent-child relationship, other features should not be referenced to the edges created by rounds or to the round surface.
2. Preferably, rounds should be added at the end of model creation. This means, the rounds should be the last features in any solid model.
3. The regeneration time of the model increases if the number of round features in it is increased.
4. Use the shortcut menus extensively while creating rounds to reduce the modeling time.

Creating Auto Rounds

Ribbon: Model > Engineering > Round drop-down > Auto Round

You can create rounds on all the edges of the model by using the **Auto Round** tool in a single step. To do so, choose the **Auto Round** tool from the **Round** drop-down of the **Engineering** group; the **Auto Round** dashboard will be displayed, as shown in Figure 6-69.

There are two check boxes available in the dashboard. By default, these check boxes are selected. The **Convex** button is used to create round on the outer edges of the model. The **Concave** button is used to create round on the inner edges of the model, if the edges are compatible with the radius value. The tabs available in this dashboard are discussed next.

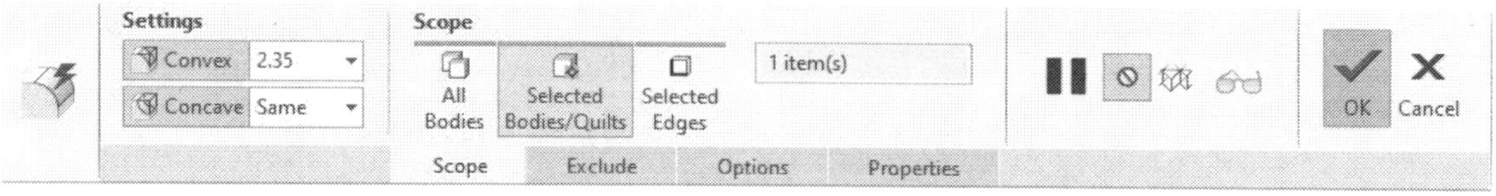

*Figure 6-69 Partial view of the **Auto Round** dashboard*

Scope Tab

When you choose the **Scope** tab, a slide-down panel will be displayed, as shown in Figure 6-70. This slide-down panel displays three radio buttons and two check boxes. These options are discussed next.

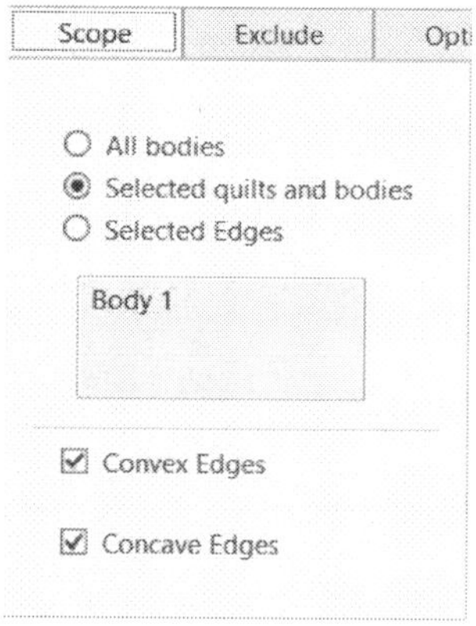

*Figure 6-70 The **Scope** slide-down panel*

All bodies

This radio button is used to create rounds in all the bodies. The solid models have property to contain weight.

Selected quilts and bodies

This radio button is used to create rounds in the selected bodies and surfaces models. The surface models do not have property to contain weight.

Selected Edges

This radio button is used to create rounds on the selected edges of the model.

Convex Edges

This check box is used to include or exclude the outer edges of the model. Select this check box to include the outside edges.

Concave Edges

This check box is used to include or exclude the inner edges of the model. Select this check box to include the inside edges.

Exclude Tab

When you choose the **Exclude** tab, a slide-down panel will be displayed. It is used to exclude those edges that you do not want to make round. By holding the **CTRL** key, select the edges you want to exclude; the selected edges will be displayed in the **Excluded Edges** area of the slide-down panel. The **Geom Checks** option will be available only when the **Auto Round** feature fails to round some of the edges. On selecting this option, the **Troubleshooter** dialog box appears displaying the reason why the edge or edge chain could not be rounded.

Options Tab

When you choose the **Options** tab, a slide-down panel will be displayed, as shown in Figure 6-71. The **Create Group of Regular Round Feature** check box is used to create a group of individual round features from the **Auto Round** dashboard. These rounds will be displayed as a grouped feature in the **Model Tree**. The **Make Round Features Dimensions Dependent** check box is used to make the dimensions of each individual round feature in a group dependent on each other. Changing any one the dimension will update the dimension of all the other round features of that round group. Select the **Preserve A-R Feature Relations and Parameters** check box to retain the attributes of the Auto Round feature such as parameters and relations in the individual round features of a round group.

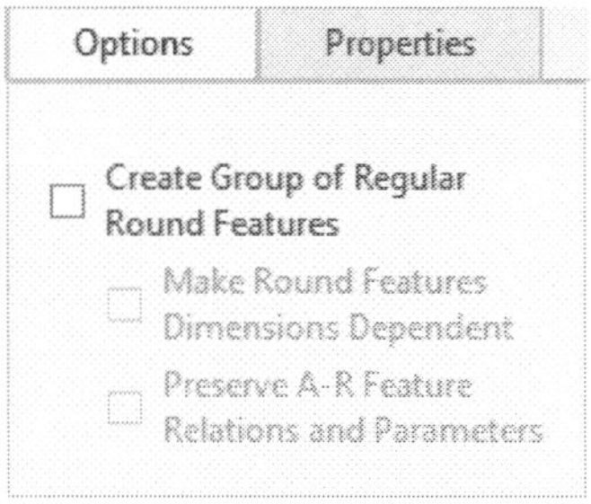

*Figure 6-71 The **Options** slide-down panel*

Tip

Rounds and chamfers are used in components to reduce the stress concentration at the sharp corners and edges. Therefore, they reduce the chances of failure of a component under a specified loading condition.

Creating Rounds Using Intent Edges

A very powerful feature of Creo Parametric is the creation of Intent objects. Intent objects are automatically created based on the model geometry. They are families of associated points, curves, edges or surfaces that logically define boundaries of geometry created or modified by a feature. These objects are regenerated for even small changes such as modifications in the sketch.

Use of intent edges or surfaces makes the selection of references quicker. You can place a round feature in a model by selecting intent edges or intent surfaces. Such a method of creating rounds is more appropriate because it prevents failing of the feature when changes are made in the model. It is because references for the rounds are tied to the features in the design model, not to the individual edge references.

To create a feature using Intent edge method, choose the **Round** tool from the **Engineering** group; the **Round** dashboard will be displayed. Next, choose the **Intent Chain** option from the Filter drop-down list. Now, hover the cursor over the edge; intent edges are highlighted, as shown in Figure 6-72. Next, click on the highlighted edges; the entities are selected and displayed in thick green color. Choose the **OK** button from the **Round** dashboard; the round is created on the selected edges, as shown in Figure 6-73.

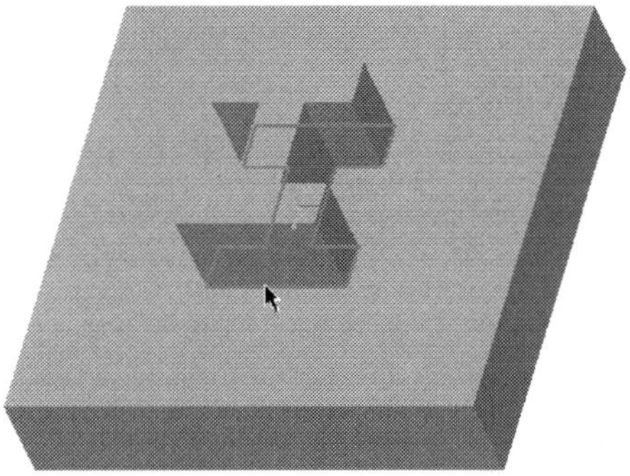

Figure 6-72 Intent edges of the model

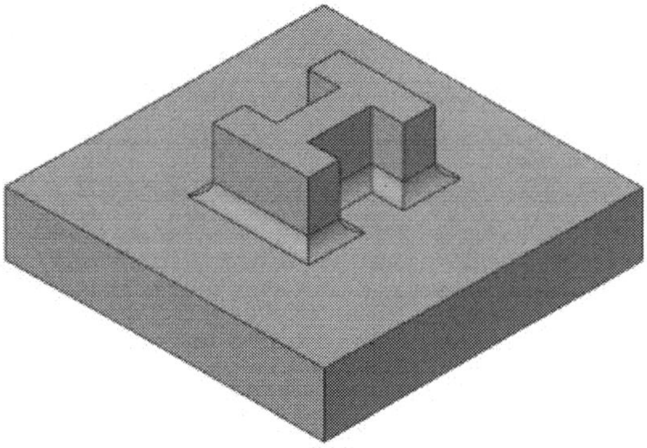

Figure 6-73 Resulting round feature

When you change the sketch for second extrude feature, the round feature preserves its attributes, refer to Figure 6-74.

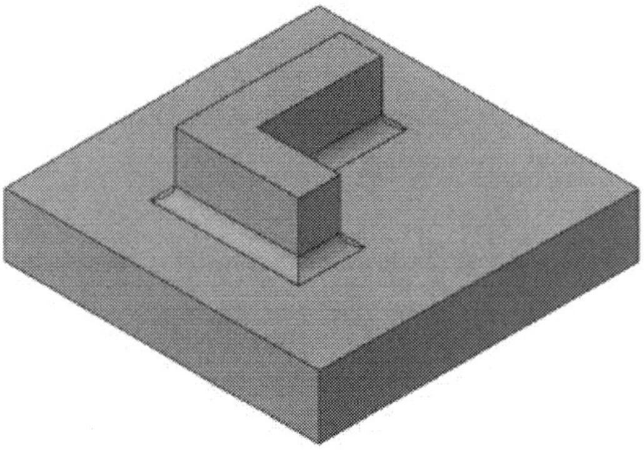

Figure 6-74 Resulting round feature after changing the sketch of the extruded feature

Another example shown in Figure 6-75 illustrates that references are bound to features and not edge references.

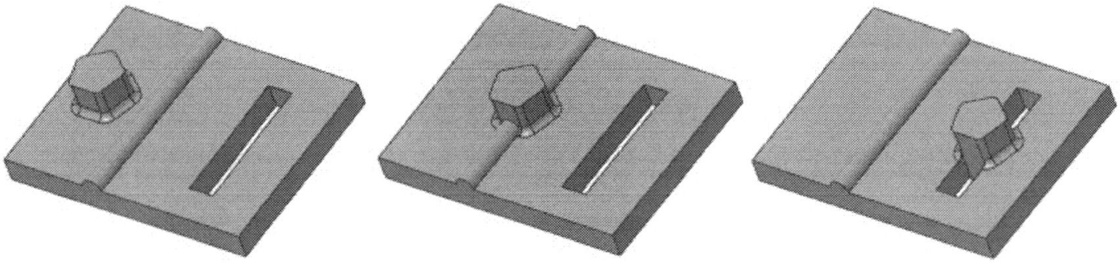

Figure 6-75 Regeneration of the round feature with respect to geometry changes

CREATING CHAMFERS

The **Chamfer** tool is used to bevel the edges and corners as per some specified parameters. In Creo Parametric you can create two types of chamfers. The first is the Corner chamfer and the second is the Edge chamfer. Figure 6-76 shows two types of chamfers. To create a chamfer on a corner, choose the **Corner Chamfer** tool from the **Chamfer** drop-down in the **Engineering** group. To create a chamfer on an edge, choose **Chamfer > Edge Chamfer** from the **Engineering** group.

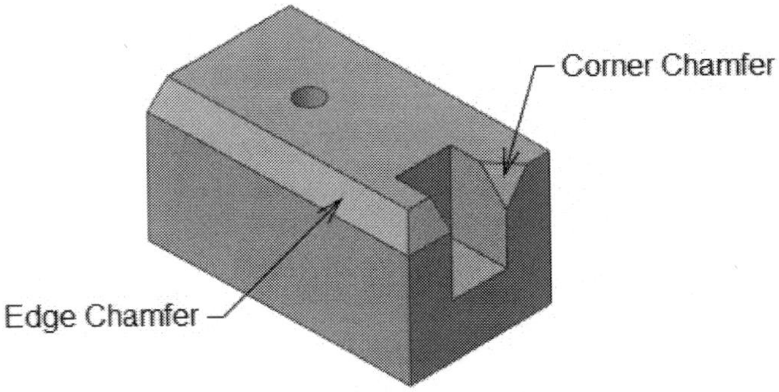

Figure 6-76 *Two types of chamfers*

Corner Chamfer

Ribbon: Model > Engineering > Chamfer > Corner Chamfer

A corner chamfer creates a beveled surface at the intersection of three edges. When you choose the **Corner Chamfer** tool from the **Chamfer** drop-down list of the **Engineering** group, the **Corner Chamfer** dashboard will be displayed, refer to Figure 6-77. Also, you will be prompted to select a corner to chamfer. Select one of the corner points, a preview of chamfer will be displayed with drag handles. Now, you need to specify the required values of **D1**, **D2**, and **D3** in their respective edit boxes in the dashboard. After specifying the values, choose the **OK** button from the dashboard to create the chamfer.

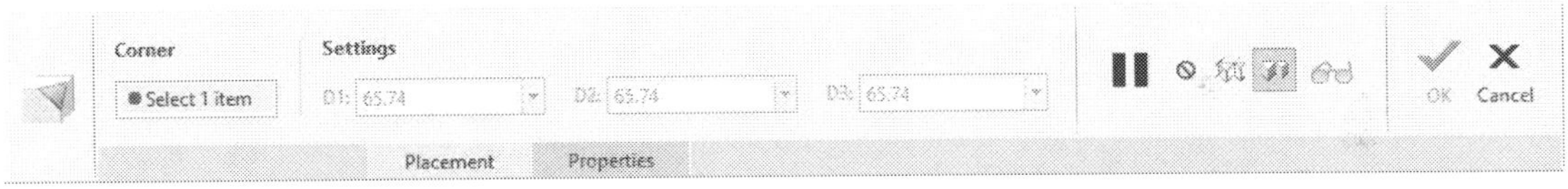

Figure 6-77 *Partial view of the* ***Corner Chamfer*** *dashboard*

Note

When you are creating the corner chamfer, notice that only the ***Vertex*** *filter is available in the Filter drop-down list in the Status Bar. As mentioned earlier, the filters in Creo Parametric narrow the entities available for selection. Therefore, you can easily select the reference entities on the model.*

Edge Chamfer

Ribbon: Model > Engineering > Chamfer > Edge Chamfer

An edge chamfer creates a beveled surface along the selected edge. When you choose the **Edge Chamfer** tool from the **Chamfer** drop-down in the **Engineering** group, the **Edge Chamfer** dashboard will be displayed, as shown in Figure 6-78 and you will be prompted to select any number of edges to create a chamfer. The options in this dashboard are discussed next.

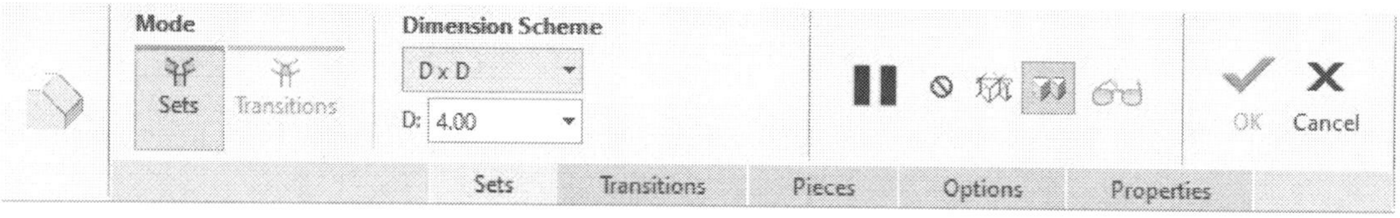

Figure 6-78 *Partial view of the* ***Edge Chamfer*** *dashboard*

Sets

When you invoke the Chamfer dashboard, the **Sets** button is chosen by default in the **Mode** area. This button is used to create chamfers by defining either one set or more than one sets. When this button is chosen, the drop-down list in the **Dimension Scheme** area of the **Edge Chamfer** dashboard is also displayed. This drop-down list is used to specify the type of chamfer. The options in the drop-down list are discussed next.

D x D

The **D x D** option creates a chamfer such that the chamfer is created at an equal distance from the selected edge on both the faces connected by the edge. In Figure 6-79, the dimension value 60 is the value of D where the distance D is measured from the corner.

D1 x D2

The **D1 x D2** option creates a chamfer at two user-defined distances from the selected edge. In Figure 6-80, the dimension value 60 is the value of D1 and the dimension value 82.78 is the value of D2. The values for D1 and D2 can be entered from the Chamfer dashboard. You can choose the **Interchange distance dimensions of the chamfer** button available on the right of the D2 edit box to interchange the two dimensions.

Figure 6-79 *Creating chamfers using the* ***D x D*** *option*

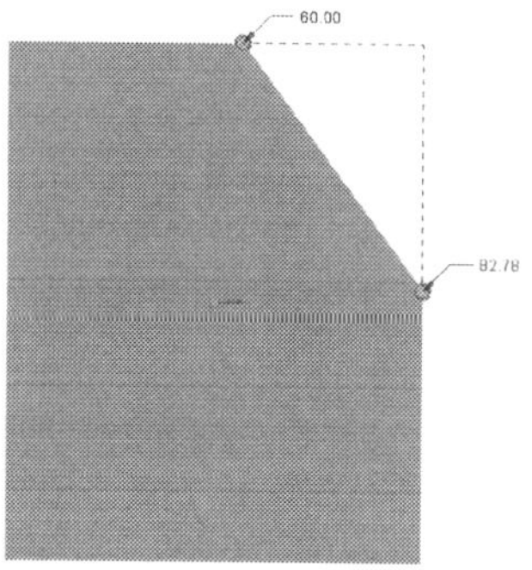

Figure 6-80 *Creating chamfers using the* ***D1 x D2*** *option*

Angle x D

The **Angle x D** option creates a chamfer at a user-defined distance from the selected edge and also at a user-defined angle measured from a specified surface. In Figure 6-81, the angle is 60 degrees and the value of D is 60.

45 x D

The **45 x D** option creates a chamfer at the intersection of two perpendicular surfaces. The chamfer created is at an angle of 45-degree from both the surfaces and at a distance D from the edge along each surface. Once the chamfer is created, the distance D can be modified. Figure 6-82 shows this type of chamfer where the value of D is 60.

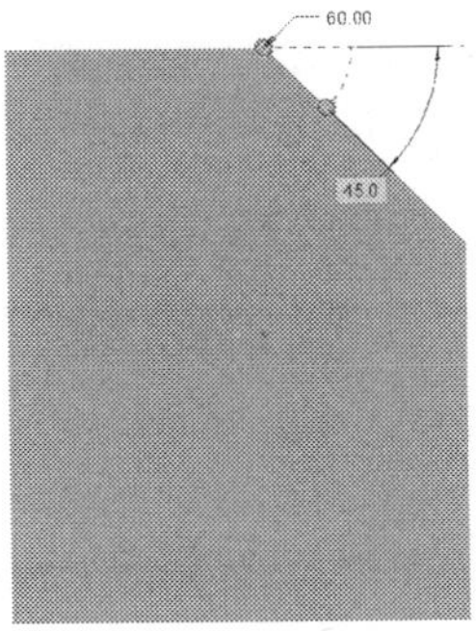

*Figure 6-81 Creating the chamfers using the **Angle x D** option*

*Figure 6-82 Creating the chamfers using the **45 x D** option*

O x O

The **O x O** option is used when **Offset Surfaces** is selected from the slide-down panel that is displayed on choosing the **Sets** tab of the Chamfer dashboard. In this case, the chamfer is created by offsetting the neighboring surfaces and you need to define the offset distance from the edge.

O1 x O2

The chamfer is created by offsetting the neighboring surfaces independently.

Sets Tab

When you choose the **Sets** tab in the Chamfer dashboard, a slide-down panel will be displayed, as shown in Figure 6-83. All references and dimensions of the chamfer can be specified in this slide-down panel. The options in this slide-down panel are discussed next.

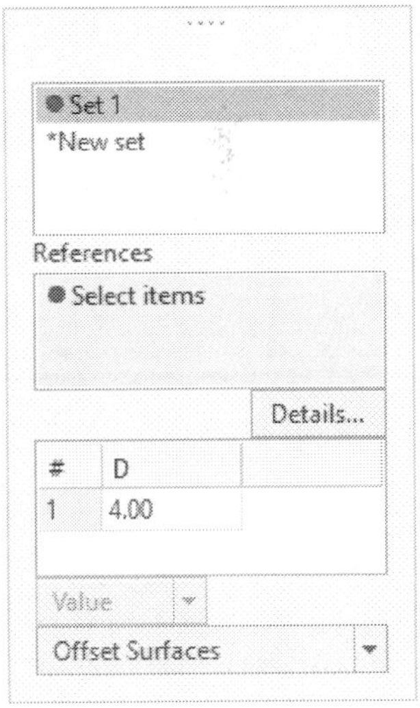

*Figure 6-83 The **Sets** slide-down panel*

New set Area

This area displays the number of sets created. To add a set, right-click to invoke a shortcut menu and choose the **Add** option from it or just click on **New set**. It is recommended that you define an additional set only when you have defined the first set completely. You can switch among the sets by selecting them from this area. The set that is highlighted in this area can be previewed in the drawing area. You can delete this set by using the **Delete** option from the shortcut menu.

References collector

The **References** collector displays the edges selected to chamfer. An edge can be removed from the selection set by right-clicking. On doing so, a shortcut menu will be invoked and you can choose the **Remove** option from it.

The area below the **References** collector is used to specify the dimension values. The dimension values can also be specified on the dashboard.

Transitions Tab

When you choose the **Transitions** tab, a slide-down panel will be displayed, as shown in Figure 6-84. The transitions of a chamfer will be displayed in the slide-down panel only when the **Transitions** button is chosen in the **Chamfer** dashboard. The options available in the Transition type drop-down list will be according to the geometrical context.

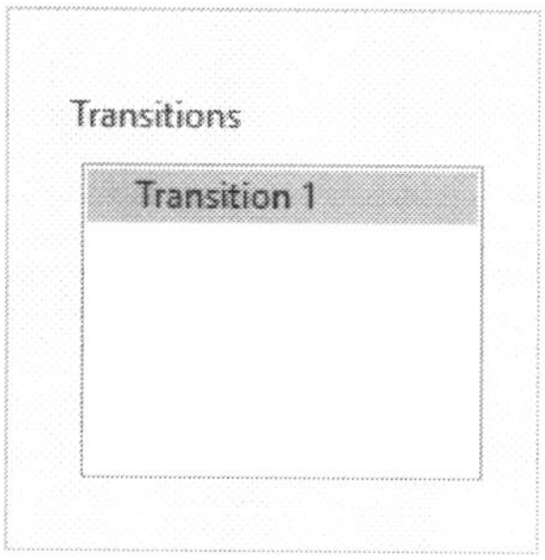

***Figure 6-84** The **Transitions** slide-down panel*

Pieces Tab

The use of **Pieces** tab is the same as discussed in the **Round** dashboard. You can use it to exclude selected pieces in case some ambiguity exists while creating chamfers.

Options Tab

When you choose the **Options** tab, the slide-down panel, which has two radio buttons will be displayed. These radio buttons allow you to change the chamfer as a surface or as a solid.

Properties Tab

When you choose this option, the slide-down panel will be displayed, which shows the feature identity.

Transitions

Transitions allow you to manage overlapping or discontinuous chamfer pieces. Default transitions are automatically applied according to the geometry of the chamfer feature created. They are mostly retained but you can modify them to obtain preferred geometry.

The **Transitions** button in the Chamfer dashboard is used in the same way as it is used in the **Round** dashboard. This button will be available only when you have defined at least one reference for the chamfer. Using this button, you can define the shape of the chamfer that is formed at a vertex of the given geometry and also define a limit up to which the chamfer will be created. Most of the transition types for a Chamfer feature are same as those that were discussed for a Round feature.

Note

*After entering into the transition mode, you cannot change the value of **D**. Therefore, it is recommended to set the value of **D** prior to specifying the transition. In case it is required to change the value of **D** after setting the transition, choose the **Sets** button to return to the set mode and change the value of **D**.*

The **Corner Plane** option can be used only when three chamfers converge at a vertex. By using this option, a plane is created at the point of merger, as shown in Figure 6-85. This plane is in the shape of a triangle and its size depends on the specified value of **D**.

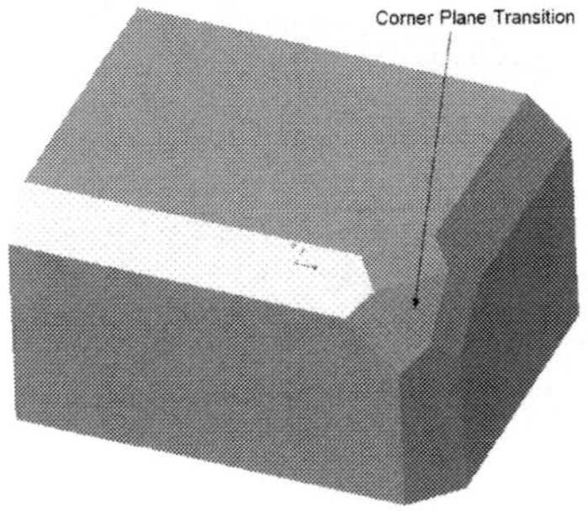

Figure 6-85 Chamfer created using the ***Corner Plane*** *option*

UNDERSTANDING RIBS

Ribs are defined as thin wall-like structures used to bind the joints together so that they do not fail under an increased load. In Creo Parametric, the section for the rib is sketched as an open section and can be extruded equally in both directions of the sketch plane or on either side. The procedure of creating a rib is similar to that of creating an extrusion.

In Creo Parametric, you can create two types of ribs: Trajectory Ribs and Profile Ribs. The steps to create both the ribs are almost same. These steps are discussed next.

Creating Trajectory Ribs

Ribbon: Model > Engineering > Rib > Trajectory Rib

Trajectory Ribs are mostly used to strengthen plastic parts that include a base and a shell or a hollow area in between the pocket surfaces. To create a trajectory rib, choose the **Trajectory Rib** tool from the **Rib** drop-down in the **Engineering** group. On doing so, the **Trajectory Rib** dashboard will be displayed, as shown in Figure 6-86.

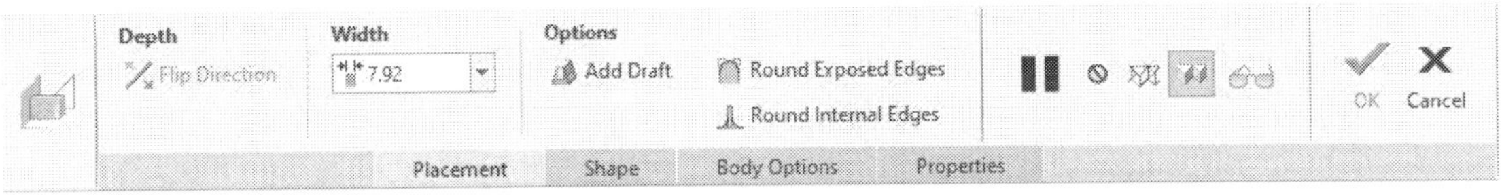

Figure 6-86 Partial view of the ***Trajectory Rib*** *dashboard*

To create a trajectory rib, first you need to have a shell feature with a solid base, or a hollow area in between the pocket surface. Next, you need to select a sketching plane and then draw the geometry of the rib feature in the sketcher environment. Generally, the sketching plane defines the top surface of the rib. The side surfaces of the rib geometry extend to the next surface encountered. The profile for the rib feature can be open, closed, or intersecting. Also, the profile must be created in between the hollow area and can include any number of segments. Note that, like base features, you can select an existing sketch or create an internal sketch for the **Trajectory Rib** tool.

You can now add draft to ribs by using the **Add Draft** button from the dashboard. When you choose this button, a drag handle for adding a draft is displayed. Using this drag handle, you can set the value of draft.

You can also add rounds at the inner edges as well as the exposed edges by using the **Round Internal Edges** and **Round Exposed Edges** buttons, respectively. Choose the **Round Internal Edges** button; the drag handle will be displayed to control radius of round.

The procedure to create a trajectory rib feature is discussed next.

1. Choose the **Trajectory Rib** tool from the **Rib** drop-down in the **Engineering** group of the **Model** tab.

2. Choose the **Placement** tab from the dashboard to display a slide-down panel and then choose the **Define** button from the group; the **Sketch** dialog box will be displayed.

3. Select the sketch plane and specify the reference in the **Sketch** dialog box. The sketch plane will define the top surface of the rib, refer to Figure 6-87.

4. Draw the sketch and exit the sketcher environment; the preview of the rib feature will be displayed with the default width. Also, an arrow will be displayed in the drawing area. This arrow shows the direction of material addition and should point toward the geometry of the model and not away from it. The direction of the arrow can be reversed, if required, by using the **Flip Direction** button available in the dashboard.

   The width of rib can be dynamically modified by dragging the handles displayed on the rib or by entering a value in the edit box available in the dashboard.

5. After specifying all parameters for the rib feature, choose the **OK** button; the rib will be created in between the pocket surfaces with the bottom of rib intersecting the base of solid, as shown in Figure 6-88.

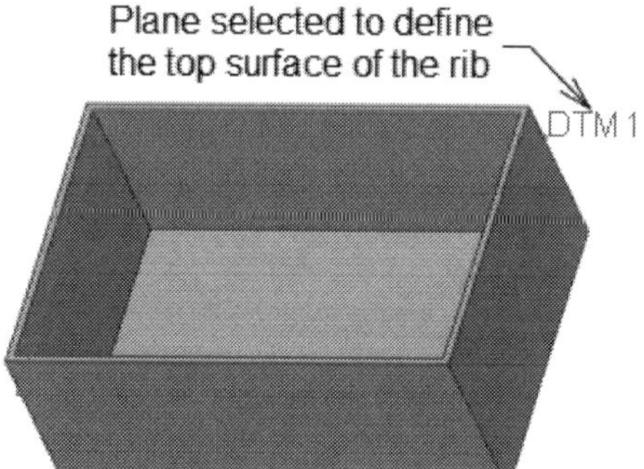

Figure 6-87 *The sketching plane selected*

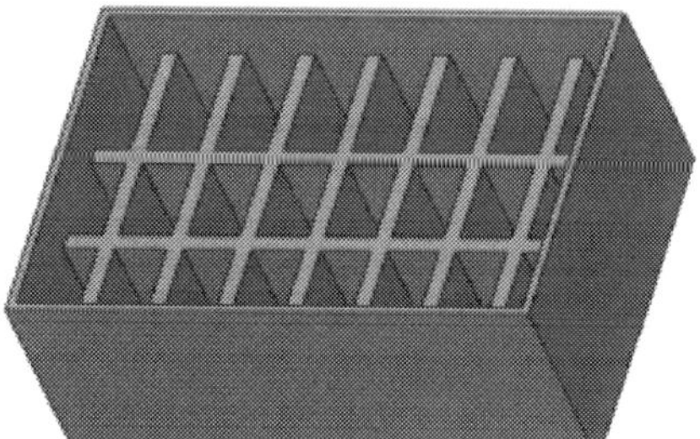

Figure 6-88 *The resultant rib feature*

Tip

In case the sketched profile is beyond the model, the rib automatically trims itself to the model boundaries.

Note

*You can change the values of rounds and drafts of the rib by using the edit boxes available in the **Shape** tab.*

Creating Profile Ribs

Ribbon: Model > Engineering > Rib > Profile Rib

To create a profile rib, choose the **Profile Rib** tool from the **Rib** drop-down in the **Engineering** group; the **Profile Rib** dashboard will be displayed, as shown in Figure 6-89.

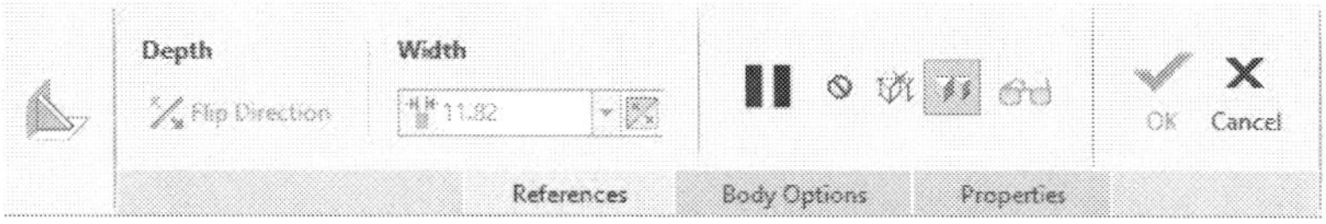

***Figure 6-89** Partial view of the **Profile Rib** dashboard*

To create a rib, first you need to select a sketching plane and then draw the geometry of the rib feature in the sketcher environment. The sketch for the rib feature should be an open entity.

The options and the tools in the **Profile Rib** dashboard are discussed next.

References Tab

When you choose the **References** tab, a slide-down panel will be displayed, as shown in Figure 6-90. In the slide-down panel, the **Sketch** collector shows **Select 1 item**. This indicates that the sketch for the rib feature is not drawn. To draw the sketch, choose the **Define** button; the **Sketch** dialog box will be displayed using which you can define the sketch plane and the references. To edit the sketch of the rib, you can choose the **Edit** button. On choosing the **Unlink** button, the association between the sketch and feature breaks and the entities and references of the sketch get copied to a new internal section.

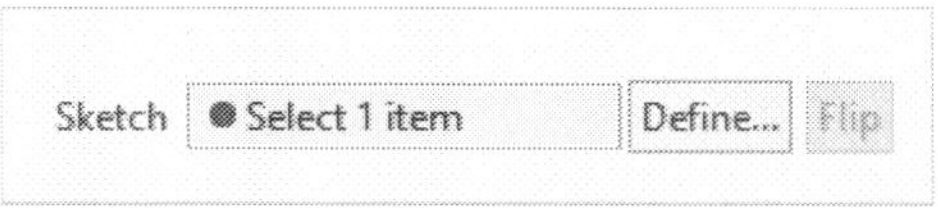

***Figure 6-90** The **References** slide-down panel*

The **Flip** button is used to flip the side where the material should be added. This button will be available only when you draw the sketch and exit the sketcher environment.

Change the thickness option between both sides, side 1, and side 2 Button

This button is available on the **Profile Rib** dashboard only when the sketch of the rib is drawn and you exit the sketcher environment. You have an option to extrude the sketch on either sides or symmetrically on both sides of the sketching plane.

Note

Since the sketch of the rib can be extruded on both sides of the sketching plane, therefore, while creating a rib, you must always select an appropriate sketching plane such that it lies at the center of the rib feature.

The procedure to create a profile rib feature using the **Profile Rib** dashboard is discussed next.

1. Invoke the **Profile Rib** dashboard by choosing the **Rib > Profile Rib** from the **Engineering** group.

2. From the **Profile Rib** dashboard, choose the **References** tab to display the slide-down panel and choose the **Define** button from it.

3. Select the sketch plane and specify the reference in the **Sketch** dialog box. Now, create an open sketch that defines the rib profile. Make sure that the endpoints of the sketch are aligned with the edges of the model.

> **Tip**
> *You can use the* ***References*** *tool from the* ***Setup*** *group of the* ***Sketch*** *tab to select the required edges for creating the Profile Rib.*

4. When you exit the sketcher environment, an arrow will be displayed in the drawing area. This arrow shows the direction of material addition and should point toward the geometry of the model and not away from it. The direction of the arrow can be reversed by using the **Flip** button available in the slide-down panel that will be displayed on selecting the **References** tab.

5. If the arrow direction is correct, then you can preview the rib feature on the graphics window. The rib thickness can be dynamically modified by dragging the handles displayed on the rib or by entering a value in the edit box available in the **Profile Rib** dashboard.

6. Now, use the **Change the thickness option between both sides, side 1, and side 2** button to change the direction of extrusion. You can create the rib feature either symmetrically about the sketching plane or on either side of it. Every successive click on this button will change the position of the rib feature with respect to the sketching plane.

7. After specifying all parameters for the rib feature, choose the **OK** button.

The profile ribs can be rotational ribs and straight ribs. Rotational ribs are constructed on cylindrical parts and straight ribs are created on planar faces. Figure 6-91 shows a rotational rib and Figure 6-92 shows a straight rib. The options for the creating these ribs are the same. The creation of these ribs depends on the geometry of the feature on which the rib is created.

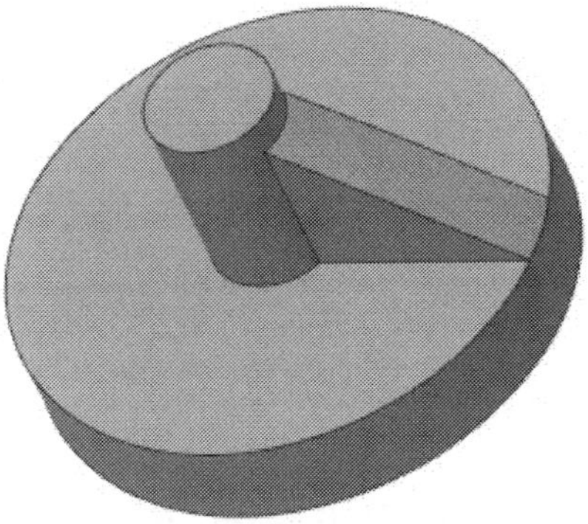

Figure 6-91 *The rotational rib feature*

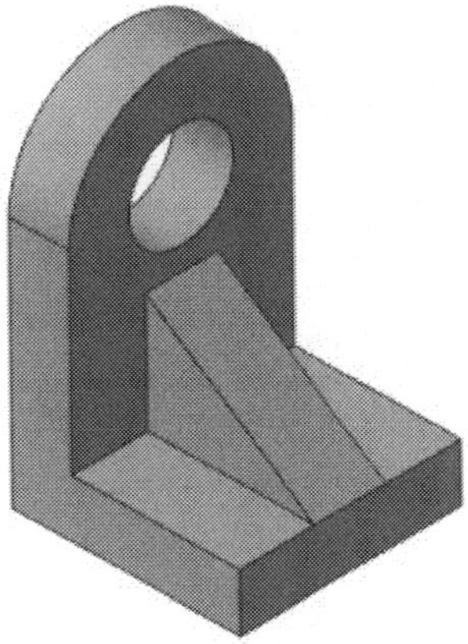

Figure 6-92 *The straight rib feature*

EDITING FEATURES OF A MODEL

Editing is one of the most important aspects of the product design. Most of the designs require editing during or after their creation. As mentioned earlier, Creo Parametric is a parametric and feature-based solid modeling software. Hence, the features constituting a model can be individually edited. Also, you can edit the datums and the features referenced to these datums. As parent-child relationship exists between the two features, therefore the child feature is also modified when the parent feature is modified. For example, if you have created a feature using a datum plane that is at some offset distance, the feature will be automatically repositioned when the offset value of the datum plane is changed. The methods to edit features and references of a model are discussed next.

Editing Definition or Redefining Features

Editing the definition of features allows you to make changes in the parameters that were used to create it. You can also modify the sketches of the sketched features by editing their definitions. There are three methods available in Creo Parametric to edit definition or redefine features. These methods are listed next.

1. Select the feature to be redefined from the drawing area; the selected feature is highlighted and a mini popup toolbar will be displayed near the cursor. Choose the **Edit Definition** option from the mini popup toolbar to redefine the selected feature.

Note

Sometimes, it is difficult to select a feature from a model to redefine. Therefore, you can use the filters in the Filter drop-down list in the Status Bar.

2. You can also use the **Model Tree** to redefine a feature. Select the feature that has to be redefined from the **Model Tree**; the selected feature is highlighted in the drawing area and a mini popup toolbar will be displayed near the cursor. Choose the **Edit Definition** option from the mini popup toolbar to redefine the selected feature.

3. To edit the definition of a feature select it from the drawing area or from the **Model Tree** and press CTRL+E keys from the keyboard.

Reordering Features

Ribbon: Model > Operations > Reorder

Reordering the features is defined as the process of changing the sequence of feature creation in a model. Sometimes after creating a model, it may be required to change the order in which the features of the model were created. While reordering features of the model, you should remember the following points.

1. You can move features forward or backward in the **Model Tree**, for regeneration.

2. You can reorder multiple features at a time.

3. You can also put features or datums in groups or take them out of these groups by reordering.

4. You can reorder the datums under a single feature or it can embed within the feature.

5. If you select a pattern member to reorder, then all the members in the pattern will be selected to be reordered.

For reordering the features, the **Model Tree** or the **Reorder** tool is used.

If you use the **Model Tree**, you just have to drag and drop the feature that you want to reorder below or above the destination feature.

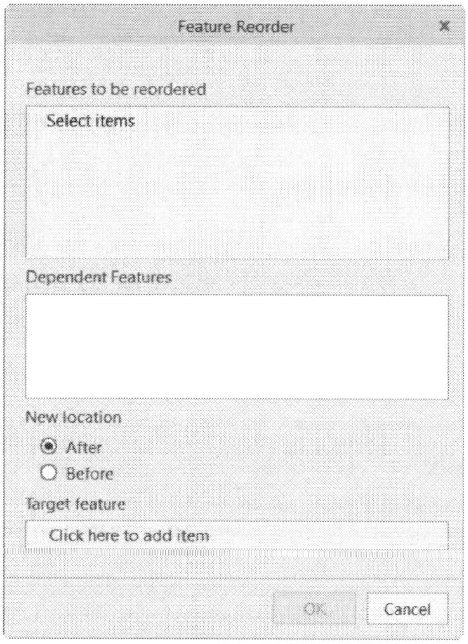

***Figure 6-93** The **Feature Reorder** dialog box*

Choose the **Reorder** tool from the expanded **Operations** group; the **Feature Reorder** dialog box will be displayed, as shown in Figure 6-93. Click in the **Features to be reordered** collector of the **Feature Reorder** dialog box and select the features to be reordered from the **Model Tree** or from the graphics window. By default, the **After** radio button is selected in the **New location** area. Now select the **Target feature** collector from the dialog box and then, in the **Model Tree**, select the feature after which the target feature is to be placed.

In the **New location** area, there are two radio buttons: **After** and **Before**. If you choose the **After** radio button then the feature to be reordered will be placed after the target feature. If you choose the **Before** radio button then the feature to be reordered will be placed before the target feature. Any dependent features that must be reordered will be automatically added to the **Dependent Features** list. These dependent features are reordered along with the features selected to reorder.

The example that is discussed here will explain why and when the need for reordering of features in a model arises. Consider the model shown in Figure 6-94. It consists of a rectangular pattern of three columns and two rows. Now, a shell feature is created on this model which removes the top and the front face, as shown in Figure 6-95.

Figure 6-94 *Model with pattern*

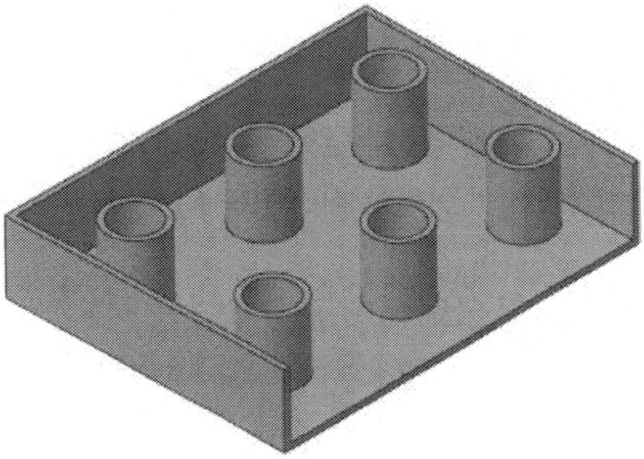

Figure 6-95 *Model after creating the shell feature*

In Figure 6-95, it is evident that the material equal to the wall thickness of the shell feature is added around the hole features. If this is not desired, the model has to be modified and here the need for reordering the feature arises. You will reorder the features such that the shell feature is inserted before the hole feature and its pattern. When you reorder the features, all features will be automatically adjusted in the new order. The model after reordering the hole feature is shown in Figure 6-96.

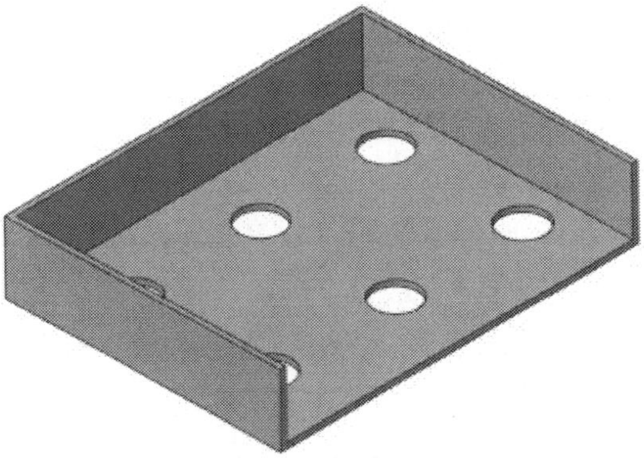

Figure 6-96 *Model after reordering the features*

Note

Usage of the ***Pattern*** *tool is discussed in the Chapter 8.*

Rerouting Features

Rerouting of features is required when there is a need of editing or replacing the references of a feature with new references. While doing so, you can also redefine the references that were selected while creating the sketch. A reference is a feature or geometry that is used (referenced) by other features. The features that use the reference are child features of that reference.

For rerouting a feature, select the required feature from the **Model Tree** or from the drawing area. Now, choose the **Edit References** or **Replace References** option from the **Operations** group.

You can note that the **Edit References** or **Replace References** option will only be available after a feature has been selected.

To edit the reference, choose the **Edit References** option from the **Operation** group of the **Model** tab; the **Edit References** dialog box will be displayed, as shown in Figure 6-97. You can also invoke this dialog box by choosing the **Edit References** option in the mini popup toolbar. Choose the reference to be replaced from the **Original references** collector. Next, click in the **New reference** collector of the dialog box, select the replacement reference, and then choose the **OK** button from the dialog box. To reset the changed reference back to the original reference, select the reference under the **Original references** collector and click on the **Reset** button next to the **New reference** collector. To edit the references of other child features of the selected reference, click ⊕ to expand the **Child handling** section of the **Edit References** dialog box. In the **Update child features** area, select the check boxes next to the child features. Click the **Roll To** button to roll the model to the recent feature from which an other reference can be selected for the selected reference. To regenerate the model with the references that were edited so far, select the **Preview** option.

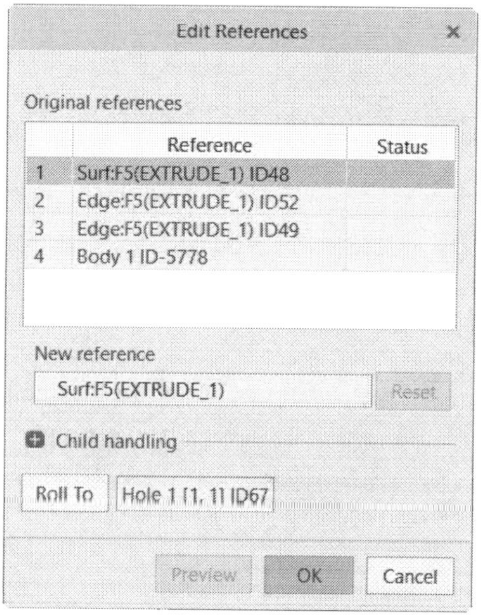

***Figure 6-97** The **Edit References** dialog box*

To replace the reference, choose the **Replace References** option from the **Operations** group of the **Model** tab; the **Replace References** dialog box will be displayed, as shown in Figure 6-98. Choose the reference to be replaced from the **Original references** collector. Now, click in the **New reference** collector in the dialog box and select the replacement reference. Next, choose the **OK** button from the dialog box. To reset the changed reference back to the original reference, choose the **Reset** button from the dialog box.

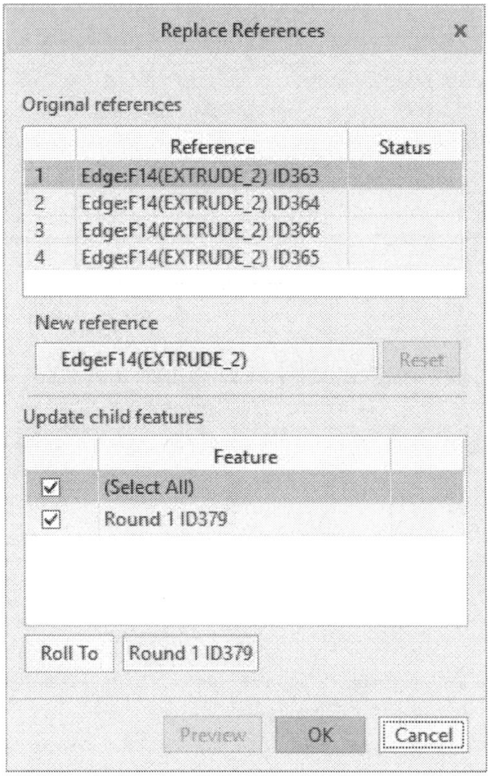

*Figure 6-98 The **Replace References** dialog box*

Suppressing Features

Once the feature is suppressed, it will neither be displayed in the drawing area nor in the drawing views. Note that on suppressing, the feature is not deleted, only its visibility and regeneration is turned off temporarily. You can resume the feature anytime by unsuppressing it using the **Model Tree** or using the **Resume** option from the **Operations** group. As soon as you unsuppress the feature, it will be displayed in the drawing area and also in the drawing views. When a model has many features, then suppressing some features decreases its regeneration time. To suppress a feature, click on it in the **Model Tree** or select it in the drawing area and then choose the **Suppress** option from the mini popup toolbar; the **Suppress** message box will be displayed prompting you to confirm the suppression of the highlighted features in the drawing area. Confirm the suppression by choosing the **OK** button in this message box. You can also suppress a feature by first selecting it from the **Model Tree** or the drawing area and then choosing **Suppress > Suppress** from the **Operations** group. To unsuppress the feature, click in the **Model Tree** or in the drawing area and choose the **Resume** option from the mini popup toolbar.

Tip

*If the suppressed features are not displayed in the **Model Tree,** then to display the suppressed features, choose **Settings > Tree Filters** from the **Model Tree;** the **Model Tree Items** dialog box will be displayed. Select the **Suppressed objects** check box and choose the **OK** button. The suppressed features will now be displayed in the **Model Tree** with a small black box displayed against their names.*

Note

If the feature being suppressed has child features, they will also be suppressed on suppressing the parent feature. However, before suppressing the features, the system prompts you to confirm the suppression of the highlighted features in the drawing area. You can suspend the suppression of child features by selecting the ***Suspend*** *option from the drop-down list of the respective child feature in the* ***Child Handling*** *dialog box. This dialog box will be displayed when you choose the* ***Options*** *button from the* ***Suppress*** *message box.*

Deleting Features

The feature that is not required can be deleted from the model. Right-click on the feature in the **Model Tree** or in the drawing area to invoke the shortcut menu. From this menu, choose the **Delete** option. Once the feature to be deleted is selected, it gets highlighted along with its child features. The system confirms you to delete the selected feature. If you confirm the deletion, the feature will be deleted along with its child features.

Modifying Features

Once a feature is created, you can still modify it by modifying its dimensions. This editing operation reflects the parametric nature of Creo Parametric. Click on the feature in the **Model Tree** or in the drawing area to invoke the mini popup toolbar. From this toolbar, choose the **Edit Dimensions** option; the dimensions of the feature will be displayed in the drawing area. These dimensions are the same as those defined while sketching or creating the feature.

In some cases, for example, rounds, holes, and so on, the dimensions of these features were defined as parameters. Therefore, these dimensions are also displayed when you select them to modify. To modify the dimensions, double-click on the dimension in the drawing area; the dimension is converted into an edit box. Type the required value for the selected dimension and press ENTER. Next, choose the **Regenerate** button from the **Quick Access** toolbar or press CTRL+G if the model is not regenerated by default.

Tip

You can double-click on a feature in the drawing area to display its dimensions. A dimension can be modified first by double-clicking on it, then entering a new value in the edit box displayed and finally by regenerating it.

Modifying the Sketch of a Feature Dynamically

If you want to modify the sketch of a feature without entering the sketcher environment, then you can modify it dynamically. To dynamically modify a sketch, follow these steps:

1. Select the feature and choose the **Edit Dimensions** button from the mini popup toolbar; the dimensions of the sketch appear on the feature.

2. Now, bring the cursor close to the sketch. You will notice that the cursor changes to a pointer. The pointer symbol represents that you can select an entity from the sketch and drag it to modify its shape.

3. Modify the shape of the sketch by dragging its entities.

4. Choose the **Regenerate** button from the **Quick Access** toolbar to apply the modifications in case the model is not regenerated by default.

Modifying the Features Dynamically

Dynamic editing of models is a new feature of Creo Parametric and it has been introduced to simplify the editing process. This feature enables you to modify the model after it has been created and can be performed any time during the creation. To edit a feature dynamically, select the feature from the drawing area or from the **Model Tree** and right-click to invoke a shortcut menu. Next, choose the **Edit Dimensions** option from the mini popup toolbar. Now you can edit the features dynamically.

The following entities displayed on the feature can be modified:

1. You can drag any vertex or edge to change the dimension of a feature.

2. The profile used for the feature is also displayed. You can drag the entities and modify the shape of the sketch to modify the feature.

3. All the dimensions that were defined while sketching the model as well as those that were entered in the dashboard are displayed on the feature. You can double-click on the dimension and enter a new value to modify the feature accordingly.

Note that in the process of dynamic modification, the feature gets modified instantaneously and you need not regenerate the model.

TUTORIALS

Tutorial 1

Create the model shown in Figure 6-99. The dimensions, and the front, top, and left-side views of the model are shown in the Figure 6-100. **(Expected time: 45 min)**

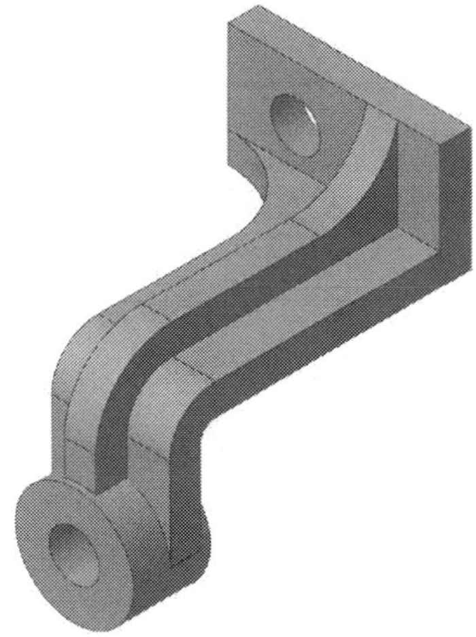

Figure 6-99 *Isometric view of the model*

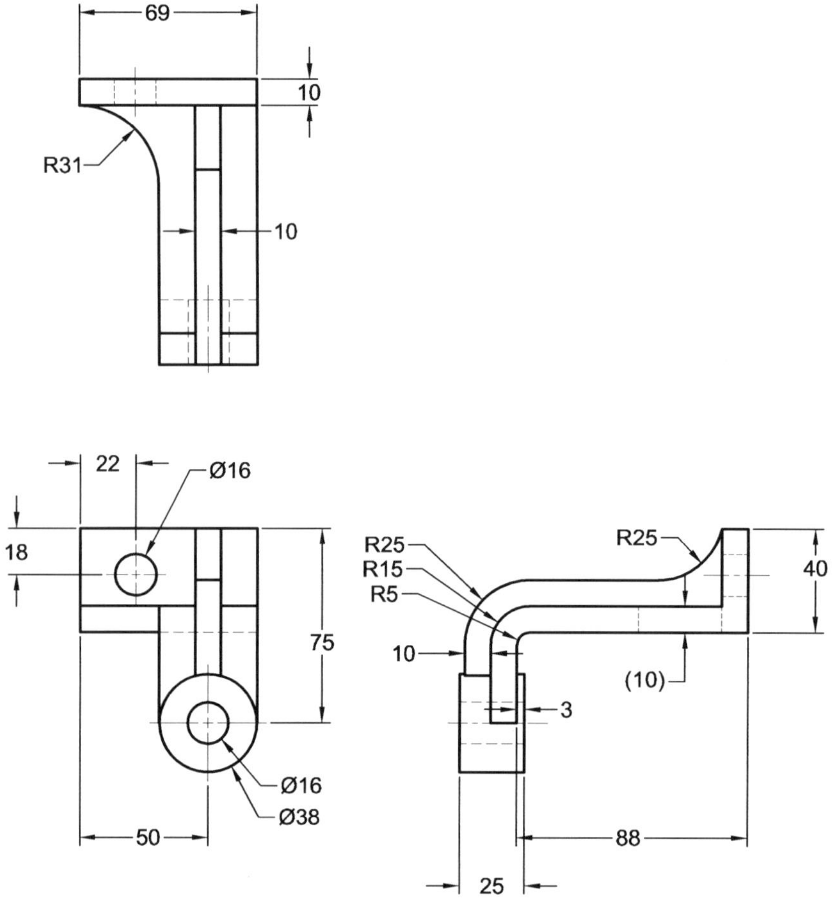

***Figure 6-100** Orthographic views of the model*

The following steps are required to complete this model:

Examine the model and determine the number of features in it, refer to Figure 6-99.

a. Create the base feature on the **FRONT** datum plane.
b. Create the second extrude feature on the **TOP** datum plane.
c. Create the third extrude feature on the front planar surface of the second feature.
d. Create the non symmetrically extruded cylindrical feature on the front face of the third feature.
e. Create a hole feature that is coaxial to the cylindrical feature.
f. Create the round features.
g. Create the last feature, which is the rib.

When a Creo Parametric session starts, the first task is to set the working directory. As this is the first tutorial of this chapter, you need to select the Working Directory first. Set *C:\ Creo-8.0\c06* as the Working Directory; the message **Successfully changed to C:\ Creo-8.0\c06 directory** is displayed in the message area.

Starting a New Object File

1. Start a new part file and name it as *c06tut01*.

   The part mode is invoked and three default datum planes are displayed in the drawing area. Also, the **Model Tree** is displayed on the left of the drawing area.

Creating the Base Feature

To create the sketch of the base feature, you first need to select the sketching plane for it. In this model, you need to draw the base feature on the **FRONT** datum plane because the direction of extrusion of this feature is normal to this plane.

1. Select the **FRONT** datum plane as the sketching plane and choose the **Extrude** tool from the **Shapes** group or from the mini popup toolbar; the sketcher environment is invoked.

2. After you enter into the sketcher environment, create the sketch of the base feature and then apply constraints and dimensions to it, as shown in Figure 6-101. Note that in the sketch, the bottom line of the rectangular section coincides with the **TOP** datum plane.

   As is evident from the sketch of the base feature shown in Figure 6-101 that the **RIGHT** datum plane is located at a distance of 50 from the left edge because later in the tutorial, the rib feature will be created on this plane. Refer to Figure 6-100 for the distance required for rib feature. As mentioned earlier, that, by default, the sketch for the rib feature is extruded on both sides of the sketching plane. Therefore, when you create the sketch for the rib feature on the **RIGHT** datum plane, it will extrude on both the sides.

3. After the sketch is completed, choose the **OK** button to exit the sketcher environment; the **Extrude** dashboard is enabled and appears above the drawing area. All model attributes that are selected by default are accepted to create the model.

4. Enter **10** as the depth in the dimension box on the **Extrude** dashboard.

5. Choose the **Saved Orientations** button from the **Graphics** toolbar; a flyout is displayed. Choose the **Default Orientation** option from the flyout; the model orients in its default orientation, that is, the trimetric view.

6. Next, choose the **OK** button from the **Extrude** dashboard to create the base feature, as shown in Figure 6-102.

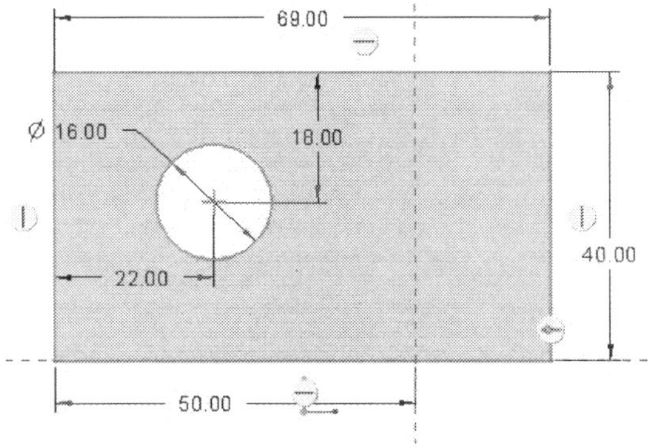

Figure 6-101 *Sketch of the base feature with dimensions and constraints*

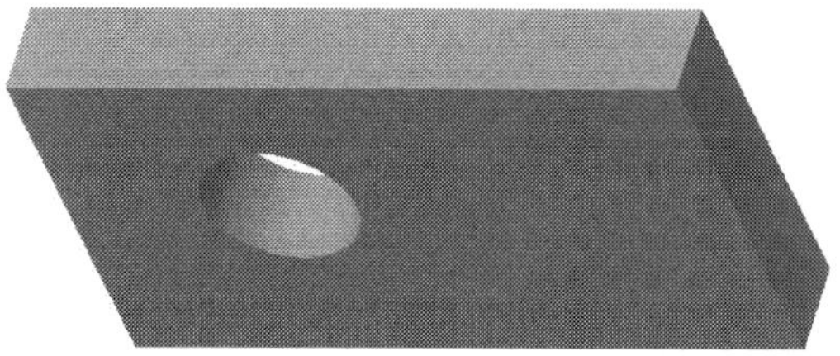

Figure 6-102 *Face of the base feature selected*

Creating the Second Feature

The second feature is also an extruded feature and it will be created on the **TOP** datum plane. Therefore, you need to define the **TOP** datum plane as the sketching plane.

1. Select the **TOP** datum plane as the sketching plane and then choose the **Extrude** tool from the **Shapes** group or from the mini popup toolbar; the sketcher environment is invoked.

2. Create the sketch for the second feature and apply constraints and dimensions to it, as shown in Figure 6-103.

   In Creo Parametric, you can draw sketches by snapping the edge references. To close the sketch, draw a line and aligned it with the bottom edge by snapping the edges of the base feature, refer to Figure 6-103.

3. After completing the sketch, turn the model display to **Shading With Edges** and choose the **OK** button; the **Extrude** dashboard is enabled above the drawing area.

4. Enter **10** in the dimension box in the **Extrude** dashboard and press ENTER.

5. Choose the **OK** button from the **Extrude** dashboard; the second feature is completed and the shaded default trimetric view is displayed, as shown in Figure 6-104.

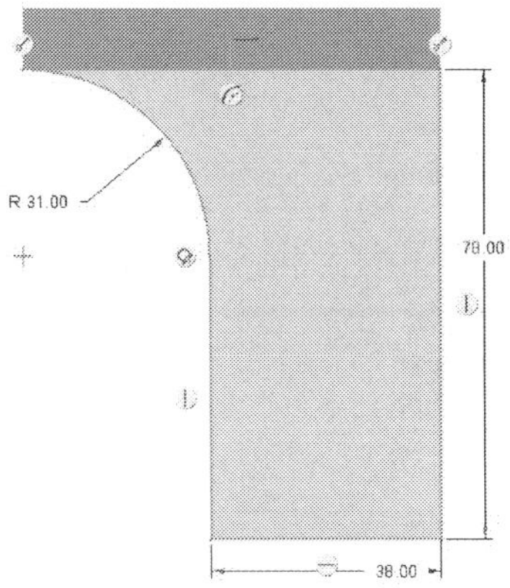

Figure 6-103 *Sketch of the second feature with dimensions and constraints*

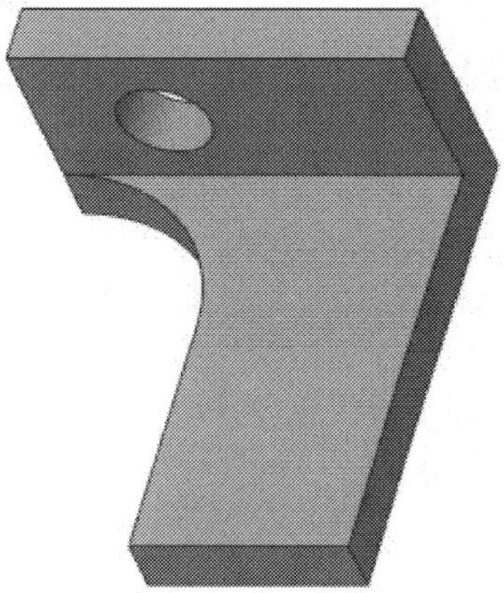

Figure 6-104 *The default trimetric view of the completed second feature and the base feature*

Creating the Third Feature

The sketch of the third feature will be drawn on the front planar surface of the second feature and it will be extruded to the given depth.

1. Select the face of the second feature as the sketching plane and choose the **Extrude** tool to invoke the sketcher environment, refer to Figure 6-105.

2. Once you enter in the sketcher environment, turn the model display to **No hidden**. Create the sketch for the third feature and apply constraints and dimensions to it, refer to Figure 6-106.

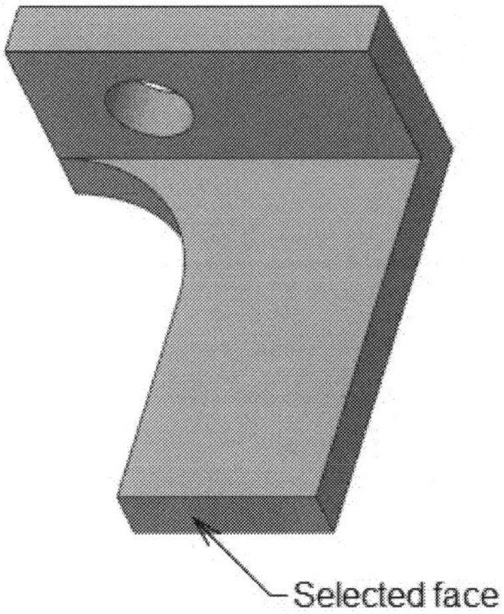

Figure 6-105 *Planar surface selected as the sketching plane for the third feature*

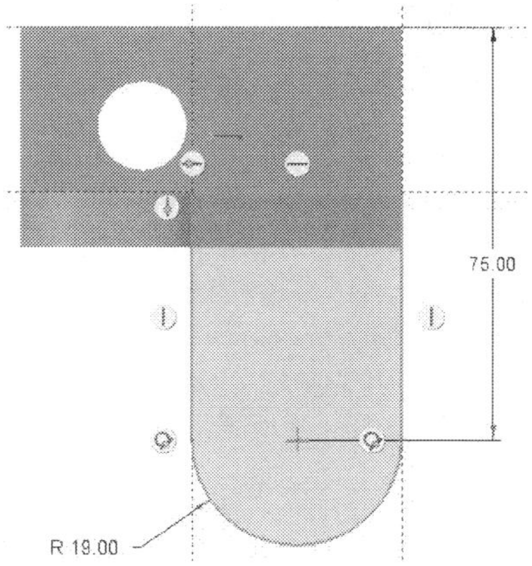

Figure 6-106 *Sketch of the third feature with dimensions and constraints*

Note

While drawing the sketch of any feature, it is recommended that you first apply the required constraints and then dimension the sketch.

3. Choose the **OK** button from the **Close** group; the **Extrude** dashboard is enabled above the drawing area.

4. Enter **10** in the dimension box in the **Extrude** dashboard.

5. Choose the **OK** button from the **Extrude** dashboard. Turn the model display to **Shading With Edges**. The third feature is completed. You can use the middle mouse button to spin the model to view its different orientations.

Creating the Fourth Feature

The fourth feature of the model is an extruded feature and its sketch is drawn on the front planar surface of the third feature. The extrusion of the cylindrical feature will be created on both sides of the front face. The attributes of the feature, like extrusion on both sides and the depth of extrusion, will be specified after the sketch of the feature is drawn.

1. Select the front planar face of the third feature, shown in Figure 6-107, as the sketching plane.

2. Choose the **Extrude** tool from the **Shapes** tab or from the mini popup toolbar; the sketcher environment is invoked.

3. Choose the **Concentric** button from the **Circle** drop-down in the **Sketching** group to draw circular section for the fourth feature, refer to Figure 6-108. Select the arc using the left mouse button. As you move the mouse, the rubber-band circle changes its size. Move the cursor close to the arc; the cursor snaps to the arc. Use the left mouse button and select a point on the arc. You will notice that the equal radius constraint is applied to the sketch. Now, exit this tool.

4. After the sketch is completed, choose the **OK** button from the **Close** group; the **Extrude** dashboard is enabled above the drawing area.

5. Choose the **Options** option from the **Extrude** dashboard. In the slide-down panel, choose the **Blind** option from both the **Side 1** and **Side 2** drop-down lists.

6. Enter **12** in the dimension box on the right of the **Side 1** drop-down list.

7. Enter **13** in the dimension box that is on the right of the **Side 2** drop-down list.

8. Choose the **OK** button from the **Extrude** dashboard to create fourth feature.

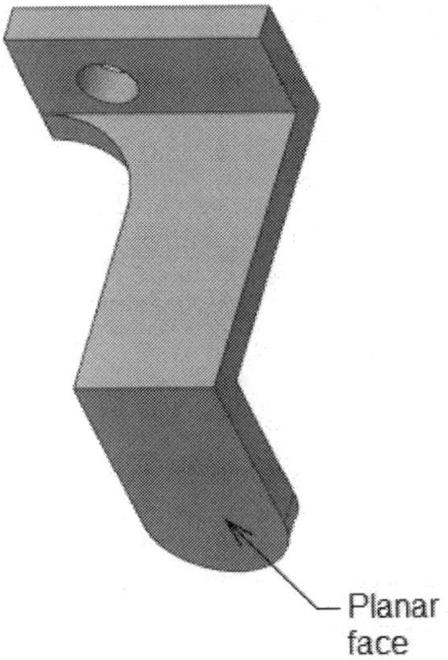

Figure 6-107 Front planar face selected

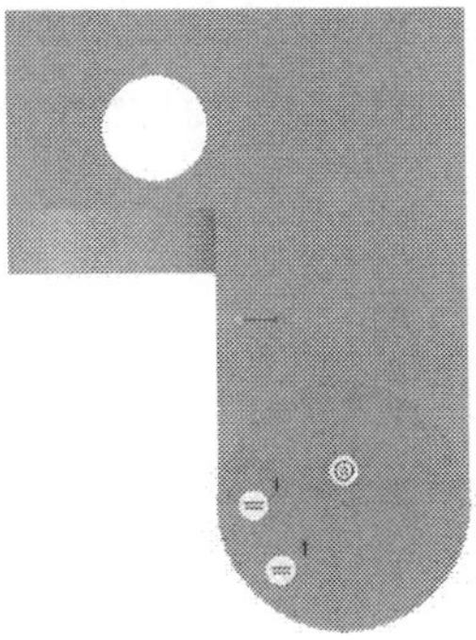

Figure 6-108 Sketch for the fourth feature

Creating the Hole Feature

The hole feature is created using the **Hole** dashboard. The hole will be placed coaxial to the fourth cylindrical feature.

Note

*For creating a coaxial hole, you need to select the axis of the circular feature. To do so, you need to turn on the display of the axis by selecting the **Axis Display** check box from the **Datum Display Filters** drop-down list in the **Graphics** toolbar, if it is not displayed.*

1. Choose the **Hole** tool from the **Engineering** group; the **Hole** dashboard is displayed and the **Simple** tool is chosen by default.

2. Choose the **Placement** tab from the **Hole** dashboard; a slide-down panel is displayed.

3. Select the front face of the cylindrical feature, as shown in Figure 6-109, to place the hole. As you select the front face of the cylindrical feature, a preview of the hole is displayed in the drawing area. Now, you need to specify reference for placing the hole.

4. Press the CTRL key and select the axis of the cylindrical feature from the drawing area.

   Note that the **Coaxial** option is automatically selected in the drop-down list of the slide-down panel of the **Placement** tab.

5. In the diameter dimension box on the **Hole** dashboard, type **16** and press ENTER.

6. In the depth flyout of the **Hole** dashboard, choose the **Drill to intersect with all surfaces** button.

7. Choose the **OK** button from the **Hole** dashboard; the hole is created and trimetric view of the shaded model with the hole is displayed, as shown in Figure 6-110.

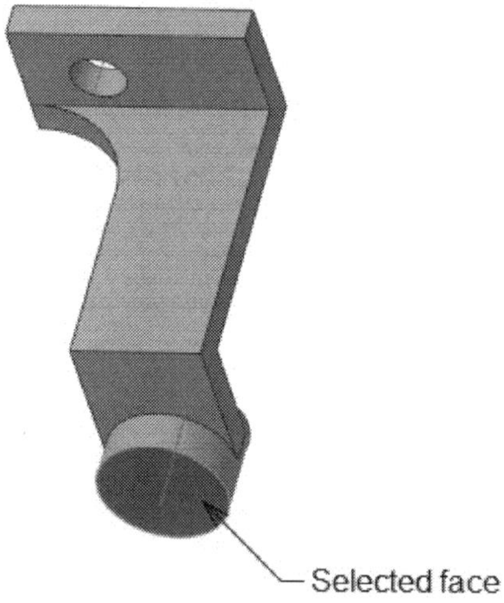

Figure 6-109 Planar face selected for creating hole

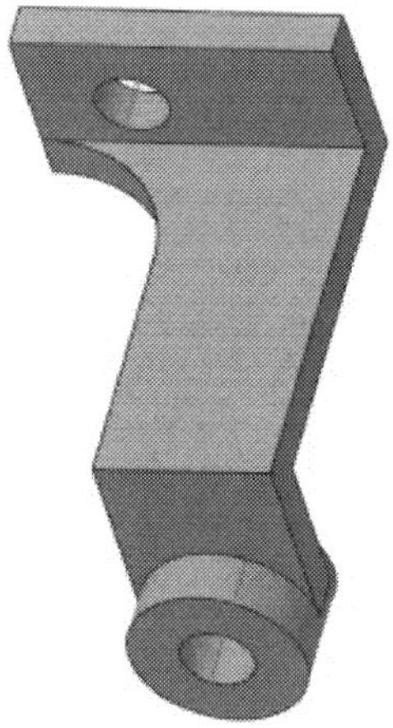

Figure 6-110 The default trimetric view

Creating Two Round Features

Now, the two round features need to be created. Both the rounds have different radii and will be created by defining two sets.

1. Choose the **Round** button from the **Engineering** group; the **Round** dashboard is displayed.
2. Choose the **Sets** tab to display the slide-down panel. Let this slide-down panel remains open so that you can view the selections made on the model.
3. Select the **Edge** option from the Filter drop-down list in the Status Bar. Spin the model and select the edge shown in Figure 6-111.

   A preview of the round is created on the edge and the default value of the radius is displayed.
4. Double-click on the default radius value that is displayed on the preview of the round. In the edit box that appears, enter the value **5** and press ENTER; the first round is created.
5. After spinning the model, select the edge shown in Figure 6-112. You will notice that in the slide-down panel, **Set 1** and **Set 2** appear. This indicates that the second set has been defined.

   A preview of the round is created on the edge and the default value of the radius is displayed on the round geometry.
6. Double-click on the default radius value that is displayed in the preview of the second round. In the edit box that appears, type the value **15** and press ENTER.

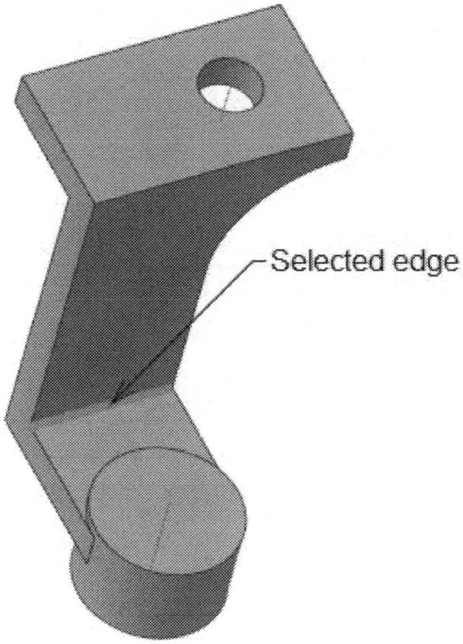

Figure 6-111 *Faces for creating a round of radius* ***5***

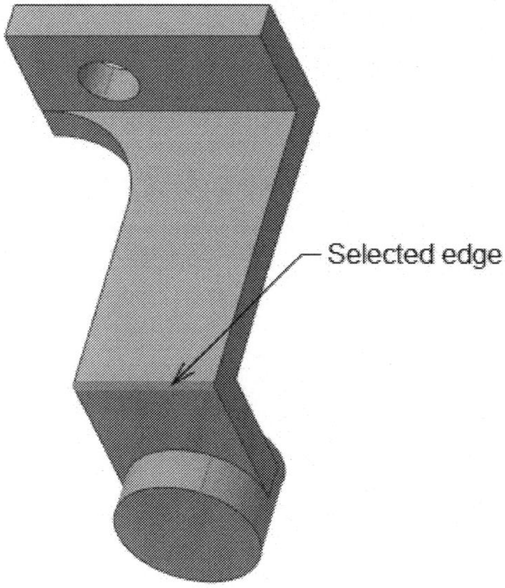

Figure 6-112 *Edges selected for creating a round of radius* ***15***

7. Choose the **OK** button from the **Round** dashboard.

   The second round is created. In the **Model Tree**, the two rounds created appear as a single feature.

Creating the Rib Feature

1. A rib feature is always sketched from the side view. Turn on the display of the datum planes from the **Datum Display** drop-down list in the **Graphics** toolbar if it is turned off.

   The location for the rib from the **RIGHT** datum plane was calculated while sketching the base feature. The section for the rib feature will be drawn on the **RIGHT** datum plane.

2. Select the **RIGHT** datum plane and then choose the **Profile Rib** tool from the **Rib** drop-down in the **Engineering** group; the sketcher environment is invoked.

3. Draw the open sketch for the rib feature and apply required constraints and dimensions, as shown in Figure 6-113.

4. After the sketch is completed, choose the **OK** button. On doing so, an arrow pointing in the direction of material addition is displayed on the sketch. As the section for the rib feature is open, Creo Parametric allows you to specify the direction where the material should be added. Therefore, you need to change the direction of the arrow toward the model, if it is not already changed.

5. Enter **10** in the dimension box in the **Profile Rib** dashboard.

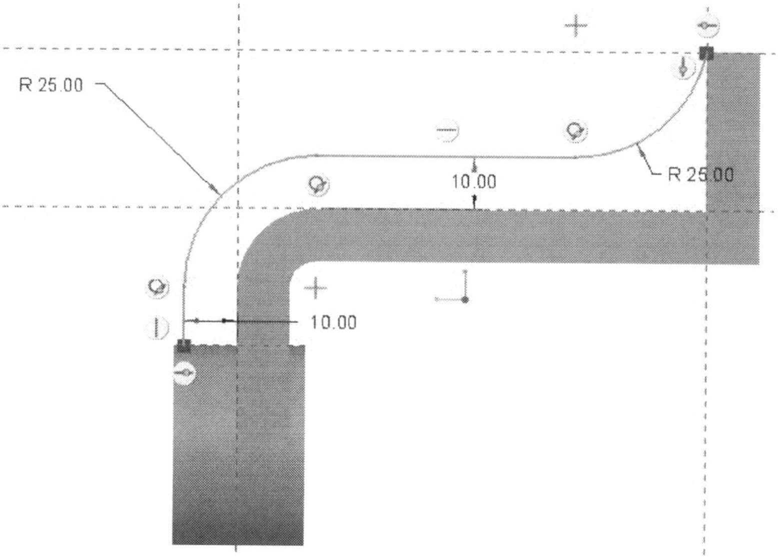

***Figure 6-113** Open sketch of the rib feature with dimensions and constraints*

6. Choose the **OK** button from the **Profile Rib** dashboard to complete creating the model and exit this feature creation tool.

   All features in the model have been created and the model is now complete. The trimetric shaded view of the completed model is shown in Figure 6-114.

Saving the Model

1. Choose the **Save** button from the **File** menu and save the model. The order of feature creation can be seen from the **Model Tree** shown in Figure 6-115.

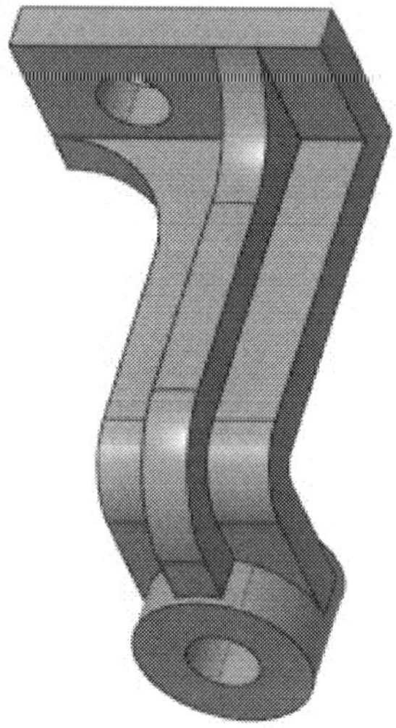

***Figure 6-114** The default trimetric view of the model*

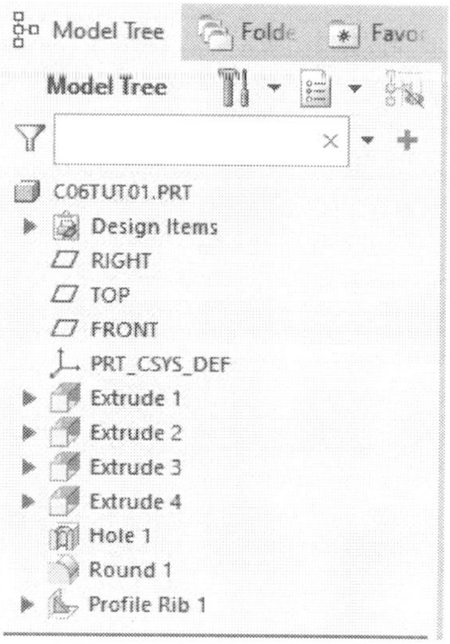

***Figure 6-115** The **Model Tree** for Tutorial 1*

Tutorial 2

In this tutorial, you will create the model shown in Figure 6-116. The dimensions of the model are shown in Figure 6-117. **(Expected time: 30 min)**

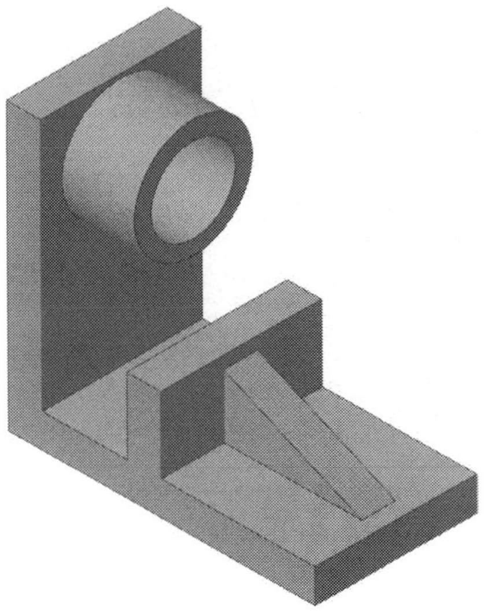

Figure 6-116 *Isometric view of the solid model*

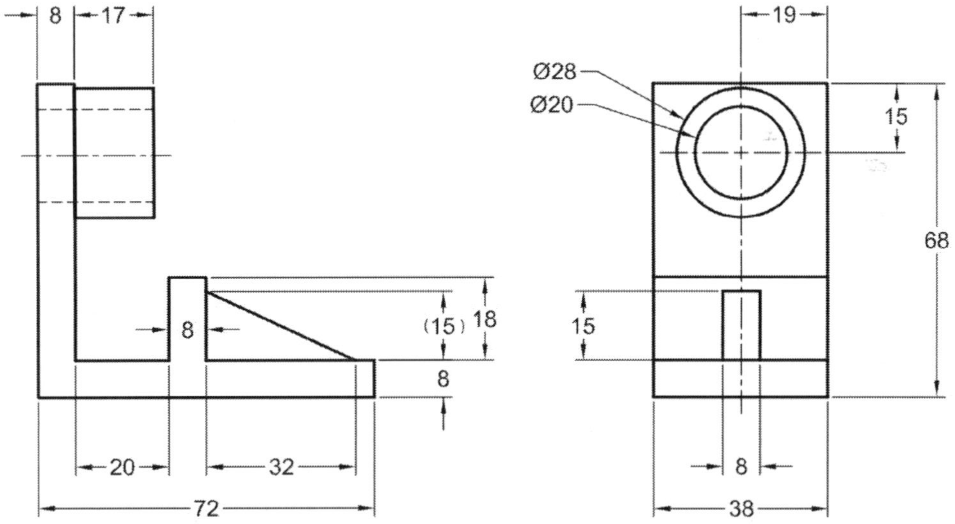

Figure 6-117 *Front and right view of the model*

The following steps are required to complete this model:

Examine the model and determine the number of features in it, refer to Figure 6-116.

a. Create the base feature.
b. Create the cylindrical feature.

c. Create the hole feature coaxially on the cylindrical feature.
d. Create the rib feature.

The working directory has already been selected in Tutorial 1 and therefore, you do not need to select it again. However, if you need to change the working directory, choose **File > Manage Session > Select Working Directory** and then select *c06* in the **Select Working Directory** dialog box.

Starting a New Object File

1. Start a new part file and name it as *c06tut02*.

   The three default datum planes and the **Model Tree** appear in the drawing area.

Creating the Base Feature

1. Select the **FRONT** datum plane from the drawing area and choose the **Extrude** tool from the mini popup toolbar; the sketcher environment is invoked automatically.

2. In the sketcher environment, create the sketch of the base feature and apply constraints and dimensions to it, as shown in Figure 6-118.

3. After the sketch is completed, choose the **OK** button and exit the sketcher environment; the **Extrude** dashboard is enabled and appears above the drawing area.

4. Enter **38** as the depth in the dimension box that is present on the **Extrude** dashboard; the default trimetric view of the base feature is displayed, as shown in Figure 6-119.

5. Choose the **OK** button from the **Extrude** dashboard to exit the feature creation tool.

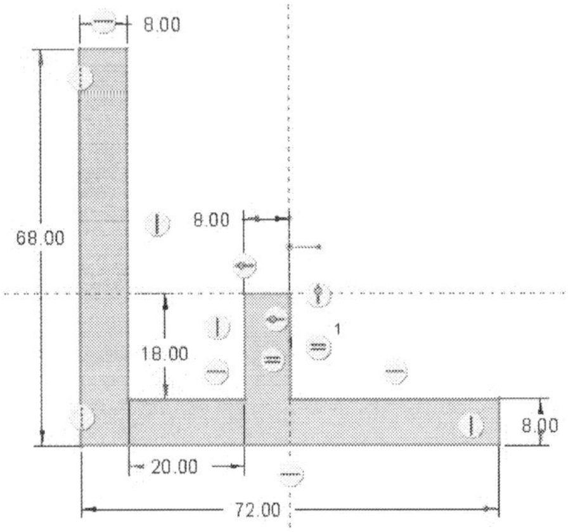

Figure 6-118 *Sketch of the base feature with dimensions and constraints*

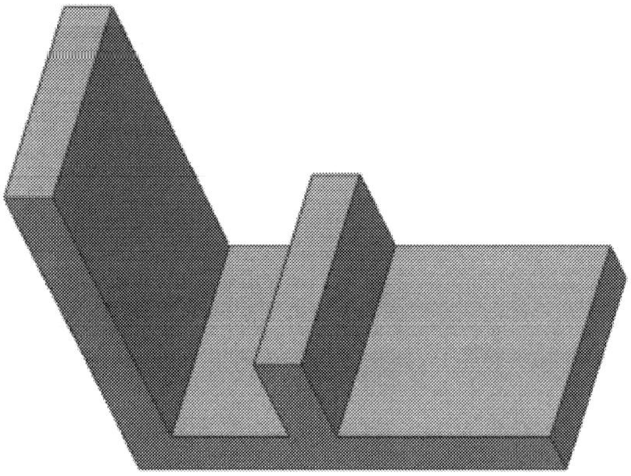

Figure 6-119 *The default trimetric view of the base feature*

Note

In this tutorial, the base feature can also be created by extruding it on both sides of the sketching plane that is the ***FRONT*** *datum plane. This would reduce the step required to create the datum plane, which in turn is used to create the rib feature. However, to familiarize you with creating embedded datum planes, the base feature is extruded on one side of the sketching plane.*

Creating the Second Feature

The second feature is a cylindrical feature that is sketched on the planar surface of the base feature which is shown in Figure 6-120.

1. Choose the **Extrude** tool from the **Shapes** group.

2. Choose the **Placement** tab and then from the slide-down panel, choose the **Define** button; the **Sketch** dialog box is displayed.

3. Select the face of the base feature as the sketching plane, refer to Figure 6-120.

4. Select the **TOP** datum plane from the drawing area and then select the **Top** option from the **Orientation** drop-down list.

5. Choose the **Sketch** button from the **Sketch** dialog box to enter into the sketcher environment.

6. Draw the sketch for the second feature. Apply and modify the dimensions, as shown in Figure 6-121.

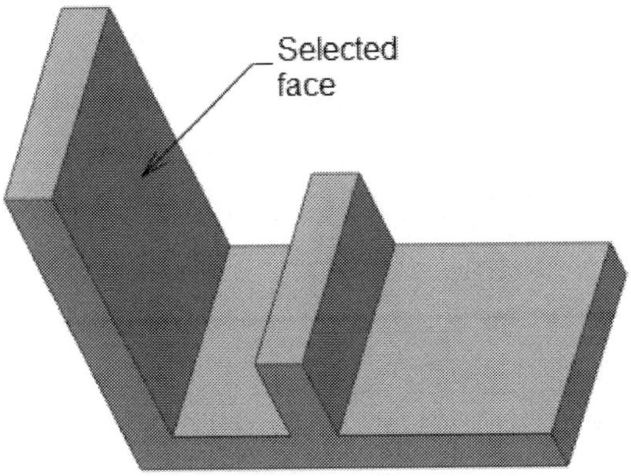

Figure 6-120 *The face of the base feature selected as the sketching plane*

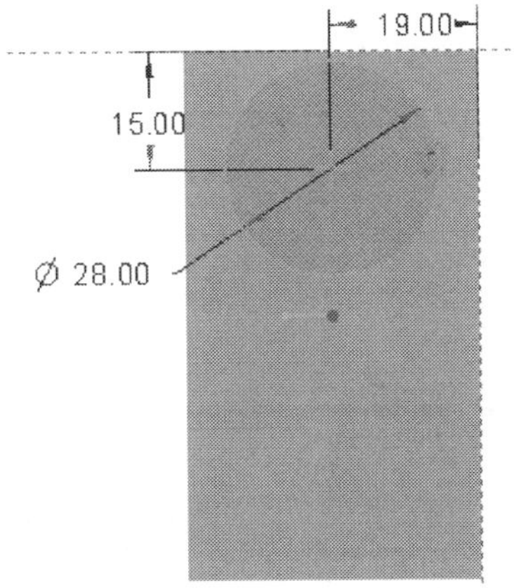

Figure 6-121 *Sketch and dimensions of the second feature*

7. Turn the display of the model to **Shading With Edges**. Exit the sketcher environment by choosing the **OK** button; the **Extrude** dashboard is enabled and appears above the drawing area.

8. Enter **17** in the dimension box in the **Extrude** dashboard.

9. Choose the **OK** button from the **Extrude** dashboard to exit the feature creation tool. The trimetric view of the model with the second feature is shown in Figure 6-122.

Creating the Hole Feature

The hole feature will be created using the **Hole** dashboard. The coaxial hole will be created on the cylindrical feature. The axis of the cylindrical feature will be used as the axial reference to create the coaxial hole.

1. Choose the **Hole** tool from the **Engineering** group in the **Ribbon**; the **Hole** dashboard is displayed. The **Simple** tool in the **Hole** dashboard is chosen by default.

2. Choose the **Placement** tab from the **Hole** dashboard; a slide-down panel is displayed.

3. Select the front face of the cylindrical feature to place the hole.

   As you select the front face of the cylindrical feature, the preview of the hole is displayed in the drawing area. Now, you need to specify the reference for the placement of hole.

4. Hold-down the CTRL key and select the axis of the cylindrical feature from the drawing area.

   The **Coaxial** option is selected automatically in the drop-down list of the slide-down panel.

5. In the diameter edit box of the **Hole** dashboard, enter **20** and press ENTER.

6. From the depth flyout on the **Hole** dashboard, choose the **Drill to intersect with all surfaces** button.

7. Choose the **OK** button from the **Hole** dashboard; the hole is created and the trimetric view of the shaded model with the hole is displayed, as shown in Figure 6-123.

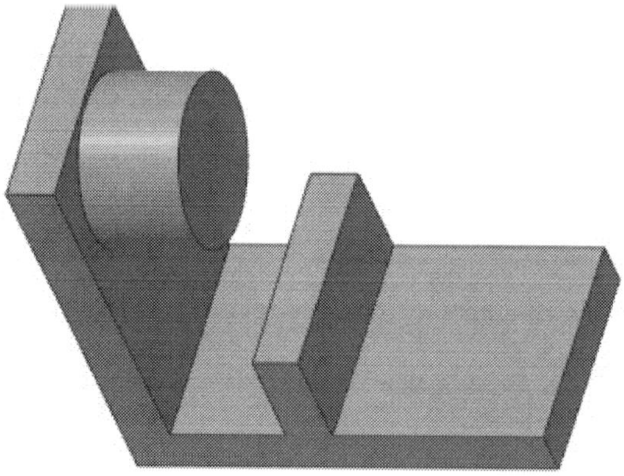

Figure 6-122 *Model with the second feature*

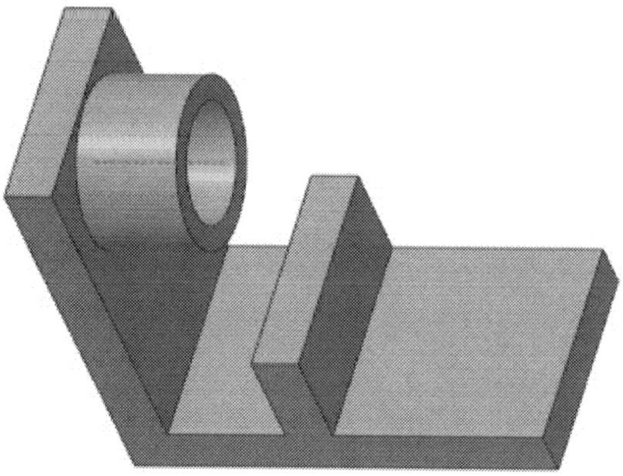

Figure 6-123 *Model with the hole feature*

Creating the Rib Feature

To create the rib feature, an internal datum plane is created. As mentioned earlier, rib features are always drawn from the side view.

1. Choose the **Profile Rib** tool from the **Rib** drop-down in the **Engineering** group.

2. Choose the **References** tab; a slide-down panel is displayed. Choose the **Define** button; the **Sketch** dialog box is displayed.

3. Choose the **Plane** tool from the **Datum** group of the **Model** tab in the **Profile Rib** dashboard; the **Datum Plane** dialog box is displayed. You may need to move the **Sketch** dialog box to bring the **Plane** button into view.

4. Select the axis of the hole, press the CTRL key, and select the **FRONT** datum plane. You can view your selections in the **References** collector.

5. Choose the **Offset** button in the **References** collector; a drop-down list appears on the right of the reference. Select the **Parallel** option from this drop-down list.

6. Choose the **OK** button from the **Datum Plane** dialog box to exit it.

   A datum plane that passes through the selected axis and is parallel to the **FRONT** datum plane will be created. This datum plane is selected automatically as the sketching plane. Now, you need to select the reference plane.

7. Select the **TOP** datum plane and then select the **Top** option from the **Orientation** drop-down list.

8. Choose the **Sketch** button; the system takes you to the sketcher environment. Choose the **No Hidden** option from the **Display Style** drop-down in the **Graphics** toolbar.

9. Draw the open sketch of the rib feature and apply dimensions, as shown in Figure 6-124. Exit the sketcher environment by choosing the **OK** button; the **Profile Rib** dashboard is enabled and appears above the drawing area.

10. Choose the **Flip** button on the **References** slide-down panel to flip the direction of the arrow; the preview of the rib is displayed on the model. Choose this button only when the preview of the rib is not displayed. Alternatively, click on the arrow to flip the direction.

11. Enter **8** as the thickness of the rib in the edit box in the **Profile Rib** dashboard. Next, choose the **OK** button.

    The trimetric view of the complete model with the rib feature is shown in Figure 6-125.

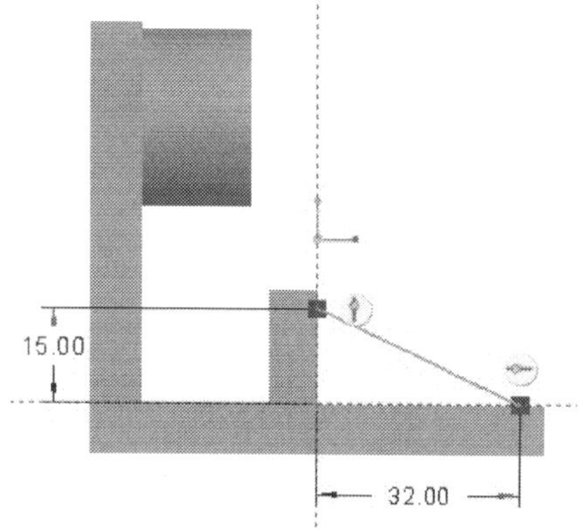

Figure 6-124 *Sketch for the rib feature with the model display set to* ***No hidden***

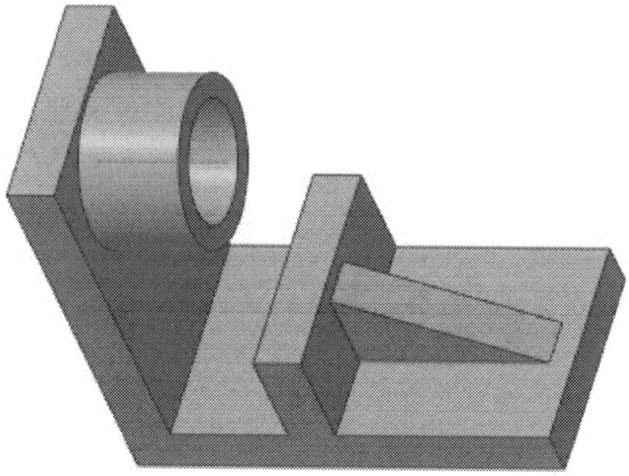

Figure 6-125 *The default trimetric view of the final model*

Saving the Model

1. Choose the **Save** button from the **File** menu to save the model and then close the active window.

Tutorial 3

In this tutorial, you will create the model shown in Figure 6-126. The solid model, dimensions, and the front and the right views are also shown in the Figure 6-127. **(Expected time: 45 min)**

Figure 6-126 *Isometric view of the model*

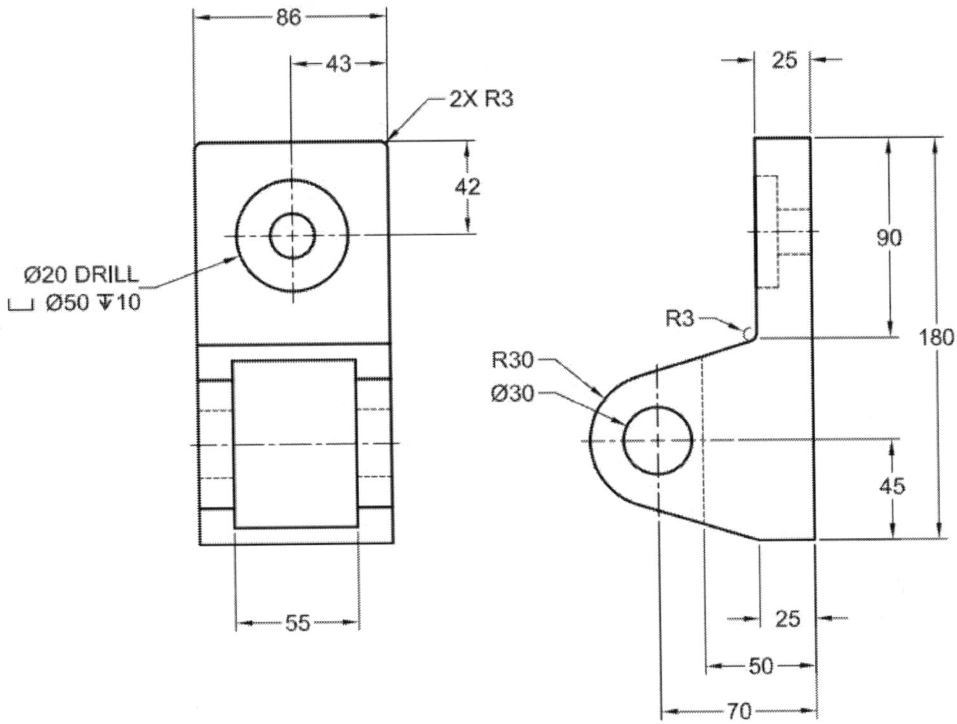

Figure 6-127 Front and right view of the model

The following steps are required to complete this model:

Examine the model and determine the number of features in it, refer to Figure 6-126.

a. Create the base feature.
b. Create the cut feature.
c. Create a counterbore hole.
d. Create the rounds.

Starting a New Object File

1. Start a new part file and name it *c06tut03*.

   The three default datum planes and the **Model Tree** appear in the drawing area.

Creating the Base Feature

1. Choose the **Extrude** tool from the **Shapes** group.
2. Choose the **Placement** tab; a slide-down panel is displayed. Choose the **Define** button; the **Sketch** dialog box is displayed.
3. Select the **RIGHT** datum plane as the sketching plane.

4. Select the **TOP** datum plane from the drawing area and then select the **Top** option from the **Orientation** drop-down list.

5. Choose the **Sketch** button from the **Extrude** dashboard to enter into the sketcher environment.

6. Once you enter into the sketcher environment, create the sketch of the base feature and apply dimensions and constraints, as shown in Figure 6-128.

7. Exit the sketcher environment by choosing the **OK** button; the **Extrude** dashboard is enabled and appears above the drawing area.

8. Select the **Blind** option from the **Side 1** drop-down in the **Options** tab. Next, enter **86** in the dimension box in the **Extrude** dashboard.

9. Choose the **OK** button to exit the feature creation tool. The base feature is completed. The default trimetric view of the base feature is shown in Figure 6-129.

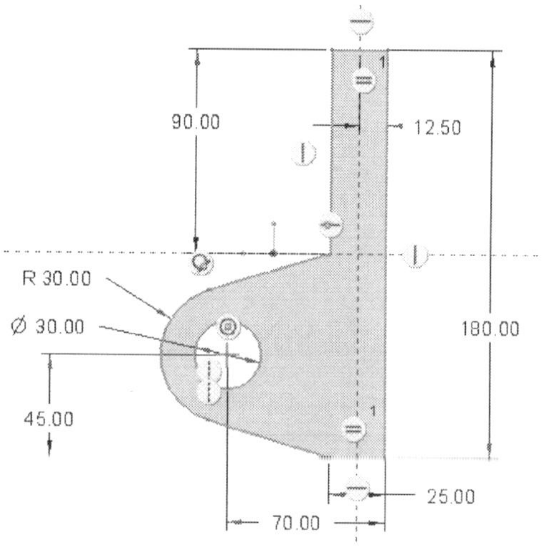

Figure 6-128 *Sketch of the base feature with dimensions and constraints*

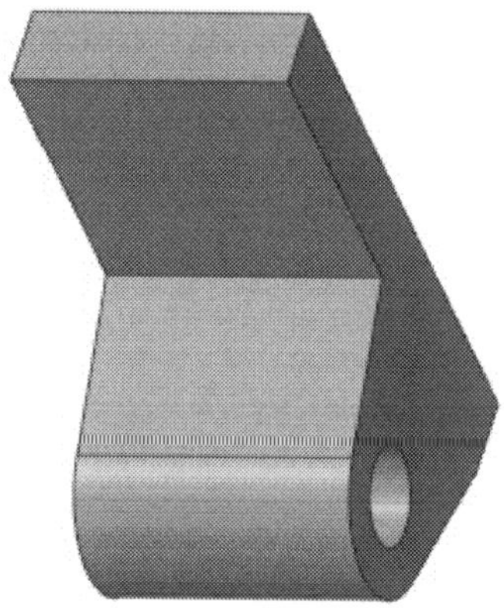

Figure 6-129 *The default trimetric view of the base feature*

Creating the Second Feature

The second feature is an extruded cut feature. This cut feature is created on a datum plane that passes through the center of the base feature.

1. Choose the **Extrude** tool from the **Shapes** group.

2. Choose the **Remove Material** button.

3. In the depth flyout, choose the **Extrude symmetrically on both sides of the sketch plane** button.

4. Choose the **Placement** tab; a slide-down panel is displayed. Choose the **Define** button; the **Sketch** dialog box is displayed.

5. Choose the **RIGHT** datum plane as the sketching plane. Next choose the **TOP** datum plane in the **Reference** collector and then select the **Top** option from the **Orientation** drop-down list.

6. Choose the **Sketch** button; you enter into the sketcher environment.

7. Draw the sketch for the cut feature and apply the required dimensions and constraints, as shown in Figure 6-130.

   In Figure 6-130, some dimensions appear light in color. These dimensions are weak dimensions and it is not important to convert them into strong dimensions because these dimensions are not important for the creation of this feature. However, the geometry for the cut should be similar to that shown in Figure 6-128.

8. After the sketch is completed, exit the sketcher environment; the **Extrude** dashboard is enabled and it appears above the drawing area.

9. Enter a depth value of **55** in the dimension box in the **Extrude** dashboard. The system will accept this depth symmetrical to the sketching plane.

10. Choose the **OK** button to exit the feature creation tool. The cut feature is completed now and the default trimetric view of the cut feature along with the base feature is shown in Figure 6-131.

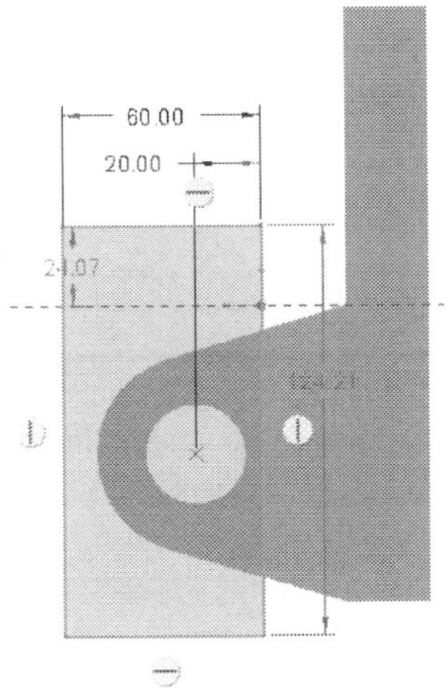

Figure 6-130 *Sketch of the cut feature with dimensions and constraints*

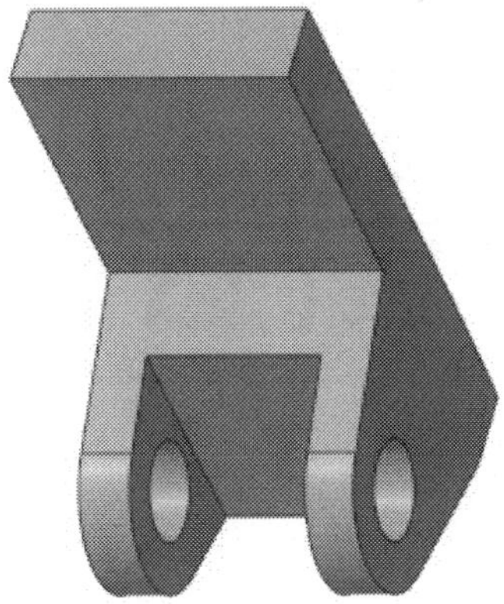

Figure 6-131 *The default trimetric view of the model*

Creating the Hole Feature

The hole is created using the **Hole** dashboard.

1. Choose the **Hole** tool from the **Engineering** group; the **Hole** dashboard is displayed.

2. In the **Hole** dashboard, choose the **Drilled** button in the **Profile** area and the **Counterbore** button in the **Settings** area.

3. Click on the **Shape** tab to open the slide-down panel and specify the attributes of the counterbore hole. For dimensions of the counterbore hole, refer to Figure 6-127.

4. Select the front face of the base feature to place the hole; a preview of the hole appears on the selected face.

5. Choose the **Placement** tab from the **Hole** dashboard; a slide-down panel is displayed.

6. Click in the **Offset References** collector; you are prompted to select two references.

7. Select the two references to position the hole feature. Refer to Figure 6-127 for dimensions of the hole placement. Note that the second reference should be selected by using the CTRL key + left mouse button.

8. Modify the dimensions of the references by entering suitable values in the **Placement** slide-down panel. The center of hole is at a distance of **42** from the top edge and **43** from the right edge. After viewing the preview of the hole, you may have to specify a negative value for any one of the distances.

9. Choose the **OK** button to completely create the hole and Figure 6-132 shows the hole created.

Creating the Round Features

In this section, you will create the round feature using the **Simple** option.

1. Choose the **Round** button from the **Engineering** group; the **Round** dashboard is displayed.

2. Using the CTRL key, select the edges that are shown in Figure 6-133. The selected edges turn green in color.

3. Enter **3** in the dimension box in the **Round** dashboard or double-click on the default value displayed on the preview of the round and enter the radius value in the edit box displayed. The round is created and its preview appears in the drawing area.

4. Choose the **OK** button to exit the round feature creation tool. The trimetric view of the final model with the round feature created is shown in Figure 6-134.

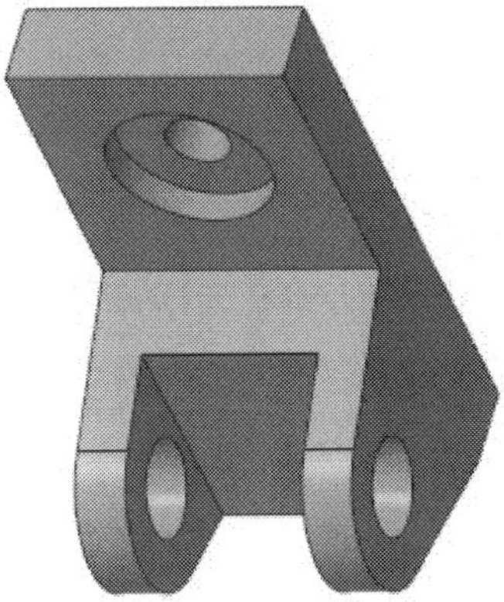

Figure 6-132 The default trimetric view

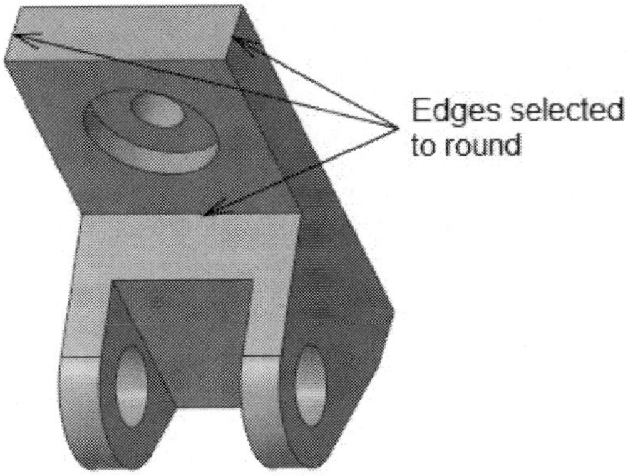

Figure 6-133 *Edges selected to create the round feature*

Saving the Model

1. Choose the **Save** button from the **File** menu and save the model.

   The order of feature creation can be seen from the **Model Tree** shown in Figure 6-135.

Figure 6-134 *Final model for Tutorial 3*

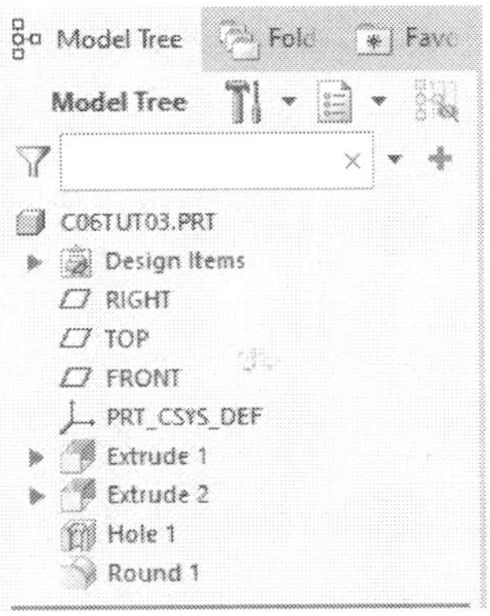

Figure 6-135 *The **Model Tree** for Tutorial 3*

Tutorial 4

Create the model shown in Figure 6-136. The dimensions, top view, and front section view of the model are shown in Figure 6-137. **(Expected time: 45 min)**

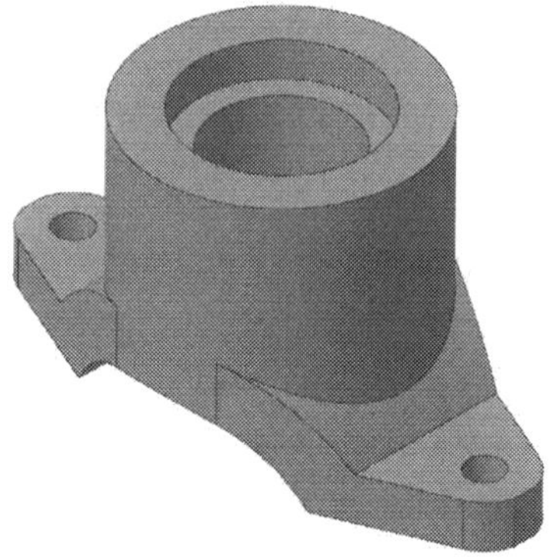

Figure 6-136 Isometric view of the model

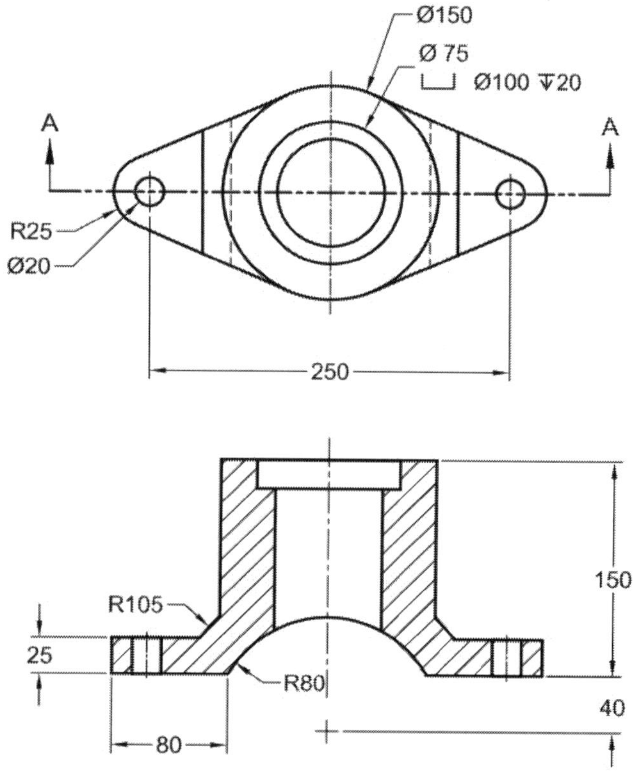

Figure 6-137 Orthographic views of the model

The following steps are required to complete this model:

Examine the model and determine the number of features in it, refer to Figure 6-136.

a. Create the base feature.
b. Create the cut feature.
c. Create the cylindrical feature on an offset datum plane.
d. Create a counterbore coaxial hole on the cylindrical feature.
e. Create straight holes on the top face of the base feature.

Starting a New Object File

1. Start a new part file and name it as *c06tut04*.

The three default datum planes are displayed in the drawing area if the **Plane Display** button is turned on.

Selecting the Sketching Plane for the Base Feature

In this model, you need to draw the base feature on the **FRONT** datum plane because the direction of extrusion of the base feature is perpendicular to the **FRONT** datum plane.

1. Choose the **Extrude** tool from the **Shapes** group.

2. Choose the **Placement** tab; a slide-down panel is displayed. Choose the **Define** button; the **Sketch** dialog box is displayed.

3. Select the **FRONT** datum plane as the sketching plane. Next, select the **RIGHT** datum plane as reference plane and set its orientation to **Right**.

4. Choose the **Sketch** button to enter into the sketcher environment.

Creating the Base Feature

From the model, the section to be extruded for the base feature is not evident. Therefore, you need to visualize the sketch for the base feature. The sketch is shown in Figure 6-138. When this sketch is extruded, it will create the base feature.

1. Draw the sketch using various sketcher tools, as shown in Figure 6-138.

2. The sketch is dimensioned automatically and some weak dimensions are assigned to it. Add the required constraints to it and modify the weak dimensions, as shown in Figure 6-138.

3. Choose the **OK** button; the **Extrude** dashboard is displayed.

4. Choose the **Saved Orientations** button from the **Graphics** toolbar; a flyout is displayed. Choose the **Default Orientation** option from the flyout; the model is set to its default trimetric orientation.

   The default view is displayed which gives you a better view of the sketch in the 3D space. Also, an arrow is also displayed on the model indicating the direction of extrusion.

5. Enter **150** in the dimension box that is present in the **Extrude** dashboard.

6. Choose the **OK** button from the **Extrude** dashboard.

   The base feature is completed, as shown in Figure 6-139. You can use the middle mouse button to spin the model to view it from various directions.

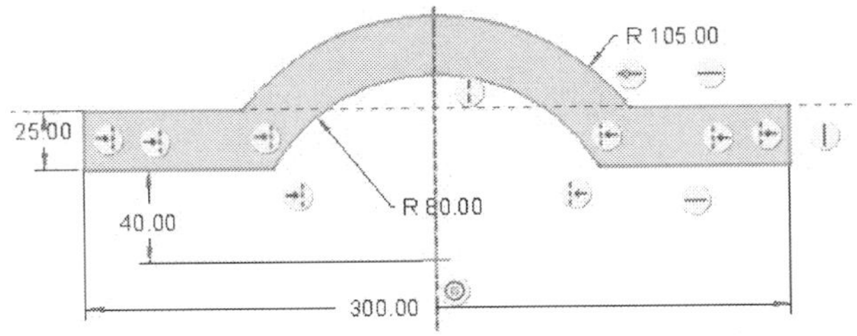

Figure 6-138 Sketch with dimensions and constraints

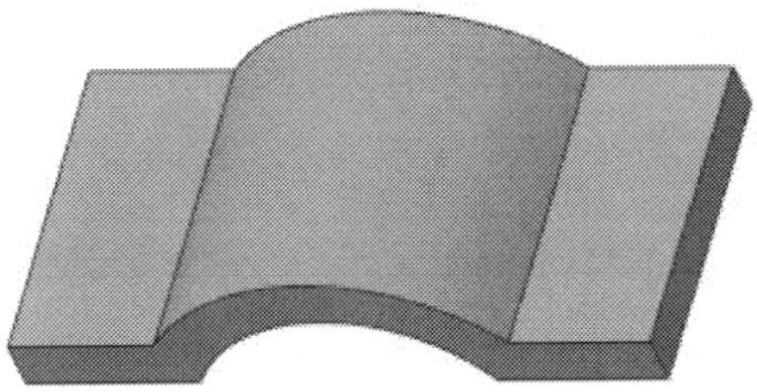

Figure 6-139 Default view of the base feature

Creating the Second Feature

The next feature is an extruded cut feature. The sketching plane for this feature is the bottom face of the base feature. To get the required shape of the base of the model, you need to cut the base feature in such a way that you get the required shape. This sketch and its dimensions can be referred from the top view of the model shown in Figure 6-137. Before drawing the sketch, change the model display to **No hidden** if required.

1. Choose the **Extrude** tool from the **Shapes** group.

2. Choose the **Remove Material** button from the **Extrude** dashboard.

3. Choose the **Placement** tab; a slide-down panel is displayed. Choose the **Define** button from the slide-down panel; the **Sketch** dialog box is displayed.

4. Select the bottom face of the base feature as the sketching plane.

5. Select the **RIGHT** datum plane and from the **Orientation** drop-down list, select the **Right** option.

6. Choose the **Sketch** button to enter into the sketcher environment.

7. Draw the sketch of the cut feature, and apply constraints and dimensions to it, as shown in Figure 6-140.

8. Choose the **OK** button to exit the sketcher environment; the **Extrude** dashboard is displayed above the drawing area.

9. Choose the **Shading With Edges** button from the **Display Style** drop-down in the **Graphics** toolbar.

10. Choose the **Saved Orientations** button from the **View** tab; a flyout is displayed. Choose the **Default Orientation** option from the flyout; the model is set to its default trimetric orientation.

11. Choose the **Change material direction of extrude to other side of sketch** button from the **Extrude** dashboard; an arrow is displayed pointing in the direction of extrusion, as shown in Figure 6-141. You can also click on the Material direction arrow to change the direction of material removal.

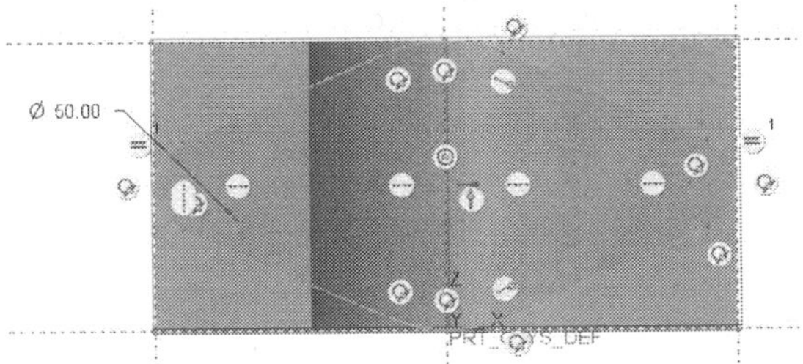

Figure 6-140 Sketch with dimensions and constraints for the cut feature

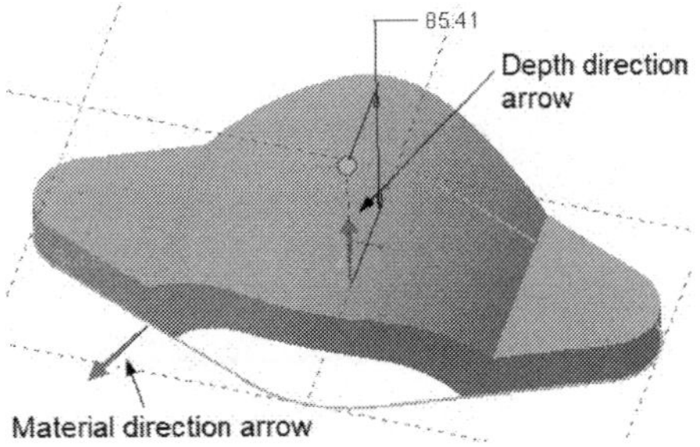

Figure 6-141 Two arrows and the preview of the cut feature

12. Choose the **Options** tab in the **Extrude** dashboard to display the slide-down panel. From the **Side 1** drop-down list, select the **Through All** option; the model can be previewed in the drawing area.

13. Choose the **OK** button from the **Extrude** dashboard to complete the feature creation. The model after creating the cut feature is shown in Figure 6-142.

Creating the Sketching Plane for the Third Feature

You need to create a datum plane to create the third feature. The datum plane will be at a distance of 150 from the bottom face of the model.

1. Choose the **Plane** tool from the **Datum** group; the **Datum Plane** dialog box is displayed.

2. Spin the model using the middle mouse button and then using the left mouse button, select the bottom face of the model.

   When you select the bottom face of the model, the **Offset** constraint is displayed in the **References** collector of the dialog box. Note that the arrow points in the opposite direction. Therefore, you need to enter a negative value of the offset distance.

3. In the **Translation** dimension box, enter **150** and press ENTER.

4. Choose the **OK** button from the **Datum Plane** dialog box.

   Datum plane **DTM1** is created, as shown in Figure 6-143, and will be selected as the sketching plane for creating the sketch.

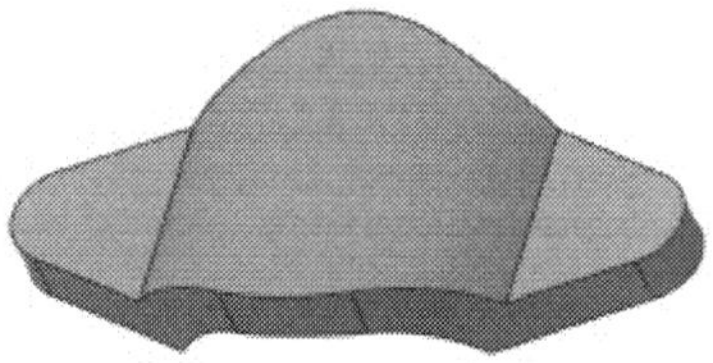

Figure 6-142 Model after creating the cut feature

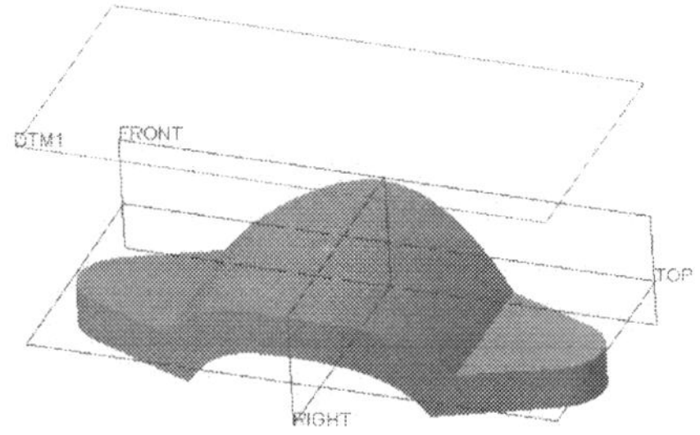

Figure 6-143 Datum plane created at an offset distance

Creating the Third Feature

The plane **DTM1** will be selected as the sketching plane and the depth of extrusion will be given from this plane. The third feature is cylindrical in shape and its outer edge is tangent to the edge of the base feature.

1. Choose the **Extrude** tool from the **Shapes** group; the **Extrude** dashboard is displayed.
2. Choose the **Placement** tab; a slide-down panel is displayed. Choose the **Define** button from the slide-down panel; the **Sketch** dialog box is displayed.
3. Select **DTM1** as the sketching plane; an arrow appears on the datum plane.
4. Select the **FRONT** datum plane and then from the **Orientation** drop-down list, select the **Top** option.
5. Choose the **Sketch** button to enter into the sketcher environment.
6. Draw a concentric circle for the cylindrical feature, as shown in Figure 6-144.
7. Choose the **OK** button to exit the sketcher environment.
8. Choose the **Saved Orientations** button from the **Graphics** toolbar; a flyout is displayed. Choose the **Default Orientation** option from the flyout; the model is set to its default trimetric orientation.
9. Choose the **Options** tab in the **Extrude** dashboard to display the slide-down panel. From the **Side 1** drop-down list, select the **To Next** option. The model can be previewed in the drawing area.
10. Choose the **OK** button from the **Extrude** dashboard to complete the feature creation. The model after creating the feature is shown in Figure 6-145.

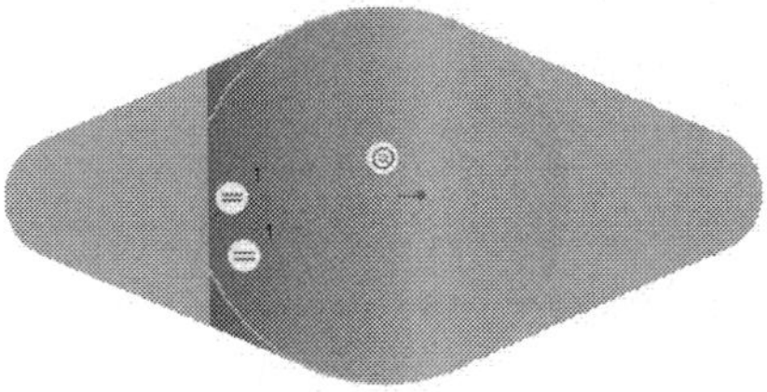

Figure 6-144 Sketch for the cylindrical feature

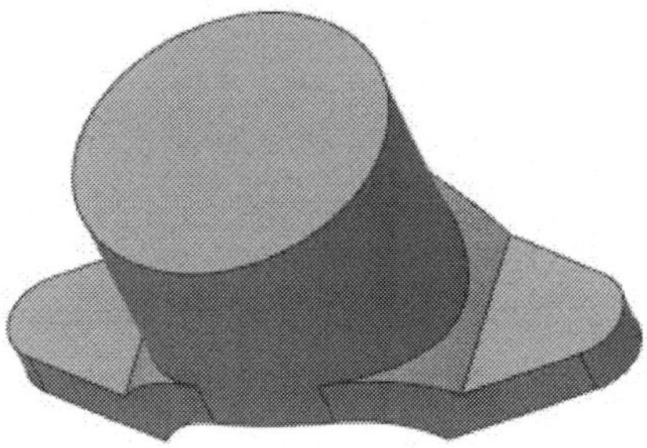

Figure 6-145 Model after creating the cylindrical feature

Creating the Counterbore Hole

The fourth feature is a counterbore hole that can be easily created using the **Hole** feature creation tool.

1. Choose the **Hole** tool from the **Engineering** group; the **Hole** dashboard is displayed.
2. Choose the **Simple** and then the **Drilled** button; some buttons are added to the dashboard.
3. Choose the **Counterbore** button from the dashboard.
4. Next, choose the **Shape** tab and enter the values, as shown in Figure 6-146.

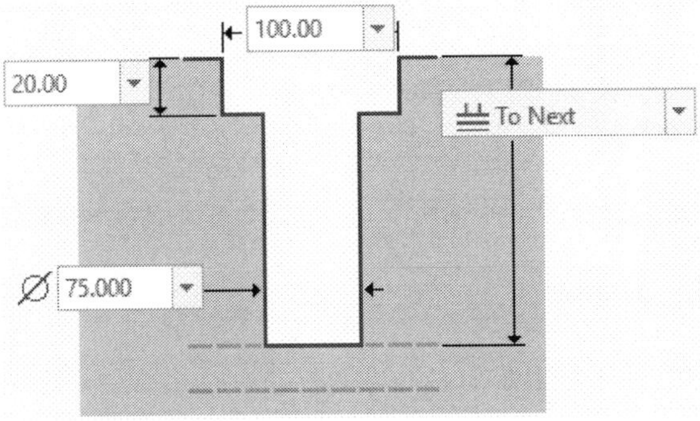

Figure 6-146 Sketch with dimensions

5. Select the **To Reference** option from the drop-down list and then select the bottom face of the base feature.
6. Choose the **Placement** tab from the **Hole** dashboard; a slide-down panel is displayed.
7. Select the top face of the third feature and then select the axis of the cylindrical feature by pressing the CTRL key; a preview of the hole can be viewed in the model.

   The **Coaxial** option is selected automatically in the **Type** drop-down list of the slide-down panel.

8. Choose the **OK** button to exit the **Hole** dashboard; the model with the counterbore hole is shown in Figure 6-147.

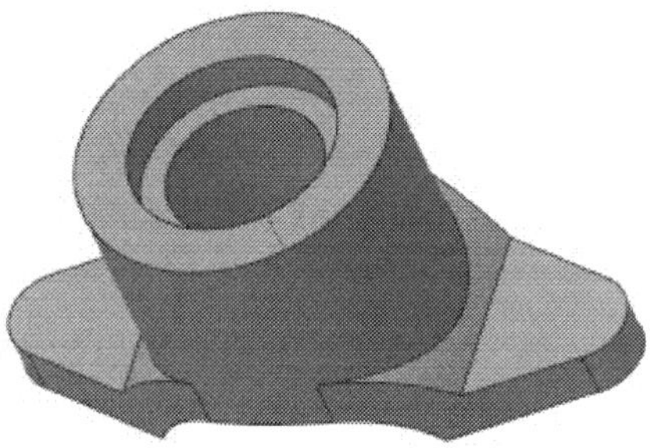

Figure 6-147 Model with counterbore hole

Creating the Datum Axis for the Hole Features

The last feature is a pair of two holes that is on the top face of the base feature. The two holes will be placed coaxially with the two axes that you need to create.

1. Choose the **Axis** button from the **Datum** group; the **Datum Axis** dialog box is displayed.
2. Select the curved surface shown in Figure 6-148 to create the datum axis.
3. Select **Through** from the drop-down list in the **Reference** tab and choose the **OK** button from the dialog box to accept the settings and exit.
4. Similarly, create a datum axis on the left of the model. Figure 6-149 shows the model after creating the two datum axes.

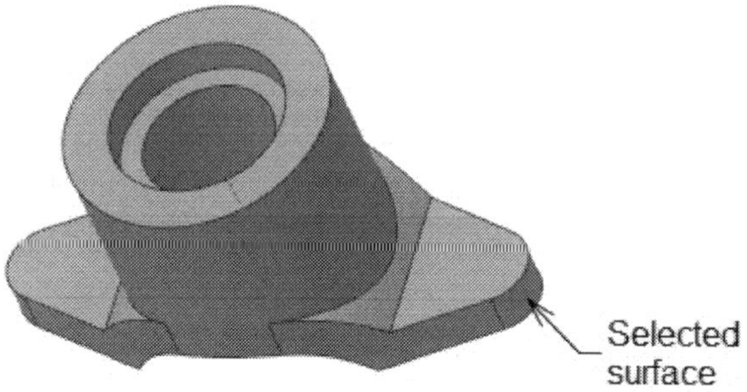

Figure 6-148 Surface to be selected for creating the datum axis

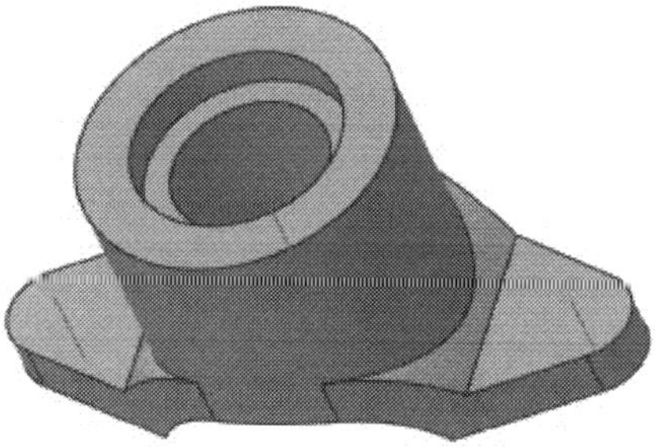

Figure 6-149 The two datum axes

Creating Holes

To create the two coaxial holes, you need to invoke the **Hole** dashboard twice. This is because only one hole can be created at a time by using the **Hole** dashboard.

1. Choose the **Hole** tool from the **Engineering** group; the **Hole** dashboard is displayed.
2. Select the top face of the base feature; a preview of the hole appears on the selected face.
3. Choose the **Placement** tab from the **Hole** dashboard; the slide-down panel is displayed.

4. Select the axis of the feature by pressing the CTRL key.

   The **Coaxial** option is selected automatically in the **Type** drop-down list of the slide-down panel.

5. Enter **20** as the diameter of hole in the dimension box of the **Hole** dashboard and choose the **Drill up to next surface** option from the dashboard.

6. Choose the **OK** button to exit the **Hole** dashboard.

7. Again, invoke the **Hole** dashboard and create the second hole similar to the first hole.

   The two holes on the base feature are shown in Figure 6-150.

Saving the Model

1. Choose the **Save** button from the **File** menu and save the model. The order of feature creation can be seen from the **Model Tree** shown in Figure 6-151.

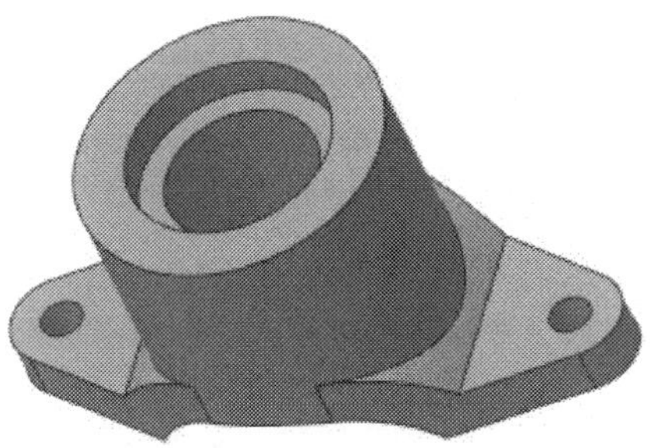

Figure 6-150 *Model after creating the two holes*

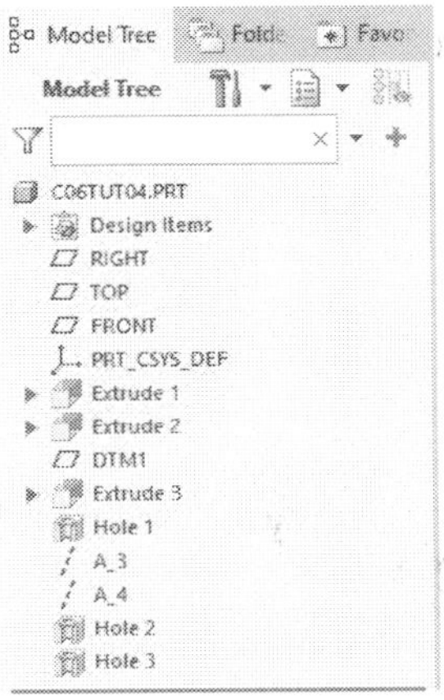

Figure 6-151 *The **Model Tree** for Tutorial 4*

Self-Evaluation Test

Answer the following questions and then compare them to those given at the end of this chapter:

1. The profile rib feature is always created from the _________ view.

2. A _________ hole is a stepped hole and has two diameters, large and small.

3. Straight holes are the holes that have a circular cross-section with a _________ diameter throughout the depth.

4. __________ are defined as thin wall-like structures that are used to bind the joints together so that they do not fail under increased load.

5. By default, the suppressed features are displayed in the __________.

6. A hole created using the **Hole** dashboard is parametric in nature. (T/F)

7. A hole cannot be created on both sides of the sketching plane or the placement plane. (T/F)

8. In Creo Parametric, you can sketch profile of a hole. (T/F)

9. The **Full Round** button in the **Sets** slide-down panel is used to enter the radius of the round. (T/F)

10. The **Model Tree** is used extensively in Creo Parametric for editing a model. (T/F)

Review Questions

Answer the following questions:

1. Which of the following is a stepped hole with two diameters?

(a) Counterbore (b) Countersink
(c) Straight (d) None

2. Which of the following options of the **REROUTE** menu is used to change the sequence in which the features are created?

(a) **Feature** (b) **Modify**
(c) **Regenerate** (d) **Reorder**

3. At the intersection of how many edges does a **Corner** chamfer create a beveled surface?

(a) One (b) Two
(c) Three (d) None

4. When you redefine a profile rib feature, the __________ dashboard is displayed.

5. The sketch of a rib feature can be extruded to __________ side(s) of the sketching plane.

6. Rounds and chamfers are used in engineering components to reduce the __________ on the corners.

7. If a feature being suppressed has some child features, then the child features will also be suppressed. (T/F)

8. While in the sketch mode, the constraints to a sketch should be applied before applying the dimensions. (T/F)

9. The chamfers created in Creo Parametric are parametric in nature. (T/F)

10. The **Model Tree** can be used to redefine a feature. (T/F)

EXERCISES

Exercise 1

Create the model shown in Figure 6-152. The dimensions and front and top views of the model are shown in Figure 6-153. **(Expected time: 45 min)**

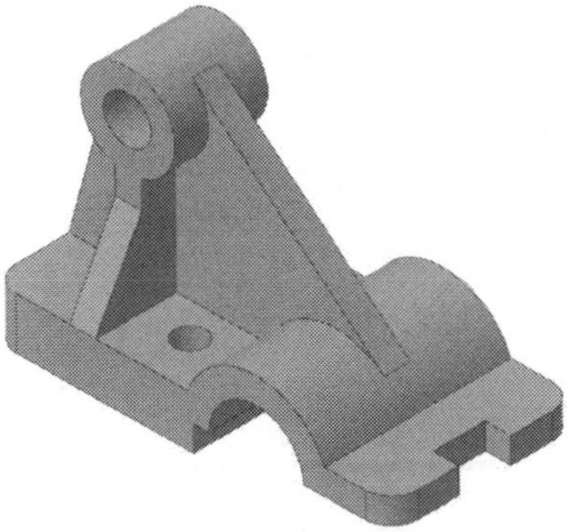

Figure 6-152 Isometric view of the model

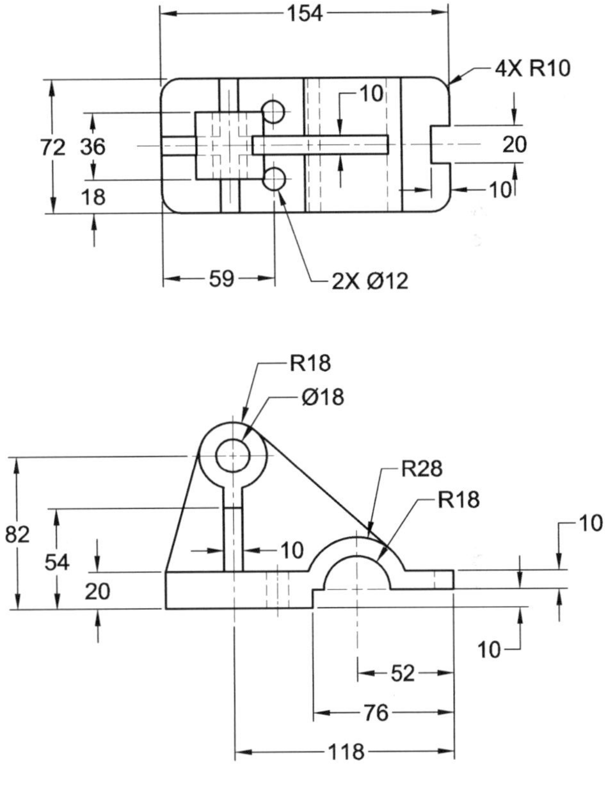

Figure 6-153 Orthographic views of the model

Exercise 2

Create the model shown in Figure 6-154. The dimensions and front and right-side views of the model are shown in Figure 6-155. **(Expected time: 30 min)**

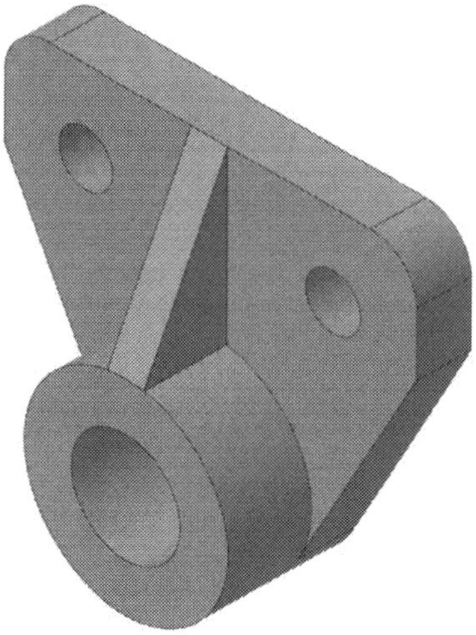

Figure 6-154 Isometric view of the model

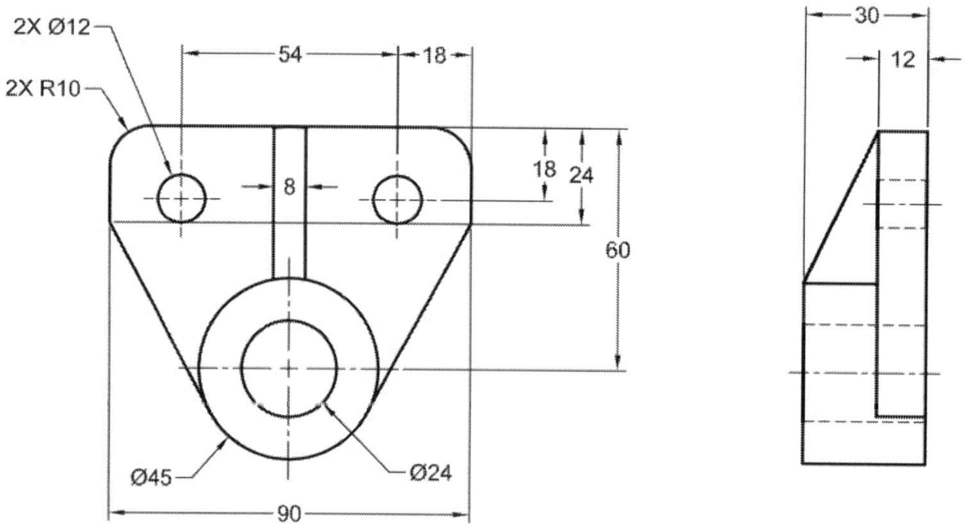

Figure 6-155 Front and right view of the model

Answers to Self-Evaluation Test

1. side, **2.** counterbore, **3.** constant, **4.** Ribs, **5.** **Model Tree**, **6.** T, **7.** F, **8.** T, **9.** F, **10.** T

Chapter 7

Options Aiding Construction of Parts-II

Learning Objectives

After completing this chapter, you will be able to:

- *Create shell and draft features*
- *Create cosmetic sketches, cosmetic threads, and cosmetic grooves*
- *Create toroidal bend and spinal bend*
- *Create warp features*

OPTIONS AIDING CONSTRUCTION OF PARTS

In the previous chapter, you learned about using the **Hole**, **Round**, **Chamfer**, and **Rib** tools. These tools are used to create engineering features in the models.

In this chapter, you will learn about the remaining feature creation tools such as **Shell** and **Draft**, provided in Creo Parametric. Also, you will learn to create cosmetic features such as cosmetic sketch, cosmetic thread, and cosmetic groove.

CREATING SHELL FEATURES

Ribbon: Model > Engineering > Shell

The **Shell** tool scoops out material from the model and at the same time removes the selected faces, leaving behind a thin model with some specified wall thickness. If you do not select a surface to remove, a closed shell is created as a part completely hollow from inside. In this case, you can later add necessary cuts or holes to complete the geometry. The **Shell** tool can be invoked by choosing the **Shell** tool from the **Engineering** group of the **Model** tab in the **Ribbon**. When you choose this tool; the **Shell** dashboard will be displayed, as shown in Figure 7-1. The options in this dashboard are discussed next.

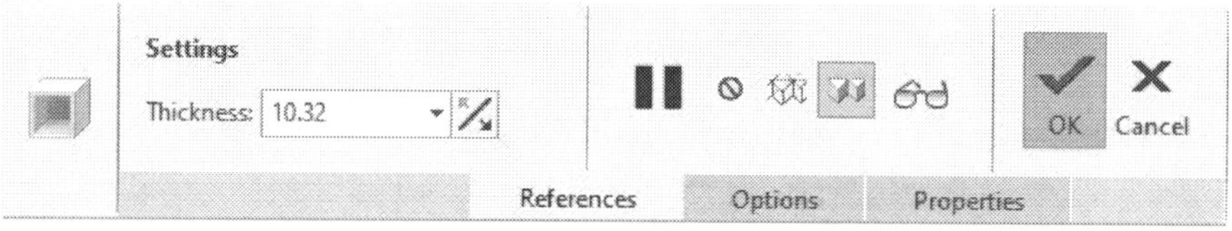

***Figure 7-1** Partial view of the **Shell** dashboard*

Change Thickness Direction Button

The **Change thickness direction** button is used to change the direction of thickness added in the model while using the **Shell** tool.

References Tab

When you choose the **References** tab from the **Shell** dashboard, the **References** slide-down panel will be displayed, as shown in Figure 7-2.

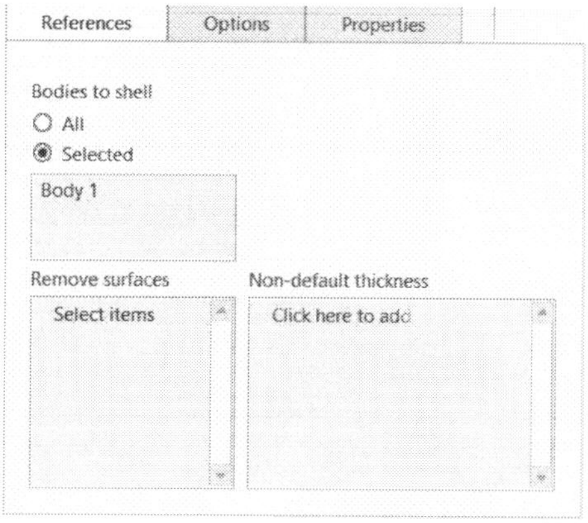

***Figure 7-2** The **References** slide-down panel*

By default, the **Remove surfaces** collector is activated in this slide-down panel. This collector displays the surfaces that are to be removed. You can also select the surfaces to remove before choosing the **Shell** tool. While creating or redefining the shell feature, you can select additional surfaces to remove or clear the selection of some of the previously selected surfaces by activating the **Remove surfaces** collector from the **References** tab. The **Non-default thickness** collector is used to specify surfaces for different thickness value. When you select a surface in this collector, a drag handle is displayed on the surface selected for different thickness and the default value of thickness is displayed with the name of the selected surface in the **Non-default thickness** collector. In Figure 7-3, the feature is created with a default thickness value and the same model with non-default thickness on the selected walls is displayed in Figure 7-4.

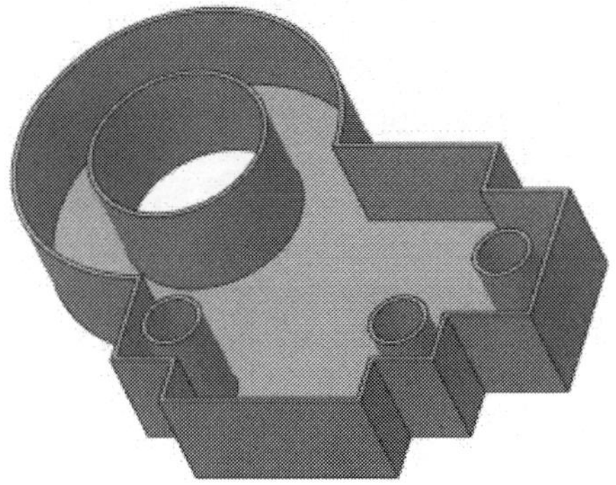

Figure 7-3 Model with constant thickness shell

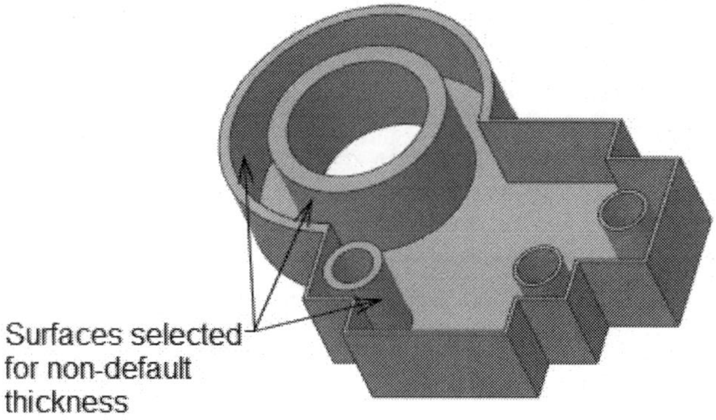

Figure 7-4 Model with non-default thickness shell

Options Tab

When you choose the **Options** tab from the **Shell** dashboard, the **Options** slide-down panel will be displayed, as shown in Figure 7-5.

The **Exclude surfaces** collector lists the surfaces excluded from being removed. If you do not select any surface from being excluded, the entire part will be shelled. To exclude surfaces from being removed, click in the **Exclude surfaces** collector and select the surfaces from the drawing area, refer to Figure 7-6; as a result the cylindrical feature in the model will be excluded during the operation, as shown in Figure 7-7. The **Details** button below the **Exclude surfaces** collector is used to invoke the **Surface Sets** dialog box. You can use the **Surface Sets** dialog box to view included surfaces or excluded surface sets and add rule-based surface sets.

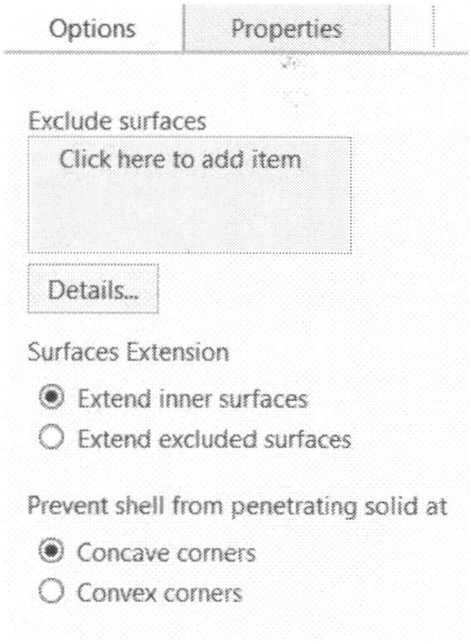

*Figure 7-5 The **Options** slide-down panel*

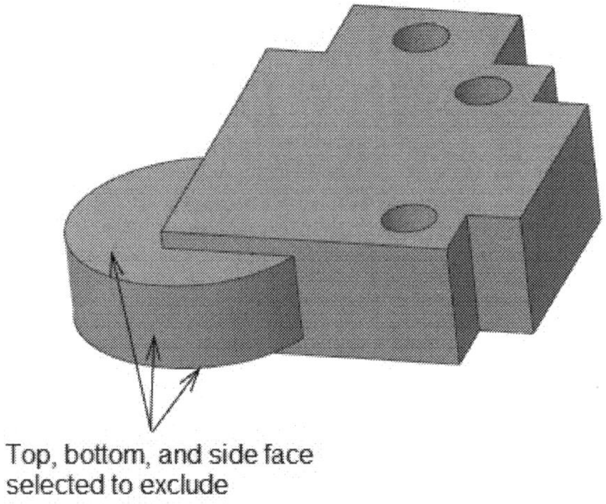

Figure 7-6 *Surfaces selected to exclude*

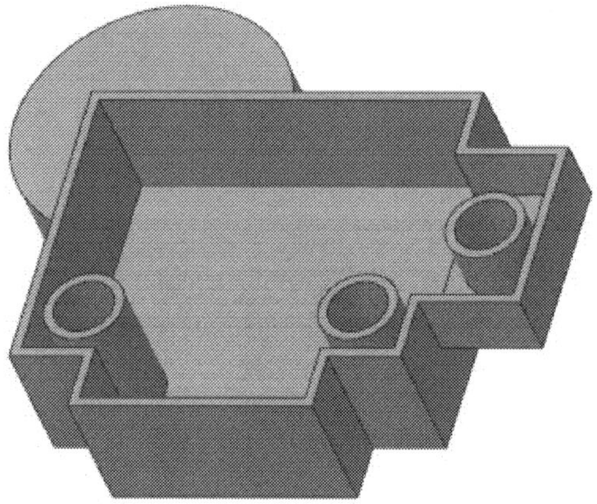

Figure 7-7 *Resulting shell feature after excluding the surfaces of cylindrical feature*

In the **Surfaces Extension** area, the **Extend inner surfaces** radio button is selected by default. As a result, the inner surfaces are capped in the model, refer to Figure 7-7. To extend the excluded surfaces, select the **Extend excluded surfaces** radio button; the excluded surfaces will be extended within the model, as shown in Figure 7-8.

Figure 7-8 *Resulting shell feature after extending the excluded surfaces*

In the **Prevent shell from penetrating solid at** area, the **Concave corners** radio button is chosen by default. As a result, the shell feature does not penetrate at concave corners when the surface is excluded from being removed, as shown in Figure 7-9.

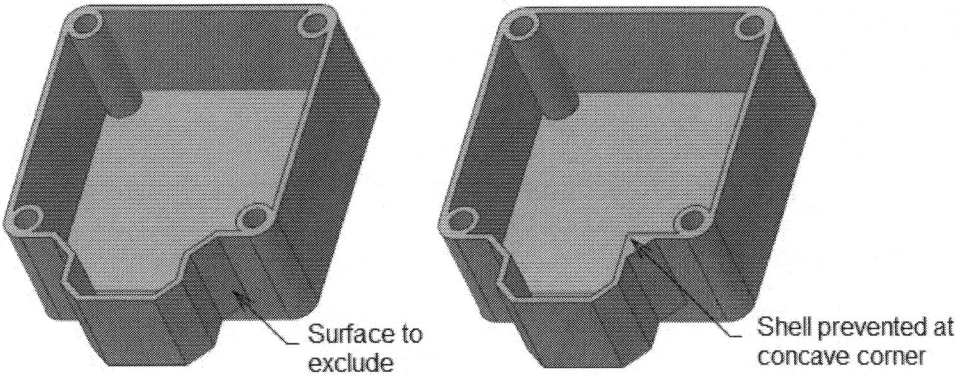

Figure 7-9 *The shell feature created with the* ***Concave corner*** *option selected*

As you choose the **Convex corners** radio button; the resulting shell feature will prevent the removal of surfaces at the convex corner, as shown in Figure 7-10.

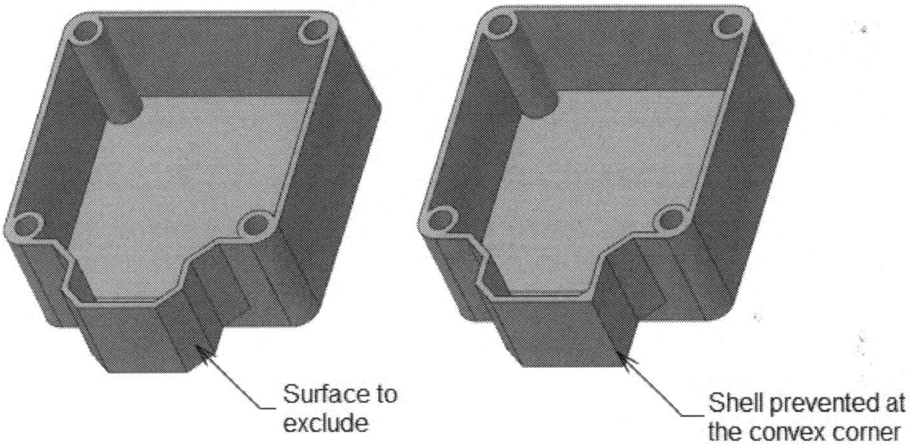

Figure 7-10 *The shell feature created with the* ***Convex corner*** *option selected*

CREATING DRAFT FEATURES

Ribbon: Model > Engineering > Draft drop-down> Draft

The **Draft** tool adds an angle to individual surfaces or to series of surfaces. The value of the angle is always between -89.9° to +89.9°. One of the applications of draft features is found in molds and castings where a taper is required to separate the casting from the mould or vice versa. To create a draft, choose the **Draft** tool from the **Draft** drop-down available in the **Engineering** group; the **Draft** dashboard will be displayed, as shown in Figure 7-11. The options in this dashboard are discussed next.

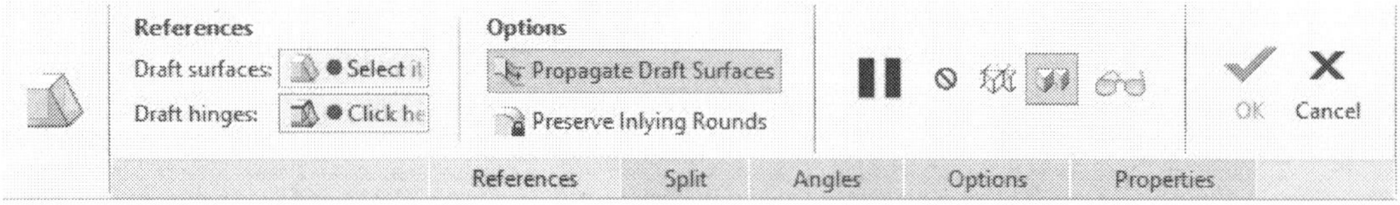

Figure 7-11 *Partial view of the **Draft** dashboard*

References Tab

On choosing the **References** tab, the **References** slide-down panel will be displayed, as shown in Figure 7-12. The options in this slide-down panel are discussed next.

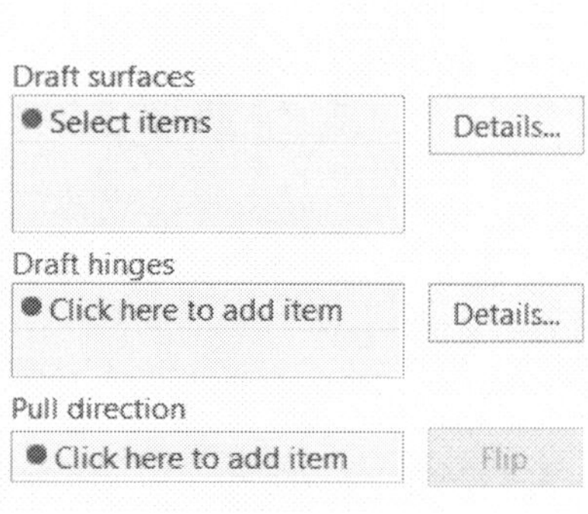

Figure 7-12 *The **References** slide-down panel*

Draft surfaces Collector

The **Draft surfaces** collector is selected by default. You can select individual surfaces as well as continuous surface chains. If this collector is not selected by default, click in this collector and then select the surface to which you need to add a draft angle. The maximum draft angle that can be added to a surface is 89.9 degrees.

Draft hinges Collector

The **Draft hinges** collector is used to select the hinge of the draft surface. The hinge that you select can be an edge, a surface, an axis, or a datum plane. The draft surface is pivoted on the hinge that you select. In other words, the draft surface is rotated about the hinge. The hinge that you select need not intersect the draft surface. For a draft hinge, you can also select a plane or a surface. Here the draft surface will be hinged at the intersection of the selected plane and surface. Figure 7-13 shows the draft surface and the hinge selected to create the draft and Figure 7-14 shows the resultant draft surface.

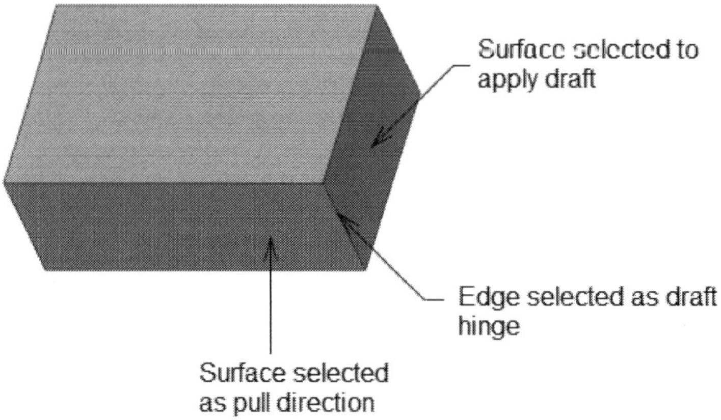

Figure 7-13 *Attributes selected to create a basic draft feature*

Pull direction Collector

The **Pull direction** collector is used to specify the direction of rotation of the draft surface. The direction of pull is indicated by a pink arrow, refer to Figure 7-14. Generally, when you select the hinge, the pull direction is selected by default. If you are selecting a plane in the **Pull direction** reference collector, then the draft surface should be normal to the selected plane. If you are selecting a straight edge or a datum axis, then the draft surface should be parallel to the selected edge or axis. The pink arrow displayed on the model is used to flip the direction of the draft feature. You can also change the direction of the draft feature by choosing the **Flip** button available next to the **Pull direction** collector in the **References** slide-down panel.

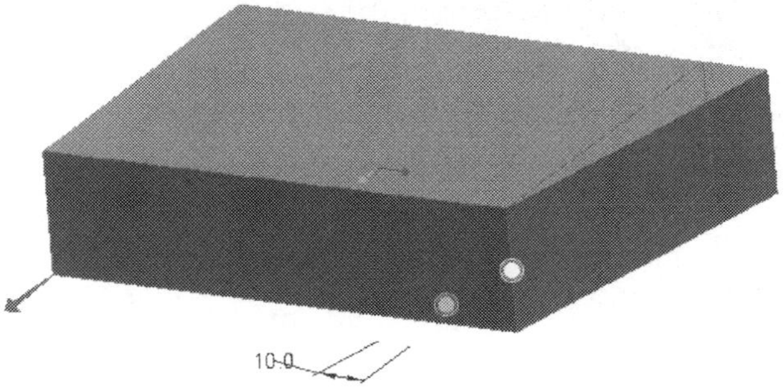

***Figure 7-14** Preview of the resultant draft feature*

Split Tab

If you choose the **Split** tab, the **Split** slide-down panel will be displayed, as shown in Figure 7-15. The options in this slide-down panel are discussed next.

Split options Drop-down List

The options in this drop-down list are used to split the surface selected to draft. Using these options, the selected surface gets split into two surfaces and different draft angles can be applied to both the surfaces. The options available in the **Split options** drop-down list are discussed next.

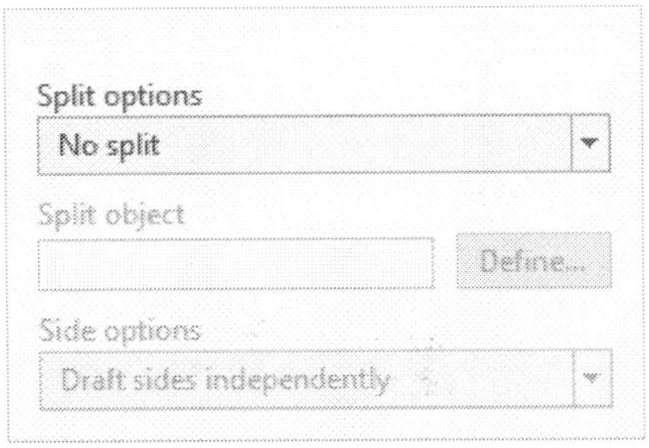

***Figure 7-15** The **Split** slide-down panel*

No split

This option is used when you do not want to split the surface selected to give the draft angle.

Split by draft hinge

This option is available only when you have selected the hinge for the draft surface. When you select this option, the selected surface is divided into two surfaces at the location on the surface where the hinge intersects it. When you choose this option, the **Side options** drop-down list gets activated with the **Draft sides independently** option selected by default. As a result, you can specify two independent draft angles for each side of the drafted surface. If you select the **Draft sides dependently** option, you can specify a single draft angle with the second side drafted in an opposite direction. Selecting the **Draft first side only** and **Draft second side only** options allow you to draft the one of the sides of the surface with the second side remaining in a neutral position. Figure 7-16 shows a model where the draft surface is split into two surfaces at the hinge.

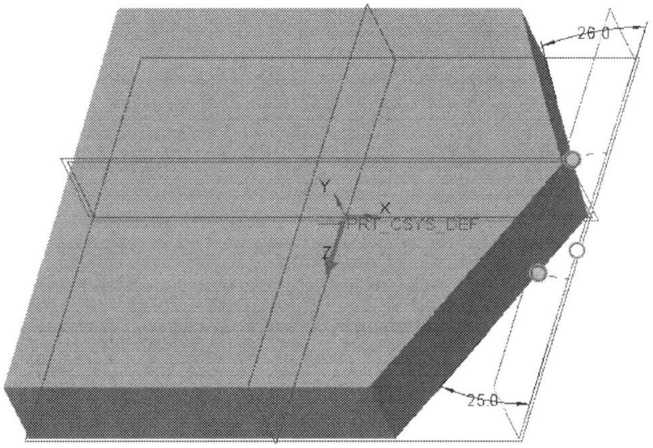

Figure 7-16 Split at draft hinge

Split by split object

This option, when chosen activates the **Split object** collector. The **Define** button next to this collector will open the Sketcher environment so that you can sketch the split curve on the draft surface or on another plane. In case, the sketch does not lie on the draft surface, the system projects it on the draft surface in a direction normal to the sketching plane. In this type of split, surface gets split into two surfaces and the sketch defines the profile of the split. Figure 7-17 shows the parameters that you need to define to create a draft by using the **Split by split object** option. Figure 7-18 shows the draft created on the cylindrical surface.

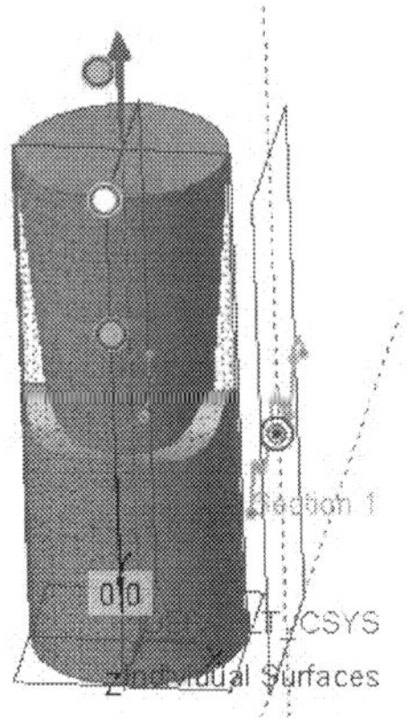

Figure 7-17 The ***Split by split object*** *option*

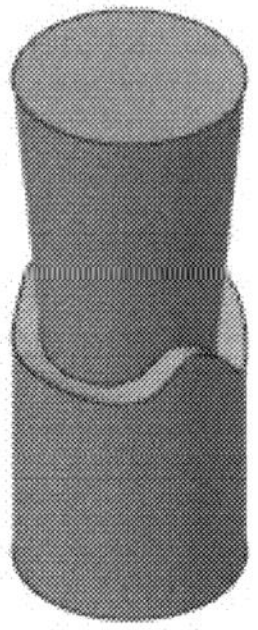

Figure 7-18 Resultant model

Angles Tab

Choose the **Angles** tab from the **Draft** dashboard to manage the angles of a variable draft feature. On choosing this option, the **Angles** slide-down panel will be displayed, as shown in Figure 7-19. For a constant draft feature, a single line containing the value of draft angle will be displayed in this slide-down panel. In case of variable draft feature, additional angles are added to this slide-down panel. To add a draft angle, right-click on a cell in the table and choose the **Add Angle** option from the shortcut menu; a line will be added to the panel. Each line contains the **Angle** edit box with the value of the draft angle, the **Reference** collector with the name of the reference, and the **Location** edit box specifying the location of the draft angle control along the reference. In the case of a split draft with both sides independent, the **Angle 1** and **Angle 2** boxes are used to specify the angles. You can also flip the angle of draft by choosing the **Flip Angle** option from the shortcut menu. The **Adjust angles to keep tangency** check box is used to force the resultant draft surface to maintain its tangency.

The procedure to create constant and variable angle draft feature is discussed next.

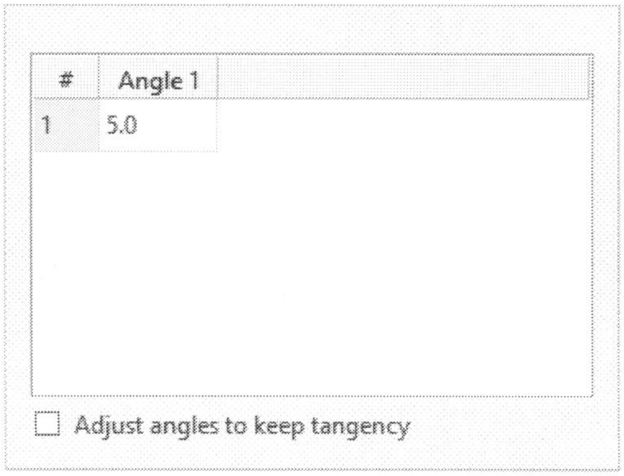

Figure 7-19 *The **Angles** slide-down panel*

Constant Angle Drafts

By default, Creo Parametric creates a constant angle draft feature. This means, the same draft angle is applied to the surface that you have selected.

Variable Angle Drafts

When you apply more than one value of the draft angle to the selected surface, it is called variable angle draft. These angles are placed using control points along the draft surface. You can modify the location of these control points using the options in the **Angles** tab or you can drag the control point that is displayed on the model. If the draft hinge is a curve, the angle control point will lie on the draft hinge. If the draft hinge is a plane or quilt, the angle control point will lie on the boundary of the draft surface.

To create the draft surface, as shown in Figure 7-20, follow the steps given below. It is assumed that you have created the base feature of the model, as shown in Figure 7-21.

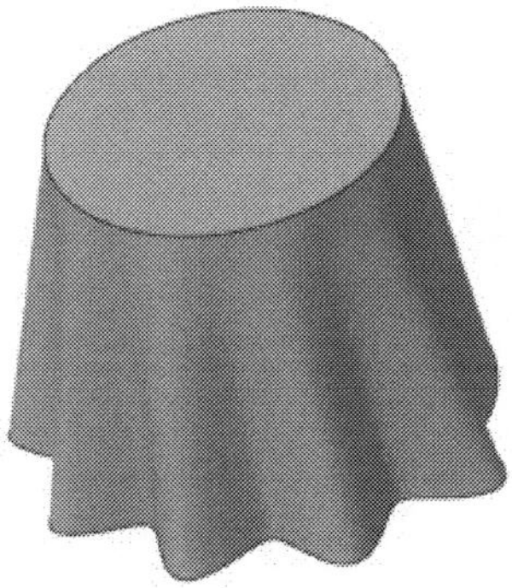

Figure 7-20 *Draft feature on the model*

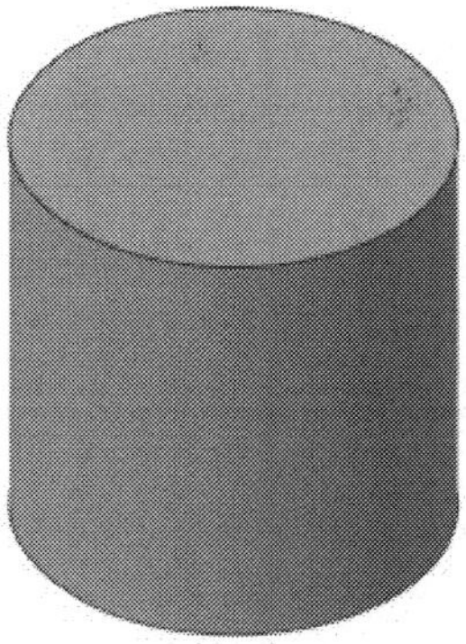

Figure 7-21 *Base feature*

1. Invoke the **Draft** dashboard and click on the first collector from the left in the **Hinge & Pull Direction** area.

2. Select the top face of the base feature. Notice that in the **Pull direction** collector, the direction of pull is selected automatically. Reverse the pull direction by choosing the **Reverse pull direction** button available on the right of the second collector.

3. Hold the right mouse button to invoke the shortcut menu shown in Figure 7-22.

4. Choose the **Draft Surfaces** option from the shortcut menu displayed and select the cylindrical surface of the base feature; a draft angle is added and a drag handle appears on the base feature. You can use the drag handle to change the draft angle. A draft angle of value **1** also appears in the drawing area. You will also notice a white ball on the edge of the top face (face selected as hinge).

5. Bring the cursor on the white ball and hold the right mouse button to invoke the shortcut menu shown in Figure 7-23.

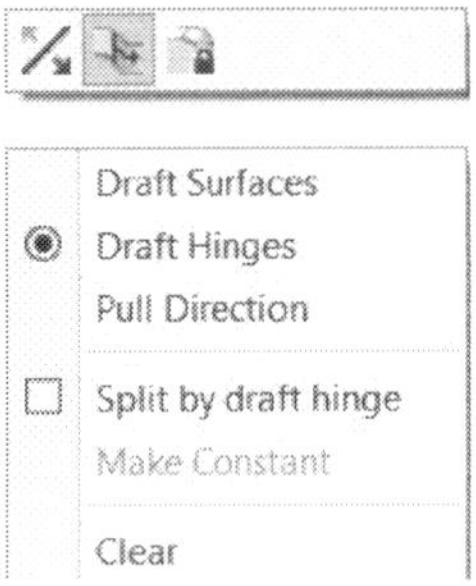

*Figure 7-22 The **Draft** shortcut menu*

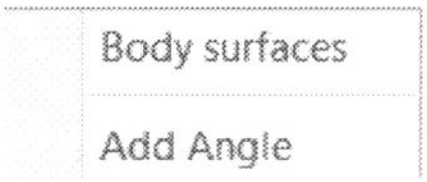

*Figure 7-23 The **Angle** shortcut menu*

6. Choose the **Add Angle** option from the shortcut menu. Notice that now there are two white balls on the edge of the top face. On the first ball, the value of **0.5** appears. This value varies from one end of the face to the other end. This value represents the location of the point that you need to use for reference in order to apply the draft angles.

7. Again bring the cursor to any one of the two white balls and invoke the shortcut menu.

8. Choose the **Add Angle** option from the shortcut menu displayed. Another point is added with a location value on the edge.

9. Add ten such points for locations varying from **0** to **0.9**.

10. After adding ten points, double-click on any of the location value that is present in the drawing area; an edit box appears. Enter **0** in this edit box. Similarly, locate all remaining ten points and increment them by 0.1. Figure 7-24 shows the model after relocating all ten points on the edge.

11. Double-click on the value of the angle that corresponds to the location **0**; an edit box appears. Enter **0** in the edit box.

12. Double-click on the value of the angle that corresponds to the location **0.1** (these values appear above the model and have a default value 0). The edit box appears; enter a value of **10**. Similarly, vary the angle at other locations. Every alternate location should have an angle of **10**. Figure 7-25 shows the preview of the model after modifying the values of the angle at all ten locations.

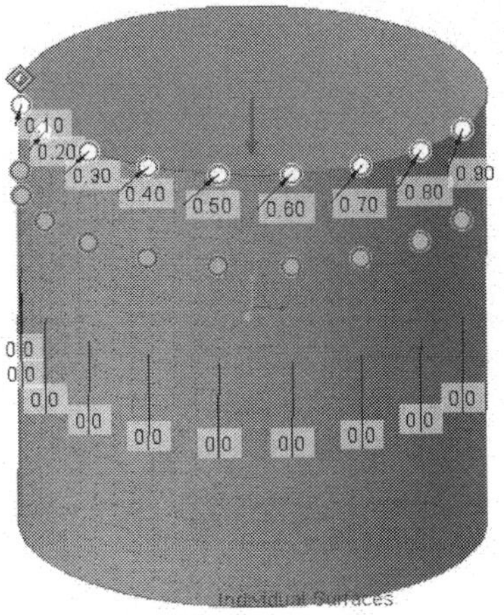

Figure 7-24 Location points on the edge

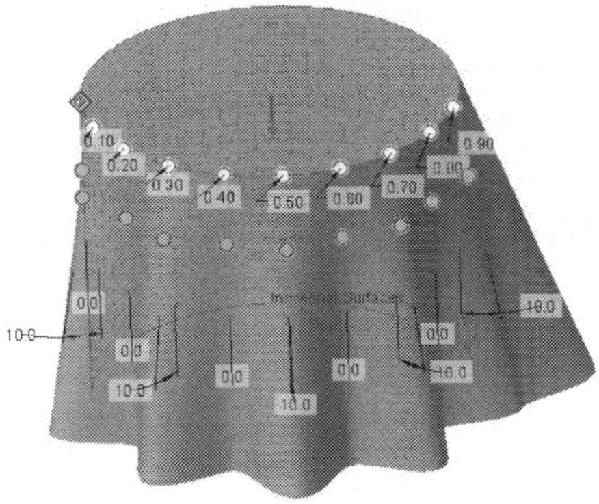

Figure 7-25 Location points with modified angle values

13. After modifying the values of the angle at all locations, choose the **OK** button from the dashboard to complete the procedure.

Options Tab

On choosing this tab, the **Options** slide-down panel will be displayed, as shown in Figure 7-26. The options in this slide-down panel are discussed next.

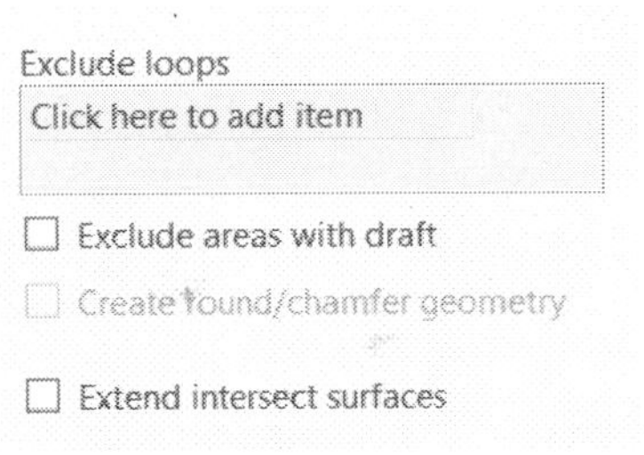

*Figure 7-26 The **Options** slide-down panel*

Exclude loops Collector

The **Exclude loops** collector is activated only when the selected surface contains more than one loop. This collector is used to select a surface to which you do not want to add a draft angle. When you select a surface to add a draft angle, all loops that are on the surface get applied the same draft angle. However, using this collector you can select the loop to which you do not want to add a draft angle.

Figure 7-27 shows an example where the surface is selected to add a draft angle. You will notice that the surface of the extruded feature is selected. If you continue with the draft feature creation and exit the **Draft** dashboard, the draft is created, as shown in Figure 7-28.

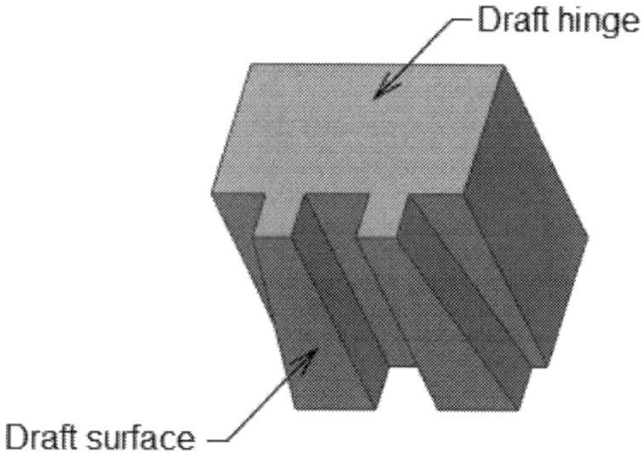

Figure 7-27 *Attributes selected for draft feature*

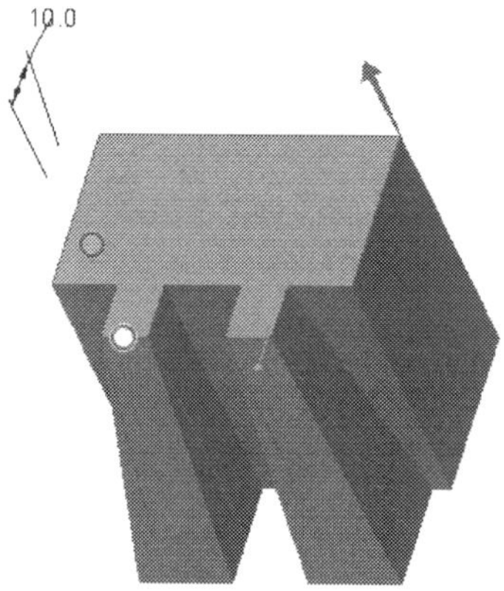

Figure 7-28 *Preview of the draft feature*

To exclude a face of the extruded feature from the loop, click in the **Exclude loops** collector to activate it. Now, bring the cursor close to the face of the feature. The boundary of the face is highlighted. Select the face to exclude it from the loop. Figure 7-29 shows the resulting draft feature after excluding the face of the extruded feature from the loop. Note that the model should be a single feature then only the **Exclude loops** collector will be active.

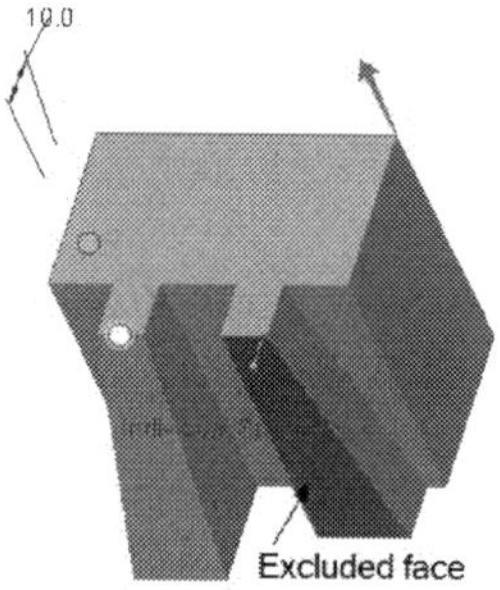

Figure 7-29 *Draft after excluding a face of the extruded feature*

Exclude areas with draft Check Box

This check box is used in such cases where the draft surface is not perpendicular to the pull direction and the angle between the draft surface and the pull direction is greater than or equal to the draft angle. Select this check box to exclude the areas from the draft feature where angle between the draft surface and the pull direction is greater than or equal to the draft angle.

Extend intersect surfaces Check Box

When this check box is selected, the draft surface extends in the direction of the feature it intersects. If the draft cannot extend to the adjacent model surface, then the model surface extends into the draft surface. If even on selecting the check box, neither of the above cases are possible then the resultant bodies will overhangs the edge of the model. Figure 7-30 shows an example of the draft that intersects the adjacent feature and extends outside the edge of the feature at the bottom. Figure 7-31 shows the feature when the draft surface gets extended.

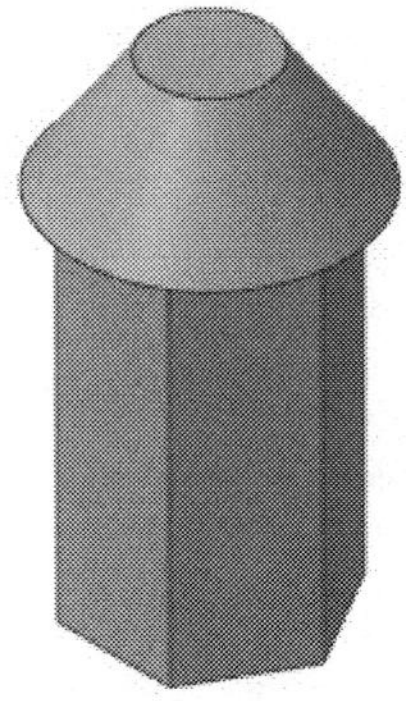

*Figure 7-30 Draft created with the **Extend intersect surfaces** check box cleared*

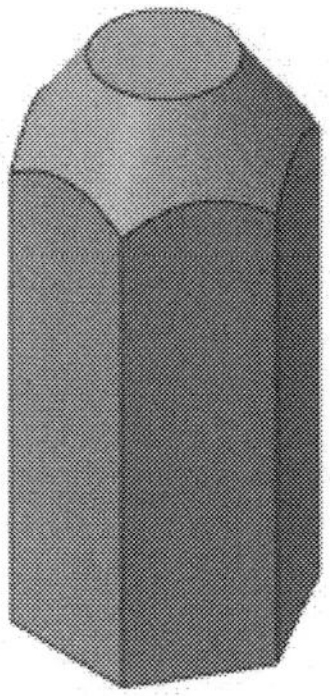

*Figure 7-31 Draft created with the **Extend intersect surfaces** check box selected*

Propagate Draft Surfaces Button

This button is used to propagate draft along the surfaces tangential to the surfaces to be drafted. By default, this button is chosen. Figure 7-32 shows an object in which draft is applied when **Propagate Draft Surfaces** button is chosen and Figure 7-33 shows same object when this button is not chosen.

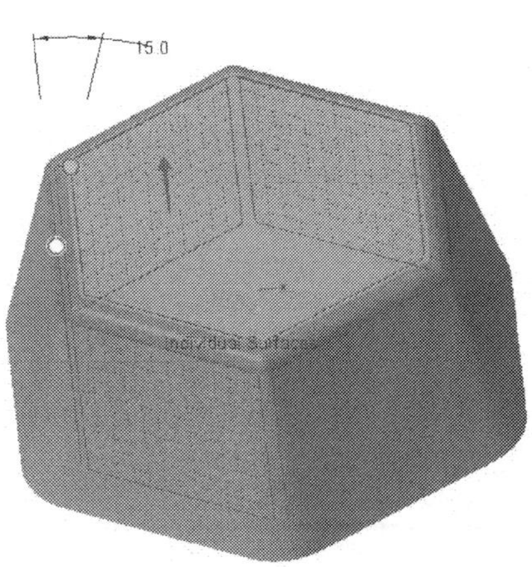

*Figure 7-32 Draft created with the **Propagate Draft Surfaces** button chosen*

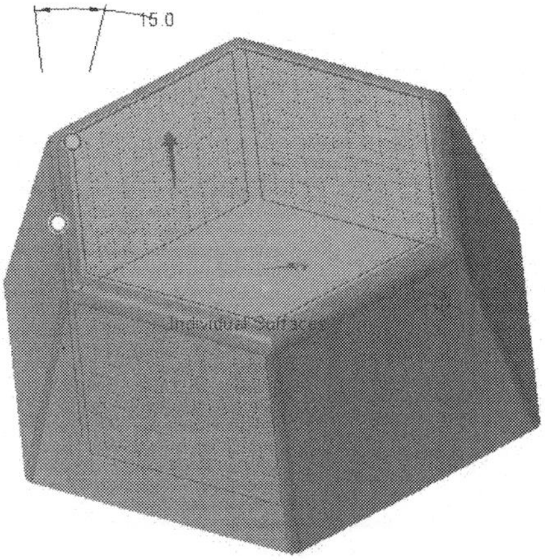

*Figure 7-33 Draft created with the **Propagate Draft surfaces** button not chosen*

Preserve Inlying Rounds in Draft

This tool is used to apply draft easily to design models containing rounds and chamfers. If the geometry selection contains inlying rounds, you can control their inclusion or exclusion from the draft operation by selecting this tool.

To do so, select the individual surfaces to draft and then select the hinge, refer to Figure 7-34. Creo parametric will automatically detect the inlying rounds and display them in green color, refer to Figure 7-34.

Figure 7-35 shows the model when inlying rounds handling tool is not selected. After selecting the tool rounds will regain their original shape, refer to Figure 7-36.

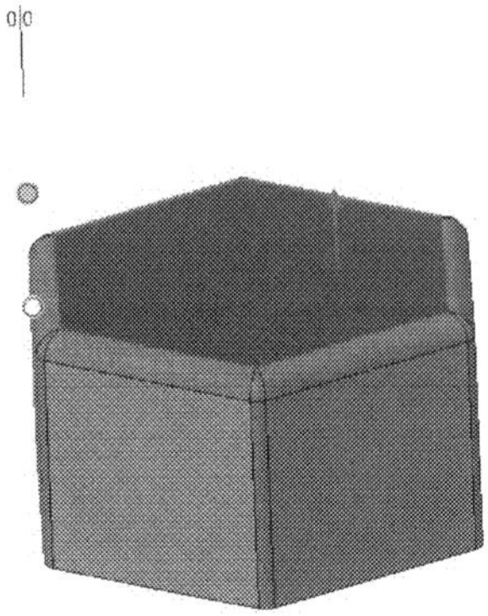

Figure 7-34 *Individual surfaces and hinge selected for draft*

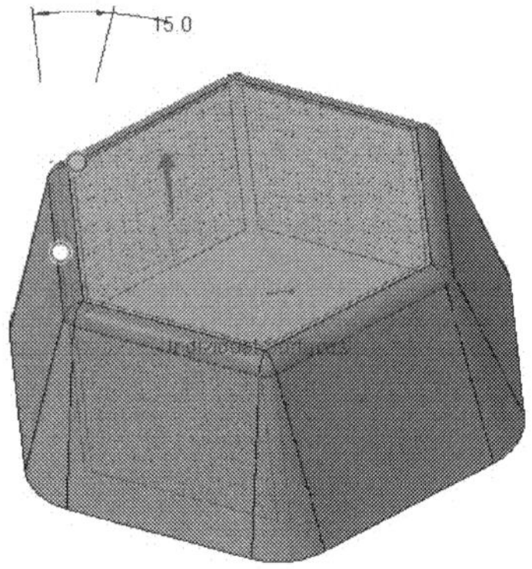

Figure 7-35 *Round handling tool is not selected*

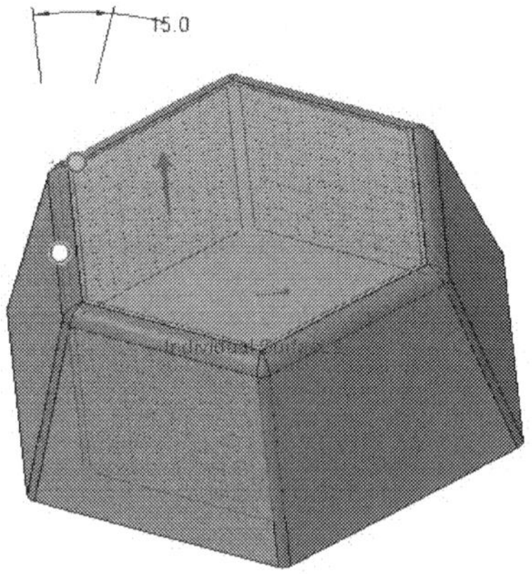

Figure 7-36 *Round handling tool is not selected*

Properties Tab

When you choose the **Properties** tab, a slide-down panel is displayed, showing the feature id of the feature you have created. You can specify a name for the feature using the **Name** edit box of this slide-down panel.

CREATING TOROIDAL BENDS

Ribbon: Model > Engineering drop-down > Toroidal Bend

The **Toroidal Bend** tool is used to provide a toroidal (revolved) shape to solids, quilts, or datum curves. The example of a plate after creating the toroidal bend is shown in Figure 7-37. To create the toroidal bend, first you will create a rectangular plate and cut the profile on it, as shown in Figure 7-38. Then, using the **Toroidal Bend** tool, you will bend the rectangular plate through an angle of 360 degrees, refer to Figure 7-37. You can also use this tool to create model of tires or wrap a logo around a revolved geometry.

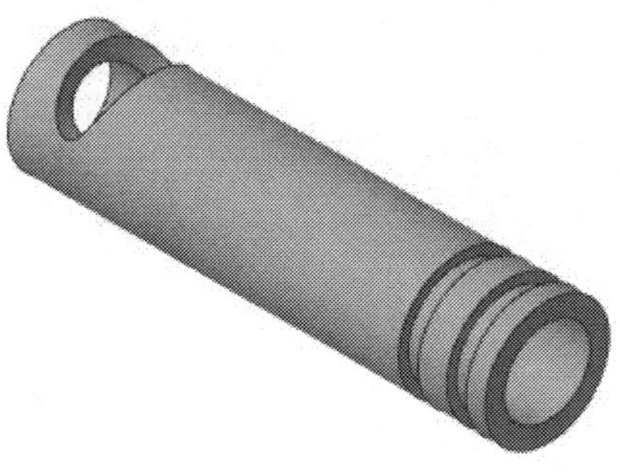

Figure 7-37 The plate after creating the toroidal bend

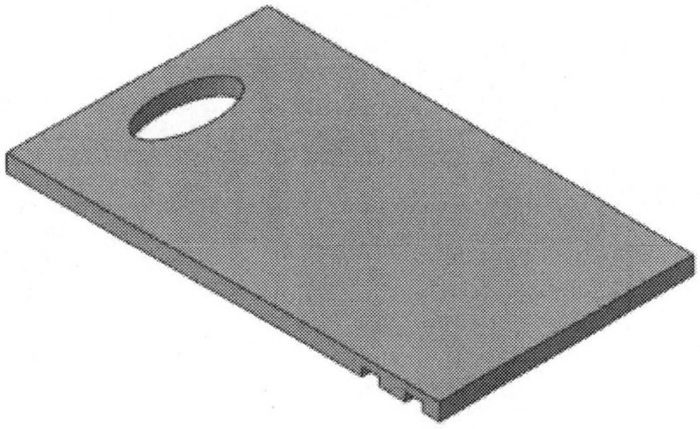

Figure 7-38 Rectangular plate with a cut profile

To create a Toroidal Bend feature, choose the **Toroidal Bend** tool from the expanded **Engineering** group of the **Model** tab; the **Toroidal Bend** dashboard will be displayed, as shown in Figure 7-39. The options available in this dashboard are discussed next.

*Figure 7-39 Partial view of the **Toroidal Bend** dashboard*

In this dashboard, the first collector from the left is the **Profile Section** collector. This collector is used to select an internal or external profile section from the model. The profile section must contain a geometry coordinate system where x-axis defines the neutral plane during the bending operation. The **Bend Radius** button in the dashboard provides options to specify the bend radius of the neutral plane. When you specify profile section for the bend geometry, the **Bend Radius** option gets automatically selected. You can specify the value of this option in the edit box next to the **Bend Radius** button. If you choose the **Bend Axis** button, you can select an axis to bend the geometry around it. To select the axis, use the collector beside the **Bend Axis** button. Note that the axis selected as bend axis must lie on the same plane on which you have created the profile section. The **360 Bend** button creates a full 360° bend of the geometry. As you choose this option, two plane collectors are displayed in the dashboard. In this case, the bend radius of the geometry will be the distance between the two planes divided by 2.

References Tab

On choosing the **References** tab, the **References** slide-down panel will be displayed, refer to Figure 7-40. The **Quilts and/or a solid body** collector is used to collect the quilts, surface pieces or solids to bend. The **Curves** collector in this slide-down panel collects all the curves that belong to the geometry being bent. The **Profile Section** collector is used to edit the attached internal or external sketch. Select the **Normals Reference Section** check box to define an external sketch as a reference for the normal vector direction of the toroidal bend.

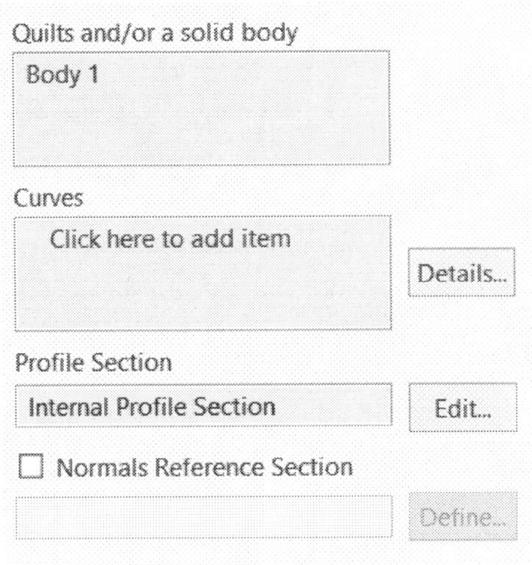

***Figure 7-40** The **References** slide-down panel*

Options Tab

When you choose the **Options** tab from the **Toroidal Bend** dashboard, the **Options** slide-down panel will be displayed, as shown in Figure 7-41. By default, the **Standard** radio button is selected in this panel. As a result, bending of the geometry takes place according to the standard of algorithm for toroidal bends.

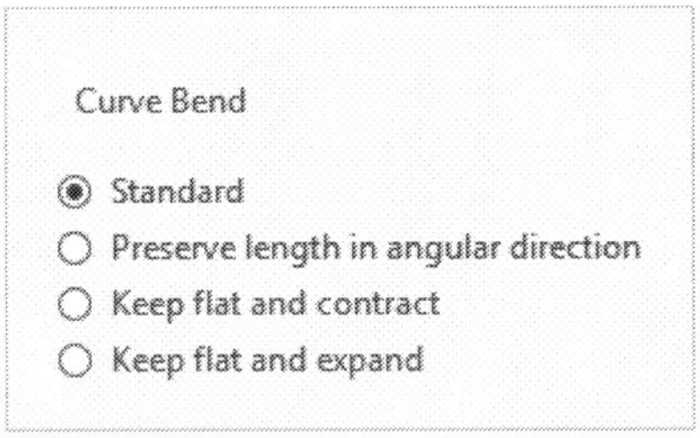

***Figure 7-41** The **Options** slide-down panel*

Select the **Preserve length in angular direction** radio button if you want to bend the curve chains in such a manner that the distance from the points on the curves to the plane of the profile section is maintained along the angular direction. If you select the **Keep flat and contract** radio button, the curve chains are kept flat in the neutral plane and the distance from the points on curves to the plane of the profile section is reduced. The **Keep flat and expand** radio button is selected to keep the curve chains flat in the neutral plane. The distance from points on the curves to the plane of the profile section is increased.

Properties Tab

The **Name** edit box under this slide-down panel is used to specify the name of the feature. You can choose the information button next to the edit box to display detailed information of the component in a browser.

The following steps explain the procedure to create the model shown in Figure 7-37. It is assumed that you have created the rectangular plate of dimensions 300 X 500 X 20.

1. Choose the **Toroidal Bend** tool from the expanded **Engineering** group to invoke the **Toroidal Bend** option; the **Toroidal Bend** dashboard will be displayed, as shown in Figure 7-39. Next,

choose the **References** tab from the dashboard; the **References** slide-down panel will be displayed. Choose the **Define** button from the **Profile Section** area.

2. Select the front face of the base feature (which is the longer face of the base plate) as the sketching plane and orient the sketch plane.

3. After entering the sketcher environment, draw the sketch, as shown in Figure 7-42. The sketch comprises of a line that is drawn from the left edge to the right edge. Also, a user-defined geometric coordinate system is placed at the midpoint of the line.

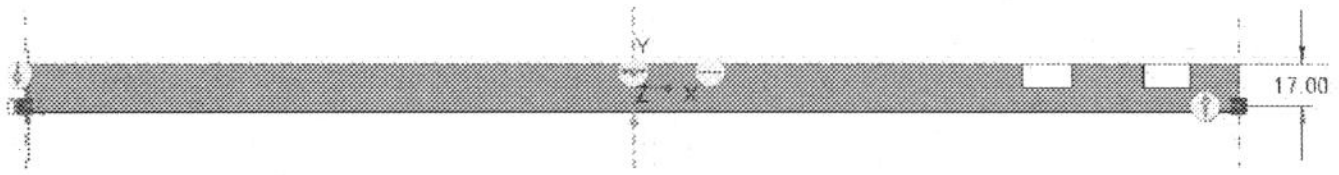

Figure 7-42 *Sketch along with the reference coordinate system*

4. Next, exit the sketcher environment.

Note

The sketch, shown in Figure 7-42, controls the bend profile of the Toroidal Bend feature. If the sketch is a straight line, the side profile of the Toroidal Bend will be a line; else it will be an arc. The user-defined coordinate system acts as a neutral point in the bend.

5. Choose the **References** tab; the **References** slide-down panel is displayed. Click on the **Quilts and/or a solid body** collector and select the required solid body from the drawing area.

6. Next, choose the **Options** tab; the **Options** slide-down panel will be displayed. Select the **Standard** radio button if not already selected.

7. Next, choose the **360 Bend** button from the **Toroidal Bend** dashboard; you will be prompted to select the two parallel planes to define the length of the bend.

8. Select the face that was selected as the sketching plane and then select the longer face which is parallel to it; preview of the toroidal bend will be displayed.

9. Choose the **OK** button from the dashboard; the Toroidal Bend will be created and a model similar to the one shown in Figure 7-37 will be displayed in the drawing area.

CREATING SPINAL BENDS

Enhanced

Ribbon: Model > Engineering > Spinal Bend

The **Spinal Bend** tool is used to bend solids or quilts about a curved spine by continuously repositioning cross-sections along the curve. Cross-sections, which are perpendicular to an axis, are repositioned perpendicular to the sketched spine without any distortion. Any kind of compression or distortion in the geometry is created along the trajectory of the spine. To create a bend geometry, select the model from the drawing area, refer to Figure 7-43, and next, choose the **Spinal Bend** tool and select a spine curve as a reference geometry; the resultant bend geometry will be created, as shown in Figure 7-44.

Figure 7-43 *Model for creating the spinal bend*

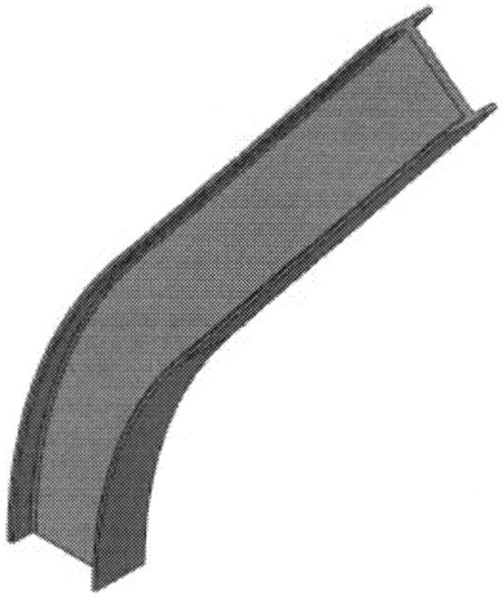

Figure 7-44 *Model with spinal bend feature*

Note

To get the shape shown in Figure 7-44, you may need to change the direction of spine curve and select the required option from the depth drop-down list.

Choose the **Spinal Bend** tool from the expanded **Engineering** group; the **Spinal Bend** dashboard is displayed, as shown in Figure 7-45. The options available in this dashboard are discussed next.

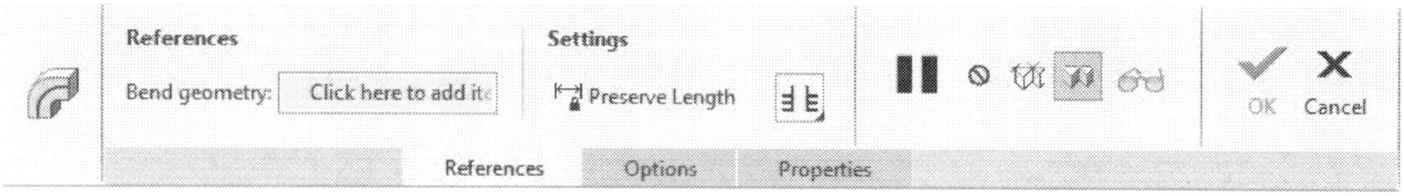

Figure 7-45 *Partial view of the* ***Spinal Bend*** *dashboard*

The first collector in the **Spinal Bend** dashboard is the **Bend Geometry** collector. This collector is used to select the geometry to bend. The **Preserve Length** button sets the length of the resulting bent geometry equal to the length of the spine curve. On choosing the **Preserve Length** button, the bend region maintains its original length, refer to Figure 7-46.

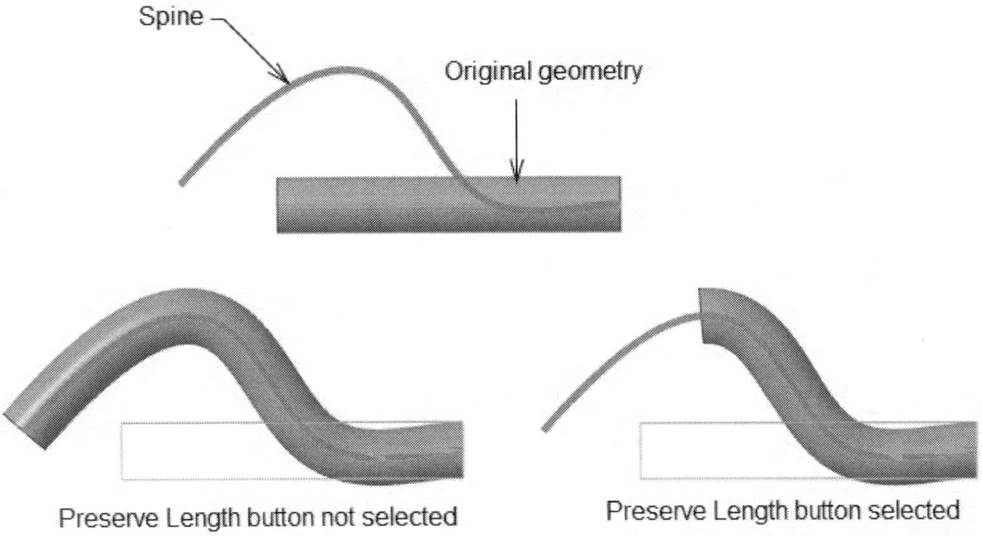

*Figure 7-46 Effect of **Preserve Length** option in geometry*

The Depth drop-down list in the dashboard is used to specify the region for the bend. The bend region is defined relative to an axis that is tangent to the spine at the start point of the spine. The bend region lies between the plane that passes through the start point of the spine. By default, the **Bend the entire selected geometry from the spine start** option is selected in the drop-down list. As a result, the geometry bends from the start of the spine to the farthest point of the geometry in the direction of the axis. The **Bend Geometry between the spine start and the specified offset** option is used to bend the geometry from the start of the spine to a specified depth in the direction of axis. Select the **Bend Geometry between the spine start and the selected reference** option to select a reference up to which the selected geometry will bend. You can also select a plane that is perpendicular to the axis, point, or a vertex.

References Tab

On choosing the **References** tab, the **References** slide-down panel will be displayed. The **Spine** collector in this slide-down panel is used to select the spine curve.

Options Tab

On choosing the **Options** tab, the **Options** slide-down panel will be displayed, as shown in Figure 7-47. By default, the **None** option is selected in the **X-Section Property Control** drop-down list. The options in this drop-down list are used to control the properties of varying cross-sections along the spine. The **Edit relation** button is used to edit the specified relations. Choose the **Sketch csys** button to sketch a coordinate system to be used in the calculation of cross-section properties. The **Linear** radio button in the **Control type** area is used to vary the section properties linearly between the start and end point of the spine. If you select the **Graph** radio button, you can vary the cross-section by using the values in the graph. You can click in the **Graph** collector to select an existing graph feature. The **Remove unbent geometry** check box is used to remove the portion of the model that is outside the bent area.

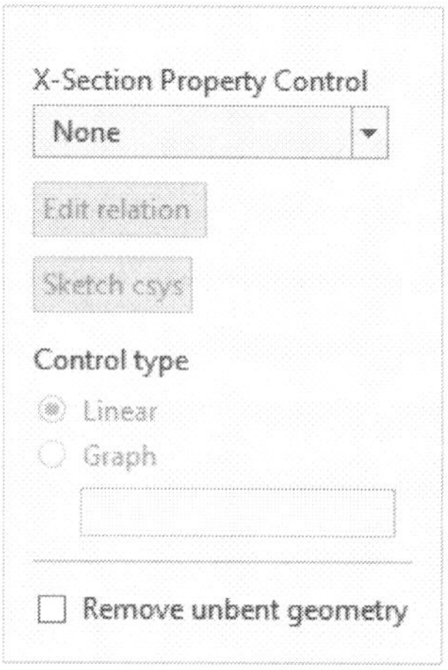

Figure 7-47 *The* ***Options*** *slide-down panel*

The following steps explain the procedure to create the model shown in Figure 7-44.

Note

It is assumed that you have created the base feature before invoking the ***Spinal Bend*** *tool.*

1. Choose the **Spinal Bend** tool from the **Engineering** group to invoke the **Spinal Bend** dashboard and then click in the **Bend Geometry** collector of the dashboard.

2. Select the solid geometry from the drawing area. Next, choose the **References** tab; the **References** slide-down panel will be displayed.

3. Click on the **Spine** collector in the slide-down panel; you will be prompted to select spine.

Note

Note that you have created the trajectory before selection. The trajectory sketched for the spinal bend should be tangent to the solid geometry, as shown in Figure 7-48.

4. Select the spine from the drawing area. You can retain the original length of the selected geometry by selecting the **Preserve Length** button. You can also define the length of the selected geometry to be bent by selecting the **Bend Geometry between the spine start and the specified offset** option from the Depth drop-down of the **Spinal Bend** dashboard. You can drag the handle or enter a value in the edit box to specify the length of the bend region.

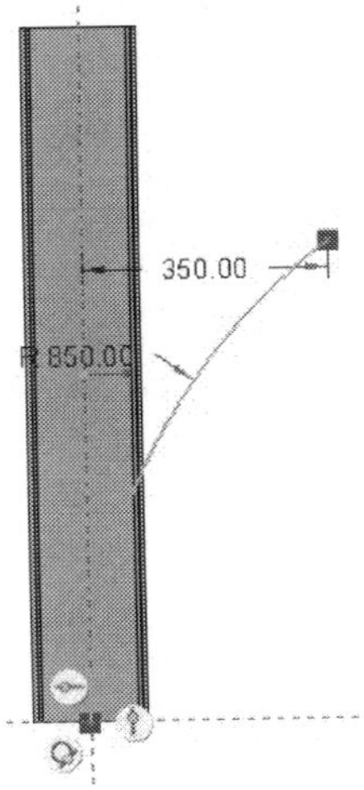

Figure 7-48 *Sketched trajectory for spinal bend*

CREATING COSMETIC SKETCHES

Cosmetic sketches are used to sketch entities on the surface of a model for visualization of stamped features such as company logos and serial numbers. The method to create cosmetic sketches is similar to the method of creating typical sketches. Cosmetic sketches cannot be used for removing or adding material to the model. You can also use imported entities as cosmetic sketch. To create a cosmetic sketch, choose the **Cosmetic Sketch** tool from the expanded **Engineering** group; the **Cosmetic Sketch** dialog box will be displayed, as shown in Figure 7-49. Notice that the attributes in this dialog box are similar to the **Sketch** dialog box.

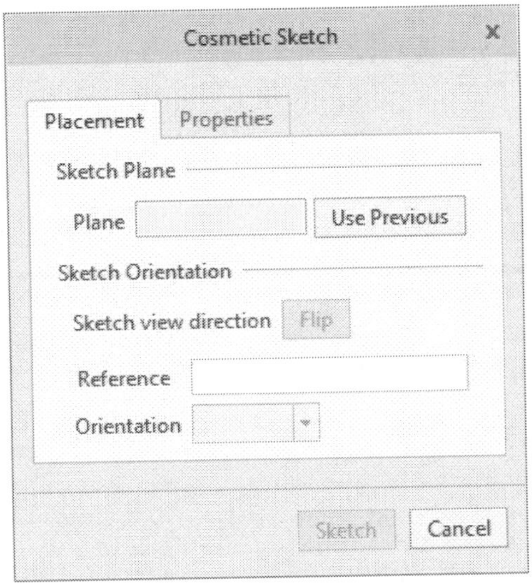

Figure 7-49 *The* ***Cosmetic Sketch*** *dialog box*

To display cross-hatching for a closed sketch, click the **Properties** tab and select the **Add hatching lines** check box. Next, specify the attributes of hatching in the **Scale** and **Angle** edit boxes and select the sketching plane and choose the **Sketch** button to create a cosmetic sketch. The cosmetic sketch created on the surface of the model is shown as an example in Figure 7-50.

Figure 7-50 *Cosmetic sketch created on the model*

CREATING COSMETIC THREADS

Cosmetic threads are used to create a lightweight representation of threads on a model. In general, a cosmetic thread is used to represent the thread symbol. Cosmetic threads can be applied to cylinders, cones, and planes. The surface selected for the placement determines whether the thread will be external or internal. To create a cosmetic thread feature, choose the **Cosmetic Thread** tool from the expanded **Engineering** group; the **Thread** dashboard will be displayed, as shown in Figure 7-51. The options available in this dashboard are discussed next.

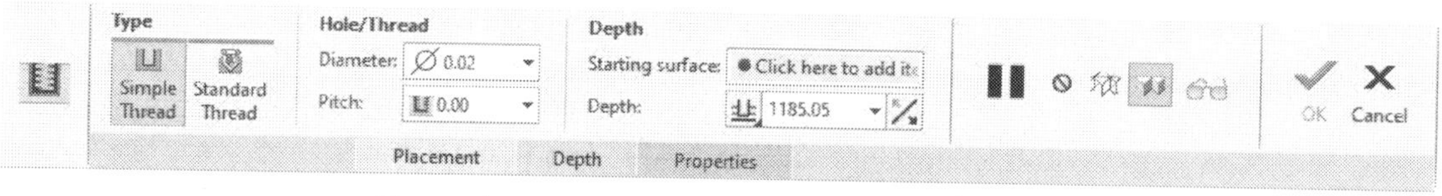

Figure 7-51 *Partial view of the* ***Thread*** *dashboard*

Creating Simple Cosmetic Thread

When you invoke the **Thread** dashboard, the **Simple Thread** button is chosen by default and you are prompted to select a surface to place the thread. As you hover the cursor over the model, you will notice that the faces only with cylindrical and conical shapes are highlighted for selection. Select a cylindrical or a conical surface; you will be prompted to select a plane, surface, or quilt to specify the start location of the thread. Select a face of the model to specify thread start location; preview of the thread geometry is displayed on the model. You can also use the **Starting surface** collector in the dashboard to select the surface for thread start location. Next, you can specify the value of thread diameter in the **Diameter** edit box. Note that for a cylindrical surface, the value of diameter is specified and for a conical surface, the perpendicular distance between the reference surface and thread center axis is specified. To specify the pitch value of thread, type a value in the **Pitch** edit box. To set the thread depth, you can select an option from the **Depth** drop-down list or specify the depth in the **Depth** edit box. You can change the direction of the thread by selecting **Changes the direction of thread surface** button from the dashboard. Next, select the **OK** button from the dashboard to create the cosmetic thread feature on the model, as shown in Figure 7-52.

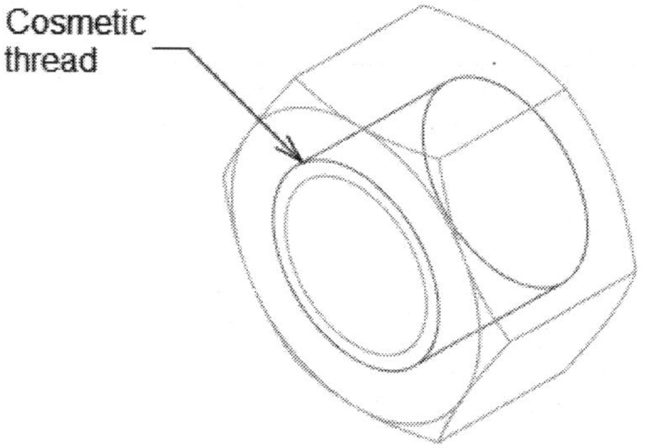

Figure 7-52 *Illustration of a cosmetic thread feature*

> **Tip**
> *You can also change the parameters of the cosmetic thread using the* ***Properties*** *tab of the* ***Thread*** *dashboard.*

Creating Standard Cosmetic Thread

To create a standard cosmetic thread, choose the **Standard Thread** button from the dashboard; the dashboard gets modified, as shown in Figure 7-53. Next, select the surface to be threaded using the **Threaded surface** collector in the **Placement** tab and then specify the thread start location by using the Thread start collector available in the dashboard. To specify the type of thread, select it from the Thread Series drop-down list and then specify size of the thread using the Thread size drop-down list.

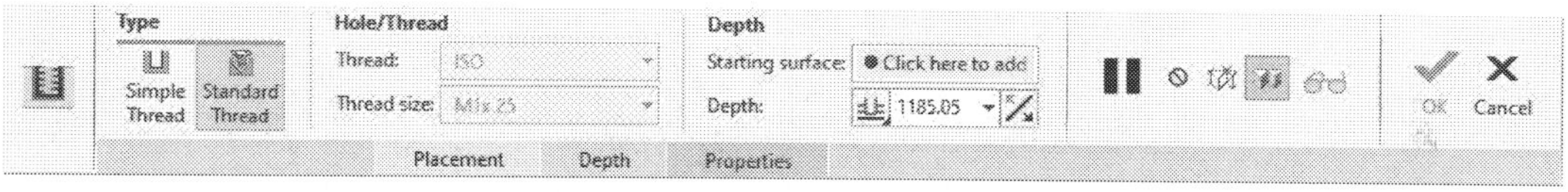

Figure 7-53 *Partial view of the* ***Thread*** *dashboard with the* ***Standard Thread*** *button chosen*

Note that when you select a cylindrical or conical reference surface on which you want to place a thread, Creo Parametric compares the size of the cylinder with the standard hole tables. If the diameter of the selected surface matches the standard hole size, a suggested value will get displayed in the **Thread Size** drop-down list. However, you can also choose any other thread size that is available in the list. Remember that if you change the size of the cylinder after defining the cosmetic thread feature, the cosmetic thread feature will not be updated according to the hole table. After specifying the attributes of the cosmetic thread, choose the **OK** button to complete the procedure.

CREATING COSMETIC GROOVES

Cosmetic groove is a projected cosmetic feature. The cosmetic groove is created by making a sketch and projecting it onto a surface. To create a cosmetic groove, invoke the **Cosmetic Groove** tool from the expanded **Engineering** group; you will be prompted to select a quilt or a set of surfaces. Select a surface from the model and then choose the **Done Refs** option in the **Menu Manager**. Next, you will be prompted to select a plane for sketching the groove section. Select the plane and then specify the direction of viewing. Choose the **Okay** option in the **Direction** rollout of the **Menu Manager** and specify the orientation of the sketching plane; the sketcher is invoked. Draw the sketch for the groove section and choose the **OK** button from the **Close** group of the **Sketch** tab; the sketch is projected on the selected surfaces, refer to Figure 7-54.

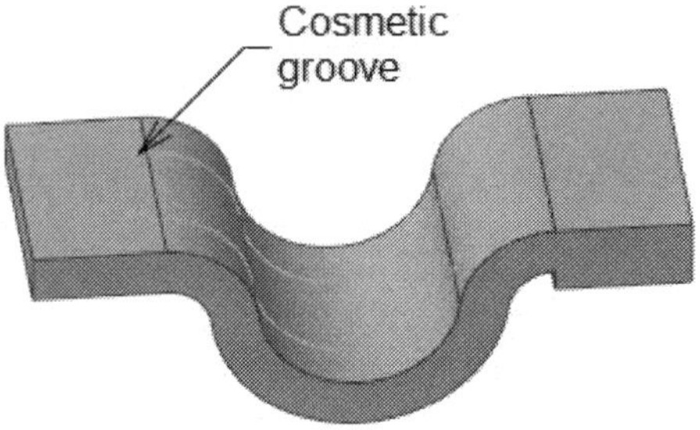

Figure 7-54 *Illustration of a cosmetic groove feature*

CREATING WARPS

Ribbon: Model > Editing > Warp

The **Warp** tool helps you study the design variations during the conceptual design stage. Using the **Warp** tool, you can manipulate the form and shape of solids, quilts, facets, and curves. The **Warp** tool can be used only in the **Part** mode. Choose the **Warp** tool from the **Editing** group; the **Warp** dashboard will be displayed, as shown in Figure 7-55. The tabs and buttons of the **Warp** dashboard are discussed next.

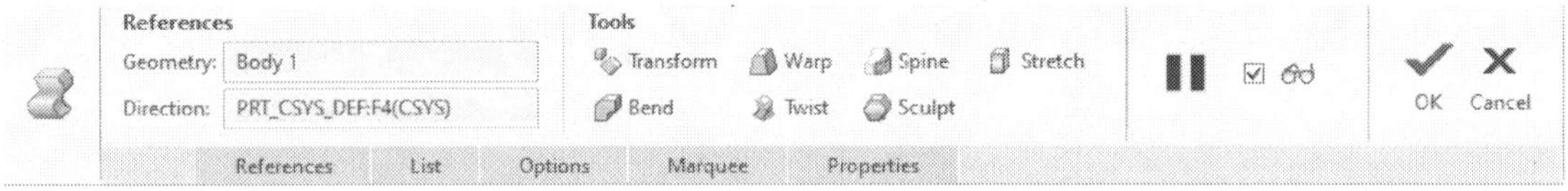

Figure 7-55 *Partial view of the **Warp** dashboard*

References Tab

When the **References** tab is chosen from the **Warp** dashboard, the **References** slide-down panel will be displayed, as shown in Figure 7-56. The **References** area indicates the geometry and direction chosen to create the warp transformation. By default, the Geometry collector gets active when you invoke the **Warp** tool. If the **Hide Original** check box is selected, the original geometry selected for the creation of the warp transformation is not displayed in the drawing area. The **Copy Original** check box, if selected, will keep a copy of the original entity chosen for creating the warp transformation. The **Copy Original** check box is not available for solid geometries. Select the **Facet Preview** check box to display a preview of the warped geometry within the feature. You can also increase or decrease the quality of the facets by dragging the slider available next to the **Facet Preview** check box.

List Tab

On choosing the **List** tab, the **List** slide-down panel is displayed showing a list of all the warp transformations performed on the selected entity. The warp operations performed on selected geometry are displayed in the sequence they were performed. You can select the required operation from the slide-down panel of the **List** tab, and then edit, preview, or delete it. The **List** slide-down panel is shown in Figure 7-57.

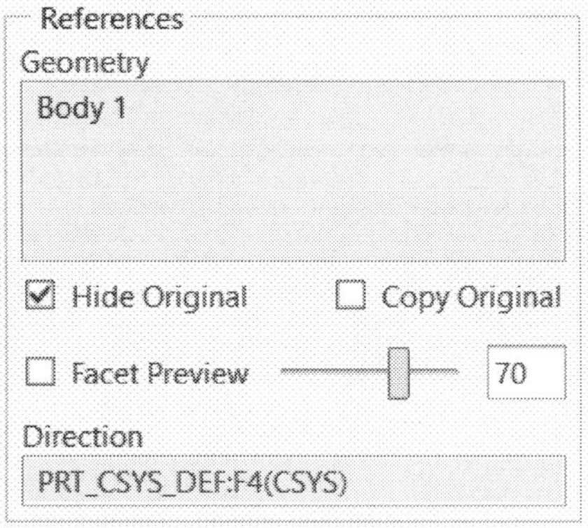

*Figure 7-56 The **References** slide-down panel*

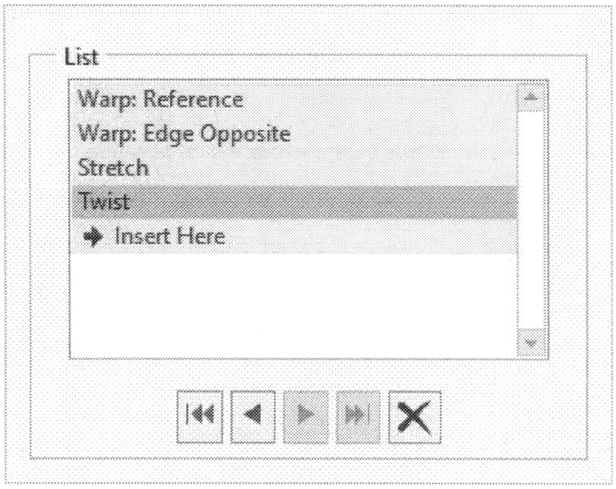

*Figure 7-57 The **List** slide-down panel*

Options Tab

The **Options** tab is used to input various parameters for the warp transformation. When the **Options** tab is chosen after invoking the **Transform** tool, the **Options** slide-down panel will be displayed, as shown in Figure 7-58. The uses of **Transform** tool will be discussed later in this chapter. The **Options** slide-down panel lists various parameters which control a particular warp transformation. In this slide-down panel, you can enter the values through which the geometry has to be transformed in the X, Y, and Z axes. The options available in this slide-down panel are context sensitive according to the transformation type selected in the **Warp** dashboard. Similarly, the parameters which control a particular warp transformation can be edited using the **Options** tab.

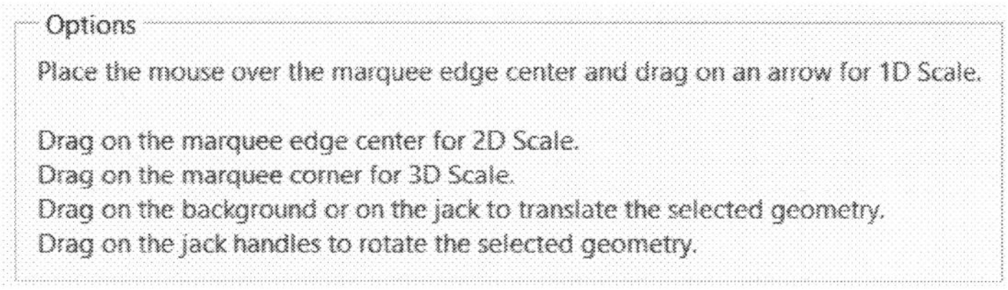

*Figure 7-58 The **Options** slide-down panel*

Marquee Tab

The marquee is a 3D bounding box that appears around the geometry in the drawing area. The **Marquee** tab is used to specify the marquee dimensions and location parameters. The options available in the slide-down panel of the **Marquee** tab are context sensitive according to the transformation type selected. Choose the **Marquee** tab after invoking the **Bend** transformation; the **Marquee** slide-down panel is displayed, as shown in Figure 7-59 The **Marquee** slide-down provides you the options to control the warping operation of an entity through primary and optional references. You can control the size of the marquee by specifying desired dimensions based on the transformation type. For transformations like bend and twist, you can use the

Center Axis option to place the marquee axis at the center of the selected references instead of at the center of the selected geometry. If required, you can restore the marquee to its default dimensions by choosing the **Reset** button available at the bottom of the panel. The remaining options available in this slide-down panel will be discussed later in this chapter.

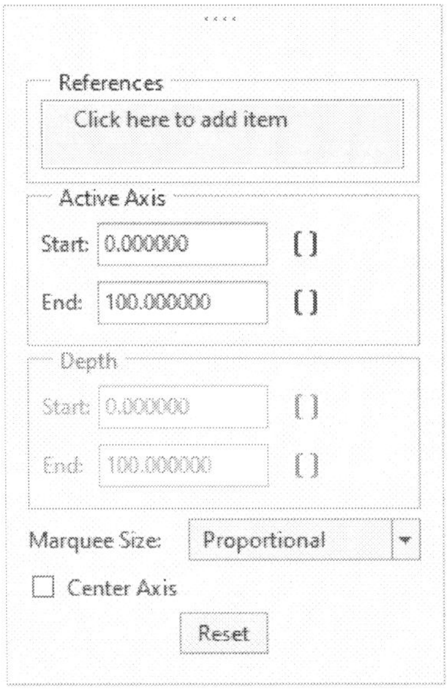

*Figure 7-59 The **Marquee** slide-down panel*

> **Tip**
> *Transforming a geometry may distort or remove the analytic surfaces such as planes, cylinders, and others. To preserve the analytic surfaces of the transformed geometry, select the **Use Boundary Tangents** option from the shortcut menu. This provides you better control tangency of surfaces thereby increasing the chances to keep analytic surfaces. Note that using this option to maintain higher tangency might cause slower regeneration of Warp feature.*

Transform Tool

The **Transform** tool can be used for various transformation operations such as translating, rotating, and scaling a geometrical entity. The translate option translates the geometry in the required direction. The rotate option rotates the geometry based on the position of the **Jack** (a blue color coordinate system which appears at the center of the drawing area). The scale option scales the geometry to the required scale factor.

The following steps explain the procedure to translate a geometry using the **Transform** tool, refer to Figure 7-60.

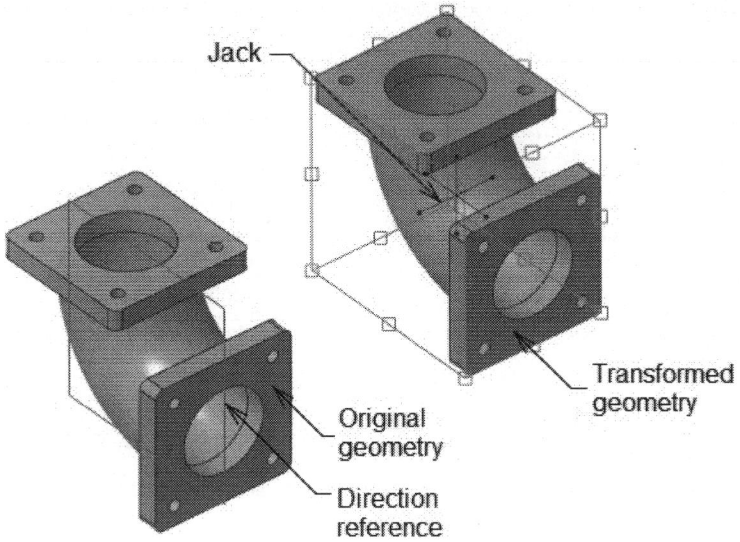

Figure 7-60 *Preview of the translated feature along with the original feature*

1. Choose the **Warp** tool from the expanded **Editing** group to invoke the **Warp** dashboard.

2. Choose the **References** tab to display the slide-down panel. Select the solid geometry from the drawing area. Now, click once in the **Direction** collector and select the **FRONT** datum plane as the direction reference. To unhide the original geometry during the operation, select the **Hide Original** check box.

3. Choose the **Transform** tool from the **Warp** dashboard.

4. Click anywhere in the drawing area and drag the mouse to place the geometry at the desired location.

Note

*1. The **Free** option in the **Drag** drop-down list is selected by default. As a result, the geometry can be translated freely anywhere in the drawing area with respect to the direction plane. The **H/V** option in this drop-down list allows you to translate the geometry horizontally or vertically with respect to the direction plane. Choose the **Normal** option if you want to translate the geometry normal to the direction plane.*

*2. By default, the **Selection** option is selected in the **Transform** drop-down list. As a result, you will be able to translate, rotate, and scale the selected geometry. **Jack** can be moved independent of the solid geometry. To do so, select **Jack** in the **Transform** drop-down list in the dashboard. Next, click on **Jack** in the drawing area and drag to place it at the required location. You can also specify x-, y-, and z-coordinate values to move **Jack**. You can also snap **Jack** to external geometries and rotate the transformed object about an external geometry. You can reposition **Jack** at center, or align it with direction plane by using the buttons available in the **Jack** area of the **Warp** dashboard. Alternatively, you can choose the **Center Jack** or **Align Jack** option from the shortcut menu to do the same.*

5. Choose the **OK** button from the **Warp** dashboard to complete the warp transformation.

The following steps explain the procedure to rotate geometry using the **Transform** tool.

1. Choose the **Warp** tool from the expanded **Editing** group to invoke the **Warp** dashboard.

2. Choose the **References** tab to display the slide-down panel. Select the solid geometry from the drawing area. Now, click once in the **Direction** collector and select the **FRONT** datum plane as the direction reference.

3. Choose the **Transform** tool from the **Warp** dashboard; **Jack** will be displayed at the center of the drawing area.

4. The rotation center is defined by the location of **Jack**. Click and drag the handles on the ends of the **Jack** to rotate the geometry to the required orientation. Hold down the SHIFT key while rotating the geometry, to perform the rotation in 15-degree increments.

5. Alternatively, click on the handles of **Jack** and then choose the **Options** tab from the **Warp** dashboard. Enter the angle of rotation in the **Angle** edit box and choose the **OK** button from the **Warp** dashboard to complete the warp transformation.

   Figure 7-61 shows the model before rotating the geometry and Figure 7-62 shows the model after rotating the geometry.

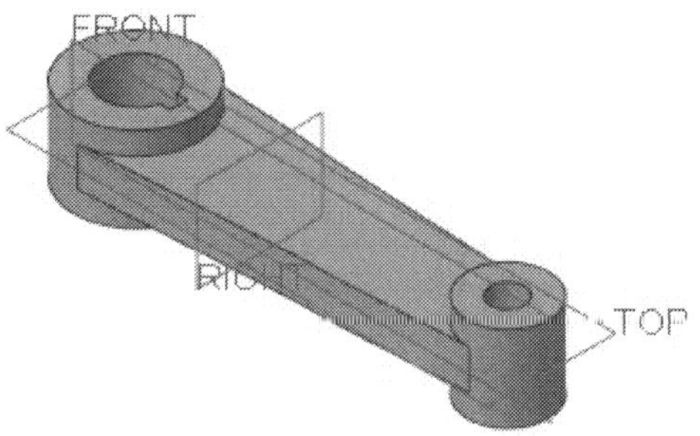

Figure 7-61 *Model before rotating the geometry*

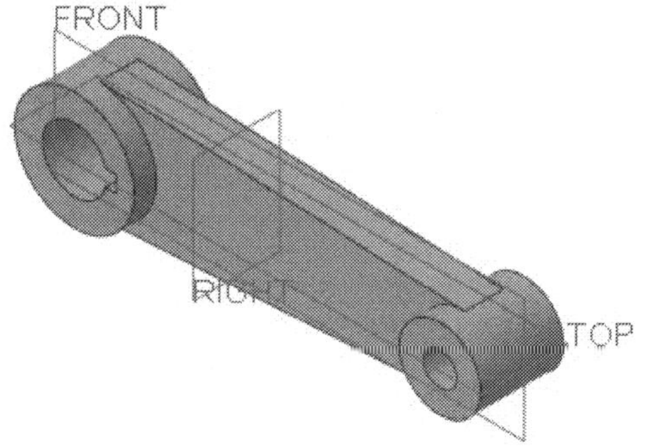

Figure 7-62 *Model after rotating the geometry*

The following steps explain the procedure to scale a geometry using the **Transform** tool.

1. Choose the **Warp** tool from the **Editing** group to invoke the **Warp** dashboard.

2. Click on the **References** tab to display the slide-down panel. Select the solid geometry from the drawing area. Now, click once in the **Direction** collector and the **TOP** datum plane as the direction reference.

3. Choose the **Transform** tool from the **Warp** dashboard.

4. Click on any corner of the marquee and choose the **Options** tab from the **Warp** dashboard; the **Options** slide-down panel will be displayed. The **Scale** area of the slide-down panel is displayed, as shown in Figure 7-63.

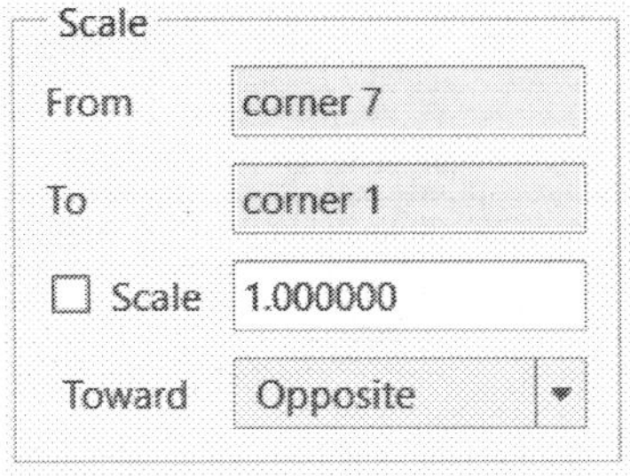

Figure 7-63 *The* ***Scale*** *area of the* ***Options*** *slide-down panel*

5. Enter the scale factor in the **Scale** edit box. The two options in the **Toward** drop-down list are **Opposite** and **Center**. The **Opposite** option scales the corner opposite to the selected reference corner and the **Center** option scales the geometry with respect to the center of the geometry. You can also drag any corner of the marquee for three-dimensional scaling about the opposite corner. Move the pointer over any edge to display the arrows and edge handle. You can drag an edge for two dimensional scale or an edge arrow for one dimension scaling.

6. Choose the **OK** button from the **Warp** dashboard to complete the warp transformation.

   Figure 7-64 shows a scaled feature previewed along with the original feature.

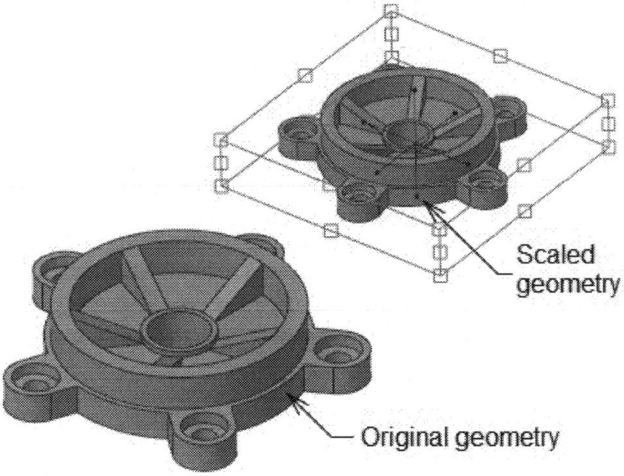

Figure 7-64 *Preview of the original geometry with the scaled geometry*

Warp Tool

The **Warp** tool can be used for changing the shape of the geometry. Some of the operations that are possible using the **Warp** tool are tapering the top or the base of the geometry, shifting the center of gravity of the object toward the base or the top of the object, or dragging a corner or an edge toward or away from the center.

The following steps explain the procedure to apply warp transformation using the **Warp** tool.

1. Choose the **Warp** tool from the expanded **Editing** group to invoke the **Warp** dashboard.

2. Choose the **References** tab to display the **References** slide-down panel. Select the solid geometry from the drawing area. Now, click once in the **Direction** collector and select the **TOP** datum plane as the direction reference.

3. Choose the **Warp** tool from the **Warp** dashboard.

4. Move the cursor toward one of the marquee points of the solid geometry to display the arrows. If you click on the arrow that is perpendicular to the edge and then choose the **Options** tab from the **Warp** dashboard; the **Options** slide-down panel will be displayed, as shown in Figure 7-65. If you click on the arrow that is perpendicular to the corner; the **Options** slide-down panel will be displayed, as shown in Figure 7-66.

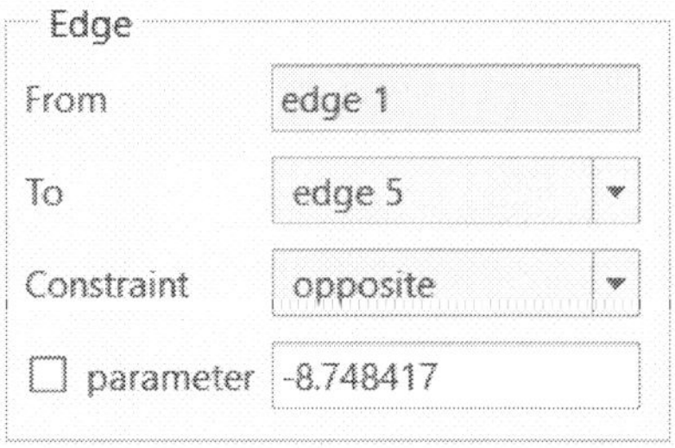

***Figure 7-65** The **Edge** area of the **Options** slide-down panel*

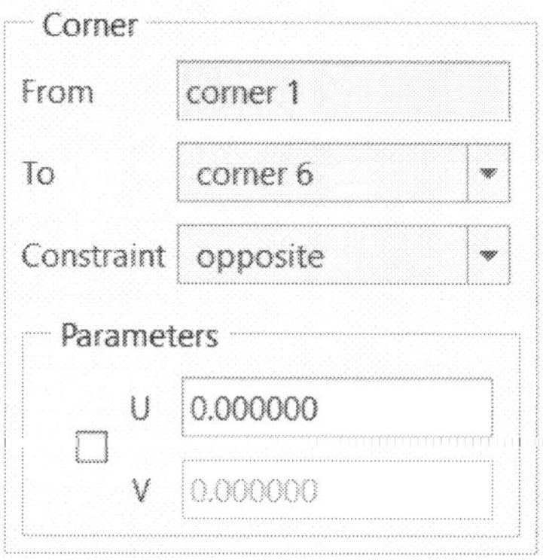

***Figure 7-66** The **Corner** area of the **Options** slide-down panel*

5. Select the **Center** option from the **Constraint** drop-down list.

6. Enter the warping length in the **Parameter** edit box; preview of the warped feature appears in the drawing area. Choose the **OK** button from the **Warp** dashboard to complete the feature creation.

7. Alternatively, move the cursor to any of the control points to display the arrows. Select one of the arrows and drag it when the arrow on the control point changes its color to orange.

   Figure 7-67 shows the model for applying the warp transformation and Figure 7-68 shows the model after applying the warp transformation.

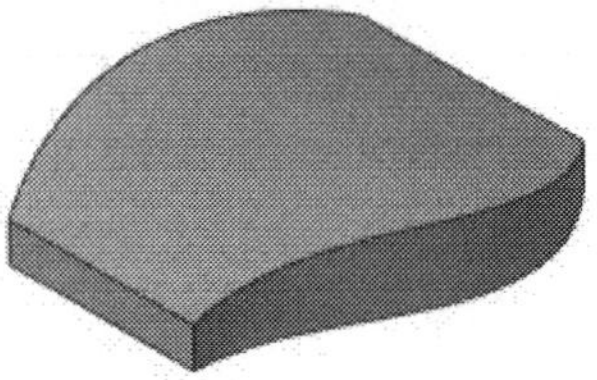

__Figure 7-67__ Model for applying the warp transformation

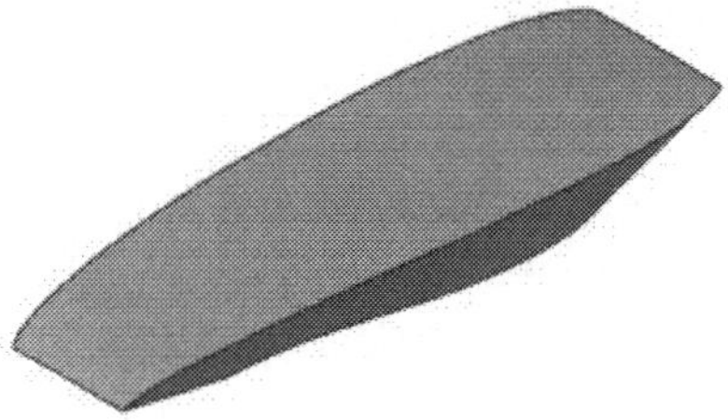

__Figure 7-68__ Model after applying the warp transformation

Note

1. When you move the cursor onto a corner point of the marquee, six arrows will be displayed; three in purple and the other three in orange. The arrows in orange are used to warp the model along the selected edge and the arrows in purple are used to warp the model by moving the selected corner within the selected marquee face. In this process, the other two corners follow symmetricity to maintain the original corner angles. If you do not require the other two corners to follow symmetricity, press and hold the ALT key and then drag the selected corner.

2. If you press and hold the SHIFT and ALT keys together while dragging the marquee points, the warp will take place symmetrically from both the corners.

3. For shifting the center of gravity, click on the orange arrow of an edge and then click the __Options__ tab. Next, select the required edge in the __Handle__ drop-down list and type a value between 0 to 1 in the __Value__ edit box. A value of 0.5 shifts the center of gravity to the midpoint of the edge.

Spine Tool

The **Spine** tool is used to reshape the geometry by manipulating the defining points of a curve (also referred to as spine). The transformation using the **Spine** tool can be created in the linear or radial direction.

The options available in the **Marquee** slide-down panel vary according to the button selected in the **Type** area of the **Warp** dashboard.

By default, the **Rectangular( )** option is selected in the dashboard. As a result, the geometry will be deformed freely. Click on the **Marquee** tab; the **Marquee** slide-down panel will be displayed, as shown in Figure 7-69. Notice that **Proportional** is selected under the **Marquee Size** drop-down list. The **Start** and **End** edit boxes in the **First** and **Second** areas are used to set the start and end points of the selected marquee. The values entered in these edit boxes are percentage of the total length of marquee. If you select the **Absolute** option in the **Marquee Size** drop-down list, you can set the start point and length of the marquee in the **First** and **Second** areas, respectively. In case, you have selected

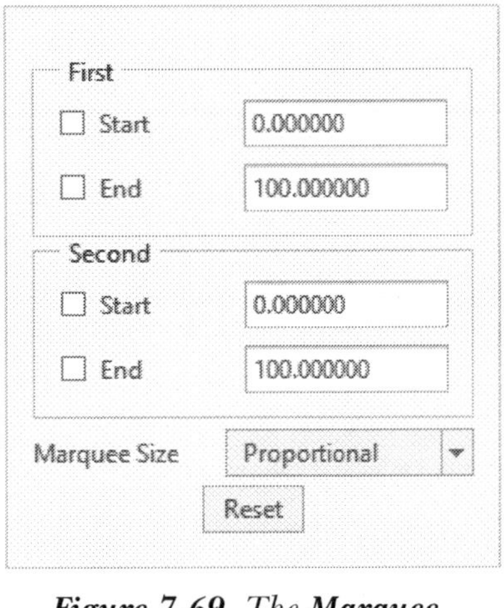

__Figure 7-69__ The __Marquee__ slide-down panel

the **Offset** option in the drop-down list, you can set the start offset and end offset points of the marquee in the **First** and **Second** areas normal to the spine.

When you select the **Radial**( ) option in the dashboard, the deformation of the geometry occurs in radial direction. You can specify the start and end values of marquee in **Radial** and **Depth** areas of the **Marquee** panel.

If you have selected the **Axial** option( ) in the dashboard, you will require to specify the start and end values only for the radial direction.

The following steps explain the procedure to apply a warp transformation using the **Spine** tool.

1. The transformation using the **Spine** tool will be applied with reference to a curve (also referred as a spine). You need to create a curve before invoking the **Spine** tool. The transformation will then be applied by dragging the control points of the spine to the required location.

Note

Remember that the spine you create needs to be sketched parallel to the plane on which you apply the warp transformation.

2. Choose **Warp** from the expanded **Editing** group to invoke the **Warp** dashboard.

3. Choose the **References** tab to display the slide-down panel. Select the solid geometry from the drawing area. Now, click once in the **Direction** collector and select the plane, which was the sketching plane for the spine, as the direction reference.

4. Choose the **Spine** tool from the **Warp** dashboard; the **Reference** slide-down panel will be displayed, as shown in Figure 7-70. Now, click once in the **Spine** collector and select the sketched spine from the drawing area, refer to Figure 7-71. In case, the spine curve is shorter than the marquee, select the **Extend Curve** button to extend the curve to the marquee faces.

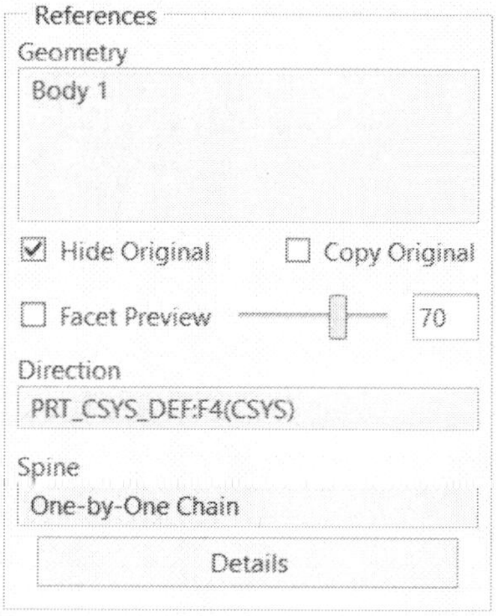

***Figure 7-70** The **References** slide-down panel*

5. Click and drag any of the control points displayed on the curve (spine) to the desired location. Choose the **OK** button from the **Warp** dashboard to complete the creation of the feature, as shown in Figure 7-72.

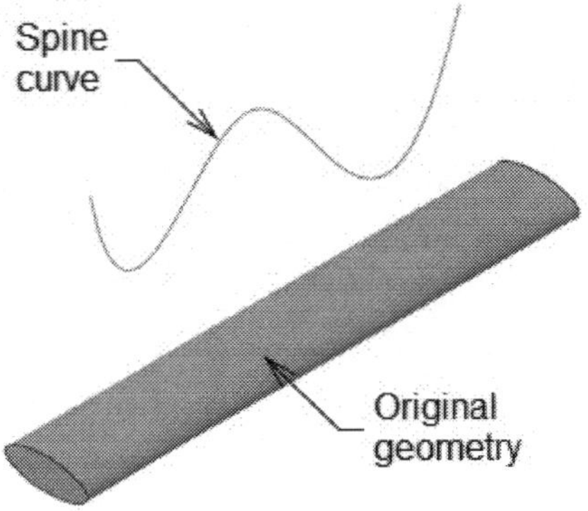

Figure 7-71 Model and spine curve selected for Spine transformation

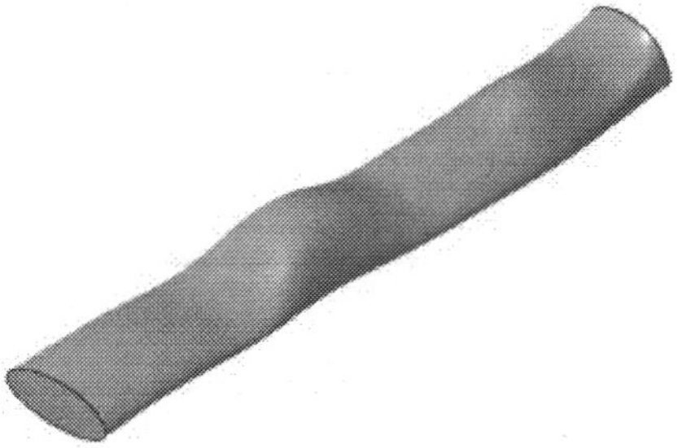

Figure 7-72 Model after applying the Spine transformation

Stretch Tool

The **Stretch** tool can be used to stretch geometry along the required direction. The range and scale of stretch can be controlled using the various options in the **Options** tab of the **Warp** dashboard.

The following steps explain the procedure to apply a warp transformation using the **Stretch** tool.

1. Choose **Warp** from the expanded **Editing** group to invoke the **Warp** dashboard.

2. Choose the **References** tab to display a slide-down panel. Select the solid geometry from the drawing area. Next, click once in the **Direction** collector and select the **FRONT** datum plane as the direction reference.

3. Choose the **Stretch** tool from the **Warp** dashboard. You can apply the transformation by moving the cursor onto the small orange dot which will be displayed in the drawing area and then dragging it to the required location.

4. Drag the small orange dot to stretch the model. Alternatively, enter a desired value of stretch in the **Scale** edit box; the model will be stretched in the direction of axis displayed and the preview of the stretched feature will be displayed in the drawing area.

5. To stretch the model in the other direction, choose the **Next Axis** button from the dashboard; the axis and a drag handle will be displayed in the other direction. Now, you can drag the rectangle to stretch the model in that direction. You can reverse the direction of stretch by choosing the **Flip Direction** button.

6. To set the length of the axis, invoke the **Marquee** tab; the **Marquee** slide-down panel will be displayed, as shown in Figure 7-73. Enter the start and end points of the axis in the **Start** and **End** edit boxes, respectively.

Figure 7-73 *The **Marquee** slide-down panel*

7. Choose the **OK** button from the **Warp** dashboard to complete the creation of the feature.

   Figure 7-74 shows the model before the stretching transformation and Figure 7-75 shows the model after the stretching transformation.

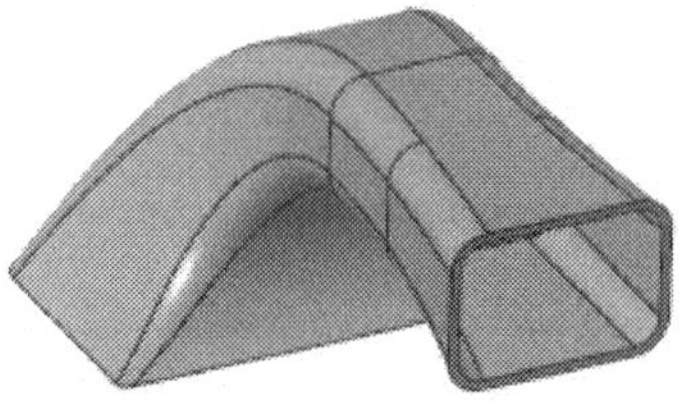

Figure 7-74 *Model before stretching*

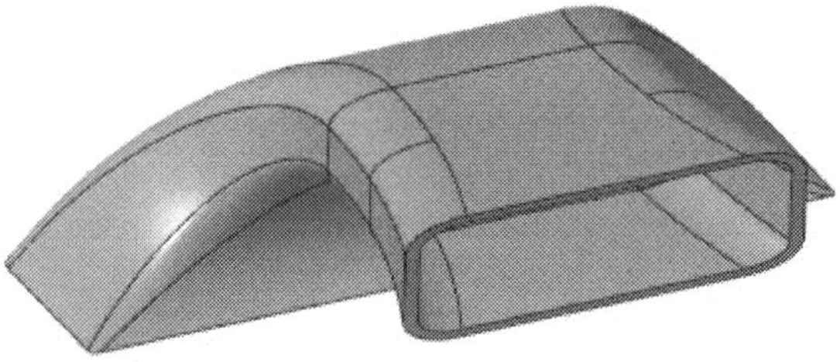

Figure 7-75 *Model after stretching*

Note

*The **Proportional** marquee size is selected by default in the **Marquee** slide-down panel, thereby stretching the model relative to its previous position. You can select **Absolute** to stretch the model to the specified length. To specify the stretch from an offset distance, select the **Offset** marquee size.*

Bend Tool

The **Bend** tool is used to bend the geometry to the required orientation. The following steps explain the procedure to create a bend using the **Bend** tool.

1. Choose the **Warp** tool from the expanded **Editing** group to invoke the **Warp** dashboard.

2. Choose the **References** tab to display the slide-down panel. Select the solid geometry from the drawing area. Now, click in the **Direction** collector once and select the **FRONT** datum plane as the direction reference.

3. Choose the **Bend** tool from the **Warp** dashboard. The model will be displayed in the drawing area with marquee and an orange line with a small circle attached at one end.

4. Move the cursor onto the small orange circle rectangle and drag it until the desired orientation is achieved. Alternatively, enter the desired orientation value in the **Angle** edit box available in the dashboard; the preview of the bend will be displayed in the model area.

5. Next, choose the **Options** tab; a slide-down panel will be displayed, as shown in Figure 7-76.

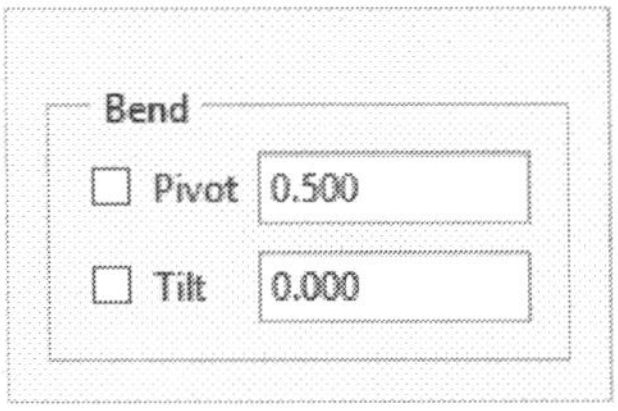

Figure 7-76 *The **Bend** area of the **Options** slide-down panel*

The **Pivot** edit box in the **Bend** area of this panel is used to specify the fixed point of the bend and the **Tilt** edit box is used to specify the angle of the bend. Enter the desired values in the respective edit boxes.

You can also set the range of bend, similar to stretch, by invoking the **Marquee** tab in the dashboard. For this model, specify 3 and 97 in the **Start** and **End** edit boxes, respectively, in the **Active Axis** area of the **Marquee** tab.

Figure 7-77 shows the geometry before creating the bend feature and Figure 7-78 shows the geometry after creating the bend feature.

Figure 7-77 *Model before bending*

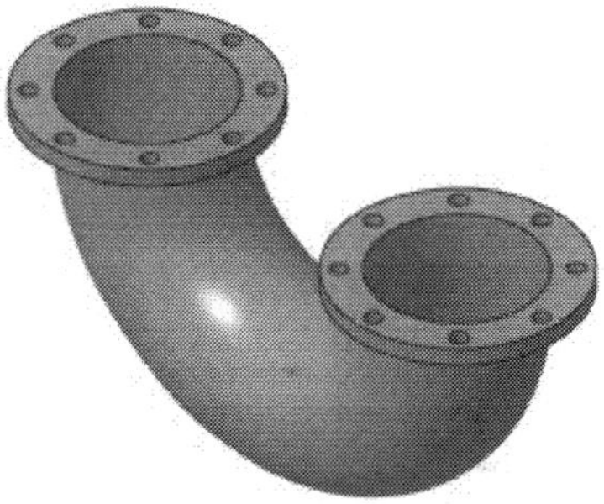

Figure 7-78 *Model after bending*

The bend operation can also be controlled by using the buttons in the **Warp** dashboard.

Next Axis

This button switches the axis of bend from one plane to another.

Flip direction

This button reverses the direction of the axis of the bend.

Increase Tilt

This button increases the tilt or the bending angle by 90 degrees from the earlier applied bend angle.

Twist Tool

The **Twist** tool is used to twist the geometry around an axis by the user specified angle.

The following steps explain the procedure to twist a geometry using the **Twist** tool.

1. Choose the **Warp** tool from the expanded **Editing** group to invoke the **Warp** dashboard.

2. Choose the **References** tab, the **References** slide-down panel is displayed. Select the solid geometry from the drawing area. Now, click once in the **Direction** collector and select the **TOP** datum plane as the direction reference.

3. Choose the **Twist** tool from the **Warp** dashboard; the model will be displayed in the drawing area along with the marquee and a small orange dot.

4. Move the cursor onto the small orange dot and drag to twist the model to the required orientation. Alternatively, enter the angle of twist in the **Angle** edit box in the dashboard; the preview of twist is displayed in the drawing area.

5. You can define the range of the twist feature similar to **Stretch** and **Bend**. To do so, choose the **Marquee** tab; the **Marquee** slide-down panel will be displayed, as shown in Figure 7-79. The **Start** and **End** edit boxes in this slide down panel are used to define the range of twist.

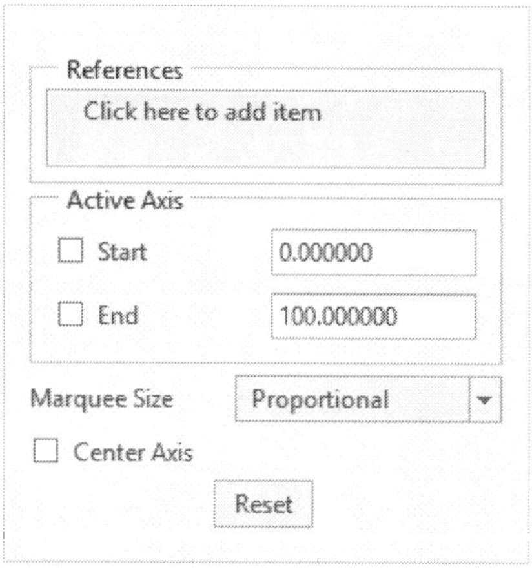

Figure 7-79 *The **Marquee** slide-down panel*

6. Choose the **OK** button from the **Warp** dashboard to complete the creation of the feature.

Figures 7-80 shows the model before twisting and Figure 7-81 shows the model after twisting.

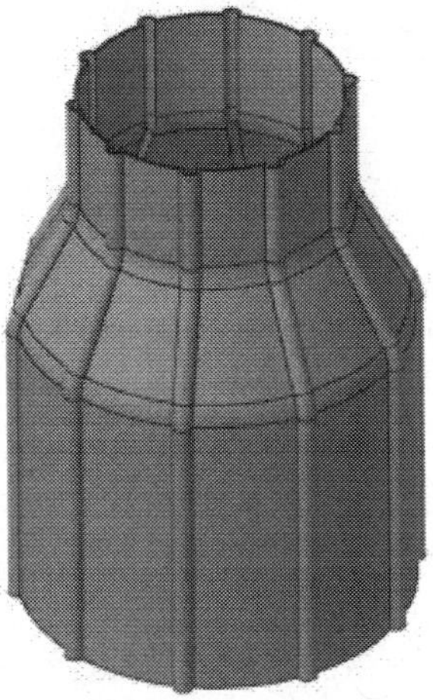

Figure 7-80 *Model before twisting*

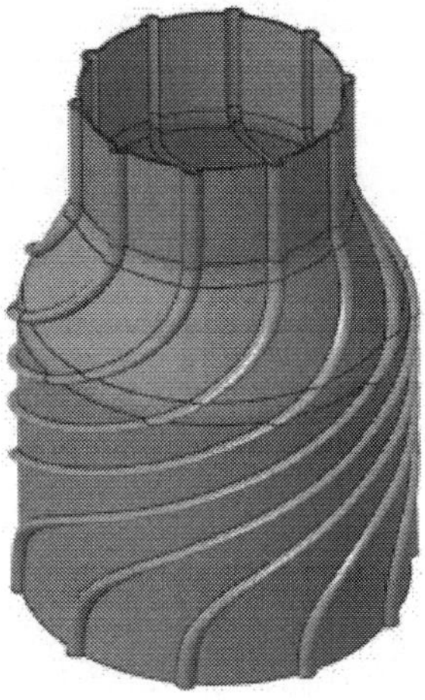

Figure 7-81 *Model after twisting*

Note
*The direction of twist and the axis of twist can be reversed or switched by using the **Flip Direction** and **Next Axis** buttons, respectively in the **Warp** dashboard.*

Sculpt Tool

The **Sculpt** tool allows you to deform the geometry through a mesh. The following steps explain the procedure to create a sculpt transformation.

1. Choose the **Warp** tool from the expanded **Editing** group to invoke the **Warp** dashboard.

2. Click on the **References** tab to display a slide-down panel. Select the solid geometry from the drawing area. Now, click once in the **Direction** collector and select the required datum plane as the direction reference.

3. Choose the **Sculpt** tool from the **Warp** dashboard.

4. Choose the **Options** tab from the **Warp** dashboard; the **Options** slide-down panel will be displayed, as shown in Figure 7-82.

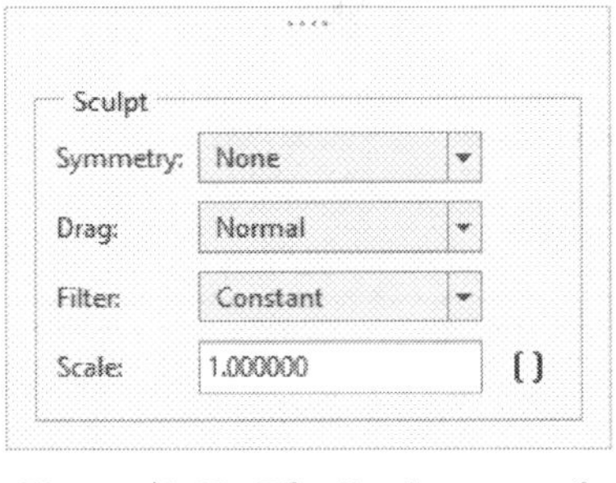

Figure 7-82 *The **Sculpt** area of the **Options** slide-down panel*

5. Select the type of symmetry from the **Symmetry** drop-down list.

Note
*The **Symmetry** drop-down list specifies the direction in which the selected group of points move while sculpting. The **None** option is selected by default in the **Symmetry** drop-down list. If the **Horizontal** option is selected from the drop-down list, then the points move symmetric to the centerline of the face parallel to the selected reference plane.*

6. Select the type of drag from the **Drag** drop-down list.

Note

*The **Drag** drop-down list specifies the constraints on the group of points being sculpted. The **Normal** option is selected by default in this drop-down list so the points move normal to the selected face when the transformation is created. If the **Free** option is selected from the **Drag** drop-down list, then the points selected for sculpting move freely in the drawing area. The **Along Row/Column** option will be used to move the points selected for sculpting towards the neighboring rows or columns.*

7. Select the type of filter required from the **Filter** drop-down list.

Note

*The **Filter** drop-down list specifies the behavior of the points being sculpted. The **Constant** filter is selected by default so the points selected for sculpting move to the same distance as the dragged point. If the **Linear** filter type is selected, then the selected points drop-off linearly with respect to the dragged point. The **Smooth** filter drops off the selected points smoothly with respect to the dragged point.*

8. Select one or more curves or points from the marquee displayed (hold down the CTRL button for multiple point selections). Drag the selected points to the required location. Choose the **OK** button from the **Warp** dashboard to complete the warp transformation.

   Figure 7-83 shows the model before applying the sculpt transformation and Figure 7-84 shows the model after applying the sculpt transformation.

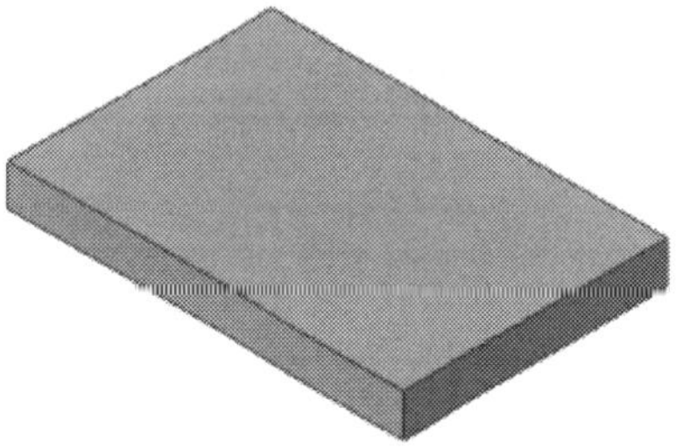

Figure 7-83 *Model before applying the sculpt transformation*

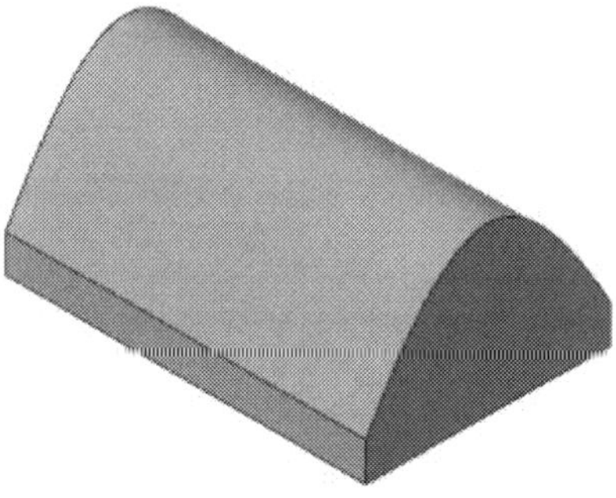

Figure 7-84 *Model after applying the sculpt transformation*

You may also specify how the geometry is sculpted by choosing the buttons that will become available when the **Sculpt** tool is chosen in the **Warp** dashboard. These buttons are discussed next.

Next Axis

This button helps you to align the mesh in the direction in which the geometry will be sculpted.

One Side

This button sculpts the selected face in only one direction and is chosen by default.

Both Sides

This button applies the motion of selected face to the opposite face.

Symmetrical

This button, if chosen, allows the geometry to be symmetric from its center.

Note

*The number of rows and columns in the marquee can be controlled by specifying the required value in the **Rows** and **Columns** edit boxes, respectively in the dashboard.*

TUTORIALS

Tutorial 1

In this tutorial, you will create the model shown in Figure 7-85. To perform this tutorial, you need to download the zipped file named as *c07_creo_8.0_input* from the **Input Files** section of the CADCIM website. The complete path for downloading the file is:

Textbooks > CAD/CAM > PTC Creo Parametric > Creo Parametric 8.0 for Designers, 8th Edition > Input Files

(Expected time: 30 min)

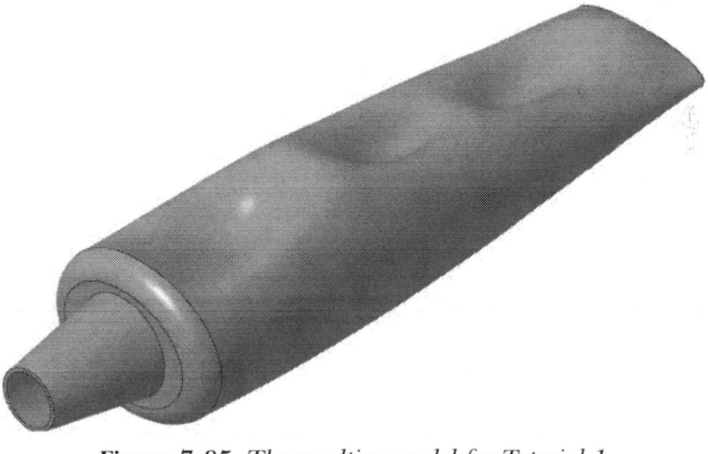

Figure 7-85 *The resulting model for Tutorial 1*

The following steps are required to complete this tutorial:

a. Create the sketch for warp feature.
b. Create the warp features using the **Spine** tool.

When you start the Creo Parametric session, the first task is to set the working directory. Make sure the required working directory is selected.

Opening the File

1. Start a new session of Creo Parametric 8.0.

2. Select the **Open** button in the **Home** tab; the **File Open** dialog box will be displayed. Browse to the location where you have stored the downloaded file and select the **Open** button in the dialog box to open the model in the Part environment.

Creating the Sketch for the Warp Feature

You need to use the **Spline** tool to create deformation in the tube. Therefore, you first need to create the curve which will be used to create the warp feature.

1. Choose the **Sketch** tool from the **Datum** group; the **Sketch** dialog box is displayed.

2. Choose the **RIGHT** plane as the sketching plane and select the **Top** option from the drop-down list in the **Orientation** tab. Next, choose the **Sketch** button to enter the sketching environment.

3. Choose the **Spline** tool from the **Sketching** group and draw the spline curve shown in Figure 7-86. Note that the dimensions of the spline are not important as this sketch will be used as reference for deciding the shape after deformation.

4. Choose the **OK** button to exit the sketcher environment.

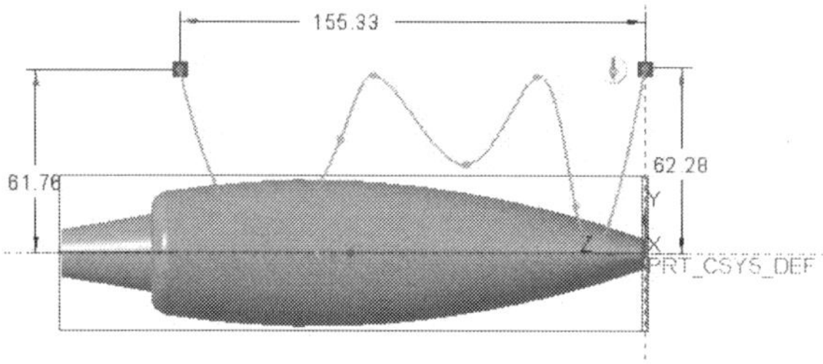

Figure 7-86 *Sketched spline curve*

Applying the Warp Transformation by Using the Spine Tool

1. Choose the **Warp** tool from the expanded **Editing** group; the **Warp** dashboard is displayed.

2. Choose the **References** tab; the **Geometry** collector is enabled by default. Select the model from the drawing area.

3. Click in the **Direction** selection box; you are prompted to select a plane or a coordinate system to define the direction of the warp.

4. Select the **TOP** plane from the drawing area; you are prompted to select the deformation tool.

5. Choose the **Spine** tool from the **Warp** dashboard; you are prompted to select a curve to define the deformation.

6. Select the sketched spine as the reference curve for the warp transformation. The model similar to the one shown in Figure 7-87 is displayed in the drawing area.

7. Select the control points on the reference curve and drag them until the required spinal deformation is obtained. The position of the control points after carrying out deformation is shown in Figure 7-88.

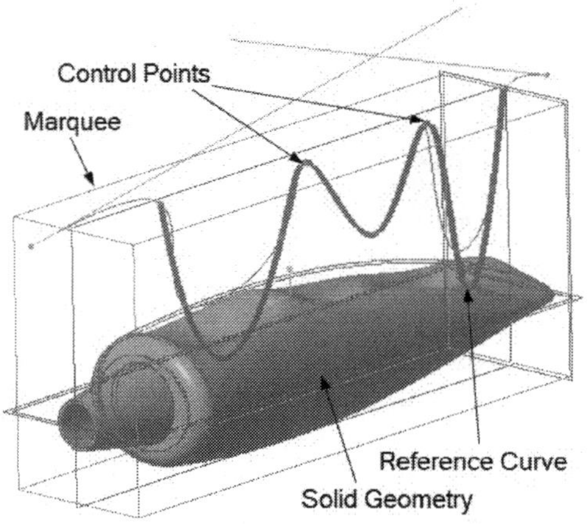

Figure 7-87 *Figure showing the parameters for the warp transformation*

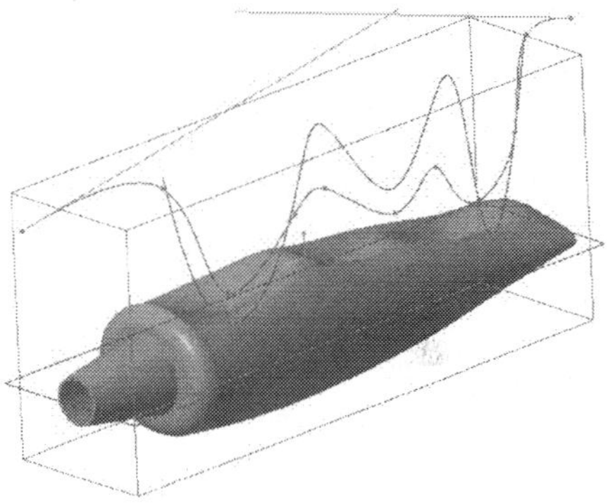

Figure 7-88 *Figure showing the deformed model*

Note

The control points for creating the deformation may be randomly selected based on the profile required.

8. Choose the **OK** button to create the feature. Figure 7-89 shows the final model after applying the warp transformation and hiding the spline created.

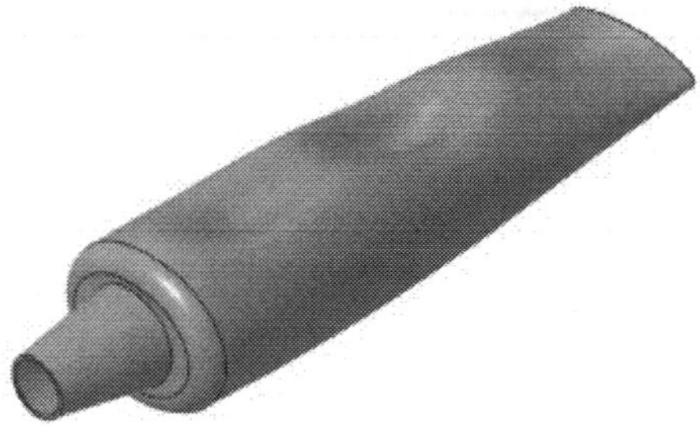

Figure 7-89 *Final model after applying the warp transformation*

Saving the Model

1. Choose the **Save** button from the **File** menu to save the model.

Tutorial 2

In this tutorial, you will create the model shown in Figure 7-90. The Figure 7-91 shows the drawing of the base feature. Angle of twist for the Warp is 120°. **(Expected time: 30 min)**

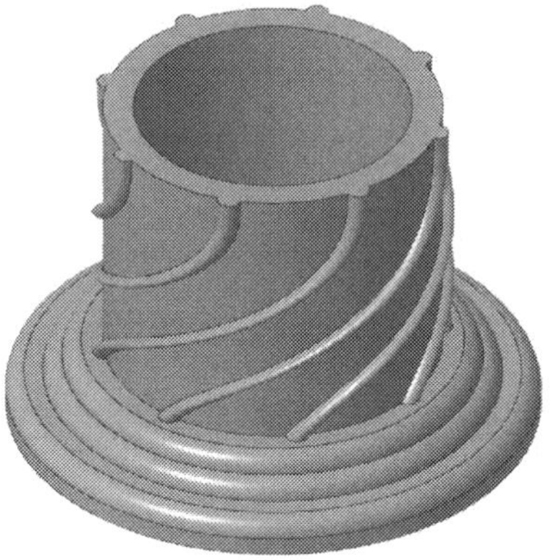

Figure 7-90 *The resulting model for Tutorial 2*

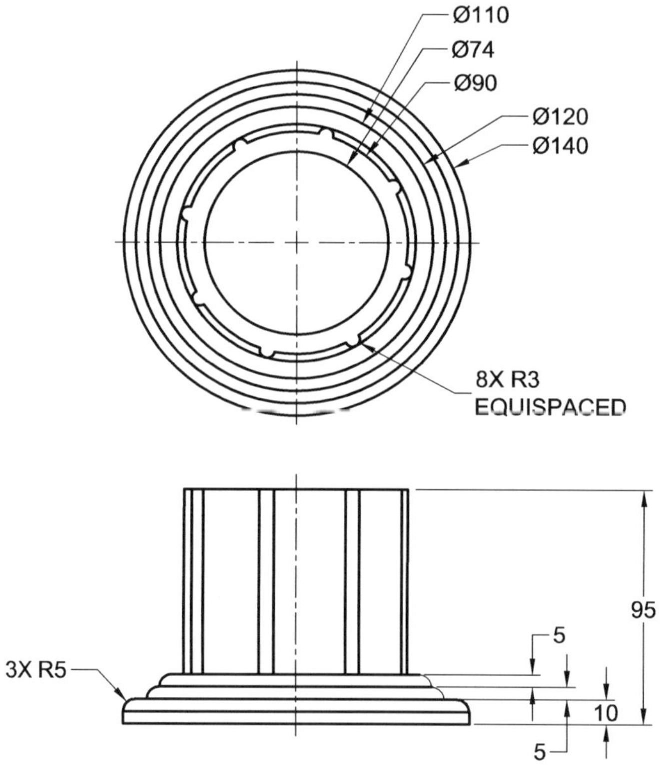

Figure 7-91 *Drawing of the base feature for Tutorial 2*

The following steps are required to complete this tutorial:

a. Create the revolve feature.
b. Create the shell feature with thickness as 8.
c. Create the extrude feature, and then pattern the extruded feature.
d. Create the round feature.
e. Create the warp transformations.

Starting a New Object File

Start a new part file and name it as *c07tut2*.

Creating the Revolve Feature

1. Draw the sketch of the revolved feature in the **FRONT** datum plane, as shown in Figure 7-92.

2. Revolve the sketch through 360 degrees to create the base feature, as shown in Figure 7-93.

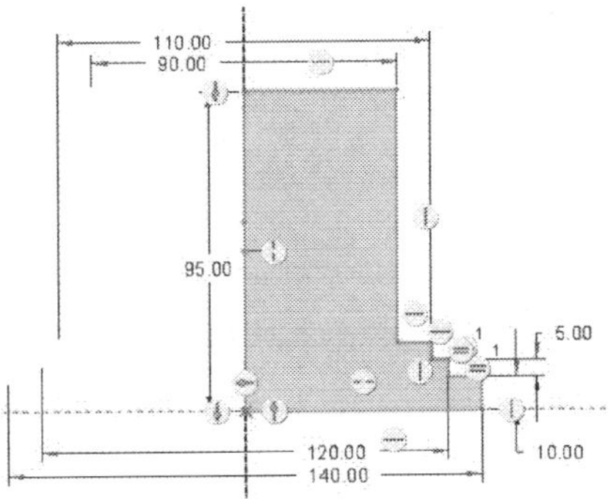

Figure 7-92 *Sketch for the revolved feature*

Figure 7-93 *Model showing the base feature*

Creating the Shell Feature

1. Create the shell feature with a wall thickness of value **8** and select the top face of the base feature. The model after creating the shell feature is shown in Figure 7-94.

Creating the Extrude Feature and its Pattern

1. Create the extrude feature similar to the one shown in Figure 7-95.

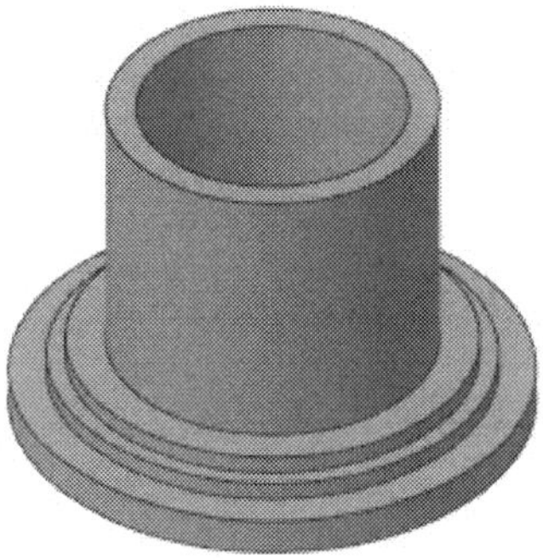

Figure 7-94 Model after creating the shell feature

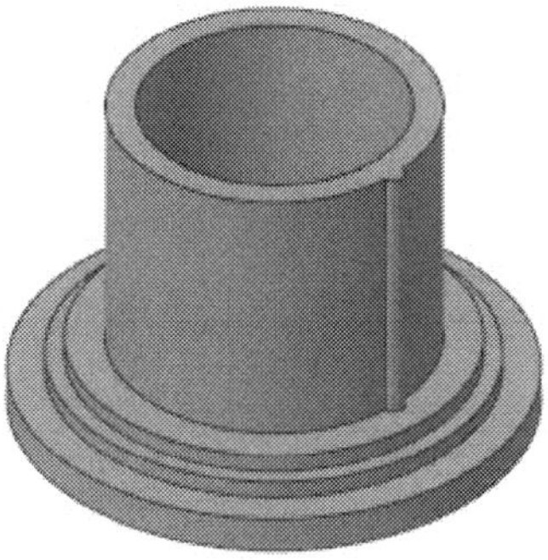

Figure 7-95 Model after creating the extrude feature

2. Create the axis pattern of the extrude feature, refer to Figure 7-96.

Creating Round Features

1. Choose the **Round** tool from the **Engineering** group; you are prompted to select the edges to create the round.

2. Choose the edges, refer to Figure 7-90.

Note

Hold the CTRL key to select multiple edges.

3. Enter **5** in the **Radius** edit box available in the **Round** dashboard and then choose the **OK** button to exit the **Round** dashboard.

   The model after creating the round feature is shown in Figure 7-97.

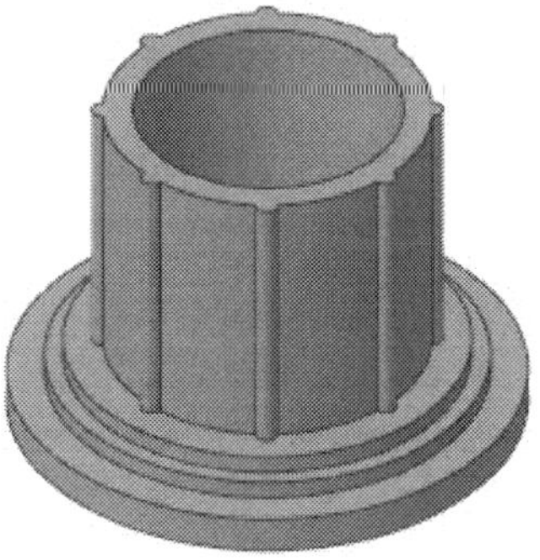

Figure 7-96 Model after creating the pattern

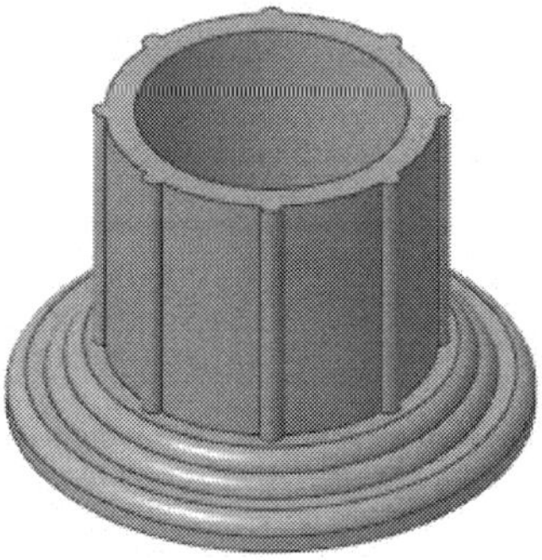

Figure 7-97 Figure showing the round feature

Applying the Warp Transformation by Using the Twist Tool

1. Choose the **Warp** tool from the expanded **Editing** group; the **Warp** dashboard is displayed and you are prompted to select the solids or quilts to warp.

2. Select the model in the drawing area and then choose the **References** tab from the **Warp** dashboard. Click in the selection area for the **Direction** collector; you are prompted to select a plane or a coordinate system to select the direction of the warp.

3. Select the **TOP** datum plane in the drawing area; you are now prompted to select the deformation tool.

4. Choose the **Twist** tool from the **Warp** dashboard.

5. Enter **120** in the **Angle** edit box in the **Warp** dashboard; preview of the warp transformation is displayed in the drawing area.

6. Choose the **OK** button from the **Warp** dashboard to complete the creation of the feature.

   The model after applying the warp transformation is shown in Figure 7-98.

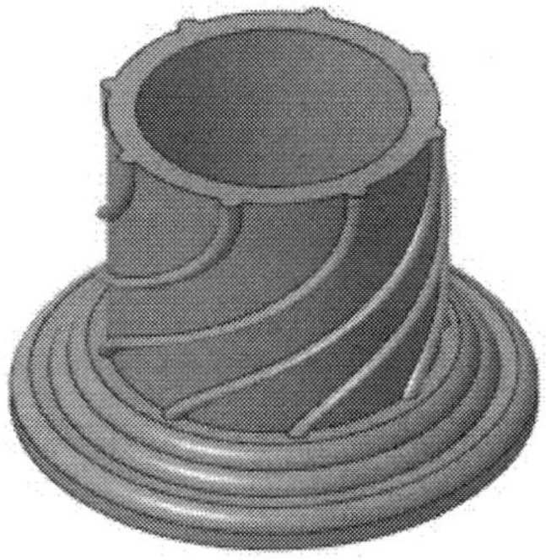

Figure 7-98 *Model after applying the warp transformation*

Saving the Model

1. Choose the **Save** button from the **File** menu to save the model.

Tutorial 3

In this tutorial, you will create the model shown in Figure 7-99. **(Expected time: 30 min)**

Figure 7-99 *Model for Tutorial 3*

The following steps are required to complete this tutorial:

a. Create the extrude feature.
b. Create the second extrude feature.
c. Create the pattern of the tread.
d. Create the toroidal bend feature.
e. Create the mirror of the model.

Starting a New Object File

Start a new part file and name it as *C07_TUT03*.

Creating the First Extrude Feature

1. Draw the sketch for the first extrude feature in the **FRONT** plane, as shown in Figure 7-100.

Figure 7-100 *Sketch of the extrude feature*

2. Extrude it to a length of **1500**, refer to Figure 7-101.

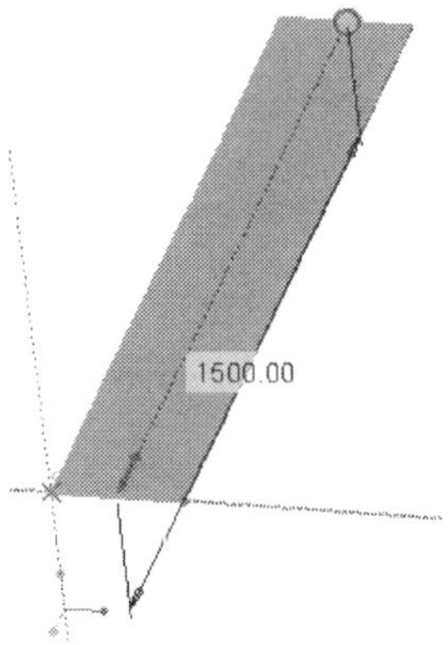

Figure 7-101 *Preview of the first extrude feature*

Creating the Second Extrude Feature

1. Select the top face of the model from the drawing area.

2. Choose the **Extrude** tool from the **Shapes** group; the **Extrude** dashboard is displayed.

3. Draw the sketch for the second extrude feature, as shown in Figure 7-102, and select the **OK** button.

4. Enter **10** in the **Depth** edit box and choose the **OK** button from the **Extrude** dashboard.

5. Specify the correct direction of extruded feature and choose the **OK** button to complete the creation of the feature.

   The model after creating the second extrude feature is shown in Figure 7-103. Constraints visibility have been turned off for clarity.

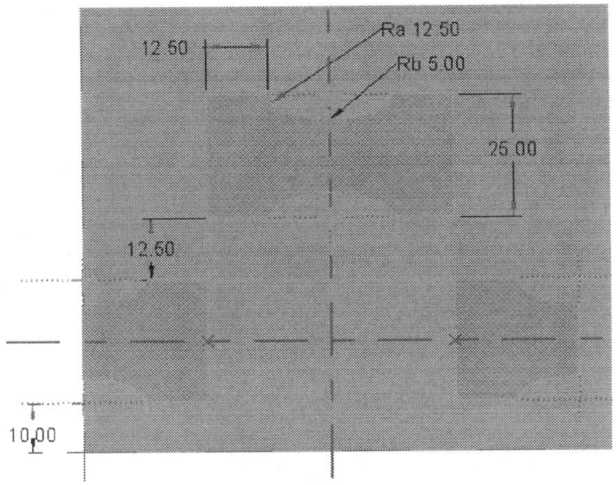

Figure 7-102 Sketch of the extrude feature

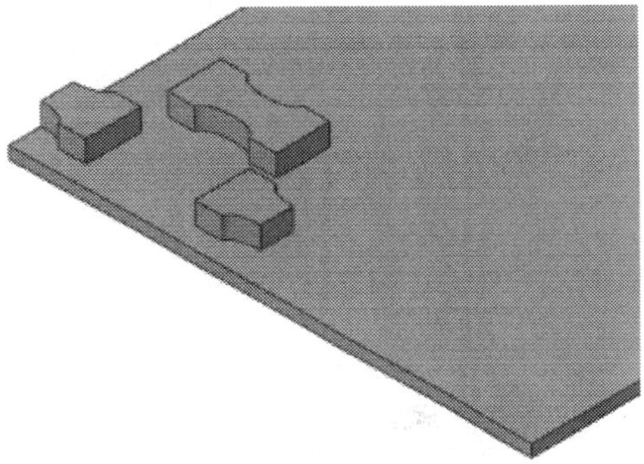

Figure 7-103 Model showing the second extrude feature

Creating the Pattern of the Tread

1. Select the extruded feature created in the previous section from the **Model Tree** or the drawing area.

2. Choose the **Pattern** tool from the **Pattern** drop-down in the **Editing** group; the **Pattern** dashboard is displayed and you are prompted to select a plane, flat face, linear curve, coordinate system axis, or axis to define the first direction.

3. Click in the Direction reference collector in the **Pattern** dashboard and select the **Front** plane from the drawing area or from the **Model Tree**.

4. Enter **75** in the **Spacing** edit box available right to the **1st direction** collector.

5. Choose the **OK** button from the dashboard to exit from the **Pattern** tool.

   The model after creating the pattern is shown in Figure 7-104.

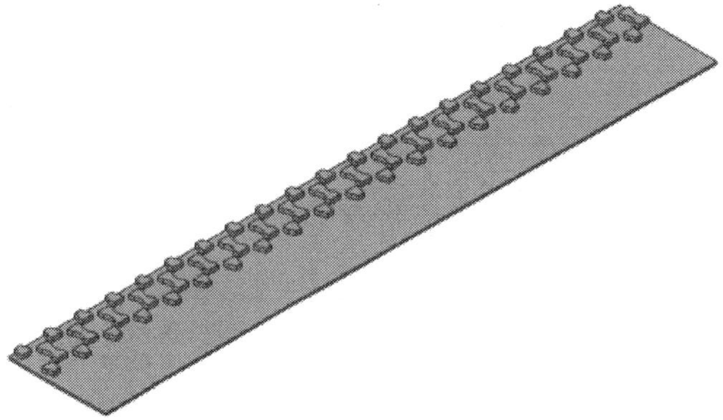

Figure 7-104 *Model after creating the pattern of the tread*

Note
You will learn about ***Pattern*** *tool in detail in next chapter.*

Creating the Toroidal Bend Feature

1. Choose the **Toroidal Bend** tool from the **Engineering** group; the **Toroidal Bend** dashboard is displayed.

2. Choose the **References** tab; the **References** slide-down panel is displayed. Select the solid body in the **Quilts and/or a solid body** area and choose the **Define** button in the **Profile Section** from the slide-down panel; the **Sketch** dialog box is displayed.

3. Select the front face of the base feature as the sketching plane. Next, choose the **Sketch** button from the **Sketch** dialog box to accept the default orientation and enter the sketcher environment.

4. Draw the section for the toroidal bend feature, as shown in Figure 7-105.

5. Choose the **Coordinate System** button from the **Datum** group and place the reference coordinate system at the origin, as shown in Figure 7-105.

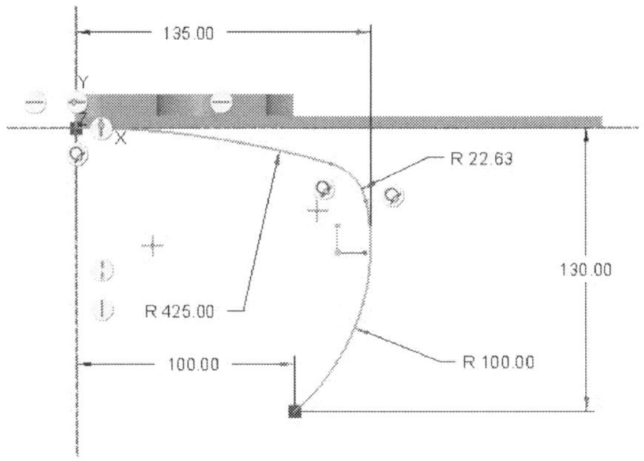

Figure 7-105 *Section created for the toroidal bend feature*

6. Choose the **OK** button from the **Sketch** dashboard to exit the sketcher mode; the **Toroidal Bend** dashboard is displayed again. Choose the **Options** tab and select the **Standard** radio button, if it is not already selected from the **Curve Bend** area of the slide-down panel.

7. Choose the **360 Bend** button in the **Toroidal Bend** dashboard; you are prompted to select two parallel planes to define the length of the bend.

8. Select the front face and then the back face of the base feature; preview of the toroidal bend feature is displayed.

9. Choose the **OK** button; the model similar to the model shown in Figure 7-106 is created in the drawing area.

Creating the Mirror of the Model

1. Select the part and then choose the **Mirror** tool from the **Editing** group; the **Mirror** dashboard is displayed.

2. Click in the **Mirror plane** collector and select the **Right** datum plane from the **Model Tree** and choose the **OK** button from the dashboard.

   Figure 7-107 shows the final model after mirroring the part.

Figure 7-106 *Model after creating the toroidal bend feature*

Figure 7-107 *Model after mirroring the part*

Saving the Model

1. Choose the **Save** button from the **File** menu to save the model.

Tutorial 4

In this tutorial, you will create the model of the pencil shown in Figure 7-108 and then change the color of the pencil. Figure 7-109 shows the front, section, and the detail views of the model. **(Expected time: 30 min)**

Figure 7-108 Solid model of the pencil

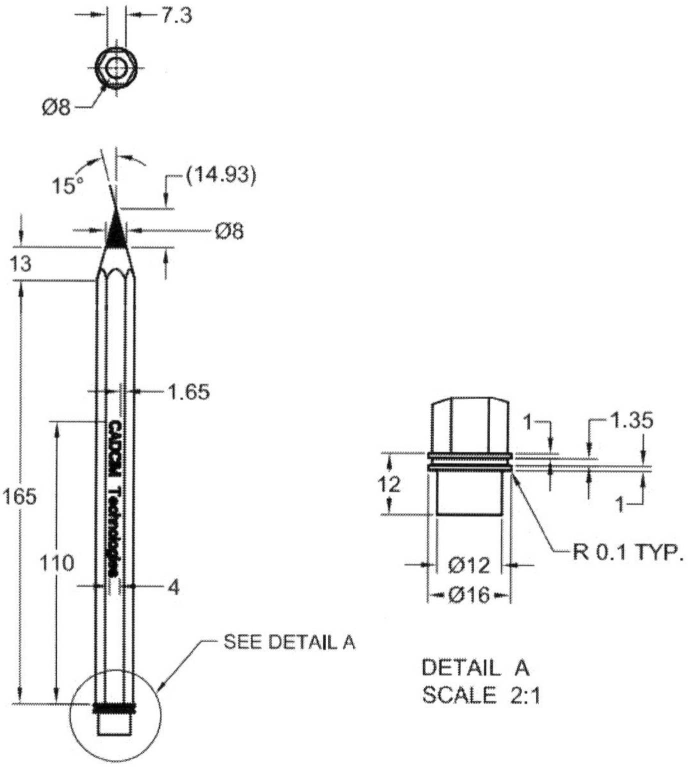

Figure 7-109 Orthographic and detail views of the model

Examine the model and determine the number of features in it. The model is composed of six features, refer to Figure 7-108.

The following steps are required to complete this tutorial:

a. Create the base feature.
b. Create the second feature that is a cylinder on the top face of the base feature.
c. Create the draft feature.
d. Create the tip of the pencil that is a revolved feature.
e. Create the tail end of the pencil that is again a revolved feature.
f. Create a cosmetic sketch on the pencil by using the **Cosmetic Sketch** tool.
g. Changing model appearances.

Make sure that the *c07* folder is the current Working Directory.

Starting a New Object File

1. Start a new part file and name it as *c07tut4*.

   The three default datum planes are displayed in the drawing area.

Creating the Base Feature

The base feature of the pencil is a protrusion feature in which an octagon is extruded to a depth of 165.

1. Choose the **Extrude** tool from the **Shapes** group to display the **Extrude** dashboard and invoke the **Sketch** dialog box.

2. Select the **TOP** datum plane as the sketching plane.

3. Select the **RIGHT** datum plane from the drawing area and then select the **Right** option from the **Orientation** drop-down list only if these options are not selected by default.

4. Draw the sketch of the hexagon by using the **Palette** button and then apply constraints and dimensions to it, as shown in Figure 7-110.

5. Exit the sketcher environment and extrude the sketch to a depth of 165. The base feature thus created is shown in Figure 7-111.

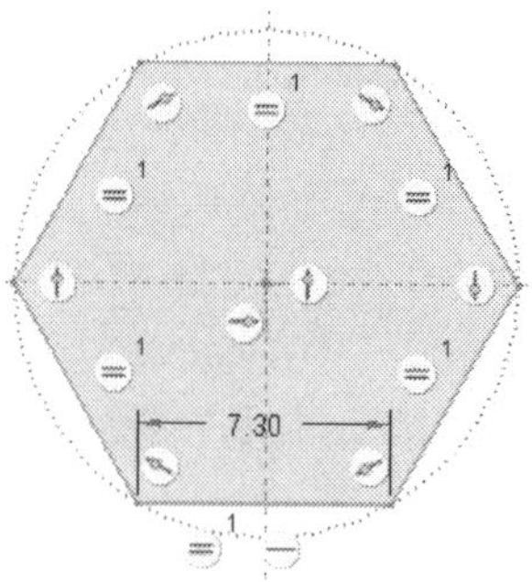

Figure 7-110 Sketch of the hexagon feature

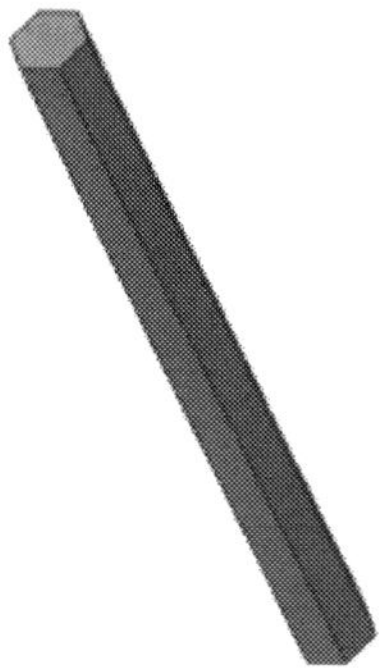

Figure 7-111 Base feature created by extruding the hexagon sketch

Creating the Second Feature

The second feature is a cylindrical feature. You need to draw the feature on the top face of the base feature.

1. Choose the **Extrude** tool from the **Shapes** group; the **Extrude** dashboard is displayed. Next, invoke the **Sketch** dialog box and select the top face of the base feature as the sketching plane.
2. Enter the sketcher environment and draw the sketch of the cylinder. Note that the diameter of the circle is 8. Next, exit the sketching mode.
3. After exiting the sketcher environment, extrude the circle to a depth of 13. The model after creating the cylindrical feature is shown in Figure 7-112.

Creating the Draft Feature

You need to create the draft feature on the cylindrical surface and the draft surface to intersect the base feature. This is because when the draft surface intersects the base feature and is extended, the required shape is obtained at the intersecting edge.

1. Choose the **Draft** tool from the **Engineering** group; the **Draft** dashboard is displayed.
2. Select the cylindrical surface of the cylinder to apply draft to it.
3. Choose the **References** tab to invoke a slide-down panel. Click in the **Draft hinges** collector and then select the top face of the cylindrical feature as the hinge; a pink arrow is displayed pointing downward.
4. Choose the **Flip** button on the right of the **Pull direction** collector to change the direction of the arrow so that it points upward.
5. Choose the **Options** tab to invoke a slide-down panel and then select the **Extend intersect surfaces** check box.
6. Enter **15** as the angle value in the **Angle 1** edit box present in the **Draft** dashboard.

7. Choose the **OK** button from the **Draft** dashboard to complete the draft feature, as shown in Figure 7-113.

Figure 7-112 *Model after creating the cylindrical feature*

Figure 7-113 *Model after creating the draft feature*

Creating the Fourth Feature

The fourth feature is the tip of the pencil which is a revolved feature.

1. Choose the **Revolve** tool from the **Shapes** group; the **Revolve** dashboard is displayed.

2. Invoke the **Sketch** dialog box and select the **FRONT** datum plane as the sketching plane and select the top face as reference.

3. After entering the sketcher environment, draw the sketch and then apply constraints and dimensions to it, as shown in Figure 7-114. Make sure that the geometry centerline is drawn. Exit the sketcher environment.

   The angle of revolution is 360 degree by default.

4. Choose the **OK** button to complete the feature creation. The default trimetric view of the model after creating the fourth feature is shown in Figure 7-115.

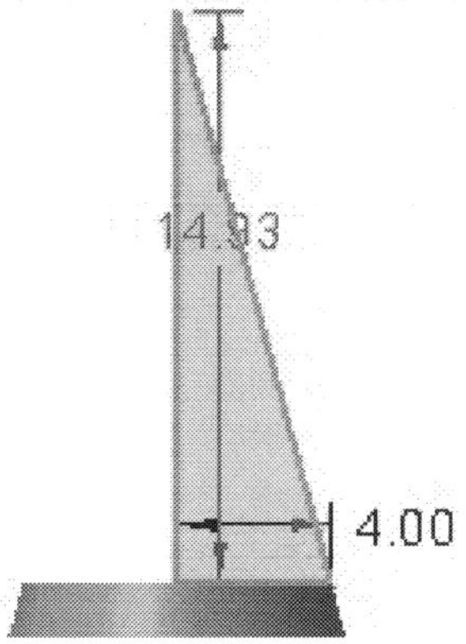

Figure 7-114 *Sketch for the tip of the pencil*

Figure 7-115 *Model after creating the tip*

Creating the Fifth Feature

The fifth feature, which is a revolve feature, is the bottom portion of the pencil.

1. Choose the **Revolve** tool from the **Shapes** group; the **Revolve** dashboard is displayed.
2. Invoke the **Sketch** dialog box and then select the **FRONT** datum plane as the sketching plane.
3. After entering the sketcher environment, draw the sketch and then apply constraints and dimensions to it, as shown in Figure 7-116. Make sure that the geometry centerline is also drawn.
4. Exit the sketcher environment; the angle of revolution is set to 360-degree by default.
5. Choose the **OK** button to complete the feature creation. The model after creating the fifth feature is shown in Figure 7-117.

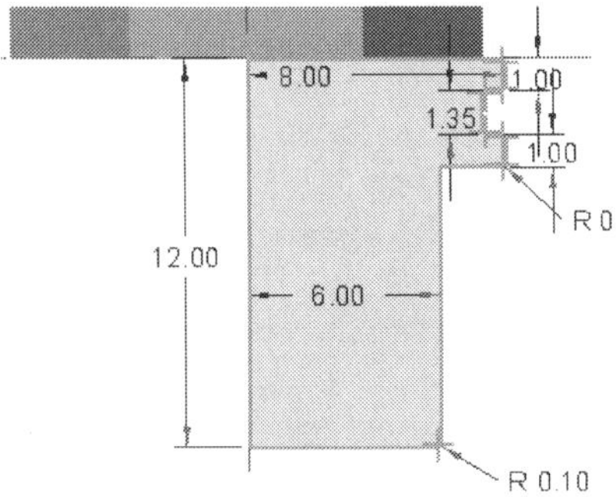

***Figure 7-116** Sketch with dimensions and constraints*

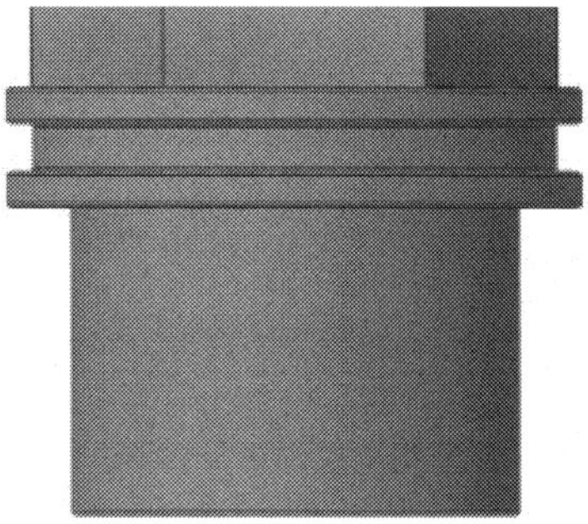

***Figure 7-117** Partial view of model after creating the top of the pencil*

Creating the Cosmetic Sketch

1. Choose the **Cosmetic Sketch** tool from the expanded **Engineering** group; the **Cosmetic Sketch** dialog box is displayed.
2. Select the front face of the base feature as the sketching plane.
3. After entering the sketcher environment, create the text by using the **Text** button. Figure 7-118 shows the text with dimensions.
4. Choose the **OK** button to exit the sketcher environment.

   The final model after writing the text is shown in Figure 7-119. The **Model Tree** of the final model is shown in Figure 7-120.

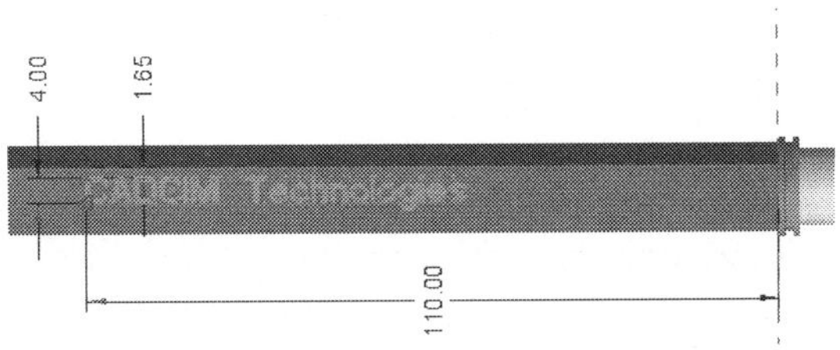

Figure 7-118 *Text with dimensions*

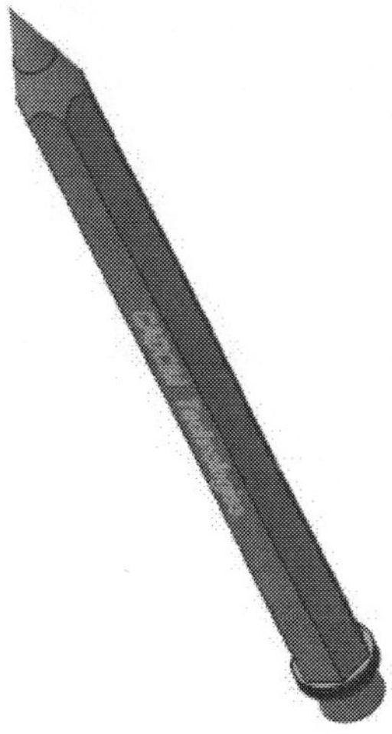

Figure 7-119 *The final model with text*

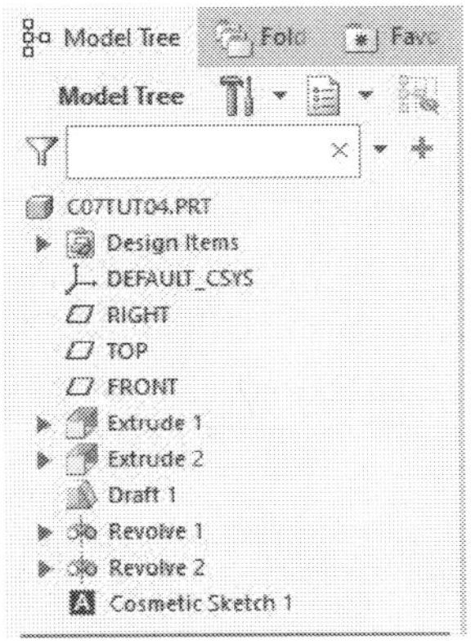

Figure 7-120 ***Model Tree*** *for Tutorial 4*

Changing the Appearance of the Model

You can apply colors and appearances to the surfaces in Creo Parametric. Remember that the colors you apply are saved with the model and remain on the model until you clear them.

1. Choose the down arrow below the **Appearances** button from the **Appearance** group in the **View** tab; a slide-down panel is displayed.

2. Click on the Library Browser button in the **Library** area to open the appearance folder tree.

3. Choose **Wood > wood-spruce.dmt**; the appearances available in the selected file gets added to the Library palette.

Note

*You can also select an appearance icon from the **My Appearances** or **Library** area. After selecting the icon, invoke the **Appearances Manager** to adjust the values as required. You can also add or remove textures to colors.*

4. Select **wood-spruce-solid-unfinished**; the cursor is displayed as a paint brush and also the **Select** dialog box is displayed. Select the surfaces to which you want to apply the selected appearance. Choose the **OK** button from the **Select** dialog box; the appearance is applied to the selected surfaces.

5. Similarly, assign different colors to other entities as desired. The final model with appearances applied is shown in Figure 7-121.

***Figure 7-121** Model with appearances applied*

Saving the Model

1. Choose the **Save** button from the **File** menu and save the model.

Note

*Colors are not features of the model and therefore, they will not appear in the **Model Tree**.*

Self-Evaluation Test

Answer the following questions and then compare them to those given at the end of this chapter:

1. While using the **Sculpt** tool, the __________ drop-down list is used to specify the direction in which the group of points selected move when sculpted.

2. To specify the length of the bend for a 360 degrees toroidal bend feature, you need to select two __________ planes.

3. The **Sculpt** tool deforms the geometry through a __________.

4. The **Next Axis** button is used to switch the axis of bend from one __________ to another.

5. The **Spinal Bend** tool is used to bend solids or quilts by continuously __________ the cross-sections along a sketched spine.

6. The **Transform** tool can be used to perform __________, __________, and __________ operations.

7. The **Spine** tool deforms the geometry in either __________ or __________ direction.

8. The **Spinal Bend** tool bends solids or quilts along a sketched spine. (T/F)

9. The warp transformations can be performed only in the part mode. (T/F)

10. The toroidal bend feature can be created only on one side of the sketching plane. (T/F)

Review Questions

Answer the following questions:

1. What is the maximum bend angle through which a geometry can be bent using the **Toroidal Bend** tool?

   (a) 90 degrees (b) 180 degrees
   (c) 270 degrees (d) 360 degrees

2. The trajectory for creating a spinal bend on geometry should be sketched __________ to the solid geometry.

3. The __________ tool is used to stretch the geometry along the required direction.

4. The __________ check box, if selected, keeps the original entity selected for the creation of warp feature in the drawing area along with the preview of the transformed feature.

5. The direction reference chosen for the creation of warp feature is any one of the __________ planes.

6. The **Spine** tool reshapes the geometry by manipulating the defining points of a __________.

7. A solid geometry can be rotated by clicking and rotating the active handles of the __________.

8. While using the **Toroidal Bend** tool, the user-defined coordinate system defines the __________ of the bend profile.

9. While creating a model using the **Toroidal Bend** tool if the sketch defining the bend is a straight line, the profile of the model is a __________.

10. The **Warp** tool modifies a model along its default __________.

EXERCISES

Exercise 1

In this exercise, you will create the model shown in Figure 7-122. The dimensions for the model are shown in Figure 7-123.

(Expected time: 20 min)

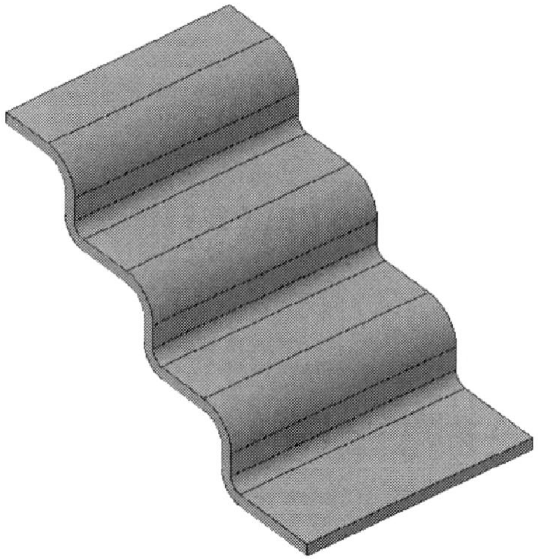

Figure 7-122 *The resulting model for Exercise 1*

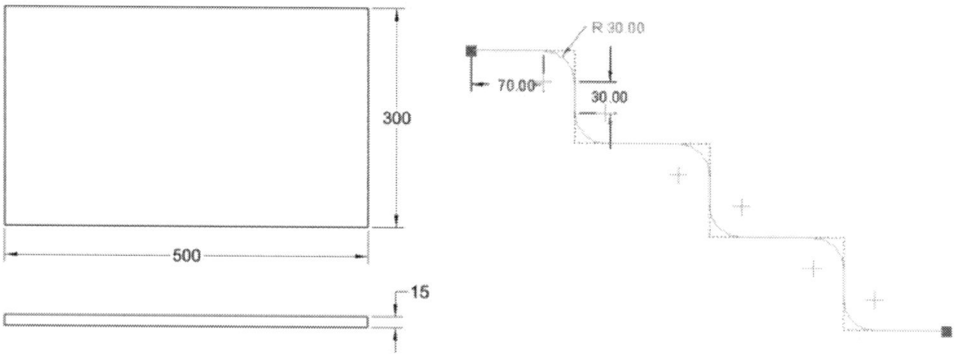

Figure 7-123 *Drawing of the base feature for Exercise 1*

Exercise 2

In this exercise, you will create the model shown in Figure 7-124. The dimensions for the base feature are shown in Figure 7-125. Angle of twist for the Warp is 100° and the thickness of the model is 2.

(Expected time: 20 min)

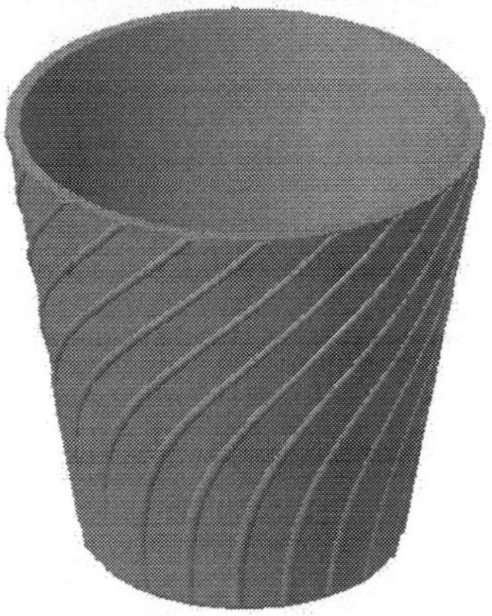

Figure 7-124 *The resulting model for Exercise 2*

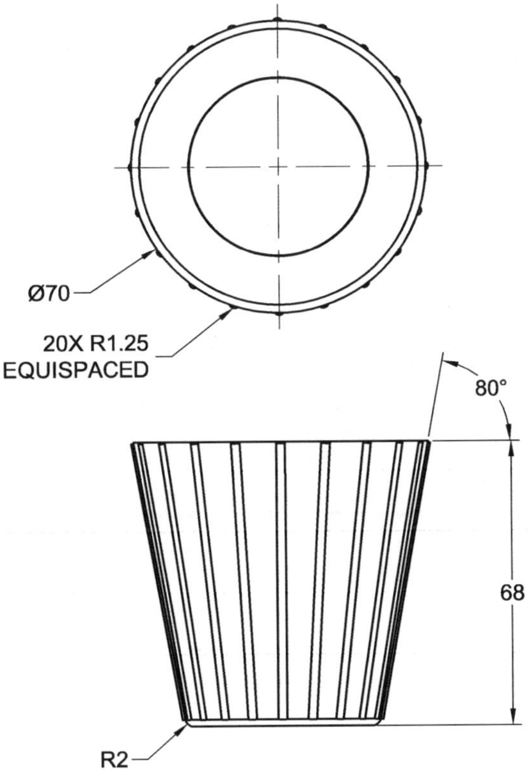

Figure 7-125 *Drawing of the base feature for Exercise 2*

Answers to Self-Evaluation Test

1. Symmetry, **2.** parallel, **3.** mesh, **4.** plane, **5.** repositioning, **6.** translate, rotate, scale, **7.** linear, radial, **8.** T, **9.** T, **10.** F

Chapter 8

Options Aiding Construction of Parts-III

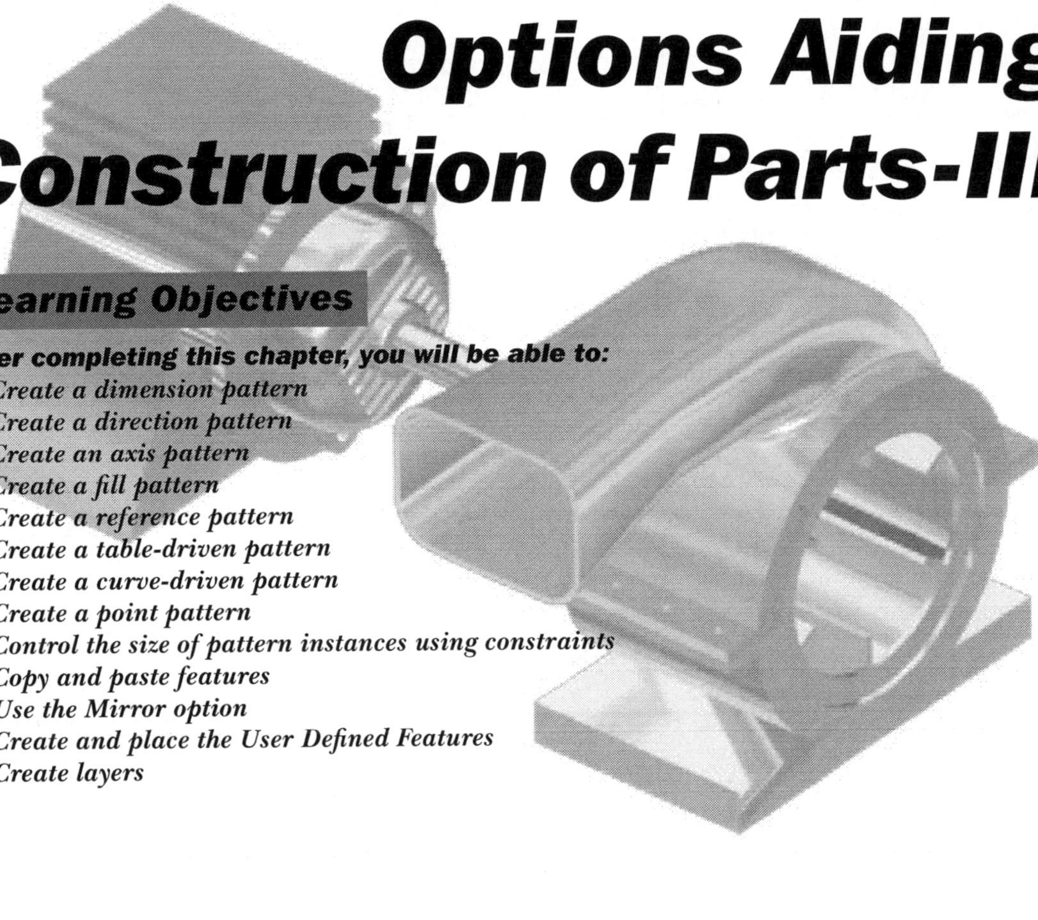

Learning Objectives

After completing this chapter, you will be able to:

- *Create a dimension pattern*
- *Create a direction pattern*
- *Create an axis pattern*
- *Create a fill pattern*
- *Create a reference pattern*
- *Create a table-driven pattern*
- *Create a curve-driven pattern*
- *Create a point pattern*
- *Control the size of pattern instances using constraints*
- *Copy and paste features*
- *Use the Mirror option*
- *Create and place the User Defined Features*
- *Create layers*

INTRODUCTION

In this chapter, you will learn about various methods for duplicating existing features. In Creo Parametric, you can duplicate a feature by using the **Pattern**, **Copy**, and **Mirror** tools.

You will also learn to create user defined features and simplified representation in solid models.

CREATING A FEATURE PATTERN

A pattern consists of multiple instances of a feature. Patterns are used to create incremental array of features in one or two directions from a single feature called the parent feature or the leader. When a pattern is created, the leader also becomes a part of the pattern. When you pattern a feature, you need to specify the total number of features to be created including the one being patterned and also the increment in the dimensions, if required. You can pattern several features at a time by creating a local group of the selected features. You can also apply move or rotational transformations to a pattern, group pattern, or to a pattern of a pattern.

Uses of a Pattern

Patterns are very helpful in solid modeling as they speed up the model creation process. The uses of patterns in solid modeling are discussed next.

1. Patterns are used to create multiple copies of a feature, and therefore, save time that would otherwise be spent in creating the features individually.

2. All instances in a pattern, including the parent feature, act as a single feature. Therefore, they can be easily suppressed or mirrored.

3. All instances in a pattern are related parametrically. Therefore, you can modify the number of instances in a pattern, the spacing between the instances, and other pattern related parameters.

4. If the dimensions of the parent feature are modified, then the dimensions of the child features are also modified.

Creating a Pattern

Ribbon: Model > Editing > Pattern drop-down > Pattern

Enhanced

In Creo Parametric, patterns are created by choosing the **Pattern** tool from the **Editing** group. This tool is an object-action tool which means the **Pattern** tool is enabled only when you have selected the feature to be patterned.

To pattern a feature, select the feature and then choose the **Pattern** tool from the **Pattern** drop-down available in the **Editing** group or from the mini popup toolbar; the **Pattern** dashboard will be displayed, as shown in Figure 8-1. Note that a reference is already selected in the **Pattern** dashboard. The options and tools in the dashboard depend on the type of pattern you create. In Creo Parametric, you can create eight types of pattern. These patterns are **Dimension**, **Direction**, **Axis**, **Fill**, **Table**, **Reference**, **Curve**, and **Point**. The options and tools in the **Pattern** dashboard are discussed next.

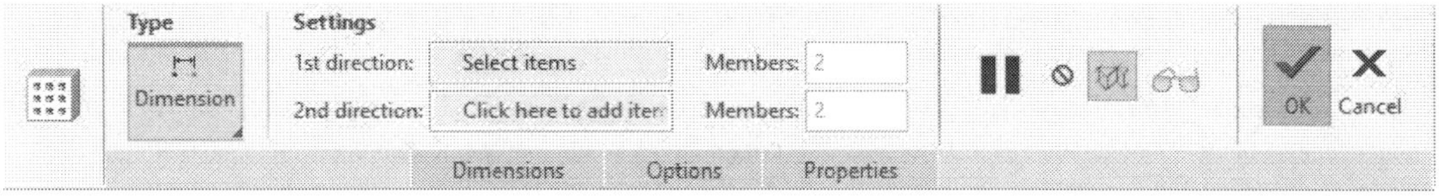

*Figure 8-1 The **Pattern** dashboard*

Type Drop-down List

In the **Type** drop-down list of the **Pattern** dashboard, the **Dimension** option is selected by default. The other options in this drop-down list are **Direction**, **Axis**, **Fill**, **Table**, **Reference**, **Curve**, and **Point**. When you select these options from this drop-down list, the options and tools in the **Pattern** dashboard change accordingly. The types of pattern that can be created in Creo Parametric are discussed next.

Dimension Pattern

In dimension patterns, the driving dimensions of the parent feature are used to create a pattern. This pattern can be created in one direction or in two directions. When you select the option to create the pattern in the second direction, all instances that were created in the first direction can also be created in the second direction. You need to specify the increment value for the instances, which can be positive or negative. Once you have specified the increment value in a direction, the system creates the specified number of instances (including the parent feature) in that direction. Figure 8-2 shows a feature to be patterned, the patterned feature is shown in Figure 8-3.

Figure 8-2 Hole to be patterned

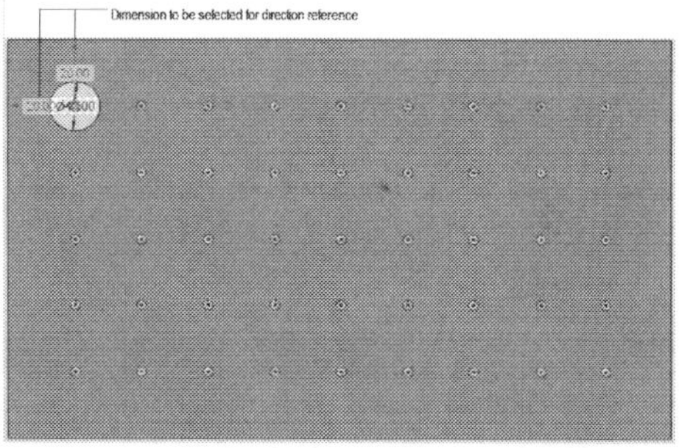

Figure 8-3 Hole patterned in two directions

Tip

1. A feature can be used as a pattern leader for a single pattern only. After the pattern is created, the leader becomes a dependent part of the pattern feature.

2. Always create dimensions of features in such a manner that the location and size of the pattern feature can be controlled easily in the later stages of design process.

*3. The number of instances in a pattern feature can affect the regeneration time. For simple type of pattern, use the **Identical** or **Variable** option to speed up the regeneration process.*

To create a dimension pattern, use the following steps:

1. Select the feature to be patterned and then choose the **Pattern** tool.

2. By default, **Select items** in the **1st direction** collector is activated. Next, select the dimension; the dimension will be added in the **1st direction** collector and an edit box will displayed. Enter the increment in the edit box where you need to place the second instance of the pattern.

3. After specifying the dimensions in the first direction, you need to specify the number of instances in the **Members** edit box

4. To create instances in the second direction, click on **Click here to add item** in the **2nd direction** collector; **Select items** in the **2nd direction** collector will be activated. Click the dimension in the second direction; an edit box will be displayed. Enter the increment in the edit box.

5. Enter instances in the **Members** edit box for the second direction. On doing so, you can view the black dots where the instances of the pattern will be placed.

6. Choose the **OK** button to exit the dashboard and to view the pattern.

You can also add additional dimensions to **1st direction** and **2nd direction** reference collectors to create a varying pattern. The procedure to create a varying pattern feature is discussed later in this chapter.

Direction Pattern

The **Direction** option is used to create the pattern in the specified direction. The direction is specified by selecting an edge, plane, axis, coordinate system or linear curve. This pattern is different from the Dimension pattern. This pattern does not use the dimensions of the feature to be patterned. When you select this option from the **Type** drop-down list, you will be prompted to select the reference for the first direction. After specifying the first direction, an arbitrary dimension will be displayed in the specified direction. This dimension is used to adjust the distance between the instances. You can also vary dimensions of the patterned feature by using the **Dimensions** slide-down panel. To create pattern in other direction, click the second direction collector and select the second direction reference. You can also activate a collector by using the options in the shortcut menu.

In Creo Parametric, there are three options to define the direction reference, **Translation**, **Rotation**, and **Coordinate System**. These options are available in the **Translation** drop-down lists adjacent to the **1st direction** and **2nd direction** edit boxes. The drop-down list adjacent to **1st direction** edit box is used for defining the direction reference in the first direction and the drop-down list adjacent to **2nd direction** edit box is used for defining the direction reference in the second direction. The options in these drop-down lists are discussed next.

Translation: When this option is selected in the drop-down list, then you need to select a flat face, axis, plane, linear curve, or coordinate system axis to define the direction reference. When you select a flat face or a plane, then a linear pattern is created in the direction perpendicular to the selected face or plane. When you select an axis, linear curve or a coordinate system axis, then a linear pattern is created in the direction of that curve. Now you need to enter the spacing between two instances in the corresponding edit boxes.

Rotation: When this option is selected in the drop-down list, then you need to select a linear curve, coordinate system axis, or an axis to define the direction reference. When you select an axis, linear curve or a coordinate system axis, then a circular pattern is created around that curve. Next you need to enter the required values in the corresponding edit boxes.

Coordinate System: When this option is selected in the drop-down list, then you need to select a coordinate system to define the direction reference. Select a coordinate system; the dashboard will be displayed, as shown in Figure 8-4. Specify the required values in **X**, **Y**, and **Z** edit boxes to define the direction vector.

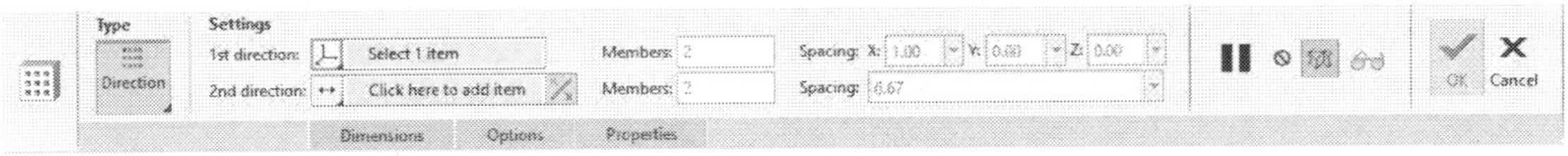

*Figure 8-4 The **Pattern** dashboard with the **Coordinate System** option selected*

Note that the direction of pattern can be reversed by choosing the **Flip** button in the **Pattern** dashboard. Figure 8-5 shows the linear curve selected to create the direction pattern. The direction is indicated by an arrow. Preview of the pattern instances are shown as black dots and the pattern created is shown in Figure 8-6.

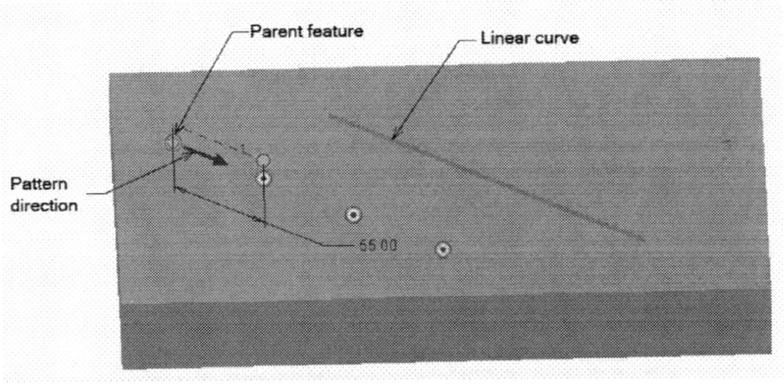

Figure 8-5 Linear curve selected to create the direction pattern

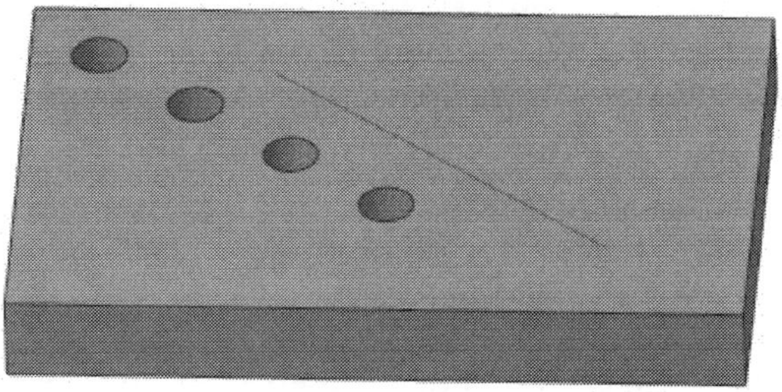

Figure 8-6 Direction pattern created

Note

The reference that you select for specifying the direction of direction pattern should belong to the feature created before the feature to be patterned otherwise the desired reference will not be highlighted for selection. However, you can create a reference using the ***Datum*** *drop-down list of the* ***Pattern*** *dashboard.*

Axis Pattern

The **Axis** option is used to create rotational patterns. To create this type of pattern, an axis is required. The axis can be a datum axis or the axis of a feature. For example, an axis formed by a revolved feature, like a hole. When you choose this option, the **Pattern** dashboard will be displayed, as shown in Figure 8-7. Note that a reference is already selected. An axis pattern allows you to create instances of a feature in two directions. The first direction for the axis pattern is the angular direction of the selected axis. The pattern members created in the first direction revolve around the axis with equal spacing. The second direction of the pattern is the radial direction of the revolution axis. The pattern members are added radially along the selected axis.

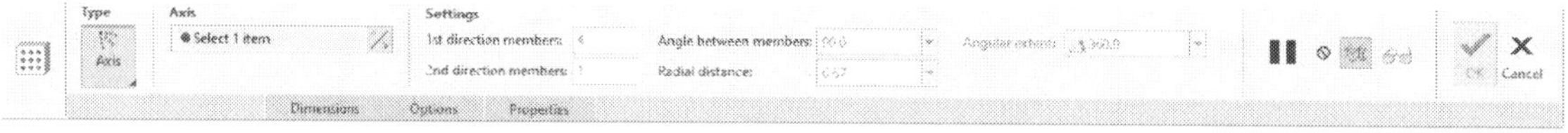

Figure 8-7 *The* ***Pattern*** *dashboard with the* ***Axis*** *option selected*

Note

Remember that for creating axis pattern of a feature, the reference axis must be created before the feature to be patterned.

The parameters and attributes that you need to specify to create an axis pattern are shown in Figure 8-8.

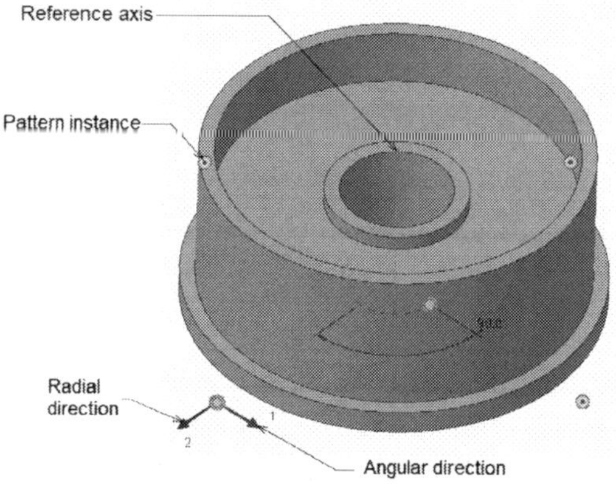

Figure 8-8 *Parameters to be specified for the axis pattern*

To create an axis pattern, use the following steps:

1. Select the feature or a feature group to be patterned and then choose the **Pattern** tool. Now, select the **Axis** option from the **Type** drop-down list in the **Pattern** dashboard; you will be prompted to select an axis about which the pattern will be created.

2. Turn on the display of the axis and then select it.

3. Enter the required angle between each instance and then the number of pattern instances in the respective edit box. Instead of specifying the angle between each instance, you can specify the total angle of pattern by choosing the **Angular extent** button from the dashboard. When you enter the value of angular extent of the pattern in the Angular extent edit box, the instances will be placed at equal distance on the arc length.

   The instances in the first direction have been specified and now you can specify the number of instances and the increment in dimension in the second direction. Note that the dimension and number of instances in the second direction should be specified only if you need to create instances in the second direction. To create the rotational pattern, as shown in Figure 8-9, you need to create instances in the first direction. The instances in the second direction, along with the instances in the first direction, are shown in Figure 8-10. To create instances in the second direction, specify a value in the **2nd direction members** edit box and then enter the radial distance between the pattern instances in the **Radial distance** edit box. The pattern is created and you can view the black dots where the instances will be placed.

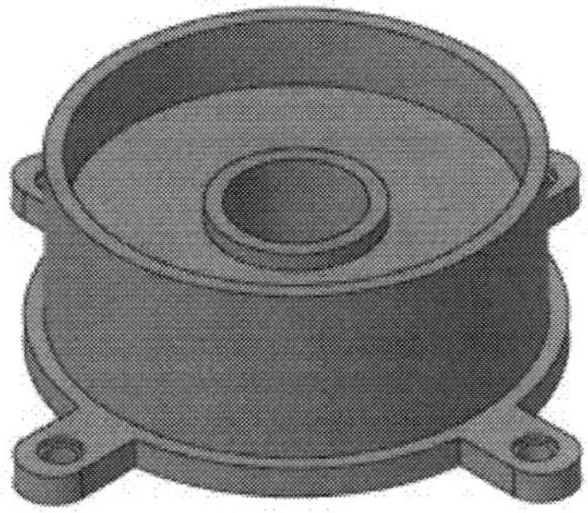

Figure 8-9 *Rotational pattern created using the **Axis** option*

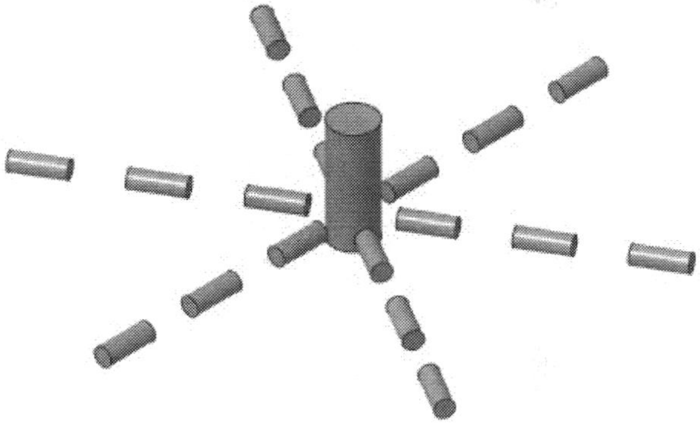

Figure 8-10 *Pattern created both in circular and linear directions*

4. Choose the **OK** button to exit the dashboard and view the pattern.

Fill Patterns

Fill patterns are used to fill the sketched area by a selected feature. This is the easiest and fastest method to create a pattern in the sketched area. The instances of the feature are positioned in a grid. You can specify the type of fill pattern using one of the predefined template. When you select the **Fill** option from the **Type** drop-down list, the **Pattern** dashboard will appear, as shown in Figure 8-11. Note that a reference is already selected.

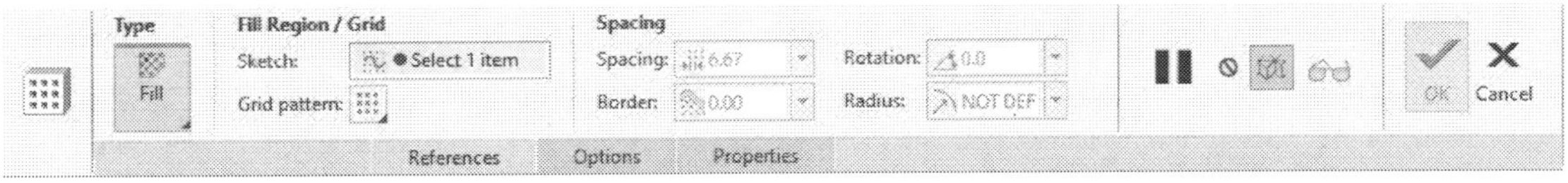

Figure 8-11 *The **Pattern** dashboard with the **Fill** option selected*

The options and tools in this dashboard are discussed next.

References Tab: When you choose this tab, the slide-down panel will be displayed in which the **Define** button is available. Choose the **Define** button to invoke the **Sketch** dialog box. This dialog box is used to sketch the area that will be filled by the pattern instances. Select the sketching plane and its orientation and then enter the sketcher environment to draw the sketch.

Grid Pattern Drop-Lown List: This drop-down list is available only when the area to be filled is defined. The options in this drop-down list are used to specify the shape of the fill pattern. Using the options in this drop-down list, you can shape the fill pattern along square pattern, diamond pattern, hexagon pattern, concentric circle pattern, spiral pattern, and along the sketched curves. Generally, it is recommended that the sketch should be similar to the shape of the fill pattern you need. Figures 8-12 through 8-17 show the patterns of various shapes.

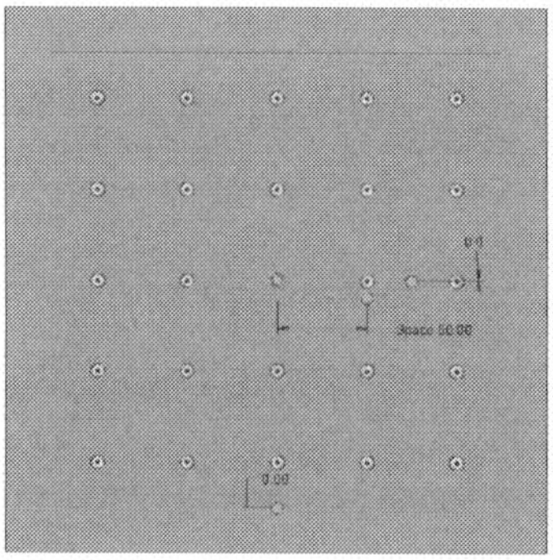

Figure 8-12 *The square pattern fill in the sketched square*

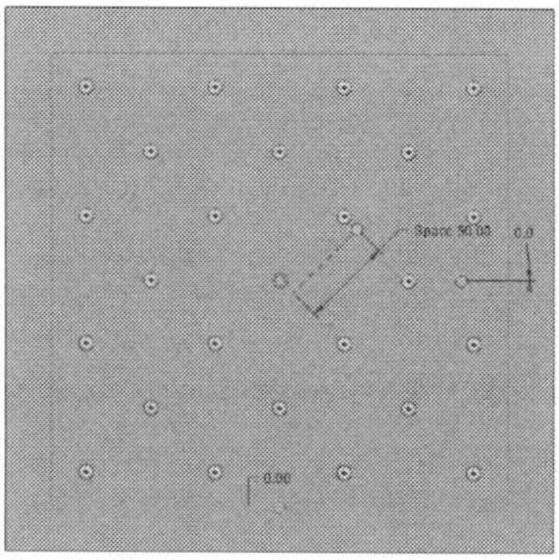

Figure 8-13 *The diamond pattern fill in the sketched square*

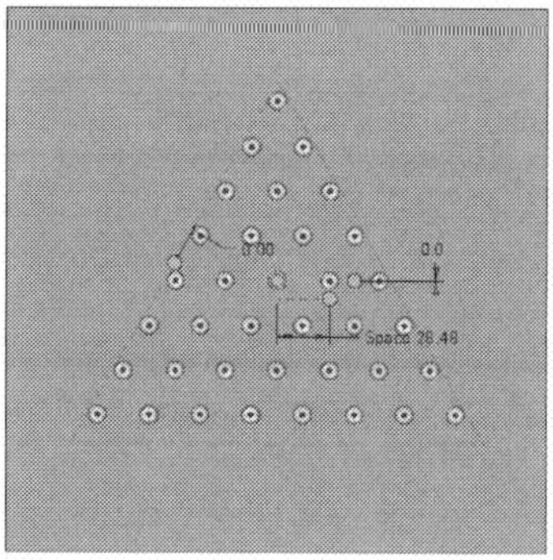

Figure 8-14 *The hexagon pattern fill in the sketched triangle*

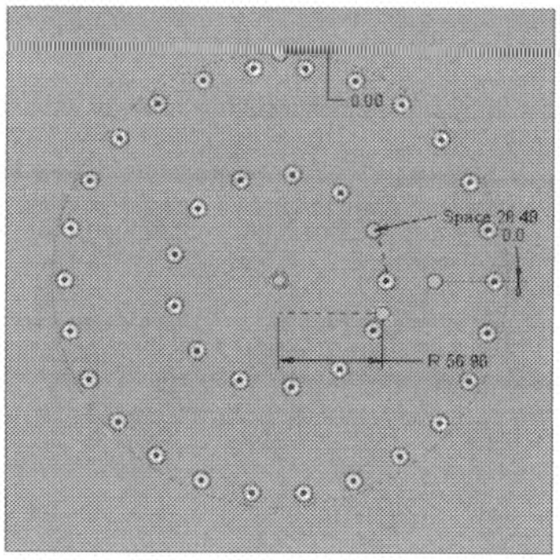

Figure 8-15 *The concentric circle pattern fill in the sketched circle*

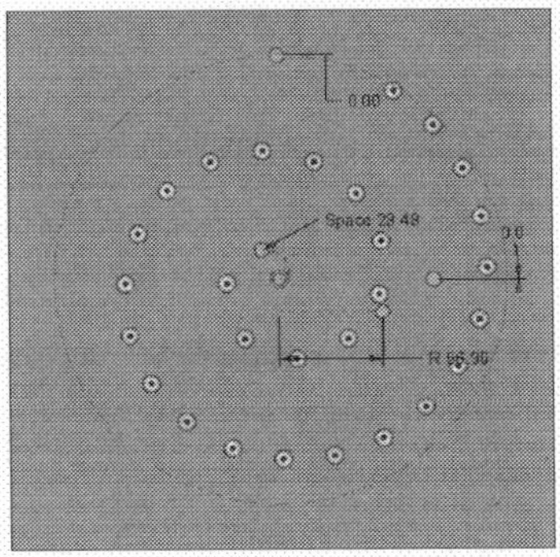

Figure 8-16 The spiral pattern fill in the sketched circle

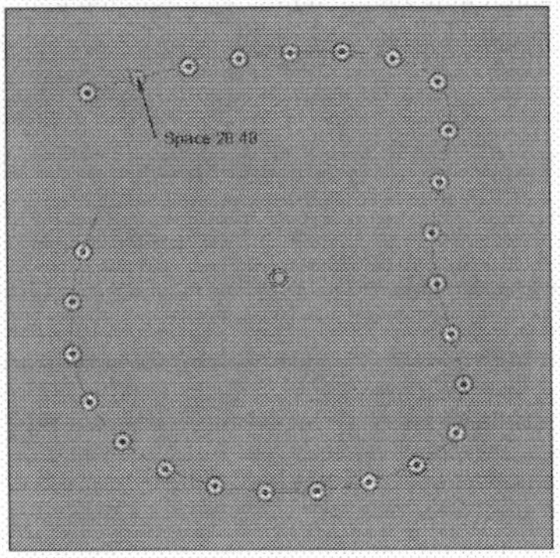

Figure 8-17 The sketched curves pattern fill in the sketched curve

Spacing Area: The **Spacing** area in the dashboard are used to specify the different parameters of the fill pattern. You can specify the gap between the two members of the pattern, the distance of all members (instances) from the sketched boundary, rotation angle of the pattern, and radial distance between instances, in case of circular and spiral shapes.

Options Tab: When you choose this tab, the slide-down panel will be displayed in which four check boxes are available, as shown in Figure 8-18. These check boxes are discussed next.

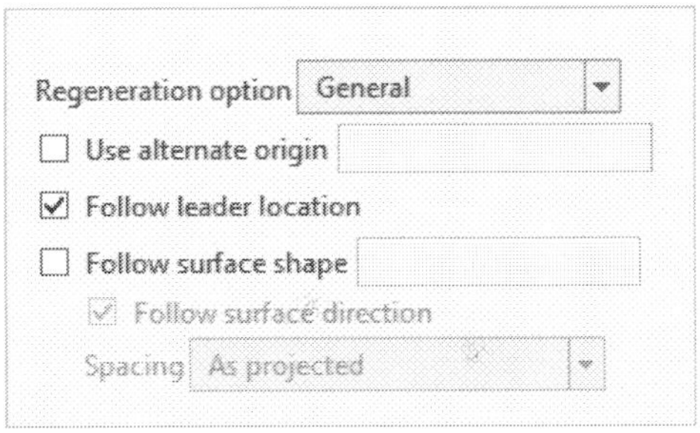

*Figure 8-18 The **Options** slide-down panel*

When you select the **Use alternate origin** check box, you will be prompted to specify the alternate origin point for the pattern. Select a point from where you want to start the pattern.

The **Follow leader location** check box is selected by default. When you clear this check box, the pattern will be created on the plane which was selected for drawing the area to be filled.

When you select the **Follow surface shape** check box, you will be prompted to select a surface. Select a surface; the pattern instances will follow the shape of the selected surface.

The **Follow surface direction** check box is active only when you select the **Follow surface shape** check box. When you select the **Follow surface direction** check box, the **Spacing** drop-down list will get activated. In this drop-down list, three options are available to define the spacing of the pattern instances on the surface. When you select the **As projected** option from the drop-down list, the instances will be projected on the surface opposite to the selected surface. When you select the **Map to Surface Space** option, the pattern leader is projected straight onto the surface, and the remaining pattern members are placed according to the uv-lines that pass through the pattern leader. When you select the **Map to Surface UV Space** option, the instances are mapped to the uv-space of the surface based on their xy coordinates. The **Follow template rotation** check box is available only for circle and spiral templates. This option is used to rotate the pattern members around an origin. The **Follow curve direction** check box in the panel allows you to place pattern members in the sketch plane to follow the curve.

Tip

While creating a pattern, you can select the member that you want to remove. This can be done before exiting the ***Pattern*** *dashboard. To remove the member of a pattern, select it in the preview of the pattern. The selected member turns white suggesting that it is removed. To resume the removed member of the pattern, select it once again.*

Reference Pattern

In the reference patterns, an existing pattern is referenced to create a new pattern. The instance number is always the same as the reference pattern. In this pattern, the parent feature of the new pattern should be referenced to the parent feature of the existing pattern. The **Reference** pattern option can be used only when the feature to be patterned is using the geometry references of the originally patterned feature.

Figure 8-19 shows an extruded cut referenced to the parent hole feature. In this figure, the parent cut feature is created on a datum plane and passes through the axis of the parent hole feature. Therefore, a relationship is built between the parent cut feature and the parent hole feature.

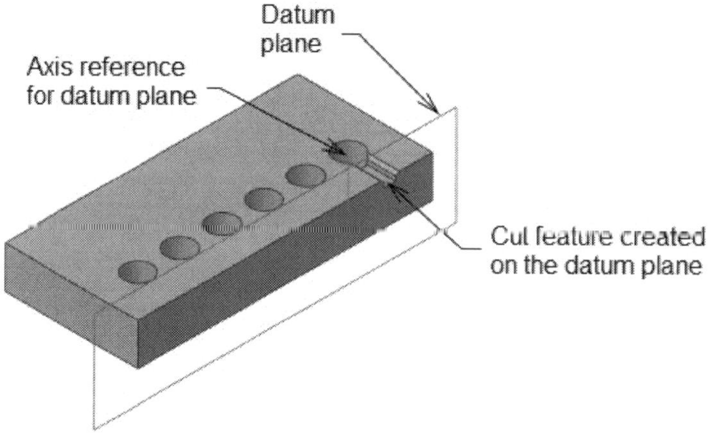

Figure 8-19 *Cut feature referenced to the parent hole feature*

To create reference pattern of the cut feature, select the feature from the **Model Tree** and choose the **Pattern** tool from the **Editing** group, or from the mini popup toolbar to invoke the **Pattern** dashboard. The **Reference** option is selected by default in the **Pattern** dashboard. Choose the **OK** button; the pattern of the cut feature is created without specifying the increment in dimensions. Figure 8-20 shows the cut feature patterned using the reference of the hole pattern.

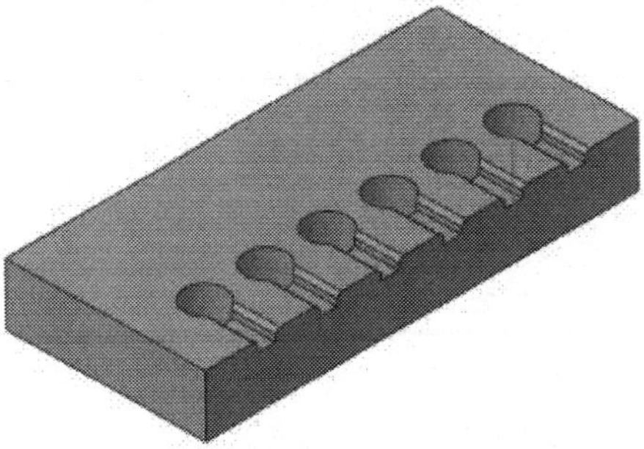

Figure 8-20 *Cut feature patterned using the* ***Reference*** *pattern*

Note

If the cut feature is not created on an embedded datum plane, the relationship between the hole feature and the cut feature will not exist. Therefore, the cut feature cannot be patterned using the ***Reference*** *option. Ensure that the feature which will be patterned using the* ***Reference*** *option has a relationship with the parent feature which in this case will be the patterned hole feature.*

Table Pattern

Table-driven patterns are created by defining a table. In this table, you need to specify the dimensions of the instances from the edge or faces from where the leader of the pattern is referenced (leader of a pattern is the feature that is selected to create a pattern). You can create complex or irregular patterns of features, components, or groups using pattern tables. Pattern tables allow you to control the location of individual instances by specifying absolute dimensions for each instance in the pattern, such as the pattern leader. When you select the **Table** option from the **Type** drop-down list, the **Pattern** dashboard will appear, as shown in Figure 8-21.

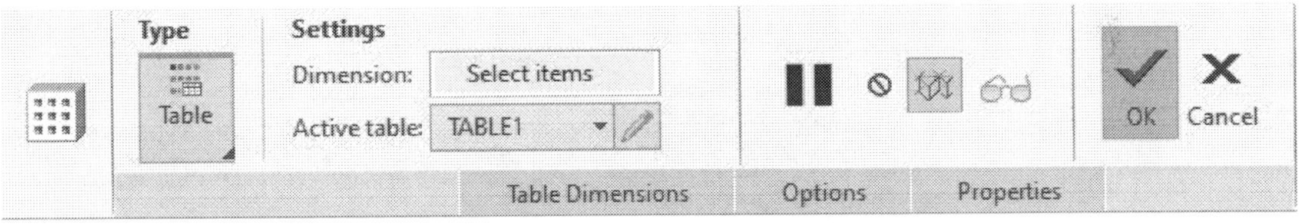

Figure 8-21 *The* ***Pattern*** *dashboard with the* ***Table*** *option selected*

In Figure 8-21, the **Active table** collector shows an active table. You can create more than one table to create a pattern. You can choose the **Edit the active table** button to display the **Pro/TABLE** window, as shown in Figure 8-22. This button will be available only after you have selected at least one dimension.

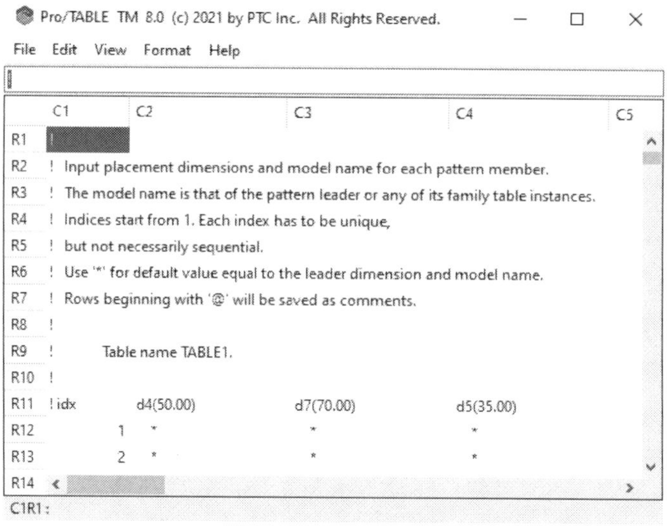

***Figure 8-22** The **Pro/TABLE** window*

To create the table-driven pattern shown in Figure 8-23, follow the given steps:

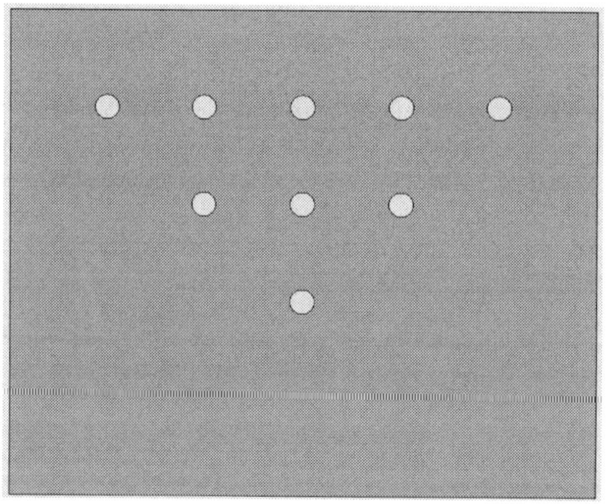

***Figure 8-23** Table-driven pattern*

1. Create a solid protrusion of dimension 120 X 100 X 15.

2. Create a through hole of diameter **5** on the top face of the base feature at a distance of **20** from the left edge and **20** from the top edge.

3. Select the hole feature and choose the **Pattern** tool from the **Editing** group; the **Pattern** dashboard will be displayed and the dimensions of the hole feature will appear on the hole.

4. Select the **Table** option from the **Type** drop-down list in the dashboard.

5. Select the dimension which is **20** from the left edge and then use CTRL+left mouse button to select the dimension which is **20** from the top edge.

6. Choose the **Edit the active table** button from the **Pattern** dashboard to display **Pro/TABLE**.

7. In column **C1**, click under **idx** (index number). Enter **1** when the cell is highlighted. The value **1** signifies the first instance.

8. In column **C2**, toward the right of **1**, enter the distance along the first dimension and in column **C3**, enter the distance along the second dimension, as shown in Figure 8-24.

9. Similarly, under the rows below **idx**, enter the dimensions of other instances of the holes, as shown in Figure 8-24. Remember that the distance of each hole is measured from where the leader is dimensioned.

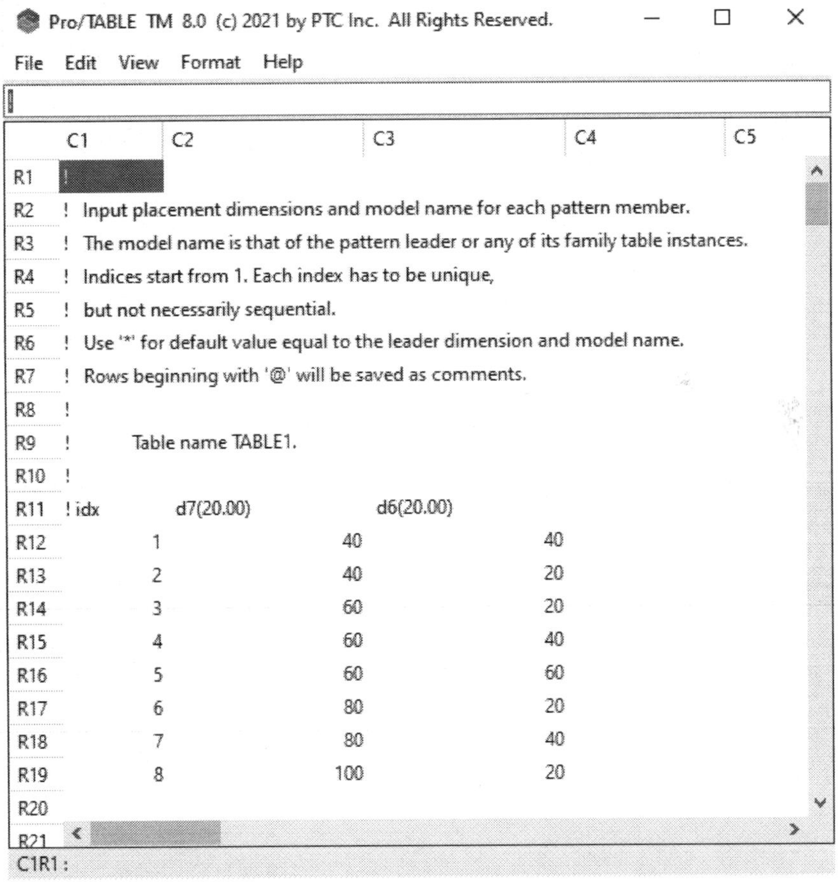

Figure 8-24 Coordinate values of instances

10. Choose **File > Save**, and then choose the **Close** button to exit **Pro/TABLE**.

11. Choose the **OK** button to exit the feature creation tool.

Tip

1. Pattern tables can be created from scratch by picking dimensions and filling in all values. In most cases, it is easier and less time-consuming to create a dimensional pattern similar to what you need and then convert it into a table.

*2. You can import a previously saved pattern table by choosing **File > Read** in the menu bar of the table editor window and typing the name of the table file. You can also save the current table for future use by choosing **File > Save** or by choosing **File > Save As** and typing the file name.*

*3. To create additional pattern tables, choose the **Add** option from the shortcut menu displayed on right-clicking in the **Tables** tab.*

Generally, table-driven patterns are used when the incremental distances between the instances of the pattern are nonuniform. This pattern is also useful when the coordinate locations of the instances are known.

Curve Driven Pattern

The **Curve** option is used to create patterns along user-defined curves. When you select the **Curve** option from the **Type** drop-down list, the **Pattern** dashboard will appear, as shown in Figure 8-25.

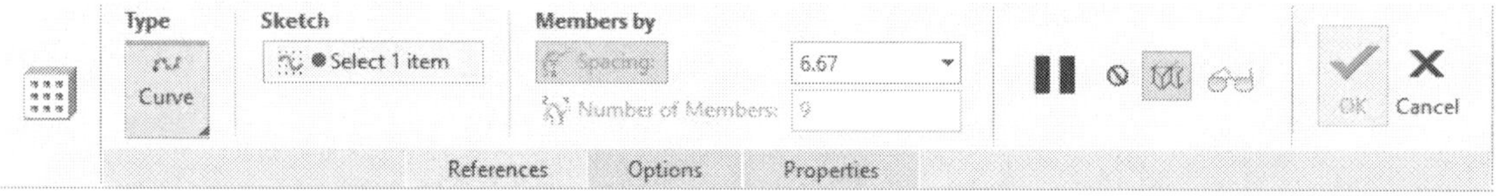

Figure 8-25 *The **Pattern** dashboard with the **Curve** option selected*

To create a curve driven pattern, first you need to sketch a curve. You can use an internal or an external sketch to draw the curve. The curve also specifies the direction for the pattern. For open sketches, the direction of the pattern is always from the start to the end point of the curve. For closed sketches, the direction of the pattern can be from either side of the selected vertex. For single entity closed sketches, you can use the **Divide** tool in the sketcher environment to specify the start point of pattern. All instances created using the **Curve** option are identical in size and geometry. The pattern instances also follow the shape of the sketched curve. After choosing the **Curve** option from the drop-down list, you have to sketch a curve to pattern. Then enter the spacing between the patterned instances in the edit box in the pattern dashboard. The preview of the patterned feature can be seen in the drawing area. Figure 8-26 shows the model with the feature to be patterned along the sketched curve, and Figure 8-27 shows the patterned feature.

Note

The start point of the curve pattern acts as the start point of the curve. The black arrow on the curve displays the start point of the curve and the direction of the pattern.

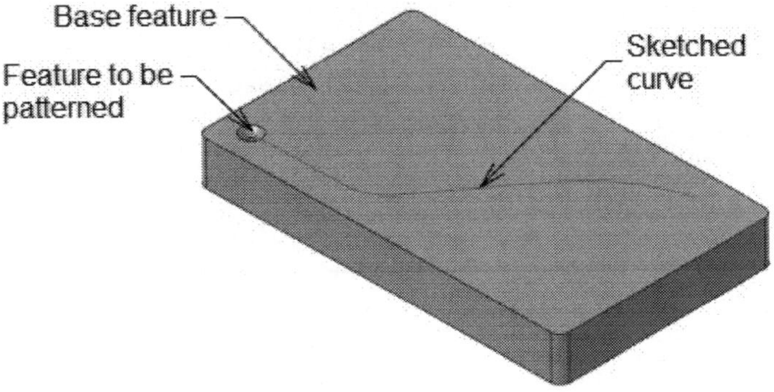

Figure 8-26 *Model with the feature to be patterned and the sketched curve*

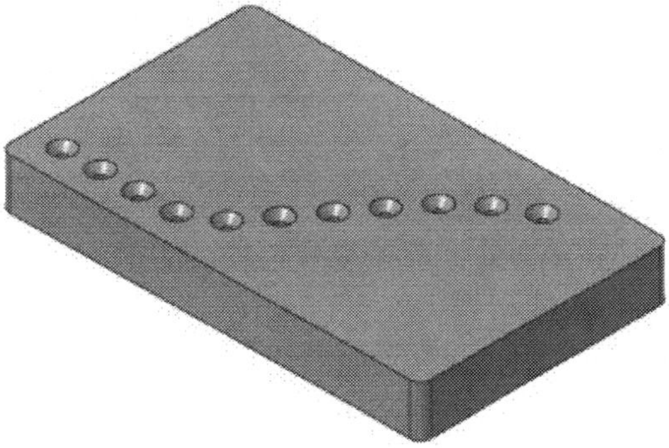

Figure 8-27 *The curve-driven pattern*

> **Tip**
> *You cannot use a combination of open and closed sections to create a curve pattern.*

You can choose the **Follow curve direction** option in the **Options** slide-down panel to rotate the pattern members about the selected curve. When you select this option, the origin of each pattern member becomes tangent to the selected curve.

Point Pattern

The **Point** option is used to create patterns at sketched points, datum points, or coordinate systems. To create a pattern, select the feature to be patterned and invoke the **Pattern** tool, as discussed earlier. Select **Point** from the **Type** drop-down list; the **Pattern** dashboard will be displayed, as shown in Figure 8-28.

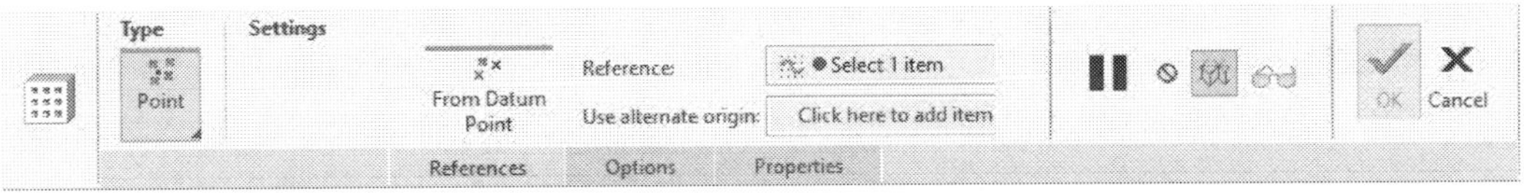

Figure 8-28 *The* ***Pattern*** *dashboard with the* ***Point*** *option selected*

The **From Datum Point** button in the dashboard enable you to select the geometry points, geometry coordinate systems, or datum points at which the pattern members will be placed. Select the required entities from the model area and then choose the **OK** button; a pattern will be created at each point or coordinate system, refer to Figures 8-29 and 8-30. Note that the points and the coordinate systems must be drawn first and then you need to invoke the **Pattern** tool. You can also use the geometric points or coordinate systems used in an internal sketch or in an imported feature.

When you create a point pattern, by default the origin is selected. However, you can change the default selected origin by using the **Use alternate origin** collector displayed on selecting the **Point** option from the **Pattern** drop-down list.

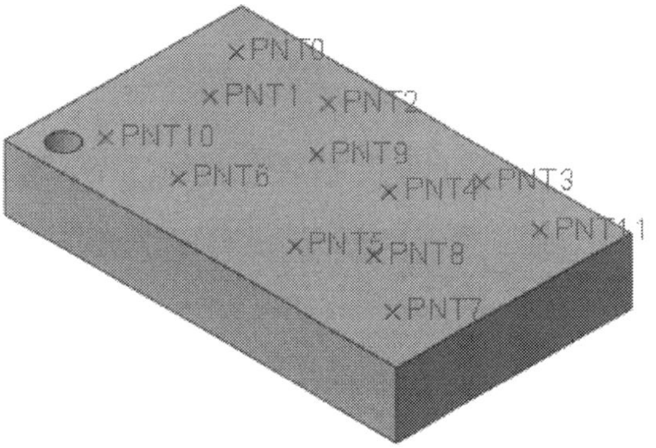

Figure 8-29 *Hole on the base feature*

Figure 8-30 *Hole patterned on the base feature*

Options Tab

When you choose the **Options** tab, a slide-down panel will be displayed. The options in this slide-down panel are discussed next.

Identical Pattern

The **Identical** option in the **Regeneration option** drop-down list is used to create an identical pattern. You need to select at least one incremental dimension to pattern the feature. Depending on the incremental dimension selected, the resulting pattern will be linear or rotational. A linear pattern is created when the driving dimension is linear and a rotational pattern is created when the driving dimension is angular. You can enter a positive or a negative value as the increment in a pattern dimension. All instances of a pattern that are created using this option are identical in size and geometry and are placed on the same surface. This is the reason the patterns created using this option are known as Identical patterns. Figure 8-31 shows a hole feature on the base feature and Figure 8-32 shows the holes patterned linearly.

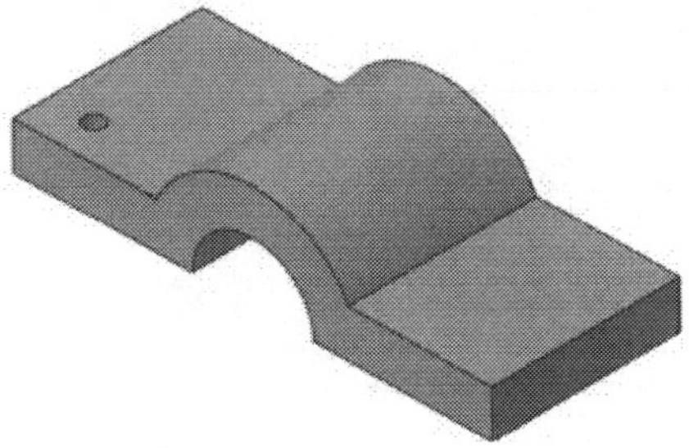

Figure 8-31 Hole on the base feature

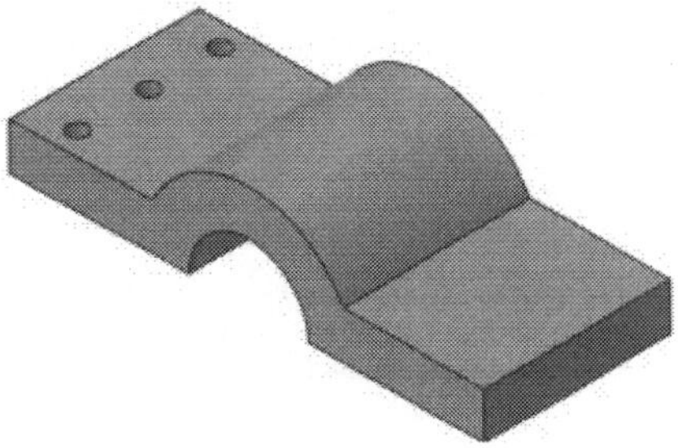

Figure 8-32 Hole patterned on the base feature

Similarly, Figure 8-33 shows a hole feature on the base feature and Figure 8-34 shows rotational pattern of the hole feature.

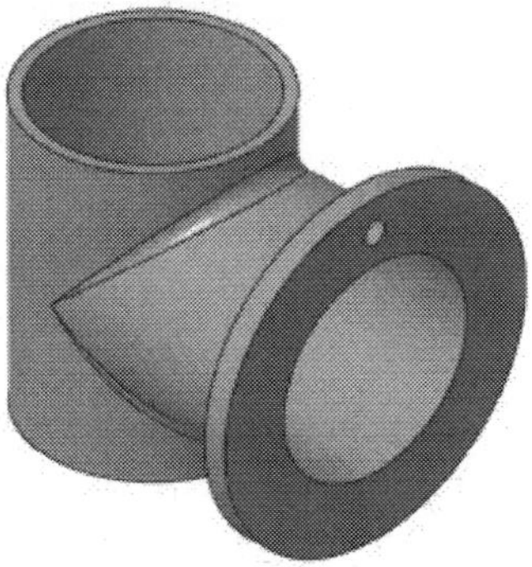

Figure 8-33 Hole on the base feature

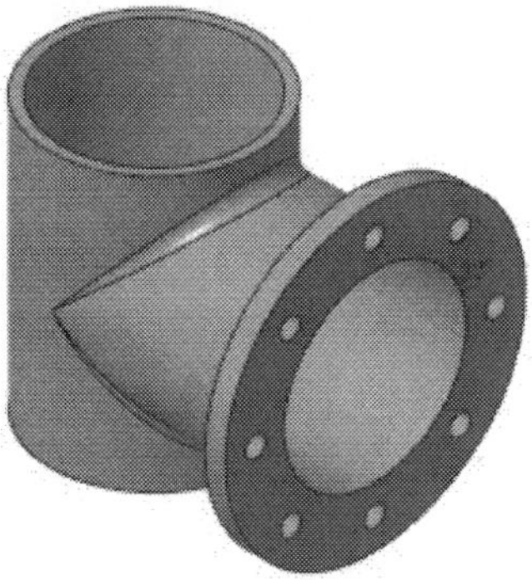

Figure 8-34 Rotational pattern of the hole feature

As evident from Figures 8-32 and 8-34, all instances in identical patterns are placed on same placement surface and no feature intersects edges of the placement surface, any other instance, or any feature other than the placement surface. Note that you cannot pattern the hole feature on the right flap shown in Figure 8-32 by using the **Identical** option. However, you can also use the **General** option to create such type of pattern feature.

Variable Pattern

The Variable pattern is used when instances vary in size. In this type of pattern, the instances can be placed on different surfaces and can also intersect with the edges of the placement surface. To create a varying pattern feature, you need to select the **Variable** option from the **Regeneration option** drop-down list in the **Options** tab and then select the dimension(s) to be varied. Next, click in the desired direction collector in the **Dimensions** slide-down panel. Select multiple dimensions of the feature by pressing and holding the CTRL key. Next, specify the value for the dimension increment and choose the **OK** button to create varying pattern of the selected feature.

Figures 8-35 shows the feature to be patterned and Figure 8-36 shows the feature after creating variable pattern.

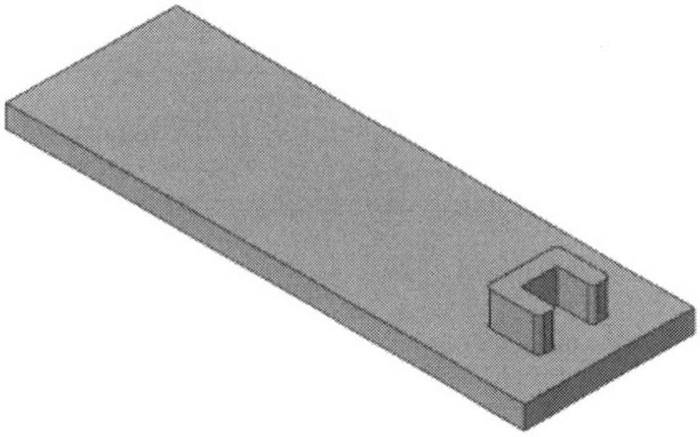

Figure 8-35 The feature to be patterned

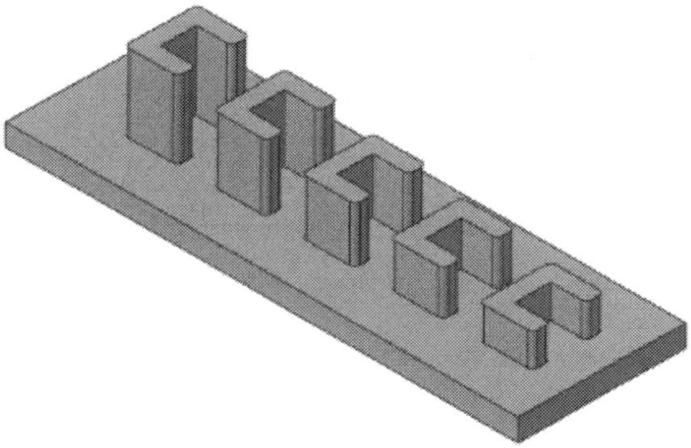

Figure 8-36 Varying pattern feature created

General Pattern

You can create the most complex patterns by using the **General** pattern. This option is used to create patterns in which the instances touch each other and intersect with other instances or the edges of a surface. This option of creating patterns is also used when instances intersect with the base feature and the intersection is not visible. Figure 8-37 shows a hole on the base feature and Figure 8-38 shows the hole pattern created using the **General** option.

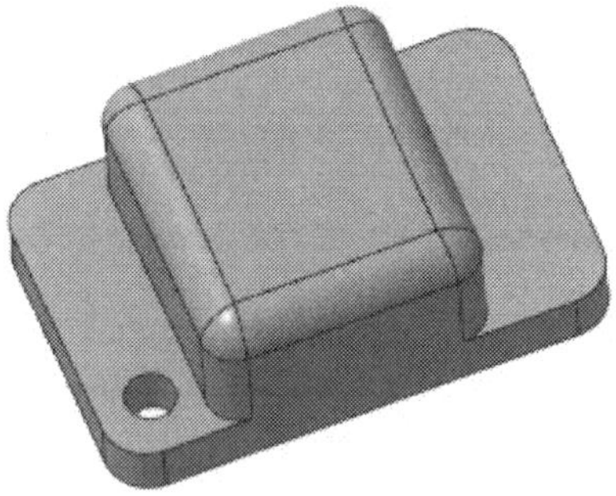

Figure 8-37 Hole on the base feature

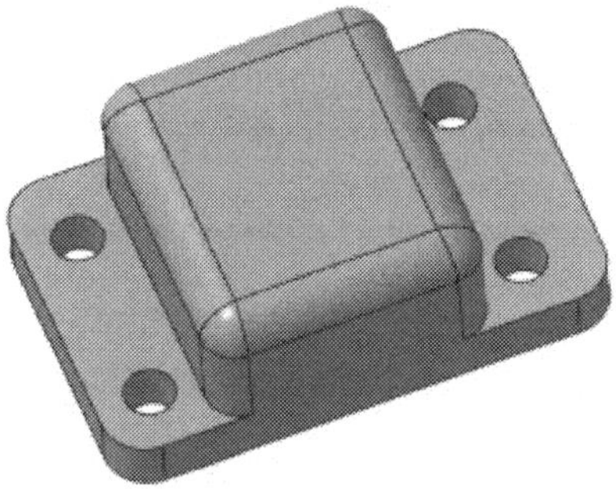

Figure 8-38 General pattern

Note

If the features that you need to pattern have formed a group and are listed in the ***Model Tree*** *as a group feature, you need to select the desired feature from the group to pattern it. A feature forms a group when it has used other features to form itself. For example, if you create an asynchronous datum plane and use it as a sketch plane, then after the feature is created, it is listed in the* ***Model Tree*** *as a group feature. This group comprises feature, datum plane, and sketch of feature.*

Creating a Geometry Pattern

Ribbon: Model > Editing > Pattern drop-down > Geometry Pattern

In Creo Parametric, the **Geometry Pattern** tool is used to pattern geometries rather than features. In the geometry pattern, the selected features need not be made consecutive. To create a geometry pattern, select the geometry from the model and then choose the **Geometry Pattern** tool from the **Pattern** drop-down in the **Editing** group; the **Geometry Pattern** dashboard will be displayed, as shown in Figure 8-39.

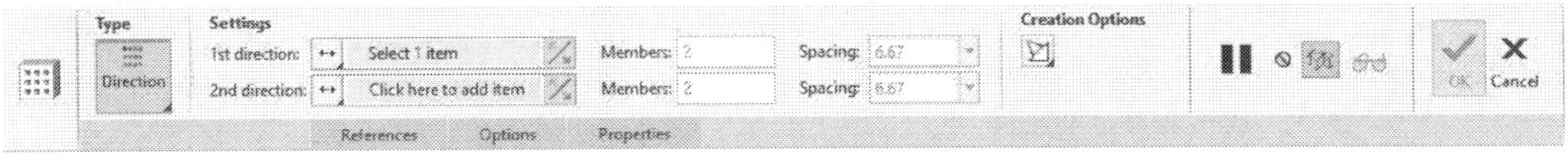

***Figure 8-39** The **Geometry Pattern** dashboard*

In Creo Parametric, you can create six types of geometry patterns. These patterns are **Direction**, **Axis**, **Fill**, **Table**, **Curve**, and **Point**. The options in the **Geometry Pattern** dashboard are same as available in the **Pattern** dashboard with some exceptions such as the **Dimension** and **Reference** patterns are not available for geometry patterns. The options displayed depend upon the type of the pattern selected. The options that are not available in the **Pattern** dashboard are discussed next.

When you click on the **References** tab, a slide-down panel is displayed. Choose the **Geometry References** button to display the **Geometry References** dialog box. This dialog box has three collectors. The **Surface Sets** collector is used to select surface sets. The **Chain** collector is used to select curves and edge chains from the model. The **References** collector in this dialog box is used to select features and groups. You can use the options in the **Creation Options** drop-down list to specify the attachment type between the base feature and patterned features. By default, the Flexible Modeling attach option is selected in this drop-down list. As a result, the patterned members are extended or trimmed until they are attached to the base geometry. If you choose the Copy geometry without attachment option, the surfaces, curves, chains, features, and groups are copied to a new location without getting attached to the base feature. The Fill option in the drop-down list is used to fill the volume defined by the quilt with new solid material without removing existing material. The Remove option is used to remove solid material from the volume defined by the quilt. You can choose the Replace option to add new material within the geometry. The addition of material depends upon the relationship of pattern members with the neighboring geometry.

Figure 8-40 shows a model for creating geometric pattern. Both the cut features are created by using one sketch. But you need to pattern only the circular cut feature. Figure 8-41 shows the model after creating the geometric pattern.

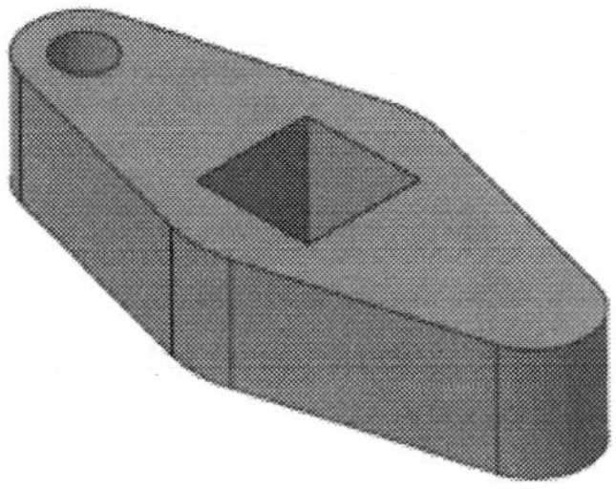

***Figure 8-40** Model with cut feature*

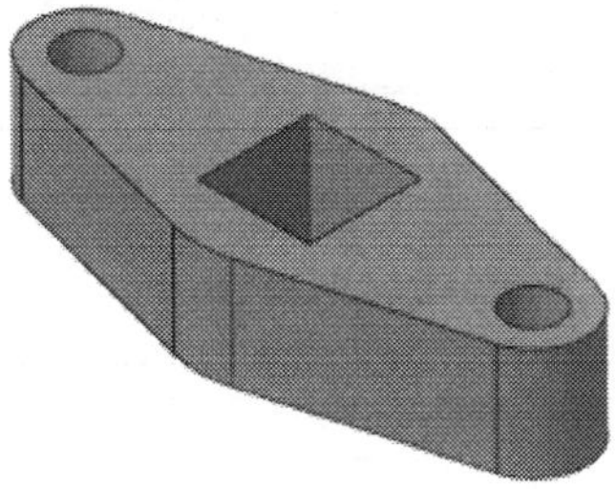

***Figure 8-41** Model after creating the geometry pattern*

Note

*Unlike the **Pattern** tool, the **Geometry Pattern** tool has the ability to pattern the feature more than twice.*

Deleting a Pattern

To delete a pattern select the pattern feature from the **Model Tree** and right-click to invoke the shortcut menu. Choose the **Delete Pattern** option from the shortcut menu. However, note that the parent feature (leader) is not deleted when you delete the pattern by using the **Delete Pattern** option even if it was selected along with other instances for deletion.

If you want to delete the pattern along with the parent feature, choose the **Delete** option from the shortcut menu. The system highlights the pattern in the drawing area and confirms the deletion of the pattern from the user.

COPYING AND PASTING FEATURES

The **Copy**, **Paste**, and **Paste Special** tools allow you to duplicate and place features, geometries, curves, and edge chains within the same model or in other models. The instances created using the **Copy** and **Paste** tools can be independent, partially dependent, or fully dependent on the original features and geometries. You can also apply move or rotate transformation to the pasted instances.

When you copy a feature or geometry, it is copied to the clipboard and it will be available for pasting with its references, settings, and dimensions, until another feature is copied to the clipboard. The features in the clipboard retain their original references, settings, and dimensions when you change the references, settings, and dimensions of any one of the instances or all the instances during a multiple paste operation. Pasting the feature in a different model also does not affect the references, settings, and dimensions of the copied features in the clipboard.

The methods to copy and paste features are discussed next.

Copying Features

Ribbon: Model > Operations > Copy

The **Copy** tool available in the **Operations** group of the **Model** tab is used to copy an object to the clipboard. To copy a feature or a set of features to clipboard, select it from the drawing area or from the **Model Tree**; the selected object gets highlighted in green color. Next, choose the **Copy** tool from the **Operations** group; the selected object will be copied to the clipboard. This tool can also be accessed from the shortcut menu of the selected object.

Pasting Features

The **Paste** and **Paste Special** tools available in the **Operations** group of the **Model** tab are used to paste the features, set of features, geometries, curves, surfaces, and edge chain that were copied to the clipboard. Note that these tools will be available only if objects for pasting are available in the clipboard. The **Paste** and **Paste Special** tools are discussed next.

Pasting Objects Using the Paste Tool

Ribbon: Model > Operations > Paste

Pasting of objects from the clipboard works in different ways. If you are pasting a single feature, the creation tool for that feature is invoked automatically. For example, if you are trying to paste a revolved feature, the **Revolve** dashboard will be invoked automatically. If you are pasting a datum plane, the **Datum Plane** dialog box will be invoked. To complete the placement of the pasted feature, you need to specify its placement references.

If you are trying to paste multiple features, first feature in the **Model Tree** will determine the creation tool invoked. As you specify references for the pasted object, the creation tools will be invoked for each feature in the group in a sequential manner. If you have selected the same primary reference as the one used by the original feature, the system places the feature using the same reference as used by the original feature. If you select a different primary reference, the system lets you place the feature using new references. Note that here the first reference should be valid for further operation.

You can also copy and paste features between two different parts that are using two different systems of measurement units. In such a case when you choose the **Paste** tool, the **Scale** dialog box will be displayed, as shown in Figure 8-42. Select the **Keep dimension values** radio button if you want to retain the dimension values in their default state. For example, 1 will remain 1 regardless of different measurement units of both parts.

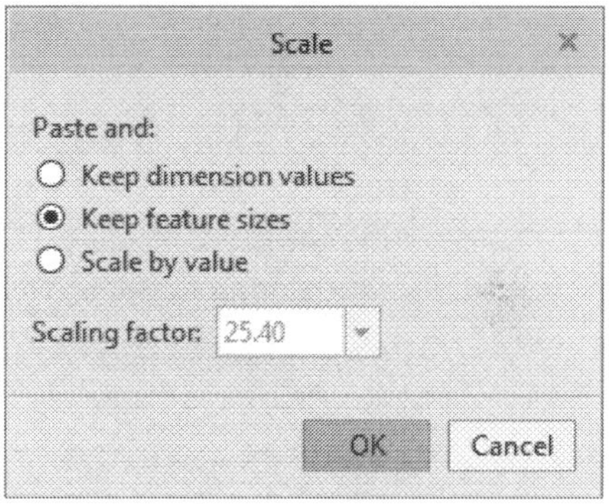

*Figure 8-42 The **Scale** dialog box*

The **Keep feature sizes** radio button is used to convert the dimension values of the copied feature to the measurement system of the part in which you are pasting. For example if the copied feature uses inches as units and the part where you want to paste the feature uses centimeters as units, a dimension value of 1 inch will be converted into 2.54 centimeters.

Select the **Scale by value** radio button and specify a value in the **Scaling factor** edit box if you want to scale the pasted item by a specified value.

After choosing the appropriate radio button from the **Scale** dialog box, choose the **OK** button; the creation tool of the feature that you are pasting will be invoked. The original feature is copied and placed according to the references specified.

Pasting Objects Using the Paste Special Tool

Ribbon: Model > Operations > Paste drop-down > Paste Special

You can use the **Paste Special** tool in case pasting of features is not possible with the help of the **Paste** tool. When you choose the **Paste Special** tool from the **Paste** drop-down in the **Operations** group of the **Model** tab, the **Paste Special** dialog box is displayed, as shown in Figure 8-43. The options in this dialog box are discussed next.

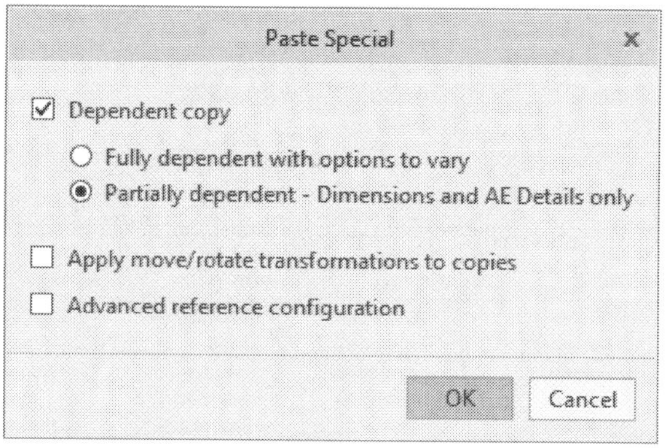

Figure 8-43 *The **Paste Special** dialog box*

Dependent copy

This check box is selected by default. As a result, the copied features are dependent on the original features. The copied features are dependent on the dimensions attributes, elements, parameters, or the sketch of the original feature. You can create independent copies of the original feature or the feature set by clearing the **Dependent copy** check box.

By default, the **Partially dependent - Dimensions and AE Details only** radio button is selected under this option. As a result, the copied objects will be dependent on the dimensions or the sketch of the original feature. The **Fully dependent with options to vary** radio button is used to create copies of the original feature that are fully dependent on all attributes, elements, and parameters of the original feature, but allows to vary the dependency of dimensions, annotations, parameters, sketches, and references. When you copy features and elements across different models, the **Make Fully Dependent copies with options to vary** radio button is available in place of the **Dependent copy** check box.

Apply move/rotate transformations to copies

By selecting this check box, you can move the pasted feature by translation or rotation, or both. You can move and rotate the pasted object by using the options available in the **Transformations** slide-down panel. Note that the copied object or feature can be fully dependent on the original feature.

Advanced reference configuration

Select this check box and choose the **OK** button from the **Paste Special** dialog box; the **Advanced Reference Configuration** dialog box will be displayed, as shown in Figure 8-44. This dialog box lists the references of the original feature and allows you to retain these references or replace them with new references in the pasted feature. It also allows you to edit the references and replace references while pasting the copied features instead of separately editing the references after pasting the feature.

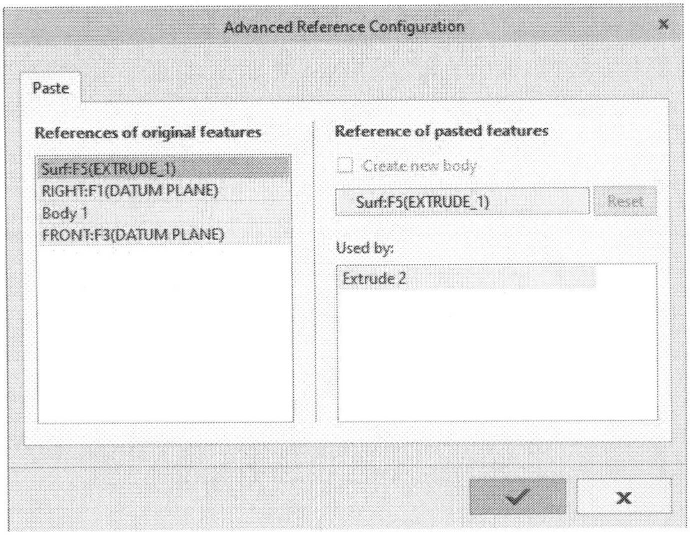

Figure 8-44 *The **Advanced Reference Configuration** dialog box*

Tip
To copy and paste objects, you can also use the CTRL+C and CTRL+V keys from the keyboard.

Cancelling the Paste Operation

You can cancel a paste operation any time during the process by choosing the **Cancel** button available in the dashboard or in the dialog box of the pasted feature. When you choose to cancel a paste operation, the **Cancel Paste** dialog box will be displayed, as shown in Figure 8-45. Note that if you are cancelling the paste of a single feature or object, the **Cancel Paste** dialog box will only ask for the confirmation to cancel the operation.

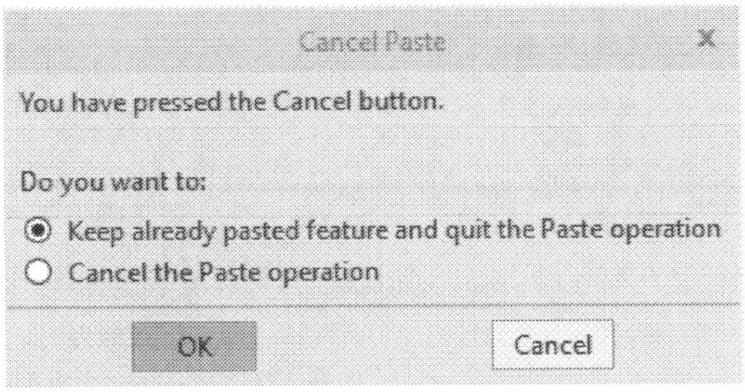

Figure 8-45 *The **Cancel Paste** dialog box*

By default, the **Quit pasting this feature and continue** radio button is selected under the **Do you want to:** area of the **Cancel Paste** dialog box. If you are pasting multiple objects or features at a time, then this option will cancel the pasting of the current feature and will continue with the pasting of the next feature available in the clipboard. Select the **Keep already pasted feature and quit the Paste operation** radio button if you want to keep the feature that is already pasted and cancel pasting of the current feature or the remaining features available in the clipboard. This option is useful when you are pasting multiple features or if the current feature is the last one to be pasted. The **Remove already pasted features and continue** radio button is used to cancel the pasting of all the features that are pasted so far and continues pasting the remaining features on the clipboard. Select this option when you are pasting multiple features. This option is not available when the current feature is the first or the last feature to be pasted. Select the **Cancel the Paste operation** radio button if you want to cancel the pasting of all features.

Note that, the radio buttons available in the **Cancel Paste** dialog box may vary according to the method of cancelling the paste operation.

MIRRORING A GEOMETRY

Ribbon: Model > Editing > Mirror

The **Mirror** tool is used to copy features and geometries about a planar face. The copy created by using the **Mirror** tool can be independent or dependent on the parent geometry. To create a mirror feature, choose the **Mirror** tool from the **Editing** group; the **Mirror** dashboard will be displayed, as shown in Figure 8-46. Note that this tool is available only when a feature or geometry has been selected. The options available in this dashboard are discussed next.

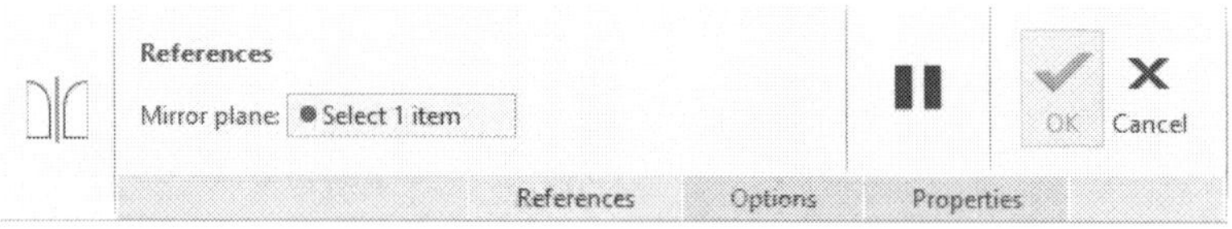

Figure 8-46 *The **Mirror** dashboard*

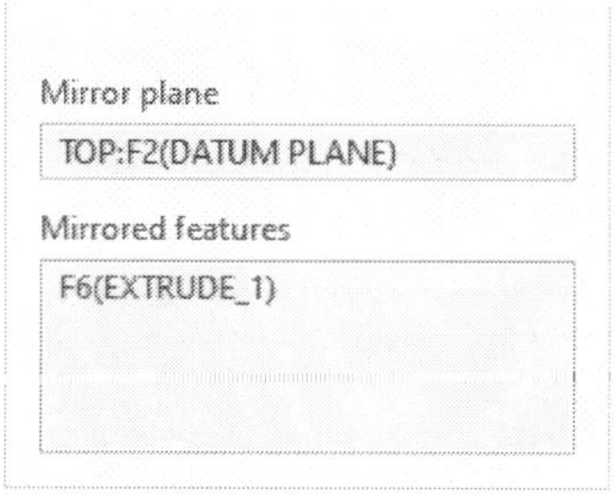

Figure 8-47 *The **References** tab*

In the Settings area of the Mirror dashboard, select the **Mirror plane** collector and then select required datum plane or planar surface as the mirroring plane. The **Mirror plane** collector is also available in the slide-down panel in the **References** tab. On selecting the mirroring plane, the preview of the mirrored feature is displayed in the graphics window. Also, the name of feature will be displayed in the **Mirrored features** collector in the **References** tab, as shown in Figure 8-47. If any source feature/s is/are missing, then the name of that feature will be highlighted in yellow color in the **Mirrored features** collector. Furthermore, the **Mirrored features** collector gives you the freedom to add or remove features while applying or editing the mirror feature. When you select the **Reapply Mirror** button, missing features are removed from the definition. Note that the **Reapply Mirror** button will be available in the **Mirror** dashboard only when you redefine a mirrored feature. To add or remove features, select the **Reapply Mirror** button, press CTRL, and select feature/s from the graphics window.

Note

1. *If source features are deleted, they are not included in the **Mirrored features** collector.*

2. *The modifications done earlier on target features are lost on reapplying mirror.*

If references of mirrored features get modified then they become visible in the **Mirrored features** collector with an error and therefore they need to be redefined. To do so, click on the feature name in the **Mirrored features** collector; a shortcut menu is displayed. From the shortcut menu, choose the **Fix** option to replace the references. It is recommended to first add or remove features from the **Mirrored features** collector before using the **Fix** option. Figure 8-48 shows the **Mirror** dashboard with the **Reapply Mirror** button.

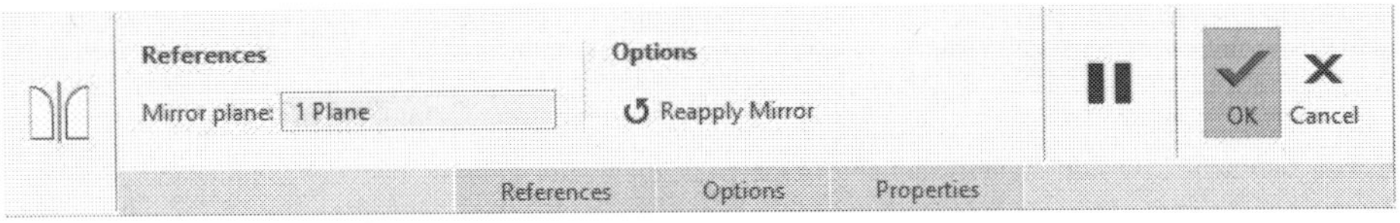

*Figure 8-48 The **Mirror** dashboard with the **Reapply Mirror** button*

To make the mirrored copy partially or fully dependent on the parent geometry, you can select the **Dependent copy** check box available in the **Options** tab. To remove or restore dependency of the mirrored feature, you can choose the options available in the shortcut menu of the **Model Tree**. These options have been discussed earlier in the **Paste** tool. Next, choose the **OK** button to create the mirrored feature. In Figure 8-49, a model is selected to mirror about the selected plane and Figure 8-50 shows the resulting model after mirroring.

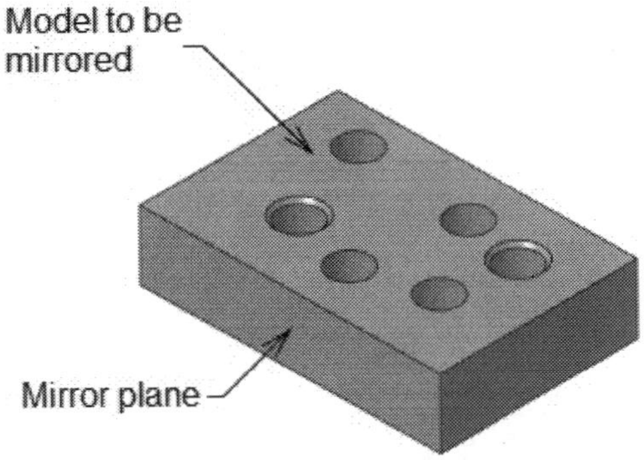

Figure 8-49 Model and datum plane

Figure 8-50 Resulting mirrored model

USER-DEFINED FEATURES (UDF)

Creo Parametric allows you to reuse existing geometry while creating new design models. This is done with the help of user defined features (UDFs). A user defined feature (UDF) is a group of selected features, all their dimensions, relations between the selected features, and a list of references for placing the UDF on a model. User defined features save considerable amount of time in designing of parts by establishing a library of commonly used geometry. Various types of UDFs that you can create in Creo Parametric are discussed next.

Subordinate UDFs

A subordinate UDF is dependent upon the original model. When a subordinate UDF is placed, it gets its values directly from the original model, so the original model must be present for the subordinate UDF to function properly. If you make any changes in the dimension values in the original model, they are automatically reflected in the UDF. You can associate more than one UDF with a model.

Standalone UDFs

A standalone UDF copies all the information of the original model into the UDF file. A standalone UDF is not dependent upon the original geometry. If you make any changes in the reference model, they are not reflected in the UDF. A standalone UDF requires more storage space than a subordinate UDF because the data and information is written in a file and then saved into your computer.

Although a reference part is not required for a standalone UDF, it is advised to have a reference part displayed when you place a UDF. The system highlights the dimensions to be entered and the reference information at appropriate time during the UDF placement. If there is no reference part, the number of UDF elements you can modify is limited.

Creating UDF

Ribbon: Tools > Utilities > UDF Library

Before creating a UDF, it is advised to create a UDF library first. By default the UDF created is saved in the current working directory but you can set any folder as the default directory for the UDF. To create a UDF library, choose the **UDF Library** tool available in the **Utilities** group of the **Tools** tab; the **Menu Manager** will be displayed, as shown in Figure 8-51. The **Create** option in the **Menu Manager** is used to add a new UDF to the library. To modify an existing UDF, you can choose the **Modify** option. You can choose the **List** option to list all the UDFs stored in the current directory. To use database management functions for the current UDF, choose the **Dbms** option. The **Integrate** option is used to resolve the differences between the source and target UDFs.

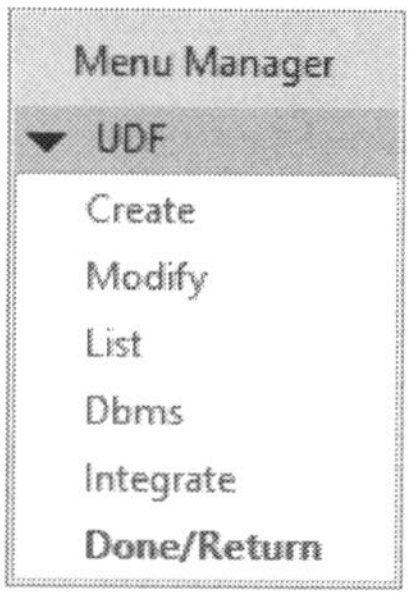

Figure 8-51** The **Menu Manager

Now, to create a new UDF, choose the **Create** option; the **UDF name** edit box is displayed above the graphics toolbar. Next, specify a name in the edit box and choose the **OK** button; the **Menu Manager** will be expanded with the **Subordinate** selected in the **UDF OPTIONS** rollout. If required, select the **Stand Alone** option from the rollout and choose the **Done** option; a **Confirmation** message box will be displayed. Choose the **Yes** button from this message box and select the reference part. Next, choose the **Done** option; the **UDF:<udf name>, Subordinate** or **UDF:<udf name>, Standalone** dialog box will be displayed along with the **Menu Manager**, refer to Figures 8-52 and 8-53, and you will be prompted to select features. Next, select the features to include in UDF and choose the **Done** option in the **Menu Manager**. To add more than one feature in the UDF, choose the **Add** option in the **Menu Manager** and then select the next feature. After selecting the features for the UDF, choose the **Done/Return** option to finish the selection of features for the UDF. Next, you are required to specify prompts for the references that are being used by the selected features. These prompts will guide you to place the UDF. You can type or select the default prompts for the references used by the selected features. Each time you enter or accept the default prompt, a reference will be highlighted.

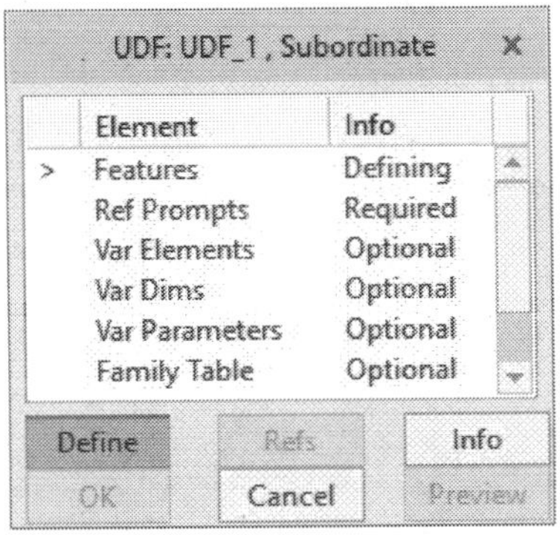

***Figure 8-52** The **UDF** dialog box*

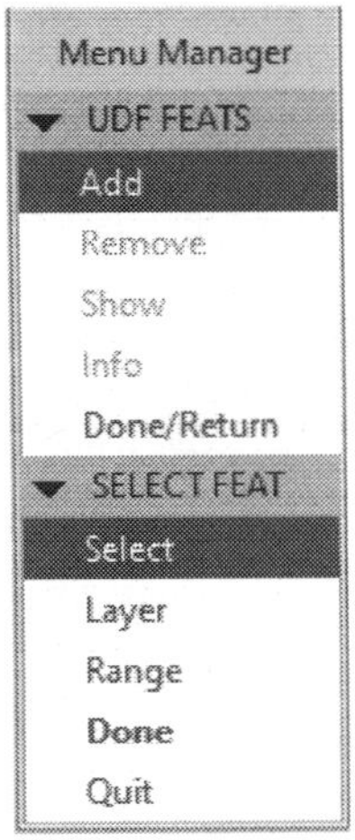

Figure 8-53** The **Menu Manager

Note that if you select a single feature for the UDF, the Prompt edit box will be displayed for the highlighted reference. If you add multiple features to include in the UDF, the **PROMPTS** rollout will be displayed in the **Menu Manager**.

The **Single** option in the rollout is used to specify a single prompt for the highlighted reference that can be used by more than one feature. When you place the UDF, the prompt appears only once, but the reference you select for this prompt applies to all the features in the group that are using the same reference. The **Multiple** option in this rollout is used to specify individual prompt for each feature using the selected reference. When the UDF is placed, each feature using the select reference is highlighted so that different prompt for each of the feature can be specified. After specifying the prompts, choose the **Done/Return** option from the **SET PROMPT** rollout. To edit the prompt, choose the **Next** and **Previous** option in the **MOD PRMPT** rollout and then choose the **Enter Prompt** option to edit the prompt. To finish, choose the **Done/Return** option from the **Menu Manager**.

The required UDF attributes are now defined and you can choose the **OK** button from the **UDF:<udf name>, Subordinate** or **UDF:<udf name>, Standalone** dialog box to save the UDF in the specified directory. Next, choose the **Done/Return** option in the **Menu Manager** to finish the procedure. However, there are also some optional attributes available in the dialog box, you can specify them as per your requirements.

To specify an optional attribute for the UDF, select it and then choose the **Define** button available in the **UDF:<udf name>, Subordinate** or **UDF:<udf name>, Standalone** dialog box. The **Var Elements** option allows you to specify the features that can be redefined when the UDF is placed. You can also select the dimensions that can be varied while placing the UDF. To do so, choose the **Var Dims** option and click the **Define** button; the **Menu Manager** will be displayed. Select the dimensions you want to make variable and choose the **Done/Return** option from the **Menu Manager**. Next, you will require to specify a prompt for the dimension. Type the prompt in the edit box and choose the **OK** button to finish defining variable dimensions for the UDF. To modify the parameters of the UDF while placing, choose the **Var Parameters** option. The

Dim Prompts option will appear only when you have specified variable dimensions for the UDF. This option lets you modify prompts for variable dimensions. You can also create family table instances for the UDF by choosing the **Family Table** option. To include external dimensions and parameters to the UDF, choose the **Ext Symbols** option.

Placing UDF

Ribbon: Model > Get Data > User-Defined Feature

In the previous section, you learned to create UDFs. A UDF is saved as *.gph* file format in the current directory. In this section, you will learn to place UDFs in the target model. The procedure to place a UDF is discussed next.

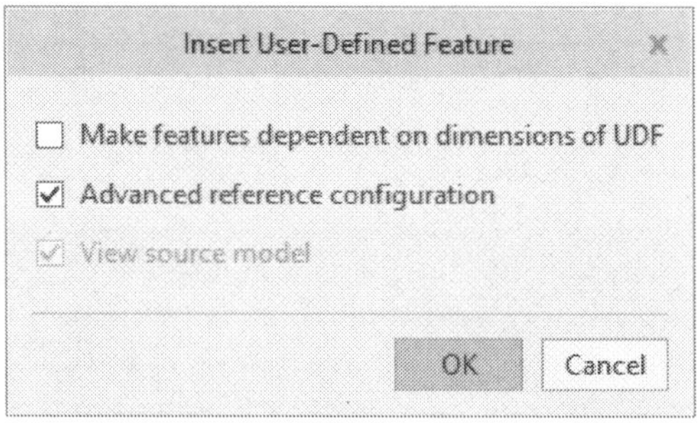

***Figure 8-54** The **Insert User-Defined Feature** dialog box*

To place a UDF, first open the target model. Next, invoke the **User-Defined Feature** tool in the **Get Data** group of the **Model** tab; the **Open** dialog box will be displayed. Browse to the location where you have stored the UDF file. Select the UDF file and choose the **Open** button in the dialog box; the **Insert User-Defined Feature** dialog box appears, as shown in Figure 8-54. By default, the **Advanced reference configuration** check box is selected in this dialog box. As a result, the UDF is placed using the references of the source model. If you clear this check box, you can manually define the UDF placement using the feature definition interface. If you select the **Make features dependent on dimensions of UDF** check box, the dimensions of the UDF will become dependent upon the source feature. Any change made in the source document will be reflected in the target model. You can select the **View source model** check box to view the source model of the UDF. Next, choose the **OK** button from the **Insert User-Defined Feature** dialog box; the **User-Defined Feature Placement** dialog box will be displayed, as shown in Figure 8-55. Also, the source model will be opened in a separate part window and preview of the UDF will be displayed on the target model. You can switch between windows by selecting the required window from the **Windows** drop-down in the **Quick Access** toolbar. The options and tabs in this dialog box are discussed next.

Placement Tab

When you select a reference from the **References of Original Features** list; the selected reference is highlighted in the source model. The **References of UDF Features** collector is used to select references for the placement of a UDF. When you select a reference from the **References of Original Features** list, the **Used By** area displays the features in the UDF that are being used by highlighted original references. By default, the **Auto Regeneration** check box is already selected. As a result, the features will be regenerated automatically when changes are made in **User Defined Feature Placement** dialog box. If you clear this check box, you will be able to manually regenerate the features of the UDF by choosing the Next or Previous buttons available on the left of the check box.

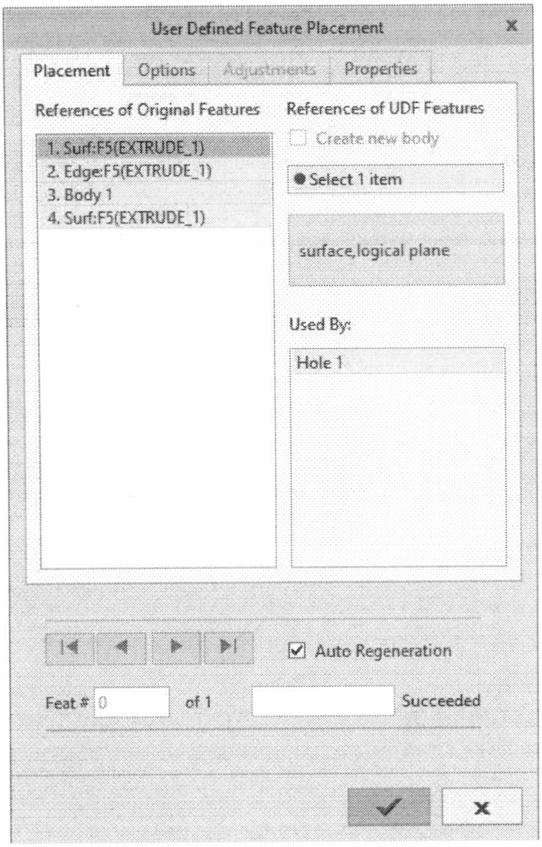

*Figure 8-55 The **User Defined Feature Placement** dialog box*

Variables Tab

The **Variables** tab in the dialog box will be displayed only when there are variable dimensions in the UDF. To change the value of the dimension, select the dimension from the list and specify the values in the respective column.

Options Tab

The options available in this tab are used to scale and redefine the features of the UDF. By default, the **Keep dimension values** radio button is selected in the **Scaling** area. As a result, the numeric values for all the dimensions are kept same. For example, the dimension of 10 inches will become 10 mm. Select the **Keep feature sizes** radio button to retain the size of the feature and convert the dimensions according to the target model units. This radio button will be available only when the source model and target model are created in different unit systems. Select the **Scale by value** radio button to enable the **Scaling factor** edit box in which you can specify the scaling value for the UDF.

The **Dimensions that may not be varied during placement of UDF** area is used to specify the behavior of non-variable dimensions. By default, the **Unlock** radio button is selected in this area. As a result, the dimensions are visible and can be modified. Select the **Lock** radio button to lock the dimensions. You can also hide the dimensions by selecting the **Hide** radio button.

The **Redefine these features** area is used to add, remove, or redefine the features in a UDF during placement. To redefine a feature, select the check box next to it and choose the **Edit Definition** button. Note that features with variable elements are checked for redefinition by default. After redefinition of a feature, the check box corresponding to that feature is automatically cleared. You can restore the changes done in the features by choosing the **Undo All** button.

Adjustments Tab

The options in this tab are used to orient the UDF during placement. The **Orientation Items** displays a list of placement orientations of the UDF. The Description list displays the alternative placement option for the selected orientation. Choose the **Flip** button if you want to flip the orientation of the feature in other direction.

Properties Tab

This tab is used to rename the UDF. You can also view the information of the UDF by selecting the **Displays information for this feature** button.

LAYERS

Layers provide you an effective way to manage and organise model or assembly items. With the help of layers, you can group items, such as features, datum planes, parts in an assembly, and even other layers so that you can perform operations on those items collectively. Layers enable you to simplify geometry selection by temporarily hiding or displaying specific model features or assembly components in the drawing area. Layers can also be used to perform actions, such as suppressing all the items in a layer at once. You can also create filters in a layer so that the items that fulfill the conditions of that filter are automatically added to the layer.

The most common operations that can be performed using layers are hiding and unhiding the items and selection of items in a layer. You can easily select multiple items rather than selecting them individually. Such kind of selection is beneficial and time saving approach to follow in special cases.

The procedure to create and manage the layers and layer states is discussed next.

Creating and Managing Layers

By default some layers are already included in every part or assembly document. These layers are called default layers. Some layers are automatically created as you start working on the model. However, you can also create your own layers and add items to them manually.

Layers in a model or assembly are managed using the **Layer Tree**. With the help of layer tree, you can manipulate layers, their items, and their display status. To toggle between the **Model Tree** and **Layer Tree**, choose the **Layer Tree** option from the **Show** drop-down list in the **Navigator** window; the layer tree will be displayed in the **Navigator** window. You can also activate the layer tree by choosing the **Layers** tool in the **Visibility** group of the **View** tab.

To create a new layer, choose the **New layer** option from the **Layer** drop-down list in the navigator; the **Layer Properties** dialog box will be displayed, as shown in Figure 8-56.

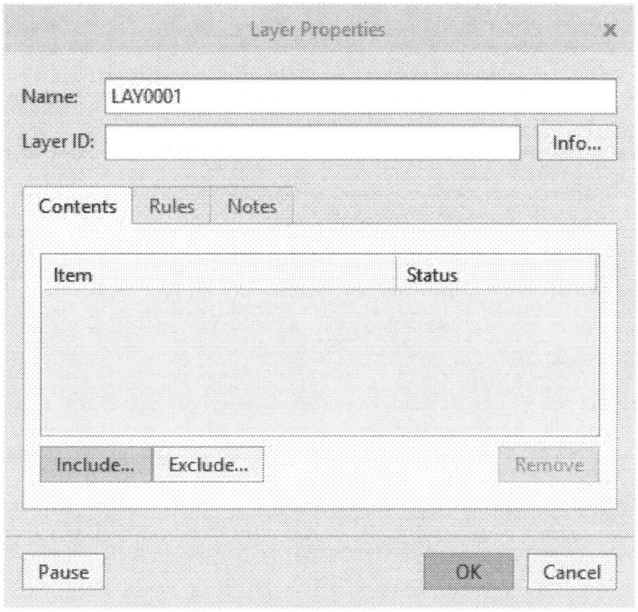

***Figure 8-56** The **Layer Properties** dialog box*

In the **Layer Properties** dialog box, specify the name of the layer in the **Name** edit box. Next, you can specify a number in the **Layer ID** edit box. The **Info** button next to the **Layer ID** edit box is used to display the information of the layer in a separate window. The **Contents** collector in this dialog box is activated by default. Also, the **Include** button below the collector is chosen by default. To add an item to the layer, select the layer from the **Model Tree** or from the drawing area; the selected item is added to the **Contents** collector. To exclude an item from the list, select the item and choose the **Exclude** button; the item will be excluded from the layer you are creating. You can remove the item by choosing the **Remove** button. To pause the process of layer creation, you can choose the **Pause** button. Next, choose the **OK** button; the layer will be added to the **Layer tree**.

You can control the layer on which the items are being placed by activating it. When the layer is activated, any drawing items created, shown, or pasted in the drawing area are automatically placed on the layer. If a layer is activated in a 3D model, all the newly created 3D features will be added to this layer. To activate a layer, select it from the **Layer Tree** and right-click to invoke a shortcut menu. Next, choose the **Activate** option from this menu; the layer is activated. An activated layer is highlighted with a green colored symbol() in the **Layer Tree**. If you are activating a layer that has the same name as another layer, then each layer with that name will be activated.

You can also create nested layers in Creo Parametric. To create a nested layer, invoke the **Layer Properties** dialog box and select the layer from the **Layer Tree** that you want to be nested. The name of the selected layer will be displayed in the **Contents** list. However, the items in a nested layer will not be displayed in the **Layer Tree**. To display the items of a nested layer, choose the **Settings** option from the **Navigator** and select the **Items on Nested Layers** check box. Alternatively, you can copy a layer and paste it into the layer that you want to be nested.

To hide or show the contents of a layer or the layer itself, choose the **Hide** or **Show** option from the layer shortcut menu.

Tip

1. Hiding an item simply removes it from the graphics window, whereas suppressing an item removes it from the regeneration cycle of the model.

2. A hidden item is included in calculations, such as mass properties analyses, whereas a suppressed item is not included in calculations.

Creating Layer States

Layer states allow you to toggle between different layer displays. You can create layer states in the view manager to save the hide/unhide status of all the layers in a model. Note that, hiding a feature or component will place it in the hidden layer. Therefore, you can use the layer states with hidden items without accessing the **Layer Tree**. You can apply layer states to any item in the drawing area, such as features, components, or drawing views and annotations. The procedure to create layer states is discussed next.

To create layer states in a model, choose the **View Manager** tool from the **Graphics** toolbar; the **View Manager** dialog box will be displayed. Next, choose the **Layers** tab and then choose the **New** option; a layer state with a default name will be added under the **Names** column. If required, add more layer states to the model by choosing the **New** option. To add items to a layer state, select the item from the **View Manager** dialog box and then choose the **Activate** option from the **Options** drop-down list. Alternatively, you can double click on the name of the layer state to activate it. Select the items from the drawing area or from the **Model Tree** and then choose the **Hide** option from the mini popup toolbar. Next, select the **Save** option from the **Edit** drop-down list; the **Save Display Elements** dialog box is displayed. Choose the **OK** button to save the changes in the selected layer state. You can repeat the above process for each layer state that you have added to the **View Manager** dialog box.

TUTORIALS

Tutorial 1

In this tutorial, you will create the model shown in Figure 8-57. The orthographic views of the model are shown in Figure 8-58. **(Expected time: 30 min)**

Figure 8-57 Solid model for Tutorial 1

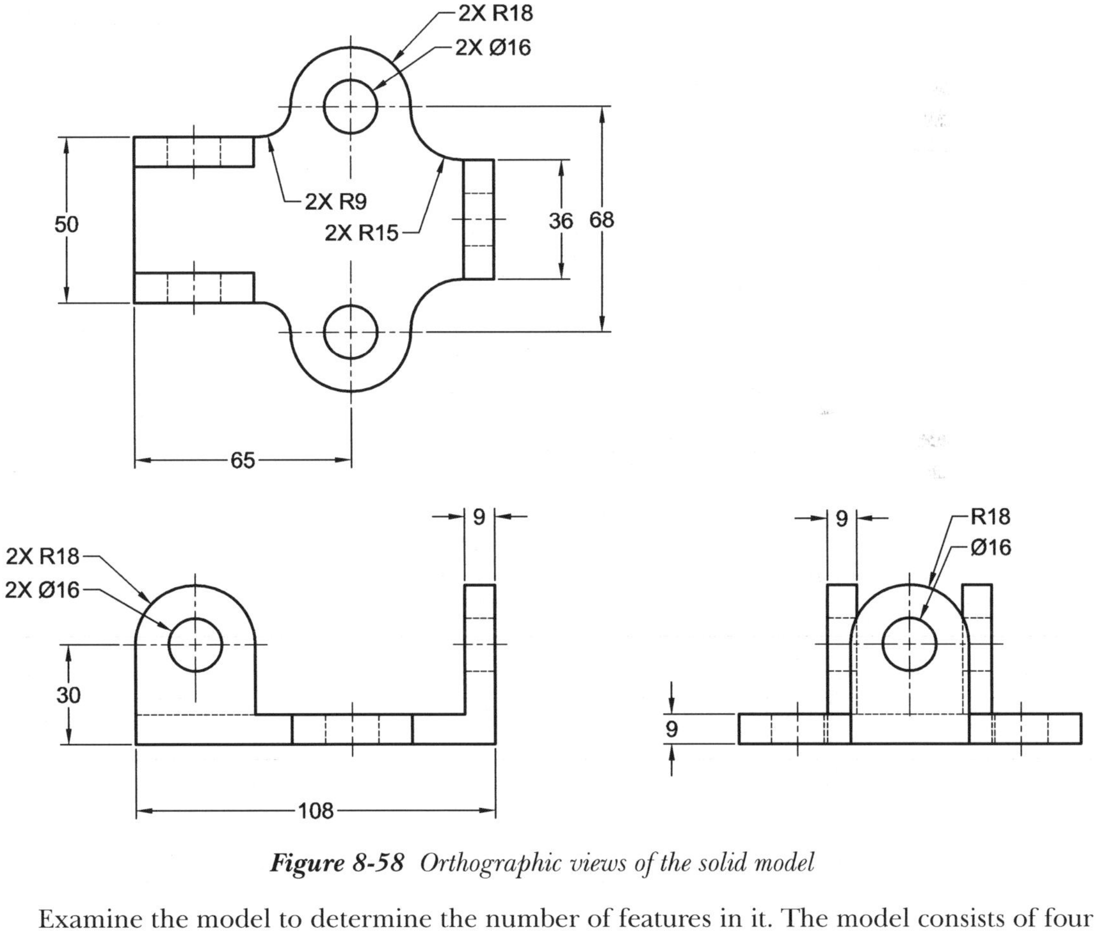

Figure 8-58 Orthographic views of the solid model

Examine the model to determine the number of features in it. The model consists of four features, refer to Figure 8-57.

The following steps are required to complete this tutorial:

a. Create the base feature on the **TOP** datum plane.
b. Create the second feature on the right face of the base feature.
c. Create the third feature.
d. Create the fourth feature by mirroring the third feature.

After starting Creo Parametric session, the first task is to set the working directory. Since it is the first tutorial of this chapter, you need to create a folder with the name *c08*, if it does not exist, and then set it as working directory.

Starting a New Object File

1. Start a new part file and then name it as *c08tut1*.

The three default datum planes are displayed in the drawing area. Also, the **Model Tree** is displayed in the drawing area.

Creating the Base Feature

To create a sketch for the base feature, you need to select the **TOP** datum plane as the sketching plane.

1. Choose the **Extrude** tool from the **Shapes** group of the **Model** tab; the **Extrude** dashboard is displayed above the drawing area.

2. Choose the **Placement** tab; the slide-down panel is displayed. From the slide-down panel, choose the **Define** button; the **Sketch** dialog box is displayed.

3. Select the **TOP** datum plane as the sketching plane. Next, select the **RIGHT** datum plane as the reference plane and set its orientation to **Right**, if not set by default.

4. Choose the **Sketch** button to enter the sketcher environment.

5. Next, create the sketch of the base feature and apply constraints and dimensions to it, as shown in Figure 8-59.

   Note that in the sketch, the bottom half of the sketch is mirrored to create the top half of the sketch. This is evident from the constraints of symmetry applied to the sketch in Figure 8-59.

6. After the sketch is completed, choose the **OK** button to exit the sketcher environment; the **Extrude** dashboard is enabled and displayed above the drawing area.

7. Enter **9** as the value of depth in the dimension box in the **Extrude** dashboard.

8. Choose the **OK** button from the **Extrude** dashboard.

Now, the base feature is completed and you need to create the second feature. The default trimetric view of the base feature is shown in Figure 8-60.

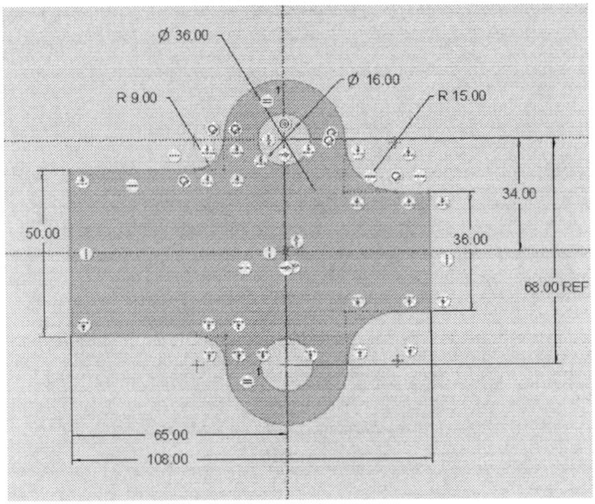

Figure 8-59 Sketch of the base feature

Figure 8-60 The default trimetric view of the base feature

Note

The two holes in the base feature are sketched while drawing the sketch for the base feature. These holes are integrated with the base feature. Therefore, the base feature is created as a single feature that includes two holes. The other method is to create the two holes separately on the base feature by using the ***Hole*** *dashboard. When you create the features separately, the total number of features created will be three.*

Creating the Second Feature

The second feature is also an extruded feature. You will create the second feature on the right face of the base feature. Therefore, you need to select the right face as the sketching plane.

1. Choose the **Extrude** tool from the **Shapes** group; the **Extrude** dashboard is displayed above the drawing area.
2. Choose the **Placement** tab to display the slide-down panel. Then, choose the **Define** button from the slide-down panel; the **Sketch** dialog box is displayed.
3. Select the right face of the base feature as the sketching plane.
4. If required, choose the **Flip** button to reverse the direction of the pink arrow.
5. Select the **TOP** datum plane as reference and then select the **Bottom** option from the **Orientation** drop-down list.
6. Choose the **Sketch** button from the **Sketch** dialog box to enter the sketcher environment.
7. Choose the **Line Chain** tool from the **Sketching** group and draw a vertical line starting from the edge of the base feature, as shown in Figure 8-61. Notice that the cursor automatically snaps the edge of the base feature while creating the vertical line.

8. Complete the sketch of the second feature and then add dimensions and constraints, as shown in Figure 8-61. After completing the sketch, choose the **OK** button from the **Close** group; the **Extrude** dashboard is enabled.

9. Enter **9** as the value of depth in the dimension box in the **Extrude** dashboard.

10. Choose the **OK** button from the **Extrude** dashboard; the default shaded trimetric view of the model after creating the second feature is shown in Figure 8-62.

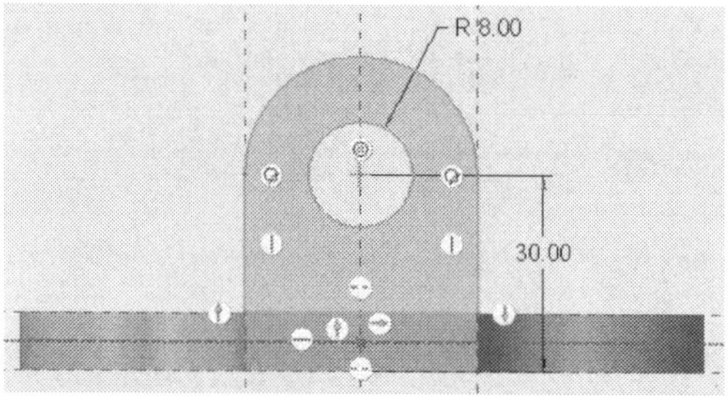

Figure 8-61 *Sketch of the second feature with dimensions and constraints*

Figure 8-62 *Model after creating the second feature*

Creating the Third Feature

Now, you need to create the third feature.

1. Select the second feature from the **Model Tree** and then choose the **Copy** tool from the **Operations** group.

2. Select the face shown in Figure 8-63 and then choose the **Paste** tool from the **Operations** group; the **Sketch** dialog box is displayed and you are prompted to select a reference, surface, or plane to define the view orientation.

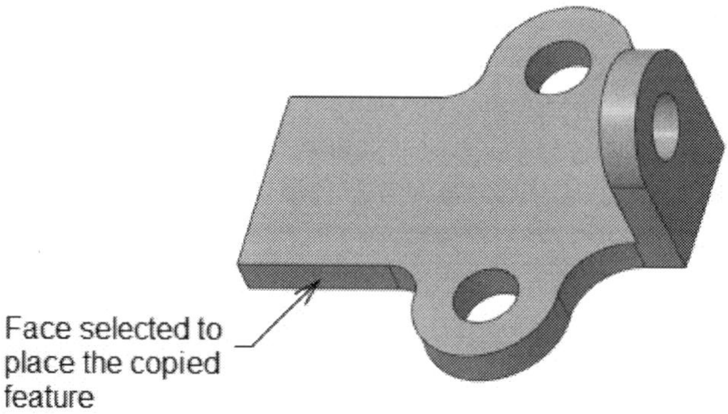

Figure 8-63 *Reference plane selected to paste the copied feature*

3. Choose the **Sketch** button; the sketch of the second feature is attached to the cursor, as shown in Figure 8-64.

4. Click anywhere in the drawing area to place the sketch. Notice that some weak dimensions and constraints are applied to the sketch.

5. Add required dimensions and constrains to the sketch and then choose the **OK** button from the **Close** group; the **Extrude** dashboard is enabled and displayed above the drawing area.

6. Click on the pink arrow in the drawing area to flip the direction of extrusion and then enter **9** as the value of depth in the dimension box in the **Extrude** dashboard. Next, choose the **OK** button from the **Extrude** dashboard. The model after creating the third feature is shown in Figure 8-65.

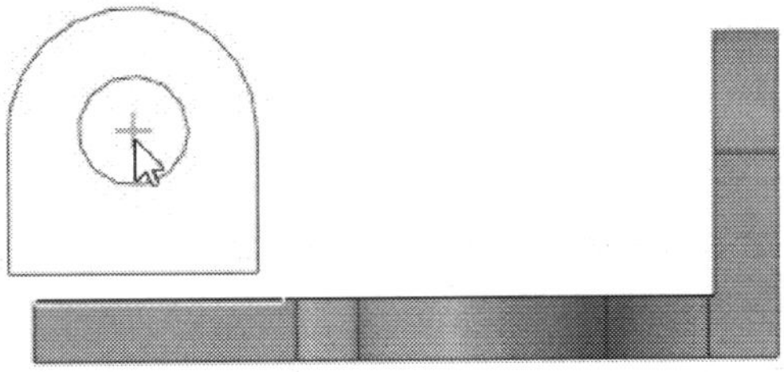

Figure 8-64 *Sketch of the second feature attached to the cursor*

Figure 8-65 *The model after creating the third feature*

Creating the Fourth Feature

The fourth feature can be created by sketching and extruding it to a given depth. You can also create this feature by placing a mirrored copy of the third feature at the required location. In this tutorial, you will use the second method because it consumes less time.

1. Select the third feature from the **Model Tree** and choose the **Mirror** tool from the **Editing** group of the **Model** tab; the **Mirror** dashboard is displayed and you are prompted to select the mirror plane.

2. Select the **FRONT** datum plane as the mirror plane.

3. Choose the **OK** button from the **Mirror** dashboard. The third feature is mirrored about the **FRONT** datum plane. The trimetric view of the final model is shown in Figure 8-66.

4. Choose the **Save** button from the **File** menu and save the model. The order of feature creation can be seen from the **Model Tree** shown in Figure 8-67. Note that the feature id numbers in your model may be different from the ones shown in this figure.

Figure 8-66 Default trimetric view of the model

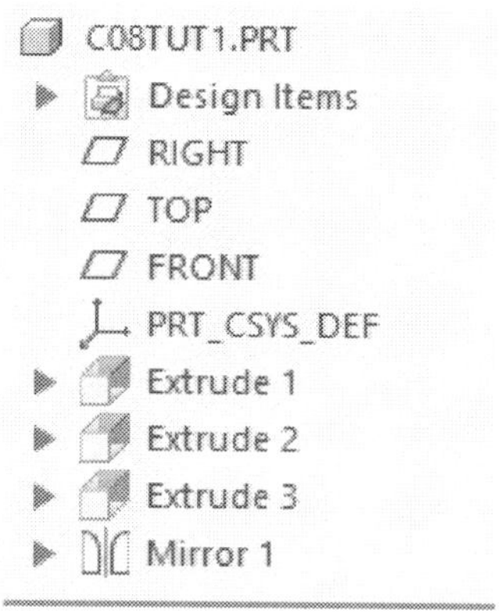

Figure 8-67 The ***Model Tree*** *for Tutorial 1*

Tutorial 2

In this tutorial, you will create the model shown in Figure 8-68. The orthographic views of the model are shown in Figure 8-69. **(Expected time: 45 min)**

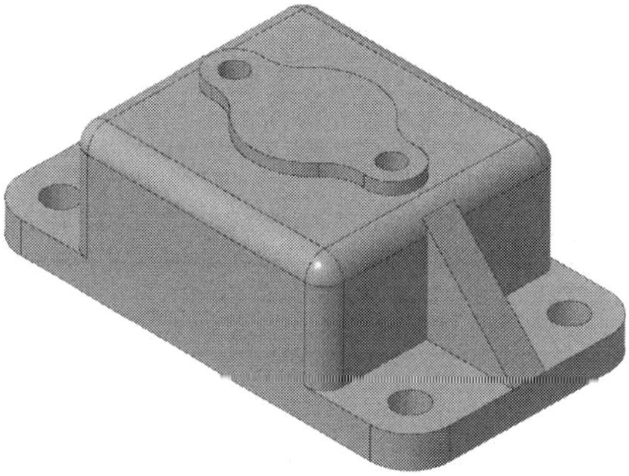

Figure 8-68 Solid model for Tutorial 2

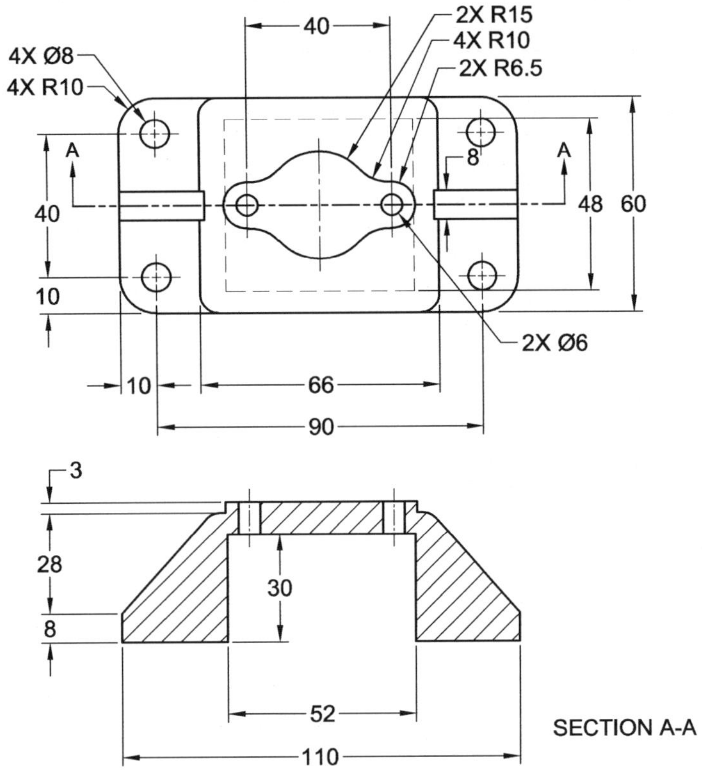

Figure 8-69 Top view and front section view of the model

Examine the model and determine the number of features in it. The model is composed of nine features, refer to Figure 8-68.

The following steps are required to complete this tutorial:

a. Create the base feature on the **FRONT** datum plane.
b. Create round features.
c. Create the hole feature and then pattern it.
d. Create the rib feature on the **FRONT** datum plane.
e. Mirror the rib feature.
f. Create the extrude feature on the top planar face of the base feature.
g. Create the cut feature on the bottom planar face of the base feature.
h. Create the hole feature on the top face of the model and then copy it.

Starting a New Object File

1. If required, set the working directory to the *c08* folder and start a new part file and then name it as *c08tut2*. The three default datum planes are displayed in the drawing area.

Creating the Base Feature

To create sketch for the base feature, first you need to select the sketching plane for the base feature. In this model, you need to draw the base feature on the **FRONT** datum plane because the direction of extrusion of this feature is perpendicular to it. The base feature will be created symmetric to the **FRONT** datum plane.

1. Choose the **Extrude** tool from the **Shapes** group; the **Extrude** dashboard is displayed above the drawing area.

2. Choose the **Placement** tab; the slide-down panel is displayed. Choose the **Define** button from the panel; the **Sketch** dialog box is displayed.

3. Select the **FRONT** datum plane as the sketch plane.

4. Select the **TOP** datum plane and then select the **Top** option from the **Orientation** drop-down list.

5. Choose the **Sketch** button to enter the sketcher environment.

6. Next, create the sketch of the base feature and apply constraints and dimensions to it, as shown in Figure 8-70.

   In the sketch, note that the base feature is symmetrical; therefore, a vertical center line is drawn and then the right half of the sketch is mirrored to create the left half. This is evident from the constraints of symmetry applied to the sketch in Figure 8-68. These constraints appear as symmetry symbols in the sketch.

   As evident from the sketch of the base feature shown in Figure 8-69, the **TOP** datum plane is aligned with the bottom line segment.

7. After the sketch is completed, choose the **OK** button to exit the sketcher environment; the **Extrude** dashboard is enabled and appears above the drawing area. Now, to create the base feature symmetrical to the **FRONT** datum plane, you need to extrude the sketch symmetrically on both sides of the sketching plane. This is because, later in the tutorial, you need to use the default datum planes as mirror planes for mirroring the features. On extruding symmetrically, you need not create the datum planes.

8. Choose the **Symmetric** button from the **Depth** drop-down in the **Extrude** dashboard.

9. Enter **60** as depth value in the dimension box in the **Extrude** dashboard.

10. Choose the **OK** button from the **Extrude** dashboard to exit the feature creation tool.

    The base feature is completed and now you need to create the second feature. The default trimetric view of the base feature is shown in Figure 8-71.

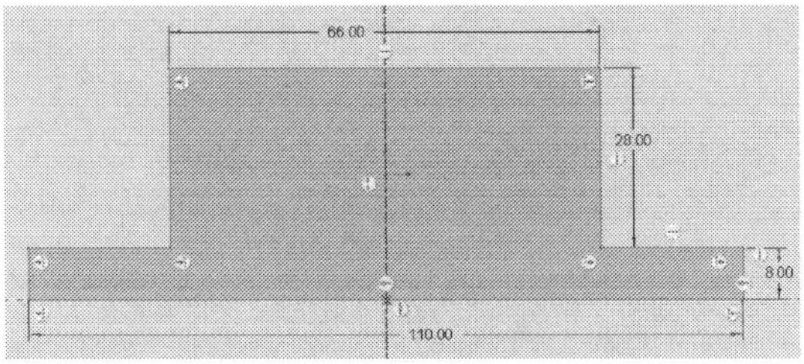

Figure 8-70 Sketch with dimensions and constraints for the base feature

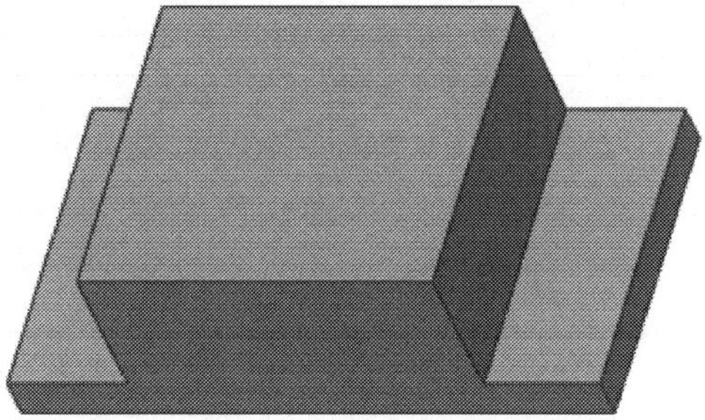

Figure 8-71 Default trimetric view of the base feature

Creating the Second Feature

The second feature is a round feature of radius 5.

1. Choose the **Round** tool from the **Engineering** group; the **Round** dashboard is displayed.

2. Choose the **Sets** tab to display the slide-down panel. Let this slide-down panel remain open so that you can view the selections made on the model.

3. Select one of the edges shown in Figure 8-72. To select the second edge and the subsequent edges, press the CTRL key and then select the edges one-by-one.

   A preview of the round is created on the selected edges; a default value of the radius is also displayed.

4. Double-click on the default radius value that is displayed in the preview of the round; an edit box appears. Enter **5** in it. The first set of rounds is created.

   Now, you need to create the second set of rounds.

5. After spinning the model, select the four vertical edges of the bottom portion of the base feature to apply round. Remember to use the CTRL key+left mouse button to select the second and subsequent edges.

   You will notice that **Set 1** and **Set 2** appear in the slide-down panel. This indicates that the second set is defined. A preview of the round is created on the edges. The default value of the radius is also displayed.

6. Double-click on the default radius value that is displayed along with the preview of the round; an edit box is displayed. Enter **10** in it.

7. Choose the **OK** button from the **Round** dashboard to create the round feature; the round feature is completed. The default trimetric view of the round feature is shown in Figure 8-73. In the **Model Tree**, the two rounds created appear as one feature.

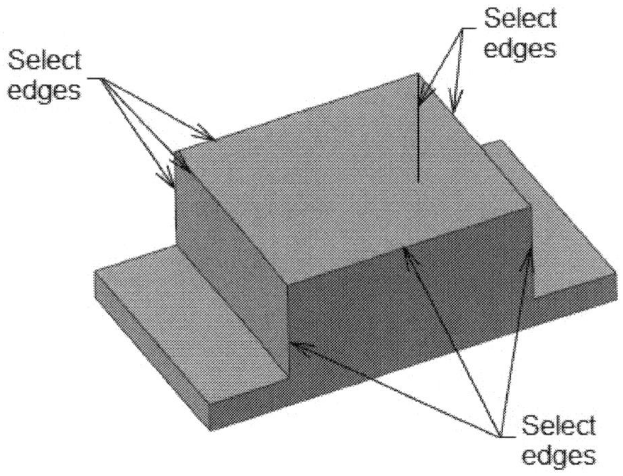

Figure 8-72 *Edges selected to create the round features*

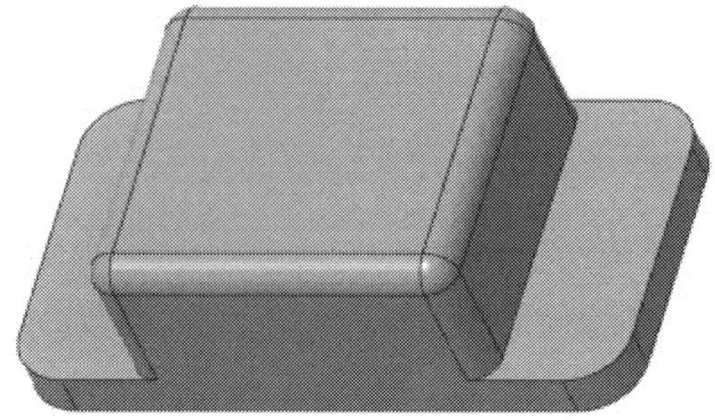

Figure 8-73 *The default trimetric view of the base feature with two sets of round feature*

Creating the Third Feature

The third feature is a through hole and needs to be created by using the **Hole** dashboard.

1. Choose the **Hole** tool from the **Engineering** group; the **Hole** dashboard is displayed. By default, the **Create simple hole** button is chosen in it.

2. Create the hole, as shown in Figure 8-74, by specifying the placement parameters. Refer to Figure 8-69 for the placement parameters.

Creating a Pattern of the Hole Feature

As evident from Figure 8-68, you need to create four instances of the hole. First instance will be created by using the **Hole** dashboard and the remaining three instances will be created by using the **Pattern** tool. You will create a rectangular pattern of the hole feature. You can also create all the holes by using the **Hole** dashboard and specifying the placement parameters for each of them. However, to save time, it is recommended that you create a pattern of the hole.

1. Select the hole feature and then choose the **Pattern** tool from the **Editing** group; the **Pattern** dashboard is displayed. Also, you are prompted to select dimensions to be changed in the first direction.

2. Make sure the **General** option is selected in the **Options** slide-down panel.

   You cannot use the **Identical** option to create rectangular pattern of the hole feature. This is because when you use the **Identical** option, the pattern cannot intersect the base feature on which the hole is created. If the top portion of the base feature is created as a separate feature, the hole can be patterned by using the **Identical** option.

3. You need to select the dimension value **10** from the drawing area. Since the two dimensions displayed in the drawing area have the dimension value 10, select the dimension that is

along the shorter side of the base feature. After you have selected the dimension in the first direction, the edit box is displayed. Enter **40** in it.

4. Hold the right mouse button to display a shortcut menu. Choose the **Direction 2 Dimensions** option from the shortcut menu.

5. Select the dimension value **10** that is along the longer side of the base feature. After you have selected the dimension in the second direction, the edit box is displayed. Enter **90** in the edit box.

   Note that the number of instances specified in the edit boxes present on the dashboard is 2 by default.

6. Choose the **OK** button from the **Pattern** dashboard; the rectangular pattern of the hole is displayed.

   You can use the middle mouse button to spin and display the model, as shown in Figure 8-75.

Figure 8-74 *The hole feature on the base feature*

Figure 8-75 *Model after creating the hole pattern*

Note

If multiple holes need to be created on a model, it is recommended that you create their pattern, if possible. This is because when you assemble bolts in these holes in the ***Assembly*** *mode, it becomes very easy to assemble them using the reference pattern.*

Creating the Rib Feature

The sketch of the rib feature will be created on the **FRONT** datum plane and the required thickness will be applied to it. This is the fourth feature of the model.

1. Choose the **Profile Rib** tool from the **Rib** drop-down in the **Engineering** group; the **Profile Rib** dashboard is displayed.

2. Choose the **References** tab to invoke the slide-down panel. Choose the **Define** button from the slide-down panel; the **Sketch** dialog box is displayed.

3. Select the **FRONT** datum plane as the sketching plane.

4. Select the **TOP** datum plane from the drawing area and then select the **Top** option from the **Orientation** drop-down list.

5. Choose the **Sketch** button to enter the sketcher environment.

6. Draw the sketch of the rib feature, as shown in Figure 8-76, and then exit the sketcher environment.

Note

In Figure 8-76, the top end of the inclined line in the sketch is aligned with the curve and the tangent constraint is applied to the line and the curve. Similarly, the bottom end of the inclined line is also aligned with the two edges. This is the reason, there are no dimensions in the sketch and the sketch for the rib feature is fully constrained.

7. Choose the **Flip** button from the **References** slide-down panel if the desired direction is not obtained.

8. Specify **8** as the rib thickness in the dimension box in the **Rib** dashboard. Choose the **OK** button to create the rib feature.

   The rib feature created is shown in Figure 8-77. You can use the middle mouse button to orient the model to view the correct placement of the rib.

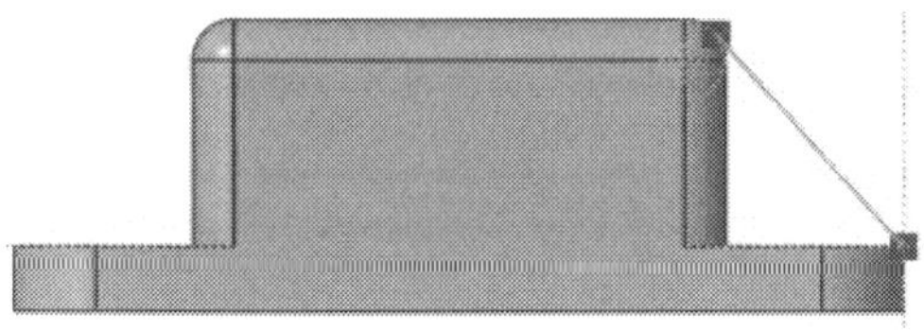

Figure 8-76 *Fully constrained sketch for the rib*

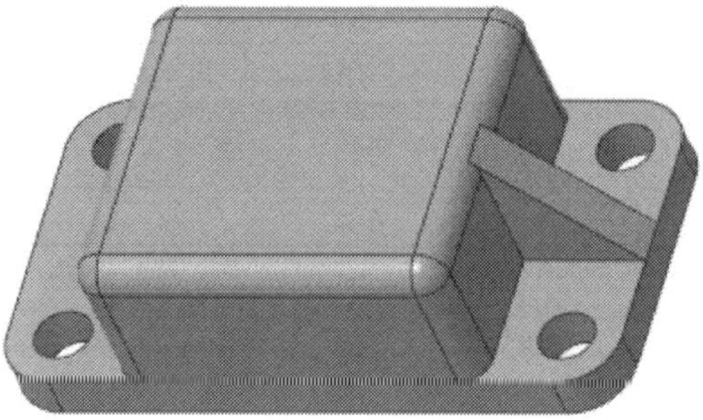

Figure 8-77 *Model after creating the rib feature*

Mirroring the Rib Feature

Now, you need to create another rib feature, as shown in Figure 8-78. You can create this feature either by mirroring or by creating the sketch of the rib feature on the sketching plane. Here, you will create the rib feature by mirroring. The rib feature will be mirrored about the **RIGHT** datum plane.

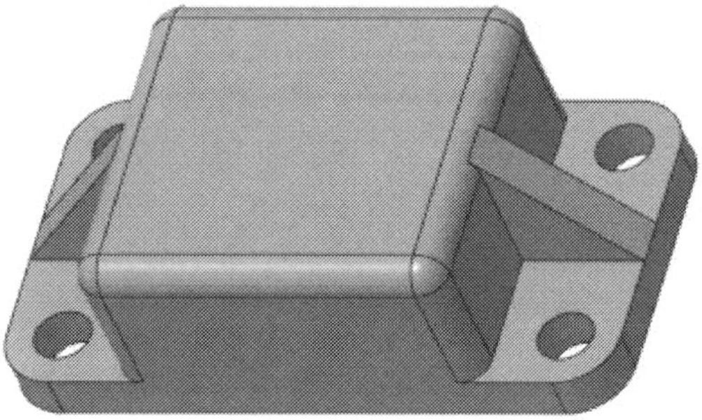

Figure 8-78 *Model after mirroring the rib feature*

1. Select the rib feature that you need to mirror. Then, choose the **Mirror** tool from the **Editing** group of the **Model** tab; the **Mirror** dashboard is displayed.

2. Choose the **References** tab to invoke the slide-down panel. In this panel, the **Mirror Plane** collector is selected by default. Select the **RIGHT** datum plane as the mirror plane.

3. Choose the **OK** button from the dashboard; the selected feature is mirrored about the **RIGHT** datum plane, refer to Figure 8-78.

Creating the Protrusion Feature

The sixth feature is an extruded feature. You need to create the extruded feature on the top face of the base feature.

1. Choose the **Extrude** tool from the **Shapes** group; the **Extrude** dashboard is displayed.

2. Invoke the **Sketch** dialog box.

3. Select the top face of the base feature as the sketching plane.

4. Select the **RIGHT** datum plane and then select the **Right** option from the **Orientation** drop-down list.

5. Choose the **Sketch** button to enter the sketcher environment.

6. Next, draw the sketch of the extruded feature, as shown in Figure 8-79. The tangent and equal radii constraints are applied to the sketch. Also, the center of the top arc and the bottom arc coincide with the intersection of the two datum planes.

7. After the sketch is complete, choose the **OK** button to exit the sketcher environment; the **Extrude** dashboard is enabled and it appears above the drawing area.

8. Enter **3** as the depth value in the dimension box in the **Extrude** dashboard. Choose the **OK** button from the **Extrude** dashboard; the extruded feature is completed and the default trimetric view is shown in Figure 8-80.

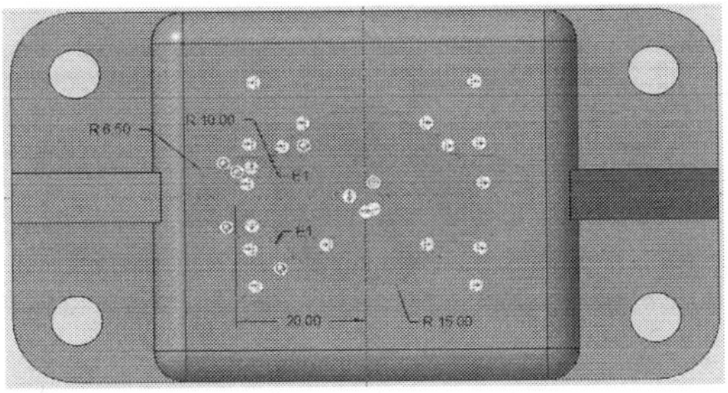

Figure 8-79 *Sketch with dimensions and constraints for the extruded feature*

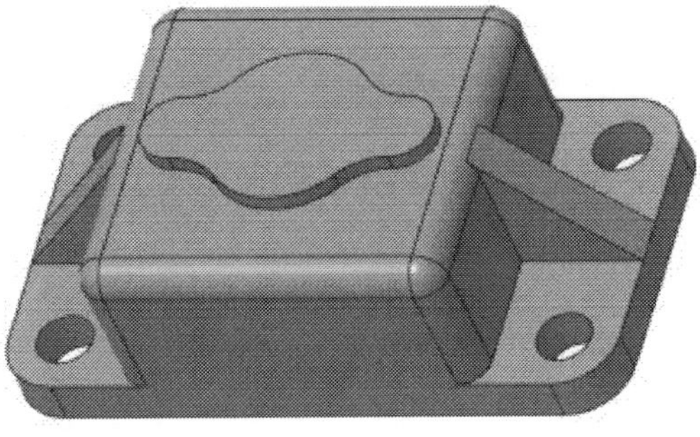

Figure 8-80 *The default trimetric view after creating the extruded feature*

Creating the Cut Feature

You need to create an extruded cut on the bottom planar surface of the base feature. This is the seventh feature of the model.

1. Choose the **Extrude** tool from the **Shapes** group.

2. Choose the **Remove Material** button from the **Extrude** dashboard.

3. Choose the **Placement** tab to invoke the slide-down panel. Choose the **Define** button from the slide-down panel; the **Sketch** dialog box is displayed.

4. Select the bottom face of the base feature as the sketching plane.

5. Select the **RIGHT** datum plane and then select the **Right** option from the **Orientation** drop-down list in the **Sketch** dialog box.

6. Choose the **Sketch** button to enter the sketcher environment.

7. Next, draw the sketch of the cut feature and dimension it, as shown in Figure 8-81.

8. After completing the sketch, choose the **OK** button; the **Extrude** dashboard is enabled and it appears above the drawing area.

9. Enter the value **30** in the dimension box of the **Extrude** dashboard.

10. Choose the **OK** button from the dashboard to create the cut feature. You can spin the model by using the middle mouse button. The cut feature is shown in Figure 8-82.

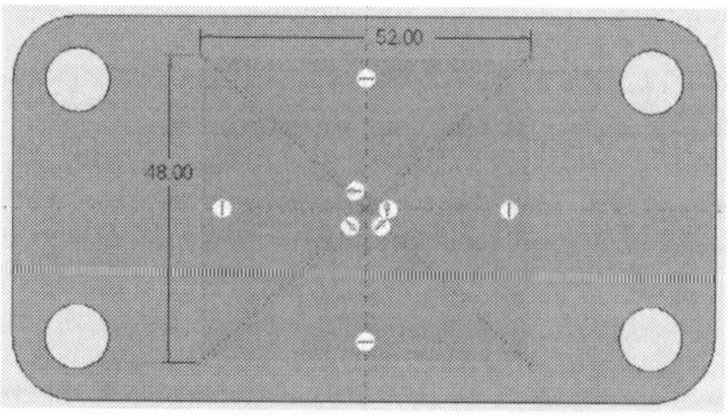

Figure 8-81 *Sketch of the cut feature with dimensions*

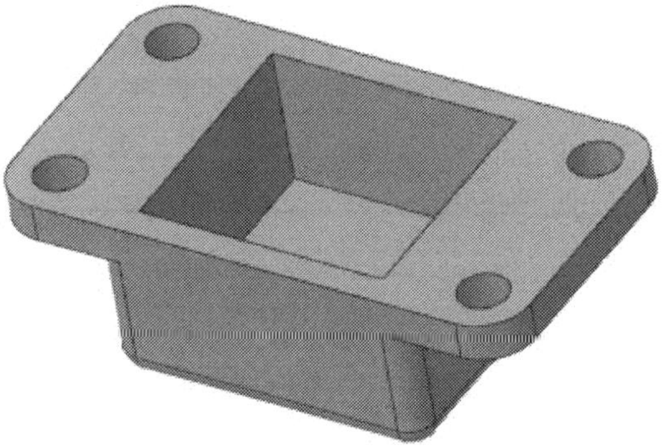

Figure 8-82 *Model after creating the cut feature*

Creating the Hole

Next, you need to create a through hole on the top extruded feature by using the **Hole** dashboard.

1. Invoke the **Hole** dashboard; the **Simple** button is chosen by default in it. Specify the placement parameters of the hole as given in Figure 8-69 and create the hole. The model after creating the hole is shown in Figure 8-83.

Mirroring the Hole Feature

Now, you need to mirror the hole feature, as shown in Figure 8-84. This is the ninth feature of the model. You will mirror the hole feature about the **RIGHT** datum plane.

1. Select the hole feature to be mirrored. Then, choose the **Mirror** tool from the **Editing** group of the **Model** tab; the **Mirror** dashboard is displayed. Choose the **References** tab to invoke the slide-down panel. Next, select the **RIGHT** datum plane as the mirror plane.
2. Choose the **OK** button from the dashboard; the selected feature is mirrored about the **RIGHT** datum plane, as shown in Figure 8-84.

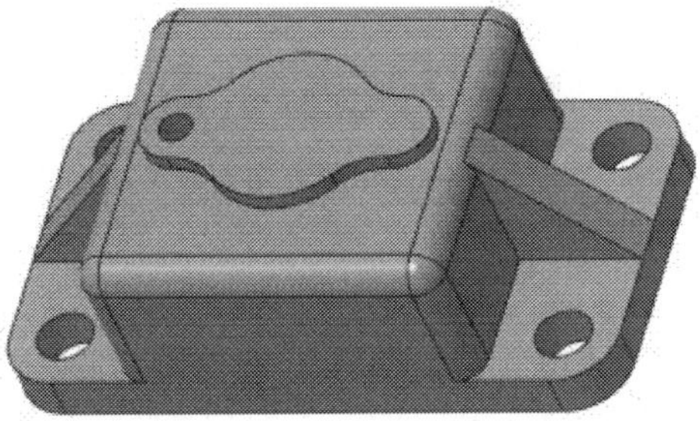

Figure 8-83 *The model after creating hole*

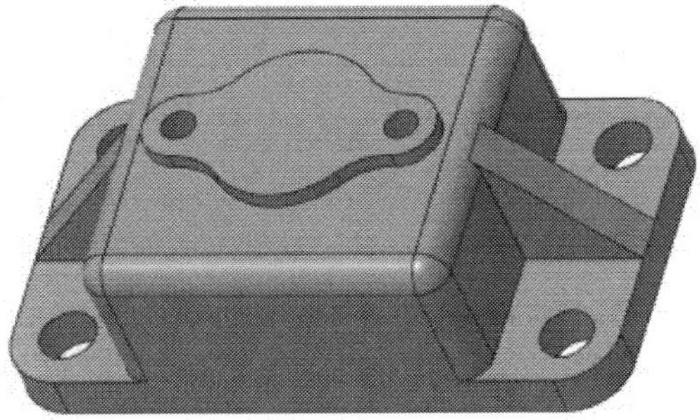

Figure 8-84 *The mirrored hole feature*

Saving the Model

You need to save the model to use it later.

1. Choose the **Save** button from the **File** menu and save the model.

The order of feature creation of the model can be seen in the **Model Tree**, as shown in Figure 8-85.

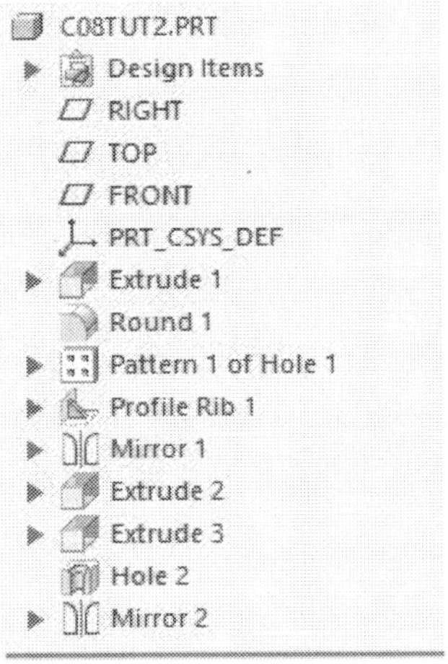

Figure 8-85 *The* ***Model Tree*** *for Tutorial 2*

Tutorial 3

In this tutorial you will create a UDF of the feature shown in Figure 8-86. Next, you will open the model shown in the Figure 8-87 and place the UDF. The resulting model after placing the UDF is shown in Figure 8-88. **(Expected time: 15 min)**

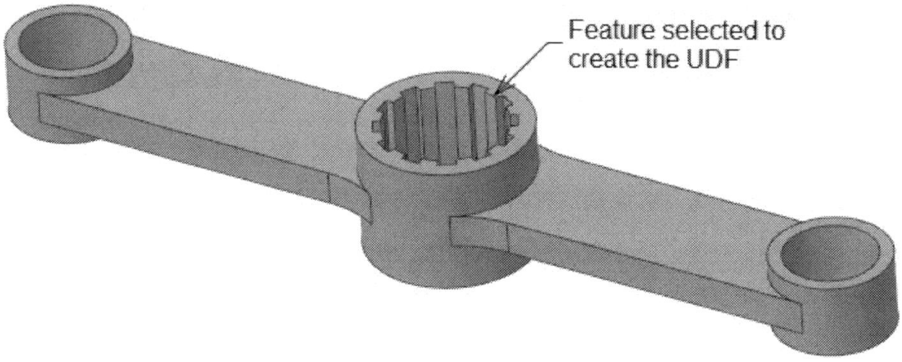

Figure 8-86 *Feature selected in the source model to create the UDF*

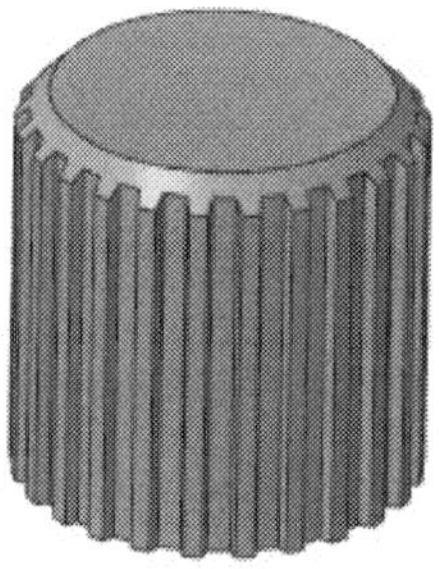

Figure 8-87 *Target model without UDF*

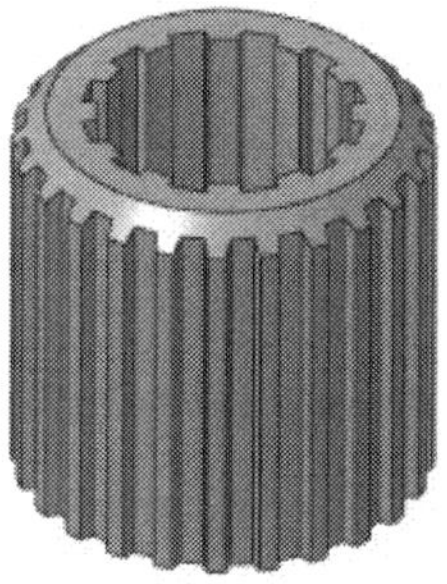

Figure 8-88 *Target model with UDF*

To perform this tutorial, you need to download the zipped file named as *c08_creo_8.0_input* from the Input Files section of the CADCIM website. The complete path for downloading the file is:

Textbooks > CAD/CAM > PTC Creo Parametric > Creo Parametric 8.0 for Designers, 8.0th Edition > Input Files > C08_input

The following steps are required to complete this tutorial:

a. Open the source model and create the UDF of the feature.
b. Specify variable dimensions in the UDF.
c. Open the target model, place the UDF, and modify the dimensions of the UDF.

When you start the Creo Parametric session, the first task is to set the working directory. Make sure the required working directory is selected.

Opening the File

1. Start a new session of Creo Parametric.

2. Choose the **Open** button in the **Home** tab; the **File Open** dialog box will be displayed. Browse to the location where you have stored the downloaded file. Select the *C08tut03_source.prt* file and choose the **Open** button in the dialog box to open the model in the part environment.

Creating the UDF

1. Choose the **UDF Library** tool in the **Utilities** group of the **Tools** tab; the **Menu Manager** will be displayed in the drawing area.

2. Choose the **Create** option in the **Menu Manager**; the UDF name edit box is displayed above the drawing area. Type C08_TUT03_UDF in the edit box and choose the **Accept value** button.

3. Choose the **Subordinate** option in the **UDF OPTIONS** rollout of the **Menu Manager** and then choose the **Done** option; the **UDF: C08_TUT03_UDF, Subordinate** dialog box along with the **Select** selection box is displayed, as shown in Figure 8-89 and Figure 8-90. Also, you are prompted to select features to add to the UDF.

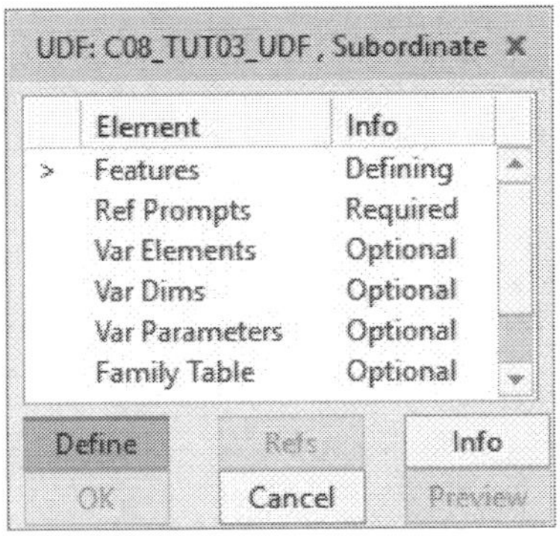

Figure 8-89 *The* ***UDF*** *dialog box*

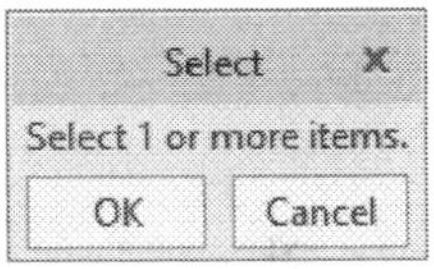

Figure 8-90 *The* ***Select*** *selection box*

4. Select the **Extrude 3** feature from the **Model tree** or from the drawing area and then choose the **Done** option in the **SELECT FEAT** rollout. Next, choose the **Done/Return** option in the **UDF FEATS** rollout; the **Enter prompt for the surface in reference color** edit box is displayed above the drawing area. Also, you will notice that the first placement reference of the feature is highlighted on the model.

5. Specify the prompts for the placement of the UDF in the respective edit boxes and choose the **Done/Return** option in the **Menu Manager**. Notice that the reference for which you are defining the prompt is highlighted in the drawing area.

6. Choose the **Var Dims** option from the **UDF: C08_TUT03_UDF, Subordinate** dialog box and then choose the **Define** button; the **VAR DIMS** rollout is activated in the **Menu Manager**.

7. Select a dimension from the drawing area, and choose the **Done/Return** option in the **ADD DIMS** rollout of the **Menu Manager**. You can add dimensions to be varied by choosing the **Add** option in the **VAR DIMS** rollout of the **Menu Manager**.

   The dimensions to be changed are shown in Figure 8-91.

8. Choose the **Done/Return** option in the **ADD DIMS** rollout to finish adding the dimensions.

9. Choose the **Done/Return** option in the **VAR DIMS** rollout; the **Enter prompt for the dimension value** edit box is displayed above the drawing area. Type the prompt in the edit box and choose the **OK** button. Note that you have to specify a prompt for each dimension that you have selected to become variable.

10. Choose the **OK** button in the **UDF: C08_TUT03_UDF, Subordinate** dialog box and then choose the **Done/Return** option in the **Menu Manager** to finish the UDF creation. Note that the UDF created will be saved in the current working directory. However, you can select any folder to save the UDF file.

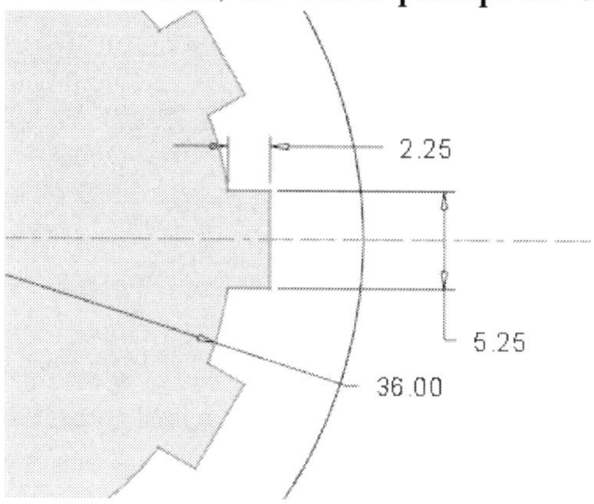

Figure 8-91 *Dimensions to be changed in UDF*

Placing the UDF

To place a UDF you must open the target model first. The original source model is also required for the placement of the UDF. The steps to place a UDF in the target model is explained next.

1. Without closing the source model, open the downloaded model file *C08tut03_target.prt* in a new window by choosing the **Open** button in the **File** menu.

2. Choose the **User-Defined Feature** tool from the **Get Data** group of the **Model** tab; the **Open** dialog box is displayed.

3. Browse to the location where you have saved the UDF file. Select the *C08tut03_udf.gph* file and choose the **Open** button in the **Open** dialog box; the **Insert User-Defined Feature** dialog box is displayed.

4. Choose the **OK** button from the **Insert User-Defined Feature** dialog box; the **User Defined Feature Placement** dialog box is displayed and preview of the UDF is highlighted in the drawing area, refer to Figure 8-92.

   The required references for the placement of the UDF are listed under the **References of Original Features** area.

5. Select the bottom face, right plane, and front plane from the drawing area by using the **References of UDF Features** collector in the **User Defined Feature Placement** dialog box; the UDF is placed on the target model, as shown in Figure 8-93.

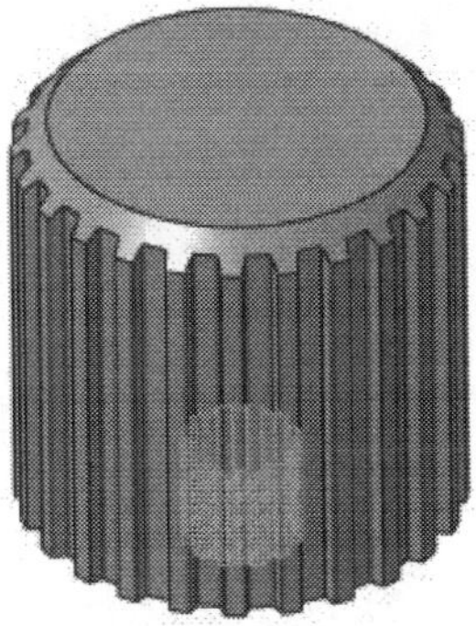

Figure 8-92 *Preview of the UDF on the target model*

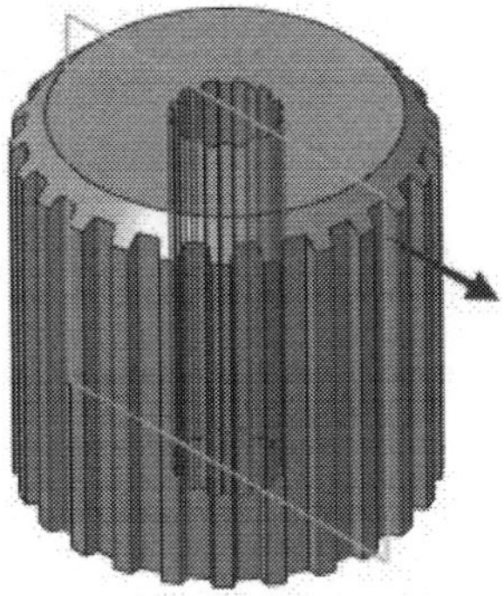

Figure 8-93 *UDF placed on the target model*

6. Choose the **Variables** tab and click in the **Value** column of the **Extrude 3** feature and specify **90** as the diameter of the feature. Next, enter **5.5** in the slot height and **15** in the slot width value column.

7. Choose the **OK** button in the **User Defined Feature Placement** dialog box; the UDF is placed on the target model. The resulting model after placing the UDF is shown in Figure 8-94. Also, a local group is added to the **Model Tree** with the same name as the UDF.

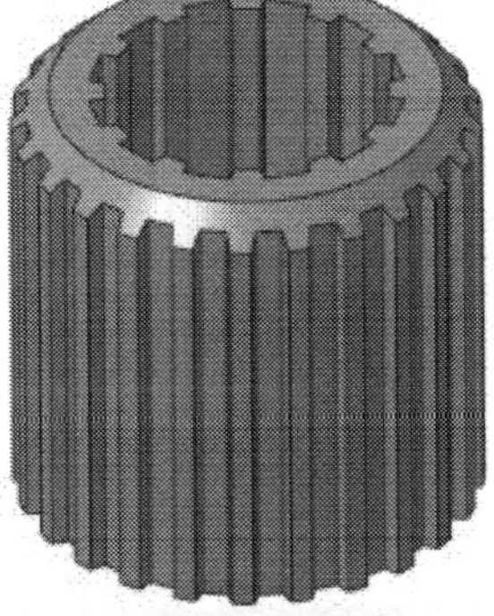

Figure 8-94 *Resulting model after placing the UDF*

8. Choose the **Save** button from the **File** menu and save the model.

Tutorial 4

In this tutorial, you will create the model shown in Figure 8-95. The orthographic views of the model are shown in Figure 8-96. **(Expected time: 30 min)**

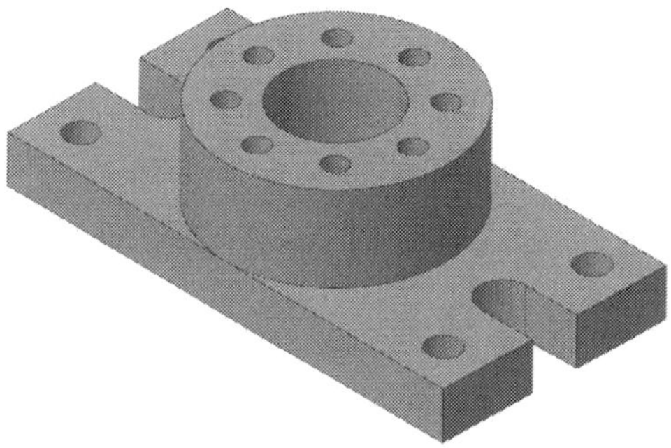

Figure 8-95 Isometric view of the solid model

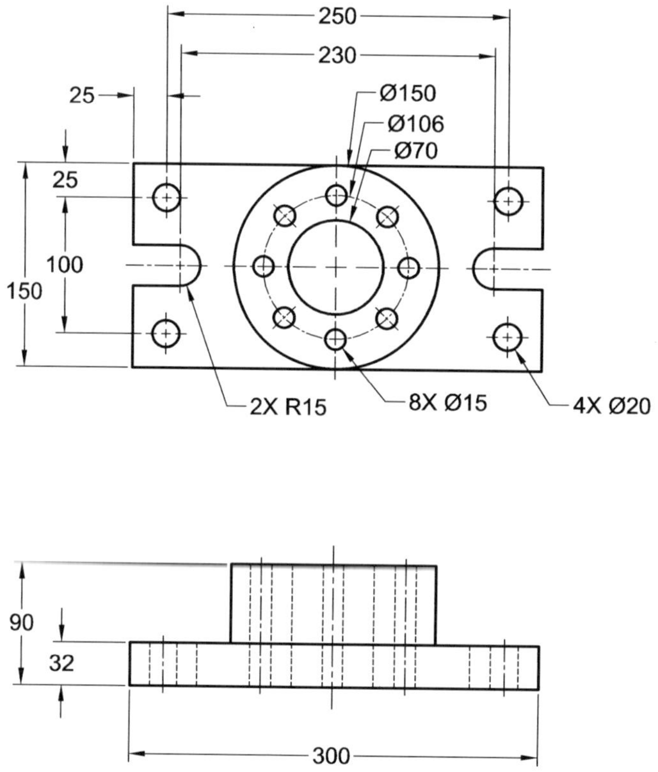

Figure 8-96 Top and front views of the solid model

Examine the model and determine the number of features in it. The model consists of five features, refer to Figure 8-95.

The following steps are required to complete this tutorial:

a. Start a new file and create the base feature on the **TOP** datum plane.
b. Create the cylindrical extrude feature.
c. Create the hole feature coaxial with the cylindrical feature.

d. Create the hole feature on the top planar surface of the base feature and then pattern it.
e. Create the hole feature on the top planar surface of the cylindrical feature.
f. Create a rotational pattern of this hole.

If required, set the working directory to the *c08* folder.

Starting a New Object File

1. Start a new part file and name it as *c08tut4*.

The three default datum planes are displayed in the drawing area.

Creating the Base Feature

To create sketch for the base feature, you first need to select the sketching plane for the base feature. In this model, you need to draw the base feature on the **TOP** datum plane. This is because the direction of extrusion of this feature is perpendicular to the **TOP** datum plane.

1. Choose the **Extrude** tool from the **Shapes** group; the **Extrude** dashboard is displayed.

2. Invoke the **Sketch** dialog box by using the **Extrude** dashboard.

3. Select the **TOP** datum plane as the sketching plane.

4. Select the **RIGHT** datum plane and then select the **Right** option from the **Orientation** drop-down list, if it is not selected by default.

5. Choose the **Sketch** button to enter the sketcher environment.

6. Create the sketch of the base feature and apply required constraints and dimensions to it, as shown in Figure 8-97. For better visibility of the sketch, you can turn off the display of constraints.

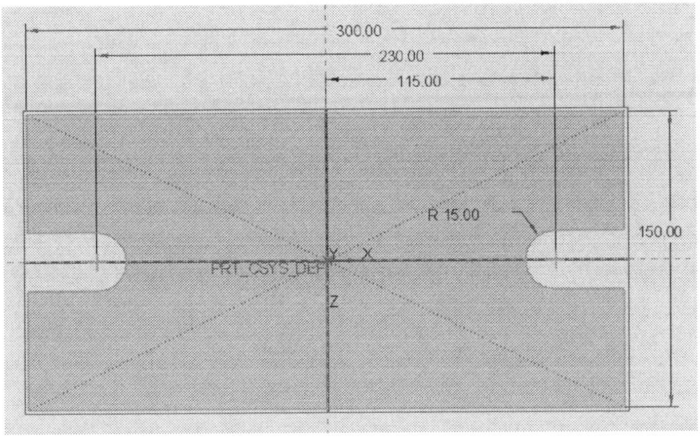

Figure 8-97 *Sketch of the base feature with dimensions and constraints*

7. After the sketch is complete, choose the **OK** button to exit the sketcher environment; the **Extrude** dashboard is enabled and displayed above the drawing area.

8. Enter **32** as the depth value in the dimension box of the **Extrude** dashboard and then choose the **OK** button from the dashboard.

   The base feature is completed and now you need to create the second feature. The default trimetric view of the base feature is shown in Figure 8-98.

***Figure 8-98** Default trimetric view of the base feature*

Creating the Second Feature

The second feature is also an extruded feature. You need to create this feature on the top face of the base feature. Therefore, you need to define the top face as the sketching plane for the second feature.

1. Choose the **Extrude** tool from the **Shapes** group; the **Extrude** dashboard is displayed.

2. Invoke the **Sketch** dialog box by using the **Extrude** dashboard.

3. Select the top face of the base feature as the sketching plane.

4. Select the **RIGHT** datum plane and then select the **Right** option from the **Orientation** drop-down list, if it is not selected by default.

5. Choose the **Sketch** button to enter the sketcher environment.

6. Create the sketch of the second feature and then dimension it, as shown in Figure 8-99.

7. After creating the sketch, choose the **OK** button; the **Extrude** dashboard is enabled.

8. Enter **58** in the dimension box of the **Extrude** dashboard and choose the **OK** button. The model similar to the one shown in Figure 8-100 is displayed in the drawing area.

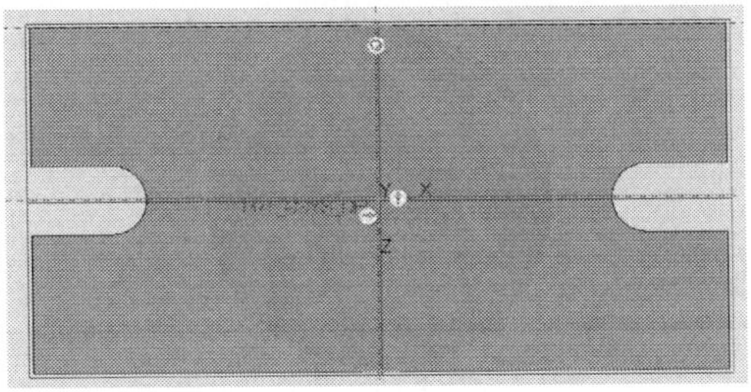

Figure 8-99 Sketch of the cylindrical feature with diameter of the cylinder

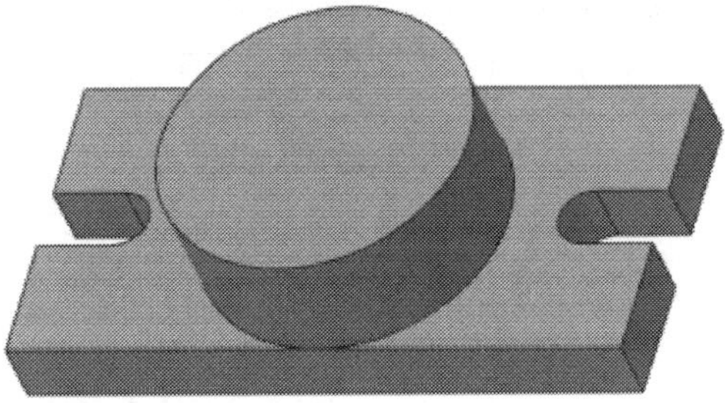

Figure 8-100 Default trimetric view of the cylindrical feature

Creating the Third Feature

The third feature is a through hole that is coaxial to the cylindrical feature. You need to create the hole feature by using the **Hole** dashboard.

1. Choose the **Hole** tool from the **Engineering** group; the **Hole** dashboard is displayed. By default, the **Simple** button is chosen in the **Hole** dashboard.

2. Create a hole of diameter **70**, as shown in Figure 8-101. Refer to Figure 8-96 for specifying the placement parameters.

Creating the Fourth Feature

The fourth feature is a through hole. You need to create this feature on the top planar surface of the base feature by using the **Hole** dashboard.

1. Choose the **Hole** tool from the **Engineering** group; the **Hole** dashboard is displayed. By default, the **Simple** button is chosen in the **Hole** dashboard.

2. Create the hole, as shown in Figure 8-102, by specifying the placement parameters (refer to Figure 8-96).

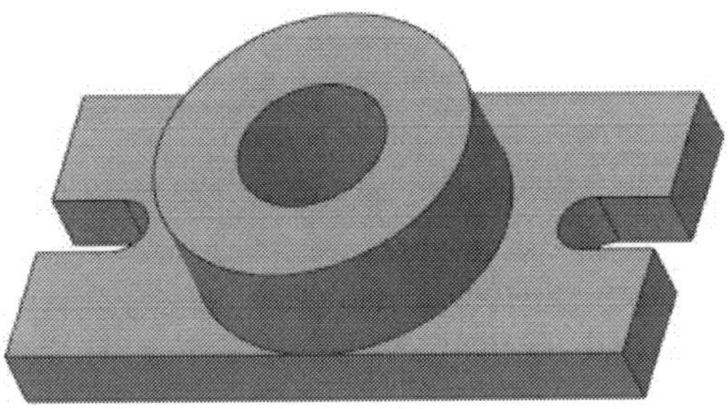

Figure 8-101 Coaxial hole on the cylindrical feature

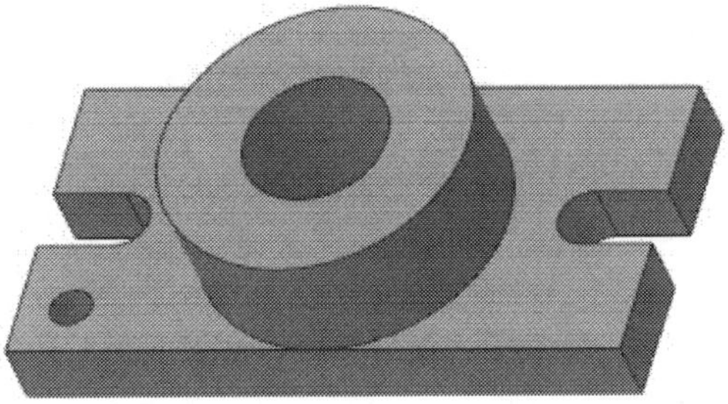

Figure 8-102 Hole on the base feature

Patterning the Hole Feature

Next, you need to create a rectangular pattern of the hole feature that is created on the base feature. You can also create individual holes but it is time-consuming and increases the number of features. Therefore, it is recommended that you create a rectangular pattern of the hole feature.

1. Select the hole feature and then choose the **Pattern** tool from the **Editing** group; the **Pattern** dashboard is displayed and you are prompted to select the dimensions to be changed in the first direction.

2. Select the **Identical** option from the **Options** slide-down panel.

   Here you need to use the **Identical** option because the feature on which the pattern is created does not intersect the pattern.

3. Select the dimension value **25** from the drawing area. Since both the dimensions displayed in the drawing area have the value **25**, select the dimension **25** that is along the shorter side of the base feature. After you have selected the dimension in the first direction, an edit box is displayed. Enter **100** in the edit box.

4. Hold down the right mouse button to display the shortcut menu. Choose the **Direction 2 Dimensions** option from the shortcut menu.

5. Select the dimension value **25** that is along the longer side of the base feature; an edit box is displayed.

6. Enter **250** in the edit box. Note that the number of instances, **2**, is specified by default in the instances edit boxes of the dashboard.

7. Choose the **OK** button from the **Pattern** dashboard; a rectangular pattern of the hole feature is displayed, as shown in Figure 8-103.

Creating a Hole on the Cylindrical Feature

You need to create a hole on the cylindrical feature diametrically by using the **Hole** dashboard.

1. Choose the **Hole** tool from the **Engineering** group; the **Hole** dashboard is displayed. By default, the **Simple** tool is chosen in the **Hole** dashboard.

2. Choose the **Placement** tab from the **Hole** dashboard to display the slide-down panel.

3. Select the top face of the cylindrical feature as the placement plane.

4. From the drop-down list in the slide-down panel, select the **Diameter** option.

5. Click in the **Offset References** collector and select the axis of the cylindrical feature.

6. Enter **106** in the second dimension box on the right of the **Diameter** option in the **Offset References** collector.

7. Use the CTRL key+left mouse button and select the **FRONT** datum plane from the drawing area. Enter the value **90** in the dimension box of the **Offset References** collector.

8. Enter **15** as the diameter of the hole in the diameter dimension box in the **Hole** dashboard.

9. Choose the **Drill to intersect with all surfaces** option from the depth flyout in the **Depth** area of the **Hole** dashboard.

10. Choose the **OK** button from the **Hole** dashboard; the hole is created, as shown in Figure 8-104.

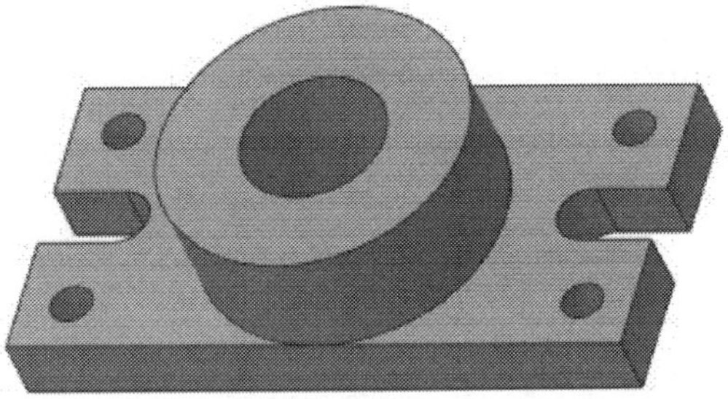

Figure 8-103 *Rectangular pattern of the hole feature*

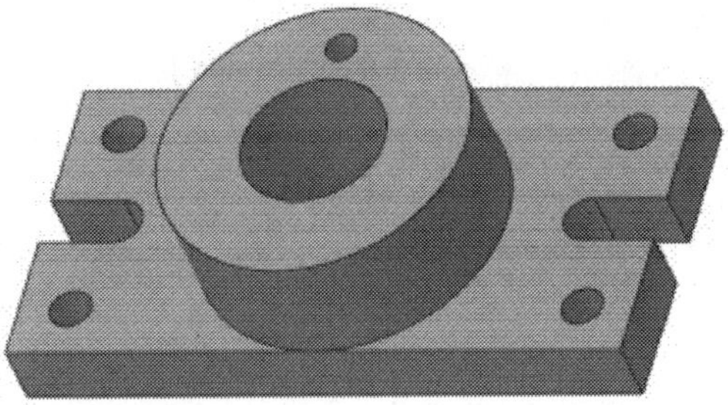

Figure 8-104 *Diametrical hole on the cylindrical feature*

Creating the Rotational Pattern of the Hole Feature

As the creation of the remaining holes individually on the cylindrical feature is time-consuming, you need to create a rotational pattern of the hole feature.

1. Select the hole feature and then choose the **Pattern** tool from the **Editing** group; the **Pattern** dashboard is displayed. Also, the dimensions of the hole feature are displayed in the drawing area and you are prompted to select dimensions to vary in the first direction.

2. Select the **Identical** option from the **Options** slide-down panel.

3. Select the angular dimension **90** from the model; an edit box is displayed.

4. Enter **45** in the edit box and press ENTER. Now, you need to specify the number of instances of the hole feature in the pattern.

5. Enter **8** in the **Members** edit box of **1st direction** and press ENTER. Choose the **OK** button from the dashboard. The rotational pattern is created and the model is completed, as shown in Figure 8-105. You can see the order of feature creation in the **Model Tree**, as shown in Figure 8-106.

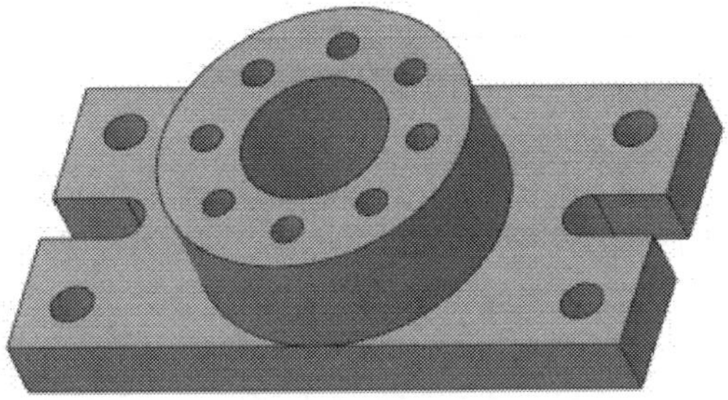

Figure 8-105 *The complete model*

Figure 8-106 *The* ***Model Tree*** *for Tutorial 4*

Saving the Model

You need to save the model because you may need it later.

1. Choose the **Save** button from the **File** menu and save the model.

Tutorial 5

In this tutorial, you will create the model of the cylinder head shown in Figure 8-107. Figure 8-108 shows the top view and the front section view of the model. **(Expected time: 45 min)**

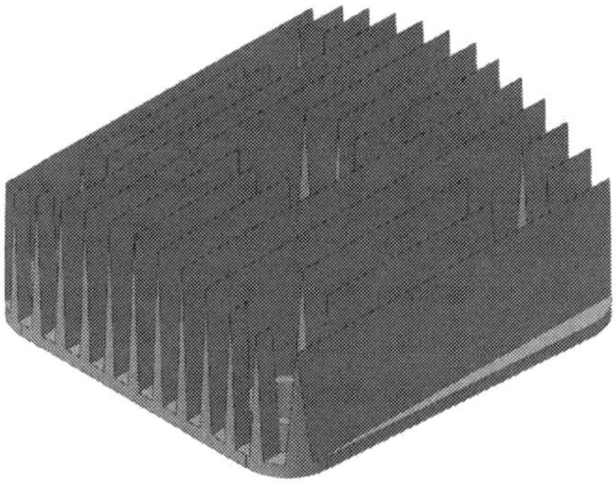

Figure 8-107 *Isometric view of the model*

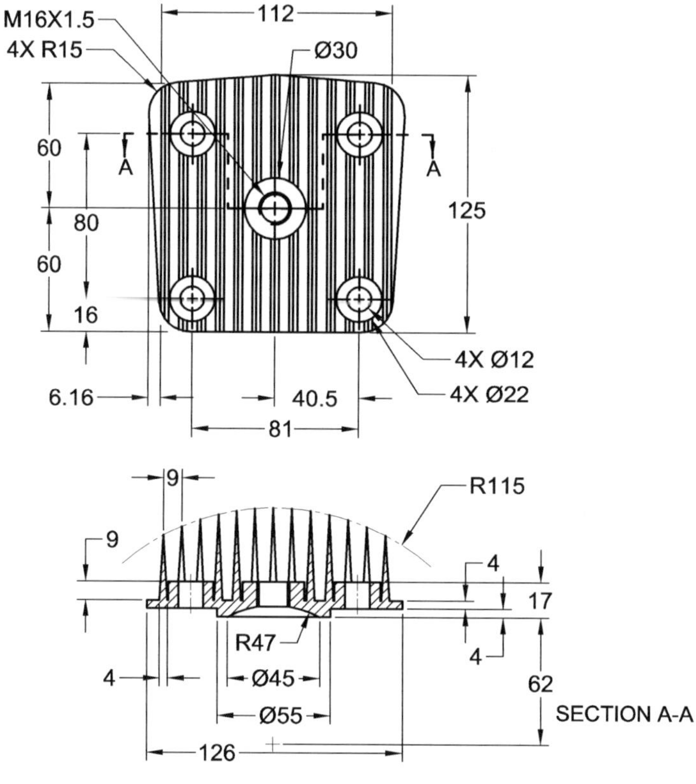

Figure 8-108 *Top and front section views of the model*

Examine the model and determine the number of features in it. The model consists of twelve features, refer to Figure 8-108.

The following steps are required to complete this tutorial:

a. Create the base feature on the **TOP** datum plane.
b. Create the round features on the vertical edges of the base feature.
c. Create the cylindrical feature on the bottom face of the base feature.
d. Create the revolve cut feature on a plane passing through the center of the cylindrical feature.
e. Create the fifth feature of the fin that will be patterned later on.
f. Create the cut feature that will remove the protrusions of the fins projecting out of the base feature.
g. Create a circular cut feature on the top face of the base feature.
h. Create a cylindrical feature on the top face of the base feature.
i. Create the ninth feature, which is a cut feature. This feature is reference patterned to create its other instances.
j. Create the tenth feature as a cylindrical feature on the top face of the base feature.
k. Create the referenced pattern of the tenth feature.
l. The eleventh feature is a coaxial hole. After creating the feature, create a reference pattern of this hole.
m. Create the last feature as the coaxial hole that will be created on the eighth feature.

If required, set the working directory to the *c08* folder.

Starting a New Object File

1. Start a new part file and name it as *c08tut5*.

   The three default datum planes are displayed in the drawing area.

Creating the Base Feature

In this model, you need to draw the sketch of the base feature on the **TOP** datum plane. The sketch is a polygon. First you need to draw the right half of the polygon and then mirror it about the centerline to create the complete polygon.

1. Choose the **Extrude** tool from the **Shapes** group.

2. Choose the **Placement** tab to display a slide-down panel. From this panel, choose the **Define** button; the **Sketch** dialog box is displayed.

3. Select the **TOP** datum plane as the sketching plane.

4. Select the **RIGHT** datum plane and then select the **Right** option from the **Orientation** drop-down list, if it is not selected automatically.

5. Choose the **Sketch** button to enter the sketcher environment.

6. Next, create the sketch of the base feature and apply constraints and dimensions to it, as shown in Figure 8-109.

7. After the sketch is complete, choose the **OK** button to exit the sketcher environment.

   The **Extrude** dashboard is enabled and appears above the drawing area.

8. Enter **4** as the depth value in the dimension box at the **Extrude** dashboard and then choose the **OK** button from the dashboard.

   The base feature is completed and now you need to create the second feature. The default trimetric view of the base feature is shown in Figure 8-110.

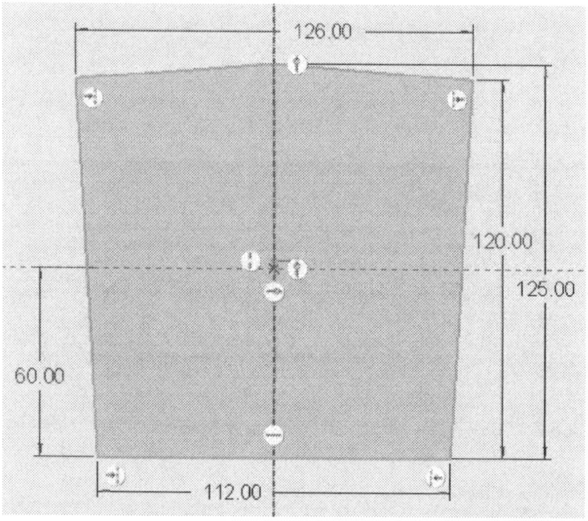

Figure 8-109 *Sketch of the base feature with dimensions and constraints*

Figure 8-110 *Default trimetric view of the base feature*

Creating the Second Feature

The second feature is a round feature of radius 15.

1. Choose the **Round** tool from the **Engineering** group; the **Round** dashboard is displayed.

2. Choose the **Sets** tab to display the slide-down panel.

3. Select one of the four vertical edges at the corners of the base feature. To select the second edge and the subsequent edges, press the CTRL key and then select the edge.

   Preview of the round is created on the selected edges. Also, a default value of the radius is displayed.

4. Double-click on the default radius value displayed along with preview of the round; an edit box appears. Enter **15** in the edit box. You can also enter the radius value in the **Radius** edit box of the **Round** dashboard.

5. Choose the **OK** button to create the round feature.

   The default trimetric view of the model after creating the round feature is shown in Figure 8-111.

Creating the Third Feature

The third feature is a cylindrical feature. You need to create the feature on the bottom face of the base feature.

1. Invoke the **Extrude** dashboard and using the **Sketch** dialog box, select the bottom face of the base feature as the sketching plane.
2. Enter the sketcher environment and draw the sketch of the cylindrical feature. Refer to Figure 8-108 for drawing the sketch.
3. Exit the sketcher environment and enter **4** as the depth of extrusion in the dashboard.
4. Exit the **Extrude** dashboard.

   The cylindrical feature is created and is shown in Figure 8-112.

Figure 8-111 *Model with round*

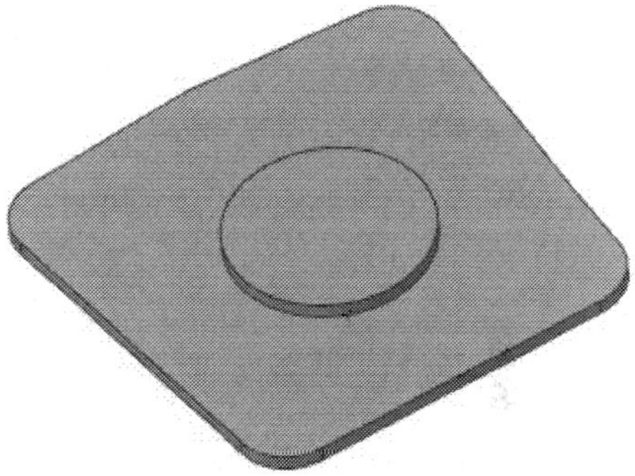

Figure 8-112 *Model after creating the cylindrical feature on the bottom face*

Creating the Fourth Feature

The fourth feature is a revolve cut that will be created on the datum plane passing through the center of the cylindrical feature.

1. Invoke the **Revolve** dashboard and choose the **Remove Material** button.
2. Choose the **Placement** tab; the slide-down panel is displayed. Choose the **Define** button from the slide-down panel; the **Sketch** dialog box is displayed. Select the **FRONT** datum plane as the sketching plane.
3. Select the **RIGHT** datum plane and then select the **Right** option from the **Orientation** drop-down list. Next, choose the **Sketch** button.

4. Once you enter the sketcher environment, create sketch of the revolve cut feature, draw a vertical geometric centerline, and apply constraints and dimensions to it, as shown in Figure 8-113.

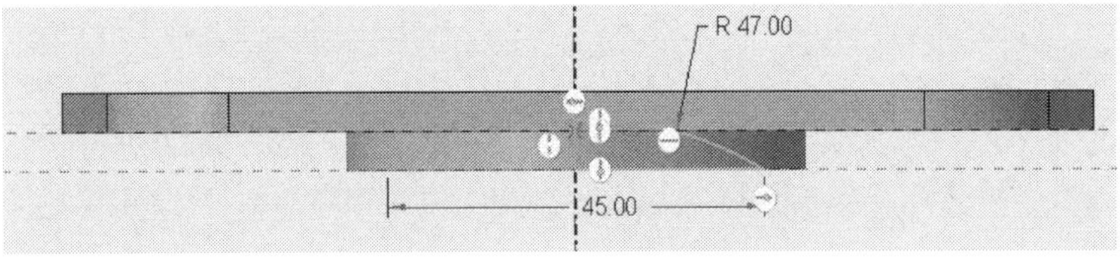

Figure 8-113 *Sketch of the revolved feature with dimensions and constraints*

5. After completing the sketch, exit the sketcher environment.

   The **Revolve** dashboard is enabled and displayed above the drawing area. The value for the angle of revolution is **360** by default.

6. Choose the **OK** button from the dashboard.

   The revolved cut feature is created and shown in Figure 8-114.

Figure 8-114 *Revolved cut feature*

Creating the Fifth Feature

Fifth feature consists of a fin and its pattern. You need to create sketch of the fin on the **FRONT** datum plane and the pattern of the fin feature will be created using the **Variable** option.

This type of pattern can only be created if dimensions of the feature to be patterned are appropriate. In the sketch of the feature, a construction entity must exist. This construction entity can be a circle, an arc, or a line. The construction entity is used as the driving entity for the pattern to be created later. The pattern instances are created in such a way that they follow the geometry of the construction entity.

1. Invoke the **Extrude** dashboard and using the **Sketch** dialog box, select the **FRONT** datum plane as the sketching plane.

2. Select the **RIGHT** datum plane and then select the **Right** option from the **Orientation** drop-down list, if it is not selected by default.

3. Choose the **Sketch** button to enter the sketcher environment.

4. Next, draw the arc, refer to Figure 8-115, by using the **Center and Ends** button.

Note

In the sketch of the fin feature, you need to draw the sketch such that the size of the instances of the pattern can be controlled later. You will use the dimension 6.16 to create the pattern. Also, the vertical height of the fin feature is not specified in the sketch because in that case, the fin feature will not slide along the construction arc but will be locked with the top construction arc.

5. Select the arc and then choose the **Toggle Construction** button from the mini popup toolbar; the arc is converted into a construction arc. Next, create the sketch of the fin feature and apply the required constraints and dimensions to it, as shown in Figure 8-115.

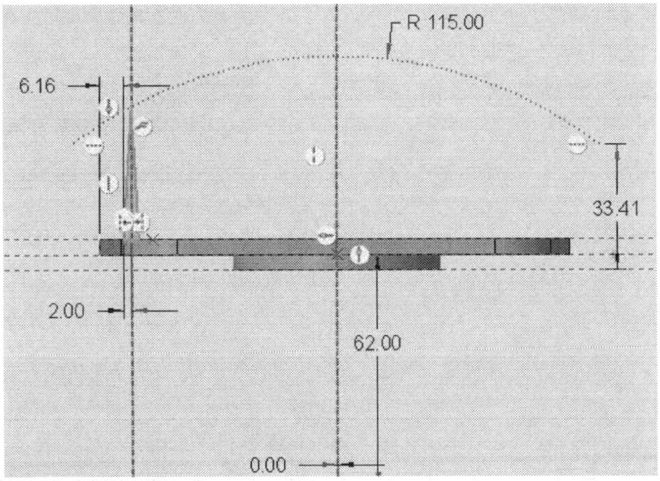

Figure 8-115 *Sketch for the fin feature with two weak dimensions*

6. After the sketch is completed, exit the sketcher environment. Note that the sketch must be closed.

7. Choose the **Symmetric** option from the **Depth** flyout in the **Extrude** dashboard.

8. Enter **135** in the dimension box of the dashboard.

9. Choose the **OK** button from the dashboard.

The default trimetric view of the fin feature is shown in Figure 8-116. Now, you need to pattern the fin feature to create the remaining 12 fins.

10. Select the fin and invoke the **Pattern** dashboard; the dimensions of the fin feature are displayed in the drawing area.

11. Select the dimension value **6.16**; an edit box appears. Enter **9** in this edit box.

12. Enter **13** in the **Members** edit box in **1st direction**. These are the number of instances of the fin feature.

13. Choose the **OK** button to exit the **Pattern** dashboard. The model after patterning the fin feature is shown in Figure 8-117.

Figure 8-116 *Fin created on the base feature*

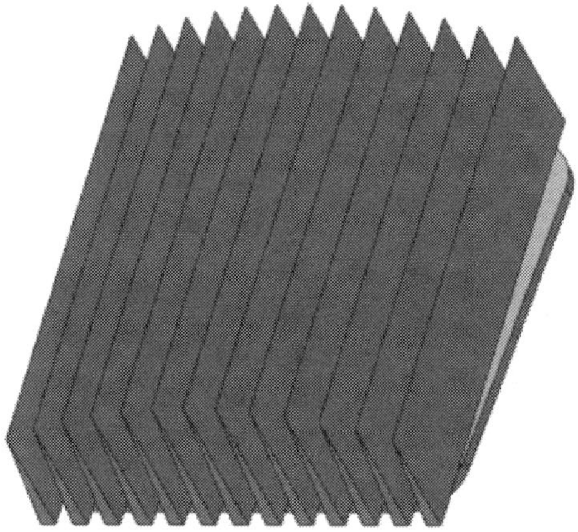

Figure 8-117 *Pattern of the fin feature*

Creating the Sixth Feature

The sixth feature is an extrude cut. You need to create this feature on the bottom face of the base feature. This cut will remove the fins that are projecting out of the base feature.

1. Invoke the **Extrude** dashboard and then choose the **Remove Material** button.

2. Select the bottom face of the base feature as the sketching plane.

3. Select the **RIGHT** datum plane and then select the **Right** option from the **Orientation** drop-down list, if they are not selected by default.

4. Once you enter the sketcher environment, use the edges of the base feature to create the sketch that will be extruded later to create the cut.

5. After the sketch is complete, exit the sketcher environment. The **Extrude** dashboard is enabled and it appears above the drawing area.

6. Choose the **Through All** button from the depth flyout.

7. If required, choose the **Change depth direction of extrude to other side of sketch** button from the dashboard to reverse the direction of the pink arrow. This arrow points in the

direction from where the material will be removed. Note that the arrows displayed on the preview should be pointing upward and outward. You can click on the arrows to change the direction of the cut.

8. Choose the **OK** button from the dashboard. The default trimetric view of the model after creating the cut feature is shown in Figure 8-118.

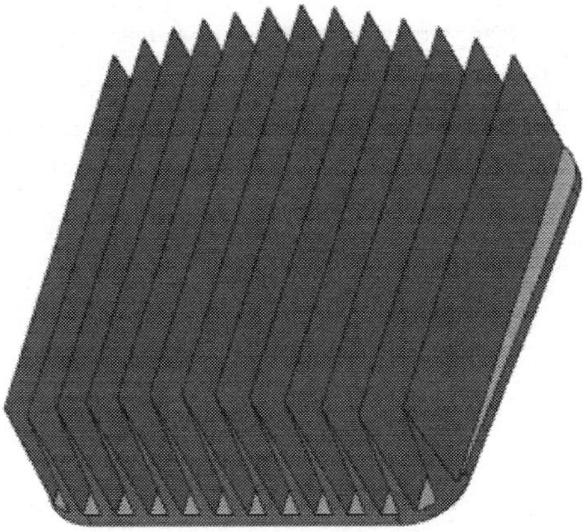

Figure 8-118 *Model after creating the cut feature*

Creating the Seventh Feature

The seventh feature is an extrude cut feature. You need to create the feature on the top face of the base feature.

1. Invoke the **Extrude** dashboard and then choose the **Remove Material** button from it.

2. Select the top face of the base feature as the sketching plane.

3. Select the **RIGHT** datum plane and then select the **Right** option from the **Orientation** drop-down list, if they are not selected by default. Next, choose the **Sketch** button.

4. Draw the sketch of the circular cut feature, as shown in Figure 8-119.

5. After the sketch is completed, exit the sketcher environment.

   The **Extrude** dashboard is enabled and it appears above the drawing area.

6. Choose the **Through All** button from the depth flyout.

7. If required, change the depth direction of extrude such that the fins of the model are cut.

8. Choose the **OK** button from the dashboard. The top view of the model after creating the circular cut is shown in Figure 8-120.

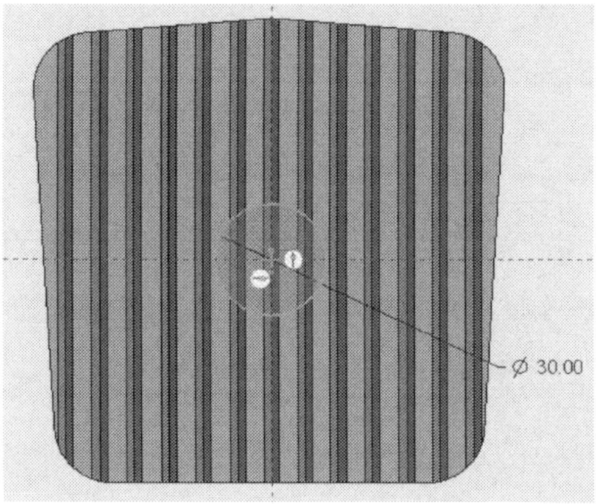

Figure 8-119 Sketch of the circular cut feature

Figure 8-120 Model after creating the circular cut

Creating the Eighth Feature

The eighth feature is a protrusion feature. You need to create this feature on the top face of the base feature.

1. Invoke the **Extrude** dashboard and select the top face of the base feature as the sketching plane.

2. Select the **RIGHT** datum plane and then select the **Right** option from the **Orientation** drop-down list, if they are not selected by default.

3. Once you enter the sketcher environment, use the edge of the cut feature to create the sketch of the protrusion feature.

4. After the sketch is completed, exit the sketcher environment.

5. Enter **9** in the dimension box of the **Extrude** dashboard.

6. Choose the **OK** button from the dashboard. The model after creating the cylindrical feature is shown in Figure 8-121.

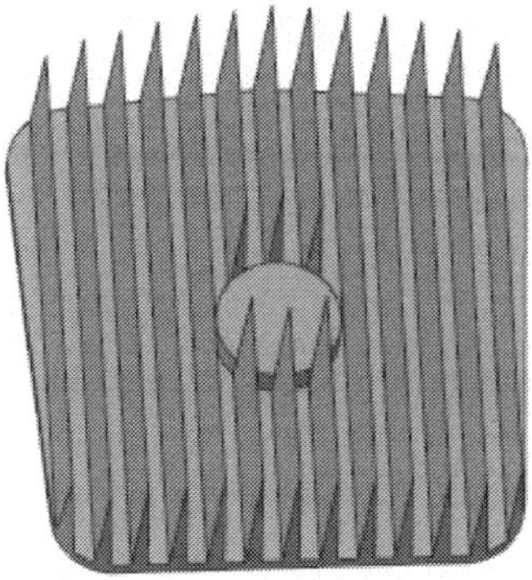

Figure 8-121 Model after creating the cylindrical feature

Creating the Ninth Feature

The ninth feature is a cut feature. You need to pattern this feature to create its other instances. The cut feature is created on the top face of the base feature.

1. Invoke the **Extrude** dashboard and then choose the **Remove Material** button.

2. Select the top face of the base feature as the sketching plane.

3. Select the **RIGHT** datum plane and then select the **Right** option from the **Orientation** drop-down list.

4. Once you enter the sketcher environment, draw the sketch of the circular cut feature, as shown in Figure 8-122.

5. After the sketch is completed, exit the sketcher environment. The **Extrude** dashboard is enabled and it appears above the drawing area.

6. Choose the **Through All** button from the depth flyout.

7. Change the depth direction of extrude such that the fins of the model are cut.

8. Choose the **OK** button from the dashboard. The model after creating the circular cut is shown in Figure 8-123.

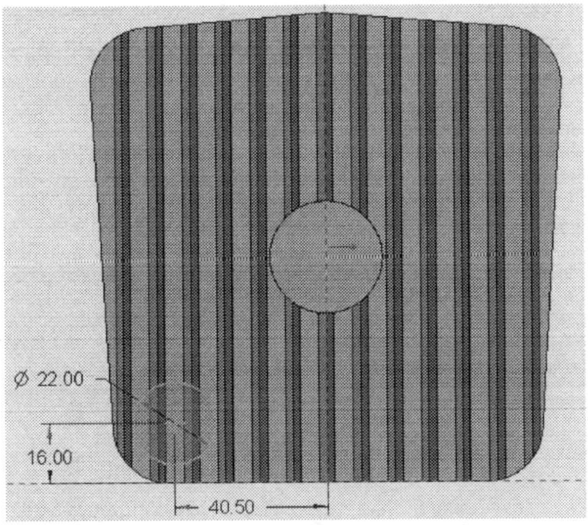

Figure 8-122 *Sketch of the cut feature*

Figure 8-123 *Model after creating circular cut*

Now you need to pattern the cut feature to create its other instances.

9. Select the cut feature from the **Model Tree** and then choose the **Pattern** tool from the **Editing** group or from the mini popup toolbar. The **Pattern** dashboard is displayed and you are prompted to select dimensions to be changed in the first direction. Select the **Dimension** option from the **Type** drop-down list, if it is not selected by default.

10. Select the **General** option from the **Options** slide-down panel, if it is not selected.

11. Select the dimension value **40.5** from the drawing area. After you have selected the dimension in the first direction, an edit box is displayed.

12. Enter **-81** in the edit box.

13. Hold the right mouse button to display the shortcut menu. Choose the **Direction 2 Dimensions** option from the shortcut menu.

14. Select the dimension **16**. After you have selected the dimension in the second direction, an edit box is displayed.

15. Enter **80** in the edit box. Note that in the instances edit boxes in the dashboard, the number of instances **2** is specified by default.

16. Choose the **OK** button from the **Pattern** dashboard. The rectangular pattern of the cut feature is displayed, as shown in Figure 8-124.

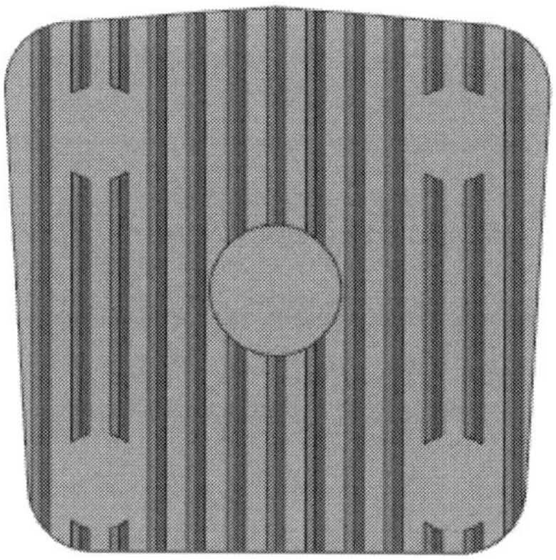

Figure 8-124 *Model after patterning the cut feature*

Creating the Tenth Feature

The tenth feature is an extrude feature created on the top face of the base feature. You need to pattern this feature to create other instances.

1. Invoke the **Extrude** dashboard.

2. Select the top face of the base feature as the sketching plane.

3. Select the **RIGHT** datum plane and then select the **Right** option from the **Orientation** drop-down list, if they are not selected by default.

4. Once you enter the sketcher environment, draw the sketch of the protrusion feature using the edges of the cut feature.

5. After the sketch is completed, exit the sketcher environment.

6. Enter **9** in the dimension box of the dashboard.

7. Choose the **OK** button from the dashboard. The model after creating the cylindrical protrusion feature is shown in Figure 8-125.

   Now you need to create a reference pattern of the protrusion feature to create its other instances.

8. Select the extrude feature from the **Model Tree** and then choose the **Pattern** tool from the **Editing** group.

   The **Reference** option is selected by default in the drop-down list of the **Pattern** dashboard.

9. Choose the **OK** button from the **Pattern** dashboard. The model after creating the pattern is shown in Figure 8-126.

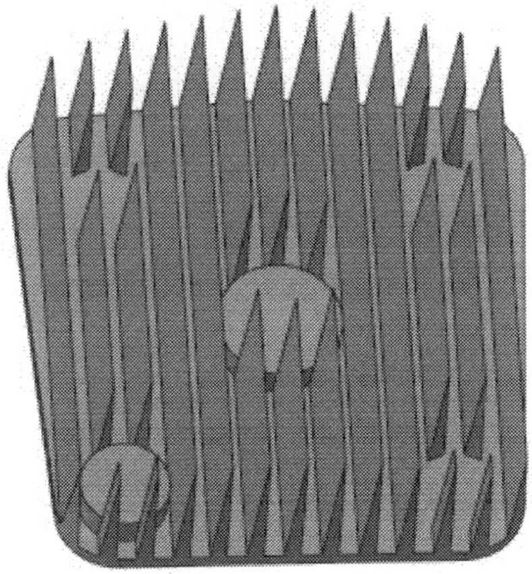

Figure 8-125 Model after creating the cylindrical protrusion feature

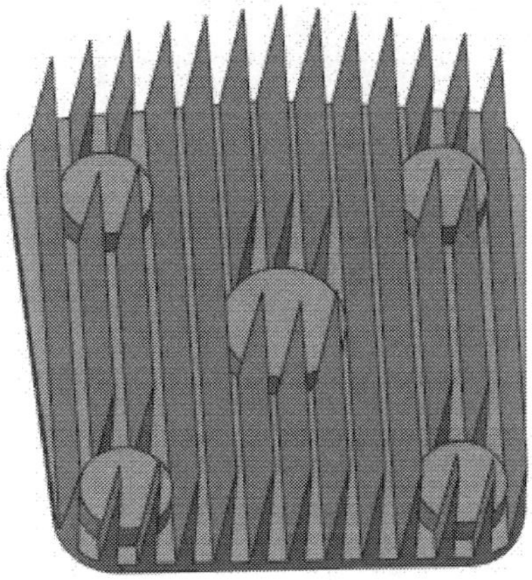

Figure 8-126 Model after creating the reference pattern of the cylindrical feature

Creating the Eleventh Feature

The eleventh feature is a coaxial hole and is created on the top face of the tenth feature. Later it is reference-patterned to create its other instances.

1. Create a hole of diameter **12** on the top face of the protrusion feature created at the lower left corner. Make sure the bottom face of the base feature is selected. You must create a hole that passes through all faces.

Note

While creating the coaxial hole on the top face of the tenth feature, make sure that the axis of the coaxial hole and the axis of the tenth feature are same.

2. Pattern the feature by using the **Reference** option. The model after creating the pattern of the hole feature is shown in Figure 8-127.

Creating the Last Feature

The last feature is a hole of dimension **16x1.5** and is created on the top face of the eighth feature. This hole is coaxial with the cylindrical feature on which it is being created. The model after creating the hole feature is shown in Figure 8-128.

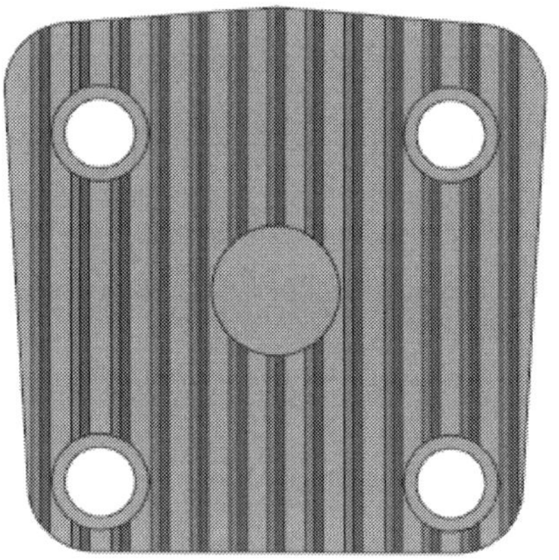

Figure 8-127 *Model after creating the reference pattern of the hole feature*

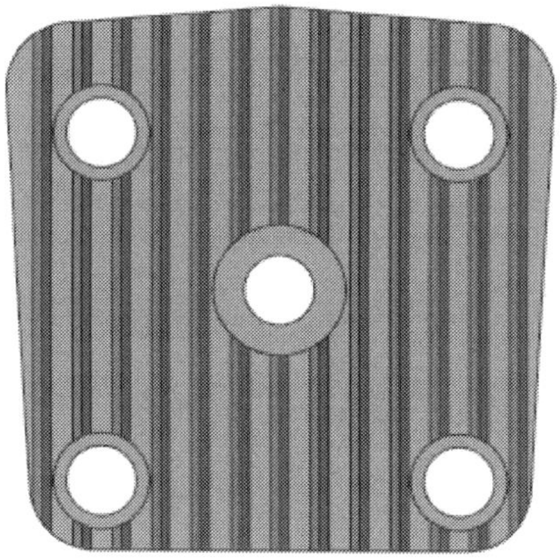

Figure 8-128 *Model after creating the hole on the cylindrical feature*

Saving the Model

You need to save the model because you may need it later.

1. Choose the **Save** button from the **File** menu and then save the model. The order of feature creation can be seen from the **Model Tree** shown in Figure 8-129.

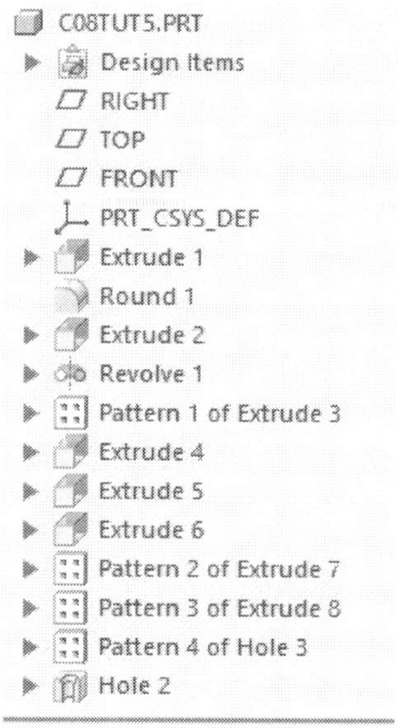

Figure 8-129 *The* ***Model Tree*** *for Tutorial 5*

Self-Evaluation Test

Answer the following questions and then compare them to those given at the end of this chapter:

1. The _________ option is used to delete a pattern except the leader feature.

2. The _________ tool is used to mirror the entire geometry about a plane.

3. The _________ option from the **Pattern** dashboard is used to create a pattern in which the dimensions of instances can be varied.

4. The first feature in a pattern is called _________.

5. The _________ menu is used to select the dimensions that you want to vary from the leader while copying.

6. When a pattern is created, the leader or the parent feature also becomes a part of the pattern. (T/F)

7. Once a pattern is created, all instances in the pattern, including the parent feature, act as a single feature. (T/F)

8. In the **Reference** pattern, an existing pattern is referenced to create a new pattern. (T/F)

9. Using the **Pattern** dashboard, you can only create linear patterns. (T/F)

10. If you select a pattern feature from the **Model Tree** and right-click to display the shortcut menu and then choose the **Delete** option, the whole pattern is deleted including the leader. (T/F)

Review Questions

Answer the following questions:

1. Which of the following options in the **Pattern** dashboard is used to create patterns in which instances touch each other and intersect with other instances or edges of a surface?

   (a) **Identical** (b) **Variable**
   (c) **General** (d) None of these

2. Which of the following options in the **Pattern** dashboard is used to create a pattern on the specified path?

   (a) **Direction** (b) **Curve**
   (c) **Axis** (d) **Point**

3. Which of the following options in the **Pattern** dashboard cannot be used to create a pattern that intersects an edge of a feature on which the pattern has to be created?

   (a) **Identical** (b) **Variable**
   (c) **General** (d) None of these

4. Which of the following options in the **Pattern** dashboard is used to create a pattern that has all instances of different sizes?

   (a) **Identical** (b) **Variable**
   (c) **General** (d) None of these

5. You can mirror features by using datum planes or planar surfaces. (T/F)

6. To create a rotational pattern, you should specify an angular increment. (T/F)

7. The dimension pattern can be created in two directions. (T/F)

8. In the **Pattern** dashboard, the **Dimension** option is selected by default. (T/F)

9. The **Direction** option in the **Pattern** dashboard is used to create the pattern in the specified direction. (T/F)

10. Fill pattern is used to fill the sketched area by a selected feature. (T/F)

EXERCISES

Exercise 1

Create the model shown in Figure 8-130. The top, front section, right, detailed, and sectioned views of the model are shown in Figure 8-131. **(Expected time: 45 min)**

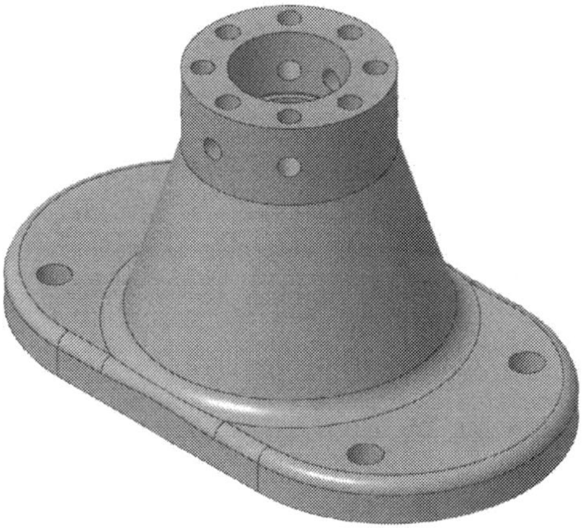

Figure 8-130 *Solid model for Exercise 1*

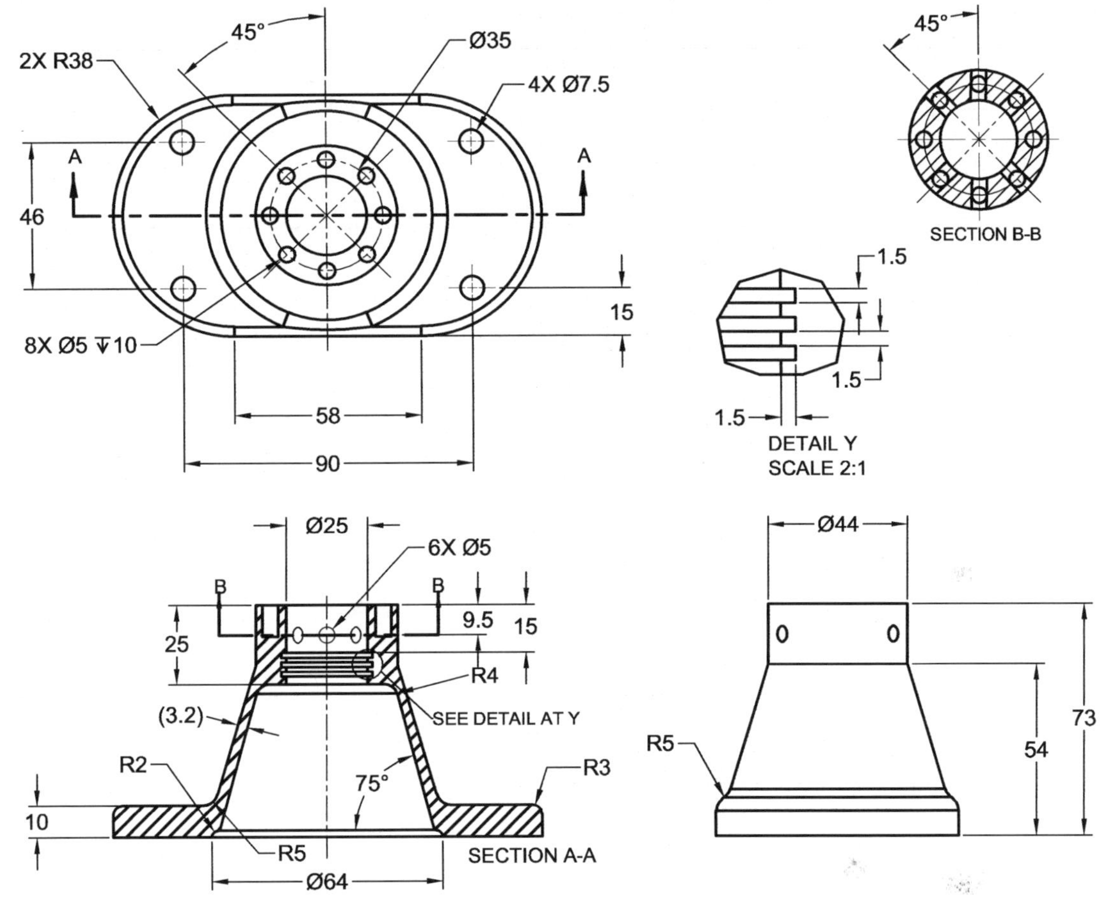

Figure 8-131 Different views of the model

Exercise 2

Create the model shown in Figure 8-132. The orthographic views of the model are shown in Figure 8-133. **(Expected time: 30 min)**

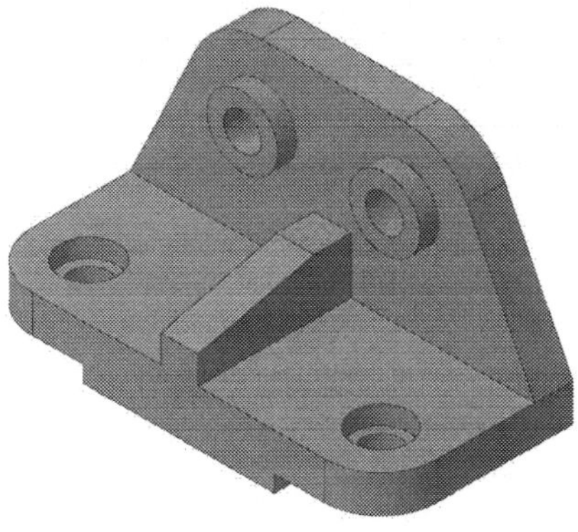

Figure 8-132 Isometric view of the model

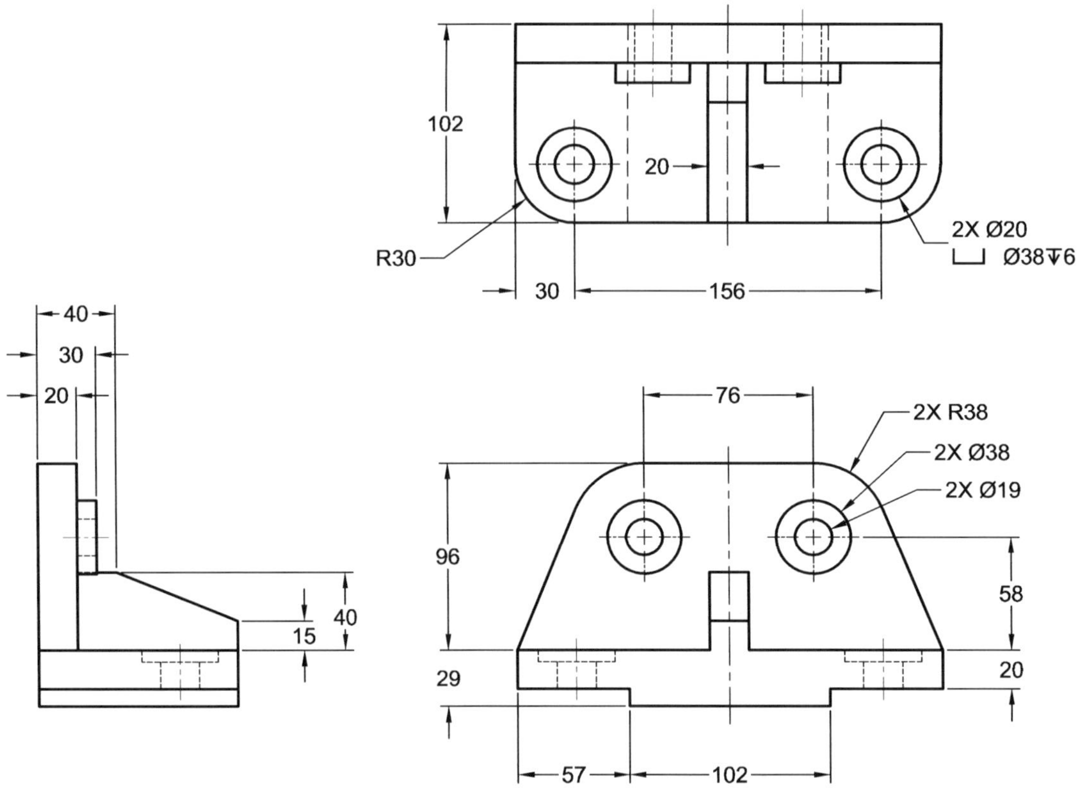

Figure 8-133 *Orthographic views of the solid model*

Exercise 3

Create the model shown in Figure 8-134. The top, front, and isometric views are shown in Figure 8-135. **(Expected time: 30 min)**

Figure 8-134 *Isometric view of the solid model*

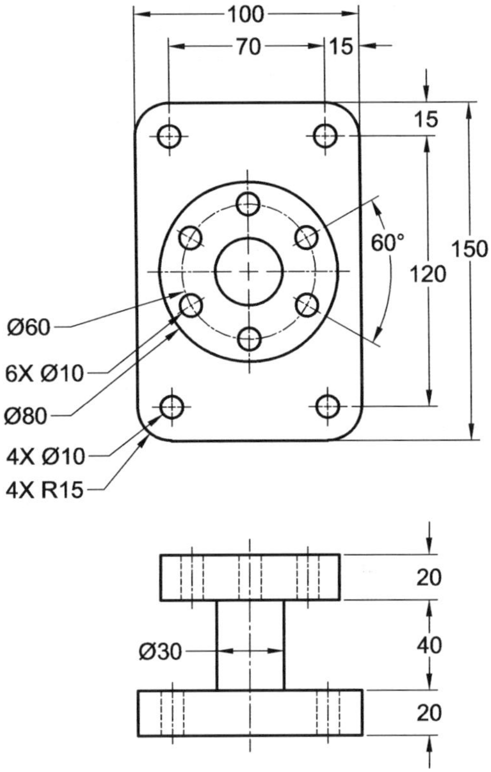

Figure 8-135 *Orthographic views of the solid model*

Exercise 4

Create the model shown in Figure 8-136. The top and sectioned views of the model are shown in Figure 8-137. **(Expected time: 1hr)**

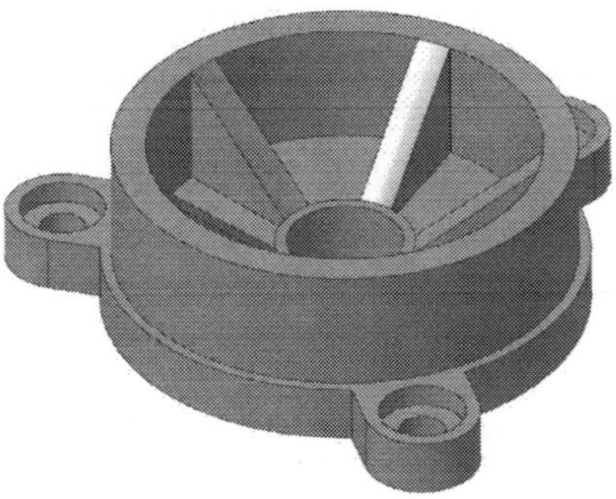

Figure 8-136 *Solid model for Exercise 4*

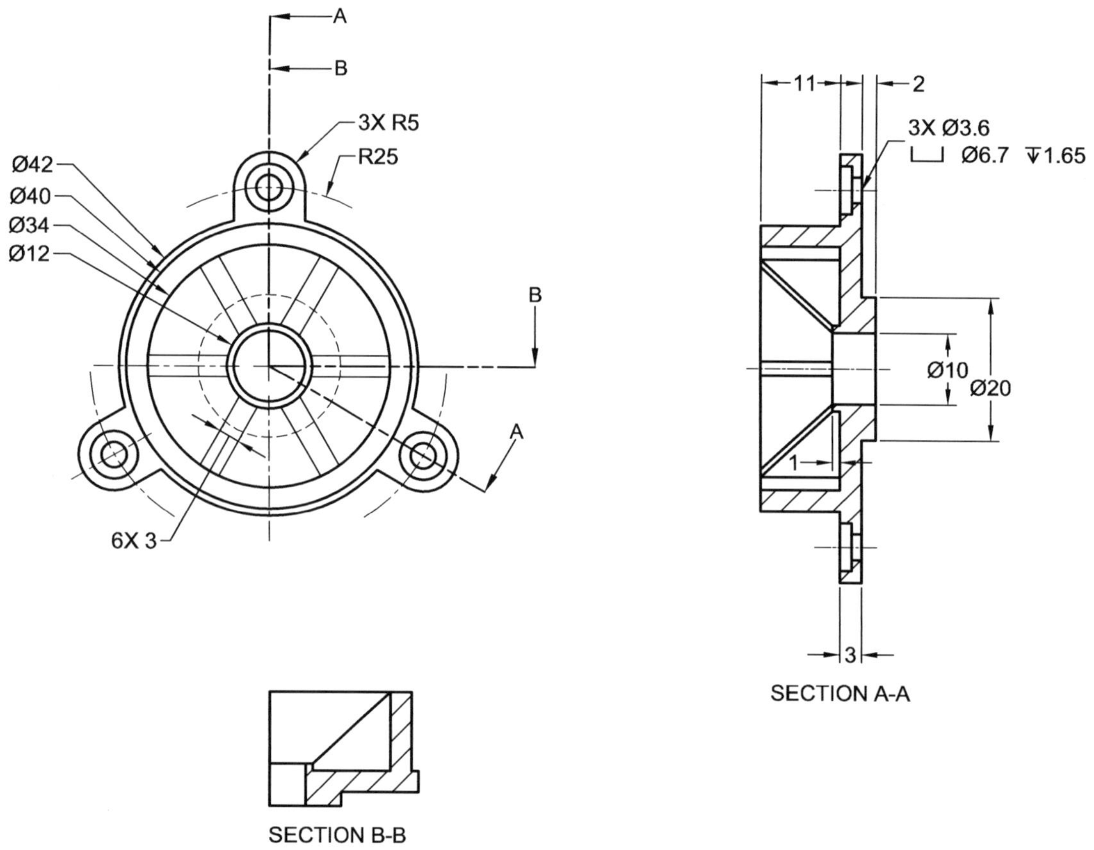

Figure 8-137 *Top and sectioned views of the model*

Answers to Self-Evaluation Test

1. Delete Pattern, **2. Mirror**, **3. Variable**, **4.** leader, **5. GP VAR DIMS**, **6.** T, **7.** T, **8.** T, **9.** F, **10.** T

Chapter 9

Advanced Modeling Tools

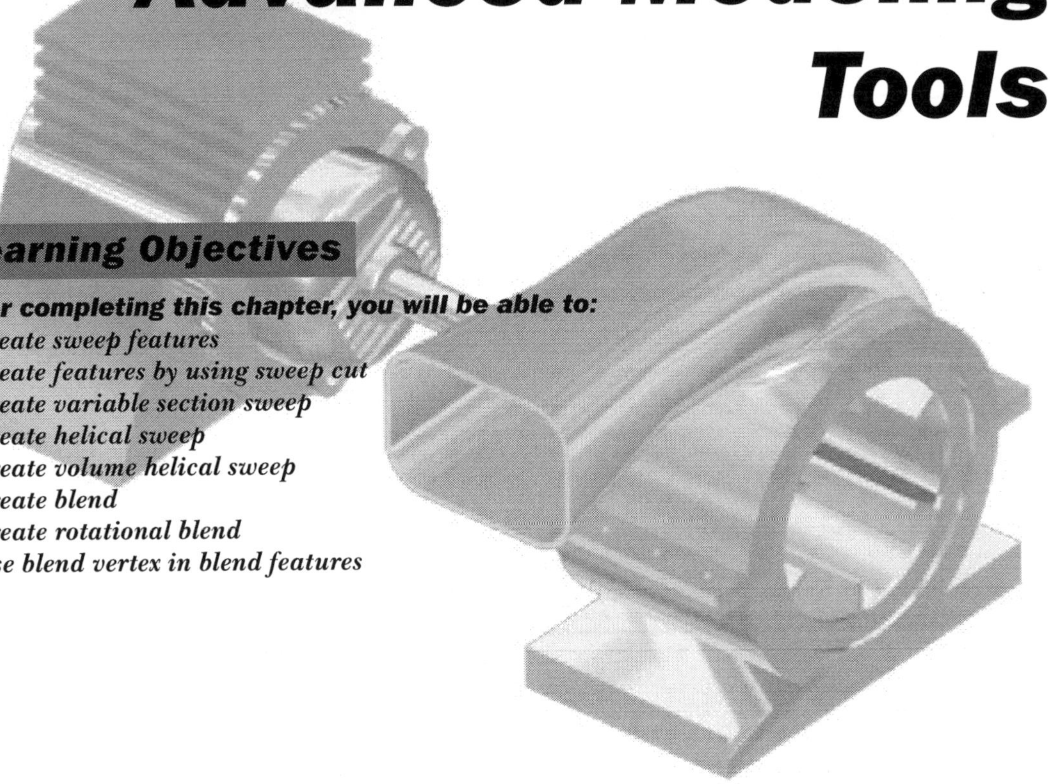

Learning Objectives

After completing this chapter, you will be able to:

- *Create sweep features*
- *Create features by using sweep cut*
- *Create variable section sweep*
- *Create helical sweep*
- *Create volume helical sweep*
- *Create blend*
- *Create rotational blend*
- *Use blend vertex in blend features*

ADVANCED MODELING TOOLS

In this chapter, you will learn about the tools that are used to create complex features. These tools are called advanced modeling tools. The advanced modeling tools available in Creo Parametric are **Sweep**, **Helical Sweep**, **Volume Helical Sweep**, **Blend**, **Swept Blend**, and **Rotational Blend** which are discussed next.

SWEEP FEATURES

Sweeping is a process of creating solid and surface geometries by moving an open or a closed section along an open or closed trajectory. A sweep feature consists of a single constant or variable section which is swept along one or more trajectories. The section can vary in its shape and orientation as it is swept in different directions along different trajectories. Trajectory is the path along which a section is swept. The trajectory for a sweep feature can be either sketched or selected. The order of operation is to first create a trajectory and then a section. In Creo Parametric, two type of sweeps, Sweep and Helical sweep, can be created and the tools to create them are discussed next.

Sweep

Ribbon: Model > Shapes > Sweep drop-down > Sweep

Enhanced

The **Sweep** tool is similar to the **Extrude** tool. The only difference is that in case of the **Extrude** tool, the feature is extruded in a direction normal to the sketching plane, but in case of the **Sweep** tool, the section is swept along the sketched or selected trajectory. The sketching tools available in the sketcher environment are used for sketching the trajectory.

To invoke the **Sweep** dashboard, choose the **Sweep** tool from the **Shapes** group of the **Model** tab; the **Sweep** dashboard will be displayed, as shown in Figure 9-1. The options available in this dashboard are discussed next.

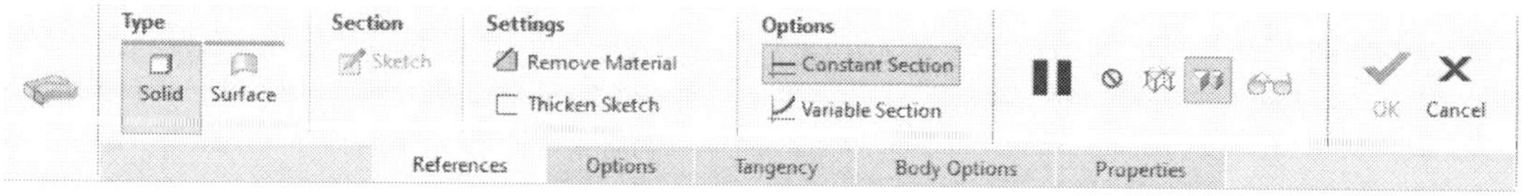

***Figure 9-1** Partial view of the **Sweep** dashboard*

Solid

In the dashboard, the **Solid** button is chosen by default and is used to create solid by sweeping the section along the selected trajectory. The section that you create should be a single closed loop.

Surface

The **Surface** button is used to create surface by sweeping the section along a trajectory. The sketch that is drawn for the section need not be a closed loop.

Sketch

The **Sketch** button in the **Section** area is used to open the sketcher environment to create or edit sweep section. The sketch created using this button is an internal sketch of the feature. This button will be activated only when you have specified trajectory(ies) for the sweep feature.

Remove Material

The **Remove material** button is available in the **Sweep** dashboard only after the base feature is created. So this button is used to remove material from an existing feature.

Thicken Sketch

The **Thicken Sketch** button is used to add thickness to the feature. Using this option, you can create thin solid, thin solid cut, or a thin surface trim. When you choose this button, the **Change direction to thicken between one side, other side, or both sides of sketch** button and a dimension edit box appear on the right of this button. This option is similar to the **Extrude** tool that was discussed in Chapter 4.

Constant Section

The **Constant Section** toggle button is used to create a sweep feature whose section is constant throughout the trajectory.

Variable Section

The **Variable Section** toggle button is used to create a sweep feature whose section varies along the trajectory. When you select multiple trajectories for sweeping, this button is chosen automatically in the **Sweep** dashboard. After choosing this button, you can also use section relations and parameters to make a variable sketch.

Note

The options and procedure to create constant and variable section sweep will be discussed later in this chapter.

References Tab

Choose the **References** tab from the dashboard; the **References** slide-down panel will be displayed. Select the trajectory, the slide-down panel will be modified, refer to Figure 9-2.

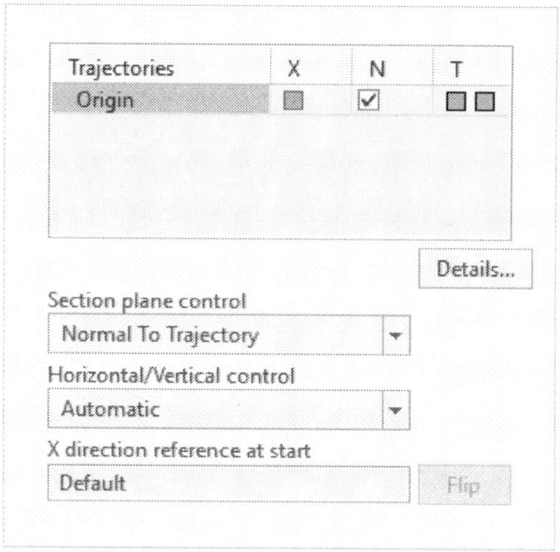

Figure 9-2 *The **References** slide-down panel*

Trajectories Collector

This collector is used for selecting the trajectories. The first trajectory that you select is by default listed as the origin trajectory. This collector has three columns for each selected trajectory. The first column is **X**. When the check box in this column is selected, the trajectory corresponding to that check box is converted into a X trajectory. This means that the x-axis of the section plane is oriented normal to the selected trajectory. The second column is **N**. When the check box in this column is selected, the trajectory corresponding to that check box is converted into a normal trajectory. The first selected trajectory cannot be used as X trajectory. The third column is **T**. When this check box is selected, the trajectory is converted into a tangent trajectory. For a tangent trajectory, a centerline is added to the sketch of the sweep section. This centerline will be tangent to the neighboring surfaces at the intersection point of the trajectory and the sketch plane. The tangency is maintained as you sweep the section. Hence, the angle of the centerline changes depending on the position of the section and the neighboring surfaces.

There are two check boxes available in **T** column in each row to specify Side 1 and Side 2 tangency for the trajectory. When you select any of these check boxes, the name of the trajectory is filled in that cell and the value T=0 appears on the trajectory in the drawing area. This value is displayed on both the ends of the trajectory. Initially, this value is 0. When you enter a value, the point of alignment of the section with the trajectory varies.

Note that first you need to select the normal and X trajectories and then draw the sketch of the section. This is because the sketch plane will be oriented only if you have defined the normal trajectory. Also, since the sketch needs to pass through the endpoints of the X-trajectory, it is better to define the X-trajectory before drawing the section. This way you will have the reference point from which the sketch needs to pass.

Note

*1. Except the origin trajectory, all other trajectories are auxiliary trajectories by default until you select the **X**, **N**, or **T** check box.*

2. Only one trajectory can be used as an X trajectory.

3. Only one trajectory can be used as an N trajectory.

4. A single trajectory can be selected as an X trajectory and an N trajectory at the same time.

5. Any trajectory with a neighboring surface can be used as a tangent trajectory.

Section plane control Drop-down List

The options in this list are used to specify the orientation of the plane on which the sweep section is drawn. By default, the **Normal To Trajectory** option is selected in this drop-down list. As a result, the z-axis of the section plane remains normal to the origin trajectory throughout its length, refer to Figure 9-3.

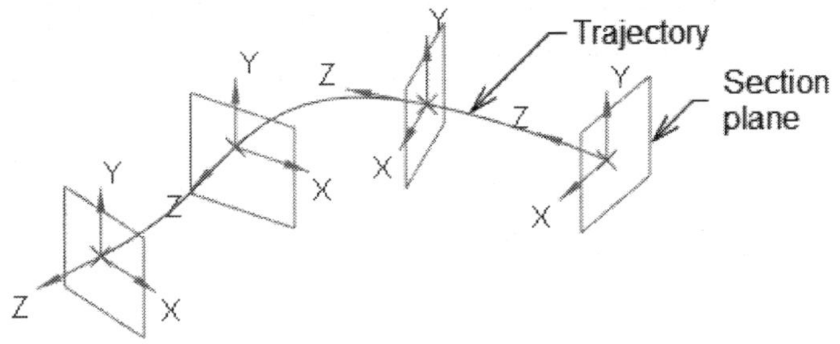

Figure 9-3 *Section plane oriented normal to trajectory*

If you choose the **Normal To Projection** option, the z-axis of the section plane remains normal to the origin trajectory as it is viewed along the projection direction. Also, the **Direction reference** collector gets activated and you are prompted to select a plane, an axis, a coordinate system axis, or a straight entity to define the projection direction. In Figure 9-4, the z-axis of the section plane is tangent at all points to the projected curve along the projection direction. Note that the y-axis of the section plane will always be normal to the plane selected in the **Direction reference** collector.

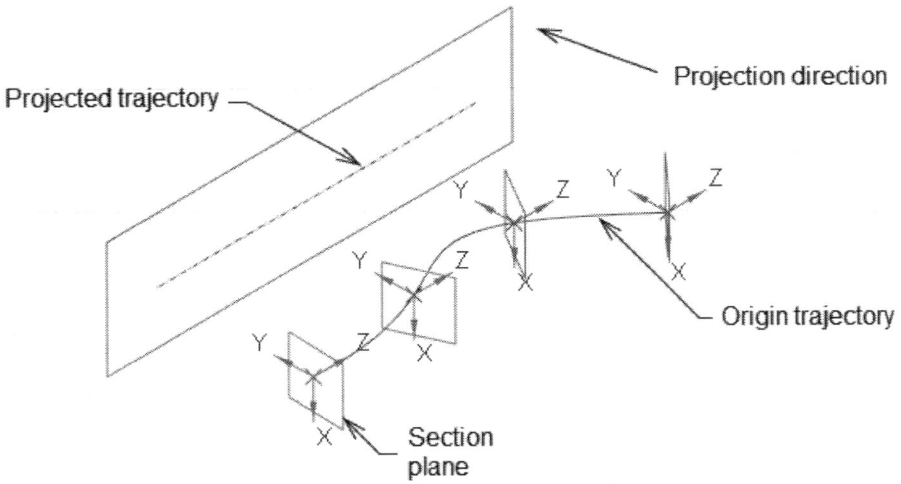

Figure 9-4 *Orientation of the section plane normal to projection*

The **Constant Normal Direction** option is used for setting the z-axis of the section plane parallel to the specified direction vector. When you choose this option, the **Direction reference** collector gets activated. Also, you will be prompted to select a plane, an axis, a coordinate system axis, or a straight entity to define the section plane in the normal direction. In Figure 9-5; the z-axis of the section plane is oriented along the direction defined by the normal direction.

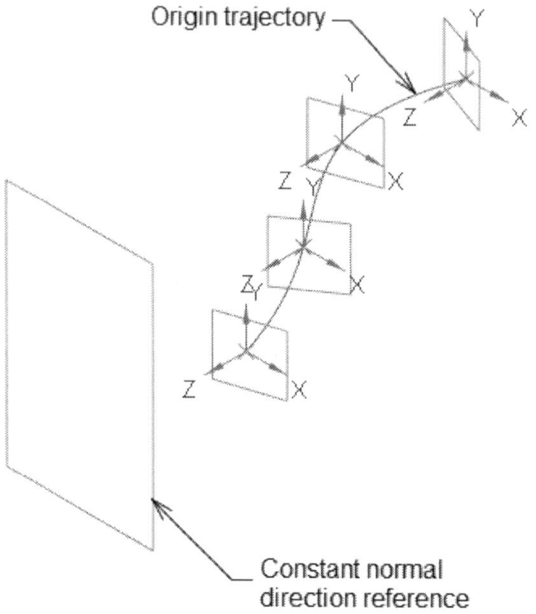

Figure 9-5 *Orientation of the section plane constant and normal to the direction reference*

Horizontal/Vertical control Drop-down List

The options available in this drop-down list are used to control the rotation of the frame normal to the section plane. The frame is essentially a coordinate system that slides along the origin trajectory and carries the sweep section with itself. Axes of this coordinate system are defined by auxiliary trajectories and other references.

By default, the **Automatic** option is selected in the **Horizontal/Vertical control** drop-down list. As a result, section plane is oriented in the xy-direction. The direction of the x-vector is calculated automatically so that the twisting produced in the swept geometry is minimum. The **X direction reference at start** collector is used to specify the x-direction reference for the section at the start. This way you can sketch the section in a preferred orientation.

The **Normal To Surface** option will be available only when the origin trajectory is created as a reference curve on a surface, one-sided edge of a surface, two-sided edge of a surface or solid edge, curve created though intersection of surfaces, or two projection curves. If this option is selected, the y-axis of the section plane becomes normal to the surface on which the origin trajectory is lying. The **Next** option is used to move the section to the next normal surface.

The **X-Trajectory** option in this drop-down list allows you to orient the x-axis of the section plane to pass through the intersection point of the specified x-trajectory and the section plane along the sweep.

Options Tab

Choose the **Options** tab; the **Options** slide-down panel will be displayed. The options available in this slide-down panel are discussed next.

Capped ends

The **Capped ends** check box is used to close the ends of a sweep feature. This option will be available only when you have created a surface sweep using a closed section and open trajectory.

Merge ends

The **Merge ends** check box is used to connect the ends of a solid sweep feature to a nearby solid surface without leaving gaps, refer to Figures 9-6 and 9-7. This option is available only when you are creating a constant section sweep. Note that to use this option, at least one solid feature should be present close to the trajectory.

Figure 9-6 *Sweep feature created with the* ***Merge ends*** *check box cleared*

Figure 9-7 *Sweep feature created with the* ***Merge ends*** *check box selected*

Sketch placement point Collector

This collector allows you to select the start point for the sweep feature. In this collector **Origin** is selected, by default, as the start point of the section. As a result, the section is drawn at the start point of the trajectory. However, if you activate the **Sketch placement point** collector by clicking and then select a datum point that lies on the origin trajectory, then the section will start from that point. This means that the selected datum point becomes the start point of the trajectory.

Tangency Tab

Choose this tab; a slide-down panel will be displayed, as shown in Figure 9-8. The options available in this slide-down panel are used to specify trajectories as tangent trajectories. This will make the surface of the variable section sweep to be tangent with existing geometry. You can specify a trajectory to be tangent either by selecting the **T** check box in the **References** tab or by using the options in the **Tangency** tab.

The **Trajectories** area in the slide-down panel lists the trajectories selected for the sweep feature. The options in the **References** drop-down list are used to specify a trajectory tangent to a surface. If you choose the **None** option from the **References** drop-down list, the tangency will get remove from the selected trajectory. The **Default1** option makes the centerline of the sweep section tangent to surfaces on side 1 of the trajectory. Choosing the **Default2** option will make the centerline of the sweep section tangent to surfaces on side 2 of the trajectory. The **Selected** option allows you to specify the surfaces for tangent centerlines in the sweep section. The usage of these options is discussed briefly later in this chapter.

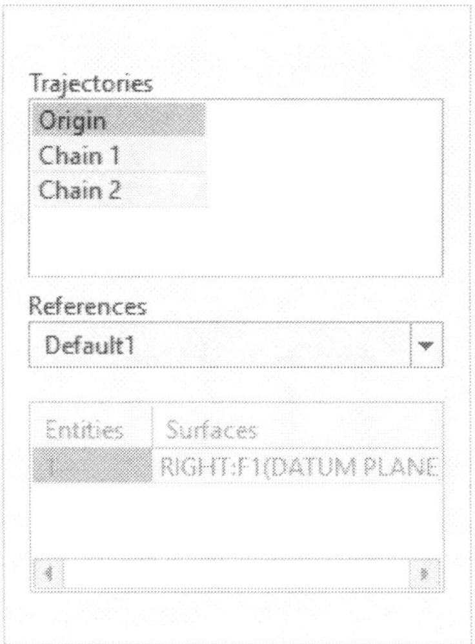

Figure 9-8 *The* ***Tangency*** *slide-down panel*

Body Options Tab

This tab will be available when the feature is created as solid. While adding material, the **Add geometry to body** option will get displayed under this tab. Select the **Create new body** check box; the created feature will get added to **Body2 (new body)**. While removing material, the **Cut geometry from body** option with two radio buttons will get displayed in the tab. The **All** radio button is used to cut the geometry from all the bodies that the feature passes through and the **Selected** radio button is used to cut the geometry from the selected bodies only.

Properties Tab

The options in this tab are used for specifying the name of the sweep feature. You can also display detailed information of components in a browser by choosing the Information button next to the **Name** edit box.

Creating Constant Section Sweep

To create a sweep feature, you need to specify a trajectory and a section. When you use only one trajectory to create a sweep feature, a constant section sweep is created. The cross-section of such sweep feature remains constant throughout the sweep. To create a constant section sweep, first select a trajectory and then invoke the **Sweep** tool or draw the trajectory using the **Sketch** tool in the **Datum** flyout of the **Sweep** dashboard. Ensure that the **Constant Section** ( ) button is chosen in the dashboard and then choose the **Sketch** button to invoke the sketcher environment. Draw the section for sweep and choose the **OK** button in the **Close** group of the sketcher environment; the **Sweep** dashboard is activated again and preview of the feature is displayed in the drawing area. Next, choose the **OK** button in the **Sweep** dashboard to finish the process.

Figures 9-9 and 9-10 show the section and closed trajectory, and the resulting swept surface.

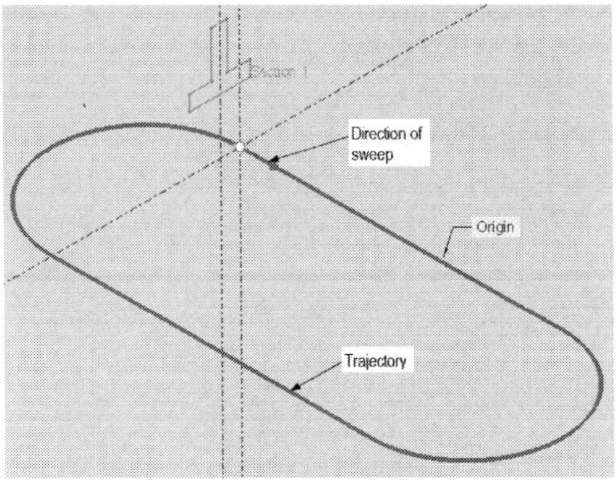

Figure 9-9 Section and close trajectory

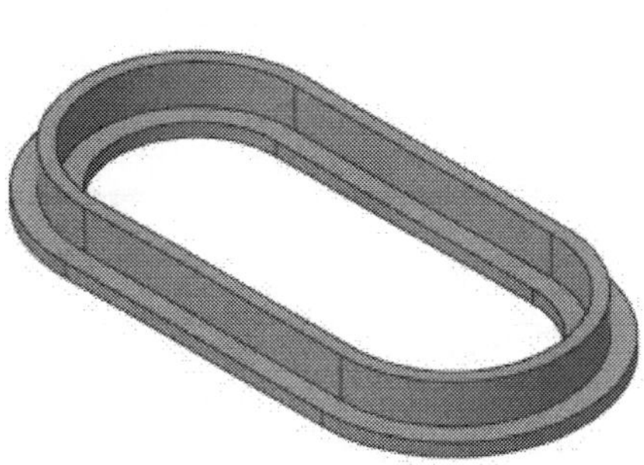

Figure 9-10 Resulting swept geometry

Figures 9-11 and 9-12 show a sweep created using an open trajectory.

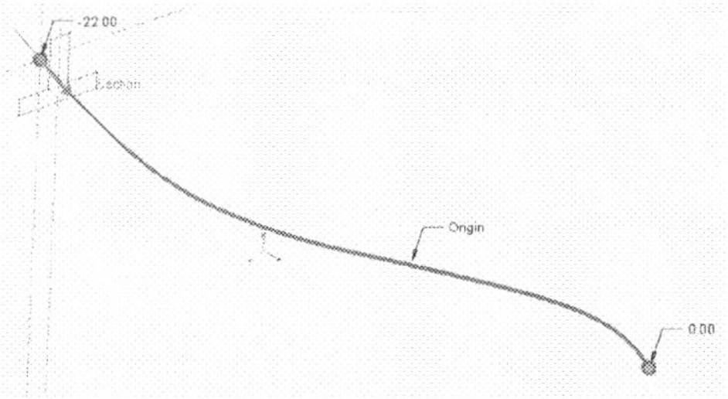

Figure 9-11 Section and an open trajectory

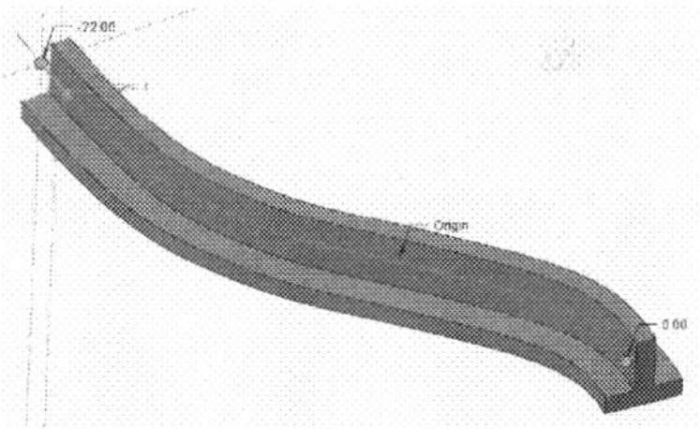

Figure 9-12 Resulting swept geometry

You can use a combination of trajectories and sections for creating a sweep feature. The combinations that can be used are as follows:

1. Closed section and open trajectory.
2. Closed section and closed trajectory.
3. Open section and open trajectory for surface sweep feature.
4. Open section and closed trajectory for surface sweep feature.

Tip

1. *Similar to other sketched features, the trajectory of the sweep feature is also sketched after selecting a sketching plane.*
2. *At bends in a trajectory, the radius of the bend should be proportionate to the cross-section to be swept to avoid overlapping. If the section size is large and the radius of the curve or bend is small, overlapping takes place and the sweep feature will not be created. Therefore, make sure that the ratio of the size of the section to the size of the trajectory is appropriate.*
3. *You can also use the edges of an existing base feature as trajectory.*

Creating a Sweep Cut

The procedure to create a sweep cut feature is similar to that of creating a sweep protrusion. The only difference between them is that in case of cut features, the material is removed from an existing feature. The **Remove Material** button in the **Sweep** dashboard is used to create cuts in the solid geometries. The cut can be a solid sweep cut or thin sweep cut. Figure 9-13 shows the part edge selected as a closed trajectory and the section created for swept cut. Figure 9-14 shows the resulting geometry after creating the sweep cut.

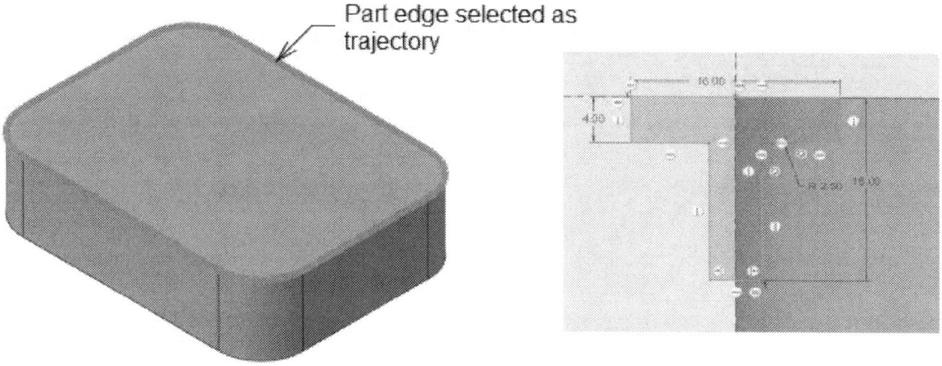

Figure 9-13 *Trajectory and section for swept cut*

Figure 9-14 *Resulting sweep cut*

Figure 9-15 shows a slot created using swept cut using an open trajectory. This is quite helpful and time saving method for creating slots and keyways in solid models.

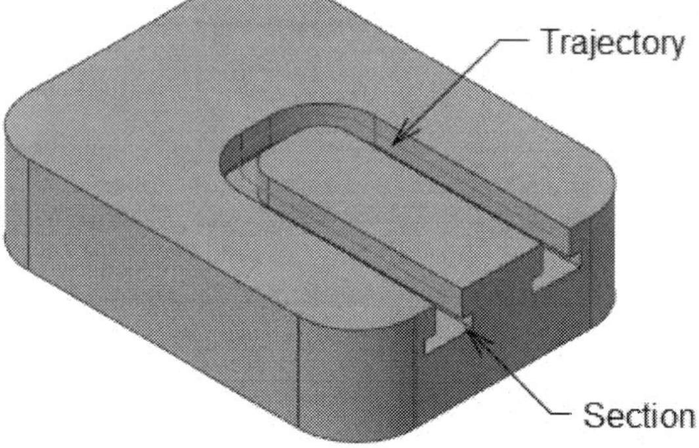

Figure 9-15 *Swept cut created using an open trajectory*

Creating Variable Section Sweep

A variable section sweep consists of one or more trajectories and a single section whose shape and orientation changes along the geometry. Depending upon the variety and complexity of the geometry, you can use the **Sweep** tool to add or remove material while sweeping a section along one or more selected trajectories. The section is swept along the origin trajectory and the variation of the section is controlled by the X-trajectory and other trajectories. Remember that the origin trajectory is the first selected trajectory. You can sketch the sweep trajectory using various sketcher environment tools. You can also select a previously created entity as the trajectory for sweep.

The following points should be remembered while creating a variable section sweep feature:

1. You should define the origin trajectory to which the section is normal.

2. After creating the origin trajectory, you have to create an X-trajectory that defines the horizontal vector of the section.

3. You can also define additional trajectories to facilitate the creation of a complex profile.

4. When you draw the section for the variable section sweep feature, the section should be aligned to the endpoints of the X-trajectory and the other trajectories. If the section is not aligned, then the feature will be created without following the trajectory path and it will be a uniform section sweep.

To create a variable section sweep feature, invoke the **Sweep** tool and then choose the **Variable Section** button () in the dashboard. Remember that you should have trajectories in the drawing area so that they can be selected for the creation of variable section sweep feature. You can also draw trajectories using the **Sketch** tool available in the **Datum** flyout in the dashboard. While selecting the trajectories for variable section sweep, select the origin trajectory and then select the remaining trajectories by pressing CTRL key, refer to Figure 9-16. After you have selected the trajectories, choose the **Sketch** button from the **Section** area of the **Sweep** dashboard and draw the sweep section in the sketcher environment. Choose the **OK** button to close the sketcher environment; a preview of the sweep feature will be displayed in the drawing area. Next, choose the **OK** button in the **Sweep** dashboard to finish the process, refer to Figure 9-17.

You can create complex geometries and features using the options provided with the **Sweep** tool. The options for creating complex (variable) sweep features are discussed next.

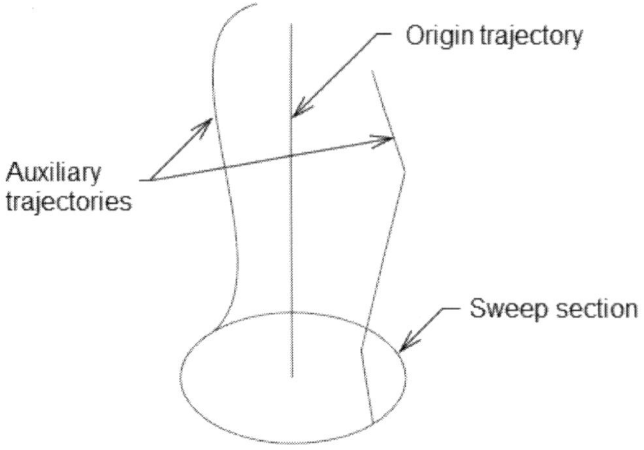

Figure 9-16 Attributes for variable section sweep

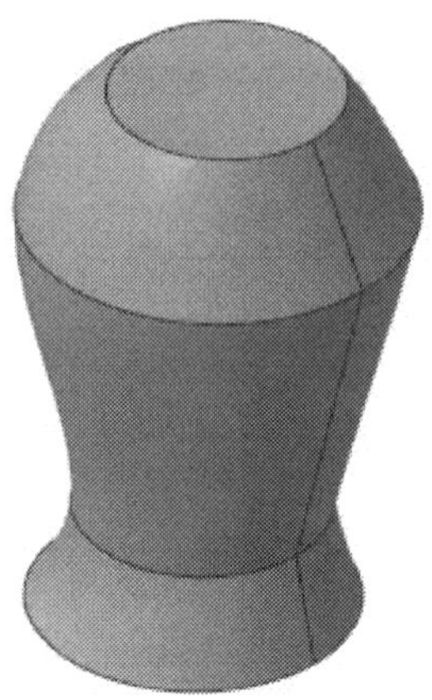

Figure 9-17 Resulting sweep feature

Creating Variable Section Sweep Normal to Trajectory

When you choose the **Normal To Trajectory** option in the **Section plane control** drop-down list, the frame moves in a direction normal to the specified trajectory. By default, the frame moves along the origin trajectory, but you can create a variable section sweep normal to any additional trajectory. To select any other trajectory as the normal trajectory, select the **N** check box corresponding to the specified curve in the **References** slide-down panel. Figures 9-18 and 9-19 show the preview of the geometry created with the origin trajectory selected as normal trajectory.

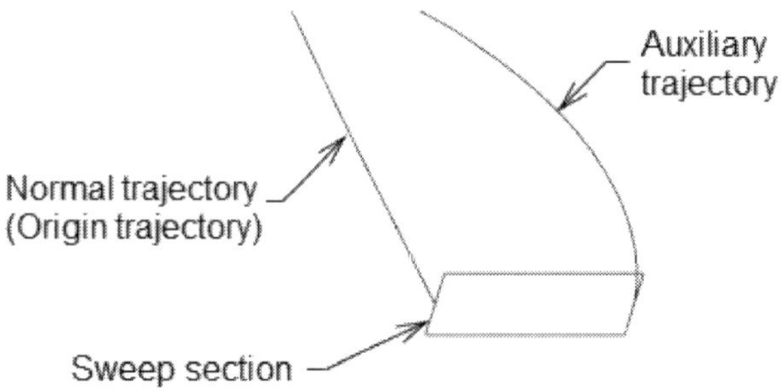

Figure 9-18 Origin trajectory selected as normal trajectory

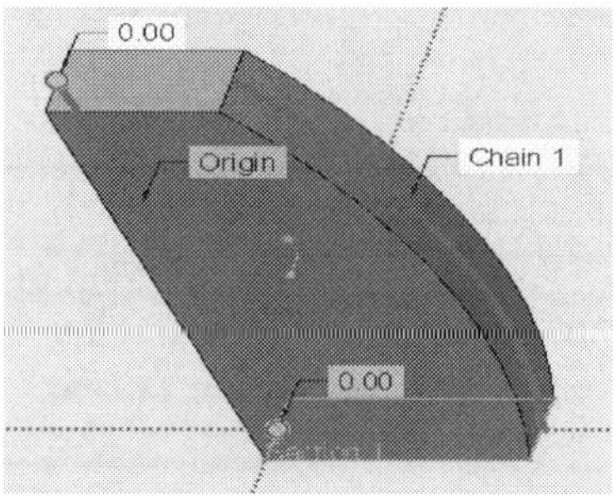

Figure 9-19 Preview of the sweep feature

When you select the **N** check box of the second trajectory in the **References** slide-down panel; the sketching plane on which you draw the sweep section, is oriented normal to specified curve, refer to Figures 9-20 and 9-21.

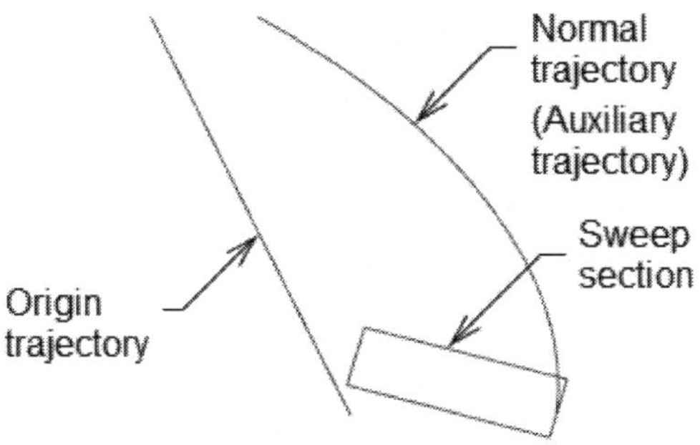

Figure 9-20 Auxiliary trajectory selected as normal trajectory

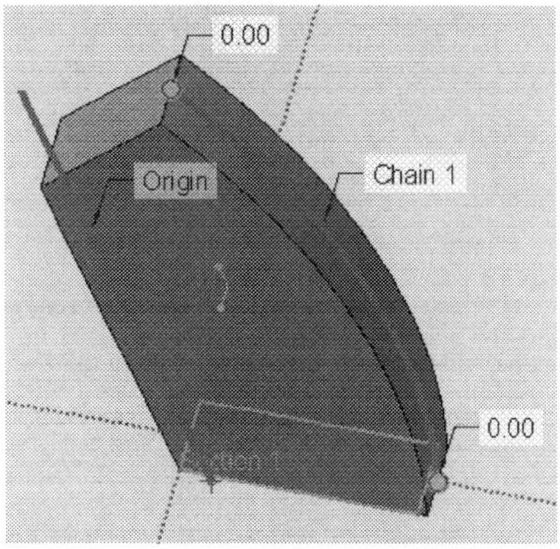

Figure 9-21 Preview of the sweep feature

You can specify the x-direction reference to set initial orientation of the frame on which the sweep section is created. Specifying the x-direction reference is similar to specifying the orientation of a sketch that is discussed in Chapter 4. In general, the orientation direction of a sketch is always set to Right. When you select a reference in the **X direction reference at start** collector; the x-axis of the frame on which the sweep section is drawn gets oriented in the specified direction, refer to Figure 9-22.

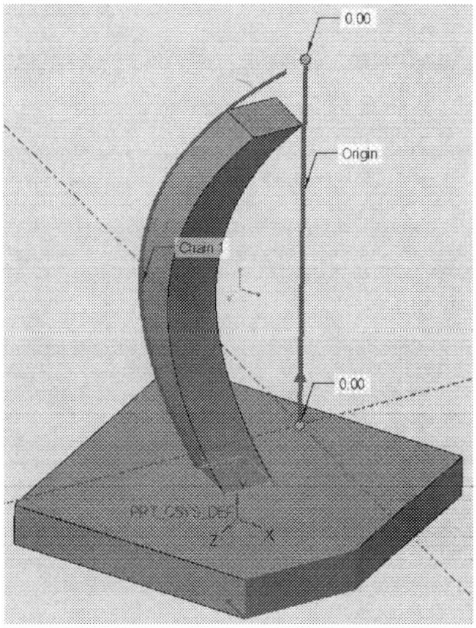

Figure 9-22 Orientation of the sweep section with respect to x-direction

Creating Variable Section Sweep Normal to Projection

When you choose the **Normal To Projection** option in the **Section plane control** drop-down list, the y-axis of the frame becomes parallel to the specified direction reference. Also the z-axis of the frame becomes tangent to the projection of the origin trajectory along the specified direction.

Figure 9-23 shows the datum plane selected as the direction reference for creating the sweep feature. Notice the changes in the swept geometry when you select the datum plane in the

Direction reference collector of the **References** slide-down panel. The y-axis of the sketch plane (frame) becomes normal to the projection plane and the z-axis becomes tangent to the projection of origin trajectory, refer to Figure 9-24.

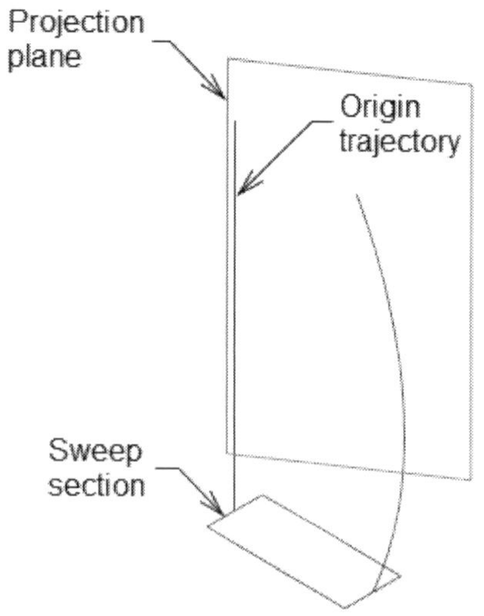

Figure 9-23 *Datum plane selected as the direction of projection*

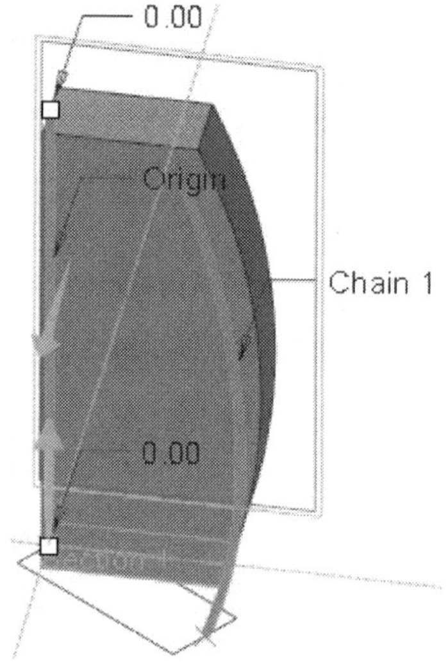

Figure 9-24 *Sweep section oriented normal to the projection plane*

Creating Variable Section Sweep Using Constant Normal Direction

When you choose the **Constant Normal Direction** option in the **Section plane control** drop-down list, the **Direction reference** collector is activated in the **References** slide-down panel and you are prompted to select a plane, an axis, a coordinate system axis, or any straight entity to define the projection direction. Figure 9-25 shows the datum plane selected as the normal direction reference for the sweep. Figure 9-26 shows the z-axis of the frame oriented parallel to the specified direction.

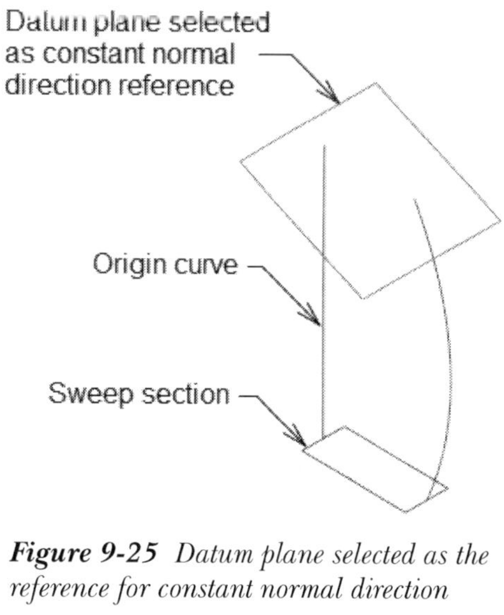

Figure 9-25 *Datum plane selected as the reference for constant normal direction*

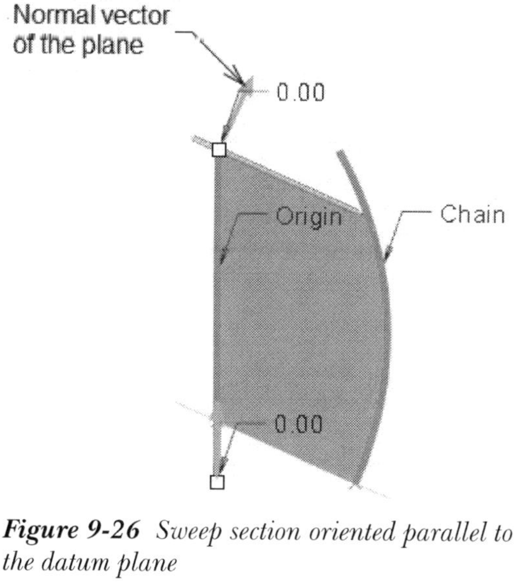

Figure 9-26 *Sweep section oriented parallel to the datum plane*

Creating Variable Section Sweep Using Tangent Trajectories

You can create variable section sweep by specifying trajectories as tangent trajectories. You can specify a trajectory as a tangent trajectory by selecting the **T** check box of the respective trajectory in the **References** slide-down panel or by using the **Tangency** tab of the **Sweep** dashboard. Figure 9-27 shows the datum curve selected as origin trajectory and the edges of the model are selected as auxiliary trajectories. In the **Tangency** slide-down panel, select the **Default1** or **Default2** option from the **References** drop-down list to specify the tangency of selected trajectory with neighboring surfaces; the corresponding surface is highlighted in the drawing area. For each tangent trajectory, a centerline appears in the sketcher environment while creating sweep section, refer to Figure 9-28. You can use these centerlines to apply tangent constraint to the entities in the sweep section thereby making the entire sweep feature tangent to the surfaces of the existing geometry.

After drawing the sketch in the sketcher environment, choose the **OK** button to close the sketcher environment; a preview of the sweep feature will be displayed in the drawing area, as shown in Figure 9-29.

Note

*You can use any option in the **Section plane control** drop-down list with the tangent trajectories specified.*

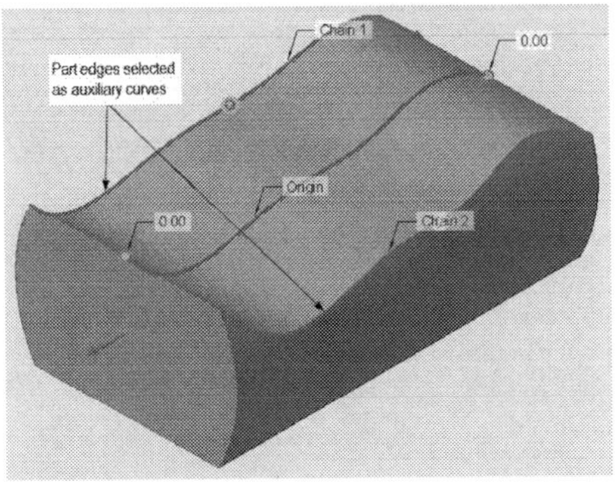

Figure 9-27 *Trajectories selected as tangent trajectories*

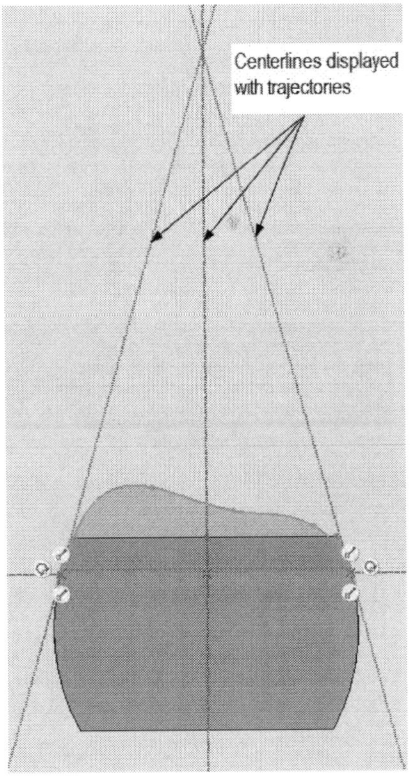

Figure 9-28 *Centerlines displayed in sketcher environment*

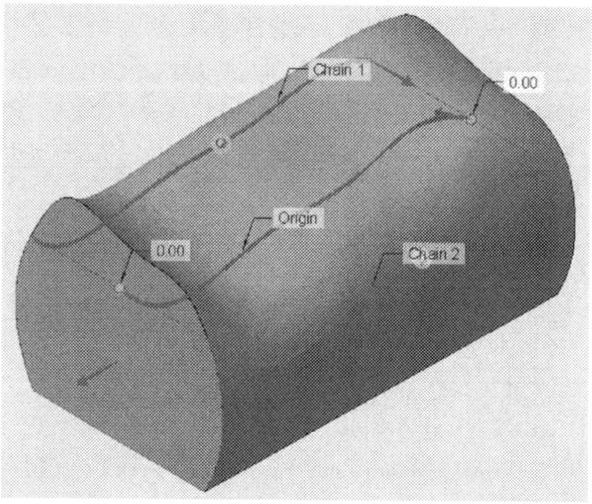

Figure 9-29 *Preview of the resulting sweep feature with tangent trajectories*

Creating Variable Section Sweep Using Trajectory Parameters

Trajectory parameters also known as Trajpar, are relations used with the sweep section to increase or decrease the dimensions along the length of the sweep feature. The value of Trajpar parameter ranges between 0 to 1 and can be used with variable section sweeps, helical sweeps, and composite curves. For a sweep feature, the value of parameter is 0 at the start point, 0.5 at the mid point, and 1 at the end point of the trajectory.

Figure 9-30 shows a model is created using the Trajpar parameter with the dimension of sweep section.

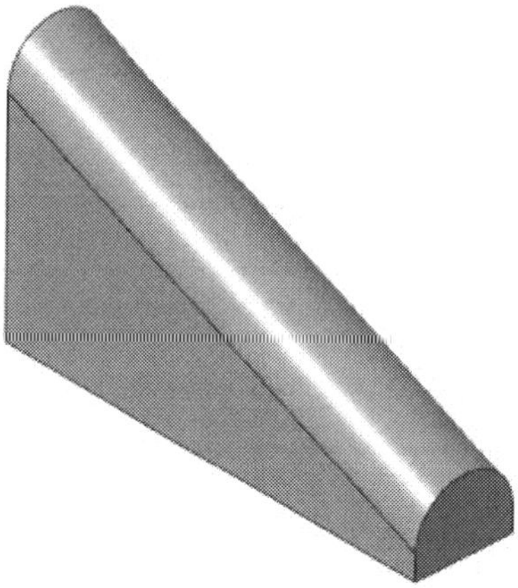

Figure 9-30 *Variable section sweep created using Trajpar parameter*

The following steps are required to create the model.

1. Draw a line on **TOP** plane in the sketcher environment, refer to Figure 9-31, and exit the sketcher environment. Use this line as a trajectory for sweep feature.

2. Select the trajectory and choose the **Sweep** tool from the **Shapes** group of the **Model** tab.

3. Choose the **Variable Section** button from the **Sweep** dashboard.

4. Choose the **References** tab and select the front plane in the **X direction reference at start** collector to orient the sweep section.

5. Choose the **Sketch** button to invoke the sketcher environment and draw a sketch for the sweep section, as shown in Figure 9-31.

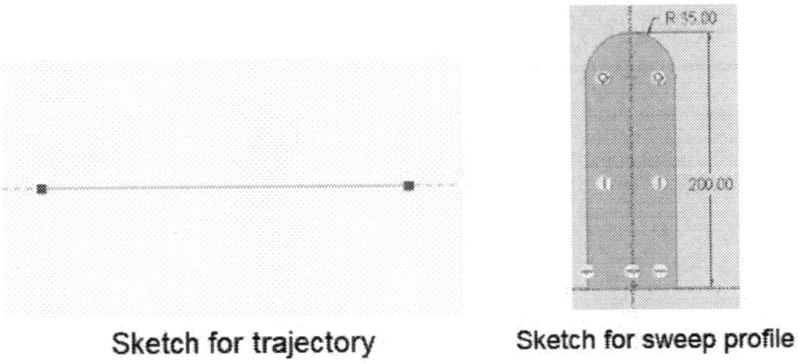

Figure 9-31 *Sketches to be drawn for the* ***Sweep*** *tool*

6. Choose the **Tools** tab and then choose the **Relations** tool from the **Model Intent** group; the **Relations** dialog box will be displayed, as shown in Figure 9-32. Also the values of the dimensions in the drawing area will change into names of the dimensions when you invoke the **Relations** dialog box. You can toggle between the name and value of the dimension by choosing the **Toggle between dimension values and names** button in the **Relations** rollout of the **Relations** dialog box.

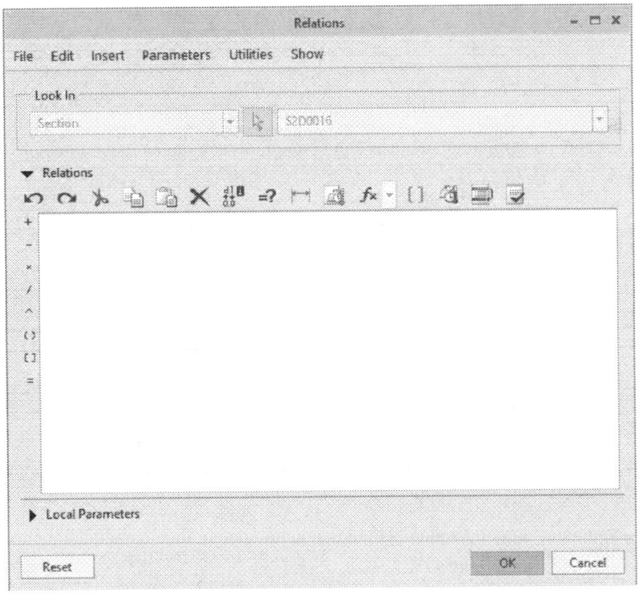

Figure 9-32 *The* ***Relations*** *dialog box*

7. Next, select the dimension that you want to make variable. For the model shown in Figure 9-30, the height of the section is varied using the Trajpar parameter. Click on the name/value of the dimension; the dimension name sd# (# is the dimension number) appears in the text area of the **Relations** dialog box.

8. Type the relation in the text box, refer to Figure 9-33. Since the value of Trajpar parameter varies 0 to 1, the value of the section height will vary 0 to 200, plus 50, which means that the height of the section will vary from 50 to 250 resulting in the creation of sweep feature displayed in Figure 9-30.

9. Choose the **OK** button in the **Relations** dialog box and choose the **Sketch** tab. To close the sketcher environment, choose the **OK** button in the **Close** group of the **Sketch** tab; a preview of the sweep feature will be displayed in the drawing area, as shown in Figure 9-34.

You can use the Trajpar parameter combined with mathematical equations to create a variable section sweep, as shown in Figure 9-35. To do so, repeat the steps 1 to 8 as described earlier, with the exception that you need to type the equation in the text box, as shown in Figure 9-36.

In the equation shown in Figure 9-36, sd10 is the section height controlled by Trajpar parameter. Because value of Trajpar varies from 0 to 1, the sine wave starts at 0 and has a magnitude of 10 with 5 cycles, plus 200. As a result, the height of the sweep section varies in sine wave form from the start point of the trajectory to the end point.

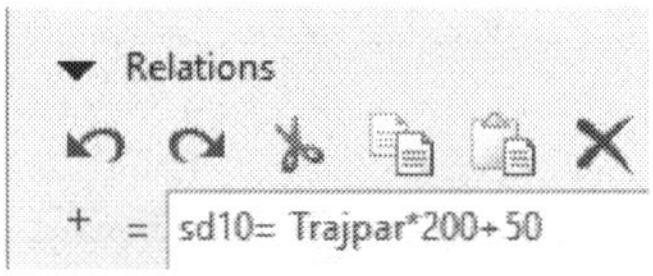

Figure 9-33 *Relation for the dimension*

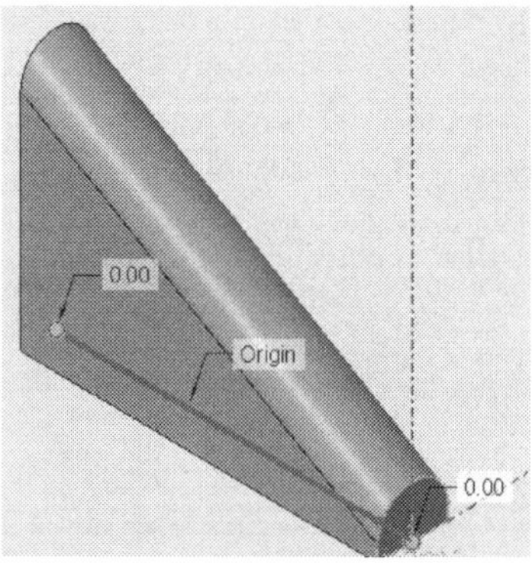

Figure 9-34 *Variable section sweep created using the Trajpar parameter*

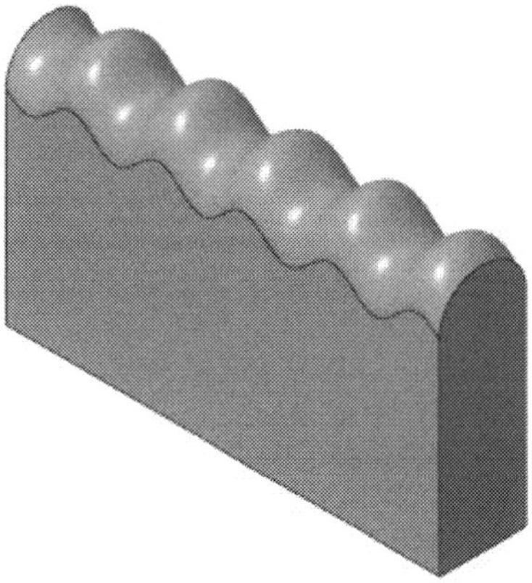

Figure 9-35 *Variable section sweep created using Trajpar parameter*

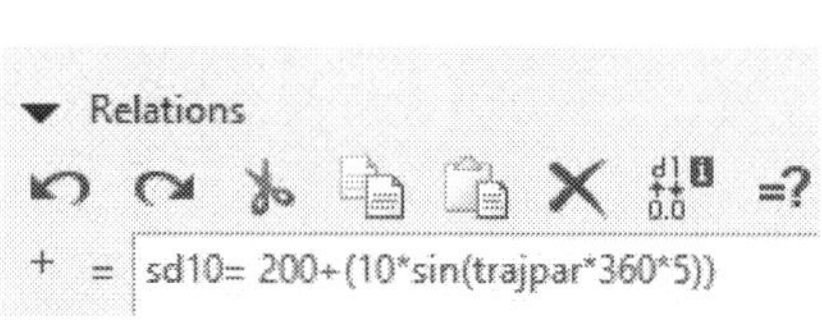

Figure 9-36 *Equation used for variable section sweep*

Helical Sweep

Ribbon: Model > Shapes > Sweep drop-down > Helical Sweep

The **Helical Sweep** tool is used to create helical sweep features by sweeping a section along a helical trajectory, refer to Figure 9-37. The main use of this tool is to create the helical springs and threads. To create a helical sweep feature, you have to define a trajectory that will specify the shape and height of the helix, a pitch value, and a cross-section which will be swept along the trajectory with the specified pitch value. The distance of the trajectory from the center line defines the radius of the helical path. The resulting helical sweep feature after specifying the attributes is shown in Figure 9-38.

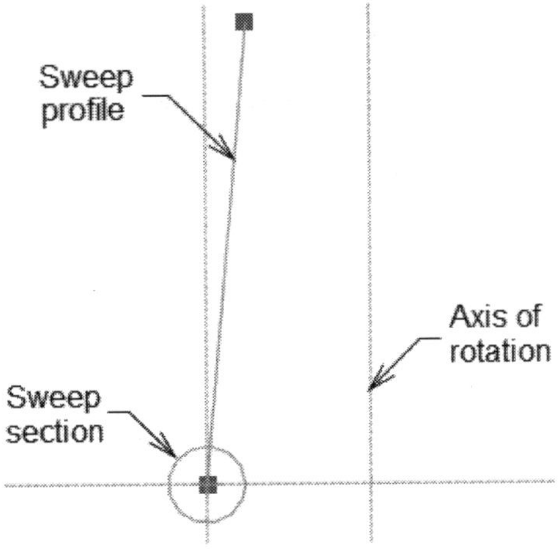

Figure 9-37 *Attributes for creating a helical sweep*

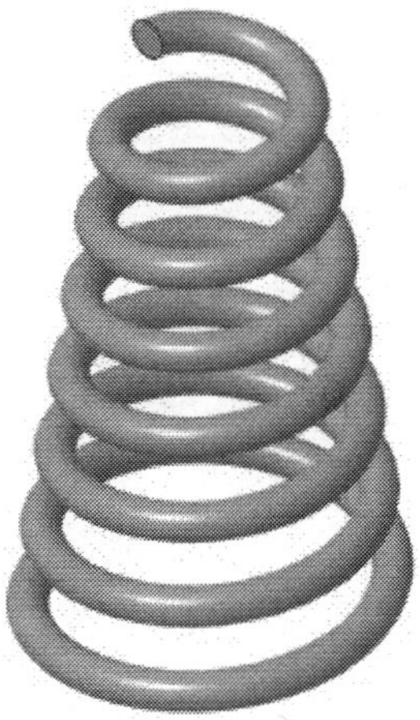

Figure 9-38 *Resulting feature*

When you choose the **Helical Sweep** tool from the **Sweep** drop-down in the **Shapes** group, the **Helical Sweep** dashboard is displayed, as shown in Figure 9-39. The options in this dashboard are discussed next.

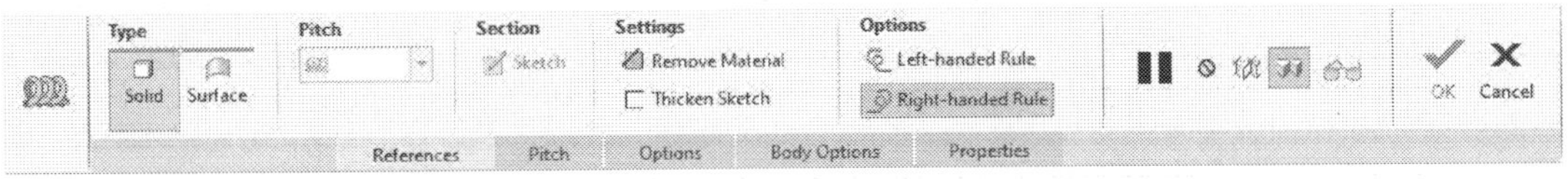

Figure 9-39 *Partial view of the* ***Helical Sweep*** *dashboard*

Few options in the dashboard are same as those in the **Extrude**, **Revolve**, and **Sweep** dashboards and have already been explained. The remaining options in the **Helical Sweep** dashboards are discussed next.

The **Pitch** edit box ( ) in the dashboard is used to specify the pitch value for the helical sweep. The **Right-handed Rule** button is selected by default in the dashboard. As a result, the helical sweep is created in a counterclockwise direction. To create a helical sweep feature in which the section is swept in the clockwise direction from the start sketch, choose the **Left-handed Rule** button in the **Helical Sweep** dashboard.

The options available in the tabs of the **Helical Sweep** dashboard are discussed next.

References Tab

Choose the **References** tab; a slide-down panel will be displayed, as shown in Figure 9-40. The **Helix profile** collector is used to select the profile (trajectory) for helical sweep. If you have not created a profile then choose the **Define** button to create the helix profile in the sketcher environment. This sketch will be an internal sketch for the feature and you can edit it, if required, by choosing the **Edit** button. In case of external sketch, you can break the association between the feature and selected sketch by choosing the **Unlink** button. Note that, if you choose the **Unlink** option, then a copy of the selected sketch will be created as an internal sketch for the helical sweep feature.

The **Flip** button next to **Start point** is used to toggle the start point of the helical sweep between the two ends of the helix sweep profile.

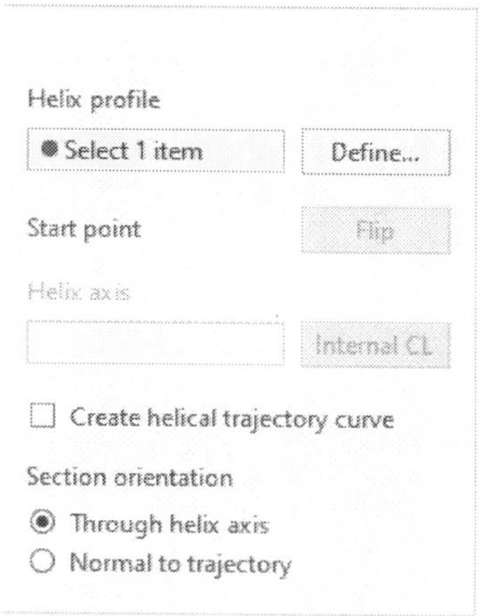

***Figure 9-40** The **References** slide-down panel*

The **Helix axis** collector is used to select the revolution axis for the helical sweep feature. The **Internal CL** option is used to select the centerline created in the sketcher environment with the helix profile as the revolution axis for the helical feature. The procedure of defining the axis of revolution for the **Helical Sweep** tool is similar to that discussed in the **Revolve** tool.

The options in the **Section orientation** area of the **References** slide-down panel are used to specify the orientation of the internal sketch plane. By default, the **Through helix axis** radio button is selected, making the sweep section to lie on a plane that passes through the axis of revolution. Select the **Normal to trajectory** radio button if you want to make the sweep section to be oriented normal to the helical sweep profile (trajectory).

Pitch Tab

The options in this tab are used to create a helical sweep feature with varying pitch values. Choose the **Pitch** tab; a slide-down panel will be displayed, refer to Figure 9-41. The **#** column displays a numbered list of pitch points in the table. The **Pitch** column displays the pitch value at selected point. The **Location Type** column displays the type of method by which the pitch point placement is determined. The location type for the pitch point can be specified after adding pitch values for the start and end points, refer to Figure 9-41. By default, the **By Value** option is selected in the drop-down list in the **Location Type** column after you add the third pitch point. As a result, the location for the selected pitch point is specified by a distance value from the start point of the helix profile (trajectory). Selecting the **By Reference** option in the drop-down list will allow you to select a point, vertex, plane, or surface for determining the pitch point location. The **By Ratio** option can be used to set the pitch point location using a ratio of the helix profile (trajectory) length from the start point. The **Location** column in the **Pitch** slide-down panel displays the location to the value for the pitch points added to the table.

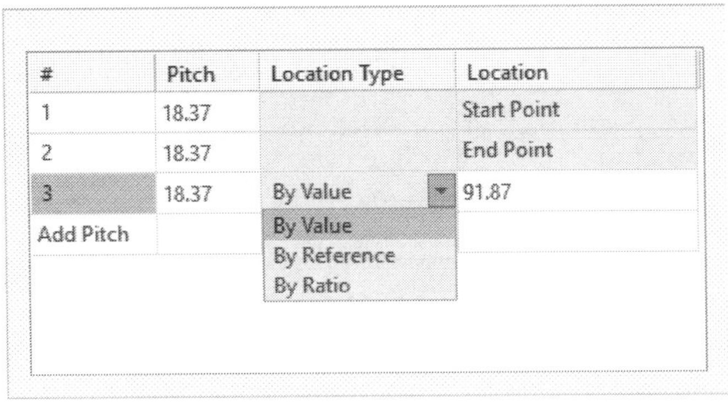

***Figure 9-41** The **Pitch** slide-down panel*

Options Tab

When you choose this tab, a slide-down panel is displayed, as shown in Figure 9-42. The **Capped ends** check box is activated only when you choose the **Surface** button in the **Helical Sweep** dashboard. On using this option, the ends of the sweep feature which are created using an open helix profile (trajectory) are closed. By default, the **Constant** radio button is selected in the **Along trajectory** area resulting a constant section helical sweep. To vary the cross- section of the helical sweep feature using parameters and relations, select the **Variable** radio button.

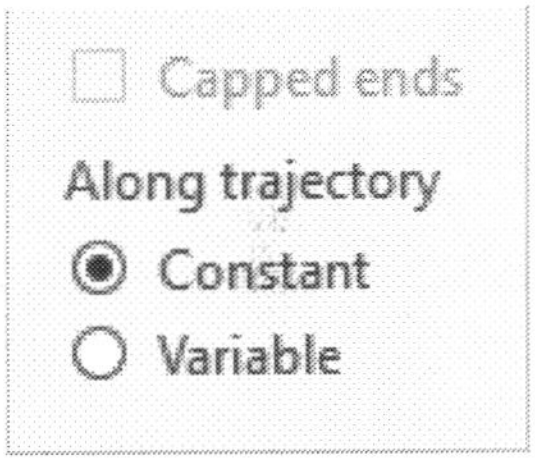

***Figure 9-42** The **Options** slide-down panel*

Note

You can use the Trajpar parameter to vary the dimensions of the sweep section along the helix profile.

Creating a Helical Spring with Variable Pitch

The brief procedure to create a variable pitch spring is discussed next.

1. Choose the **Helical Sweep** tool from the **Sweep** drop-down in the **Shapes** group of the **Ribbon** to invoke the **Helical Sweep** dashboard.

2. Select the **Through helix axis** radio button from the **References** tab if not selected by default.

3. Choose the **Define** button from the slide-down panel of the **References** tab to invoke the **Sketch** dialog box.

4. Select the sketching plane and then orient it. Draw the sketch of the sweep trajectory and the center axis, as shown in Figure 9-43.

5. Choose the **OK** button in the **Close** group of the **Sketch** tab and exit the sketcher environment and specify the value of the pitch in the **Pitch** edit box of the **Pitch** tab in the dashboard, refer to Figure 9-44.

6. After entering the pitch value at the start point, choose the **Add Pitch** option in the **Pitch** tab to specify the pitch value at the end point, refer to Figure 9-44.

7. After specifying the pitch value at the two ends, choose the **Add Pitch** option again; a drag-handle is displayed on the centerline of trajectory. Specify the value of location and the pitch value of that point, refer to Figure 9-44.

8. Similarly, you can specify the pitch values at other points by using the **Add Pitch** button, as shown in Figure 9-44.

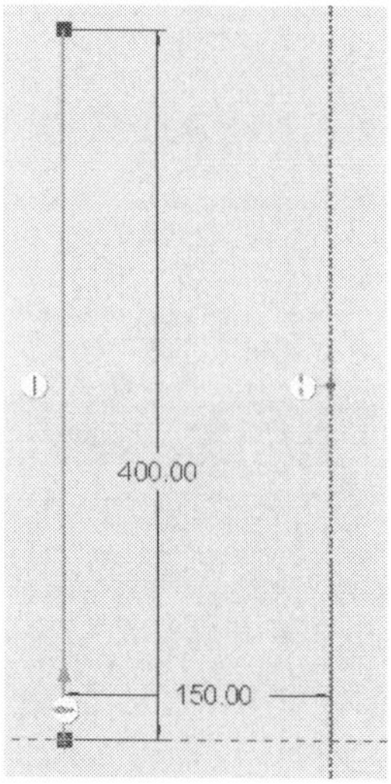

Figure 9-43 *Trajectory for helical sweep*

#	Pitch	Location Type	Location
1	18.00		Start Point
2	18.00		End Point
3	40.00	By Value	80.00
4	65.00	By Value	160.00
5	90.00	By Value	240.00
Add Pitch			

Figure 9-44 *Pitch values specified for helical sweep*

9. After all the pitch values are entered, select the **Sketch** option to enter the sketcher environment and draw the section of the spring at the intersection of the two infinite dotted lines, refer to Figure 9-45.

10. Exit the sketcher environment and choose the **OK** button from the **Close** group in the dashboard. The model of the spring will be displayed in the drawing area, as shown in Figure 9-46.

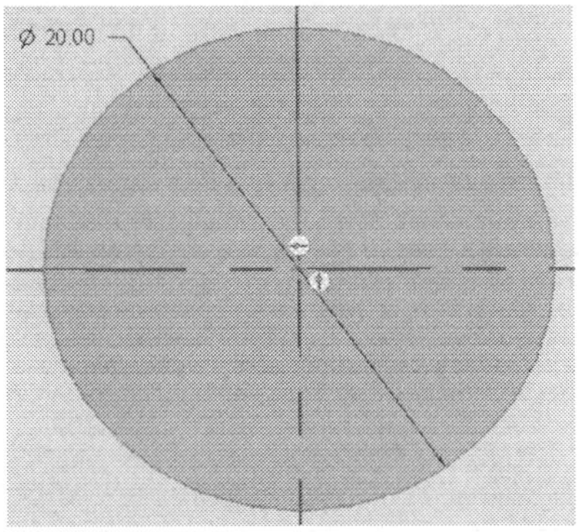

Figure 9-45 *Section for helical sweep*

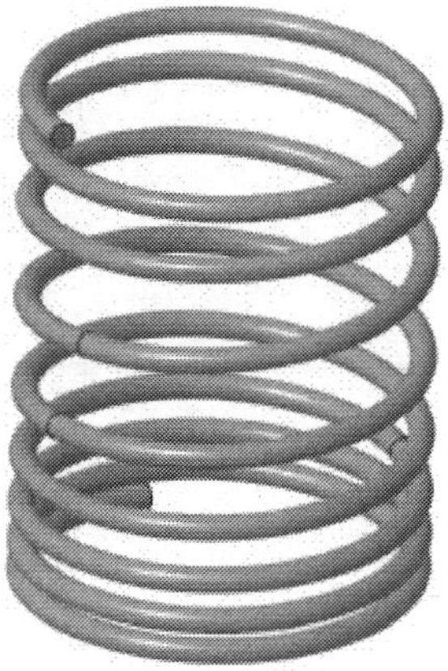

Figure 9-46 *Final model of the spring*

Note

1. You can use any geometric entity such as spline, line, or arc to create the sweep trajectory.

2. If you do not specify any additional pitch values, a constant pitch helical sweep is created.

Volume Helical Sweep

Ribbon: Model > Shapes > Sweep drop-down > Volume Helical Sweep

The **Volume Helical Sweep** tool is used to create a volume helical cut. The process of this tool resembles machining processes, particularly turning and milling. Using this tool, a rotating 3D object formed by a revolved sketch, moves along a helix axis and removes material from the part. This tool creates accurate 3D geometric representation for parts that are machined with cutting tools.

To create a volume helical sweep, choose the **Volume Helical Sweep** tool from the **Sweep** drop-down in the **Shapes** group; the **Volume Helical Sweep** dashboard is displayed, as shown in Figure 9-47. This tool can be used only if a base feature is already present in the graphics area. Assume that the component shown in Figure 9-48 is the base feature and it is selected to display dimensions.

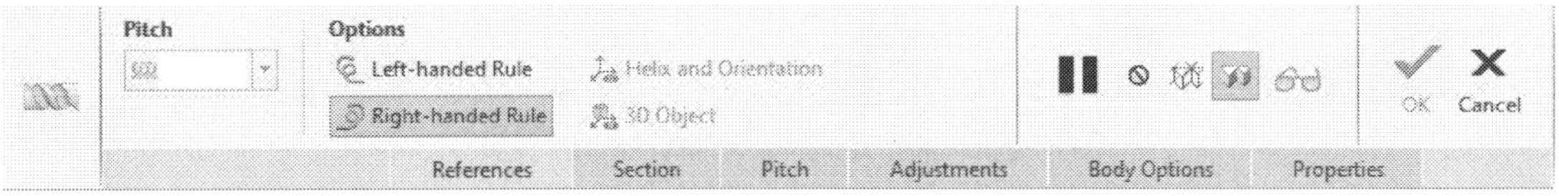

Figure 9-47 *Partial view of the **Volume Helical Sweep** dashboard*

Some options in the dashboard are same as those in the **Extrude**, **Revolve**, and **Sweep** dashboards and have already been explained. The remaining options in the **Volume Helical Sweep** dashboard are discussed next.

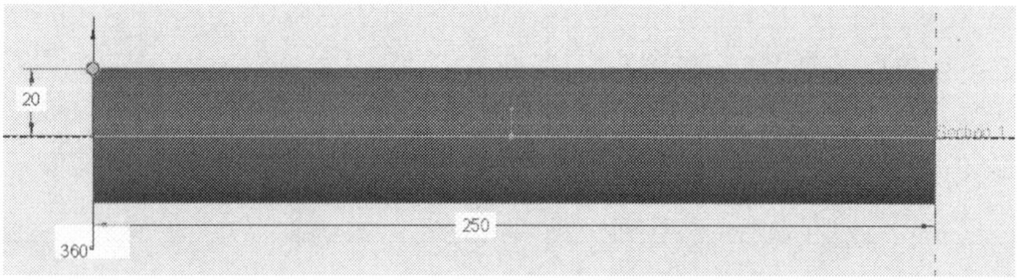

Figure 9-48 *Base feature for volume helical sweep*

Pitch

This edit box is used to specify the pitch value.

Left-handed Rule

This button is used to set the sweep direction using the left-hand rule.

Right-handed Rule

This button is used to set the sweep direction using the right-hand rule.

Helix and orientation

This button is used to show the helix and orientation of the 3D object with 3D dragger. This button gets' activated only when the sketch for the helix direction is drawn.

3D object

This button is used to show the 3D object. This button gets activated only when the **3D dragger** button is turned on.

Creating a Volume Helical Sweep

1. Invoke the **Volume Helical Sweep** dashboard and choose the **References** tab, as shown in Figure 9-49.

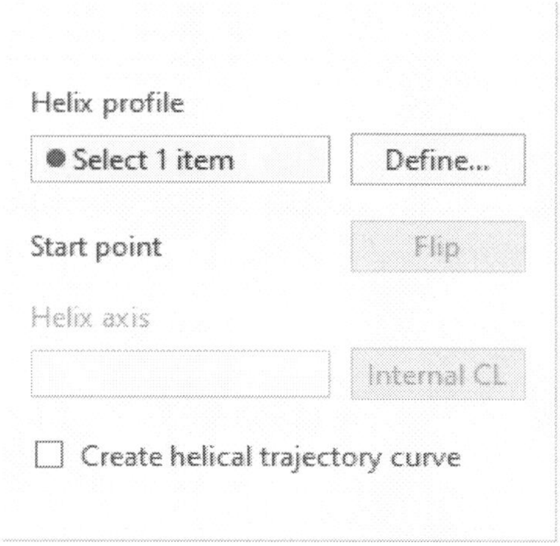

Figure 9-49 *The **References** tab*

2. Next, choose the **Define** button; the **Sketch** dialog box is displayed. Select a plane to sketch an open helix profile, refer to Figure 9-50, and choose **OK**. You can also select a predefined open sketch.

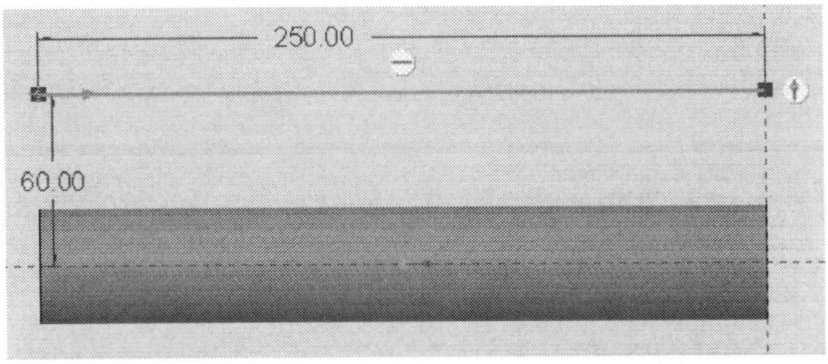

Figure 9-50 *The sketch for open helix profile*

3. Now, select the helix axis for revolution if not drawn in the sketch, refer to Figure 9-51. Next, choose to the **Section** tab and select the **Sketched section** radio button. Next, choose the **Create/Edit section** button to create a sketch for the tool section, refer to Figure 9-52. After defining the sketch choose the **OK** button. The **Selected section** radio button is used to select the pre-sketched section.

Note

1. Section profile should contain a line attached to the y-axis (colored green). y-axis serves as the axis of revolution of the 3D object.

2. Section should not overlap the y-axis and must contain lines and circular arcs. Section should be closed and convex and the resulting 3D object must be convex.

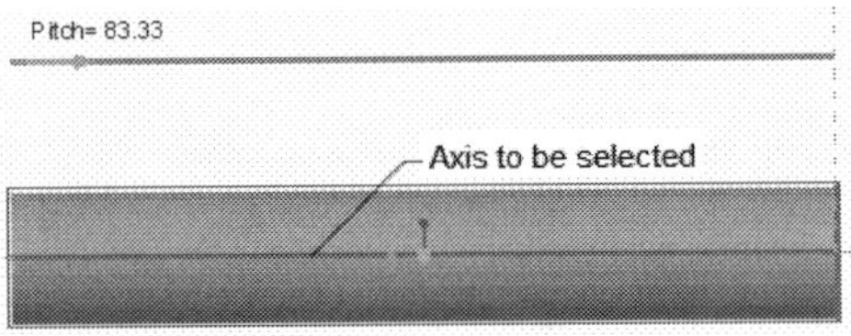

Figure 9-51 *The axis to be selected*

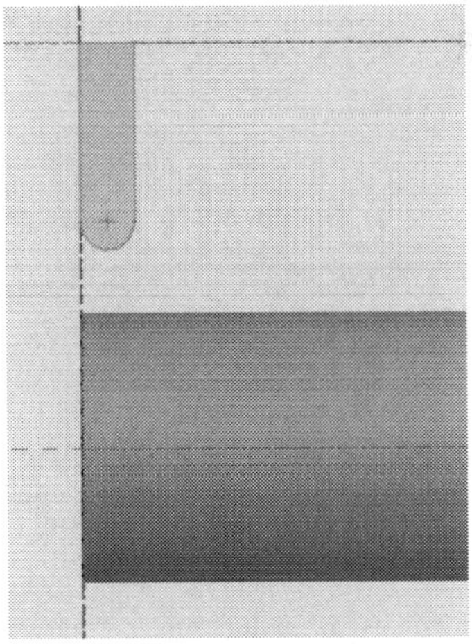

Figure 9-52 *Sketch for tool section*

Note that for better visibility of sketch, dimensions and constraints are hidden in Figure 9-52.

Tip

In some cases, the sketch automatically snaps to entities in the model, which could change the resulting geometry. To prevent the sketch from automatically snapping to the model, perform one of these actions:

- *To disable snapping as the default, click* ***File > Options > Sketcher****. Under* ***Sketcher references****, clear the* ***Allow snapping to model geometry*** *check box.*

- *To disable snapping for the current action, while the* ***Sketch*** *tab is open, hold down the* ***SHIFT*** *key while you sketch*

4. Enter the required pitch value in the **Pitch** edit box and drag the 3D object downwards with the help of axis dragger upto the required penetration distance, refer to Figure 9-53. Note that you can also dimension the tool section for accurate penetration distance.

   The 3D object can be tilted at a constant angle around the x- or z-axis. To do so, follow the steps given below:

   a. Choose the **Adjustments** tab.
   b. Choose either the **X-axis** or **Z-axis** radio button to tilt the 3D object.
   c. Enter a value from -90° through 90° in the **Tilt angle** edit box.

5. Choose the **OK** button; the volume helical sweep is created, as shown in Figure 9-54.

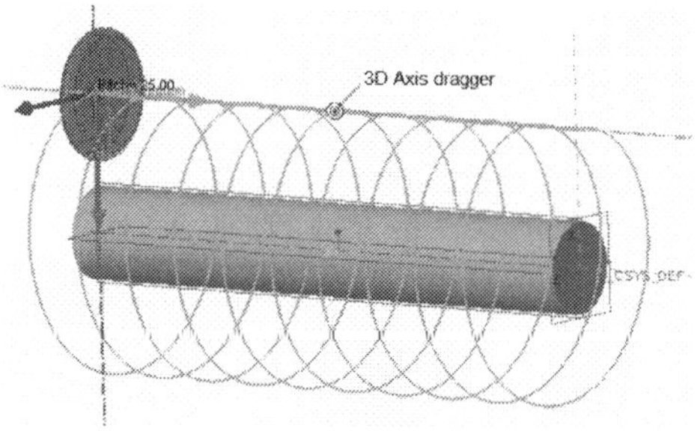

Figure 9-53 *The 3D axis dragger*

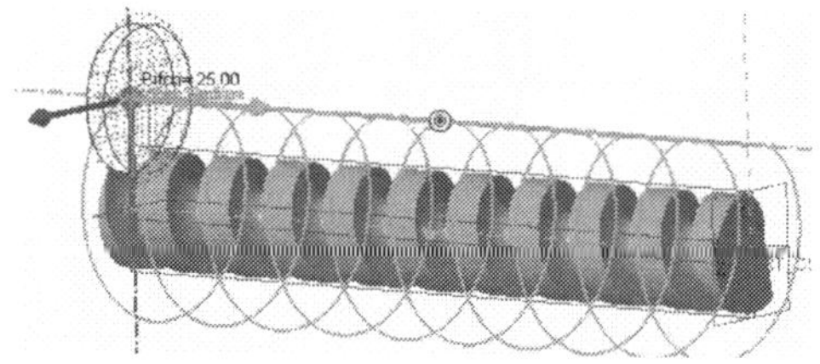

Figure 9-54 *The volume helical sweep created*

BLEND FEATURES

A blend feature is created using a minimum two planar sections joined at their vertices with transitional surfaces to form a continuous feature. Depending upon the direction of blending, three types of blend features can be created in Creo: Parallel blend, Swept blend, and Rotational blend. The tools to create blend features are discussed next.

Blend

Ribbon: Model > Shapes drop-down > Blend

Enhanced

The **Blend** tool is used to create a feature that has varying cross-sections. To create a blend feature using the **Blend** tool, all sections should lie on parallel planes.

To invoke this option, choose **Blend** from the expanded **Shapes** group; the **Blend** dashboard will be displayed, as shown in Figure 9-55. Options and tabs available in the dashboard are discussed next.

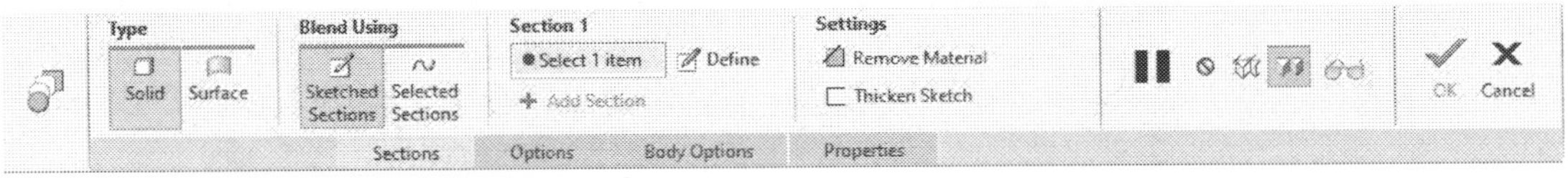

*Figure 9-55 Partial view of the **Blend** dashboard*

Solid

The **Solid** button is chosen by default and is used to blend two or more sketches in a solid form and create solid feature. Note that the sketch that is drawn for creating solid feature need to be a closed loop.

Surface

The **Surface** button is used to blend two or more sketches in a surface form and create surface feature. The sketch that is drawn for the surface model need not be a closed loop.

Sketched Sections

The **Sketched Sections** button is used to create sketches to blend. The options for creating sketches are available in the **Sections** tab that is discussed further.

Selected Sections

The **Selected Sections** button is used to select sketches available in the drawing area. The options for inserting sketches are available in the **Sections** tab.

Remove Material

The **Remove Material** button is used to remove a material from an existing feature. Hence it is displayed in the **Blend** dashboard only after base feature is created.

Thicken Sketch

The **Thicken Sketch** button is used to create thin parts with specified thickness. When you choose this button, the **Change direction to thicken between one side, other side, or both sides of sketch** button and a dimension edit box appear on the right side of this button.

Sketch Plane Location

The Sketch plane location option appears after you have specified first section for the blend feature. It is used to specify the location of a sketch plane by providing an offset distance from another sketch plane. The drop-down list adjacent to Sketch plane location option displays the section from which the current sketch is offset. To specify the offset value, choose the edit box next to the **Section** drop-down list.

This option is used to specify section plane location using a reference. You can select the reference that defines the location of the current sketching plane in the reference collector.

Sections Tab

When you choose the **Sections** tab from the dashboard, the **Sections** slide-down panel will be displayed, as shown in Figure 9-56. This panel has two radio buttons that are discussed next.

*Figure 9-56 The **Sections** slide-down panel*

Sketched sections

This radio button is used to create sketches to form a blend. When you select this radio button, the **Define** button will be displayed in the panel. It is used to enter the sketcher environment. You can also use the **Insert** button displayed in the panel to add more sketches to the blend. When you insert second section for blending, the **Sketch plane location defined by** option is displayed in the slide-down panel. By default, the **Offset dimension** radio button is selected in the panel. As a result, you can specify the offset distance of the current section plane in the Offset distance edit box. You can choose the **Reference** radio button if you want to specify the distance of the plane using a reference.

Selected sections

This radio button is used to select sketches from the drawing area to form the blend. You can use the **Insert** button displayed in the panel to add more sketches to the blend. When you choose this button, the **Move Up** and **Move Down** buttons will be displayed in the panel. You can use these buttons to arrange the inserted sketches.

Add Blend Vertex

This button is used to insert a blending vertex in the current section.

Options Tab

The **Options** tab contains the parameters related to the blend feature. These options are discussed next.

Straight

This radio button is used to connect the vertices of all sections using straight lines, refer to Figure 9-57.

Smooth

This radio button is used to connect the vertices of all sections using smooth curves, refer to Figure 9-58.

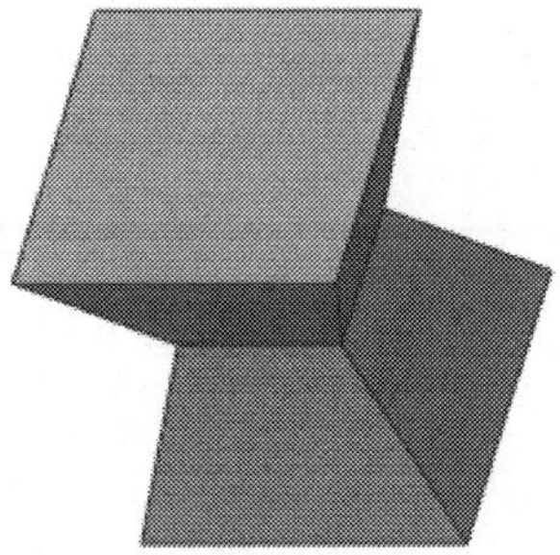

Figure 9-57 *Blend with straight edges*

Figure 9-58 *Blend with smooth edges*

Capped ends

The **Capped ends** check box is available in the **Start and End Sections** area and is used to close the ends of the blend feature. This check box will be activated when you choose the **Surface** option from the dashboard.

Tangency Tab

Choose the **Tangency** tab to display the slide-down panel, as shown in Figure 9-59. The **Tangency** tab will only be available when you select the **Smooth** radio button from the **Options** tab. This tab is used to set the boundary of created or selected sketches in the **Free**, **Tangent**, or **Normal** condition. By default, the **Free** option is selected in the **Condition** column. As a result, the sketch is not influenced from its side references, as shown in Figure 9-60.

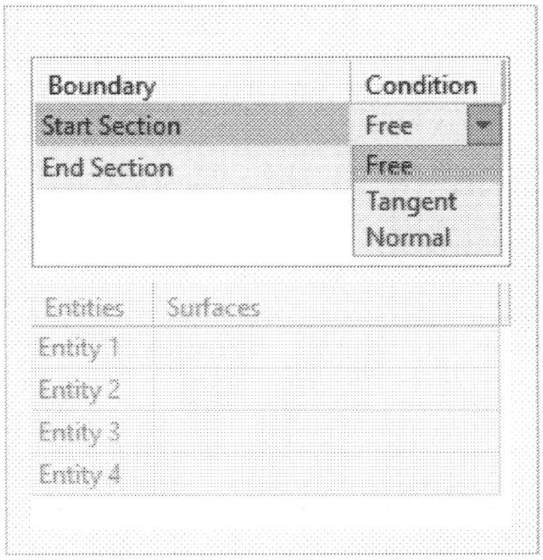

Figure 9-59 Tangency *slide-down panel*

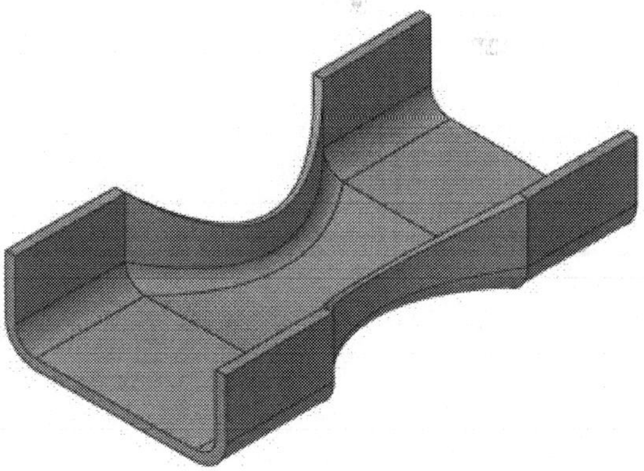

Figure 9-60 *Free condition applied to the sections*

When you select the **Tangent** option from the drop-down list corresponding to the **Start Section** or **End Section** options the entity of that section is highlighted in the drawing area and you are prompted to select a surface that lies on the highlighted boundary component. Select the surfaces of the model as they are highlighted one by one in the drawing area; the surfaces of the resulting feature are blended tangentially to the adjacent surfaces of the feature, as shown in Figure 9-61. Choose the **Normal** option to blend the section plane and the geometry of the resulting feature in normal direction, as shown in Figure 9-62.

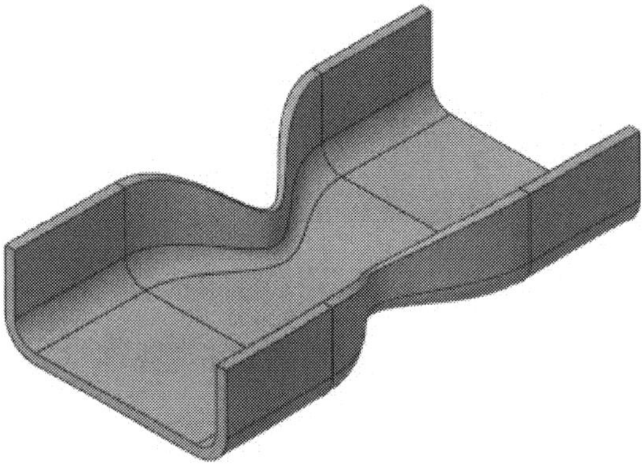

Figure 9-61 *Model created using the* ***Tangent*** *option*

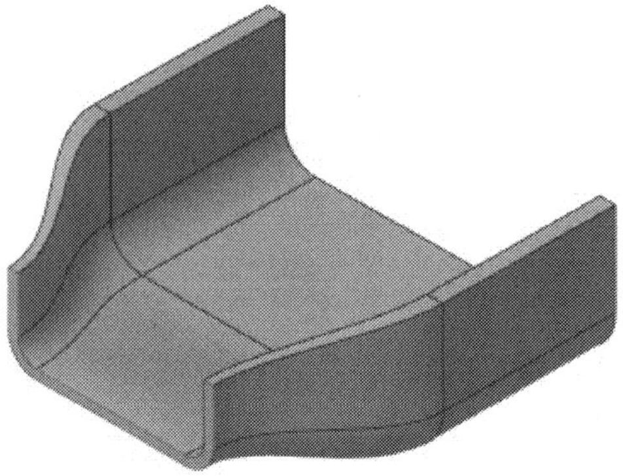

Figure 9-62 *Blend created using the* ***Normal*** *option*

Tip

You can right-click on a tangency tag located at both ends of the feature to set the tangency condition for that end.

Note

1. All sections in a blend feature must have equal number of entities. However, you can blend a point with any section irrespective of the number of entities.

2. While drawing a section, the start point of all sections should be in same direction in order to avoid twisted blend features. By default, the start point of any entity drawn to define a section is considered to be the start point of the section. To change the start point of a section, select the point to be defined as the start point and hold the right mouse button to display the shortcut menu. Choose the ***Start Point*** *option from this shortcut menu to change the start point.*

3. You can also drag the start point handle on the section to change the start point of the selected section.

Using Blend Vertex

As mentioned earlier, each section of the blend feature must have equal number of blending vertex. If you are using the **Selected sections** option to select the sketched sections, you can use the **Add Blend Vertex** option in the **Sections** slide-down panel to make the number of blending vertex equal for each section.

When you are using the **Sketched Sections** option to create sections, you can add a blend vertex while drawing the sketch. To do so, select the point where you want to place the blend vertex; the selected point is highlighted in red. Next, choose the **Blend Vertex** option from the shortcut menu or from the **Feature Tools** flyout in the **Setup** group of the sketcher environment; the blend vertex will be placed at the selected point. After adding a blending vertex; a circle appears at the vertex. Note that more than one blend vertex can be created at the same point. Each additional vertex will create a concentric circle of bigger diameter on the selected point.

For example, to create a blending between a square and a triangle, add blend vertex on a point other than the start point of the triangle. The two vertices of the triangle will be blended with the two vertices of the rectangle and the blend vertex in the triangle will be blended with the remaining two vertices in the rectangle, as shown in Figure 9-63.

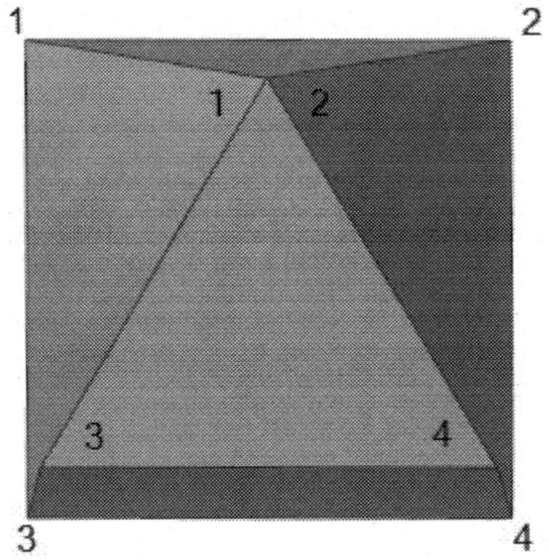

***Figure 9-63** Blending a square with a triangle by using the blend vertex*

Another method of adding a blending vertex to the section is to divide the sketched entities using the **Divide** tool located in the **Editing** group of the sketcher environment.

Swept Blend

Ribbon: Model > Shapes > Swept Blend

The **Swept Blend** tool is used to create a blend that is a combination of sweep and blend features. In this type of blend, multiple sections are placed along the main trajectory and the geometry is created by blending the sections while sweeping along the trajectory. The sections can vary in shape and size along the swept blend. A swept blend can have only two trajectories: origin trajectory and secondary trajectory. To define a trajectory of the swept blend, you can select a sketched curve, a chain of datum curves, or edges.

To create a swept blend feature, choose the **Swept Blend** tool from the **Shapes** group; the **Swept Blend** dashboard will be displayed, as shown in Figure 9-64. Note that the trajectories should be present in the drawing area so that they can be selected for the creation of the swept blend feature. However, you can draw trajectories using the **Sketch** tool available in the **Datum** flyout of the **Swept Blend** dashboard. The options in the **Swept Blend** dashboard are discussed next.

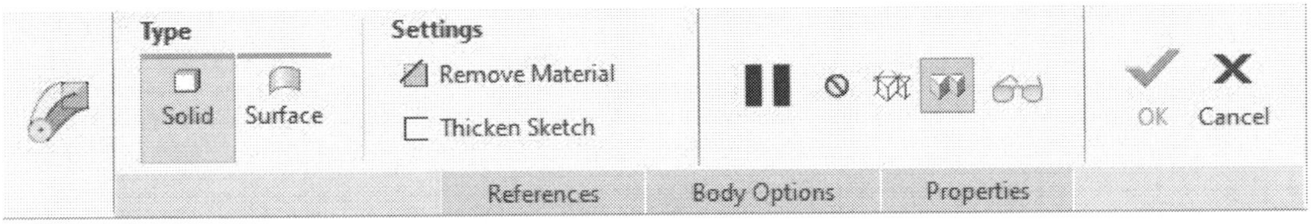

***Figure 9-64** Partial view of the **Swept Blend** dashboard*

References Tab

When you choose the **References** tab, the **References** slide-down panel will be displayed, as shown in Figure 9-65. The options available in slide-down panel are discussed next.

Trajectories Collector

This collector is used for selecting the trajectories. The trajectories that you select will be listed in this collector. By default, the first trajectory that you select is listed as the origin trajectory (also referred to as normal trajectory) and the second trajectory is listed as the secondary trajectory, refer to Figure 9-66. Note that you need to select the origin trajectory first and then draw the sketch of the section. This is because the sketch plane will be oriented only if you have defined the origin trajectory. The direction of the sweep along the origin trajectory is displayed by an arrow. You can click on the arrow to reverse the direction of swept blend creation.

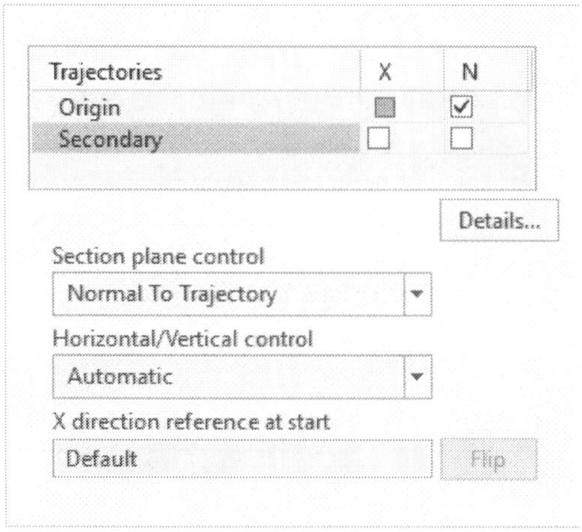

***Figure 9-65** The **References** slide-down panel*

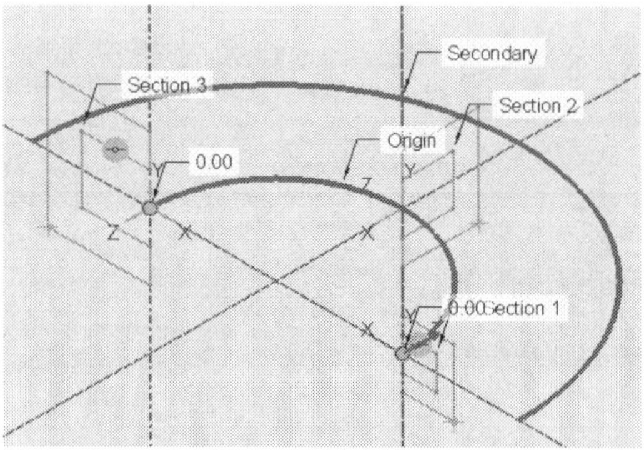

***Figure 9-66** Trajectories selected for swept blend*

The **Trajectories** collector has two columns and the selected trajectories are listed in rows. The first column is **X**. The second column is **N**. Note that only one trajectory can be selected as X-trajectory and only one trajectory can be selected as N-trajectory.

The options in the **References** slide-down panel are the same as the options in the **References** slide-down panel of the **Sweep** tool and have been discussed earlier.

Sections Tab

When you choose the **Sections** tab, the **Sections** slide-down panel will be displayed, as shown in Figure 9-67. The options in the slide-down panel are discussed next.

Sketched Sections

This radio button is used to sketch the section that is used to create the swept blend.

Selected Sections

This radio button is used to select the section that is used to create the swept blend.

The **Sections** column displays a list of sections specified for the swept blend and the **#** column displays the number of blending vertices in the sketched section. The **Insert** button is used to add a section for swept blend and will be activated only when you select the **Selected Sections** radio button. To remove a section from the **Sections** list, choose the **Remove** button. If you choose the **Sketch** button when the **Sketched Sections** radio button is selected then the sketcher environment will be invoked where you can draw a section for swept blend.

The **Section Location** collector is used to select a vertex, datum point, or end point of a curve to locate the section. By default, the start point of the origin trajectory is selected in this collector. The **Rotation** edit box is used to specify the rotation of a section about the z-axis. The value of rotation angle ranges from -120° to +120° for each section selected in the **Sections** list.

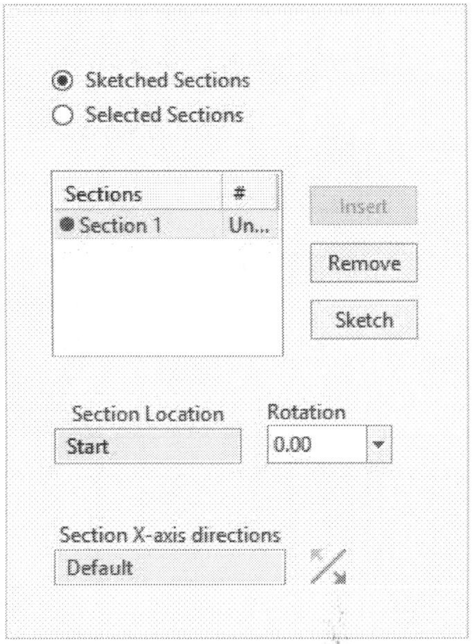

Figure 9-67 *The **Sections** slide-down panel*

The **Section X-axis directions** collector will be activated only when you select the **Automatic** option from the **Horizontal/Vertical control** drop-down list in the **References** slide-down panel. This collector is used to select x-direction reference for the selected section in the **Sections** list.

The **Add Blend Vertex** option gets activated when you select the **Selected Sections** radio button. You can use this button to add vertices to the pre-sketched sections. Remember that the blend vertex can be added to the start and end sections only.

Tangency Tab

The options in the **Tangency** tab are used to control tangency between the start and end sections of the swept blend feature. These options are same as the options explained in the **Blend** tool. Note that, the **Tangency** tab will get disabled when you choose the **Create a thin feature** button in the dashboard.

Options Tab

When you choose the **Options** tab, the **Options** slide-down panel will be displayed, as shown in Figure 9-68. The **Capped ends** check box in this panel is used to cap both the end faces of the feature. This check box will be activated only when you select the **Create a Surface** option in the dashboard. By default, the **Adjust to keep tangency** check box is selected and is used to maintain the tangency of the surfaces created between blended sections. The **No blend control** radio button is selected by default. As a result, no blend control is applied to the resulting geometry. If you select the **Set perimeter control** radio button, the perimeter

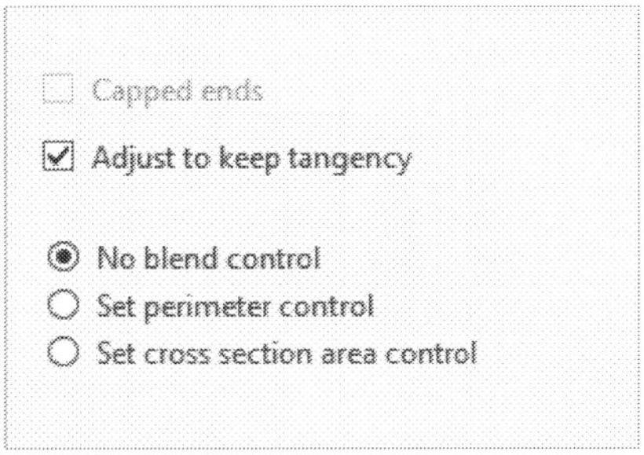

Figure 9-68 *The **Options** slide-down panel*

of the blend geometry varies linearly between sections. As you select this radio button, the **Create curve through center of blend** check box is enabled in the panel. Select this check box if you want to place the curve at the center of the swept blend. The **Set cross section area control** radio button is used to specify the area of cross-section at specific locations of the swept blend. Select this radio button to enable the Cross-sections table which allows you to select points on the trajectory and specify the area of cross-section at selected points. The **Location** column in the table displays the location of the section and the **Area** column displays the area specified at the location.

Figure 9-69 shows origin and secondary trajectory selected for swept blend. Figure 9-70 shows the completed feature created using two sections. Both sections are connected to each other at their vertices through the trajectories.

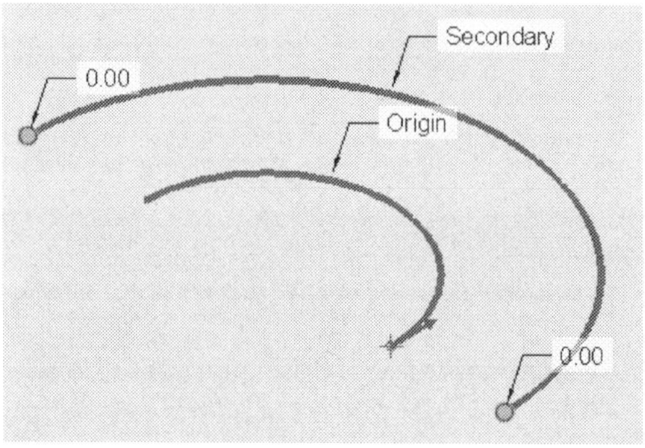

Figure 9-69 *Trajectories selected for swept blend*

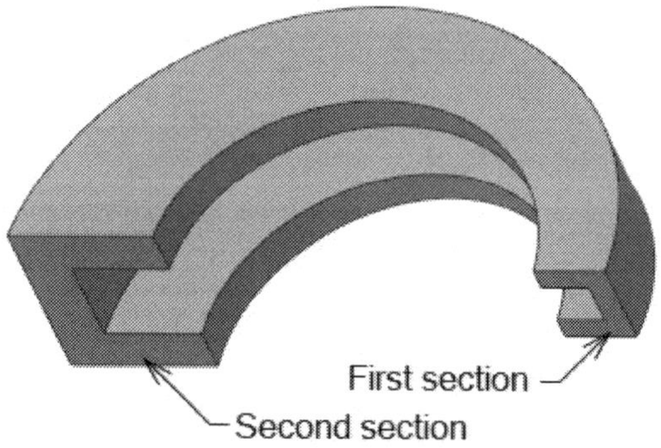

Figure 9-70 *Resulting swept blend feature*

Figure 9-71 shows three sections and a trajectory used to create the feature. Note that the number of entities should be same in all the sections.

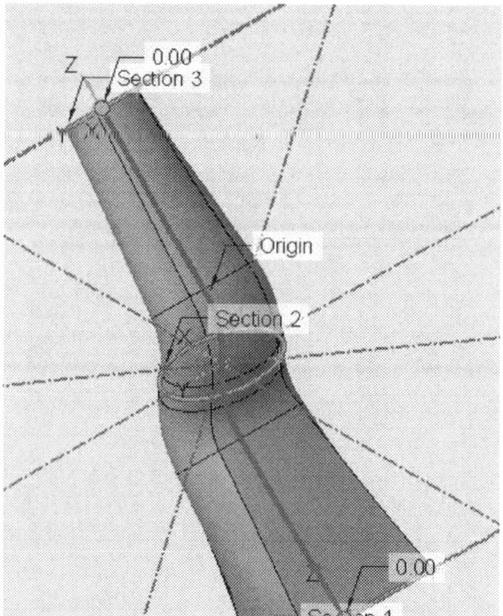

Figure 9-71 *Preview of the swept blend feature with three sections and a trajectory*

Rotational Blend

Ribbon: Model > Shapes drop-down > Rotational Blend

The rotational blends are created by blending the sections that are rotated about an axis of revolution. If the first section contains an axis of revolution or a centerline, it will be automatically selected as the axis of revolution. If the first sketch does not contain an axis of revolution or centerline, you can select a datum axis, part edge, or sketched entity as the axis of revolution. To create a rotational blend feature, all the sections must lie in planes that intersect the axis of rotation. For sketching the sections, you can define the sketch plane by using an offset angle from another section in the blend or by selecting a reference.

To create a rotational blend feature, choose the **Rotational Blend** tool from the **Shapes** group drop-down; the **Rotational Blend** dashboard will be invoked. Next, choose the **Sections** tab; a slide-down panel will be displayed. If you want to create sketch for the rotational blend then choose the **Define** button and draw the sketch in sketcher environment. If you want to use a pre sketched section, then select the **Selected sections** radio button and select the section. Remember that the first section should contain an axis of revolution in order to complete the feature, or you can select an axis of revolution using the **Internal CL** collector. After you have specified the first section, choose the **Insert** button in the **Sections** slide-down panel to insert more sections in the rotational blend feature. Next, specify the angle of offset in the **Offset dimension** edit box or select a reference in the **Through** collector by selecting the **Reference** radio button, refer to Figure 9-72. You can also use the drag handle to specify the rotation of the section. Next, choose the **OK** button in the dash board to complete the procedure, refer to Figure 9-73.

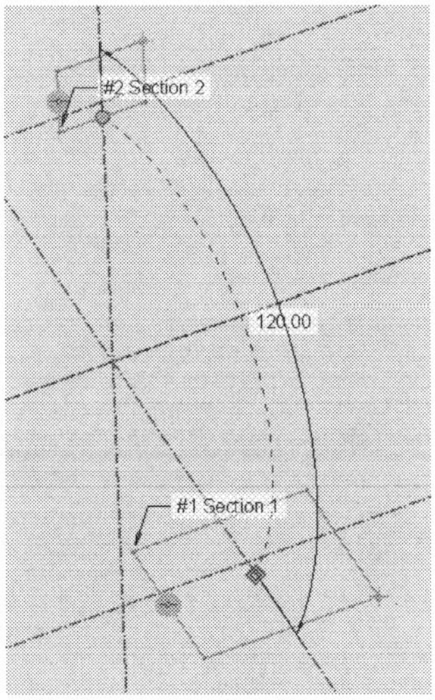

Figure 9-72 The two sections used to create the rotational blend feature

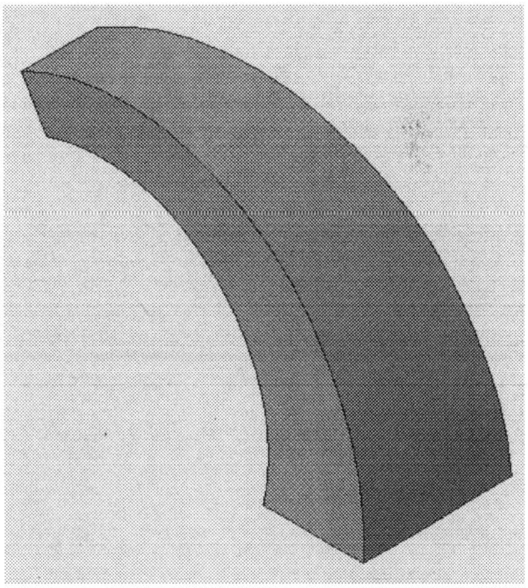

*Figure 9-73 Model created using the **Rotational Blend** tool*

Note

1. The value of the offset angle can not exceed 120°.

*2. You can select the **Connect end and start sections** check box in the **Options** tab to create a closed blend feature. But for creating a closed blend feature, at least three sections are required.*

TUTORIALS

Tutorial 1

In this tutorial, you will create the model shown in Figure 9-74. The orthographic views of the model are shown in Figure 9-75. **(Expected time: 25 min)**

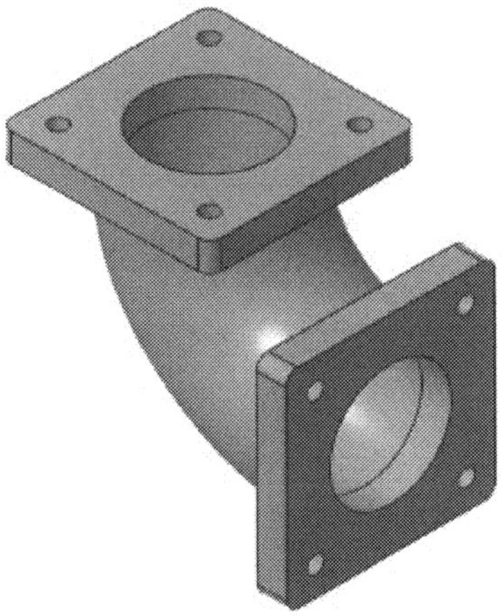

Figure 9-74 *Isometric view of the model*

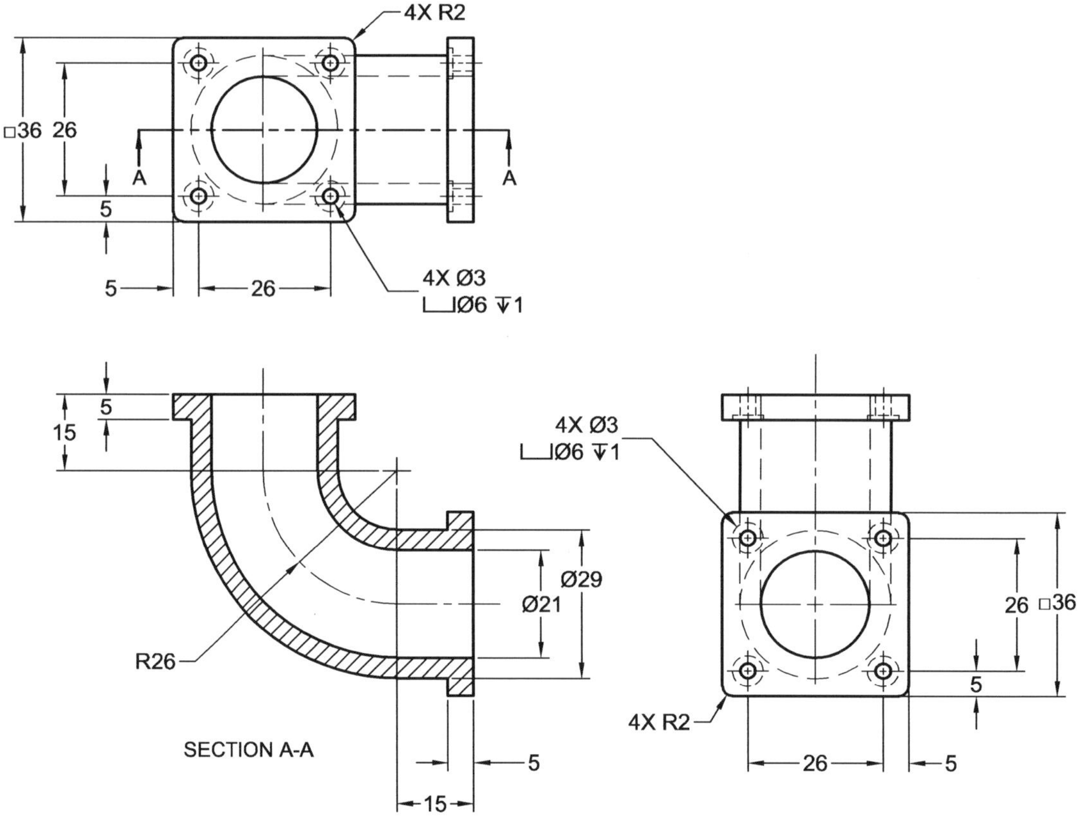

Figure 9-75 *Orthographic views of the model*

Examine the model and determine the number of features in it. The model consists of six features, refer to Figure 9-74.

The following steps are required to complete this tutorial:

a. Start a new file and create the base feature.
b. Create the shell feature of given thickness.
c. Create the third and fourth extrude features on the two ends of the sweep feature respectively.
d. Create the counterbore hole on the third and fourth features.
e. Pattern the counterbore holes.

When the Creo Parametric session starts, the first task is to set the working directory. As this is the first tutorial of the chapter, you need to create a folder *c09* if it has not been created earlier and set it as the Working Directory.

Starting a New Object File

1. Start a new part file and name it as *c09tut1*.

   The three default datum planes as well as the **Model Tree** are displayed in the drawing area.

Sketching the Trajectory

You need to sketch the trajectory of the sweep feature on the **FRONT** datum plane.

1. Invoke the **Sketch** tool from the **Datum** group. Next,select the **FRONT** datum plane as the sketching plane.
2. Select the **Top** datum plane as the reference plane from the **Reference** collector.
3. Select the **Top** option from the **Orientation** drop-down list and then choose the **Sketch** button from the **Sketch** dialog box.

   After you have selected the planes for orientation, the system takes you to the sketcher environment.
4. Draw the sketch of the trajectory and apply dimension, as shown in Figure 9-76.

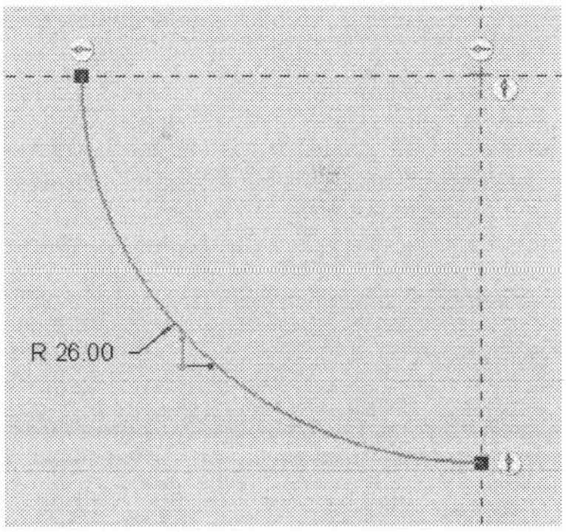

Figure 9-76 *The trajectory drawn for sweep feature*

Invoking the Sweep Tool

You need to invoke the **Sweep** tool in the **Shapes** group from the **Ribbon**.

1. Choose the **Sweep** tool from the **Sweep** drop-down in the **Shapes** group of the **Ribbon**; the **Sweep** dashboard is displayed.
2. Click on the trajectory from the drawing area; the trajectory gets highlighted and an arrow is displayed on it. This arrow displays the direction of sweep.

Tip

*To change the start point on the trajectory, click on the point where you want the start point. When the point is highlighted in red color, press and hold the right mouse button to invoke a shortcut menu. From the shortcut menu, choose the **Start Point** option.*

Drawing the Section for the Base Feature

1. After selecting the trajectory from the drawing area, choose the **Sketch** button from the dashboard; your sketching plane is automatically oriented. Also you are prompted to draw the cross-section for the sweep feature.

2. Choose the **Center and Point** button from the **Sketching** group of the sketcher environment and create a circle such that the center of the circle lies at the intersection of the two infinite perpendicular lines.

   Note that, when you draw the circle, the cursor snaps to the intersection point of the cross.

Modifying the Dimensions of the Section

1. Double-click on the dimension and modify the diameter dimension to **29**. As you enter the dimension; the sketch gets modified and refits on the screen.

2. Choose the **OK** button to close the sketcher environment.

Creating the Base Feature

After closing the sketcher environment, preview of the base feature will be displayed in the drawing area.

1. Press CTRL+ D keys to orient the model in default orientation, refer to Figure 9-77. The default orientation in Creo Parametric is **Trimetric**.

2. Now, choose the **OK** button from the **Sweep** dashboard to exit it.

Creating the Shell Feature

After creating the sweep feature, you need to create the next feature called the shell feature.

Note

*Instead of using the **Shell** tool, two concentric circles can be drawn while drawing the section for the sweep feature in order to obtain the desired hollow feature. Alternatively, the **Sweep > Thicken Sketch** option can be used to obtain the same hollow feature. In this tutorial, you will use the **Shell** option.*

1. Choose the **Shell** tool from the **Engineering** group; the **Shell** dashboard is displayed. Also, you are prompted to select the surfaces to be removed.

2. Select one end surface of the sweep feature and then by using CTRL+left mouse button, select the other end surface. The two selected surfaces are highlighted in green.

3. Enter **4** as the thickness value of the shell in the **Thickness** edit box present on the **Shell** dashboard and then press ENTER.

4. Choose the **OK** button to create the shell feature and exit the **Shell** dashboard.

   The default trimetric view of the shell feature is shown in Figure 9-78. You can use the middle mouse button to view the model from various angles.

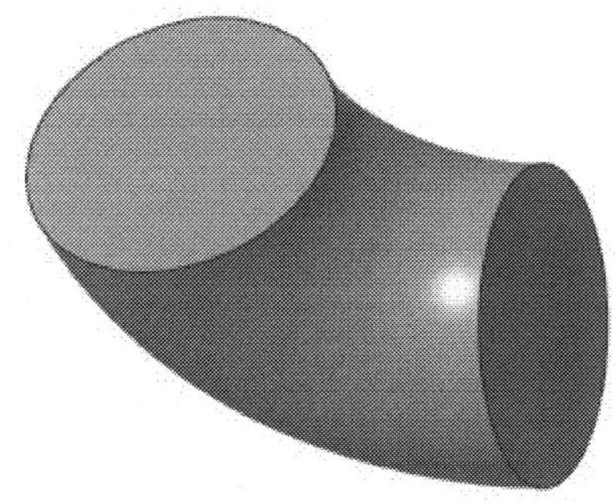

Figure 9-77 *The default view of the sweep feature*

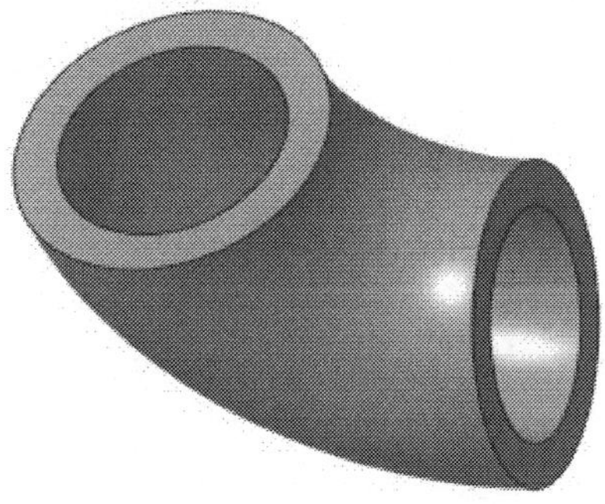

Figure 9-78 *Model after applying the shell feature*

Creating Extrude Features

The next feature is a protrusion feature with a depth of 5 and need to be created at both ends of the sweep feature. While drawing the circle for the sketch of the extrude feature, remember to use the edge of the shell in order to create a hole in the extruded feature as well.

1. Choose the **Extrude** tool from the **Shapes** group; the **Extrude** dashboard is displayed.

2. Choose the **Placement** tab from the dashboard to display the slide-down panel and invoke the **Sketch** dialog box. Select the top face of the sweep feature as the sketching plane.

3. Select the **RIGHT** datum plane and then select the **Right** option from the **Orientation** drop-down list. Then, enter the sketcher environment.

4. Draw the sketch of the extrude feature and then apply constraints and dimensions to it, as shown in Figure 9-79.

5. Create the extruded feature having an extrusion depth of 5. Similarly, create the next extruded feature at the other end of the sweep feature.

   Alternatively, you can use the **Copy** and **Paste** tools to save the time for creating the second protrusion.

   The protrusion features created at both ends of the sweep feature are shown in Figure 9-80.

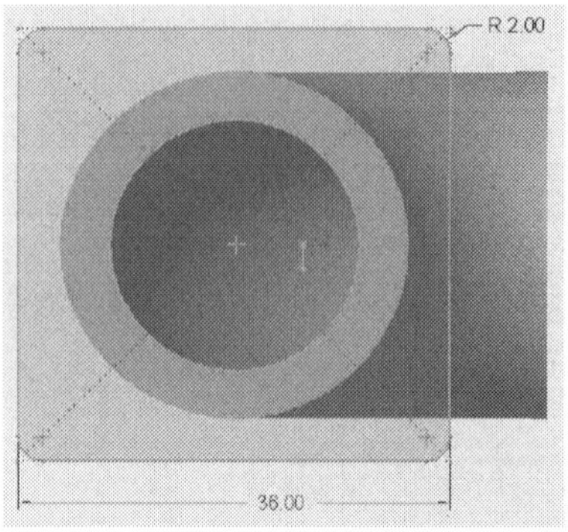

***Figure 9-79** Sketch with dimensions*

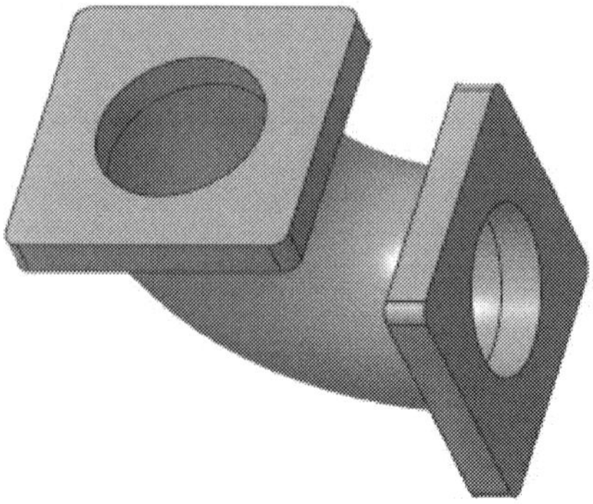

***Figure 9-80** Two extruded features created at both the ends of the sweep feature*

Creating the Hole Feature

After creating the extruded features at both ends of the sweep feature, you need to create the counterbore holes. One hole is to be created on each extruded surface and then they will be patterned on individual planes separately to create the remaining three instances.

1. Choose the **Hole** tool from the **Engineering** group; the **Hole** dashboard is displayed.

2. Choose **Simple** in the **Type** area and **Drilled** in the **Profile** area from the **Hole** dashboard; two buttons are added to the dashboard.

3. Next, choose the **Counterbore** button from the dashboard to make the hole counterbore. Also, choose **Drill up to next surface** from the depth flyout.

4. Choose the **Shape** tab from the dashboard to enter the counterbore parameters of the hole, refer to Figure 9-81.

5. Select the back face of the second extruded feature for the placement of hole; a preview of the hole appears in the drawing area.

   Now, you need to specify the placement parameters for the hole. For this purpose, refer to Figure 9-75 and look for dimensions that can help in placing the hole. The two edges are used to dimension the hole.

6. Choose the **Placement** tab and click in the **Offset References** collector to turn it green.

7. Select the front edge of the back face and then use CTRL+left mouse button to select the left edge of the third feature for specifying linear references. The hole is at a distance of **5** from both the edges. The default dimensions appear on the hole.

   You can also drag the green handles and place them on the two edges.

8. Double-click on the dimensions and modify the linear distance to **5**.

9. Choose the **OK** button from the **Hole** dashboard; a hole is created on the selected face.

10. Create another hole on the fourth feature by following the same procedure as discussed earlier.

Tip
You can also create a counterbore hole by sketching its profile. To do so, choose the ***Use sketch to define drill hole profile*** *button from the dashboard. Next, choose the* ***Activates Sketcher to create section*** *button to activate the sketcher environment.*

The trimetric view of the model completed so far is shown in Figure 9-82.

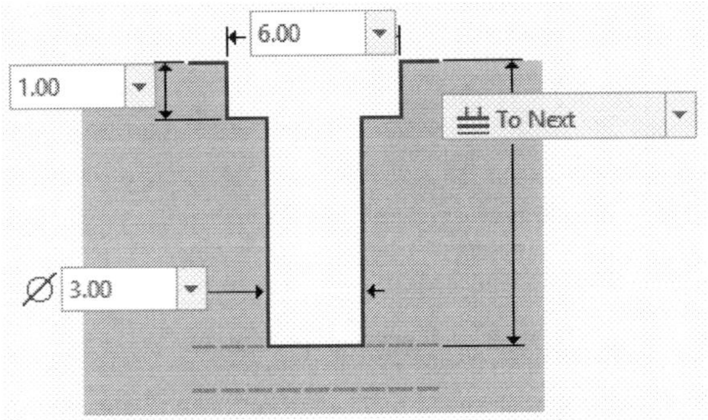

Figure 9-81 *Parameters of the counterbore hole*

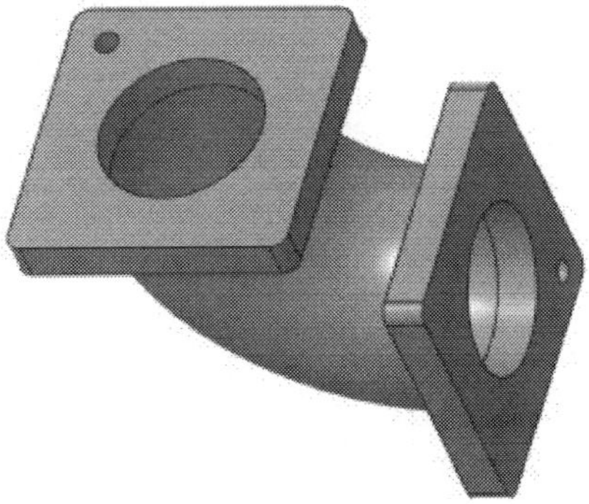

Figure 9-82 *One hole each on the two extruded features*

Creating the Pattern of Holes

After one hole is placed on each of the two faces, you need to pattern the holes.

1. Select the hole feature from the **Model Tree** or from the drawing area and hold the right mouse button to invoke the shortcut menu. Choose the **Pattern** option from the shortcut menu; the **Pattern** dashboard is displayed.

2. Specify the required parameters in the **Pattern** dashboard and create the pattern of the hole.

   Similarly, create the pattern of the hole on the other extruded feature as well. The default trimetric view of the complete model is shown in Figure 9-83.

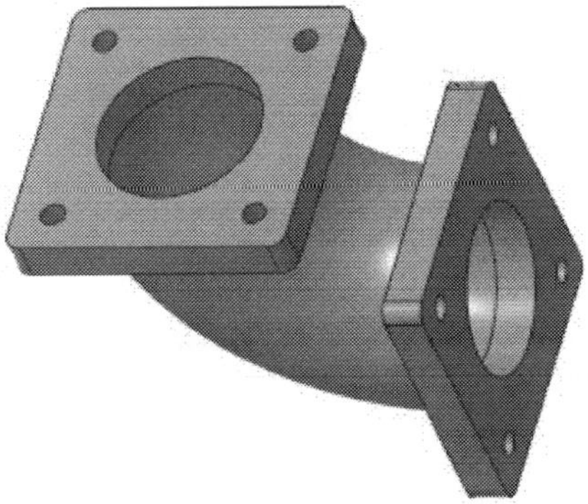

Figure 9-83 *The trimetric view of the complete model*

Saving the Model

1. Choose the **Save** option from the **File** menu and then save the model.

Note
As discussed in the earlier chapters, the ***Model Tree*** *is used to get an idea about the order of feature creation. Therefore, while creating a model, the order of feature creation will remain same but the id numbers of the features may be different.*

Tutorial 2

In this tutorial, you will create the model shown in Figure 9-84. The orthographic views of the model are given in Figure 9-85. **(Expected time: 30 min)**

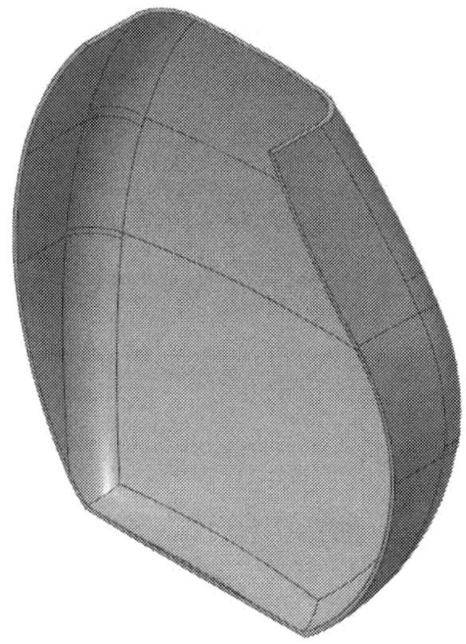

Figure 9-84 *Isometric view of the model*

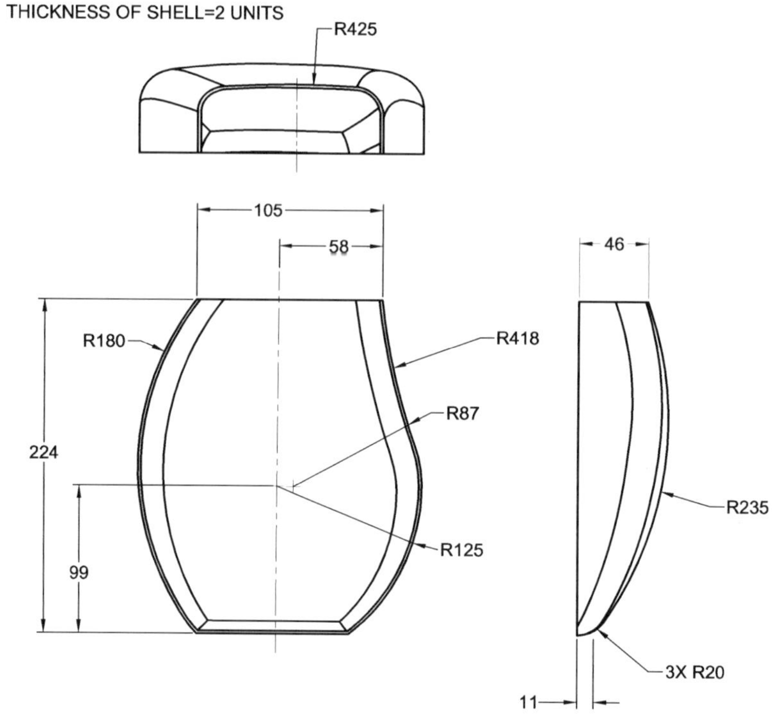

Figure 9-85 *Orthographic views of the model*

Examine the model and determine the number of features in it. The model is composed of five features, refer to Figure 9-84. The following steps are required to complete this tutorial:

a. Create the first feature as a set of datum curves that will be drawn on the **FRONT** datum plane.
b. Create the second feature. This feature is the datum curve that will be drawn on the **RIGHT** datum plane.
c. Create the variable section sweep.
d. Create the round feature.
e. Create the shell feature of thickness 2.

Make sure that the required working directory is selected.

Starting a New Object File

1. Start a new part file and name it as *c09tut2*.

The three default datum planes are displayed in the drawing area.

Creating the Trajectories

In this section, you need to draw the datum curves that will be selected as trajectories to create the variable section sweep feature. After selecting the sketching plane, draw sketch for the origin trajectory and dimension it. Then, you need to draw the sketch for the X trajectory on the same plane and dimension it. The X trajectory defines the horizontal vector of a section. If the section is aligned with the X trajectory, the section varies along the path of the X trajectory. Similarly, two additional trajectories will be drawn to guide the section throughout the sweep.

1. Choose the **Sketch** button from the **Datum** group; the **Sketch** dialog box is displayed.

2. Select the **FRONT** datum plane from the drawing area as the sketching plane.

3. Select thc **TOP** datum plane from the drawing area and then select the **Top** option from the **Orientation** drop-down list. Next, choose the **Sketch** button from the **Sketch** dialog box.

4. Once you enter the sketcher environment, create the sketch of the origin trajectory. The origin trajectory is a straight line segment aligned to the **RIGHT** datum plane and its start point is aligned to the **TOP** datum plane. The sketch of the origin trajectory is shown in Figure 9-86.

Note

1. In the sketcher environment, you will draw all the three datum curves because they are on the same datum plane.

2. You can also draw these datum curves using the ***Sketch*** *tool available in the* ***Datum*** *flyout of the* ***Sweep*** *dashboard.*

5. After the sketch of the origin trajectory is complete, draw the sketch of the X trajectory.

   The X trajectory is drawn using three arcs. It is recommended that you start the sketch from the top arc. After drawing the first arc, dimension it and then modify its dimension.

6. Draw the second arc and then modify the dimension, as shown in Figure 9-87.

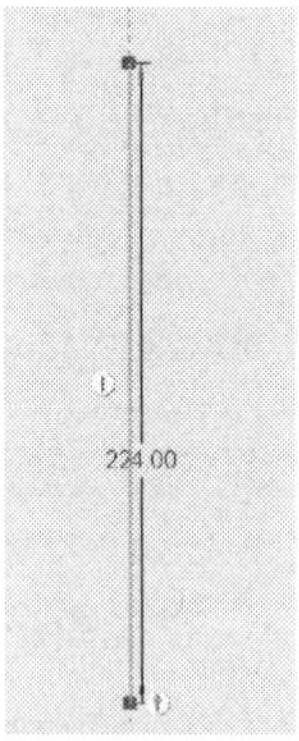

Figure 9-86 *Sketch with dimension of the origin trajectory*

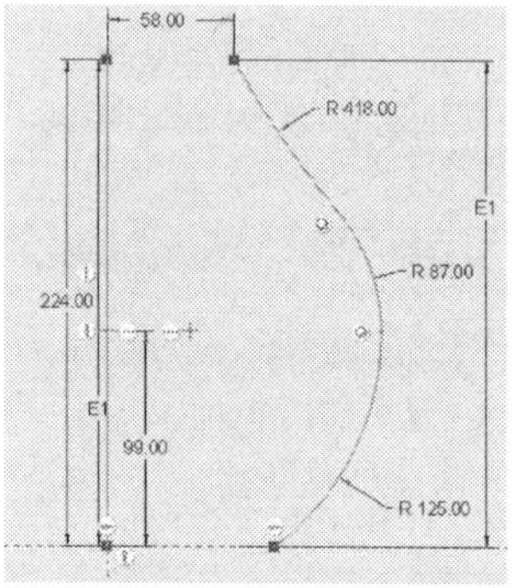

Figure 9-87 *Sketch with dimensions of the X trajectory*

7. Next, draw the third arc. As evident from the sketch of the X trajectory, the center points of the bottom and the second arcs are aligned horizontally.

8. Create the sketch of the third trajectory, as shown in Figure 9-88. This trajectory is created using a single arc.

9. After the sketch is complete, choose the **OK** button to exit the sketcher environment.

10. Similarly, create another trajectory on the **RIGHT** datum plane and keep the **TOP** datum plane at the left. The sketch of the trajectory is shown in Figure 9-89.

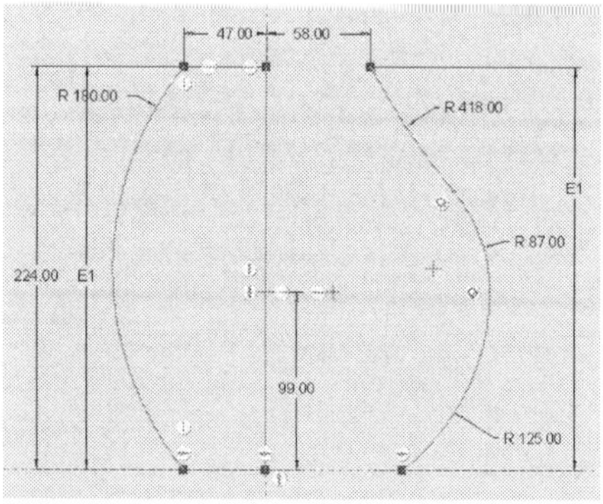

Figure 9-88 *Sketch with dimensions of the third trajectory*

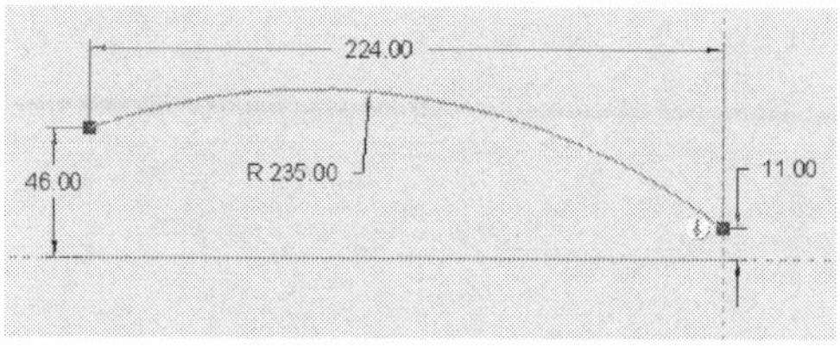

Figure 9-89 *Sketch with dimensions of the additional trajectory*

11. After the sketch is complete, choose the **OK** button to exit the sketcher environment.

12. Choose the **Default Orientation** option from the **Saved Orientations** flyout to view the trajectories that you have created, as shown in Figure 9-90. This view gives a better display of trajectories in 3D space.

Creating the Variable Section Sweep Using Sweep Feature

The four trajectories that will be used to create the variable section sweep have been created. Now, you need to invoke the **Sweep** tool and select these trajectories to create the feature.

1. Choose the **Sweep** tool from the **Shapes** group; the **Sweep** dashboard is displayed.

2. Choose the **References** tab to invoke the slide-down panel. Select the origin trajectory that is the first datum curve drawn.

3. Use CTRL+left mouse button to select the X trajectory (second datum curve). Note that as you are selecting the trajectories, they are being listed in the **Trajectories** collector.

4. Use CTRL+left mouse button to select the remaining two trajectories. Select the check box under the X column for the **Chain 1** row. This makes the second selected trajectory in the row as the X trajectory.

5. Now, choose the **Sketch** button from the dashboard to enter the sketcher environment.

6. Draw the closed sketch of the section and apply the dimensions, as shown in Figure 9-91.

Figure 9-90 *Default view of the sweep trajectories*

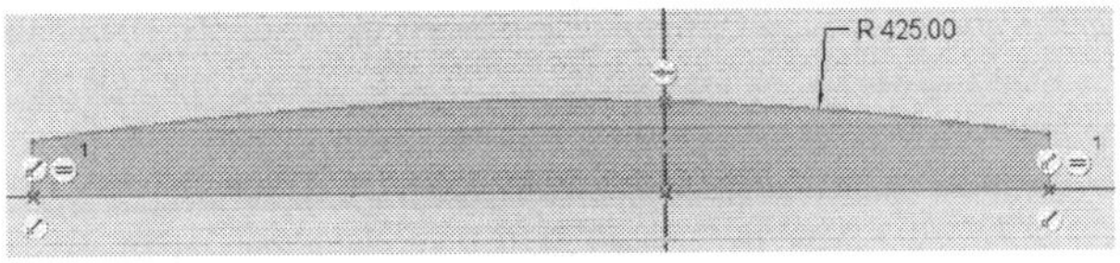

Figure 9-91 *Sketch of the section with dimensions*

Note that the section is aligned with the start points of all the trajectories that have cross marks displayed on them. The cursor will automatically snap to these points as you move it close to these points. In order to vary the section with the trajectories, it is necessary to align the section with the trajectories.

7. After the sketch is complete, choose the **OK** button to exit the sketcher environment.

   A preview of the variable section sweep feature is displayed in the drawing area.

8. Choose the **OK** button from the dashboard.

Now, you need to hide the datum curves.

9. Select the datum curves from the **Model Tree** or from the drawing area; a mini popup tool bar appears. Choose the **Hide** button from the mini popup toolbar or press CTRL+H keys from the keyboard to hide the selected curves.

The variable section sweep feature is completed, as shown in Figure 9-92. You can use the middle mouse button to orient the feature as shown in the figure.

Creating the Round Feature

Now, you need to create round on the edges of the base feature.

1. Choose the **Round** tool from the **Engineering** group; the **Round** dashboard is displayed.

2. Select the edges shown in Figure 9-93. Remember that you need to use CTRL+left mouse button to select more than one edge.

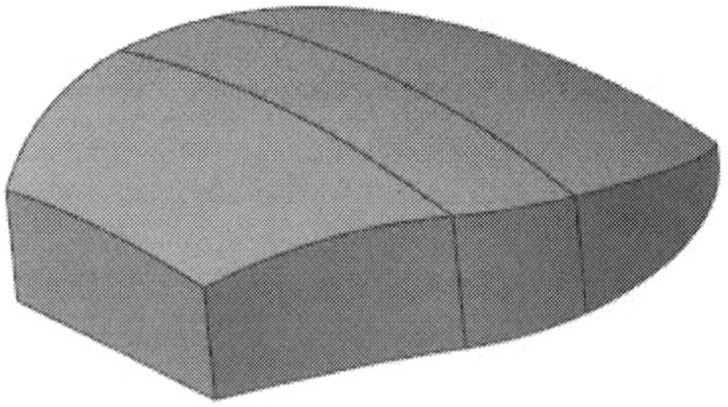

Figure 9-92 *Variable section sweep feature*

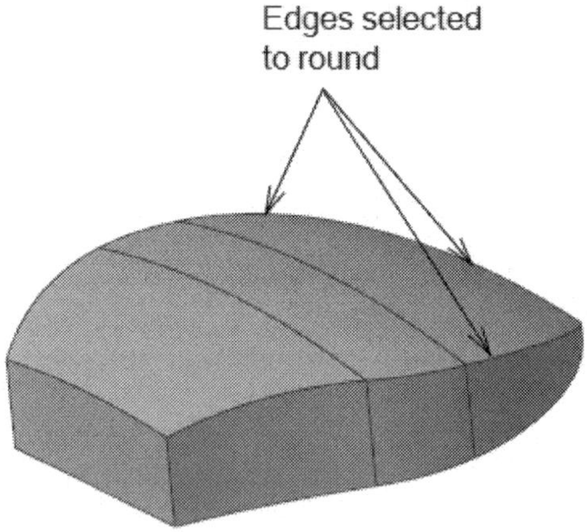

Figure 9-93 *Edges to be selected to create round*

3. Enter **20** in the **Radius** edit box present on the dashboard. Choose the **OK** button from the dashboard. The base feature after creating the round is shown in Figure 9-94.

Creating the Shell Feature

Now, you need to create the shell and remove the front and bottom faces of the base feature.

1. Choose the **Shell** tool from the **Engineering** group; you are prompted to select the surfaces to be removed.

2. Select the surfaces, as shown in Figure 9-95. Remember that you need to use the CTRL+left mouse button to select more than one face.

3. Enter **2** in the **Thickness** edit box. Then, choose the **OK** button from the dashboard.

The model after creating the shell feature is shown in Figure 9-96.

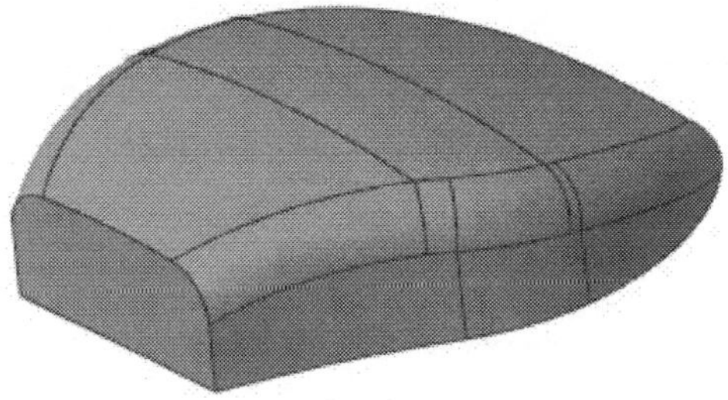

Figure 9-94 Model after creating the round

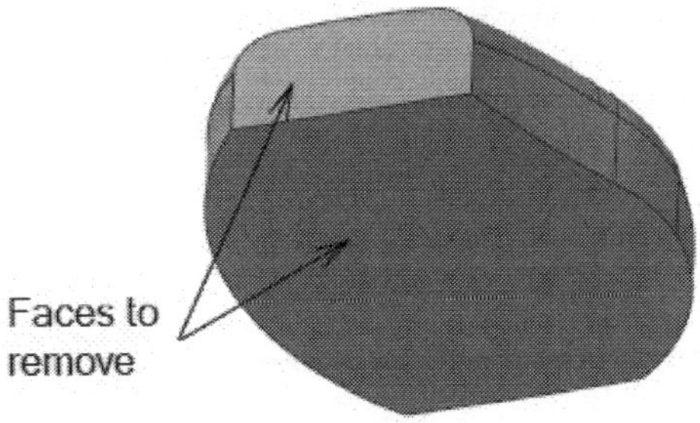

Figure 9-95 Surfaces selected to be removed

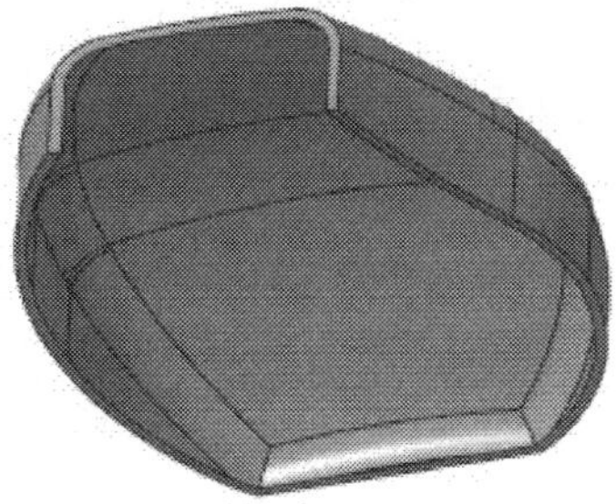

Figure 9-96 Final model after creating shell

Saving the Model

Next, you need to save the model.

1. Choose the **Save** button from the **File** menu and save the model.

Tutorial 3

In this tutorial, you will create the spring shown in Figure 9-97. Figure 9-98 shows the front view of the spring with its dimensions. **(Expected time: 10 min)**

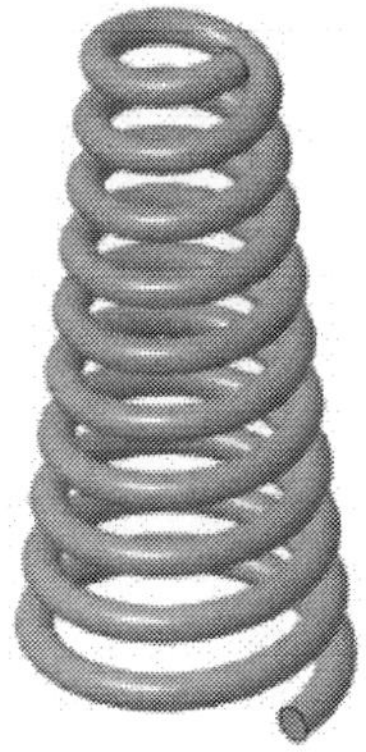

Figure 9-97 Isometric view of the spring

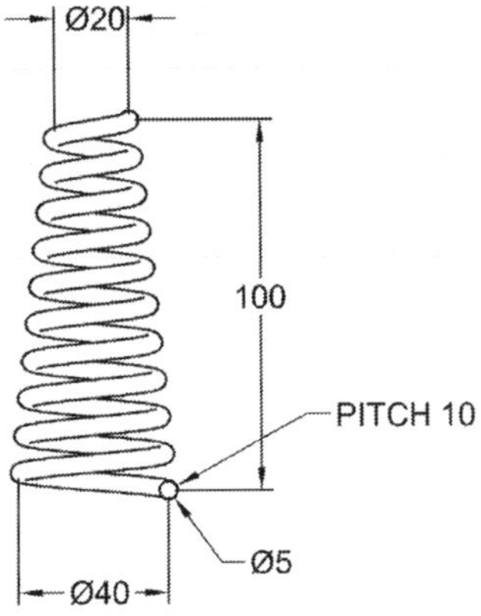

Figure 9-98 Front view with dimensions of the spring

Examine the spring and determine the specifications of the spring, refer to Figure 9-98. The following steps are required to complete this tutorial:

a. Draw the trajectory using the sketcher environment tools and apply dimensions to it.
b. Specify the pitch of the spring.
c. Draw the section of the spring.

Make sure that the required working directory is selected.

Starting a New Object File

1. Start a new part file and name it as *c09tut3*.

The three default datum planes and the **Model Tree** are displayed in the drawing area.

Creating the Helical Sweep Feature

The spring that you have to create is a right-handed spring of constant pitch.

1. Choose **Helical Sweep** from the **Sweep** drop-down of the **Shapes** group; the **Helical Sweep** dashboard is displayed. Select the **Through helix axis** radio button from the **References** tab if not selected by default.

2. Make sure that the **Right-handed Rule** button is selected in the dashboard.

3. Select the **Constant** radio button from the **Options** tab.

4. Select the **Define** radio button in the **Helix profile** area from the **References** tab; the **Sketch** dialog box is displayed and you are prompted to select a sketching plane.

5. Select the **FRONT** datum plane from the drawing area and then choose the **Sketch** button from the **Sketch** dialog box.

6. Once you enter the sketcher environment, draw the sketch of the trajectory and dimension it, as shown in Figure 9-99. As is evident from Figure 9-99, the endpoint of the trajectory is aligned with the **TOP** datum plane.

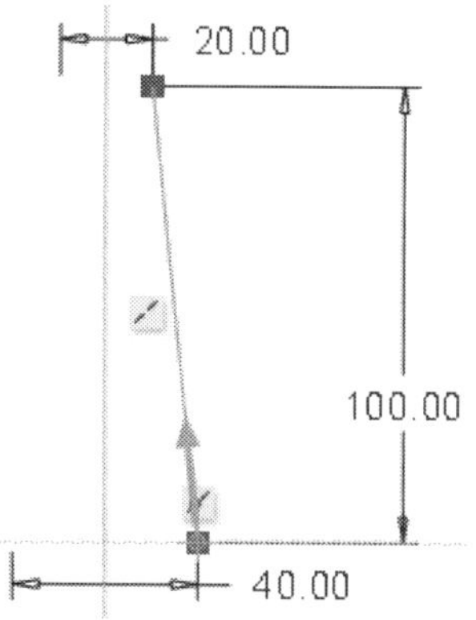

Figure 9-99 *Sketch of the trajectory*

You need to draw a geometry center line in the sketch about which the spring will be rotated. This is the axis of the spring.

7. After you complete the sketch of the trajectory, choose the **OK** button from the dashboard; the **Helical Sweep** dashboard is displayed and a default pitch value of spring is displayed on the trajectory.

8. Enter **10** in the **Pitch** edit box and choose the **Sketch** button from the dashboard. You will notice that two pink lines of infinite length crossing each other perpendicularly appear on the screen. The intersection point of these lines is the start point of the trajectory.

9. Draw the section of the spring such that the center of the circle coincides with the intersection of the two perpendicular lines. Then, dimension it, as shown in Figure 9-100.

10. After completing the sketch, choose the **OK** button to exit the sketcher environment.

11. Choose the **OK** button from the **Helical Sweep** dashboard; the spring is created, as shown in Figure 9-101.

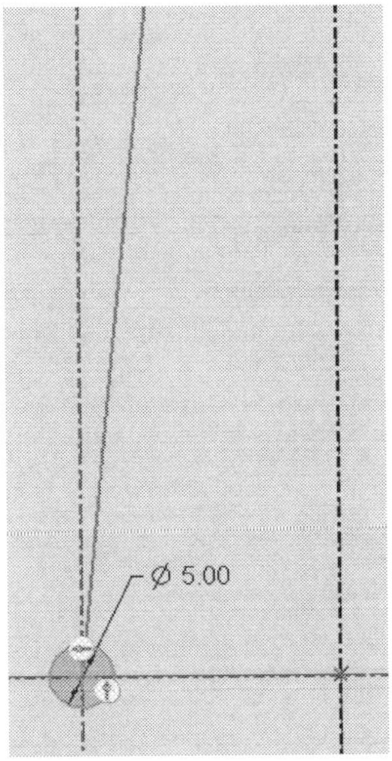

Figure 9-100 *Sketch of the section*

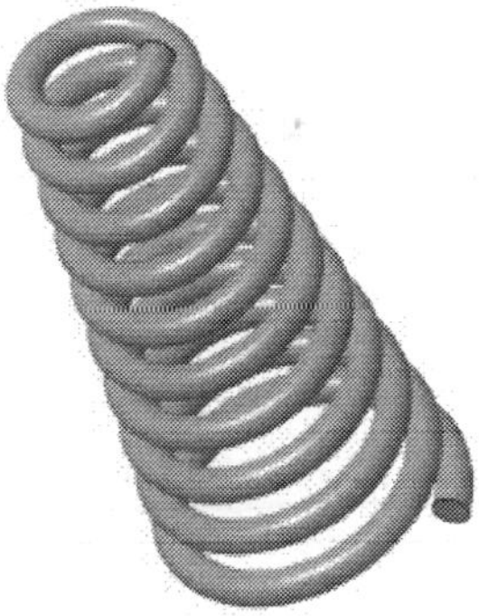

Figure 9-101 *Trimetric view of the helical spring*

Saving the Model

1. Choose the **Save** button from the **File** menu and save the model.

Tutorial 4

In this tutorial, you will create a solid model of the Upper Housing of a motor blower assembly shown in Figure 9-102. Figure 9-103 shows the left view, top view, front view, and the sectioned left view of the model. **(Expected time: 1 hr)**

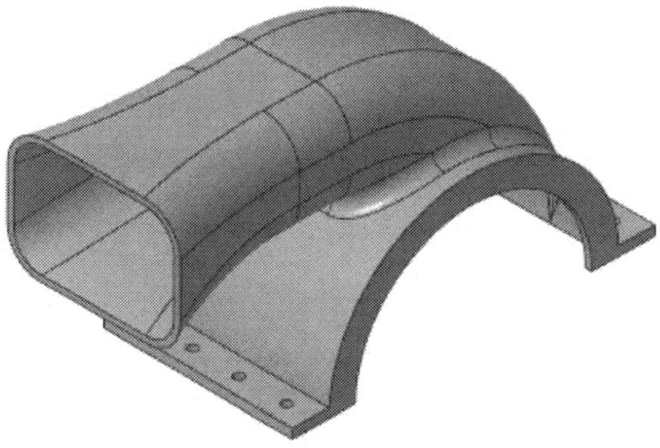

Figure 9-102 *Solid model of the Upper Housing*

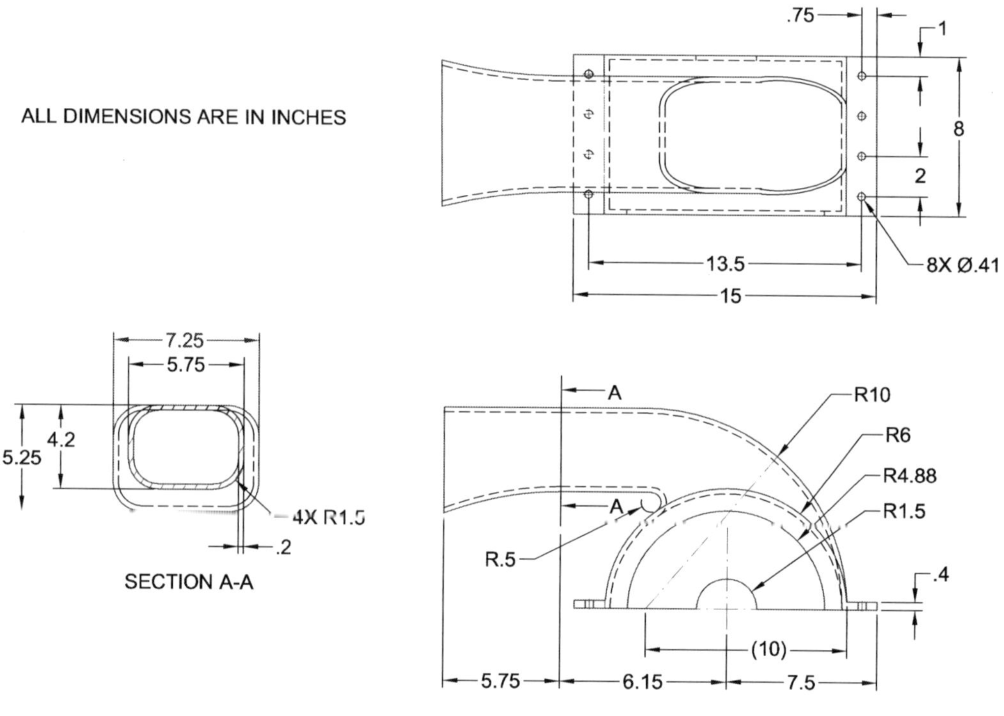

Figure 9-103 *Orthographic views of the model*

Examine the model and determine the number of features in it. The model is composed of ten features, refer to Figures 9-102 and 9-103.

The following steps are required to complete this tutorial:

a. Create the base feature.
b. Create the swept blend feature in which the section is normal to the origin trajectory.
c. Create the round feature of radius 1.5.

d. Create the round of radius 0.5.
e. Create the shell feature of thickness 0.25.
f. Create an extruded cut.
g. Create another extruded cut.
h. Create the eighth feature as an extruded feature.
i. Create a mirror copy of the eighth feature.
j. Create the holc fcature, and pattern it.

Starting a New Object File

1. Start a new part file and name it as *c09tut2*. Also, set the template to **inlbs_part_solid_abs**.

   The three default datum planes and the **Model Tree** are displayed in the drawing area.

Creating the Base Feature

To create the sketch for the base feature, you first need to select the sketching plane for the base feature. In this model, you need to draw the base feature on the **RIGHT** datum plane.

1. Choose the **Extrude** tool from the **Shapes** group. Next, invoke the **Sketch** dialog box.

2. Select the **RIGHT** datum plane as the sketching plane.

3. Select the **TOP** datum plane from the drawing area and then select the **Top** option from the **Orientation** drop-down list.

4. Choose the **Sketch** button from the **Sketch** dialog box to enter the sketcher environment.

5. Create the sketch of the base feature and apply required dimensions, as shown in Figure 9-104.

6. Choose the **OK** button to exit the sketcher environment.

   The **Extrude** dashboard gets activated and the **Extrude from sketch plane by specified depth value** button is chosen by default in the dashboard.

7. Choose the **Extrude symmetrically on both sides of the sketch plane** option from the **Depth** flyout.

8. Enter the depth as **8** in the **Depth** edit box present in the **Extrude** dashboard. Choose the **OK** button from the **Extrude** dashboard.

   The base feature is completed. The default trimetric view of the base feature is shown in Figure 9-105.

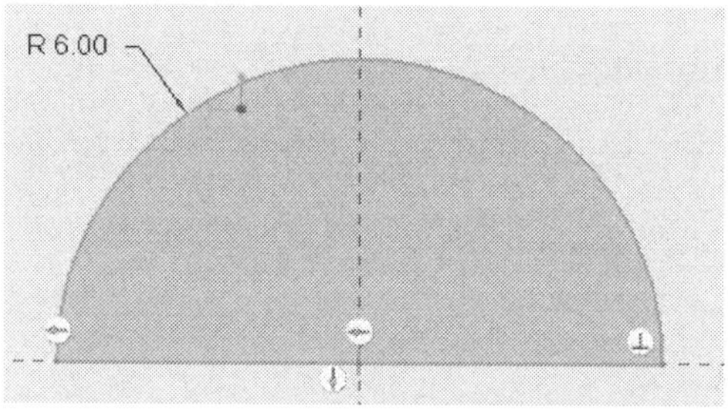

Figure 9-104 Sketch of the base feature

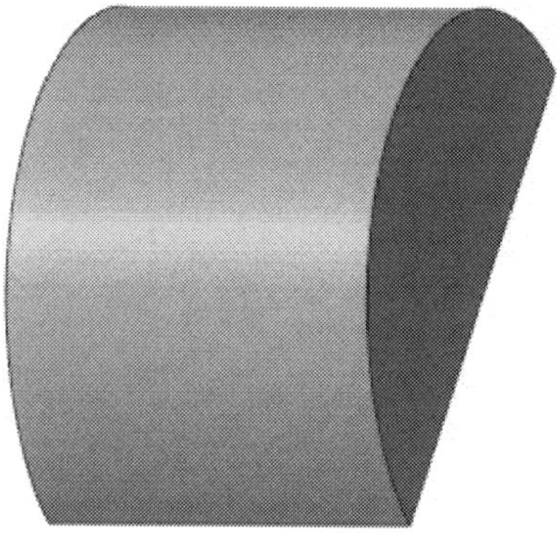

Figure 9-105 Default trimetric view of the base feature

Creating the Trajectories for the Swept Blend

You can use the **Sketch** button to create the sketch for the trajectories of the swept blend feature before or after invoking the **Swept Blend** dashboard. However, it is recommended that you create the trajectory before invoking the **Swept Blend** dashboard.

1. Choose the **Sketch** tool from the **Datum** group; the **Sketch** dialog box is displayed.

2. Choose the **Use Previous** button from the **Sketch** dialog box to enter the sketcher environment.

3. Draw the sketch, as shown in Figure 9-106, and exit the sketcher environment.

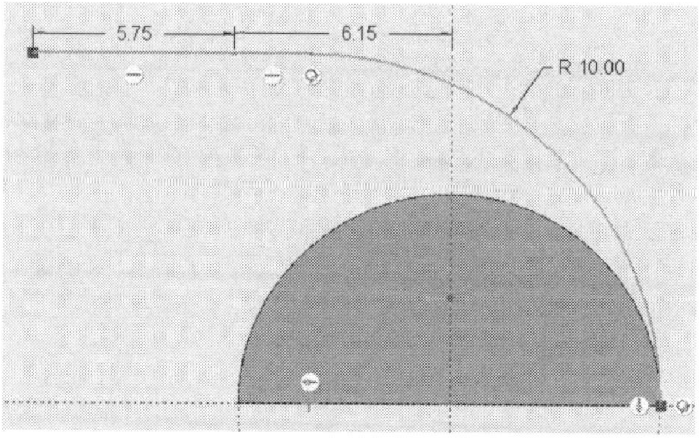

Figure 9-106 Sketch of the origin trajectory

Creating the Swept Blend Feature

1. Choose the **Swept Blend** tool from the **Shapes** group; the **Swept Blend** dashboard is displayed.

2. Choose the **References** tab to display the slide-down panel.

3. Select the trajectory from the drawing area.

Note that as you select the trajectory from the drawing area, it gets added to the **Trajectories** collector. Also note that the first trajectory you select from the drawing area is considered as the normal trajectory.

4. Next, choose the **Sections** tab to display the slide-down panel. Select the start point of the arc as the start point of the first cross-section to be created; the **Sketch** button is enabled in the **Sections** slide-down panel. Choose the **Sketch** button to enter the sketcher environment.

5. Create the sketch for the first cross-section, as shown in Figure 9-107.

6. Choose the **OK** button to exit the sketcher environment. The **Swept Blend** dashboard is enabled and is displayed above the drawing area.

7. Now, choose the **Insert** button from the **Sections** slide-down panel and select the second point on the trajectory which is at a distance of **6.15** from the **FRONT** datum plane. Choose the **Sketch** button from the **Sections** slide-down panel and create sketch for the second cross-section, as shown in Figure 9-108 and exit the sketcher environment.

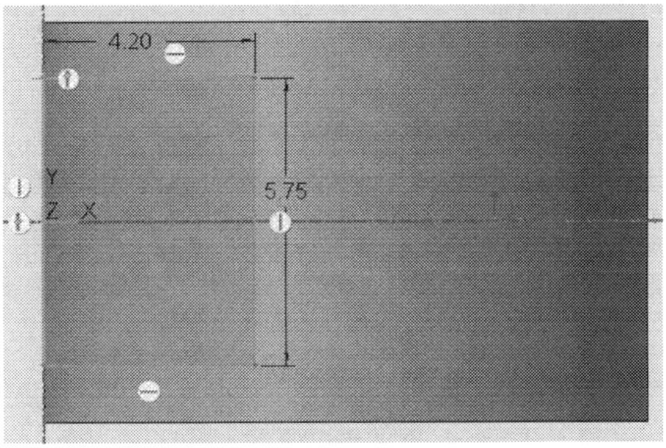

Figure 9-107 *Sketch for the first cross-section with dimensions*

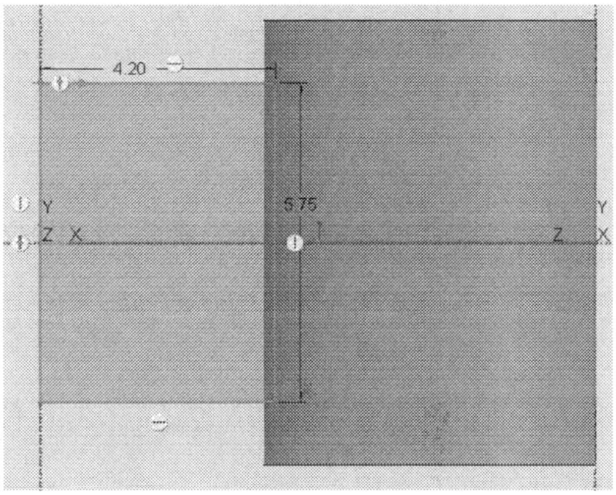

Figure 9-108 *Sketch for the second cross-section with dimensions*

8. Now, choose the **Insert** button and select the other end point of the trajectory to create the third cross-section, as shown in Figure 9-109.

9. Exit the sketcher environment after sketching the third cross-section; the **Swept Blend** dashboard is enabled and displayed above the drawing area. Choose the **Solid** button from the **Swept Blend** dashboard; a preview of the feature is displayed in the drawing area. Choose the **OK** button from the **Swept Blend** dashboard to exit the feature creation tool.

The model, similar to the one shown in Figure 9-110, is displayed in the drawing area.

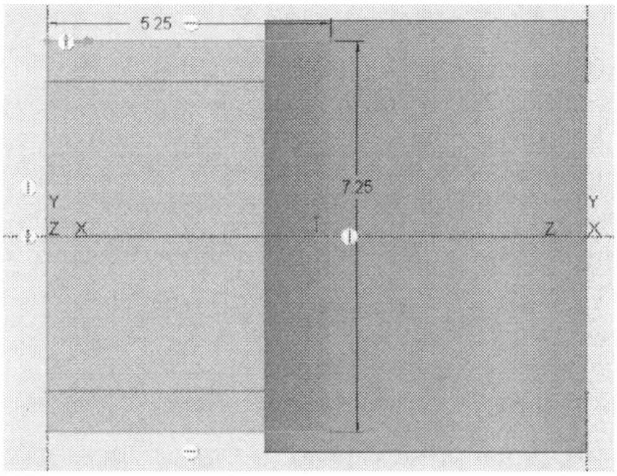

Figure 9-109 Sketch of the third cross-section with dimensions

Figure 9-110 Default view of the swept blend feature

Creating the Round Features

In this section, you need to create two round features. These features must be created before shelling because the rounds are also shelled along with the faces of the model. The third feature and the fourth feature consist of round feature of radius 1.5 and 0.5, respectively.

1. Choose the **Round** tool from the **Engineering** group; the **Round** dashboard is displayed.
2. Select all four longer edges of the swept feature, as shown in Figure 9-111 to round them. Remember that the first edge can be selected by using the left mouse button and the remaining edges can be selected by using the CTRL+left mouse button. A preview of the round is displayed in the drawing area along with the default radius value.
3. Click on the radius value, enter **1.5** in the **Radius** edit box that appears, and then press ENTER.
4. Choose the **OK** button from the **Round** dashboard. The default view of the model after creating a round feature of radius **1.5** is shown in Figure 9-112.

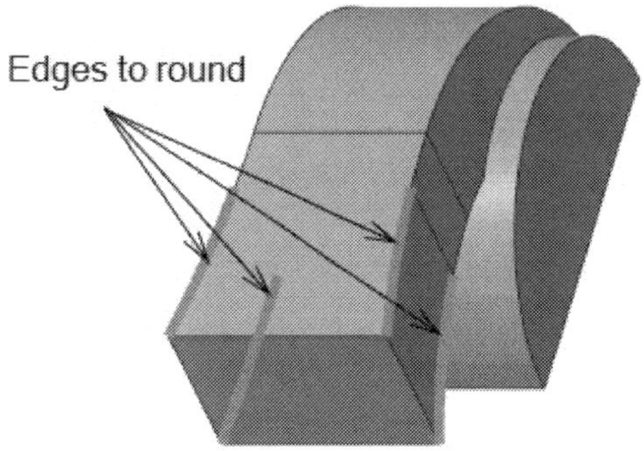

Figure 9-111 Edges to round

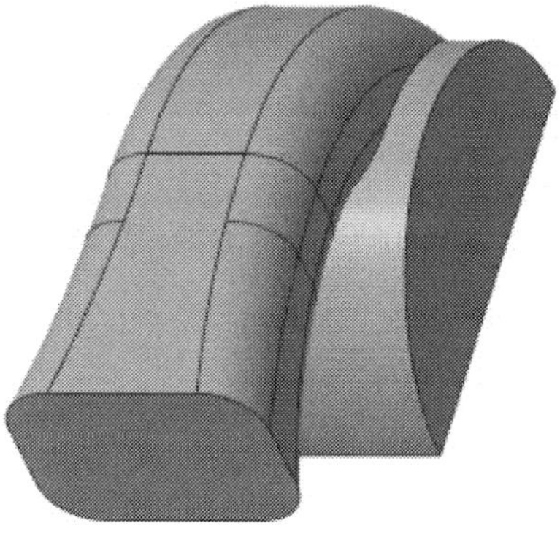

Figure 9-112 Default view of the model after creating a round feature of radius **1.5**

Note

A round of radius 0.5 cannot be created by adding a round set. This is because this round needs to be created on the geometry of the previous round. Therefore, until the first round with the radius 1.5 is created, you cannot create the second round.

The third feature of the model is completed and now you need to create the fourth feature which is also a round feature of radius 0.5. To do so, you need to invoke the **Round** dashboard.

5. Choose the **Round** tool from the **Engineering** group; the **Round** dashboard is displayed.
6. Select the edge where the base feature and the swept feature join, as shown in Figure 9-113.

   A preview of the round is displayed in the drawing area along with the default radius value.

7. Click on the radius value, enter **0.5** in the **Radius** edit box that appears, and then press ENTER.
8. Choose the **OK** button from the **Round** dashboard. The default view of the model after creating the round feature of radius **0.5** is shown in Figure 9-114.

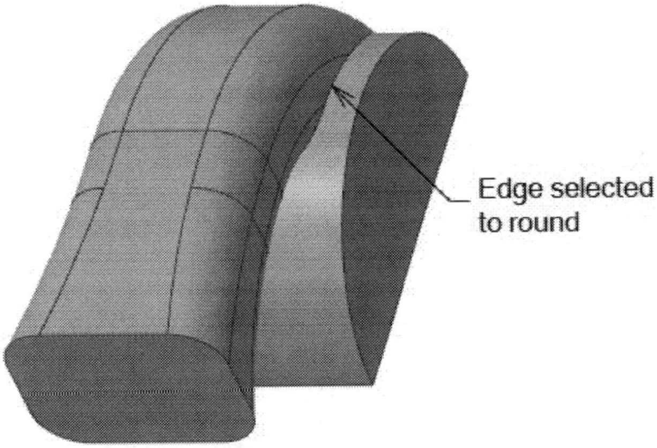

Figure 9-113 *Edge selected to round*

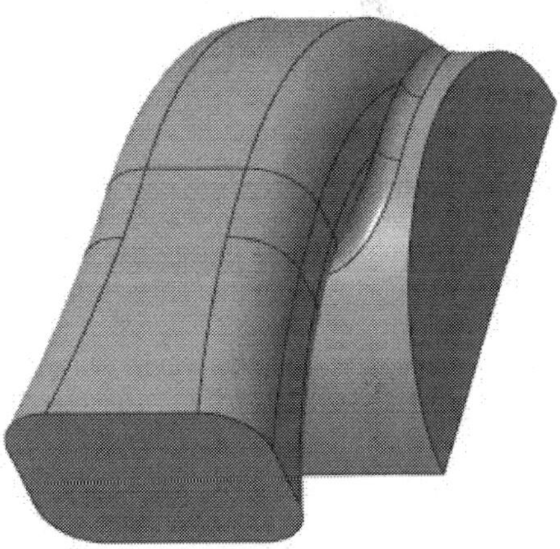

Figure 9-114 *Model after creating a round feature of radius* ***0.5***

Creating the Shell Feature

In this section, you need to create a shell of thickness **0.25** on the swept blend feature.

1. Choose the **Shell** tool from the **Engineering** group; the **Shell** dashboard is displayed and you are prompted to select the surface to be removed.
2. Using the left mouse button, select the front face of the swept blend feature and then use the CTRL+left mouse button to select the bottom face of the base feature.
3. Enter **0.2** in the **Thickness** edit box available on the **Shell** dashboard and then press ENTER.
4. Choose the **OK** button from the **Shell** dashboard. The shell feature is created as shown in Figure 9-115.

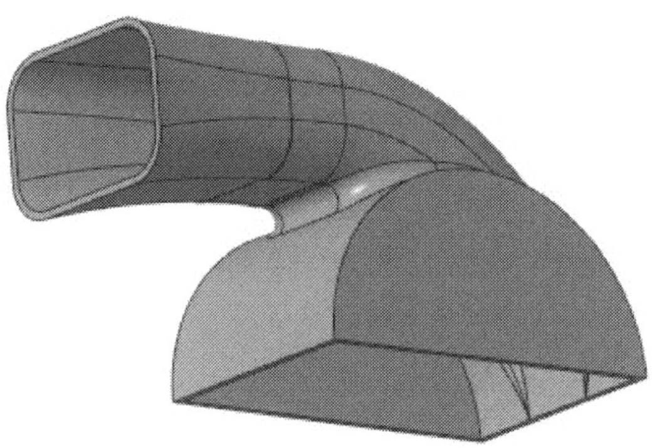

Figure 9-115 The model after shelling

Creating the Extruded Cut

The extruded cut is the sixth feature of this model. You need to create the cut on the front face of the base feature.

1. Choose the **Extrude** tool from the **Shapes** group; the **Extrude** dashboard is displayed.
2. Choose the **Remove Material** button from the dashboard.
3. Select the front face of the base feature as the sketching plane.
4. Select the **TOP** datum plane from the drawing area and then select the **Top** option from the **Orientation** drop-down list.
5. Choose the **Sketch** button to enter the sketcher environment.
6. Draw sketch of the cut feature using the **Concentric** tool from the **Circle** drop-down in the **Sketching** group and apply dimension to it, as shown in Figure 9-116.
7. Choose the **OK** button to exit the sketcher environment; the **Extrude** dashboard reappears above the drawing area. The **Extrude from sketch plane by specified depth value** button is chosen by default.
8. Choose the **Extrude upto next surface** button from the depth flyout.
9. Choose the **OK** button from the **Extrude** dashboard. The cut feature is created, as shown in Figure 9-117.

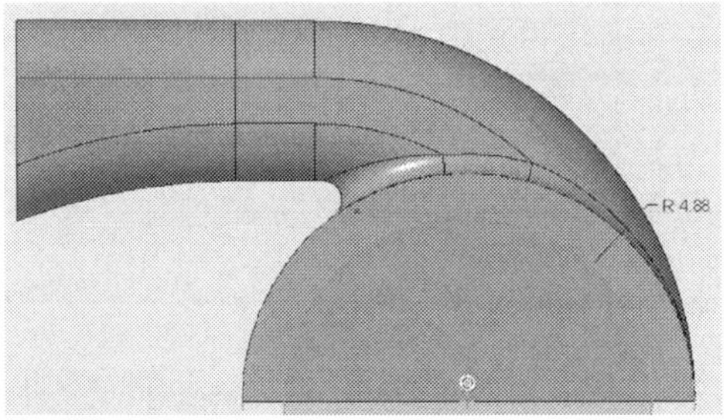

Figure 9-116 Sketch with dimension for cut feature

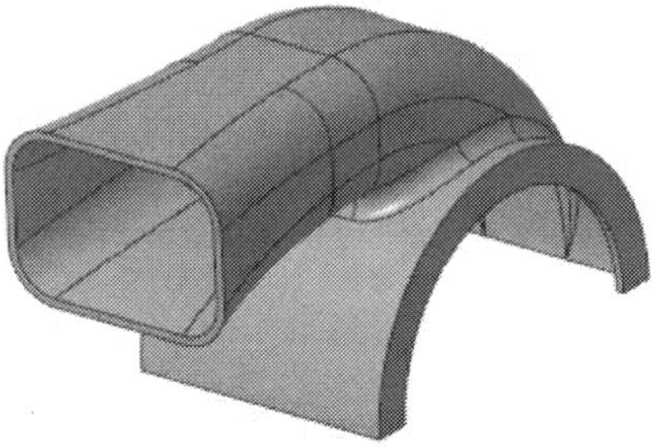

Figure 9-117 Model with cut feature at the front face

Similarly, create the next extruded cut feature at the back face of the base feature. Select the sketch plane, draw the sketch, apply dimension, and extrude the sketch using the **Extrude upto next surface** button. The sketch for the cut feature is shown in Figure 9-118.

The seventh feature of the model, which is a cut feature, is created, as shown in Figure 9-119.

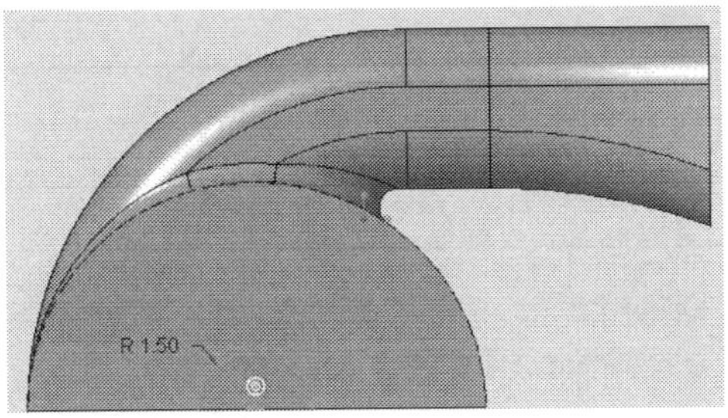

Figure 9-118 *Sketch with dimension for cut feature*

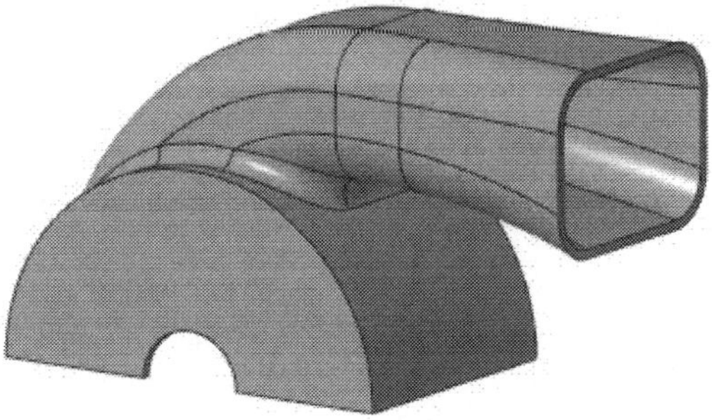

Figure 9-119 *Model with cut feature at the back face*

Creating the Extruded Feature

The eighth feature of the model is an extruded feature and it will be created on the front face of the base feature.

1. Invoke the **Extrude** tool and select the front face of the base feature as the sketching plane for the extrude feature. Once you enter the sketcher environment, draw the sketch and then apply constraints and dimensions to it, as shown in Figure 9-120.

2. After the sketch is complete, choose the **OK** button to exit the sketcher environment.

3. From the **Depth** flyout, choose the **Extrude up to selected surface, edge, vertex, quilt, body, curve, plane, axis or point** option and then select the back face of the base feature.

4. Choose the **OK** button from the **Extrude** dashboard. The extruded feature is completed and is shown in Figure 9-121.

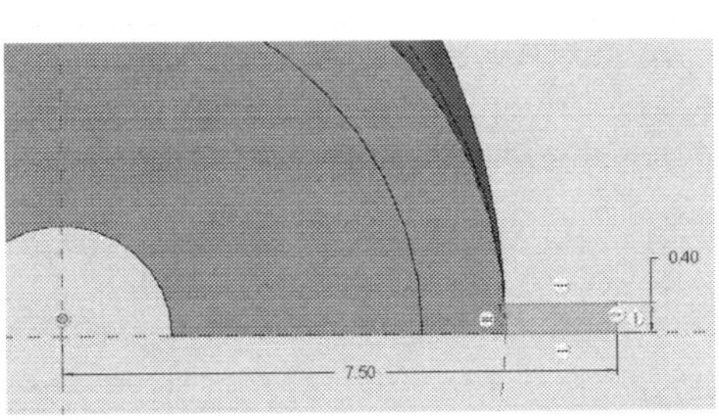

Figure 9-120 *Sketch for the extruded feature*

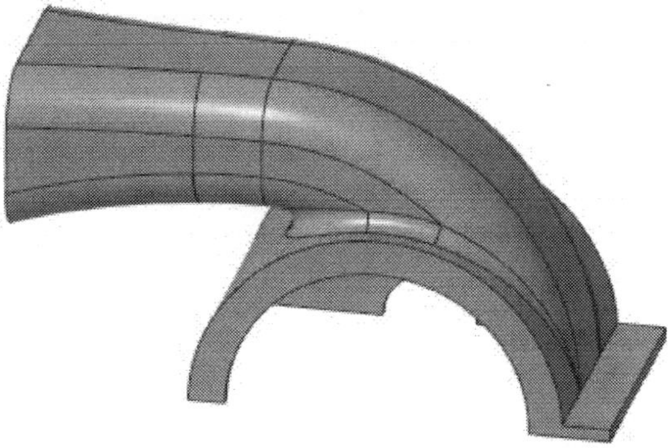

Figure 9-121 *Model after creating the extruded feature*

Creating a Mirror Copy of the Eighth Feature

You need to create a mirror copy of the extruded feature, as shown in Figure 9-122 and this will be the ninth feature of the model. The extruded feature will be mirrored about the **FRONT** datum plane.

1. Select the eighth feature of the model and then choose the **Mirror** tool from the **Editing** group of the **Model** tab; the **Mirror** dashboard is displayed.

2. Select the **FRONT** datum plane as the mirror plane from the **Mirror plane** collector and choose the **OK** button from the dashboard; the extruded feature is mirrored about this datum plane.

Creating the Tenth Feature

The tenth feature is a through hole and it will be created using the **Hole** dashboard.

1. Choose the **Hole** tool from the **Engineering** group; the **Hole** dashboard is displayed in which the **Create simple hole** button is chosen by default.

   Create the hole, as shown in Figure 9-123, by specifying the placement parameters. For placement parameters, refer to Figure 9-103.

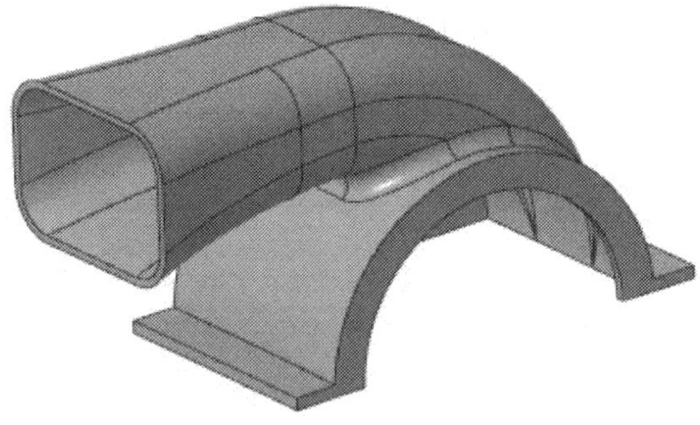

Figure 9-122 *Model after mirroring the extrude*

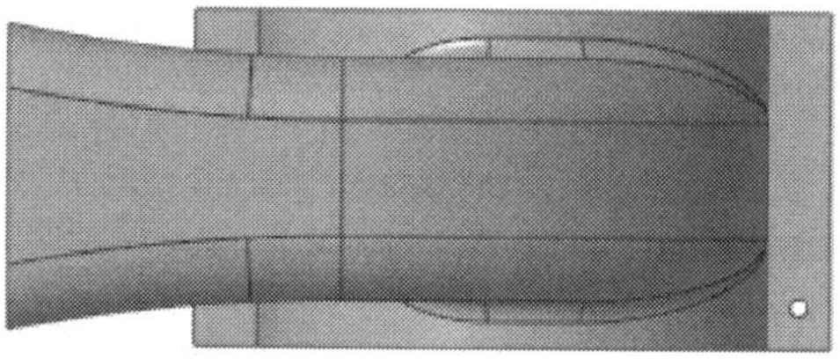

Figure 9-123 *Model after creating the hole feature*

Creating a Pattern of the Hole Feature

As evident from Figure 9-103, you need to create eight instances of the hole. The first instance is created by using the **Hole** dashboard and you can use the **Pattern** option to create remaining seven instances. You need to create a rectangular pattern of the hole feature. You can also create the holes by using the **Hole** dashboard and by specifying the placement parameters for each of them. But to save time, you should create a pattern of the hole.

1. Select the hole feature from the **Model Tree** or from the model. Hold down the right mouse button to invoke the shortcut menu.

2. Choose the **Pattern** option from the shortcut menu; the **Pattern** dashboard is displayed and the dimensions are displayed on the hole feature.

3. Select the **General** option from the **Regeneration option** drop-down list in the **Options** slide-down panel.

4. Click on the dimension value **1** from the drawing area. Enter **2** in the edit box and press ENTER.

5. In the **1st direction** area, enter **4** in the **Members** edit box and press ENTER. Now, you need to enter dimension increment in the second direction.

6. Click on the collector in the **2nd direction** area. Next, click on the dimension value **0.75** from the drawing area. Enter **13.5** in the edit box displayed, and press ENTER.

7. In the **2nd direction** area, enter **2** in the **Members** edit box and press ENTER.

8. Choose the **OK** button from the dashboard. You can use the middle mouse button to spin the model and display it, as shown in Figure 9-124.

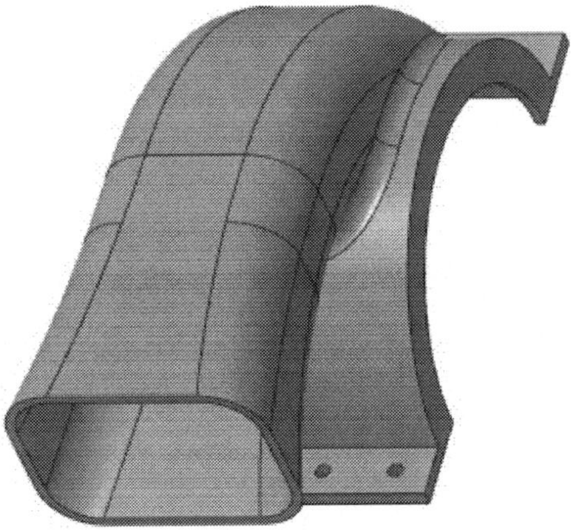

***Figure 9-124** Trimetric view of the completed model*

Saving the Model

1. Choose the **Save** button from the **File** menu and save the model.

Tutorial 5

In this tutorial, you will create the model shown in Figure 9-125. All dimensions for the three sections are shown in the Figure 9-126. Note that the model will be created by using the **Rotational Blend** tool. **(Expected time: 20 min)**

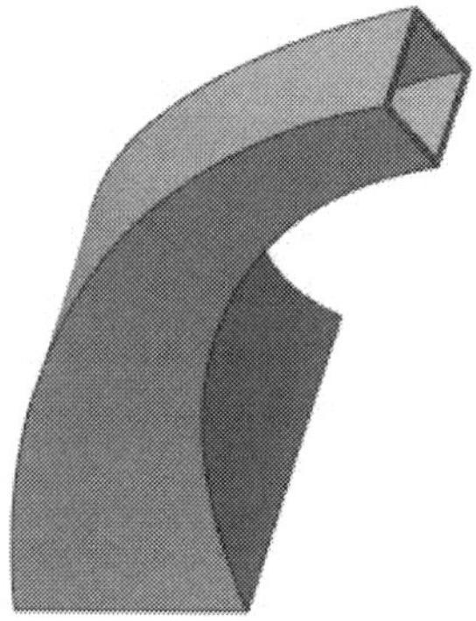

Figure 9-125 *Isometric view of the model*

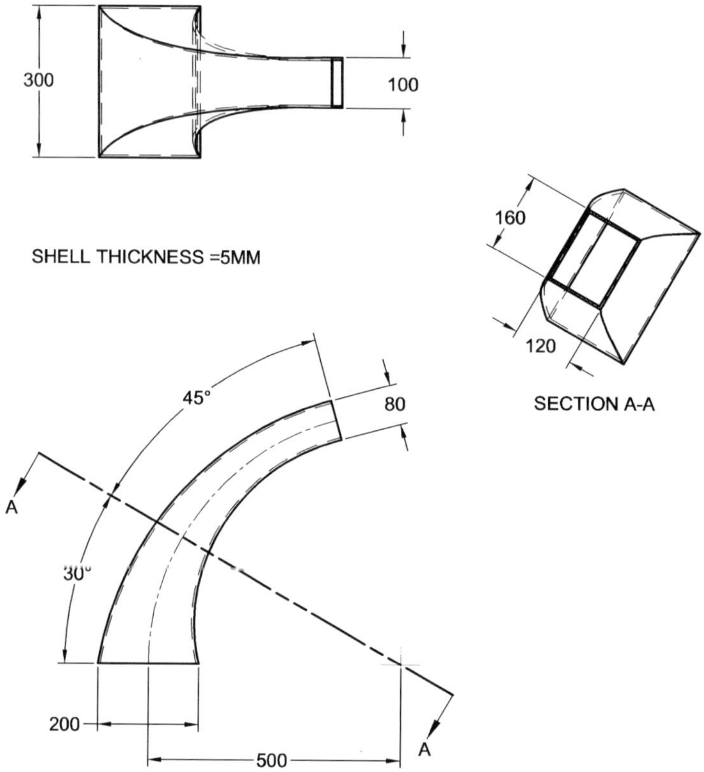

Figure 9-126 *Orthographic views of the model*

Examine the model and determine the number of sections in the blend feature, refer to Figure 9-125.

The following steps are required to complete this tutorial:

a. Start a new file and create the first section for the blend feature.
b. Create the second section for the blend feature.

c. Create the third section for the blend feature.
d. Create the Shell feature.

Starting a New Object File

1. Start a new part file and name it as *c09tut5*. The three default datum planes appear in the drawing area if they were not turned off in the previous tutorial.

Invoking the Rotational Blend Option

The first sketch of the rotational blend is **300x200** rectangle.

1. Choose **Rotational Blend** from the expanded **Shapes** group of the **Model** tab; the **Rotational Blend** dashboard is displayed.

2. Choose **Solid** and then choose the **Options** tab to select the **Smooth** radio button, if it is not selected by default. Choose the **Sections** tab from the dashboard to open the slide-down panel. Select the **Sketched sections** radio button and choose the **Define** button to define the sketching plane.

Selecting the Sketching Plane

1. Select the **TOP** datum plane as the sketching plane.

2. Select the **FRONT** datum plane as the reference plane and then select the **Bottom** option from the **Orientation** drop-down list. Choose the **Sketch** button to enter the sketching environment.

Drawing the Sketch of the First Section of the Rotational Blend

The blend consists of three sections.

1. Draw the sketch of the first section. The sketch is a rectangle of **300x200** dimension. Also, draw geometric centerline on the right side of the sketch.

2. Apply constraints and modify the dimensions, as shown in Figure 9-127.

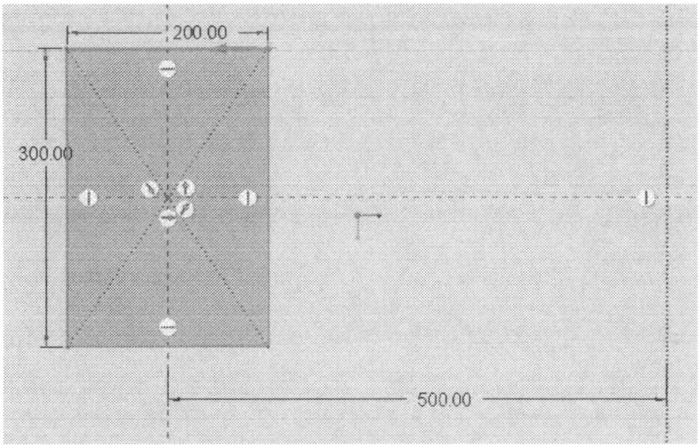

Figure 9-127 *Sketch with dimensions and constraints for the first section*

3. Choose the **OK** button from the dashboard; an angle offset value edit box with a default value becomes available in the dashboard and in the **Sections** tab. Change the angle offset value to **30** for second section and then choose the **Sketch** button from the **Sections** tab to enter the sketching environment.

The sketcher environment is invoked and you are allowed to draw the sketch for the second section. Notice that the second section does not need geometric centerline since it will rotate around the geometric centerline drawn in the first sketch.

Drawing the Sketch for the Second Section

1. Draw the sketch for the second section, apply constraints and dimensions to the sketch, and modify the dimensions, as shown in Figure 9-128.

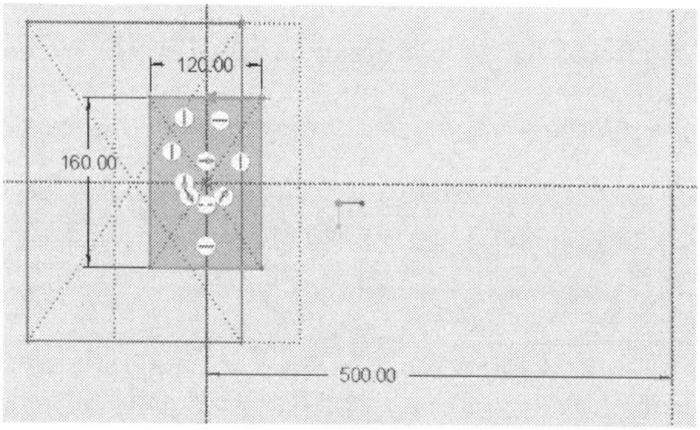

Figure 9-128 *Sketch with dimensions and constraints for the second section*

2. After completing the sketch, choose the **OK** button; a preview of the rotational blend feature is displayed. Now, choose the **Insert** button from the **Sections** tab to add the third section. Enter **45** in angle offset value box for the third section and choose the **Sketch** button from the **Sections** tab.

Drawing the Sketch of the Third Section

1. Draw the sketch for the third section, apply constraints and dimensions to the sketch, and modify the dimensions, as shown in Figure 9-129.

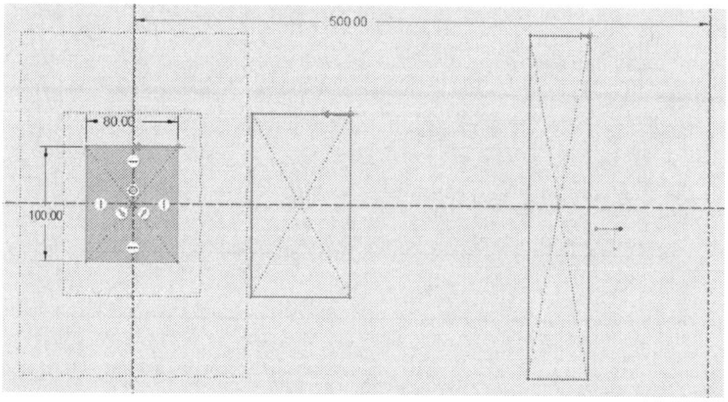

Figure 9-129 *Sketch with dimensions and constraints for the third section*

2. After completing the sketch, choose the **OK** button; a preview of the rotational blend feature is displayed.

3. If the preview is twisted at any sectioned sketch. Choose the sketch from the **Sections** tab and choose the **Sketch** button to enter the sketching environment. If the start point is not at the desired point, then select a point where you need the start point. Hold the right mouse button to invoke a shortcut menu, and then choose the **Start Point** option from it. Next, choose the **OK** button to exit the sketching environment.

4. Choose the **OK** button to complete feature creation. The default trimetric view of the model is shown in Figure 9-130.

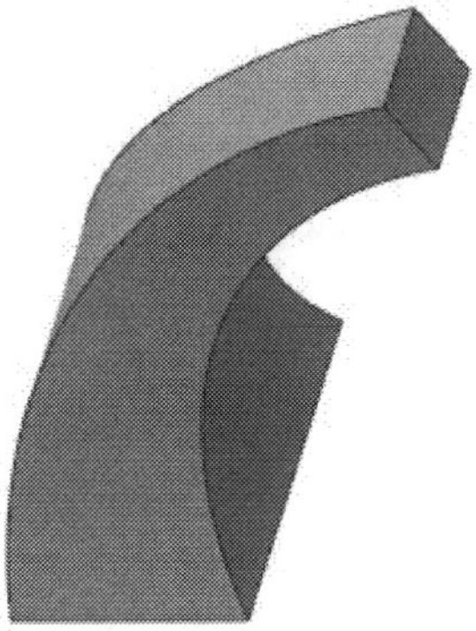

Figure 9-130 *The default trimetric view of the blend feature*

Creating the Shell Feature

1. Choose the **Shell** tool from the **Engineering** group; the **Shell** dashboard is displayed and you are prompted to select the surfaces to be removed from the part.

2. Select an end surface of the sweep feature by using the left mouse button and then select the other end surface using the CTRL+left mouse button; the two selected surfaces are highlighted.

3. Enter **5** as the thickness value of the shell in the **Thickness** edit box in the **Shell** dashboard and press ENTER.

4. Choose the **OK** button to create the shell feature.

   The default trimetric view of the model is shown in Figure 9-131. You can use the middle mouse button to change the orientation of the model.

Saving the Model

1. Choose the **Save** button from the **File** menu and save the model.

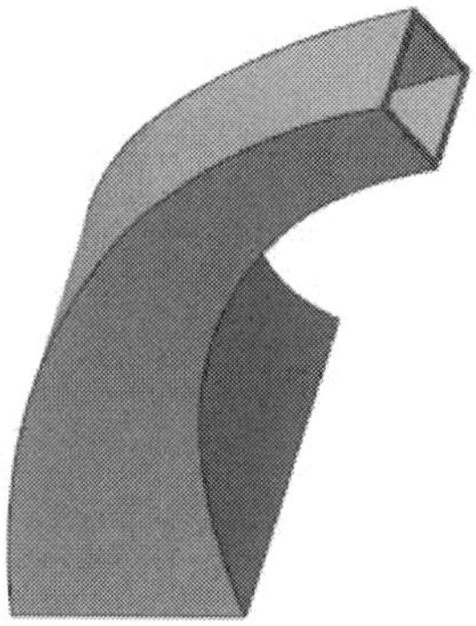

Figure 9-131 *Trimetric view of the model for Tutorial 5*

Tutorial 6

In this tutorial, you will create the blend feature shown in Figure 9-132. The orthographic detailed views of the model are shown in Figure 9-133. After creating the model, you will redefine it such that the straight blending is changed into a smooth blending.

(Expected time: 20 min)

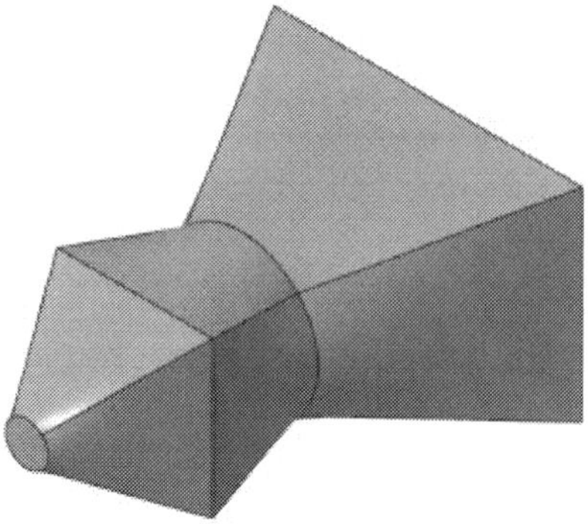

Figure 9-132 *Isometric view of the model*

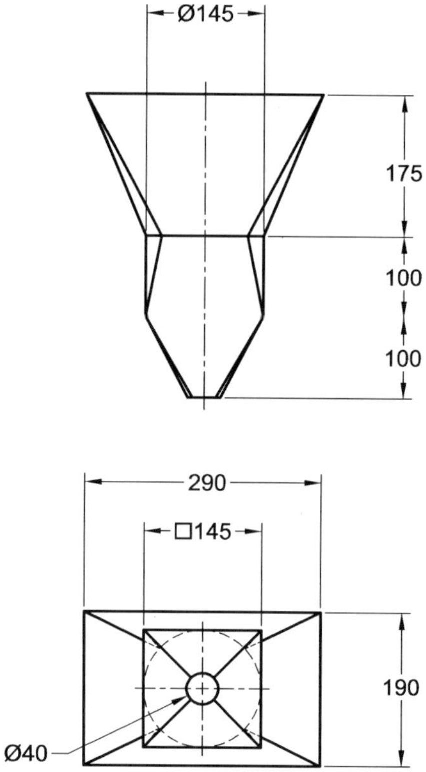

Figure 9-133 *Top and front views of the model*

Examine the blend feature and determine the number of sections in this feature. The blend consists of four sections.

The following steps are required to complete this tutorial:

a. Create the sketch for the first section of the blend feature.
b. Create the sketch for the second section of the blend feature. Define the depth values between section numbers 1 and 2.
c. Create the sketch for the third section of the blend feature and define the depth values between section numbers 2 and 3.
d. Create the sketch for the fourth section. Define the depth values between section numbers 3 and 4.
e. Redefine the model to change the straight blending into a smooth blending.

Make sure that the *c09* folder is the current Working Directory.

Starting a New Object File

1. Start a new part file and name it as *c09tut6*. The three default datum planes are displayed in the drawing area.

Invoking the Blend Option

1. Choose the **Blend** tool from the expanded **Shapes** group of the **Model** tab; the **Blend** dashboard is displayed.

2. Choose the **Solid** button.

3. Choose the **Sections** tab from the dashboard to open the slide-down panel. Next, select the **Sketched sections** radio button and choose the **Define** button to define the sketching plane.

Selecting the Sketching Plane

1. Select the **FRONT** datum plane as the sketching plane.

2. Select the **RIGHT** datum plane as the reference plane and then select the **Right** option from the **Orientation** drop-down list if not selected by default.

Drawing the First Section

1. Draw the sketch of the rectangular section of **290x190** units and then add constraints and dimensions to it, as shown in Figure 9-134.

2. Choose the **OK** button to exit the sketcher environment. Next, choose the **Options** tab and select the **Straight** radio button and then enter **175** in the offset value edit box for the second section.

3. Choose the **Sketch** button from the **Sections** tab to enter the sketcher environment.

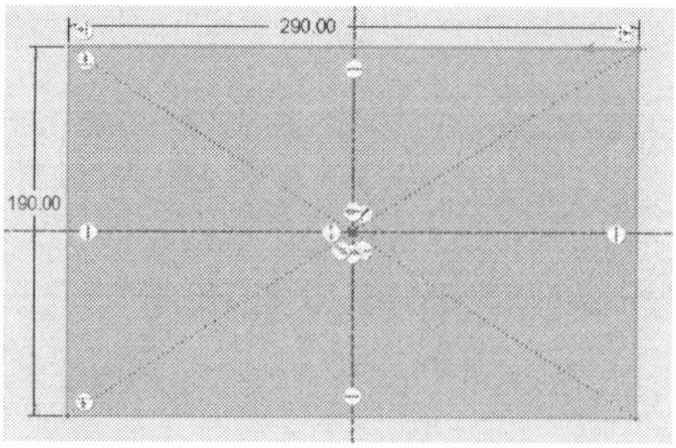

Figure 9-134 *First rectangular section*

Note

While drawing the sections for the blend feature, the start point is very important. The start point should be similar to those shown in Figures 9-134, 9-135, and 9-136. If the start point is not at the desired point then select the point where you need the start point. Next, hold the right mouse button and choose the ***Start Point*** *option from the shortcut menu invoked.*

Drawing the Next Section

The next section is a circle.

1. Draw the sketch of the circular section, refer to Figure 9-135. Modify the diameter of the circle to **145**. As discussed earlier,in a blend feature, the number of entities per section must be equal. Since a circle is a single entity, it should be divided at four points.

Dividing the Circular Section

The circular section should be divided at four points because both rectangle and square have four entities. When you divide a circle at four points, the number of entities becomes four.

1. Choose the **Divide** tool from the **Editing** group.

2. Select four points on the circle, refer to Figure 9-135.

As you select the points on the circle to divide the circle, some weak dimensions appear on the circle. Next, you need to apply constraints on the four points that were selected to divide the circle.

Applying Constraints on the Four Points

1. Choose the **Vertical** button from the **Constrain** group and select the two divisional points on the left of the circle to lie in a vertical line. Similarly, select the two points on the right to apply the constraint.

2. Now, choose the **Horizontal** button from the **Constrain** group of the **Sketch** tab and select the two division points on the upper half and the lower half to lie in a horizontal line.

3. Modify the vertical dimension of the upper right division point, as shown in Figure 9-135. After the section is completed, the two sections with dimensions will be similar to the sections shown in Figure 9-135.

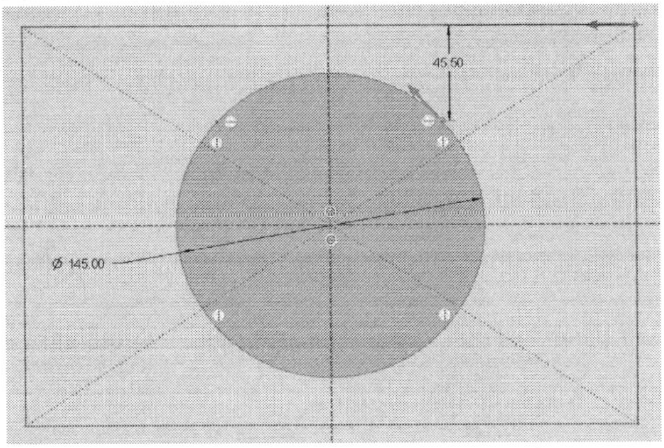

Figure 9-135 Sketch of the second section

4. Choose the **OK** button to exit the sketcher environment. Now, choose the **Insert** button from the **Sections** tab to add the third section. Enter **100** in the offset value edit box to create depth between second and third sections. The next section to be drawn is a square of **145** units.

5. After drawing the square section, choose the **Insert** button to add the fourth section. Enter **100** in the offset value edit box to create depth between third and fourth sections. Draw the circular section. Divide the circular section into four entities similar to second section and then constrain and dimension it. Figure 9-136 shows all completed sections. Choose the **OK** button to exit the sketcher environment.

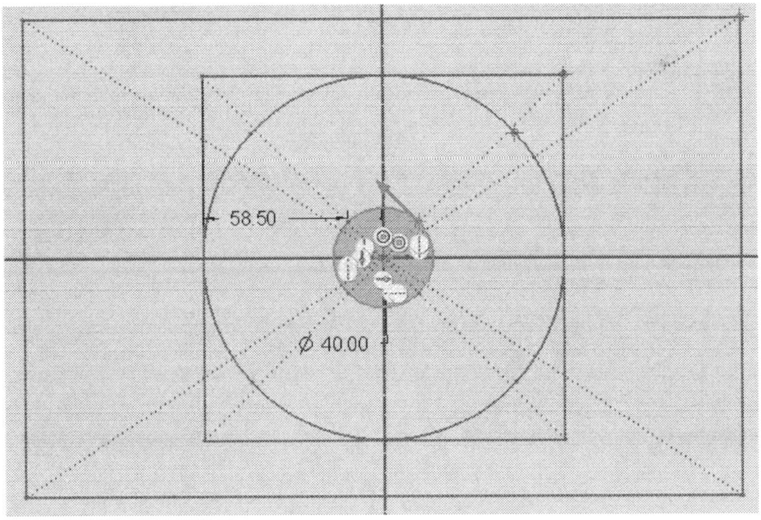

Figure 9-136 All four completed sections

Note

In Figure 9-136, note that the direction of the start points is indicated by arrows. These arrows are important as they help you avoid creating a twisted feature.

Creating the Blend Feature

Now, you need to create the blend feature. As you exit the sketcher environment after defining the sections; a preview of the blend feature is displayed in the drawing area.

1. Choose the **Default Orientation** option from the **Saved Orientations** flyout; the model orients in the drawing area, as shown in Figure 9-137.

2. Now, choose the **OK** button from the dashboard; the trimetric view of the model is displayed in the drawing area.

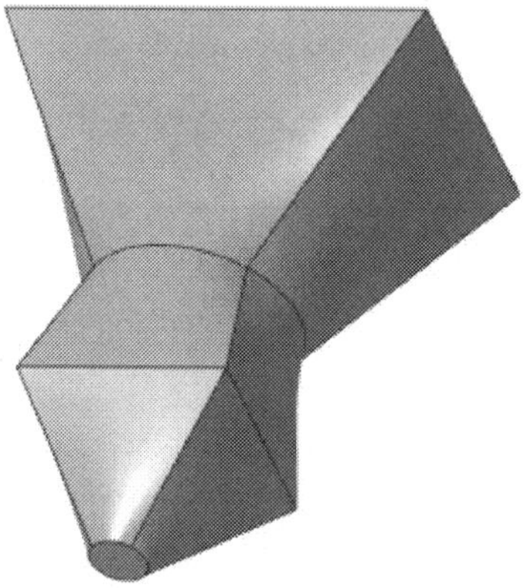

Figure 9-137 *Trimetric view of the model*

Redefining the Blend Feature

After saving the straight blend feature, you need to redefine this feature so that it can be converted into a smooth blend.

1. Select the model in the drawing area, or select the blend feature from the model tree; the edges of the model turn green and a mini popup toolbar is displayed near the cursor.

2. Click on the **Edit definition** button from the mini popup toolbar; the **Blend** dashboard is displayed.

3. Choose the **Options** tab and then select the **Smooth** radio button.

4. Now, choose the **OK** button; the straight blend feature is converted into a smooth blend, as shown in Figure 9-138.

Saving the Model

1. Choose the **Save** button from the **File** menu and save the model.

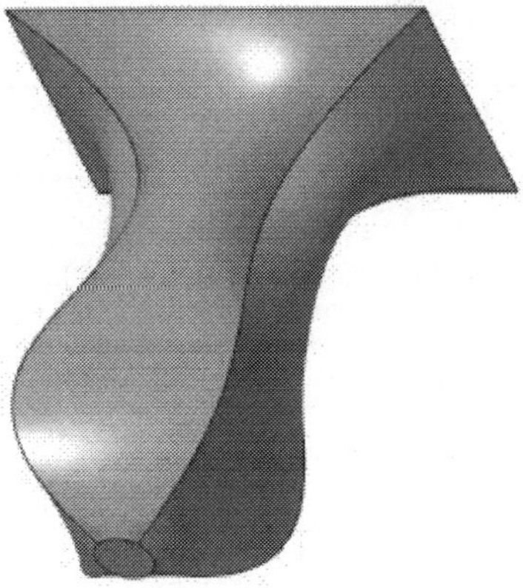

Figure 9-138 Smooth blend feature

Self-Evaluation Test

Answer the following questions and then compare them to those given at the end of this chapter:

1. What is the maximum permissible angle for the rotation of sections in a rotational blend?

   (a) 120 degree (b) 90 degree
   (c) 180 degree (d) 45 degree

2. What is the maximum possible draft angle that can be applied to a solid in Creo Parametric?

   (a) 10 degree (b) 89.9 degree
   (c) 60 degree (d) 90 degree

3. What is the minimum number of sections required for a blend feature?

   (a) one (b) two
   (c) three (d) None of these

4. While using the **Sweep** tool, the contour of the sweep feature depends on the shape of the __________.

5. While using the **Sweep** tool, when you draw section for the sweep feature, the section should be __________ to the endpoints of the trajectories.

6. When you use the **Normal To Trajectory** option to create the variable sweep, the section is created__________ to the origin trajectory and is swept along the X trajectory.

7. The **Swept Blend** tool is used to create a model that is a combination of __________ and __________ features.

8. You can create helical sweep features with a constant as well as a variable pitch. (T/F)

9. The **Helical Sweep** tool is used to create cut as well as protrusion features. (T/F)

10. You can add as well as delete points while specifying the pitch values during the creation of a variable pitch helical sweep. (T/F)

Review Questions

Answer the following questions:

1. Which of the following dashboards is displayed when you choose the **Helical Sweep** option from the **Shapes** group in the **Ribbon**?

   (a) **Sweep** (b) **Revolve**
   (c) **Swept Blend** (d) None of these

2. Which of the following options is used to create a helical feature with constant pitch throughout the sweep?

   (a) **Right Handed** (b) **Constant**
   (c) **Thru Axis** (d) None of these

3. Which of the following options in the **Swept Blend** dashboard is used to create a sweep in which the swept section is normal to the trajectory defined as the origin trajectory?

   (a) **Constant Normal Direction** (b) **Normal To Projection**
   (c) **Normal To Trajectory** (d) None of these

4. The **Sweep** tool extrudes a section along a __________.

5. The cross-section of a sweep feature remains __________ throughout the sweep.

6. The sketching plane that you select becomes __________ to the screen when you draw the trajectory.

7. The first trajectory selected for creating a sweep feature is known as __________.

8. You can use the **Helical Sweep** tool to create a spring design. (T/F)

9. The **Sweep** tool can be used to create a sweep in which the section of the sweep is constant throughout the sweep. (T/F)

10. The X trajectory is used to guide the section of the sweep feature along it. (T/F)

11. The trajectory of a sweep feature can be modified independent of the geometry of the section. (T/F)

12. A geometric centerline is required for creating a rotational blend feature. (T/F)

13. You need to have at least three sections for creating rotational blend using the **Connect end and start sections** option. (T/F)

14. Each section must have equal number of vertices for blending. (T/F)

EXERCISES

Exercise 1

Create the model shown in Figure 9-139. The front and top views of the model are shown in Figure 9-140. **(Expected time: 30 min)**

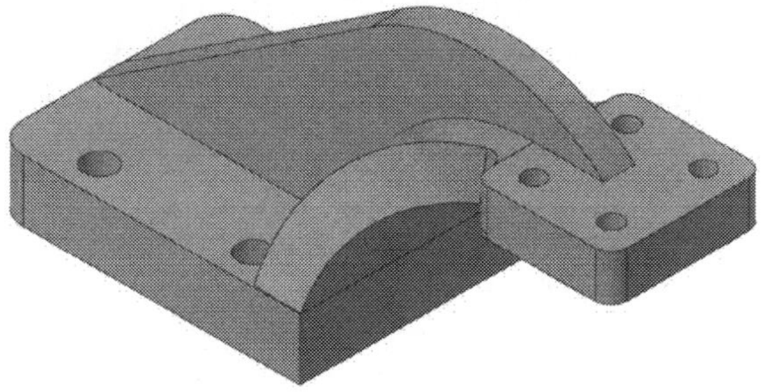

Figure 9-139 Solid model for Exercise 1

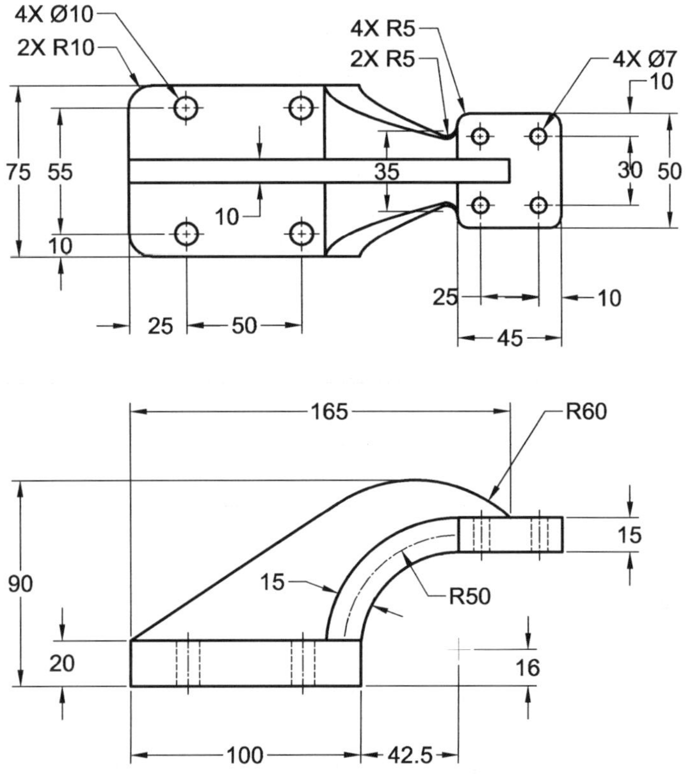

Figure 9-140 Orthographic views of the model

Hints to create the swept blend feature:

1. Select the **FRONT** datum plane as the sketching plane and select the top face of the base feature to be at the top while sketching.
2. Draw an arc of radius 50 and locate its center from the base feature.
3. Exit the sketcher environment by choosing the **OK** button.
4. Choose **Shapes > Swept Blend** from the **Ribbon**; the **Swept Blend** dashboard is displayed.
5. Choose the **Solid** button from the dashboard.
6. Choose the **Reference** tab from the dashboard and select the arc drawn.
7. Choose the **Sections** tab and select the endpoint of the arc lying on the base feature, Choose the **Sketch** button from the **Sections** slide-down panel to enter the sketcher environment. Next, draw a rectangle of dimension 15 x 75. Exit the sketcher environment.
8. Choose the **Insert** button from the **Sections** slide-down panel and then select the other end point of the arc. Then, choose the **Sketch** button from the slide-down panel; this will again take you to the sketcher environment. Draw another rectangle with the dimensions 15 x 30. Exit the sketcher environment by choosing the **OK** button.

Exercise 2

Create the model of the Helical Gear shown in Figure 9-141. Its views and dimensions are shown in Figure 9-142. **(Expected time: 45 min)**

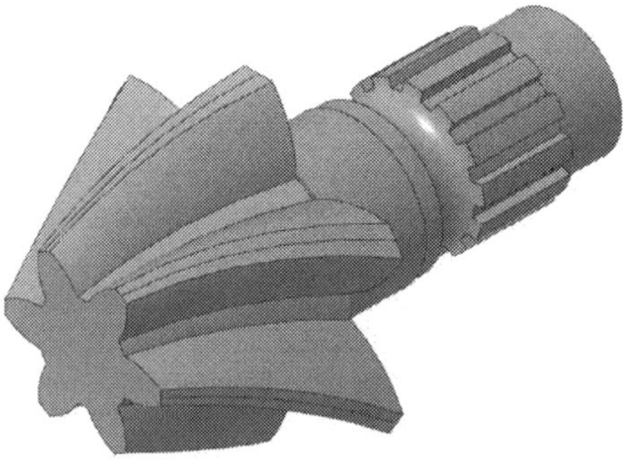

Figure 9-141 *Model of the Helical Gear*

NOTE:

1. CREATE A BLEND FEATURE WITH 3 SKETCHES. SIZE OF THE SECOND SKETCH IS 75% OF THE FIRST SKETCH AND, SIZE OF THE THIRD SKETCH IS 50% OF THE FIRST SKETCH.

2. THE PLANE FOR CREATING THE SECOND SKETCH IS 45 UNITS FROM THE FIRST SKETCH. THE PLANE FOR CREATING THE THIRD SKETCH IS 95 UNITS FROM THE FIRST SKETCH.

3. ROTATION ANGLE FOR SECOND AND THIRD SECTION IS 15° AND 30° FROM THE FIRST SECTION RESPECTIVELY.

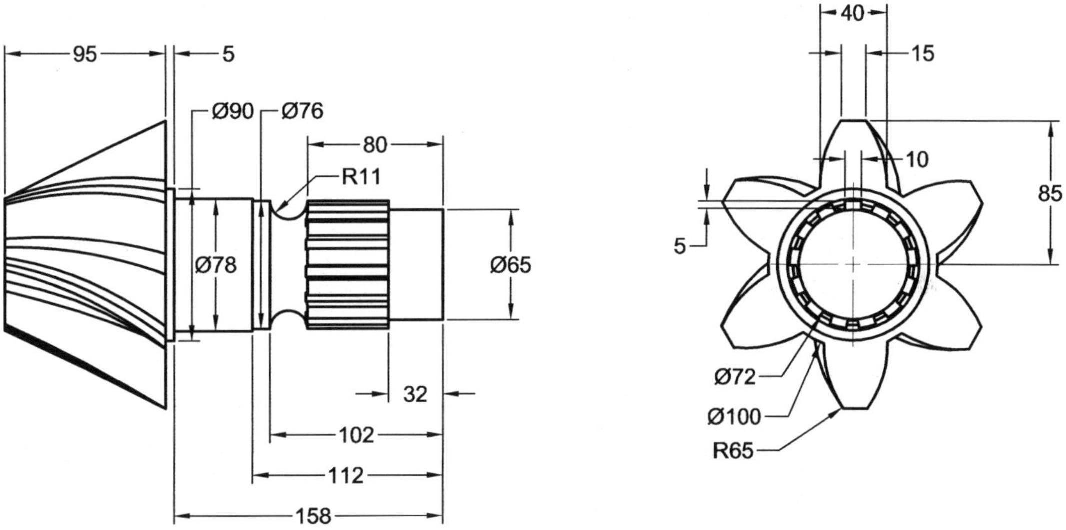

Figure 9-142 *Views and Dimensions for the model*

Hints to create a Helical Gear is discussed next:

1. Choose the **Sketch** tool from the **Datum** group of the **Model** tab in the **Ribbon** and draw the sketch, as shown in Figure 9-143. Next, choose the **OK** button to exit the sketching environment.

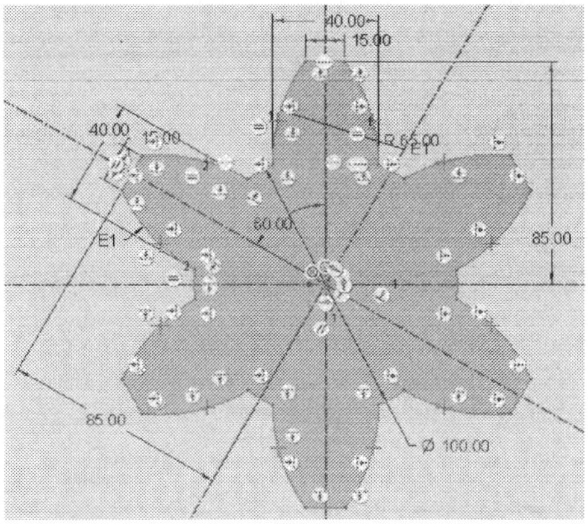

Figure 9-143 *Sketches for the Swept Blend feature (Section 1)*

2. Next, choose the **Sketch** tool from the **Datum** group and choose the **Project** tool from the **Sketching** group and choose the **Loop** option from the **Type** menu. Next, select Section 1 from the drawing area; the section will be projected.
3. Next, select the projected section and choose the **Rotate Resize** tool from the **Editing** group and scale the selected sketch by 75% and rotate it by 15 degrees.
4. Similarly, draw the third sketch at an offset distance of 50 and scale the sketch by 50% and rotate it by 15 degrees from the second sketch.
5. Now, choose the **OK** button; a blend feature is created, as shown in Figure 9-144.

Figure 9-144 *Preview of the Blend feature*

Exercise 3

Create the model of the Crane hook shown in Figure 9-145. Its views and dimensions are shown in Figure 9-146. **(Expected time: 45 min)**

Figure 9-145 *The model of the Hook*

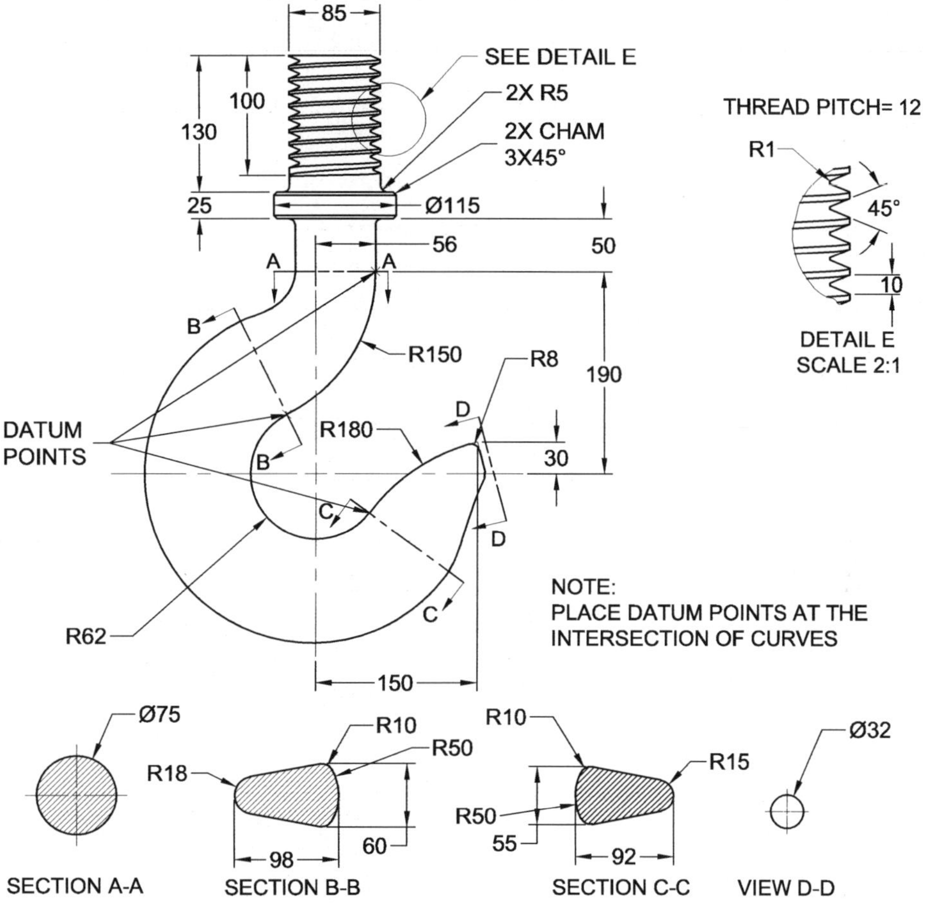

Figure 9-146 *Views and dimensions of the model*

Hints to create the Swept Blend feature

1. The model in the Figure 9-145 is created using the **Swept Blend** tool with 5 sections and a trajectory.
2. Invoke the sketcher environment, select the **FRONT** datum plane as the sketching plane for sketching a trajectory.
3. Sketch a trajectory by using the **3-Point / Tangent End** tool from the **Sketching** group. Draw the sketch of the trajectory, refer to Figure 9-146.
4. After drawing the sketch, choose the **Point** tool from the **Datum** group of the sketcher environment and place points at the intersection of the arcs, as shown in Figure 9-146.
5. Exit the sketcher environment by using the **OK** button.
6. Choose the **Swept Blend** tool from the **Shapes** group of the **Ribbon**, select the trajectory, and invoke the **Sections** slide-down panel.
7. Choose the **Insert** button in the **Sections** slide-down panel and draw first circular section. Other sections for the swept blend feature are having 6 vertices, hence you need to divide the circle into 6 parts.
8. Choose the **Divide** tool from the **Editing** group of the sketcher environment and apply dimensions as shown in Figure 9-147.

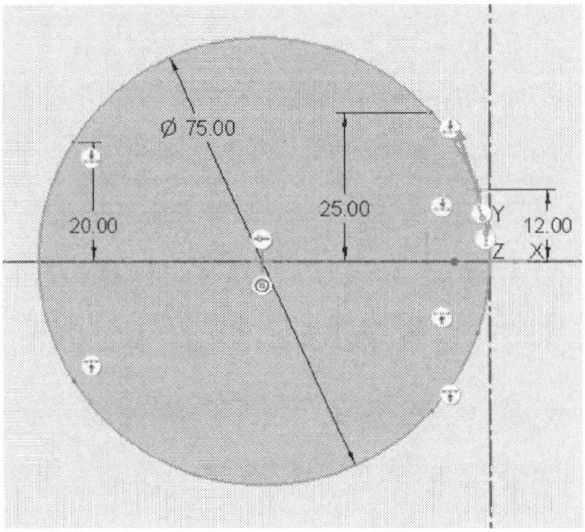

Figure 9-147 *Dimensions for the first section of the swept blend feature*

9. Close the sketcher environment after drawing the first section and choose the **Insert** button in the **Sections** panel. Select the datum point in the **Section Location** collector and choose the **Sketch** button to invoke the sketcher environment. Draw the sketch for the second section that is same as the first section.
10. After drawing the sketch for the second section, close the sketcher environment and insert the remaining sections for the **Swept Blend** feature.

Exercise 4

In this exercise, you will create the model of a carburetor cover shown in Figure 9-148. Figure 9-149 shows the top view, sectioned front view, sectioned right view, and the detailed view with dimensions. **(Expected time: 45 min)**

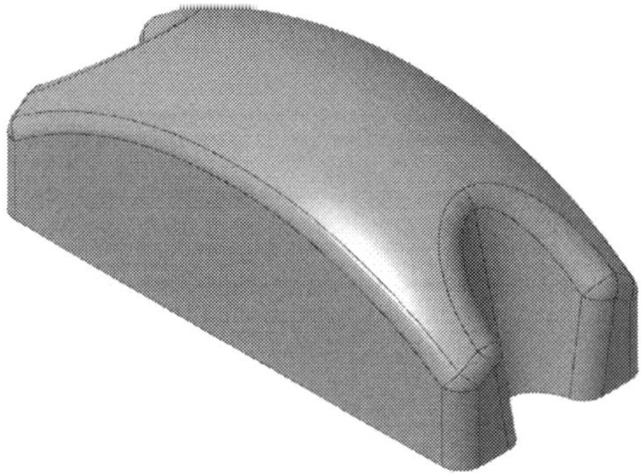

Figure 9-148 *Isometric view of the carburetor cover*

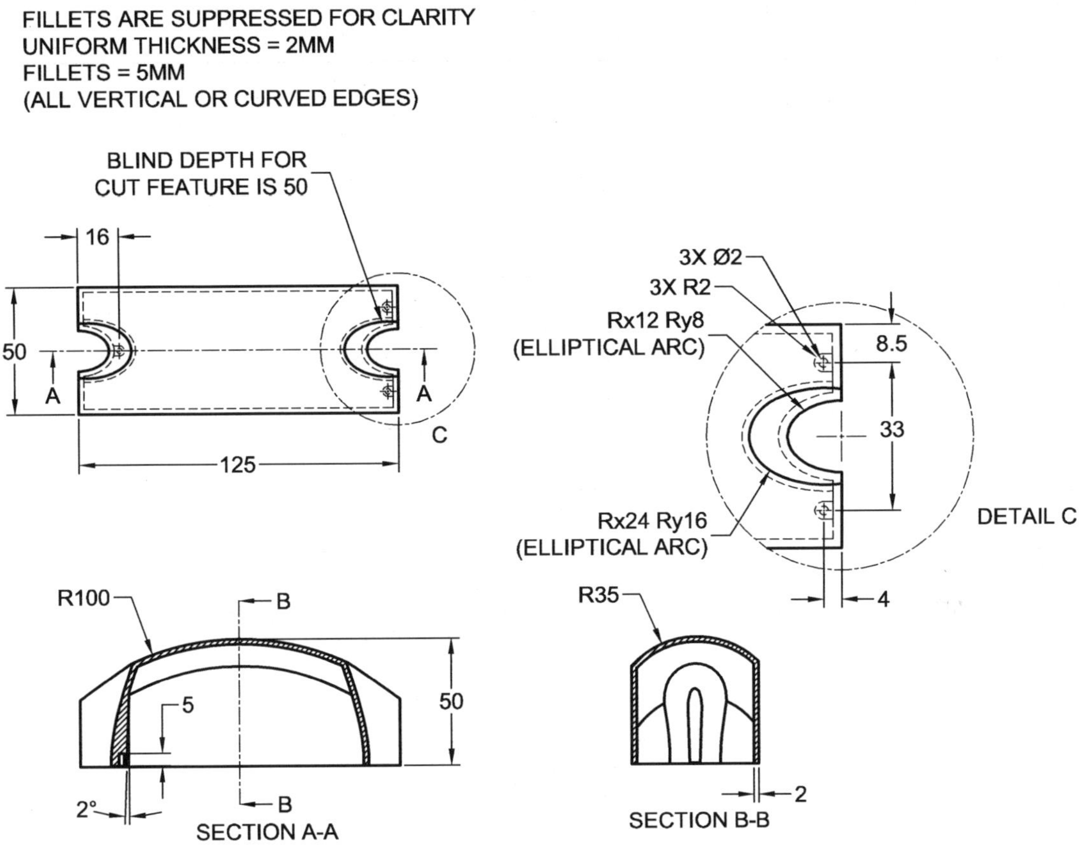

Figure 9-149 Orthographic views of the carburetor cover

Hints

1. Invoke the sketcher environment, select the **FRONT** datum plane as the sketching plane for sketching a trajectory.
2. Sketch the first trajectory by using the **3-Point / Tangent End** button from the **Sketching** group. The start point and the endpoint of the arc should be at a distance of 100 units.
3. Exit the sketcher environment by using the **OK** button.
4. Again select the **FRONT** datum plane as the sketching plane for sketching a second trajectory by drawing the line coinciding horizontal axis with 125 units distance.
5. Choose the **Sweep** tool from the **Shapes** group of the **Ribbon**, select the first sketch, and then the second sketch by keeping the CTRL button pressed.
6. Choose the **Sketch** option to create the section for the sweep feature. Create a sketch as shown in figure 9-150 and choose the **OK** button.
7. Choose the **OK** button from the **Sweep** dashboard.
8. Select the bottom face of the base feature as the sketching plane.
9. Select the **RIGHT** datum plane as the reference plane and then select the **Right** option from the **Orientation** drop-down list.
10. Create an ellipse of Rx12 and Ry8 by using the **Ellipse** button.
11. Choose the **OK** button to exit the sketcher environment.
12. Create datum plane at an offset of **50** from the bottom face and select new datum for sketching.

13. Create another ellipse of Rx24 and Ry16.
14. Exit the sketcher environment.
15. Choose the **Blend** tool in the **Shapes** group of the **Ribbon** and choose the **Remove Material** button from dashboard.
16. Choose the **Sections** tab from the dashboard. Choose the **Selected sections** radio button and select the first section.
17. Click on the **Insert** button to second secion and select the second section. Click on **OK**.
18. Mirror the cut feature about the **RIGHT** datum plane.
19. Create a round feature on all edges except the edges enclosing the bottom planar surface of the base feature.
20. Choose the **Shell** tool from the **Engineering** group. Remove the bottom face of the base feature.
21. Using the bottom face of the base feature as the sketching plane, create three protrusion features that are the supporting structures for the screws. Extrude these features by using the **Extrude up to next surface** button.
22. Create holes on three protrusion features.

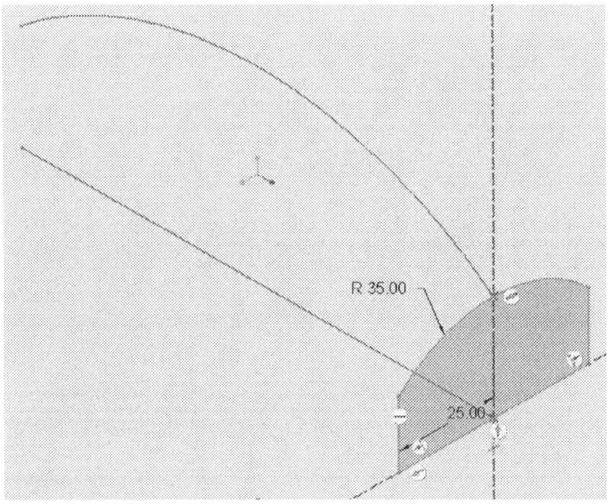

***Figure 9-150** Dimensions for the section of the sweep feature*

Answers to Self-Evaluation Test

1. a, **2.** b, **3.** b, **4.** X-trajectory, **5.** aligned, **6.** normal, **7.** sweep, blend, **8.** T, **9.** T, **10.** T

Chapter 10

Assembly Modeling

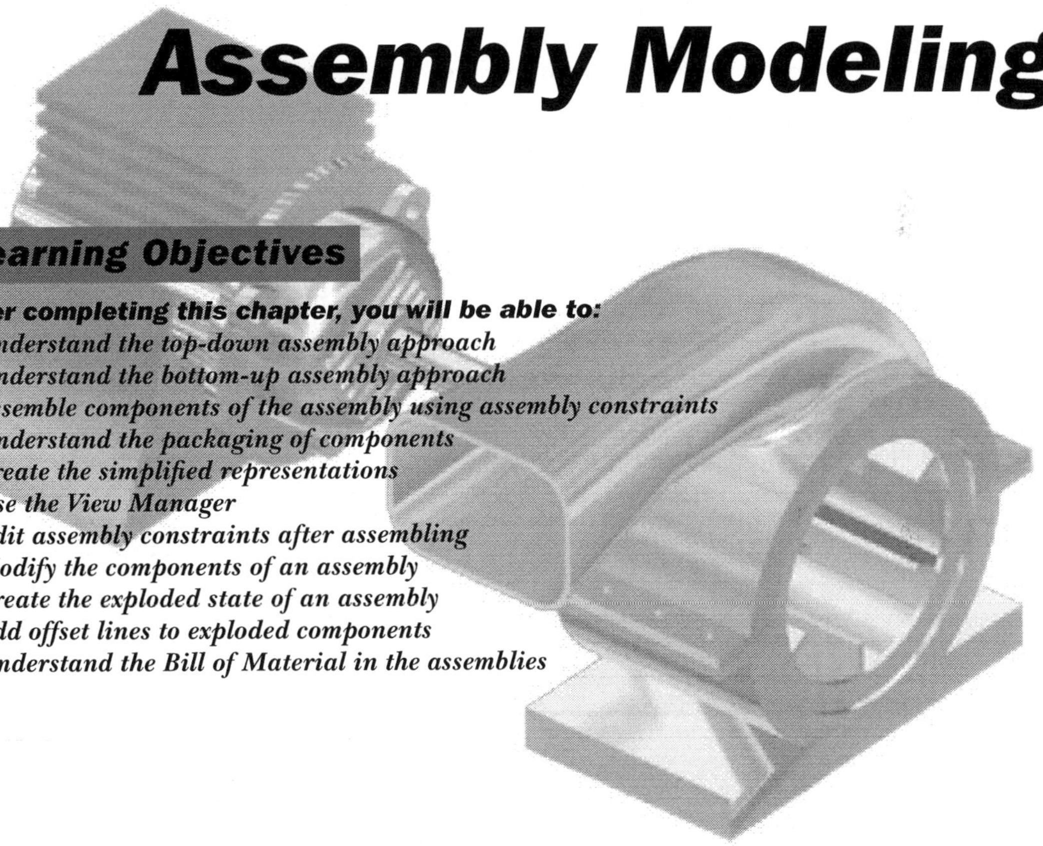

Learning Objectives

After completing this chapter, you will be able to:

- *Understand the top-down assembly approach*
- *Understand the bottom-up assembly approach*
- *Assemble components of the assembly using assembly constraints*
- *Understand the packaging of components*
- *Create the simplified representations*
- *Use the View Manager*
- *Edit assembly constraints after assembling*
- *Modify the components of an assembly*
- *Create the exploded state of an assembly*
- *Add offset lines to exploded components*
- *Understand the Bill of Material in the assemblies*

ASSEMBLY MODELING

An assembly is defined as a design consisting of two or more components bonded together at their respective working positions using assembly constraints. These assembly designs are created in the **Assembly** mode of Creo Parametric. To invoke the **Assembly** mode, choose the **New** button from the **Quick Access** toolbar; the **New** dialog box will be displayed. In this dialog box, select the **Assembly** radio button from the **Type** area and then select the **Design** radio button from the **Sub-type** area, as shown in Figure 10-1. Specify the name of the assembly in the **File name** edit box and choose the **OK** button; the **Assembly** mode will be activated. Also, the initial screen appearance of the **Assembly** mode in Creo Parametric will be displayed, as shown in Figure 10-2.

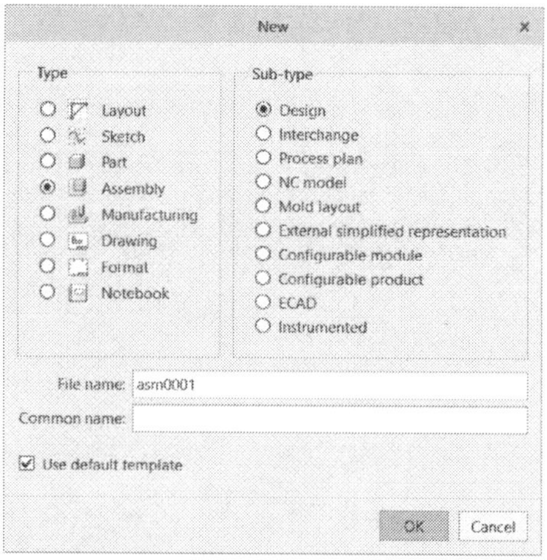

***Figure 10-1** Invoking the **Assembly** mode using the **New** dialog box*

IMPORTANT TERMS RELATED TO THE ASSEMBLY MODE

Before proceeding further in this chapter, it is very important for you to understand the following terms.

Top-down Approach

This is the method of assembling the components in which the components of the assembly are created in the same assembly file. In this type of assembly modeling approach, the components are created in the assembly file and then assembled using the assembly constraints. Note that the parts you create in the **Assembly** mode are saved as separate *.prt* files.

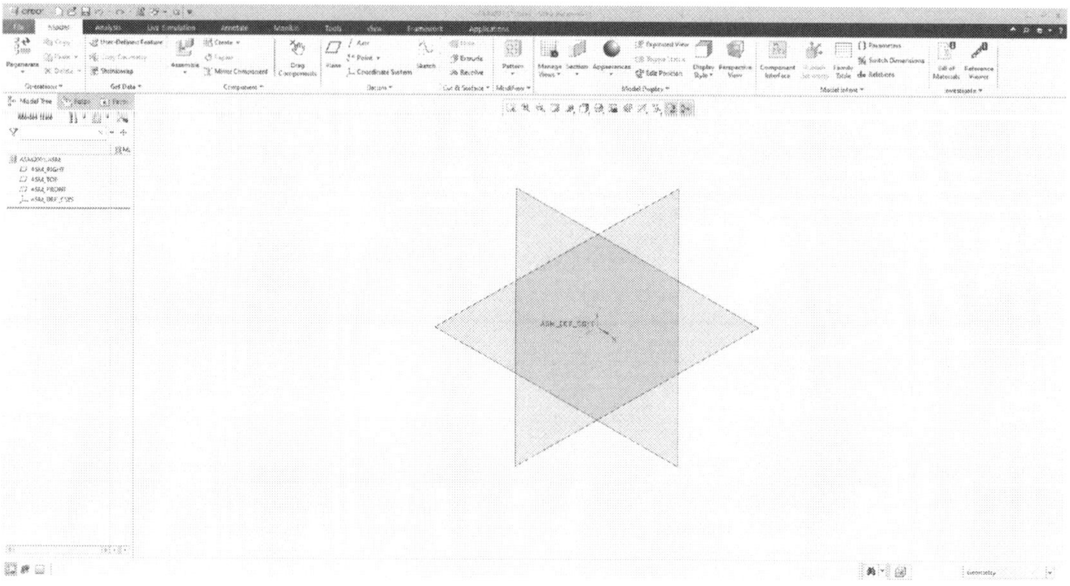

***Figure 10-2** Initial screen appearance of the **Assembly** mode*

Bottom-up Approach

This is the method of assembling the components that are created as separate parts in the **Part** mode and are saved as *.prt* files. Once all parts of an assembly have been created, you will create a new assembly file (*.asm*) and then assemble the parts using the assembly constraints available in the **Assembly** mode. Since the assembly file has information related only to the assembling of components, its size is small and therefore requires less hard disk space.

Remember that if any of these assembly components is moved from its original location on the hard disk or renamed after saving the assembly file, the assembly will not regenerate when you open it next time. This is because Creo Parametric searches for the assembly component in the folder in which it was originally saved. If it does not find the name of the component in that folder, it will show an error while opening the assembly file. The missing component will be displayed in red color in the **Model Tree**, and also a notification will be displayed in the notification area. To complete the assembly, either move the required files to the original folder or retrieve the files. To retrieve a file, right-click on the missing component in the **Model Tree**; a flyout will be displayed. Choose the **Retrieve Missing Component** option from the flyout; the **File Open** dialog box will be displayed and you will be prompted to browse to the missing file. Select the required file and then choose the **Open** button; the missing component will be assembled on the basis of previously defined relations. Alternatively, you can also use the **Notification Center** dialog box in the same manner to retrieve the missing component, refer to Figure 10-3.

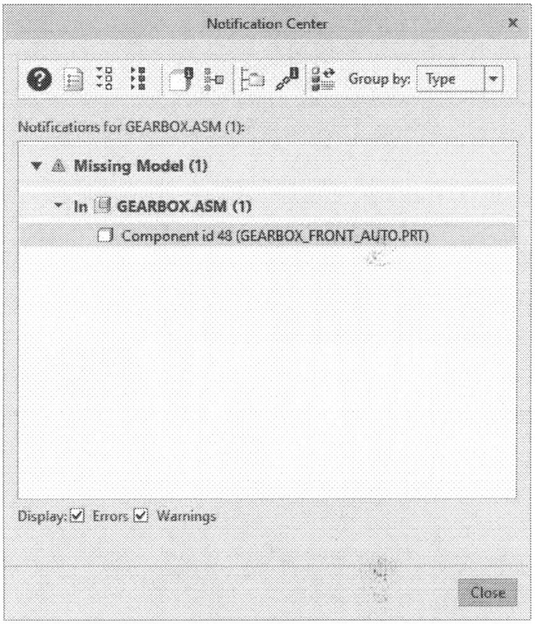

***Figure 10-3** The **Notification Center** dialog box*

> **Tip**
> 1. *It is recommended that you create separate folders for all assemblies that you create in Creo Parametric.*
>
> 2. *If you are opening an assembly with the part name same as that in the previous session then the part of the previous session will open instead of the part in the current session. To avoid this, it is recommended that you use the **Erase Not Displayed** button before opening an existing assembly. For example, if Nut was opened in the previous session and you open a component with same name in the current session then Creo Parametric will open the previously opened Nut rather than the new one.*

Placement Constraints

The placement constraints are the constraints that are used to rigidly bind the components of the assembly to their respective positions in the assembly. These constraints are also called as assembly constraints. Generally, these constraints are used in combinations and you can constrain upto six degrees of freedom of a component.

Package

Packaged components are unconstrained components in an assembly. It is the state in which the component being assembled is not fully constrained and therefore it is not rigidly placed at its actual location.

CREATING TOP-DOWN ASSEMBLIES

As mentioned earlier, in the top-down assemblies, all components are created in the **Assembly** mode and in the same assembly file. The components created in the **Assembly** mode can be saved as separate *.prt* files. As you already know that to create components, you need to invoke the **Part** mode. You can invoke the **Part** mode within the **Assembly** mode and create components using the tools available in the **Part** mode. The method to create the component from within the **Assembly** mode is discussed next.

Creating Components in the Assembly Mode

Ribbon: Model > Component > Create

To create a component in the **Assembly** mode, choose the **Create** tool from the **Component** group; the **Create Component** dialog box will be displayed, as shown in Figure 10-4. Select the **Part** radio button from the **Type** area and the **Solid** radio button from the **Sub-type** area to create a *.prt* file. Enter the name of the component in the **File name** edit box and then choose the **OK** button; the **Creation Options** dialog box will be displayed, as shown in Figure 10-5. The options available in this dialog box are discussed next.

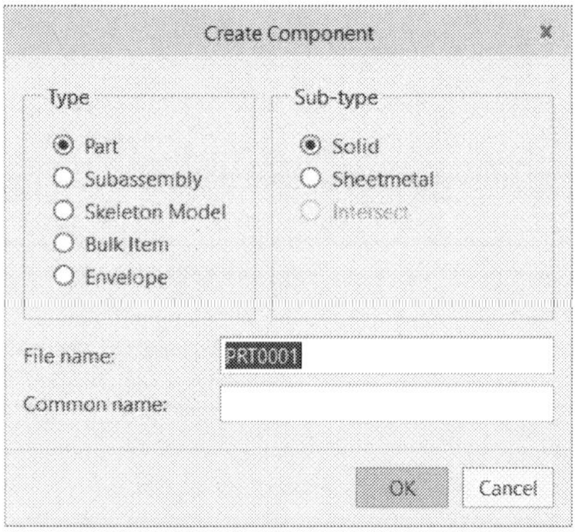

*Figure 10-4 The **Create Component** dialog box*

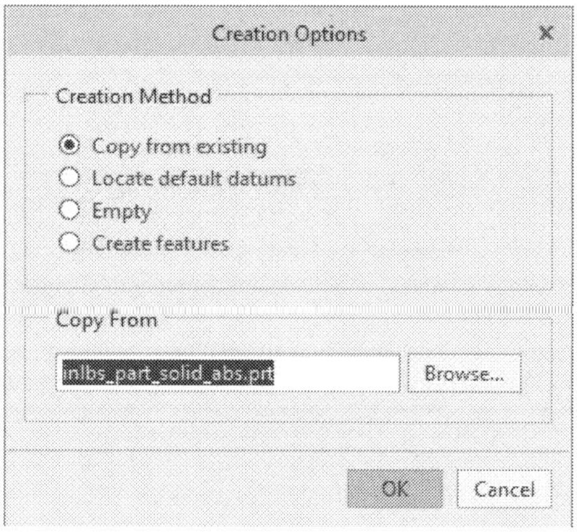

*Figure 10-5 The **Creation Options** dialog box*

In the **Creation Options** dialog box, the **Copy from existing** radio button is selected by default in the **Creation Method** area. As a result, the copy of an existing part component will be created by default. To select an existing part, click the **Browse** button next to the **Copy From** edit box; the **Choose template** dialog box will be displayed. Next, browse for the part document that you want to copy in the assembly and choose the **Open** button; the **Choose template** dialog box will close and the name of the selected part document will be displayed in the **Copy From** edit box of the **Creation Options** dialog box. Next, choose the **OK** button; the **Component Placement** dashboard will get activated. Now you can define the part placement using various assembly constraints.

The **Locate default datums** radio button in the **Creation Options** dialog box is used to create a component that assembles itself relative to the assembly references. When you select this radio button, the **Locate Datums Method** area gets activated below the **Creation Method** area. By default, the **Three planes** radio button is selected in this area. Choose the **OK** button from the **Creation Options** dialog box. As a result, you will be prompted to select planes for creating the part. After you select the planes for the part creation, a new part is created and the **Part** mode gets activated. The new part becomes activated in the assembly and remains activated until you activate another part or assembly.

The **Empty** radio button is used to create an empty part within the assembly. The component you create using this option is included in the assembly as an unplaced component.

The **Create features** radio button allows you to create the first feature of a new part. The initial features of the part are dependent on the assembly. On selecting this option, the **Part** mode will be activated, where you can create the features of the new part. The new part remains active model in the assembly until you do not activate other component or assembly.

CREATING BOTTOM-UP ASSEMBLIES

As discussed earlier, the bottom-up assemblies are created by inserting the part files one by one in the assembly file and then assembling them using the assembly constraints. When the first component is inserted in the **Assembly** mode, its three default datum planes are placed in the same orientation as that of the default datum planes of the **Assembly** mode. The method of inserting a component in the assembly is discussed next.

Inserting Components in an Assembly

Ribbon: Model > Component > Assemble drop-down > Assemble

When you enter the assembly environment, the three default assembly datum planes will be displayed in the drawing area. If you do not want the default assembly datum planes in the assembly file, then while opening a new file, clear the **Use default template** check box from the **New** dialog box and then choose the **OK** button; the **New File Options** dialog box will be displayed. Choose **Empty** from the **Template** list in this dialog box. It is recommended that you use the assembly datum planes as the first feature of the assembly and assemble the components of the assembly taking the reference of assembly datum planes. Using the datum planes as the first feature will help you during the modification of the assembly.

Some of the advantages of using the assembly datum planes are:

1. The components of the assembly can be redefined in such a way that the component placed later can be made the first component.

2. The first component can be replaced by some other component.

3. The placement constraint of the first component can be modified.

To insert a component, choose the **Assemble** tool from the **Assemble** drop-down in the **Component** group, or press the **A** key from the keyboard; the **Open** dialog box will be displayed. This dialog box is the same as the **File Open** dialog box that is used to open an existing file.

Select the part that you need to insert into the assembly. Choose the **Open** button to open the part. You can also preview the component in the **Preview** rollout in the **Open** dialog box before adding the component in the assembly. After the components are displayed in the drawing area in the **Assembly** mode, you need to specify constraints in order to assemble them by using the **Component Placement** dashboard. Even if you are placing the first component, you need to constrain it using the assembly constraints. Generally, the first component is assembled with the datum planes. The process of assembling components in an assembly is discussed next.

Assembling Components

The components in an assembly can be placed parametrically or non-parametrically. If the components are placed using the assembly constraints, it is called parametric assembly. In a parametric assembly, components are placed relative to their neighboring components and the position of the components is updated when its neighboring components are moved or changed.

On the other hand, if the components are packaged or placed without assembly constraints then the assembly is called a non-parametric assembly.

To assemble the components parametrically, choose the **Assemble** tool; the **Open** dialog box will be displayed. Select the required component and choose the **Open** button from the **Open** dialog box; the **Component Placement** dashboard will be displayed, as shown in Figure 10-6.

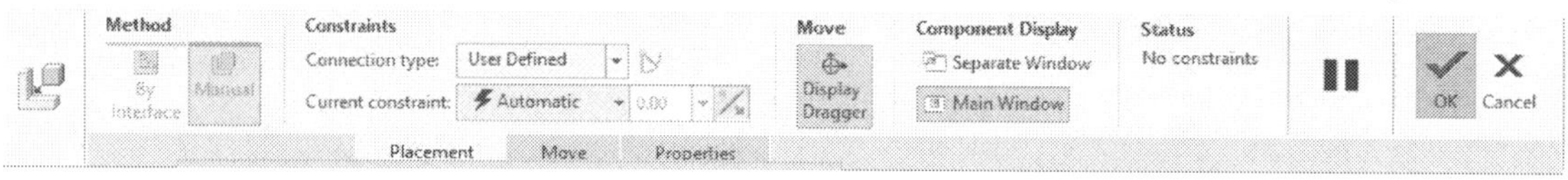

Figure 10-6 *Partial view of the* ***Component Placement*** *dashboard*

Using the **Component Placement** dashboard, you can apply constraints to assemble the components and also control the display of the component to be assembled in the same or separate windows. The options in the **Component Placement** dashboard are discussed next.

Displaying Components in a Separate Window

The **Separate Window** button is used to display the component to be assembled in a separate window. When you choose this button, a new window will open displaying the components to be assembled. Note that viewing the component to be assembled and the assembly, in separate windows, prevents cluttering of the components in the assembly window.

However, if the components are in the same window, they change their orientation as you apply the constraints, thus giving you an idea of the next constraint to be applied. In other words, you can easily find out which degree of freedom of the component is constrained and which needs to be constrained.

Displaying Components in the Same Window

The **Main Window** button is used to display the component to be assembled in the assembly window itself. This button is chosen by default when you invoke the **Component Placement** dashboard. As you apply constraints, the component in the assembly window changes its position accordingly.

Tip
*You can use the **Separate Window** and **Main Window** buttons at the same time.*

3D Dragger

The 3D Dragger is used to move or rotate the component to be assembled. The 3D Dragger can be made visible or hidden by using the **Display Dragger** button available in the dashboard. By using the 3D Dragger, you can move or rotate a component to set its orientation. It has three arrow handles and 3 circular handles, refer to Figure 10-7. The arrow handles are used to move the component and the circular handles are used to rotate the component. If the movement or rotation of a component is restricted in a particular direction then the respective handle of that direction in the 3D Dragger will not be enabled for the movement. The 3D dragger also helps you in determining the restricted direction of a component.

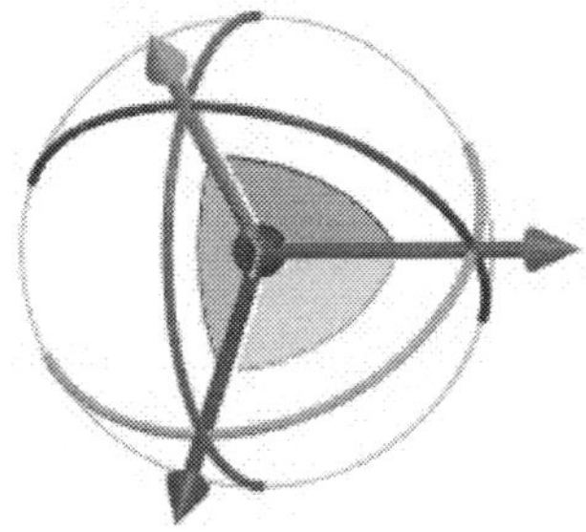

Figure 10-7 *The 3D Dragger*

Applying Constraints

As discussed earlier, the components are assembled using the assembly constraints. The assembly constraints are also called placement constraints. These constraints help in placing and positioning a component precisely with respect to the other components in the assembly. The **Constraint Type** drop-down list in the **Component Placement** dashboard contains different types of placement constraints, as shown in Figure 10-8. You can select the required constraints from this drop-down list to assemble components in the assembly. These constraints are also available when you choose the **Placement** tab of the **Component Placement** dashboard.

Automatic

When you choose this constraint, Creo Parametric assumes the constraint and applies it according to the type of entity selected. For example, if you select faces of two components to assemble, Creo Parametric will assume that you want to apply the Coincident constraint and the Coincident constraint will be applied to the two components.

Distance

The Distance placement constraint is used to constrain the distance between the two selected axes, planes, datum planes, faces, or a combination of a datum plane and a face. The selected faces or datums may or may not be in contact with each other. When you select the Distance constraint, you are prompted to select one component reference and one assembly reference. Select the references as discussed earlier. When you select the references, the distance between them is displayed in the screen and a handle is attached to the component, refer to Figure 10-9. Using this handle, you can dynamically change the distance. You can also use the **Offset** edit box available in the dashboard or the **Placement** tab to enter the distance value. If you want to flip the direction of offset then you need to choose the **Flip** button next to the **Offset** edit box.

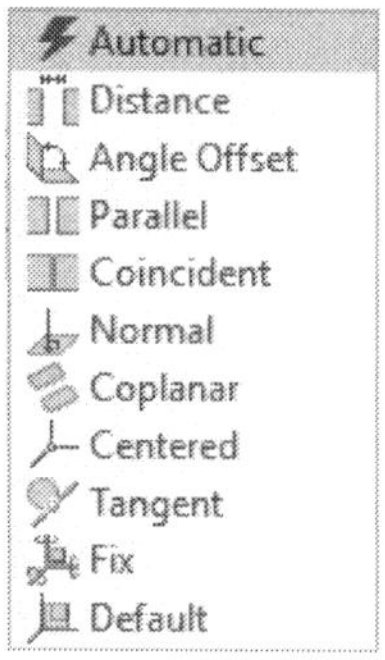

***Figure 10-8** The **Constraint Type** drop-down list*

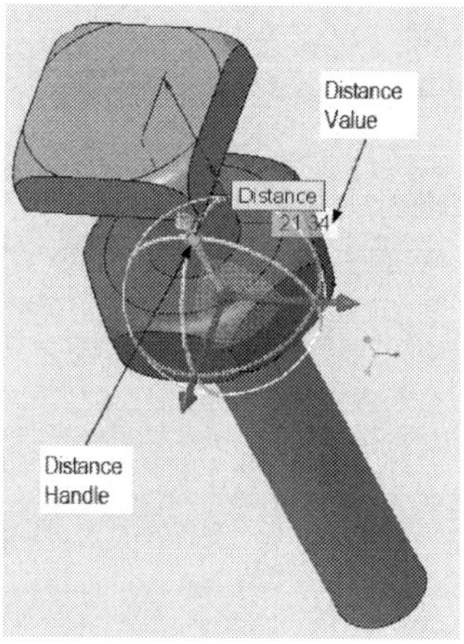

***Figure 10-9** Assembling components by using the Distance constraint*

Angle Offset

The Angle Offset placement constraint allows you to constrain the angular distance between two selected axes, planes, datum planes, faces, or a combination of a datum plane and a face. The selected faces or datums may or may not be in contact with each other. On selecting this constraint, when you select the references of the component and assembly, the component gets constrained by a default value. You can change the angle value by using the **Offset** edit box. Figure 10-10 shows the top face of the component constrained with the bottom face of the assembly using the Angle Offset constraint.

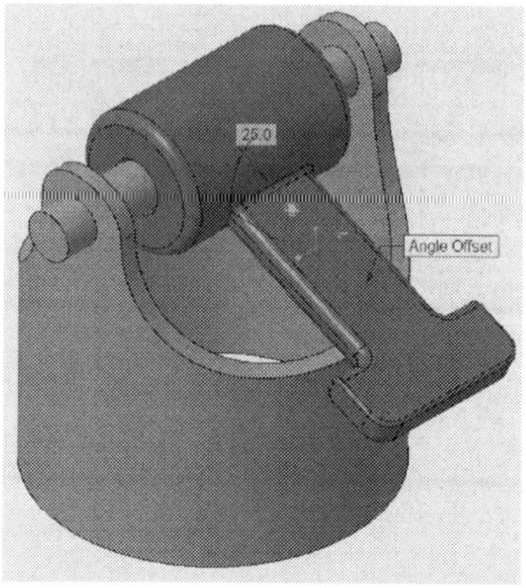

***Figure 10-10** The Angle Offset constraint applied to components*

Parallel

The Parallel placement constraint is used to make two selected axes, planes, datum planes, faces, or a combination of a datum plane and a face parallel to each other. The selected faces or datums may or may not be in contact with each other. Figure 10-11 shows the faces selected for applying the Parallel constraint. Figure 10-12 shows the components after assembling.

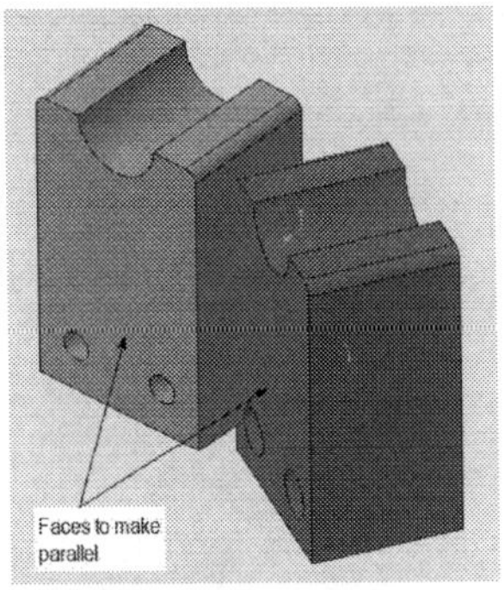

Figure 10-11 Components before applying the Parallel constraint

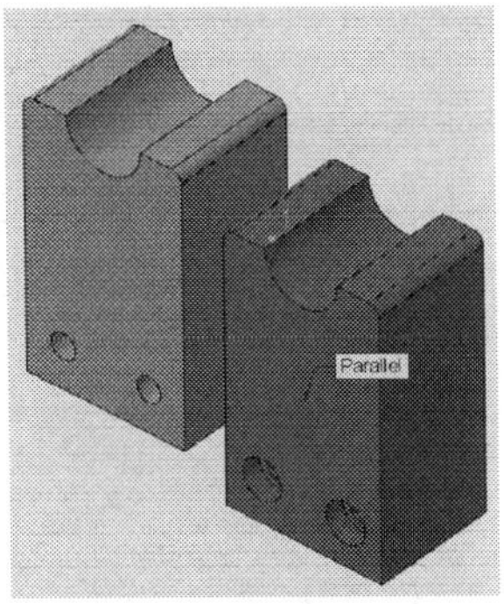

Figure 10-12 Components after applying the Parallel constraint

Coincident

The **Coincident** constraint is used to make two selected planes or faces or two axes coincident to each other. Figure 10-13 shows the axes to be made coincident. Figure 10-14 shows the components after assembling.

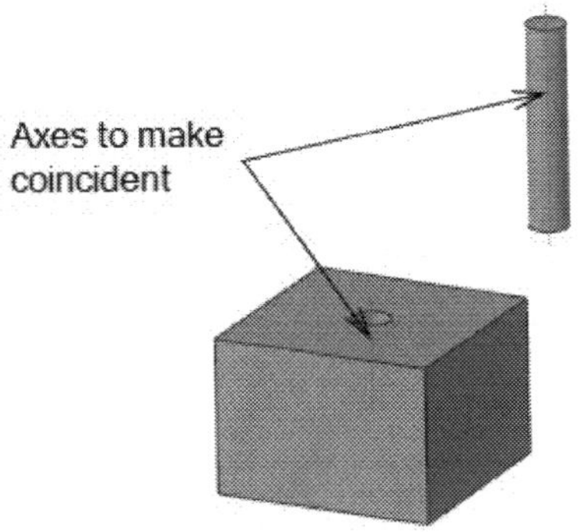

Figure 10-13 Axes selected for the Coincident constraint

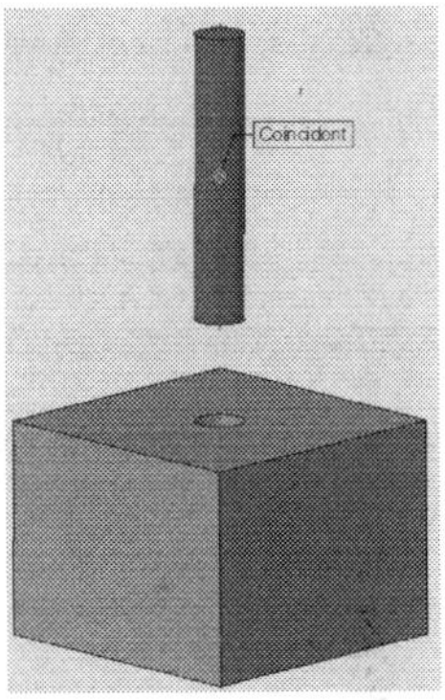

Figure 10-14 Components after applying the coincident constraint

Normal

The **Normal** constraint is used to make two selected planes or faces, or two axes perpendicular to each other. Figure 10-15 shows the faces to be constrained. Figure 10-16 shows the components after assembling.

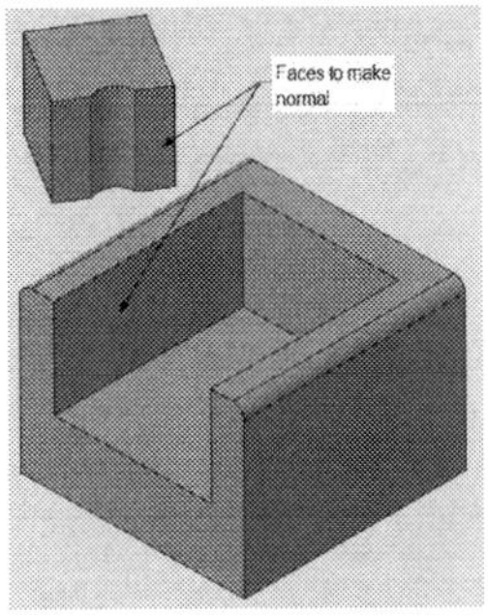

Figure 10-15 Faces selected for the Normal constraint

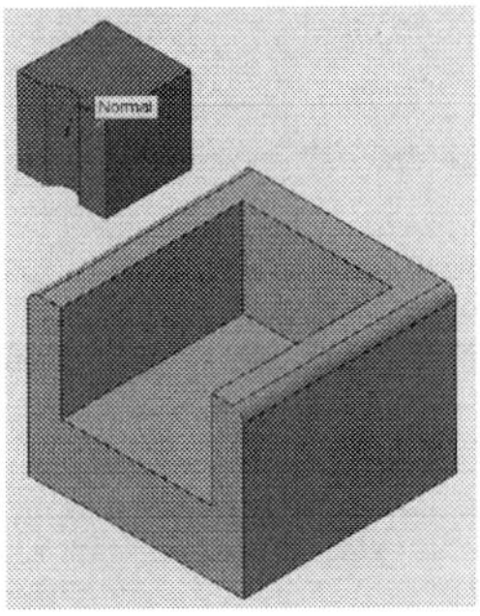

Figure 10-16 Components after applying constraint

Coplanar

The Coplanar constraint allows you to position a component edge, axis, datum axis, or surface oriented on the same plane and parallel to a similar assembly reference. Figure 10-17 shows the faces to be constrained. Figure 10-18 shows the components after assembling.

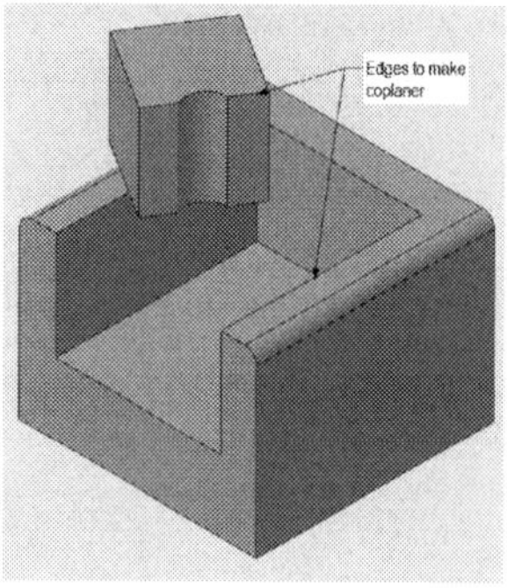

***Figure 10-17** Faces selected for the Coplanar constraint*

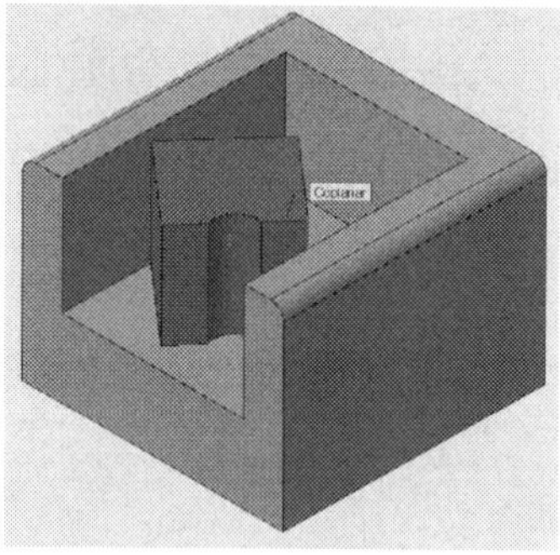

***Figure 10-18** Components after assembling*

Centered

The Centered constraint is used to assemble the revolved components. On applying this constraint, the revolved components, holes, or a combination of both will share the same orientation with respect to the central axis. Figure 10-19 shows the faces selected to be inserted and Figure 10-20 shows the components after assembling. This constraint can also be used to align the coordinate system of the first component with the coordinate system of the second component but the rotation of component will not be constrained.

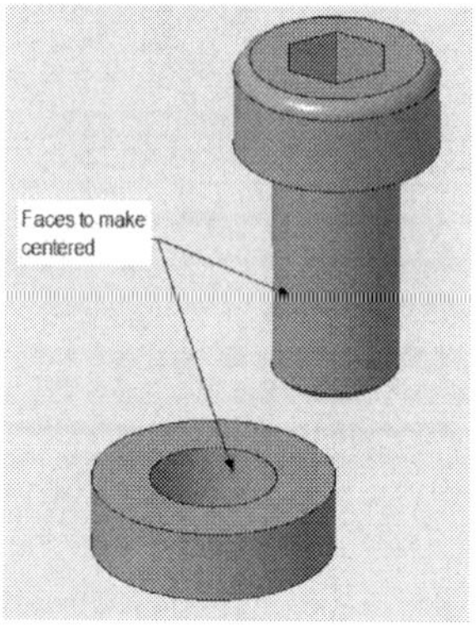

***Figure 10-19** Components to be assembled using the Centered constraint*

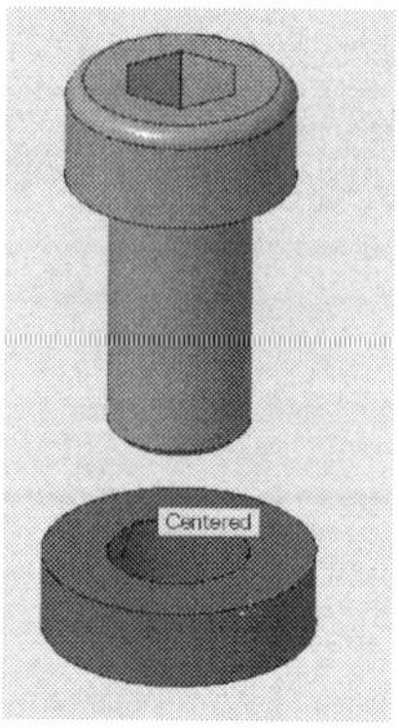

***Figure 10-20** Components after assembling*

Tangent

The Tangent constraint is used to make the selected circular face tangent to the other selected face or plane. As shown in Figure 10-21, the faces of the two components are selected to apply the Tangent constraint. Figure 10-22 shows the component after applying the tangent constraint between the circular faces of both components.

Fix

The Fix constraint is used to fix the current location of the component.

Default

The Default constraint is used to assemble the component in the assembly by aligning the default coordinate system of the component with the default coordinate system of the assembly. Also, the part datum planes are aligned with the assembly datum planes.

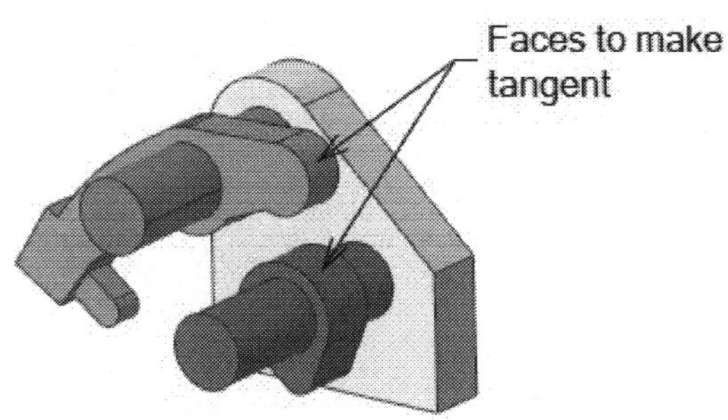

Figure 10-21 *Faces of the components selected to make tangent*

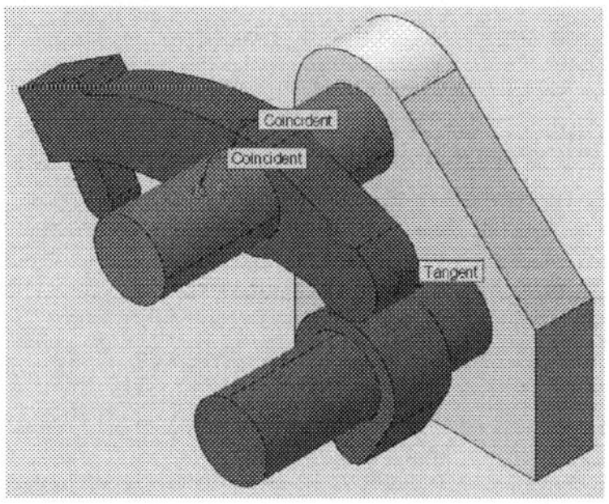

Figure 10-22 *Components after assembling*

Tip

To display the constraint sets of a component in the ***Model Tree****, choose the* ***Tree Filters*** *option from the* ***Settings*** *( ) flyout to display the* ***Model Tree Items*** *dialog box. Next, select the* ***Placement folder*** *check box in the* ***Display*** *area and choose the* ***OK*** *button. The constraints defined with the component will appear in the* ***Placement*** *folder in the* ***Model Tree****.*

Status Area

The **Status** area in the **Component Placement** dashboard displays the placement status of the component in the assembly. If you choose the **OK** button from the **Component Placement** dashboard when the placement status is displayed as **Partially Constrained**, then the component assembled will be displayed in the **Model Tree** with a small square on it. Also, the component is packaged.

The dashboard consists of five tabs: **Placement**, **Move**, **Options**, **Flexibility**, and **Properties**. The options available in these tabs are discussed next.

Placement Tab

The options in this tab are used to apply as well as edit constraints of the components to be assembled. When you choose the **Placement** tab, a panel will be displayed, as shown in Figure 10-23.

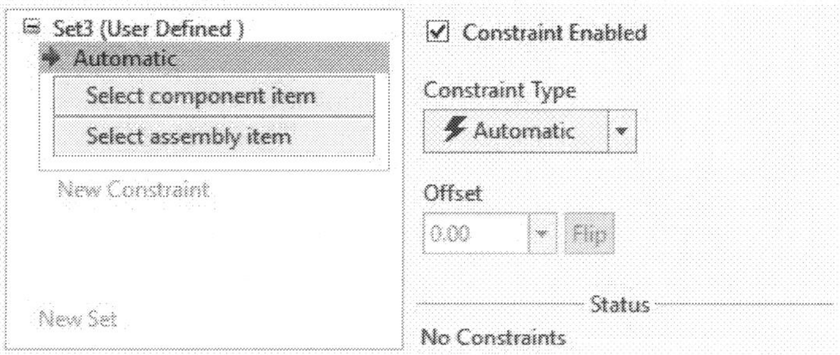

Figure 10-23 *The* ***Placement*** *panel*

The **Constraint Type** and **Offset** drop-down lists have already been discussed. Rest of the options in the panel are discussed next.

New Constraint

This option is used to apply a new set of constraints to the assembly components.

Select component Item Collector

This collector displays the entities of a component that have been chosen to assemble it with the parent component.

Select assembly Item Collector

This collector displays the entities of the parent component that have been chosen to create an assembly.

Note

Most of the options displayed on the dashboard are also available in the shortcut menu, which is displayed on right-clicking when a command is active.

Status Area

This area displays the placement status of the component in the assembly. Sometimes, after adding constraints, a few placement constraints are required to be assumed. This is the reason sometimes the **Allow Assumptions** check box will be displayed in the **Status** area.

Move Tab

The options in the **Move** tab can be used when the component to be assembled is displayed in the assembly itself. This tab is used to move or rotate a component along a degree of freedom that is not constrained. The **Move** slide-down panel is shown in Figure 10-24. The options in this tab are discussed next.

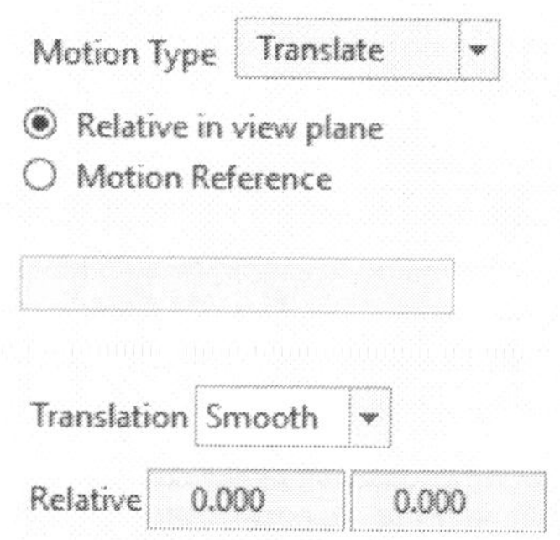

***Figure 10-24** The **Move** slide-down panel*

Motion Type

The options in the **Motion Type** drop-down list are discussed next.

Orient Mode

This option, if selected, orients the part in the assembly about its spin center.

Translate

This option, if selected, moves the component from its current location in the assembly. However, remember that the component can be moved only along the degrees of freedom that are not constrained.

Rotate

This option, if selected, causes the component to rotate in the assembly around its available degrees of freedom.

Adjust

This option is used to pack a component with reference to the assembly. When you select this option, you will be prompted to select a surface on the packaged component to adjust the reference plane. The component is assembled according to the surface specified.

> **Tip**
> *It is recommended that you use the CTRL+ALT+left mouse button to translate a component and CTRL+ALT+middle mouse button to rotate a component. It is always easier to use the above combinations to move the component rather than switching to a separate tab in the* ***Component Placement*** *dashboard and using its options to move a component. Moving the components using the keyboard shortcuts can speed up the process of locating the component correctly in the assembly and establishing constraints.*

Relative in view plane

This radio button is selected by default in the **Move** slide-down panel. As a result, the component moves relative to the viewing plane.

Motion Reference

When this radio button is selected, you need to select a reference plane through which the component will be moved. After selecting the reference plane, the **Normal** and **Parallel** radio buttons will be activated.

Normal

On selecting this radio button, the component moves in a plane normal to the selected reference plane.

Parallel

On selecting this radio button, the component moves in a plane that is parallel to the selected reference plane.

Relative

This option displays the current position of the component relative to its position before the move operation.

Options Tab

This tab will be available for the components with defined interface.

Flexibility Tab

This tab gets activated when the components with defined flexibility are placed in the assembly. Choose the **Varied Items** option in the panel to invoke the **Varied Items** dialog box. The options in the **Varied Items** dialog box are used to add variable dimensions to a flexible component.

Properties Tab

This tab is used to specify the name of the component. You can choose the i button to display the detailed information of the component in a browser window.

PACKAGING COMPONENTS

Ribbon: Model > Component > Assemble > Package

Consider a situation when you do not know the exact location of a component in the assembly or you want to specify the exact location later. In such cases, you can place the component in the assembly non-parametrically. The method of non-parametrically placing the components in the assembly is called packaging. You can package a component using the **Package** option. To package a component, choose the **Package** option from the **Assemble** drop-down in the **Component** group in the **Ribbon**; the **PACKAGE** rollout in the **Menu Manager** will be displayed, refer to Figure 10-25. The options in this menu are discussed next.

Figure 10-25 *The* ***Menu Manager***

Add

The **Add** option is used to add the component to be packaged. On choosing this option, the **GET MODEL** rollout will be displayed. Using the options of this rollout, you can open a new component, select the component from the assembly, or select the last component from the assembly to assemble again.

Choose the **Open** option from the **GET MODEL** rollout; the **Open** dialog box will be displayed. When you select the component from the **Open** dialog box, the **Move** dialog box will be displayed, as shown in Figure 10-26. The options in the **Move** dialog box are similar to those in the **Move** tab of the **Component Placement** dashboard. After you have specified the location of the component in the assembly file, choose the **OK** button to place the component and close the **Move** dialog box. Remember that you can move only the packaged component using this option.

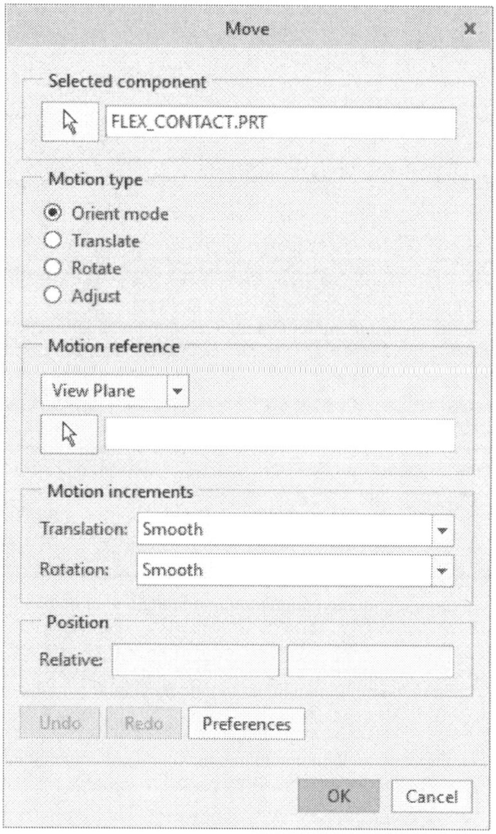

Figure 10-26 *The* ***Move*** *dialog box*

Move

The **Move** option in the **PACKAGE** rollout is used to move the packaged component to a new location. When you choose this option, the **Move** dialog box will be displayed, as shown in Figure 10-26.

Fix Location

The **Fix Location** option is used to fix the location of an existing packaged component. When you choose this option, you are prompted to select the component to be fixed. Select the component from the **Model Tree** or from the drawing area to fully constrain the packaged component in its current location.

Alternatively, you can select a component from the **Model Tree** or from the drawing area and choose the **Fix Location** option from the mini popup toolbar to fully constraint the component. Similarly, you can unfix the location of a component by choosing the **Unfix Location** option from the mini popup toolbar.

Finalize

As mentioned earlier, the packaged components are not assembled parametrically. Hence, packaging does not relate the component with its neighboring components in the assembly. Once the location of the component is decided, then it can be finalized using the **Finalize** option. When you choose this option, you will be prompted to select the packaged component. Selected the packaged component; the **Component Placement** dashboard will be displayed. You can parametrically assemble the component by adding the placement constraints to the component by using this dashboard.

Note

*When you exit the **Component Placement** dashboard before fully constraining the component, then the component is assembled as package.*

SIMPLIFIED REPRESENTATIONS

As you know the assembly designs consist of a number of parts and subassemblies. The complexity of individual components and number of parts can reduce the regeneration, retrieval, and display time of the assembly. To speed up the regeneration and display of the assembly, you can temporarily remove a complicated and unrelated component or subassembly from the assembly. You can also designate the way that individual components are represented in the assembly. This process of removing some components or changing the type of display from the current display is called Simplified Representation. A simplified representation consists only of components that have been selected for specific representation types. For example, in an engine, once the cover is assembled, the components inside it are not visible. In such a case, you can temporarily remove certain components that are not desired at some point of time from the current display or force them to be displayed in the wireframe so that you can see through them.

Types of Simplified Representations

The type of simplified representations that can be created in Creo Parametric are discussed next.

Exclude

The exclude representation is used to exclude the selected component or assembly from the drawing area. You can select more than one component or assembly at a time to exclude from the representation. However, the components that are excluded from the representation remain in the temporary memory of the system.

Master

A master representation displays a fully detailed assembly and contains all geometries and features. All components of the assembly are listed as included, excluded, or substituted in the Model Tree. For an assembly, generally the Default representation and Master representation are identical unless you do not update the Default representation to create a variation in the Master representation.

Automatic

An automatic representation displays the minimum required data for the assembly in the most accurate way. Creo determines the data to be displayed based on the actions taken by the user. This type of representation retrieves the assemblies as fast as possible. You can also use this representation to perform actions as measuring distance between two points on a light surface without retrieving the part geometry.

Assembly Only

This type of representation is used to hide all components of an assembly except the components of the subassembly.

Geometry

In this type of simplified representation, the complete geometry of the components will be displayed in the current model display style. This type of simplified representation takes a long time for regeneration. The components that are not selected in this type of representation cannot be modified. However, these components can be redefined. Generally, this type of representation is used to remove hidden lines, obtain measurement information, calculate mass properties, and reference other assembly components.

Graphics

This type of representation is generally used for assembling the components inside some other components. This type of simplified representation also reduces the regeneration time for large assemblies because it contains only information for display of the components. Note that graphical representations cannot be modified or referenced to other components.

Light Graphics

Light graphics representation utilizes less system resources than a standard representation and therefore, making the assembly to load faster. This type of representation allows you to load assembly with 3D thumbnails of assembly components. To work on a specific component, you can replace any thumbnail with a standard representation.

Symbolic

The symbolic representation enables you to simplify an assembly by representing it as a datum point and symbol. Such representation contains minimal geometry which improves system performance while loading assemblies. To define a symbolic representation, a 2D symbol must be previously defined to replace components.

Boundary Box

The boundary box representation displays only a wireframe bounding box identifying the overall size and location of the assembly. The boundary box representation is best used to simplify the representation of a selected part or sub-assembly within another simplified representation.

Default Envelope

An envelope is defined by features created in context to simplify an assembly. The default envelope representation enables you to represent an assembly with an envelope part. You can also select a pre-existing envelope as default envelope.

User Defined

A user defined representation is used to substitute the representation of a component with custom simplified representation. This type of representation can be a representation that reduces the feature count of a component and only includes selected surfaces of a geometry, or contains a work region cutout.

> **Tip**
> *1. To assemble a simplified representation of a part or subassembly into an assembly, the top-level assembly must also be a simplified representation.*
>
> *2. You can create multiple simplified representations for an assembly. Each simplified representation can correspond to an area or level of detail of the assembly in which individual designers or groups are working.*

Creating Simplified Representations

Simplified representations can be created using the cascading menu or by using the **View Manager**. The only difference between them is that by using the **View Manager**, you can give a name to the representation.

The Cascading Menu

The options available in the cascading menu are used to create a simplified representation without saving it. The simplified representation reduces the regeneration time of the assembly, especially the complex assemblies. To create a simplified representation by using the cascading menu, select a component or assembly from the **Model Tree** and then choose the **Set Representation to** option from the **Manage Views** drop-down in the **Model Display** group of the **Model** tab; a cascading menu will be displayed, as shown in Figure 10-27. To apply a representation, select the assembly or components from the **Model Tree** or from the drawing area and choose the desired representation type from the cascading menu. To save the representation, invoke the **View Manager** dialog box by choosing the **View Manager** button from the **Graphics** toolbar and right-click on the active representation; a shortcut menu will be displayed. Choose the **Save** option from the shortcut menu; the **Save Display Elements** dialog box will be displayed. Specify the name of the representation and choose the **OK** button; a new representation will be added to the list of simplified representations in the **View Manager** dialog box.

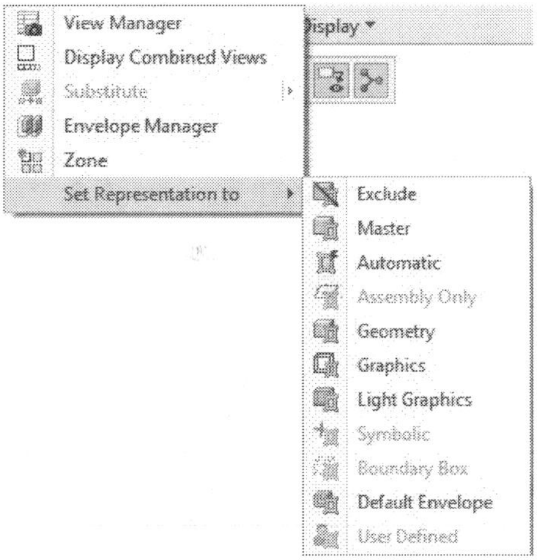

***Figure 10-27** Options in the **Set Representation to** cascading menu*

The View Manager

Ribbon: Model > Model Display > Manage Views > View Manager

You can use the **View Manager** to create, modify, and switch between the simplified representations, exploded views, and orientation of the assembly. To invoke the **View**

Manager, choose the **View Manager** button from the **Model Display** group; the **View Manager** dialog box will be displayed, as shown in Figure 10-28. Alternatively, you can invoke this dialog box by choosing the **View Manager** button in the graphics toolbar. The various uses of this dialog box and operations that can be performed on an assembly using this dialog box are discussed next.

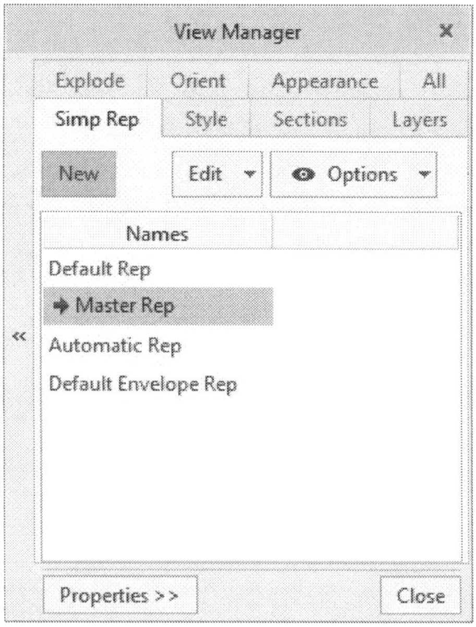

*Figure 10-28 The **View Manager** dialog box*

When the **View Manager** dialog box is displayed, the **Simp Rep** tab will be chosen by default. The options under this tab are similar to those that were available in the cascading menu. The green arrow on the left of **Master Rep** indicates that this representation is currently active. To set any of the listed representation types, select the representation and right-click to invoke the shortcut menu. Choose the **Activate** option from the shortcut menu to set the display of the assembly to that type of representation or double-click on any one of the representation types to activate it.

To create a user-defined representation, choose the **New** button from the **View Manager** dialog box. Specify a name for the representation and press ENTER; the **EDIT** dialog box will be displayed, as shown in Figure 10-29.

Using this dialog box, you can select the components that you want to include, exclude, and substitute from the display. Select the check box adjacent to the component to include or exclude the component from the representation. Next, you can choose the type of representation from the representation type column. The components that are included in the representation will be displayed in the **Model Graphics** area of the dialog box. After defining the representation, choose the **Apply** button; the new representation will be added to the list of simplified representations in the **View Manager**.

To redefine a representation type, select a representation from the list and choose the **Edit** button from the **View Manager** dialog box; a shortcut menu will be displayed. Next, choose the **Edit Definition** option; the **Edit** dialog box will be displayed where you can redefine the representation.

Note

By default, the representation state of all components is derived by the representation state of the top-level assembly.

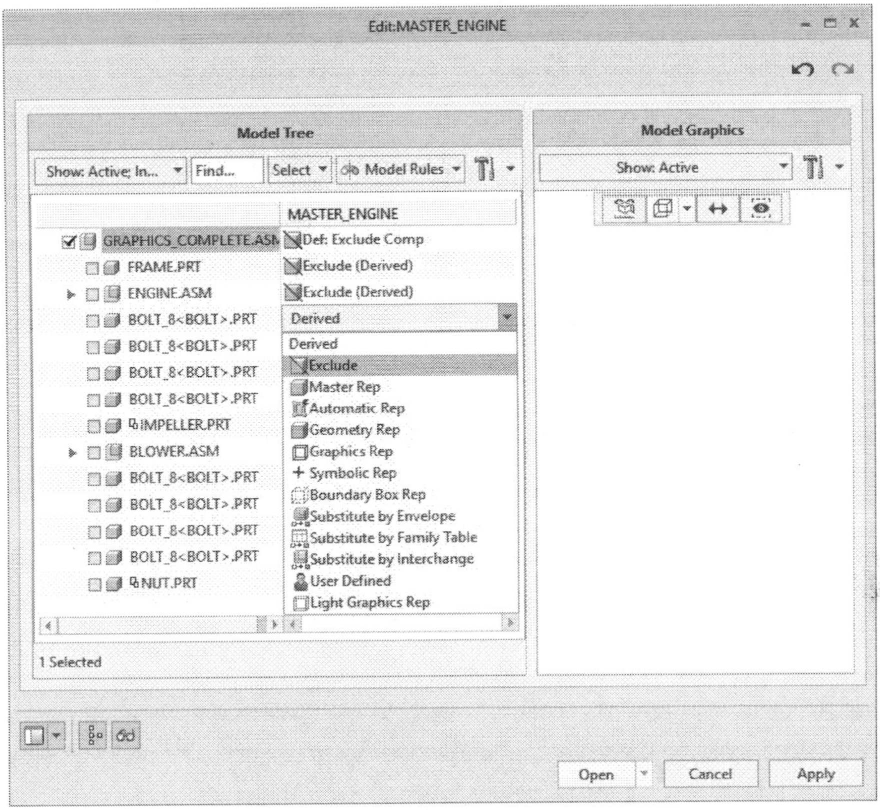

*Figure 10-29 The **Edit** dialog box*

Retrieving Simplified Representations

You can open an assembly in an existing simplified representation or create a new simplified representation on the fly. To do so, choose the **Open** option from the menu bar; the **File Open** dialog box will be displayed. Select the assembly and choose the **Open Representation** option from the **Open** flyout in the dialog box; the **Open Representation** dialog box will be displayed, listing the representations saved with the selected assembly, as shown in Figure 10-30. Select a representation from the list and choose the **Open** button to open the assembly in the selected representation. To define a new simplified representation on the fly, you can choose the **Define** button.

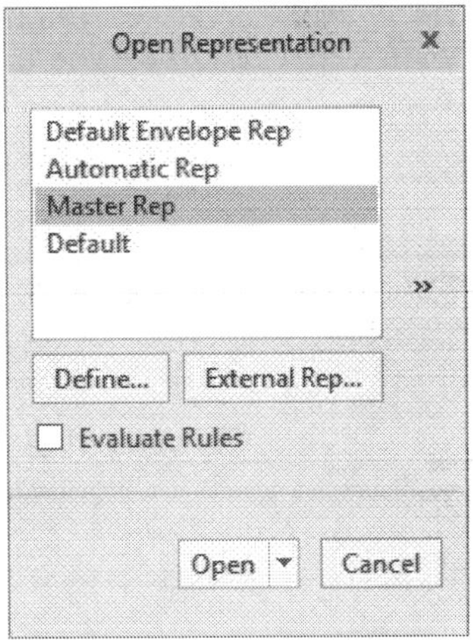

*Figure 10-30 The **Open Representation** dialog box*

Creating External Simplified Representations

The purpose of creating external simplified representations is to store the simplified representations of assemblies without modifying the original assembly. This is advantageous because the reference assembly need not to be in session while working with external simplified representations. External simplified representations are saved as new models of a special assembly type. You can create multiple external simplified representations to correspond to different assembly areas and levels of detail.

All components in an external simplified representation are same as those in the reference assembly. Therefore, it is not necessary to propagate modifications made to the external simplified representation or reference assembly. All modifications to external simplified representations are automatically reflected in the reference assembly.

You can create an external simplified representation by using the **View Manager** dialog box or the **New** dialog box.

To create an external simplified representation of an open assembly, invoke the **View Manager** dialog box and choose the **Simp Rep** tab. Next, select a simplified representation and choose the **Copy As External** option from the shortcut menu or from the **Edit** drop-down; the **Create External Simplified Representation** dialog box will be displayed. You can specify the name of the representation in the **Name** edit box. The **Open in a separate window** radio button is selected by default. As a result, the external simplified representation will open in a new active window after choosing the **OK** button from the dialog box. If you select the **Save the model** radio button, the external simplified representation will be saved but the saved model will not be displayed in the drawing area. Select the **Dependent** check box, if you want to make the external simplified representation dependent on the selected representation in the **View Manager** dialog box.

To create an external simplified representation using the **New** dialog box, choose the **External simplified representation** option in the **Sub-type** of the new assembly; the external simplified representation will be created.

Tip

*You can specify whether the components within an external simplified representation are modifiable or not. To do so, open an external simplified representation and select a top level component or assembly from the **Model Tree**. Next, right-click to invoke the shortcut menu and choose the **Set For Ref. Only** option to make the components non-modifiable. You can make the component modifiable again by choosing the **Set Modifiable** option in the shortcut menu.*

OTHER USAGES OF THE VIEW MANAGER DIALOG BOX

In assembly environment, you can use the **View Manager** dialog box to control the display of the assemblies and components. The options in this dialog box are discussed next.

Creating a Display Style

You can assign different types of display styles to the components of an assembly. To set the display style of a component, select the component and choose the desired style from the **Display Style** flyout of the **Model Display** group in the **Model** tab. The information of the display style is saved with the assembly not with the component.

To create a new display style from the **View Manager** dialog box, choose the **Style** tab. By default, two styles are available in the **Names** list of the dialog box. To create a new style, choose the **New** button; a new style will be added in the **Names** column. Specify the name of the style and press ENTER; the **EDIT** dialog box will be displayed, as shown in Figure 10-31.

By default, the **Blank** tab is selected in this dialog box and you are prompted to select the component to be made blank. Select the component to be made blank and choose the **OK** button; the selected component will be removed from the current display. You can use options in the **Show** tab to specify the type of display for the selected components in the assembly. You can also specify the display of sub-assembly components in a specified display style. To do so, choose the **By Display** tab and then select a sub-assembly from the **Model Tree**; the styles in the sub assembly will be displayed in the **Select State** area. Select a style from the list and choose the **OK** button.

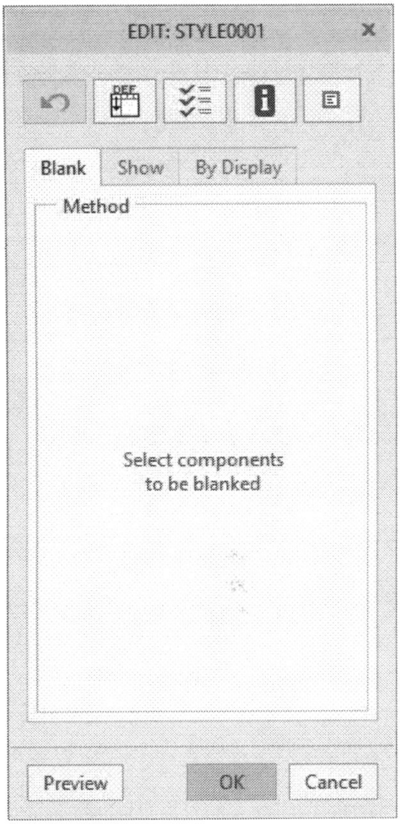

***Figure 10-31** The **EDIT** dialog box*

To revert the change, choose the **Undo Last Operation** button. To reset changes made on a component, you can choose the **Reset Selected Components** button. Choose the **Show Components With An Action Applied Only** button to display all components or blanked components in the **Model Tree** whose display style is changed.

Setting the Orientation of an Assembly

If you choose **Saved Orientations** button from the **Graphics** toolbar, a flyout will be displayed. Various orientations given in this flyout are used to display the assembly as viewed from different directions. Using the **View Manager**, you can invoke the **View** dialog box to set a user-defined orientation. To invoke the **View** dialog box, choose the **Orient** tab and create a new orientation by choosing the **New** button from the **View Manager** dialog box. Name the view and then redefine it to display the **View** dialog box. Use this dialog box to set the orientation of the assembly and then save the orientation.

Creating Sections of an Assembly

You can create the section view of the assembly using the **View Manager** dialog box. Choose the **Sections** tab from the **View Manager** dialog box and then choose the **New** button; a flyout will be displayed, as shown in Figure 10-32. Select any option from flyout and enter the name of the section in the text box displayed. Next, press the middle mouse button; the **Section** dashboard will be displayed, as shown in Figure 10-33.

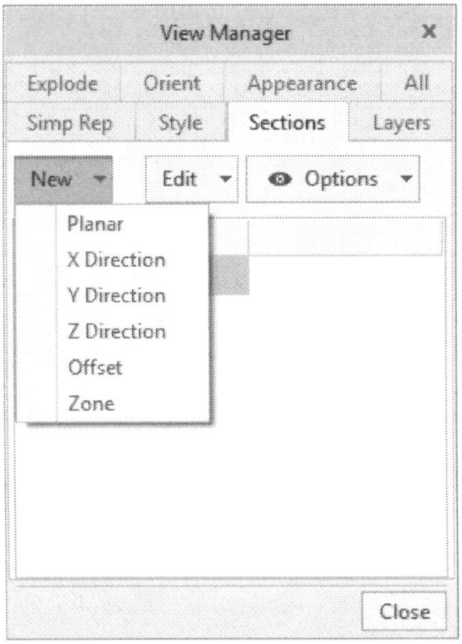

*Figure 10-32 The options in the **New** flyout*

*Figure 10-33 Partial view of the **Section** dashboard*

X Direction

This option is used to create a section along the X direction of the selected coordinate system. When you choose this option, the default coordinate system is selected by default. You can select a user defined coordinate system also.

Y Direction

This option is used to create a section along the Y direction of selected coordinate system. The method of creating a section using this option is similar to that of the **X Direction** option.

Z Direction

This option is used to create a section along the Z direction of selected coordinate system. The method of creating a section using this option is similar to that of the **X Direction** option.

Offset

The **Offset** option is used to create a section by using a user defined sketch. When you select this option, the **Sketch** tab is displayed on the **Section** dashboard. Choose the **Define** button available in the **Sketch** tab to define the sketching plane. In the sketcher environment, draw a line to define the sectioning element.

Zone

This option is used to create a 3D cross-section. You will learn about this option in later chapter of the book.

The options available in the **Section** dashboard are discussed next.

In the dashboard, the **X**, **Y**, and **Z** drop-down list is used to specify the direction of the referenced coordinate system to create the cross-section. By default, the **Offset** option is selected in the **Placement** drop-down. As a result, the cross-section is created at specified distance from the selected reference. You can specify distance in the **Distance** edit box available next to the **Placement** drop-down, or you can drag the Section dragger in the graphics area to specify the offset distance.

If you choose the **Through** option in the **Placement** drop-down list, the cross-section will be created passing through the selected reference. When you choose the **Through** option with a coordinate system selected in the **Section References** collector of the **References** tab, the **Direction** drop-down list will display the **XY**, **YZ**, and **ZX** direction for creating the cross-section. However, you can select any plane or planar surface to create the cross-section.

To reverse the direction of cross-section, choose the **Flip the clipping direction of the cross section** button. The **Preview Capped Section** button is used to cap the clipped surfaces of the model with planar surfaces. This button is selected by default in the dashboard; as a result, the color palette button is enabled. Click on the color palette button to expand the palette and specify a color; the specified color is applied to the surfaces of the cross-section in the drawing area. You can revert to the original color by choosing the **Use Model Color** option in the color palette.

To display hatching on the cross-section, choose the **Show Hatch Pattern** button. The **3D Dragger** button allows you to translate and rotate the clipping plane. The **Auxiliary Display** button is used to view the required section in a separate **2D Section Viewer** window.

Tip

Creo Parametric automatically creates a default coordinate system for generating the cross-section if the model does not have one.

The other options available in the tabs of the **Section** dashboard are discussed next.

References Tab

The **Section reference** collector in the slide-down panel of the **References** tab is used to select the reference selected to create the cross-section.

Models Tab

The slide-down panel displayed on choosing the **Models** tab is shown in Figure 10-34. By default, the **Create the section for the entire assembly** option is selected in this panel. As a result, the cross-section of the complete assembly is displayed in the drawing area. You can select the **Create the section for a single part** option from the drop-down to display the cross-section of a single part model.

By default, the **Include all items** radio button is selected in this panel. As a result, all the models in the assembly are included in the cross-section. Select the **Include selected items** radio button to select the models in the assembly that you want to include in the cross-section. Select the **Exclude selected items** radio button to exclude the selected models from the cross-section. The **Include quilts** check box is used to include all quilts of the model in the cross-section.

Options Tab

The slide-down panel displayed on choosing the **Options** tab is shown in Figure 10-35. You can select the **Show Interference** check box to display the interferences between parts or bodies by selecting **Between parts** or **Between bodies** radio button. The color button next to this check box is used to apply the desired color in the interference zone. Click in the **Select a quilt to include** collector and select a surface from the drawing area to include in the cross-section.

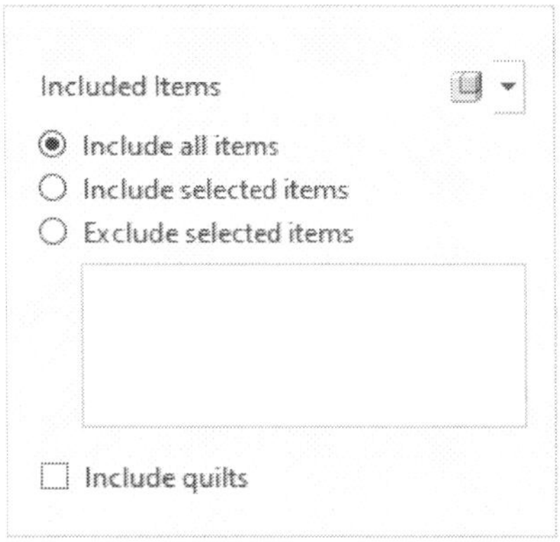

***Figure 10-34** The **Models** slide-down panel*

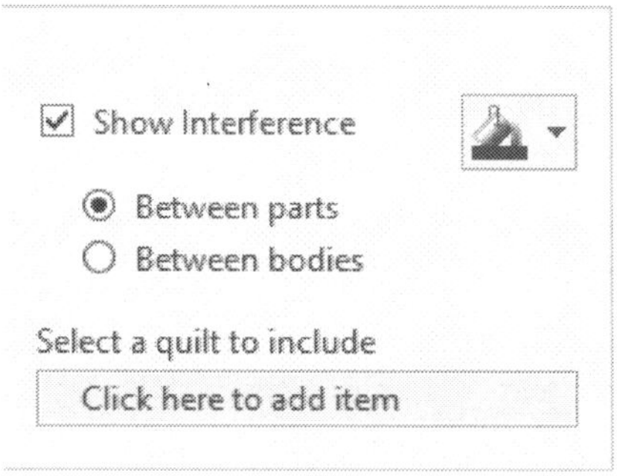

***Figure 10-35** The **Options** slide-down panel*

REDEFINING THE COMPONENTS OF AN ASSEMBLY

After you have assembled the components of an assembly, you may need to modify or redefine the location constraint or the placement constraint of the components. To do so, select the component or sub-assembly from the **Model Tree** or from the drawing area and choose the **Edit Definition** option from the mini popup toolbar; the **Component Placement** dashboard will get activated. Using the options of this dashboard, you can redefine the existing placement constraints or add new placement constraints to the selected component. All components assembled after the selected component will become invisible and the **Component Placement** dashboard will be displayed.

REORDERING COMPONENTS

Ribbon: Model > Component > Component Operations

To change the order of assembling the components, choose the **Component Operations** option from the **Component** group; the **COMPONENT** rollout of the **Menu Manager** will be displayed, as shown in Figure 10-36. Choose the **Reorder** option from this menu; you will be prompted to select the component to be reordered. When you select a component to reorder from the **Model Tree** or from the drawing area and choose the **OK** button followed by **Done**, a message will be displayed in the **Message Area** mentioning those components before which you can insert the selected component. Next, choose the **Before** or **After** option from the **REORDER** rollout and then select that component before or after which you want to place the selected component.

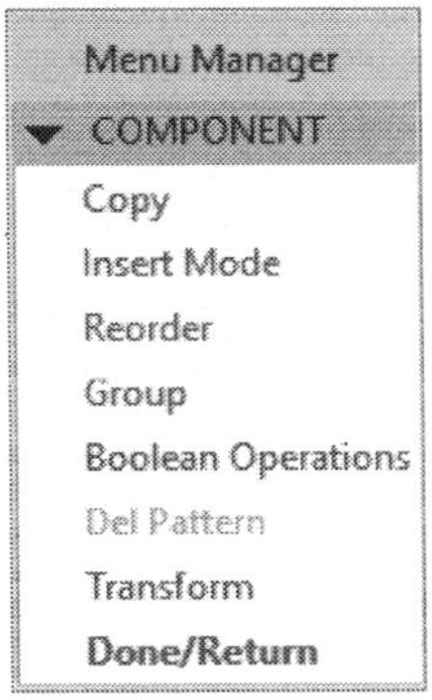

***Figure 10-36** The **COMPONENT** rollout*

Tip
*You can also reorder the components in an assembly by using the **Model Tree**. This can be done by selecting the component in the **Model Tree** and then dragging it to the position where it can be placed. Note that this method of reordering does not inform you where the component can or cannot be placed. However, the parent component of the selected child component gets highlighted when you start dragging the child component. This means if a component has a parent, then the parent is also reordered with the child. Remember that in the **Model Tree**, the parent component will always be placed before its child component.*

SUPPRESSING/RESUMING COMPONENTS

If you do not want certain components of the assembly to appear in the current display or in the drawing views, you can suppress them. To suppress the components, select them from the **Model Tree** and choose the **Suppress** option from the mini popup toolbar. Alternatively, you can choose the **Suppress > Suppress** from the **Operations** group to suppress a selected component.

Similarly, the suppressed components can be resumed by choosing the **Resume > Resume** from the same group or from the mini popup toolbar.

If the suppressed features are not displayed in the **Model Tree**, choose the **Tree Filters** option in the **Settings** drop-down of the **Model Tree**; the **Model Tree Items** dialog box will be displayed. Next, select the **Suppressed objects** check box and choose the **OK** button; the suppressed objects and features will be displayed in the **Model Tree** as a filled square symbol with the name of the component.

REPLACING COMPONENTS

Ribbon: Model > Operations > Replace

The existing components of an assembly that are assembled using the assembly constraints can be replaced with some other components, if necessary. To replace a component, choose the **Replace** tool from the **Operations** group; the **Replace** dialog box will be displayed, as shown in Figure 10-37 and you will be prompted to select the components to be replaced.

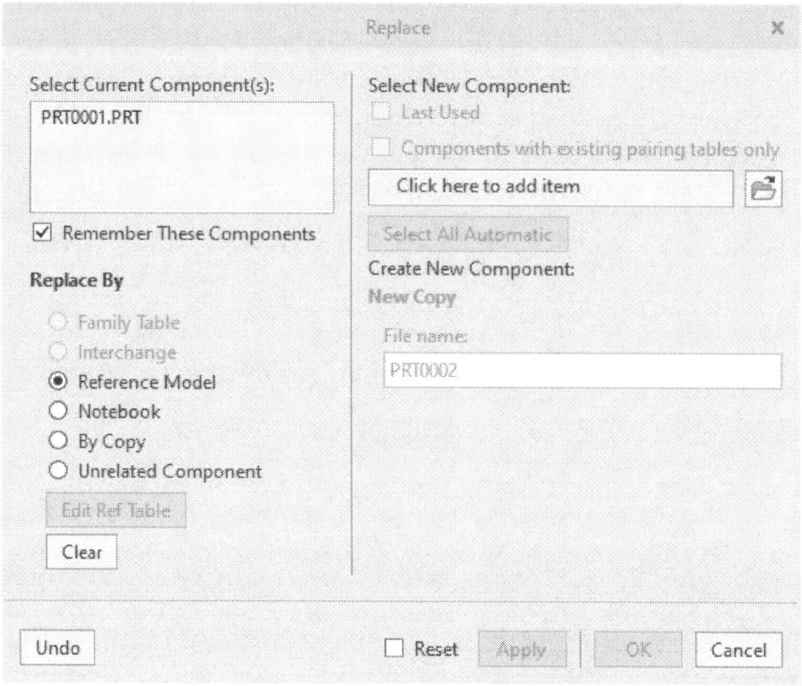

***Figure 10-37** The **Replace** dialog box*

Select the component to be replaced from the drawing area or from the **Model Tree**. The new component can be assembled using the **Family Table**, **Interchange**, **Module or Module Variant**, **Reference Model**, **Notebook**, **By Copy**, or **Unrelated Component** option available in the **Replace By** area. After you have selected one of these options, for example, the **Unrelated Component** radio button, click on the **Select New Component** collector; you will be prompted to select a model to replace the component. Choose the **Open** button available on the right of the **Select New Component** collector; the **Open** dialog box will be displayed. Select the required model from it and choose the **Open** button; the name of the selected model will be displayed in the **Select New Component** collector in the **Replace** dialog box. Now, choose the **Apply** button and then the **OK** button from the **Replace** dialog box; the **Component Placement** dashboard will be displayed. Now, add the required placement constraints to the model using this dashboard.

ASSEMBLING REPEATED COPIES OF A COMPONENT

Sometimes, the assembly design demands the assembling of a particular component more than once. One option is that you should assemble the component every time and add the placement constraints all over again. However, this is a very tedious and time-consuming process, especially in the assemblies that have a large number of similar components. To solve this problem, Creo Parametric provides you an option of assembling multiple copies of the components. To repeat a component in an assembly, select the component to repeat from the **Model Tree** or from the drawing area and then choose the **Repeat** tool from the **Component** group or from the shortcut menu; the **Repeat Component** dialog box will be displayed, as shown in Figure 10-38. The options in the **Repeat Component** dialog box are discussed next.

Component Area

The button in this area is used to select the component whose multiple copies are to be assembled. The name of the selected component will be displayed in the display box provided in this area.

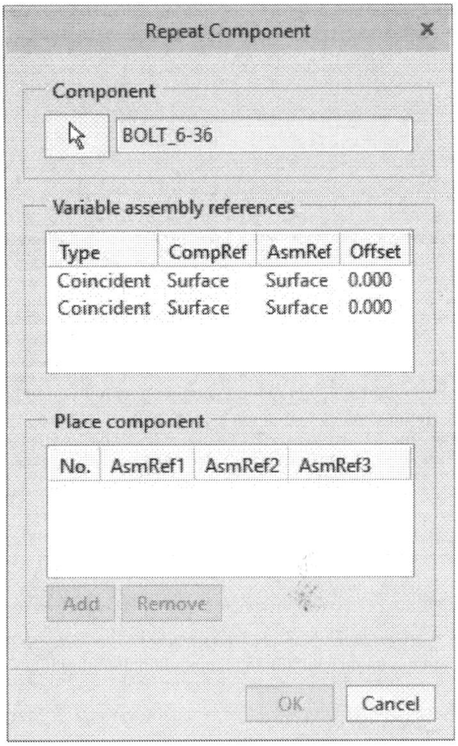

*Figure 10-38 The **Repeat Component** dialog box*

Variable assembly references Area

The **Variable assembly references** area displays the placement constraints that are applied on the selected component. You can keep the placement constraints that need to be present for the new component and modify the remaining constraints by selecting them from this area. The field in this area consists of the following columns.

Type

The **Type** column displays the type of constraints that are applied to the component.

CompRef

The **CompRef** column displays the references that were used to assemble the component. If you want to change this for the new component, select this constraint and then choose the **Add** button. Now, select a reference on the assembly. A copy of the selected component will be added to the assembly at a new location. Similarly, to add another copy, repeat this procedure.

AsmRef

The **AsmRef** column displays the reference on the assembly that was used to assemble the selected component.

Offset

The **Offset** column displays the deviation from location of the new copy to the main part which is repeated.

Place component Area

The **Place component** area displays the new references added for the newly placed copies of the selected component.

Add

The **Add** button is chosen to add the required references for the new copy of the selected component. This button will be available only when you select one of the constraints from the **Variable assembly references** area.

Remove

The **Remove** button is chosen to remove the selected reference from the new component. This button is available only when you select one of the references from the **Place component** area.

To create a repeated component in an assembly, select the required component and then invoke the **Repeat Component** dialog box. Next, select the constraint to be copied from the **Variable Assembly References** area in the dialog box and then choose the **Add** button. Now, select the new reference for the new copied component from the drawing area; the selected component will be copied to the new reference. Choose the **OK** button from the **Repeat Component** dialog box to accept the repeated copy of the component.

MIRRORING COMPONENTS INSIDE AN ASSEMBLY

You can create mirror copies of components and assemblies in the assembly environment. You can also control the dependency of the new component on the original. A dependent mirrored component is automatically updated when the original component geometry, its placement, or both are modified.

To create a mirrored copy of a component or subassembly, choose the **Mirror Component** tool from the **Component** group of the **Model** tab; the **Mirror Component** dialog box will be displayed, as shown in Figure 10-39. Alternatively, you can select a component or subassembly and right-click to invoke the shortcut menu and then choose the **Mirror Component** option from the shortcut menu to invoke the **Mirror Component** dialog box. Next, select the component or subassembly in the **Component** collector and select the plane in the **Mirror plane** collector. For mirroring a component, it is a good practice to mirror about a plane belonging to the component itself.

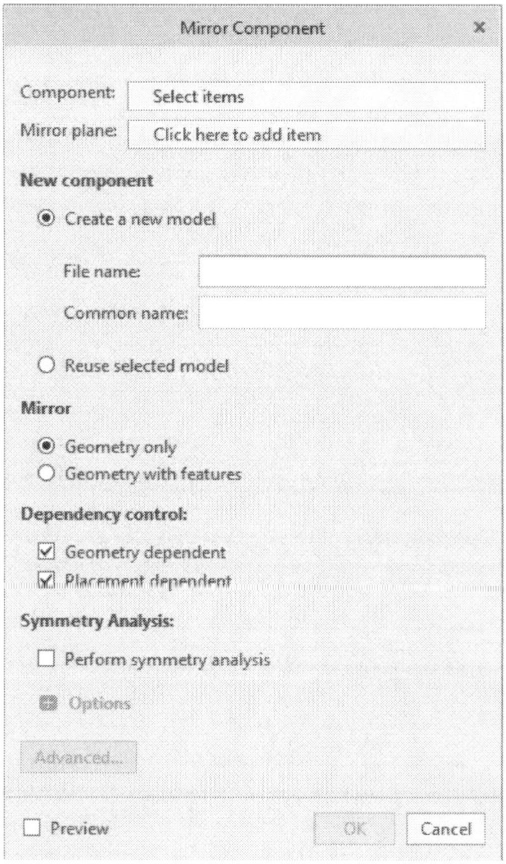

*Figure 10-39 The **Mirror Component** dialog box*

By default, the **Create a new model** radio button is selected in the **New component** area. As a result, you can specify a different name for the mirrored component in the **File Name** edit box. Select the **Reuse selected model** radio button if you want to reuse the component for creating the mirrored copy.

In the **Mirror** area of the dialog box, the **Geometry only** radio button is selected by default. As a result, the resulting mirror copy contains only the geometry of the original component. Selecting the **Geometry with features** radio button will create a mirror copy of the geometry and features of the original component.

The options under the **Dependency control** area are used to specify the dependency between the mirror component and original component. Selecting the **Geometry dependent** check box will result into a mirrored copy which is updated when modifications are done in the original component. This check box will become inactive when you select the **Geometry with features** radio button in the **Mirror** area. The **Placement dependent** check box is selected by default and it is used to create a mirror copy whose placement is automatically updated when the placement of the original component is modified.

You can perform a symmetry analysis to identify and reuse symmetric and anti-symmetric or twin, mirror components. This is useful while mirroring subassemblies. Select the **Perform symmetry analysis** check box and choose the **Advanced** button; the **Mirror Subassembly Components** dialog box will be displayed. Using this check box, you can specify whether you want to exclude, mirror, or reuse the original component. You can expand the **Options** rollout to specify the elements to consider for symmetry analysis.

After you have specified the above parameters for the mirror copy, choose the **OK** button in the dialog box to finish creating the mirror copy of the selected component or subassembly.

MODIFYING THE COMPONENTS OF AN ASSEMBLY

Sometimes, during or after the assembly of a component, you may need to modify the dimensions of the component. You can even redefine the selected feature in the **Assembly** mode itself. The methods used to modify and redefine the features are discussed next.

Modifying Dimensions of a Feature of a Component

The dimensions of a selected feature of the component can be modified by selecting the component from the **Model Tree** or from the drawing area. Select the model when the outline of the model turns green. Select the **Open** option from the mini popup toolbar to open the component as a part file in a separate window and edit the features as required. After modifying the dimensions, regenerate the component, and save the part file. Next, close the part file and activate the assembly window and regenerate the assembly to incorporate the changes made in the component.

Redefining a Feature of a Component

You can redefine the features of a component in the **Assembly** mode. This saves time in opening the part in the **Part** mode and then redefining the features. To redefine the feature of a component, select the component from the **Model Tree** and choose the **Activate** option from the mini popup toolbar; a small green button appears on the activated component in the **Model Tree**. As you activate the component, the **Part** mode becomes active automatically. To display the features of the component in the **Model Tree**, choose the **Tree Filters** option from the **Settings** drop-down of the **Model Tree** to invoke the **Model Tree Items** dialog box. Next, select the **Features** check box in the **Display** area and choose the **OK** button; the features of the component will be displayed in the **Model Tree**. Now you can redefine the desired feature by selecting it and choosing the **Edit Definition** button from the mini popup toolbar.

Adding Flexibility to a Component

Flexible components can adopt the requirements of the assembly design. Components such as springs, clips, washers and so on may require a change in their geometry in their assembled condition. Such components are called flexible components. A spring, for example, may require various compression lengths throughout the assembly. This can be done by adding flexibility to the component. Note that the changes in the geometry of a flexible component are only within the assembly. The part that is stored in the computers directory, remains unchanged. You can select dimensions, features, geometric tolerances, and parameters of a component to make it flexible.

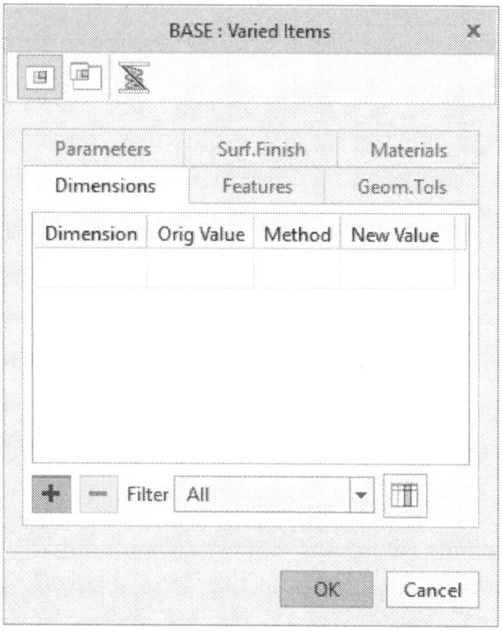

Figure 10-40 The ***Varied Items*** *dialog box*

To add flexibility to a component in assembly environment, select the component from the **Model Tree** and right-click to invoke the shortcut menu. Next, choose **Flexible Component > Make Flexible** from the shortcut menu. The **Varied Items** dialog box along with the **Select** dialog box will be displayed. Also, the **Component Placement** dashboard will become active. By default, the **Dimensions** tab is active in the **Varied Items** dialog box, refer to Figure 10-40. Select the feature of the component; the dimensions of the related feature will be displayed in the drawing area. Select the dimensions of the feature from the drawing area and click the + button; the selected dimensions will be added in the list of variable dimensions. Next, specify the new value for the dimension by clicking in the **New Value** edit box of the respective dimension and choose the **OK** button to regenerate the component. If required, you can redefine the placement constraints for the component using the **Component Placement** dashboard and choose the **OK** button to complete the procedure. You will notice that a flexible component is displayed with the symbol in the **Model Tree**.

As discussed earlier that the changes made in the geometry of a flexible component do not reflect in the original component. However, you can make the changes propagate in the original component by choosing the **Make Flexible > Propagate to > Model** option from the shortcut menu of a flexible component.

Tip

1. While redefining the placement of a flexible component, you can choose the ***Varied Items*** *option in the* ***Flexibility*** *tab of the* ***Component Placement*** *dashboard to invoke the* ***Varied Items*** *dialog box.*

2. To make an individual component flexible for assembly design, open the component in the ***Part*** *mode and invoke the* ***Model Properties*** *dialog box by choosing the* ***Prepare > Model Properties*** *from the* ***File*** *menu. Next, click the change option adjacent to the* ***Flexible*** *row in the* ***Tools*** *area; the* ***Flexibility: Prepare Varied Items*** *dialog box will be displayed where you can specify variables for the selected component.*

CREATING THE EXPLODED STATE

Ribbon: Model > Model Display > Edit Position

In assembly design, the components assembled inside other components may not be visible. This could be misleading and may confuse the viewer as the component cannot be viewed. To avoid this confusion, generally an exploded view is provided along with the assembled view. An exploded view is a state in which all the components move from their original position so that they are visible, as shown in Figure 10-41.

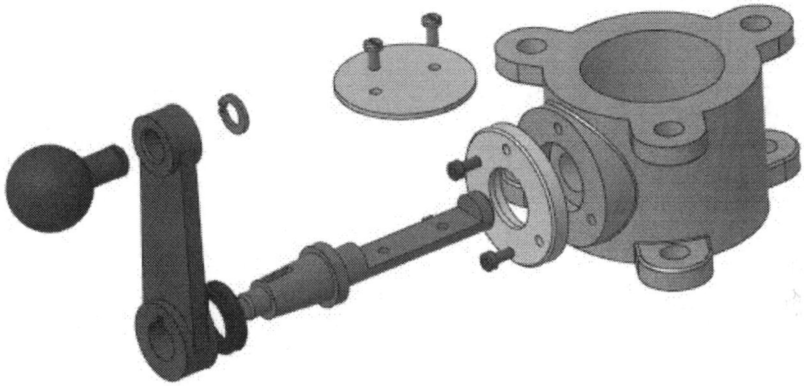

Figure 10-41 *The exploded state of an assembly*

The exploded state can be generated by invoking the **Edit Position** tool from the **Model Display** group. On choosing the **Edit Position** option, the default exploded state of an assembly will be displayed. Also, the **Explode Tool** dashboard will be displayed, as shown in Figure 10-42.

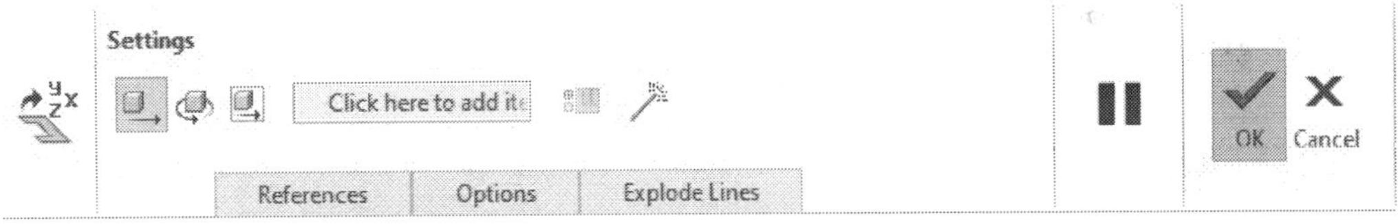

Figure 10-42 *Partial view of the* ***Explode Tool*** *dashboard*

Tip

1. You can invoke the ***Exploded view*** *tool from the* ***Model Display*** *group of the* ***View*** *tab.*

2. You can also access the options of explode tool using the ***View Manager*** *dialog box.*

Various options in this dashboard are discussed next.

Translate

This button is used to translate a component from its current location to a new location in the assembly. To move a component, choose this button; three axes will be displayed on it. Next, move the cursor on the desired axis. When the axis gets highlighted, click and drag the component.

Rotate

This button is used to rotate a component about a selected axis. To rotate a component, choose this button; you will be prompted to select a reference for rotation. Select a reference (edge, axis, planar face/datum plane, any 2 points, or coordinate system); a curved arrow will be displayed. Using this arrow, you can rotate the component.

View Plane

The **View Plane** option is used to select the current view plane (that is, the computer screen) as the reference for exploding a selected component. Choose this button; a small ball will be displayed. Click on the ball and drag to move the component to the desired location.

Toggle the explode state of selected component(s)

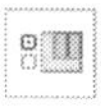

This button is used to toggle between the exploded and unexploded states of the selected components. Select a component(s) from the **Model Tree** or from the drawing area and choose the **Toggle the explode state of selected component(s)** button; the selected component(s) will be moved to their original position.

References Tab

The options in this tab are used to select the components as well as to define a reference for movement. Choose this tab; a slide-down panel will be displayed, as shown in Figure 10-43. The **Components to Move** collector allows you to select and display the name of the selected component. The **Movement Reference** collector allows you to select a reference to define the movement of the selected component. Note that you need to click in the collector so that it gets activated, thereby enabling you to select the required entities.

Options Tab

Choose this tab to display a slide-down panel, as shown in Figure 10-44. The options in this slide-down panel are discussed next.

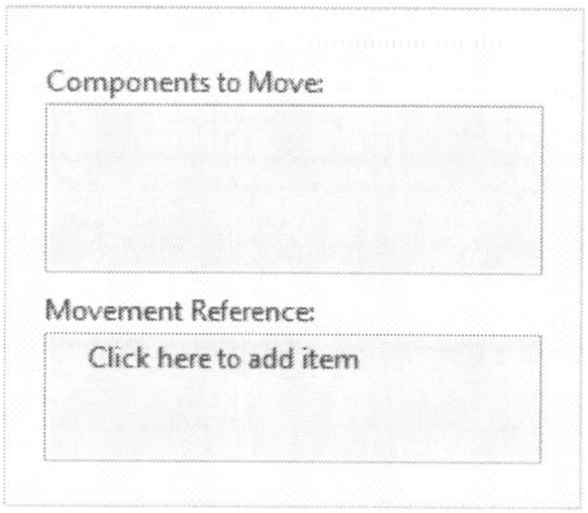

*Figure 10-43 The **References** slide-down panel*

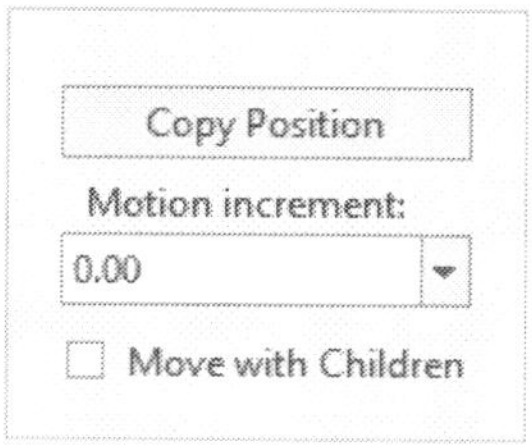

*Figure 10-44 The **Options** slide-down panel*

Copy Position

This button is used to restore the exploded component to its original position. For example, when an assembly explodes, its components move away. To restore components to their original position, choose the **Copy Position** button from the **Options** slide-down panel; the **Copy**

Position dialog box will be displayed. Click in the **Components to Move** collector to activate it and select the exploded component(s) moved away from the assembly. Next, click in the **Copy Position From** collector to activate it and then select a component from the assembly that is at a similar location in the assembly. Choose the **Apply** button from this dialog box; the selected component will be assembled at its original position in the assembly.

Motion Increment

You can specify an increment value for movement as well as define a smooth motion using this drop-down list.

Move with Children

On selecting this check box, the selected component moves along with its child components.

Note

If you have not generated the exploded view of an assembly, a default exploded state will be available on choosing ***Explode View*** *button from the* ***Model Display*** *group. To display the unexploded state of the assembly, again choose* ***Explode View*** *from the* ***Model Display*** *group.*

Alternatively, you can invoke the ***View Manager*** *and then choose the* ***Explode*** *tab. Now, double-click on the required state to activate it. Also, you can create new states, edit existing states, and switch between various states created.*

Explode Lines Tab

Explode lines are used to display the actual path and direction of the mating components. Generally, these lines are used for an easy visualization of the exploded components of the assemblies. The exploded state of an assembly with the offset lines is shown in Figure 10-45.

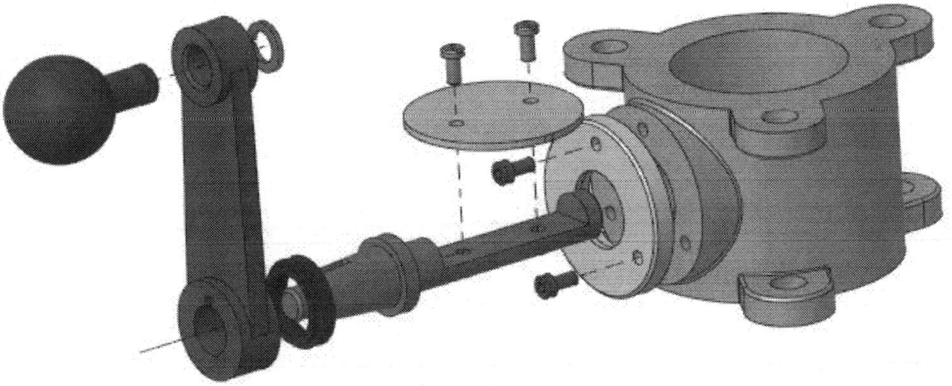

Figure 10-45 *The exploded state of an assembly with the explode lines*

Create cosmetic offset lines to illustrate movement of exploded component Button

To create explode lines, an exploded state should exist in which the explode lines will be created. Choose the **Create cosmetic offset lines to illustrate movement of exploded component** button from the panel that is displayed on choosing the **Explode Lines** tab of the **Explode Tool** dashboard; the **Cosmetic Offset Line** dialog box will be displayed, as shown in

Figure 10-46. Using this dialog box, you can select the objects for creating the explode lines. The objects that can be selected are axis, surfaces, edge, or curve.

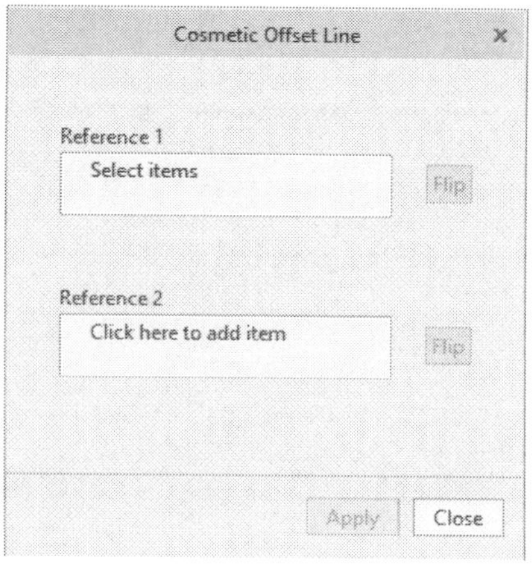

Figure 10-46 *The* ***Cosmetic Offset Line*** *dialog box*

Reference 1
Click in this collector to activate it and then select the first reference.

Reference 2
Click in this collector to activate it and then select the second reference.

Next, choose the **Apply** button to accept the creation of the explode lines and continue for the next collection. Choose the **Close** button to close this dialog box.

Note
The display of explode lines is automatically turned off once you display the unexploded state of the assembly.

Edit the selected explode line

This button will be available in the **Explode Lines** tab only when you select the explode line created earlier. To modify the explode line created, select it and then choose this button; two handles will be displayed at both ends of the line. Next, click on the handle and drag it to extend or shorten the explode line segment. To exit the edit mode, right-click and choose the **Clear** option from the shortcut menu.

Delete the selected explode line

To delete the selected explode line, select the exploded line created and then choose the **Delete the selected explode lines** button; the selected explode line will be deleted.

Edit Line Styles

You can edit the attributes of the selected exploded line by choosing the **Edit Line Styles** button. You can modify the attributes such as line style type, color of the line style, or copy the line style from an existing line type.

Default Line Style

Choose this button to set the default attributes of the explode line.

Note
To delete an explode line, select the explode line and right-click; a shortcut menu will be displayed. Choose ***Remove Explode Line*** *from it. You can also set the line style for the explode lines that you create in an exploded view by using the shortcut menu.*

THE BILL OF MATERIALS

Ribbon: Model > Investigate > Bill of Materials

The Bill of Materials or BOM is the tabular representation of all components of the assembly, along with the information associated with them. The information can be the material of the components, the additional note with the components, and so on. Creo Parametric provides you with a ready-made BOM that can be directly utilized. This BOM will be automatically updated during the assembly of the components. The BOM can be viewed by using the **Bill of Materials (BOM)** dialog box shown in Figure 10-47. This dialog box will be displayed on invoking the **Bill of Materials** tool from the **Investigate** group. The options in this dialog box are discussed next.

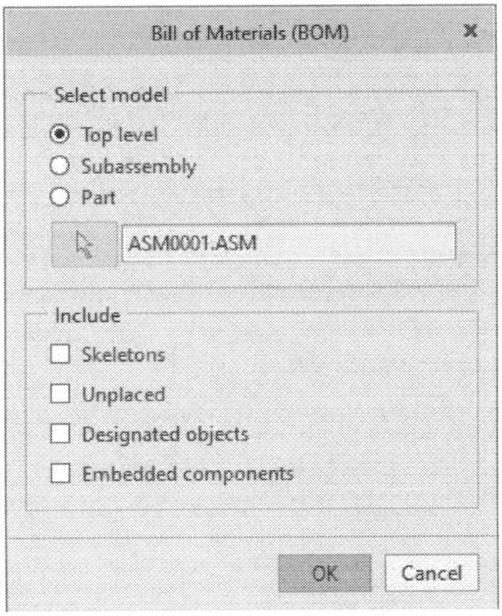

***Figure 10-47** The **Bill of Materials (BOM)** dialog box*

Select model Area

The options in the **Select model** area are used to select the type of assembly whose BOM has to be displayed. These options are discussed next.

Top Level

Using the **Top Level** radio button, you can display the top level assembly in the BOM. In this case, the subassemblies inserted in the current assembly are considered as a component.

Subassembly

The **Subassembly** radio button is used to list the components of the selected subassemblies as individual components in the BOM. When you select this radio button, you will be prompted to select the subassembly.

Part

The **Part** radio button allows you to list designated model items, including body objects. This enables you to create a part-level BOM that lists the bodies that are intended to be shown.

Include Area

The options in this area are discussed next.

Skeletons

In the Skeleton model geometry, the assembly level features such as cuts and holes do not affect the geometry of the skeleton model. The **Skeletons** check box is selected so that the skeletons, if any, in the assembly could also be included in the BOM.

Unplaced

The **Unplaced** check box is selected to include the unplaced components in the BOM.

Designated Objects

The **Designated Objects** check box is selected to include the bulk items, if any, that are used in the assembly. Bulk items are items whose solid models are not created but they are used in the assembly. In Creo Parametric, you can include the weight and other properties of these bulk items to calculate certain parameters related to the assembly like total weight of assembly. Examples of bulk items are paint, glue, cotton lace, and so on.

After you have selected the options in the **BOM** dialog box, choose the **OK** button to open the browser that displays the BOM for the selected assembly. The file of the Bill of Material is automatically saved with the name of the assembly. The extension of this file is *.bom*. You can later retrieve it in the **Drawing** mode to display the BOM in the drawing views of the assembly.

Embedded Components

This check box is selected to include the embedded components in the BOM.

GLOBAL INTERFERENCE

Ribbon: Analysis > Inspect Geometry > Global Interference drop-down > Global Interference

Global Interference is a type of analysis that can be conducted in the **Assembly** and **Drawing** modes. This analysis can be performed on the assembly involving interference fit. Typical examples of interference fit or press fit is press fitting of shaft into bearing and bearing into its housing. Using this analysis, you can check the total volume overlap of the two assembled parts. This volume can be used to calculate the compressive force working on the two assembled parts. The procedure to check the interference between the assembled parts is given next.

1. Invoke the **Global Interference** tool from the **Global Interference** drop-down available in the **Inspect Geometry** group under the **Analysis** tab; the **Global Interference** dialog box will be displayed, refer to Figure 10-48.

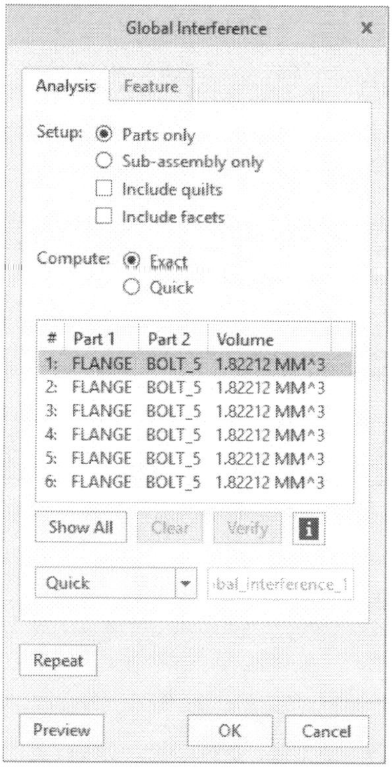

Figure 10-48 The ***Global Interference*** *dialog box*

2. Now, choose the **Preview** button from the **Global Interference** dialog box; the interfered volume will be displayed in the dialog box, refer to Figure 10-48. In this dialog box, all interfering parts will be displayed with their names and respective interfering volumes. Also, the interfering parts will be highlighted in red color in the drawing area.

PAIRS CLEARANCE

Ribbon: Analysis > Inspect Geometry > Global Interference > Pairs Clearance

The **Pairs Clearance** analysis is used to check the clearance between two assembled parts. This analysis is available in the **Part**, **Assembly**, **Piping** and **Drawing** modes. In some assemblies, a gap needs to be provided for free motion of the parts called clearance. Using the **Pairs Clearance** analysis, you can find out the clearance between the two assembled parts. Generally two edges or surfaces are required for measuring the clearance. The procedure to check clearance between the assembled parts is given next.

1. Invoke the **Pairs Clearance** tool from the **Global Interference** drop-down available in the **Inspect Geometry** group of the **Analysis** tab; the **Pairs Clearance** dialog box will be displayed.

2. Select the two surfaces or edges of the model in the drawing area.

3. As you choose the surfaces or edges, the clearance value will be displayed in the dashboard as shown in Figure 10-49.

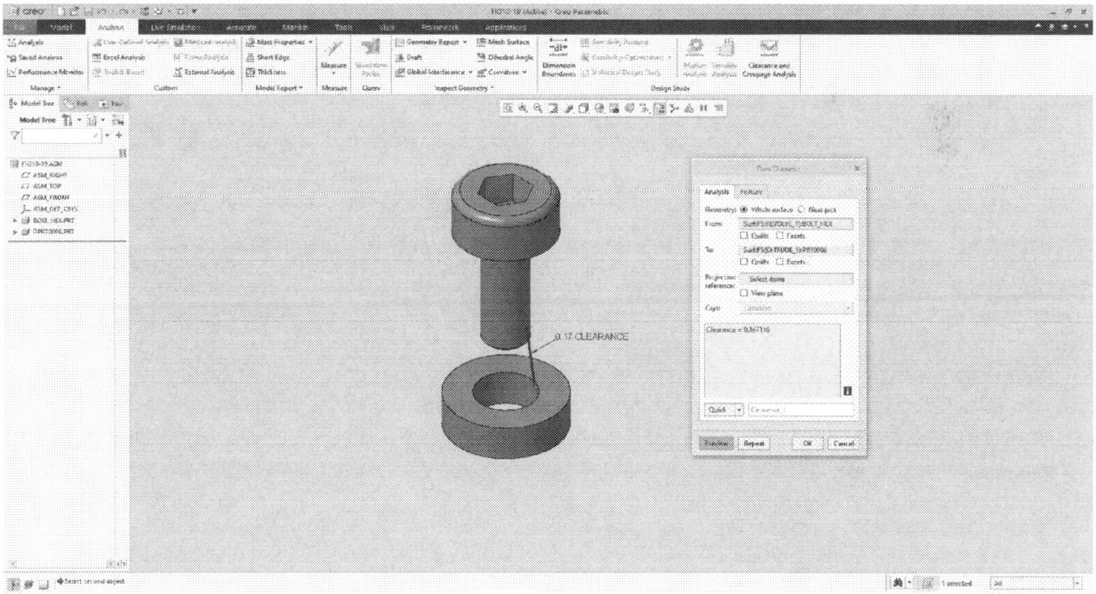

Figure 10-49 *Clearance displayed in the* ***Pair Clearance*** *dialog box and drawing area*

SHRINKWRAP FEATURES

Shrinkwrap features contain a collection of associatively copied surfaces and datums that represent the exterior shape of a referenced part or assembly. Because the surfaces are copied associatively, the Shrinkwrap feature updates when the source assembly gets modified.

Shrinkwrap models load faster than complex models which improves system performance when working with large assemblies. Shrinkwrap models usually reduce disk and memory usage by more than 90 percent. This varies depending on the complexity of the source model and the quality setting used to generate the Shrinkwrap model. Memory saving is considerable in assemblies with many hidden components.

The Shrinkwrap models provide an accurate external representation of the model for use by other design teams, suppliers, or customers, without disclosing the internal design of the assemblies. Others can visualize the product and conduct space claim studies, form and fit studies, interference checking, and so forth, while you can protect trade secrets, patented designs, and other proprietary information.

Creating a Shrinkwrap Feature

Ribbon: Model > Get Data > Shrinkwrap

To create a Shrinkwrap feature of an assembly, first create a component in the assembly environment and then activate it by choosing the **Activate** option in the mini popup toolbar, if not already activated. Next, choose the **Shrinkwrap** tool from the **Get Data** group of the **Model** tab; the **Shrinkwrap** dashboard will be displayed, as shown in Figure 10-50.

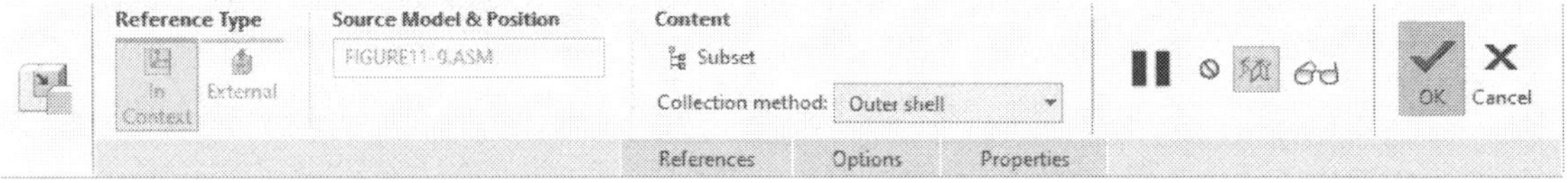

Figure 10-50 *Partial view of the* ***Shrinkwrap*** *dashboard*

In the dashboard, the **In Context** button in the **Reference Type** area is selected by default. As a result, the top level assembly is selected as reference for creating the Shrinkwrap. The **External** button is used for creating an external Shrinkwrap feature. An external Shrinkwrap feature is referenced from an external part or external model. The **Subset** button is used to invoke the **Shrinkwrap Comps** dialog box which is used to include or exclude components for creating the Shrinkwrap. The **Outer Shell** option is used to create a Shrinkwrap of the exterior shape of the model. Select the **Autocollect all solid surfaces** option to create a Shrinkwrap feature containing the external and internal solid surfaces of the model. Choose the **Manual collection** option if you want to manually select the surfaces for creating the Shrinkwrap feature.

The options in the tabs of the **Shrinkwrap** dashboard are discussed next.

References Tab

Choose the **References** tab; a slide-down panel is displayed, as shown in Figure 10-51. The **Always include surfaces** collector is used to select the surfaces that will be included in the Shrinkwrap. The **Never include surfaces** collector is used to exclude the selected surfaces from the Shrinkwrap feature. The **Chain** collector is used to select the edges and curves to include in the feature. To include the selected datums in the Shrinkwrap, you can use the **Include Datums** collector. The **Edit** button at the bottom right corner of the panel is used to include the selected annotation features in the Shrinkwrap.

Options Tab

Choose the **Options** tab to invoke the slide-down panel, as shown in Figure 10-52. By default, the **Shrinkwrap then Exclude** radio button is selected in the **Surface Copying Options** area. As a result, Creo analyzes the entire assembly to identify the external surfaces and include only those surfaces in the Shrinkwrap feature that belong to the selected components. If you select the **Exclude then Shrinkwrap** radio button, the Shrinkwrap will be created using the selected components.

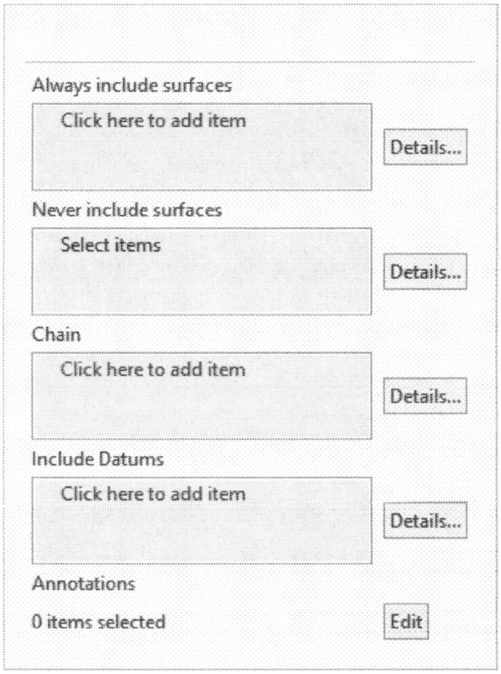

*Figure 10-51 The **References** slide-down panel*

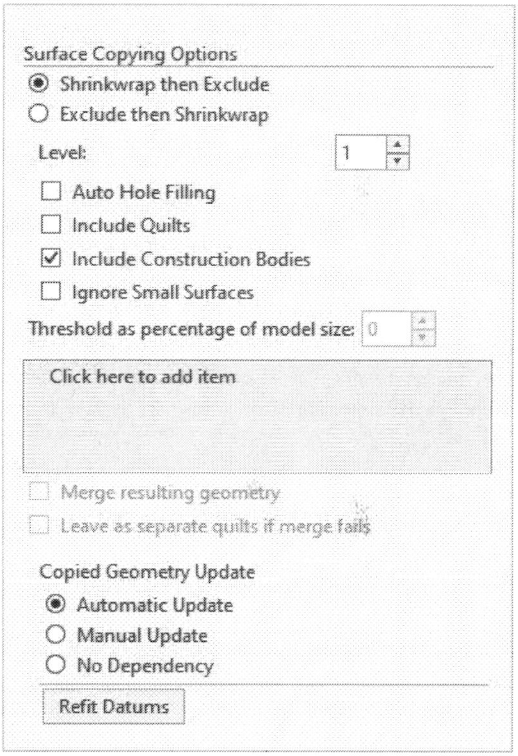

*Figure 10-52 The **Options** slide-down panel*

The **Level** edit box is used to specify the quality level of the resulting Shrinkwrap feature. The default quality is set to 1. You can specify the quality level between 1 to 10. Remember that increasing the quality level will increase the processing time of your computer.

Select the **Auto Hole Filling** check box to fill all holes and cuts that lie on a single surface. Selecting the **Include Quilts** check box will include the quilt surfaces of the assembly in the resulting Shrinkwrap feature. Generally, all quilt surfaces are considered for inclusion; however, all surfaces are not included automatically in the Shrinkwrap. The **Ignore Small Surfaces** check box is used to exclude the surfaces that are smaller than the specified value (percentage of the model's size). By default, 0 is set in the **Threshold as percentage of model size** edit box. The value in this edit box specifies the relative size of the surface to be ignored. The **Fill contours** collector is used to fill the contours of the selected surfaces. When you select a surface from this collector, the contours of the selected surface are listed in the box below the collector. The **Display curve in place of filled contour** check box is used to place a curve on the place of filled contour. The radio buttons in the **Copied Geometry Update** area are used to set geometry dependency of the Shrinkwrap feature. By default, the Shrinkwrap feature is dependent on the source components. The **Automatic Update** radio button is selected by default. As a result, the relationship between the current feature and the source component remains unchanged. Select the **Manual Update** radio button if you want to temporary suspend the relation between the current feature and the source component. Select the **No Dependency** radio button to break the relationship between the source geometry and the Shrinkwrap feature.

After you have specified the attributes for the Shrinkwrap feature, choose the **OK** button from the **Shrinkwrap** dashboard to finish the procedure. You will notice that a feature named Shrinkwrap will be added in the **Model Tree** of the component. You can use this component to provide an actual representation of the whole assembly.

TUTORIALS

Tutorial 1

In this tutorial, you will create all components of the Shock assembly and then assemble them, as shown in Figure 10-53. Also, you will create an exploded state of the assembly, as shown in Figure 10-54. The BOM is shown in Figure 10-54. The dimensions of the components are shown in Figures 10-55 through 10-60. **(Expected time: 2 hrs)**

Figure 10-53 *The Shock assembly*

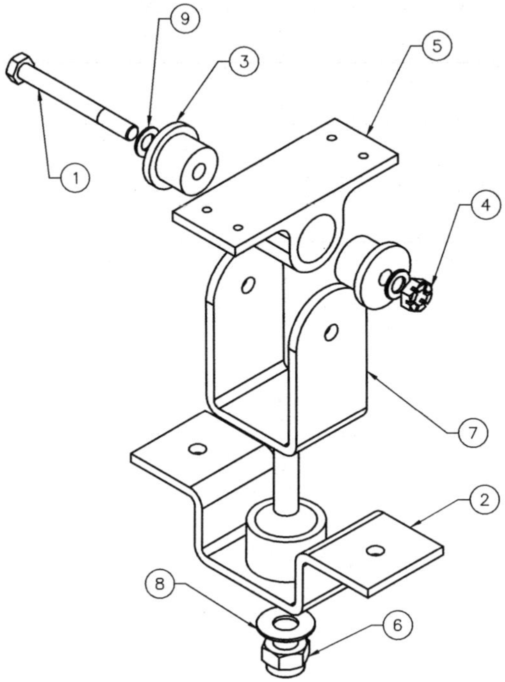

ITEM NO	PART NUMBER	QTY
1	BOLT_M48X5	1
2	BRACKET	1
3	BUSHING	2
4	CASTLE_NUT	1
5	PIVOT	1
6	SELF_LOCKING_NUT	1
7	U-SUPPORT	1
8	WASHER	1
9	WASHER_ISO7090	2

Figure 10-54 *The exploded state with Bill of Materials of the Shock Assembly*

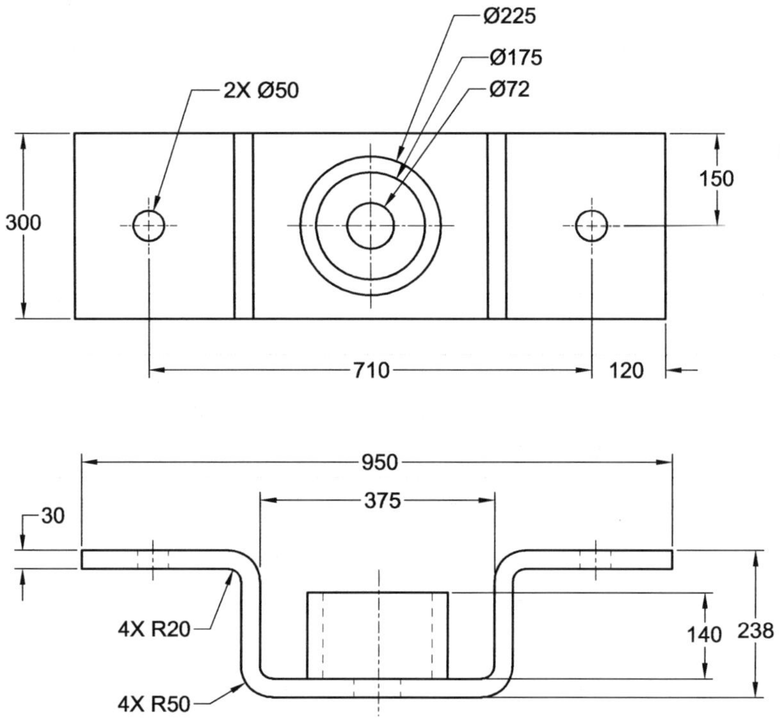

Figure 10-55 *Dimensions of the Bracket*

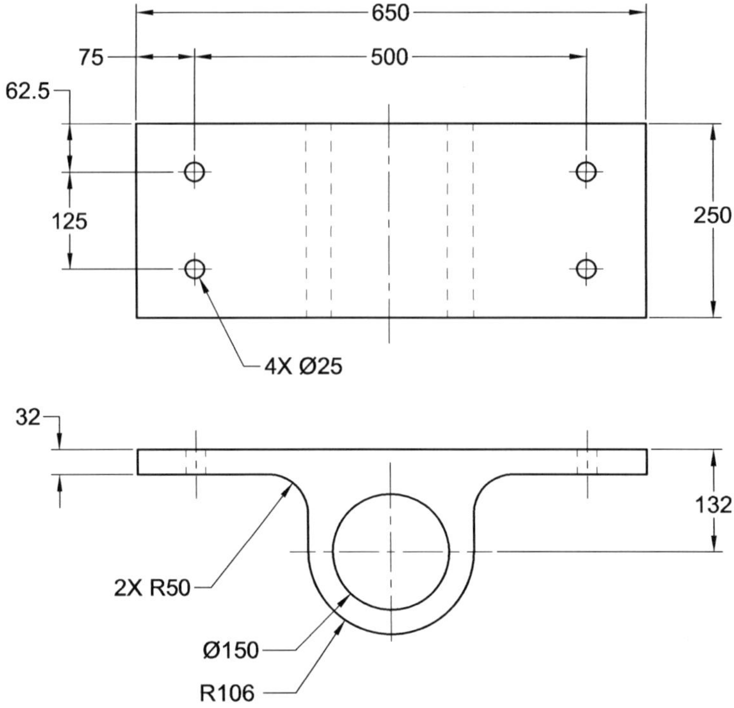

***Figure 10-56** Dimensions of the Pivot*

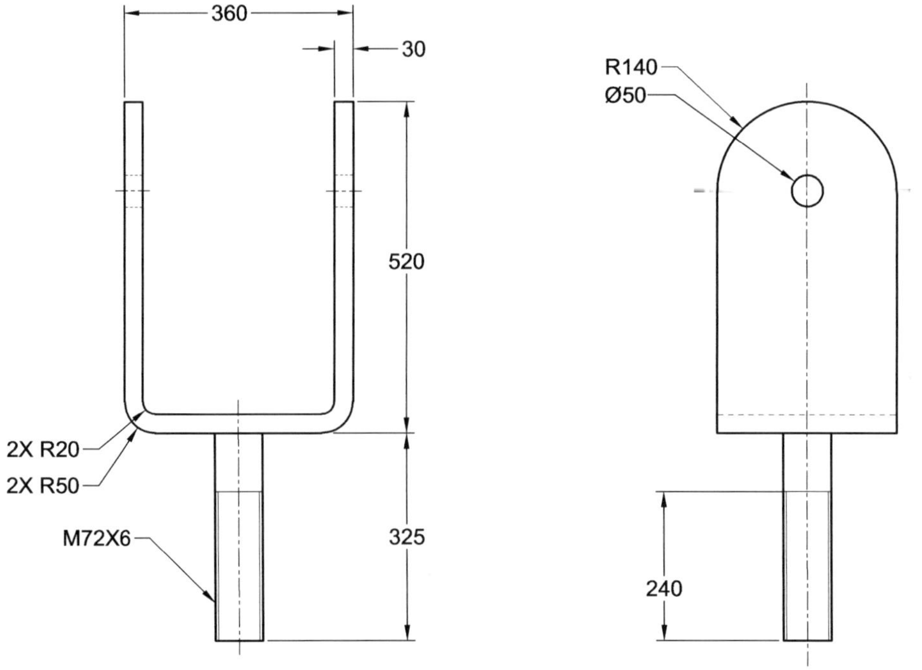

***Figure 10-57** Dimensions of the U-Support*

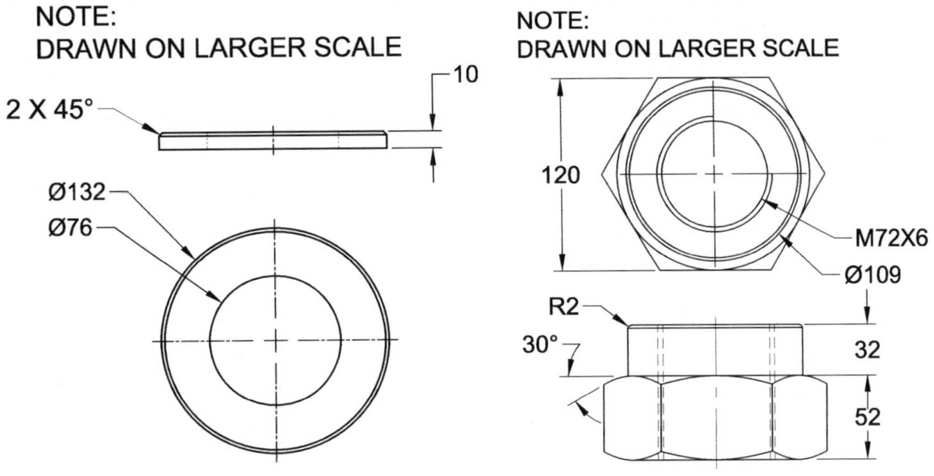

Figure 10-58 *Dimensions of the Washer and Self Locking Nut*

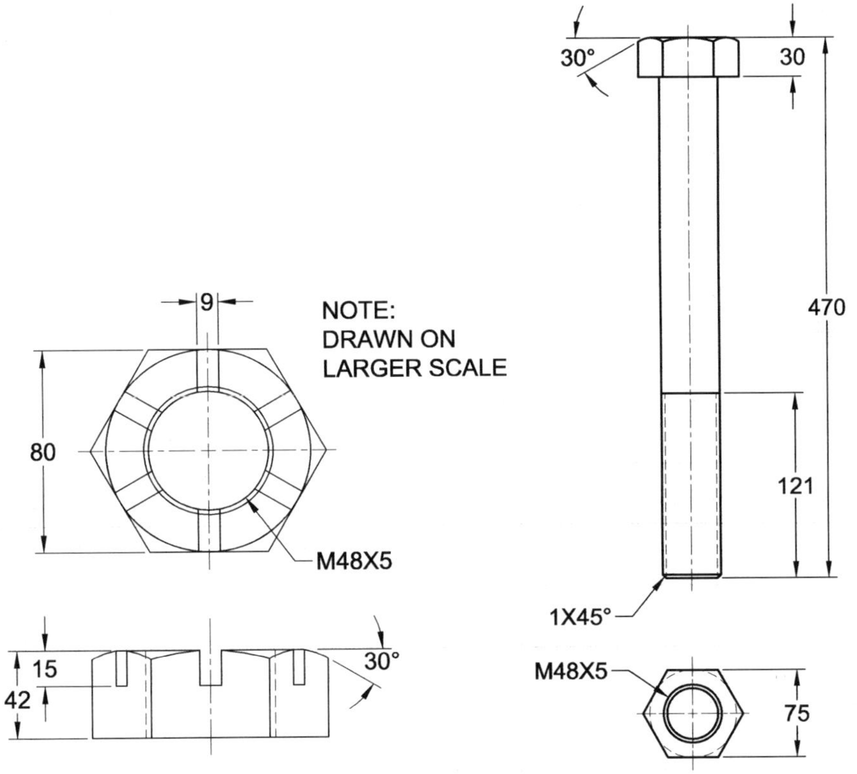

Figure 10-59 *Dimensions of the Castle Nut and Bolt*

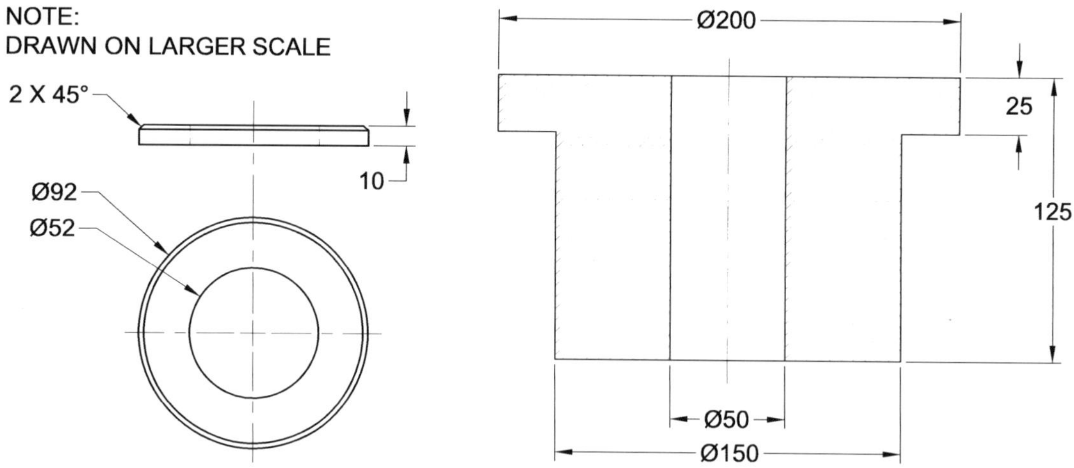

***Figure 10-60** Dimensions of the Washer and Bushing*

Note

You can download the part files of the assembly from www.cadcim.com. The complete path for downloading the files is as follows: Textbooks > CAD/CAM > PTC Creo Parametric > Creo Parametric 8.0 for Designers, 8th Edition

The following steps are required to complete this tutorial:

a. Create all components of the assembly as separate part files in the **Part** mode.
b. Create a new file in the **Assembly** mode and then assemble the U-Support with the default datum planes.
c. Assemble the Bushing with the U-Support and then repeat the Bushing in the assembly, .
d. Assemble the Pivot with the Bushing.
e. Suppress the two Bushings and the Pivot.
f. Assemble the Bracket with the assembly.
g. Assemble the Washer, Hexagonal Bolt, Castle nut, and Self locking nut in the assembly.
h. Unsuppress the suppressed components.
i. After assembling all components, create the exploded state of the assembly.
j. Generate offset lines in the exploded state.
k. Save the assembly and then exit the **Assembly** mode.

Before you start creating components, set the working directory to *C:\Creo-8.0\c10\Shockassembly*.

Starting the Components of the Assembly

The given assembly is created by using the bottom-up approach. As mentioned earlier, in the assemblies created using the bottom-up approach, the components are created as separate files and are placed in the assembly file. Therefore, first you need to create the components of the assembly.

1. Create all components of the assembly as separate part files and save them in the current working directory.

2. Close the part files if opened.

Creating a New Assembly File

You need to open a new assembly file to assemble the components.

1. Choose the **New** button from the **Data** group; the **New** dialog box is displayed.

2. Select the **Assembly** radio button in the **Type** area of the **New** dialog box. In the **Sub-type** area of this dialog box, the **Design** radio button is selected by default. Enter the name of the assembly in the **File name** edit box as **SHOCKASSEMBLY**.

3. Clear the **Use default template** check box and choose the **OK** button; the **New File Options** dialog box is displayed.

4. Select **mmks_asm_design_abs** from the **Template** area and choose the **OK** button from the **New File Options** dialog box; the assembly environment is invoked.

Assembling the U-Support with the Default Datum Planes

1. Choose the **Assemble** tool from the **Component** group or press the A key from the keyboard; the **Open** dialog box is displayed.

2. Select the **U-Support** component from the **Open** dialog box and choose the **Open** button; the **Component Placement** dashboard is displayed and you are prompted to select a reference for auto type constraining.

   In the **Component Placement** dashboard, the **Automatic** constraint type is selected by default in the **Constraint Type** drop-down list.

3. Choose the **Separate Window** button from the **Component Placement** dashboard to display the **U-Support** component in a separate window.

4. Now, select the **FRONT** plane from the drawing area or from the **Model Tree** and select the **ASM_FRONT** plane from the drawing area or from the **Model Tree** in the assembly window. Now, these two datum planes are aligned.

5. Similarly, select the other two datum planes and align them. After all the three planes are aligned, the message **Fully Constrained** is displayed in the **Status** area of the dashboard.

6. Choose the **OK** button to complete the assembly of the first part.

> **Tip**
> *Viewing the components in separate windows helps the user locate the references for the assembly constraints easily and also gives the user an independent viewing control over the two windows. You can spin, pan, and zoom the components in their respective windows.*

Note

*1. You can also assemble the first component in an assembly by selecting the **Default** option from the **Constraint Type** drop-down list in the **Component Placement** dashboard.*

*2. In an assembly, features are not displayed by default in the **Model Tree**, so you may need to display them in the **Model Tree**. To do so, choose on the **Settings** button of the **Model Tree**; a flyout will be displayed. Choose the **Tree Filters** option from the flyout; the **Model Tree Items** dialog box will be displayed. Now, select the **Features** check box from the **Display** area of the dialog box and choose the **OK** button.*

Assembling the Bushing with the U-Support

The next component you need to assemble is the **Bushing**.

1. Choose the **Assemble** tool or press the A key from the keyboard to display the **Open** dialog box.

2. Select the **Bushing** component and then choose the **Open** button from the **Open** dialog box; the **Component Placement** dashboard is displayed. Choose the **Separate Window** button from the **Component Placement** dashboard to display the **Bushing** component in a separate window, if it is not chosen by default.

3. Select the **Coincident** option from the **Constraint Type** drop-down list. Select the axis of the **Bushing** component and the axis of the **U-Support** component, as shown in Figure 10-61.

4. Next, select the planar face of the **Bushing** component and the inner planar face of the **U-Support** component, as shown in Figure 10-61. Select the **Coincident** option from the **Constraint Type** drop-down list. Choose the **OK** button; the model similar to the one shown in Figure 10-62 is displayed in the drawing area.

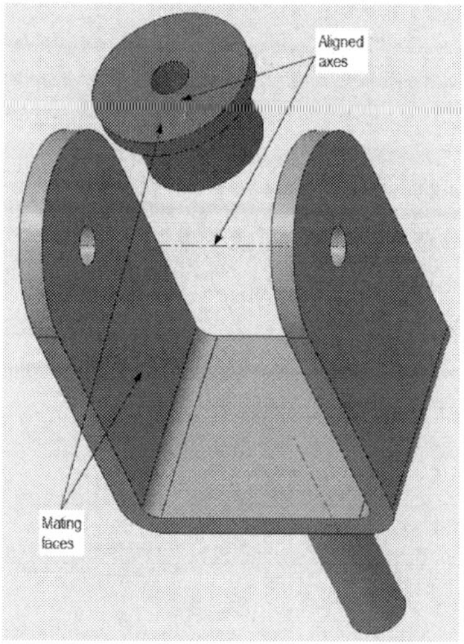

Figure 10-61 *Locations to apply the Coincident constraint*

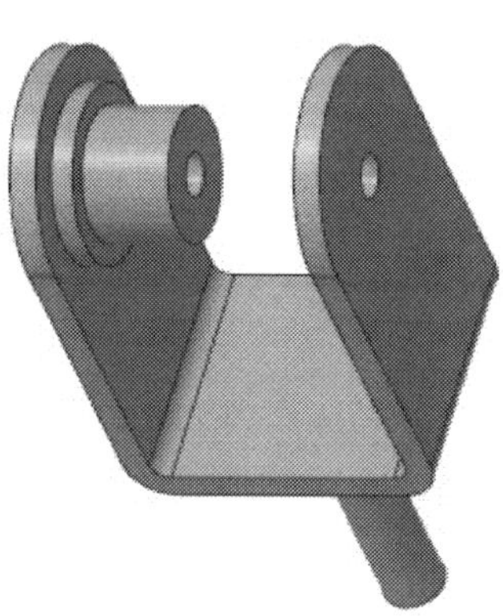

Figure 10-62 *Components after assembling*

Note

*You may need to choose the **Change orientation of constraint** button from the **Component Placement** dashboard to assemble the **Bushing** components in the required orientation.*

5. The next instance of the **Bushing** can be placed directly by using the **Repeat** tool. To repeat the **Bushing** component in the assembly, select it from the **Model Tree** or from the drawing area and then choose the **Repeat** tool from the **Component** group. On doing so, the **Repeat Component** dialog box is displayed and two **Coincident** constraints are displayed in the **Variable assembly references** area.

6. Select the second **Coincident** constraint from this area and then choose the **Add** button.

7. Select the inner right face of the **U-Support** component as the mating face. The copy of the **Bushing** component is assembled, as shown in Figure 10-63. Choose the **OK** button from the **Repeat Component** dialog box.

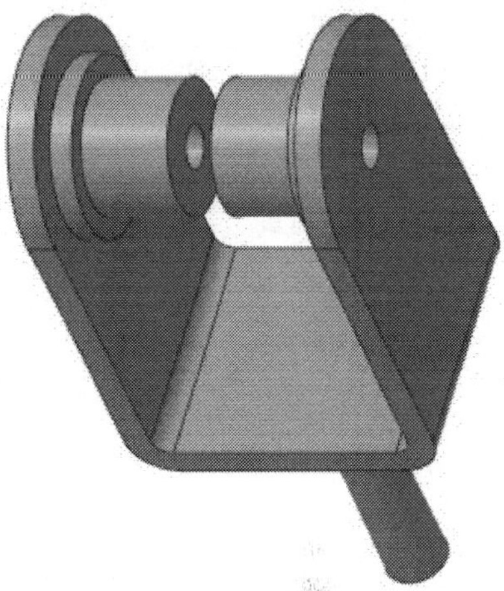

Figure 10-63 *Assembled Bushings and the U-Support*

Assembling the Pivot with the Bushing

The next component that you need to assemble is the **Pivot**.

1. Choose the **Assemble** tool to display the **Open** dialog box.

2. In this dialog box, select the **Pivot** component and then choose the **Open** button to display the **Component Placement** dashboard.

3. Choose the **Placement** tab to display the panel. Select the axis of the **Pivot** and then the axis of the hole of the **Bushing**, refer to Figure 10-64, and apply the **Coincident** constraint on them.

4. Next, click on the **New Constraint** option in the **Placement** slide-down panel. Select the **Parallel** option from the **Constraint Type** drop-down list.

5. Select the top planar face of the **Pivot** and the planar face of the **U-Support**, as shown in Figure 10-64.

6. Click on the **New Constraint** option in the **Placement** slide-down panel and select the planar face of the **Pivot** and the planar face of the **Bushing**, as shown in Figure 10-64; the **Distance** constraint is automatically applied between them and is displayed by default in the **Constraint Type** drop-down list in the **Placement** tab. Select the **Coincident** option from the drop-down list so that the selected faces are made coincident.

7. Choose the **OK** button to complete the assembly. The model similar to the one shown in Figure 10-65 is displayed in the drawing area.

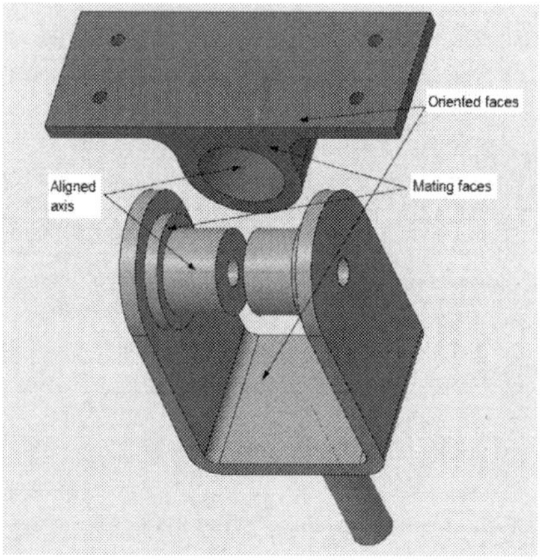

Figure 10-64 *Locations to apply constraints*

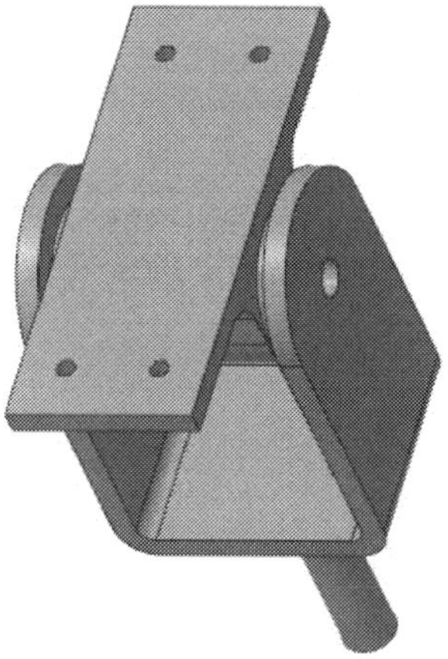

Figure 10-65 *Assembled view of the **Pivot**, **U-Support**, and **Bushing***

Suppressing Components

Next, you need to suppress both the **Bushing** and the **Pivot**. This is because when the components in the assembly increase, some components may be hidden behind the others. As a result, it gets difficult to select them. Also, more the number of components, more are the datum planes. Therefore, it becomes difficult to make selections on the components. However, when you suppress a component, its datum planes are also suppressed and the complications in the drawing area are reduced.

1. Spin the model and select the first **Bushing** from the drawing area or from the **Model Tree** and choose the **Suppress** option from the mini popup toolbar to suppress the component; the **Suppress** message box is displayed. Choose the **OK** button from the message box; the **Pivot** gets suppressed because it was dependent on the first bushing.

2. Similarly, suppress the other **Bushing** to make the assembly of the other components easy. Choose the **OK** button from the **Suppress** window to exit.

Note

*Very often you need to use the datum planes and datum axes to assemble the components. Therefore, you need to turn their display on or off from the **Graphics** toolbar, whenever required.*

Assembling the Bracket with the Assembly

The next component you need to assemble is the **Bracket**.

1. Choose the **Assemble** tool from the **Component** group to display the **Open** dialog box.

2. Select the **Bracket** component to open; the **Component Placement** dashboard is displayed.

3. Choose the **Placement** tab to display the panel and select the **Coincident** option from the **Constraint Type** drop-down list.

4. Select the axis of the **Bracket** and the axis of the **U-Support**, as shown in Figure 10-66.

5. Click on the **New constraint** option in the **Placement** panel and select the **Coincident** option from the **Constraint Type** drop-down list. You can also invoke the shortcut menu and choose the **New Constraint** option from it.

6. Now, select the faces to mate, refer to Figure 10-66; preview of the assembly is displayed.

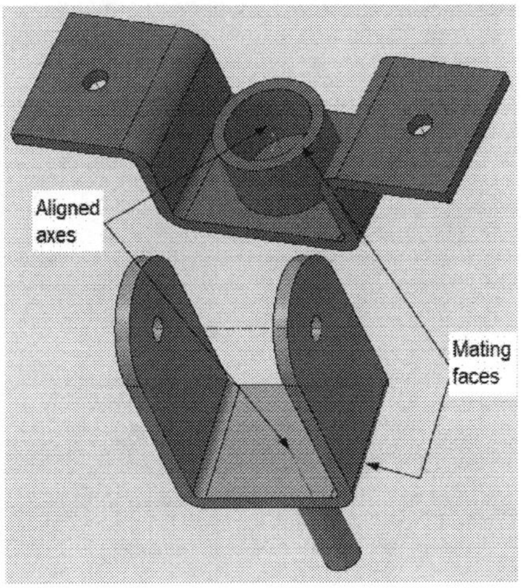

Figure 10-66 *Constraints and the location to apply them*

7. Choose the **OK** button from the **Component Placement** dashboard to complete the assembly. The model similar to the one shown in Figure 10-67 is displayed in the drawing area.

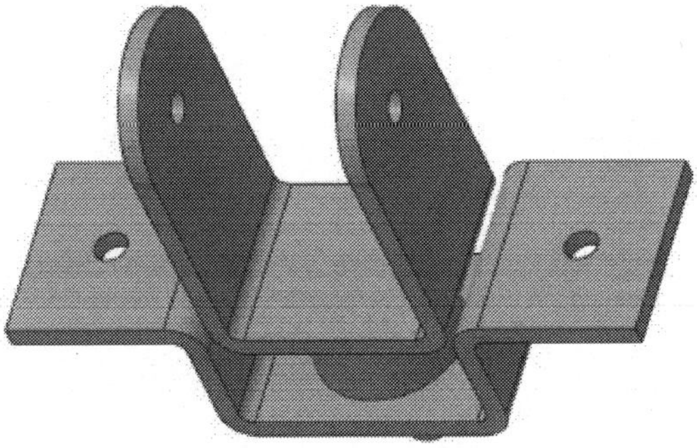

Figure 10-67 *Assembled view of the* ***Bracket*** *with* ***U-Support***

Assembling the Washer with the Assembly

The next component that you need to assemble is the **Washer**.

1. Choose the **Assemble** tool to display the **Open** dialog box.

2. Open the **Washer** component to display the **Component Placement** dashboard.

3. Choose the **Placement** tab from the dashboard to display the panel and select the **Coincident** option from the **Constraint Type** drop-down list to apply the align constraint.

4. Select the axis of the **Washer** and then select the axis of the **U-Support** from the drawing area to align the axes of the two components.

5. Now, click on the **New Constraint** option in the **Placement** panel and select the **Coincident** option from the **Constraint Type** drop-down list.

6. Select the top planar face of the **Washer** and then the bottom planar face of the **Bracket** to assemble the **Washer**; the message **Fully constrained** is displayed in the **Status** area of the dashboard.

7. Choose the **OK** button from the **Component Placement** dashboard to complete the assembly. The model similar to the one shown in Figure 10-68 is displayed in the drawing area.

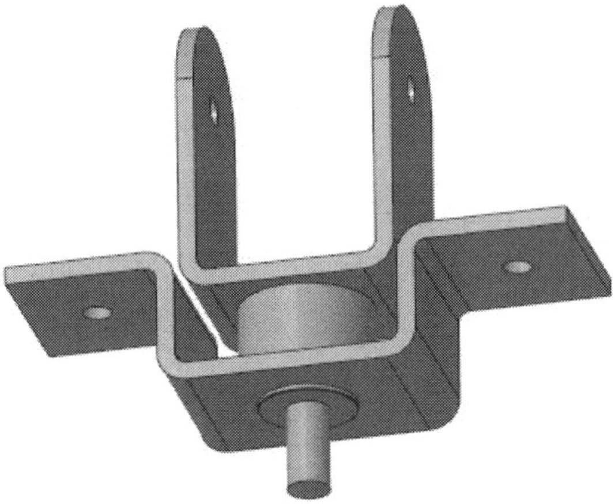

Figure 10-68 *Assembled view of the* ***Washer*** *with* ***U-Support***

Inserting the Hexagonal Bolt in the Assembly

The next component that you need to assemble is the **Hexagonal bolt**.

1. Choose the **Assemble** tool to display the **Open** dialog box.

2. Open the **Hexagonal bolt** component to display the **Component Placement** dashboard.

3. Choose the **Placement** tab to invoke a panel and select the **Coincident** option from the **Constraint Type** drop-down list to add the align constraint.

4. Select the axis or cylindrical face of the **Hexagonal bolt** and then the axis or cylindrical face of the hole on the **U-Support**. Now, the two axes are aligned. If required, you can choose the **Flip Constraint** button to flip the orientation of the **Hexagonal bolt**.

5. Click on the **New Constraint** option in the **Placement** panel and then select the **Coincident** option from the **Constraint Type** drop-down list.

6. Select the bottom planar face of the head of the **Hexagonal bolt** and then select the left outer planar face of the **U-Support**; preview of the assembly is displayed.

7. Choose the **OK** button from the **Component Placement** dashboard to complete the assembly. The model similar to the one shown in Figure 10-69 is displayed in the drawing area.

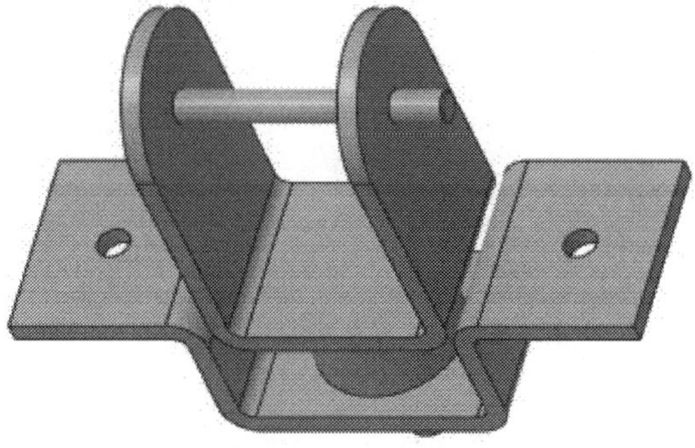

Figure 10-69 *Assembly after assembling the **Hexagonal bolt***

Assembling other Components

In this section, you need to assemble the **Washer**, **Castle nut** and the **Self locking nut** with the assembly

Unsuppressing the Components

All components are assembled at their required positions. Now, you need to unsuppress the suppressed components.

1. Choose the **Resume All** option from the **Resume** flyout of the **Operations** group; the suppressed components appear in the assembly where they were assembled. The completed assembly is shown in Figure 10-70.

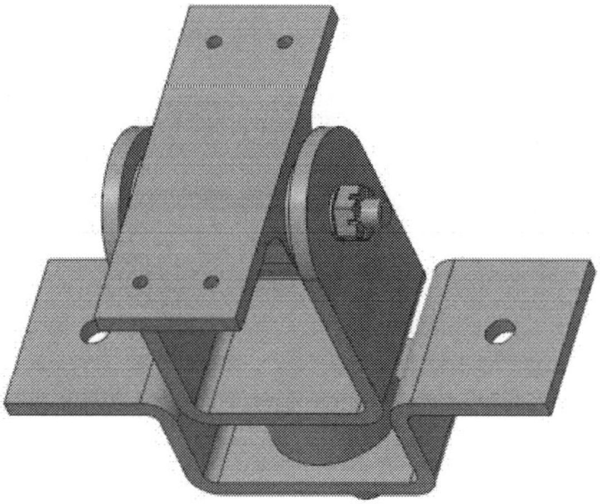

Figure 10-70 *Completed assembly with all components*

Creating the Exploded State of the Assembly

To view all the components in an assembly clearly, you need to create the exploded state of the assembly.

1. Invoke the **Edit position** tool from the **Model Display** group; the **Explode Tool** dashboard is displayed. Also, the default exploded state of the assembly is displayed.

   Now, you need to edit the position of the component in the exploded view.

2. Choose the **References** tab in the **Explode Tool** dashboard and click in the **Movement Reference** collector on the dashboard; you are prompted to select a reference to define the direction of movement.

3. Select the vertical axis of the **U-Support** from the drawing area; you are prompted to select the component to be moved.

4. Select the **Self locking** nut from the drawing area; a direction handle is displayed. Move the cursor to the required direction handle and drag to place it at the required position. Similarly, select and move the **Washer** and the **Bracket** to the required positions.

   Now, you need to move the components in the top half of the assembly. These components are moved with reference to the axis of the hole on the **U-Support**. Therefore, you need to select this axis as the motion reference.

5. Click in the **Movement Reference** collector on the dashboard; you are prompted to select the reference. Select the axis of the hole on the **U-Support** from the drawing area as the reference; you are now prompted to select the components to move.

6. Select and move the **Hexagonal bolt**, **Pivot**, **Bushings** and **Castle nut** to the required locations in the drawing area.

   The model similar to the one shown in Figure 10-71 is displayed in the drawing area.

7. Choose the **OK** button from the **Explode Tool** dashboard.

   Now, you need to save this exploded view.

8. To save this view, invoke the **View Manager** dialog box and then choose the **Explode** tab.

9. Choose the **New** button; the **Modified State Save** message box is displayed confirming to save the default exploded view. Choose the **Yes** button from the message box; the **Save Display Elements** dialog box is displayed.

10. Next, enter the **EXP1** in the **Explode** edit box and choose the **OK** button from the dialog box and close the **View Manager** dialog box.

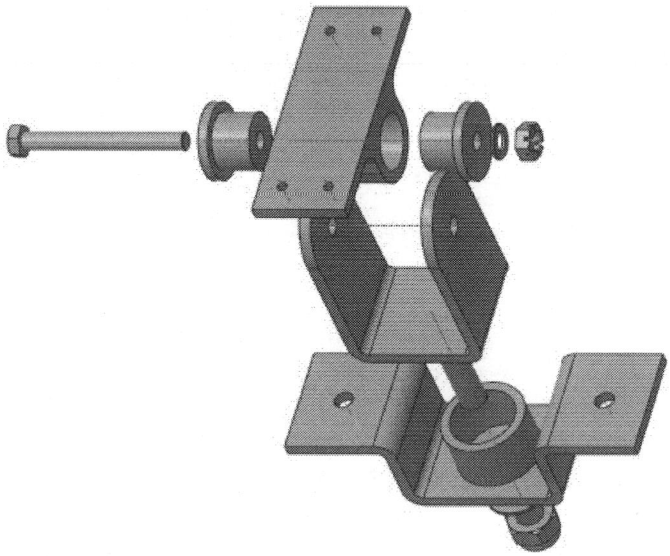

Figure 10-71 Exploded state of the assembly

Tip
*You can also view the default exploded state of an assembly by choosing the **Explode View** button from the **Model Display** group in the **View** tab.*

Creating Offset Lines

1. Invoke the **Edit position** tool from the M**odel Display** group and choose the **Create cosmetic offset lines to illustrate movement of exploded component** button from the **Explode Tool** dashboard; the **Cosmetic Offset Line** dialog box is displayed.

2. Click in the **Reference 1** collector and then select the axis of the **U-Support**.

3. Next, click in the **Reference 2** collector and then select the axis of the **Self locking nut**. Next, choose the **Apply** button from the **Cosmetic Offset Line** dialog box; the blue colored explode line is created.

4. Similarly, create the offset lines by selecting the axis of the hole on the **U-Support**, followed by the **Bushing** and then the **Castle nut**. Similarly, create offset lines by selecting the axis of the hole on the **U-Support**, followed by the **Bushing** and then the **Hexagonal bolt**. The model similar to the one shown in Figure 10-72 is displayed in the drawing area.

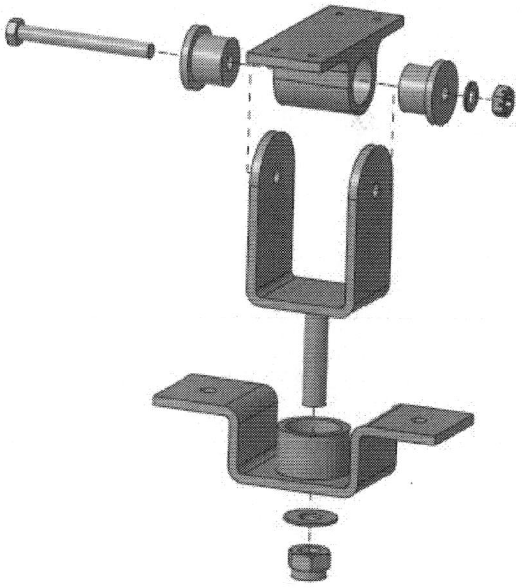

Figure 10-72 Offset lines displayed in the exploded state

5. Next, exit the **Cosmetic Offset Line** dialog box by choosing the **Close** button.

6. Choose the **OK** button from the **Explode Tool** dashboard.

Saving the Assembly File

1. Choose the **Save** button from the **File** menu and save the model.

Closing the Window

1. Choose the **Close** button from the **Quick Access** toolbar to close the assembly file.

Tutorial 2

In this tutorial, you will create all components of the Pedestal Bearing assembly and then assemble them, as shown in Figure 10-73. You will also create the exploded state of the assembly displaying the offset lines, as shown in Figure 10-74. The dimensions of the components are shown in Figures 10-75 through 10-77. **(Expected time: 2 hrs)**

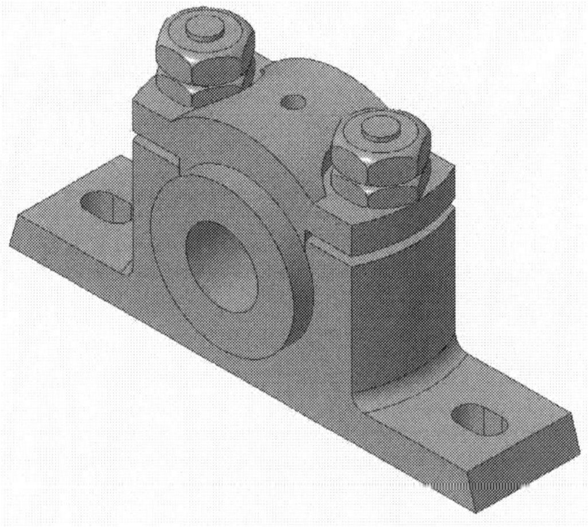

Figure 10-73 *Assembly of the Pedestal Bearing*

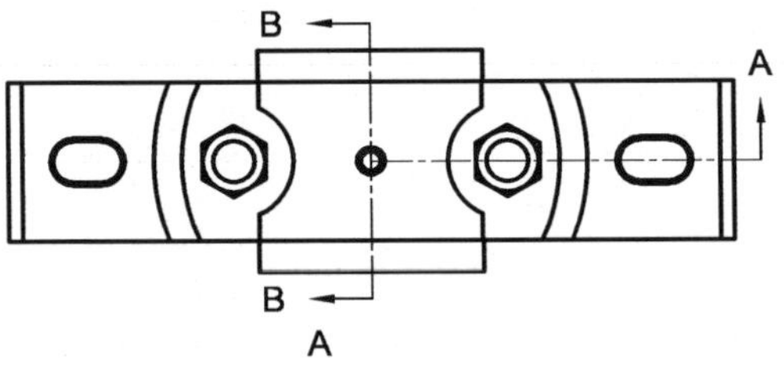

ITEM NO.	PART NO.	DESCRIPTION	QTY.
1	Casting		1
2	Cap		1
3	Bolt		2
4	Nut		2
5	Lock Nut		2
6	Brasses		1

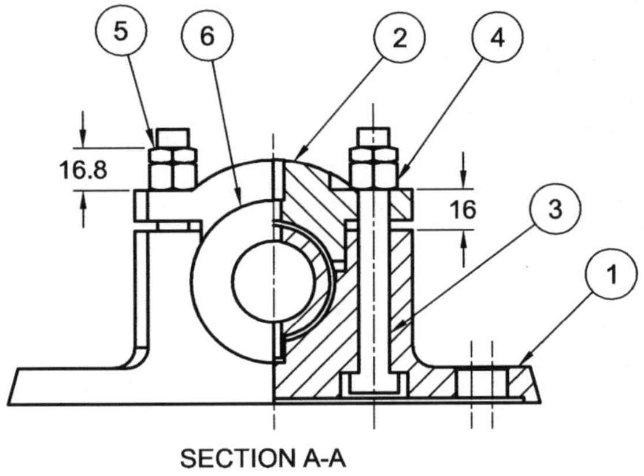

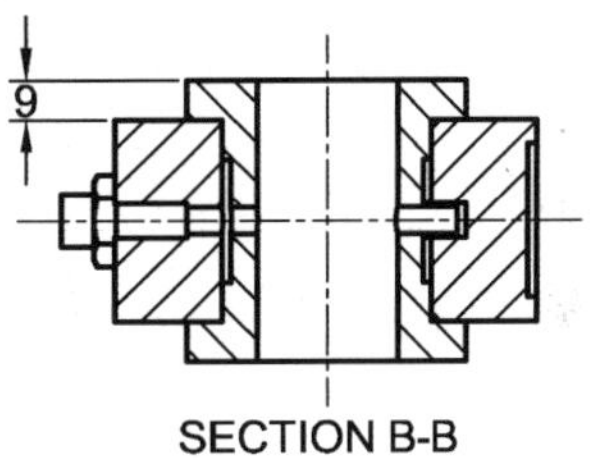

Figure 10-74 The Pedestal Bearing assembly

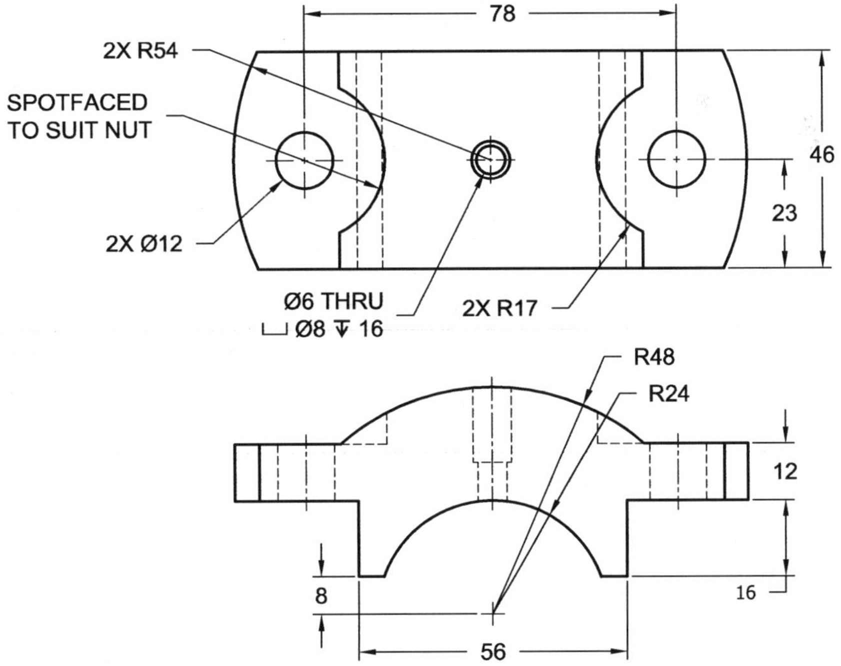

Figure 10-75 Dimensions of the Cap

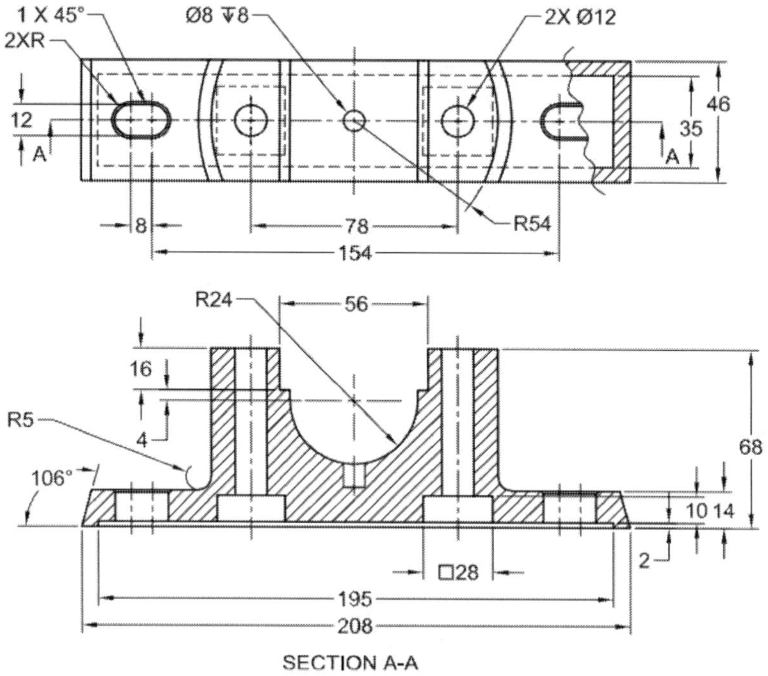

Figure 10-76 *Dimensions of the Casting*

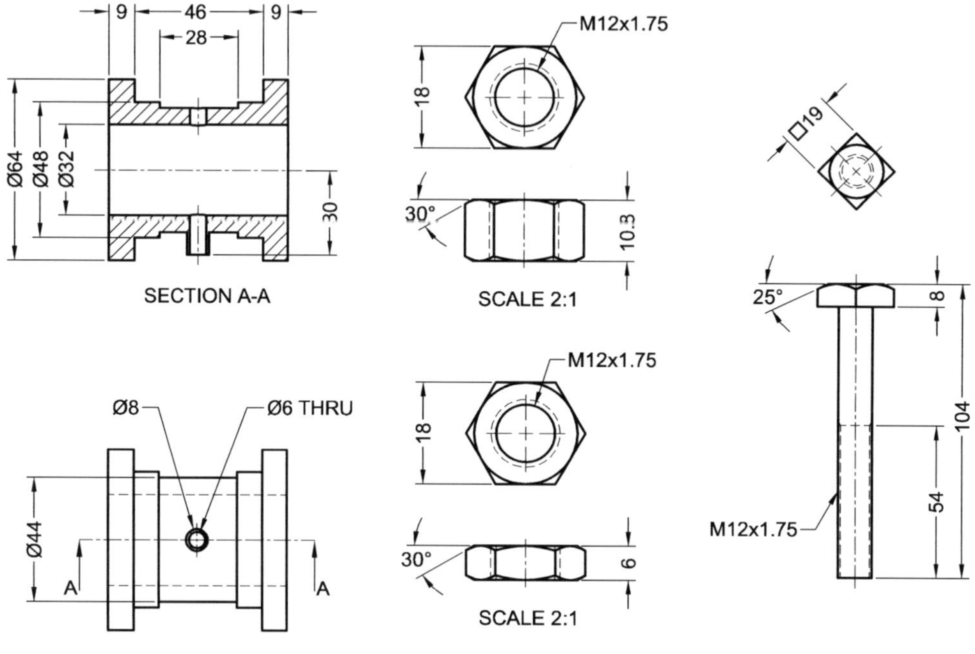

Figure 10-77 *Dimensions of the Brasses, Nut, and Locknut*

The following steps are required to complete this tutorial:

a. Create all components of the assembly as separate part files in the Part mode.
b. Create a new file in the Assembly mode and then assemble the Casting with the default assembly datum planes.
c. Assemble the Cap with the Casting.
d. Suppress the Cap from the assembly.
e. Assemble the Brasses with the Casting.
f. Insert the Square headed bolt, Nut, and Lock nut in the assembly and assemble them.
g. Resume the suppressed components.
h. Create the exploded state of the assembly.
i. Save the assembly and close the file.

Set the working directory to *C:\Creo-8.0\c10\Pedestalbearing*.

Creating Components for the Assembly

To create the assembly, all components must be created first in the **Part** mode. Also, you need to use the bottom-up approach to create the assembly.

1. Create all components of the assembly as separate part files and then save them in the current working directory.

2. Close the part files if opened.

Creating a New Assembly File

As mentioned earlier, all components created are assembled in an assembly file that has an extension *.asm*. Therefore, you need to open a new *.asm* file.

1. Choose the **New** button from the **File** menu to display the **New** dialog box.

2. Select the **Assembly** radio button in the **Type** area of the **New** dialog box. In the **Sub-type** area of the **New** dialog box, the **Design** radio button is selected by default. Enter the name of the assembly in the **File name** edit box as **PEDESTALBEARING**.

3. Choose the **OK** button to enter into the assembly modeling environment.

Assembling the Casting with the Default Datum Planes

In the new assembly file, the three default assembly datum planes are displayed in the drawing area and the **Model Tree** is displayed at the left of the drawing area. If the display of the **Model Tree** was turned off in the previous tutorial, it would not appear. Now, you can start assembling components.

1. Choose the **Assemble** tool to display the **Open** dialog box.

2. Select **Casting** from the **Open** dialog box and choose the **Open** button; the **Component Placement** dashboard is displayed.

Note
It is recommended to choose the ***Separate Window*** *button to display the component to be assembled in a new window.*

3. Select the **FRONT** datum plane of the model and then select the **ASM_FRONT** plane from the assembly to align them. Similarly, align the other two default planes with the respective assembly datum planes.

Note
If the datum planes are not displayed in the drawing area, you need to turn on their display by choosing the ***Plane Display*** *tool from the* ***Datum Display Filters*** *drop-down in the* ***Graphics*** *toolbar. After assembling the first component, you may need to turn off the display of planes. This is because the datum planes clutter the drawing area and then the selection of the references for applying assembly constraints becomes difficult. You can turn the datum planes' display on when they are required.*

Assembling the Cap with the Casting

At first, you need to assemble the **Cap** with the **Casting**.

1. Choose the **Assemble** tool to display the **Open** dialog box.

2. Select the **Cap** file and then choose the **Open** button from the **Open** dialog box to display the **Component Placement** dashboard.

3. Choose the **Placement** tab to display the panel and select the **Coincident** option from the **Constraint Type** drop-down list.

4. Select the axis of the **Cap** and then the axis of the **Casting**, as shown in Figure 10-78.

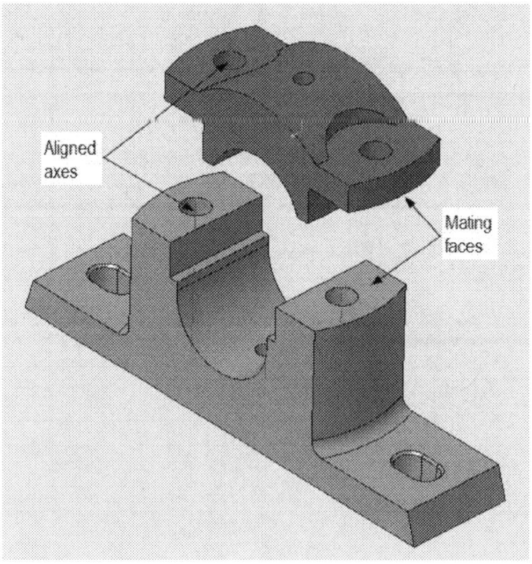

Figure 10-78 *References of the constraints used for assembling the components*

5. Choose the **New Constraint** option from the **Placement** tab and then select the **Distance** option from the **Constraint Type** drop-down list in the **Placement** panel.

6. Select the two mating faces and enter **4** in the edit box in the **Component Placement** dashboard.

7. Choose the **OK** button from the **Component Placement** dashboard to complete the creation of the assembly. The model similar to the one shown in Figure 10-79 is displayed in the drawing area.

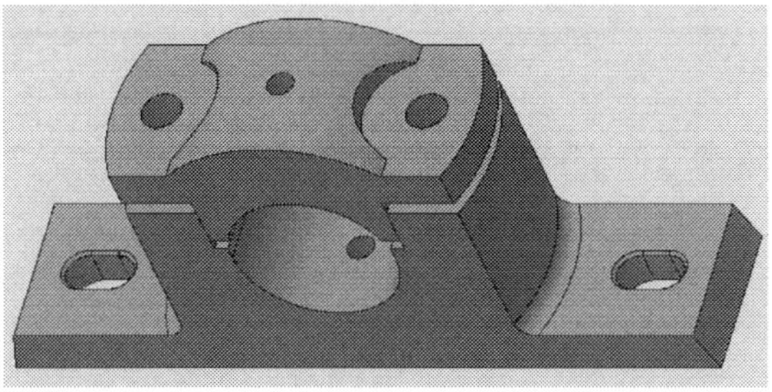

Figure 10-79 *Model after assembling the Cap with the Casting*

Suppressing the Cap from the Assembly

You need to suppress the Cap because the next component has to be assembled with the **Casting**. When you suppress the Cap from the assembly, it becomes easier to assemble the new component. The Cap will be unsuppressed later.

1. Select the **Cap** from the drawing area and right-click to display the shortcut menu.

2. Choose the **Suppress** option from the shortcut menu; you are prompted to confirm the suppression. Choose the **OK** button.

Assembling Brasses with the Casting

Next, you need to assemble the Brasses with the Casting.

1. Choose the **Assemble** tool to display the **Open** dialog box.

2. Select **Brasses** and choose the **Open** button from the dialog box; the **Component Placement** dashboard is displayed.

3. Choose the **Placement** tab to display the panel and select the **Coincident** option from the **Constraint Type** drop-down list.

4. Select the axis of the **Brasses** and then the axis of the **Casting**, as shown in Figure 10-80.

Note

*You can choose the **Separate Window** button from the dashboard to display the component in a separate window, which makes it easier to select the faces and then apply the constraints.*

5. Click on the **New Constraint** option in the **Placement** panel and select the faces of the **Brasses** and the **Casting**, refer to Figure 10-80.

As you select the faces, the **Oriented** option from the **Constraint Type** drop-down list will get selected.

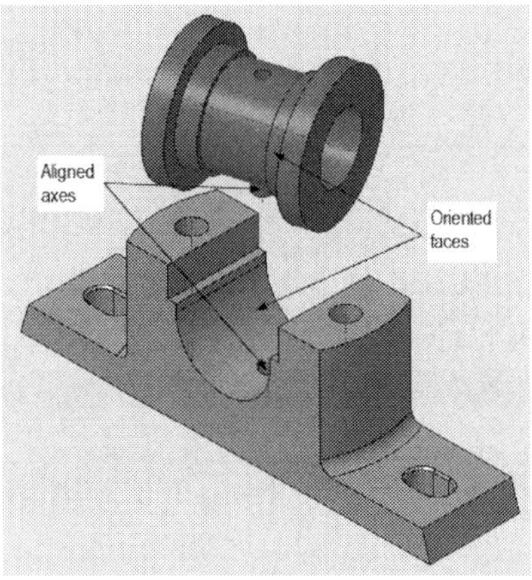

Figure 10-80 *References of the constraints used for assembling the components*

6. Choose the **OK** button from the **Component Placement** dashboard; the assembly model similar to the one shown in Figure 10-81 is displayed in the drawing area.

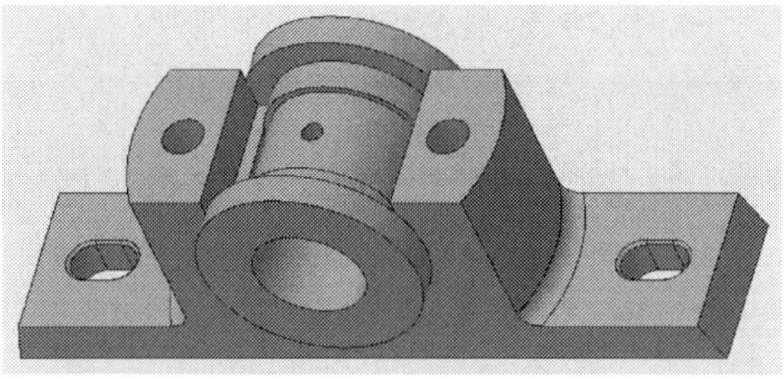

Figure 10-81 *Assembling the* ***Brasses*** *with the assembly*

7. Now, unsuppress the **Cap** by choosing **Resume > Resume All** from the **Operations** group. Similarly, assemble the remaining components. The final assembly is shown in Figure 10-82. You can choose the **Repeat** option to copy the repeated items.

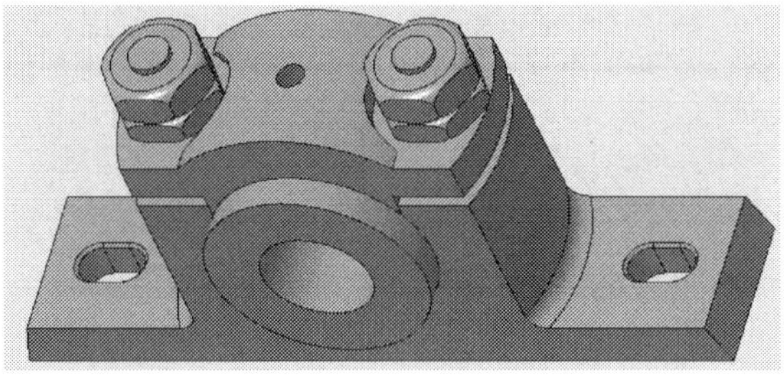

Figure 10-82 *The final Pedestal Bearing assembly*

Creating the Exploded State of the Assembly

To view all components in an assembly clearly, you need to create its exploded state.

1. Choose the **Edit Position** button from the **Model Display** tab; the **Explode Tool** dashboard and the default exploded state of the assembly are displayed.

   Now, you need to edit the position of the component in the exploded view.

2. Choose the **References** tab from the dashboard and then click in the **Movement Reference** collector; you are prompted to select a reference to define the direction of movement.

3. Select the vertical axis of the Cap from the drawing area; you are prompted to select the component to be moved.

4. Select the Casting from the drawing area; a direction handle is displayed. Move the cursor to the required direction handle and drag to place it at the required position. Similarly, place the Cap, two Square headed bolts, two Lock nuts and then two Nuts at the required location.

   Now, you need to move the Brasses to the required location in the drawing area.

5. Click in the **Movement Reference** collector again; you are prompted to select the reference. Select the axis of the Brasses from the drawing area as the reference; you are now prompted to select the component to be moved.

6. Select and move the Brasses to the required location in the drawing area.

   The model similar to the one shown in Figure 10-83 is displayed in the drawing area.

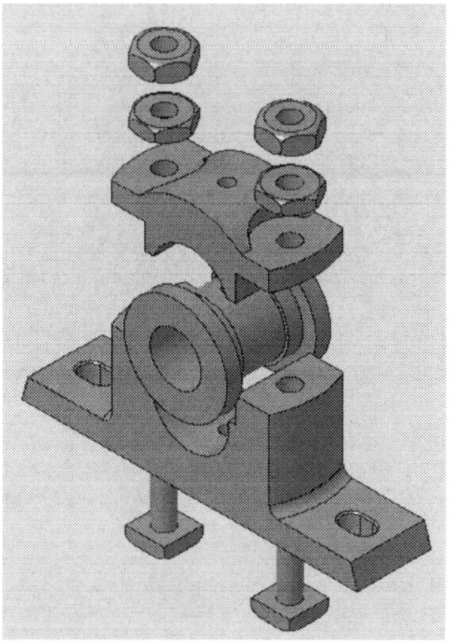

Figure 10-83 *The Exploded Pedestal Bearing assembly*

7. Now, you need to save this exploded view. To save this view, invoke the **View Manager** dialog box and choose the **Explode** tab.

8. Choose the **New** button; the **Modified State Save** message box is displayed confirming to save the default exploded view. Choose the **Yes** button from the message box; the **Save Display Elements** dialog box is displayed.

9. Next, enter the **EXP1** in the **Explode** edit box and choose the **OK** button from the dialog box and close the **View Manager** dialog box.

Creating the Offset Lines

1. Choose the **Edit Position** button from the **Model Display** tab and choose the **Create cosmetic offset lines to illustrate movement of exploded component** button from the **Explode Tool** dashboard; the **Cosmetic Offset Line** dialog box is displayed.

2. Click in the **Reference 1** collector and select the axis of the right hole on the **Casting**.

3. Next, click in the **Reference 2** collector and then select the axis of the **Square Headed bolt** on the right half of the **Casting** to display the explode line. Similarly, select the axes from the other components to create the offset lines between them.

   The model similar to the one shown in Figure 10-84 is displayed in the drawing area.

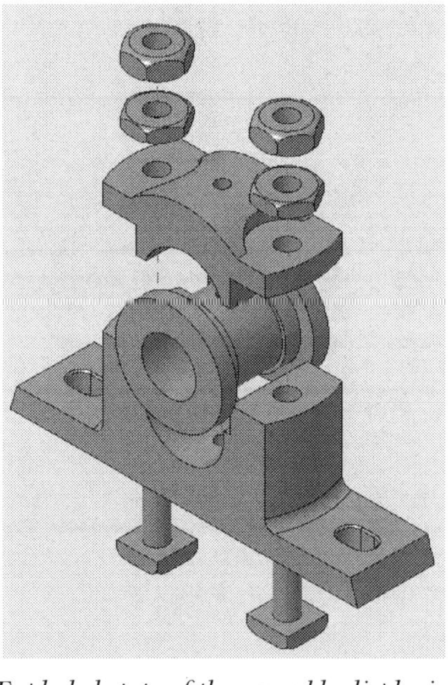

Figure 10-84 *Exploded state of the assembly displaying the offset lines*

4. Close the **Cosmetic Offset Line** dialog box and choose the **OK** button.

Saving the Assembly

1. Choose the **Save** button from the **File** menu and then save the assembly.

Closing the Window

Now, you have saved the assembly so the window can be closed.

1. Choose the **Close** button from the **Quick Access** toolbar to close the file.

Self-Evaluation Test

Answer the following questions and then compare them to those given at the end of this chapter:

1. In the ____________ type of simplified representation, the components removed from the current display are displayed again.

2. The ______________ type of simplified representation reduces the regeneration time in case of large assemblies.

3. The ____________ constraint is similar to the **Coincident** constraint with the only difference that this constraint allows you to specify some offset distance between the two coplanar faces.

4. The __________ constraint is used to assemble circular components.

5. The ___________ is a tabular representation of all components of the assembly along with the information associated with them.

6. On applying the **Coincident** constraint to two selected faces, the selected faces get connected to each other. (T/F)

7. An exploded view is a state in which all components are moved from their actual locations so that they are visible. (T/F)

8. Dimensions of a feature cannot be modified in the **Assembly** mode. (T/F)

9. You can display the unexploded state of the assembly by choosing the **Explode View** button from the **Modal Display** group. (T/F)

10. Offset lines are invisible by default in the unexploded state of the assembly. (T/F)

Review Questions

Answer the following questions:

1. Which of the following constraints is used to align the selected datum point or vertex on the first part with the selected surface or datum plane on the second part?

(a) **Orient** (b) **Pnt on Srf**
(c) **Edge on Srf** (d) **Tangent**

2. Which of the following messages is displayed in the **Placement Status** area of the **Component Placement** dashboard when the assembly is fully constrained?

   (a) **Partially constrained** (b) **Fully constrained**
   (c) **Completely constrained** (d) **None of these**

3. In which of the following types of simplified representation, the complete geometry of the components is displayed in the current model display style?

   (a) **Geometry representation** (b) **Graphics representation**
   (c) **Master representation** (d) **None of these**

4. Which of the following constraints is used to assemble two components by making the selected faces or planes coplanar such that the mating faces or planes face in the same direction?

   (a) **Insert** (b) **Orient**
   (c) **Tangent** (d) **Align**

5. Which of the following groups of the **Model** tab contains the **Explode View** button?

   (a) **Model Display** (b) **Component**
   (c) **Datum** (d) **Modifiers**

6. A component of an assembly assembled by using the assembly constraints cannot be replaced with some other components. (T/F)

7. You can hide a part from an assembly. (T/F)

8. Offset lines can only be created in the exploded state of an assembly. (T/F)

9. The order of assembling components in an assembly can be changed. (T/F)

10. Creo Parametric provides you with a ready made BOM that you can directly use. (T/F)

EXERCISE

Exercise 1

In this exercise, you will create all components of the Crosshead assembly and then assemble them, as shown in Figure 10-85. Also, you will create an exploded state of the assembly, shown in Figure 10-86, which displays the offset lines. The dimensions of the components are shown in Figures 10-87 through 10-92. **(Expected time: 2 hrs)**

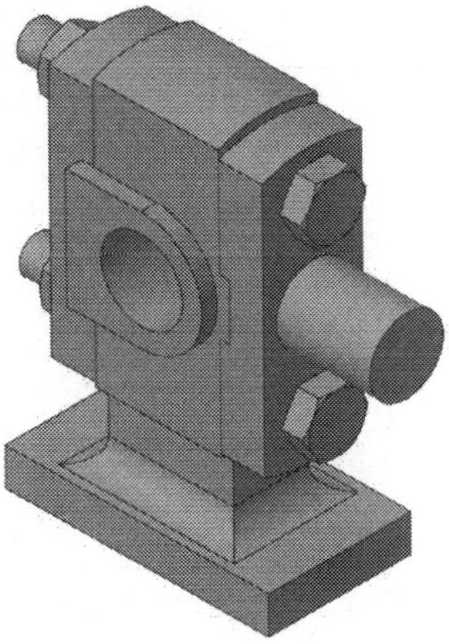

Figure 10-85 The Crosshead assembly

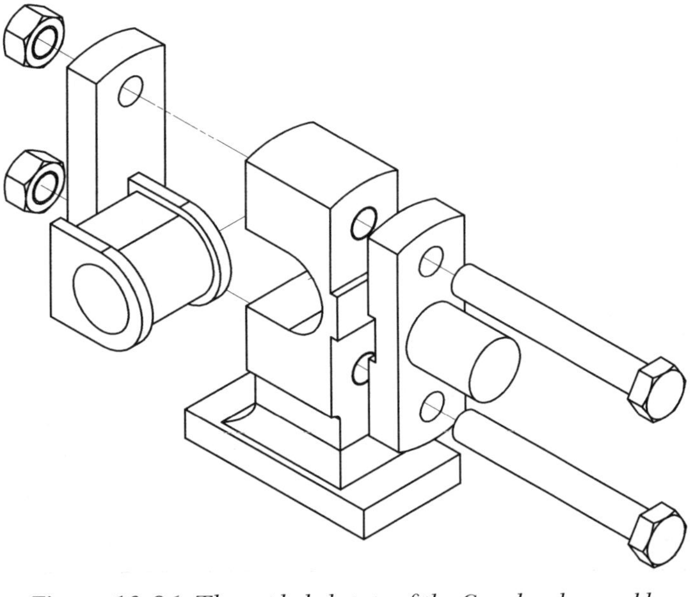

Figure 10-86 The exploded state of the Crosshead assembly

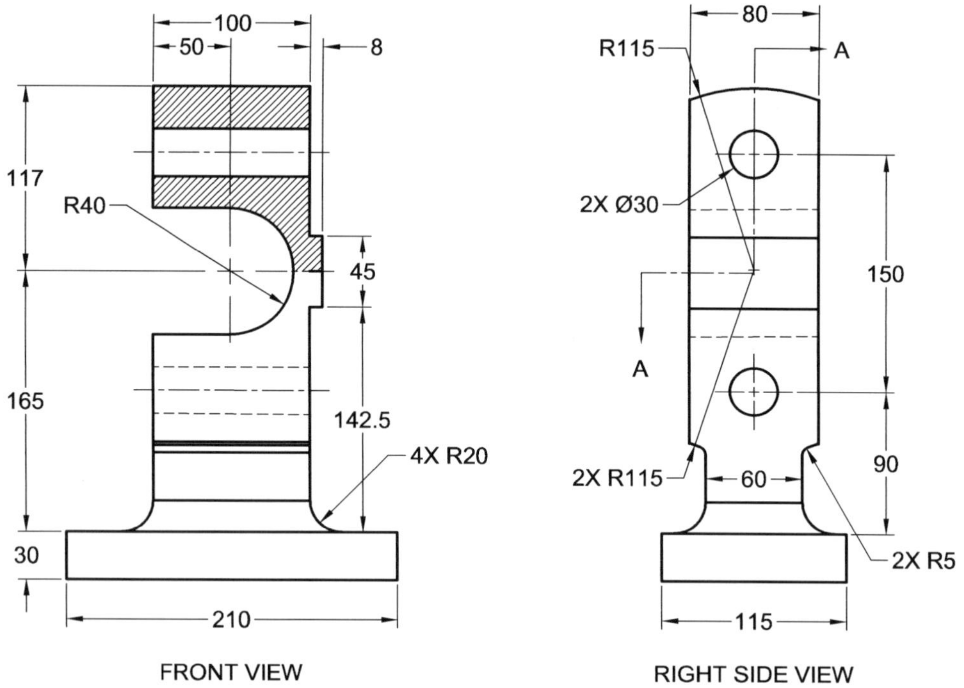

Figure 10-87 *Front view and right-side view of the Body*

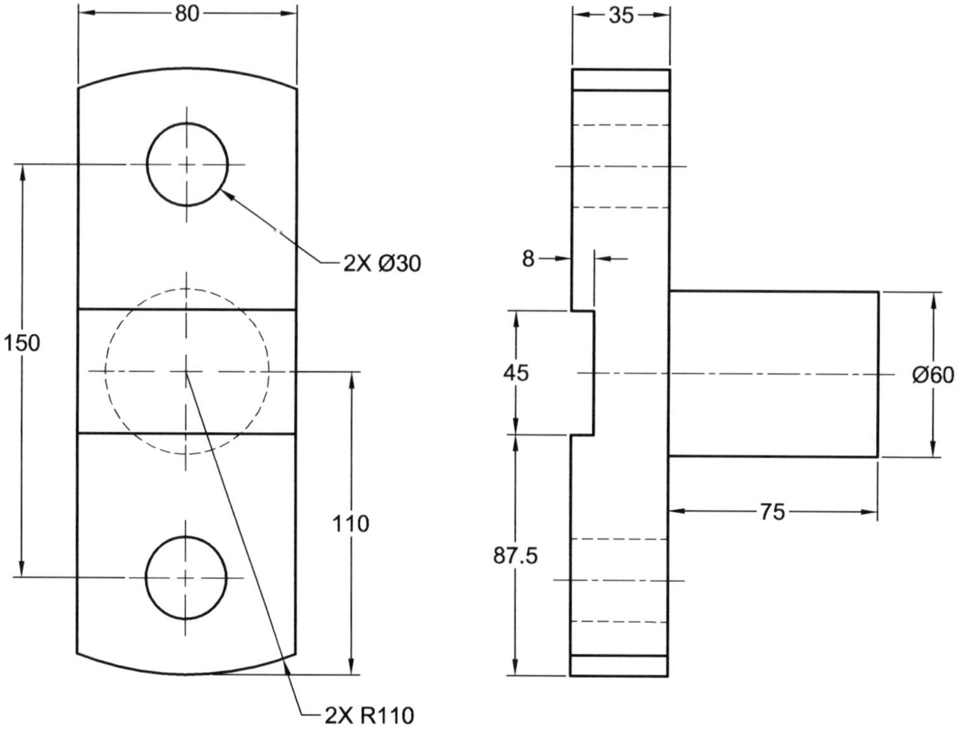

Figure 10-88 *Dimensions of the Piston Rod*

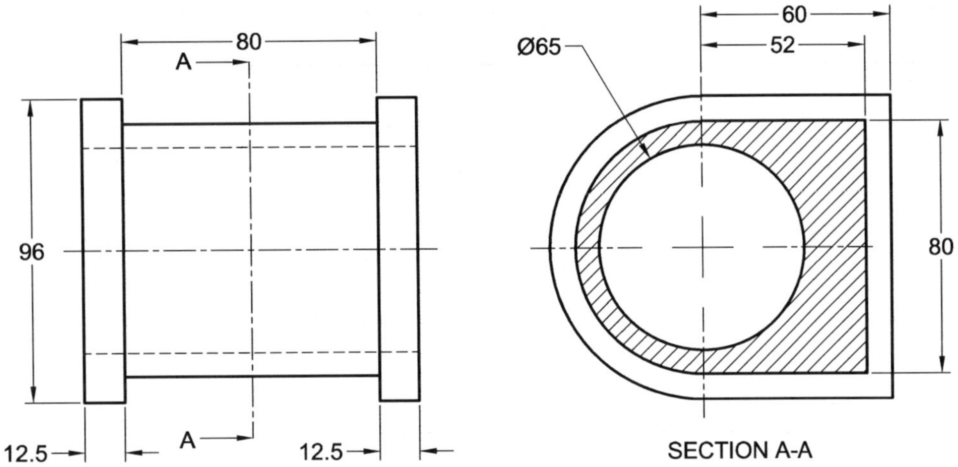

Figure 10-89 Dimensions of the Brass

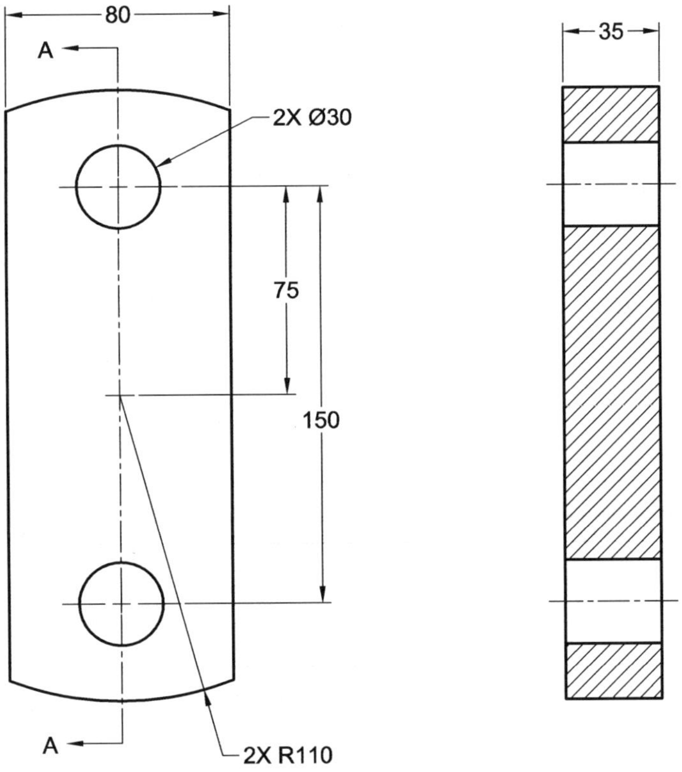

Figure 10-90 Dimensions of the Keep Plate

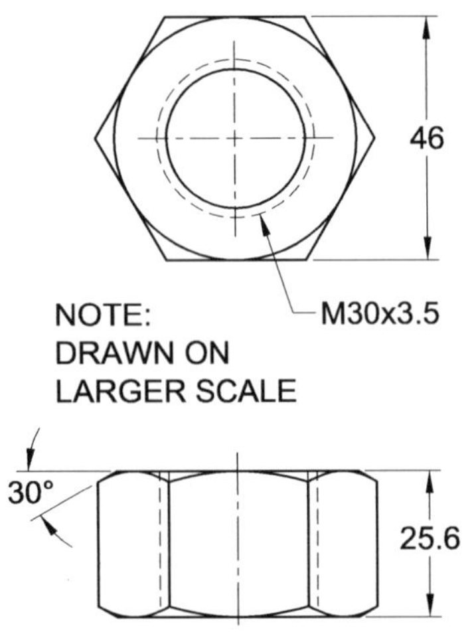

Figure 10-91 Dimensions of the Nut

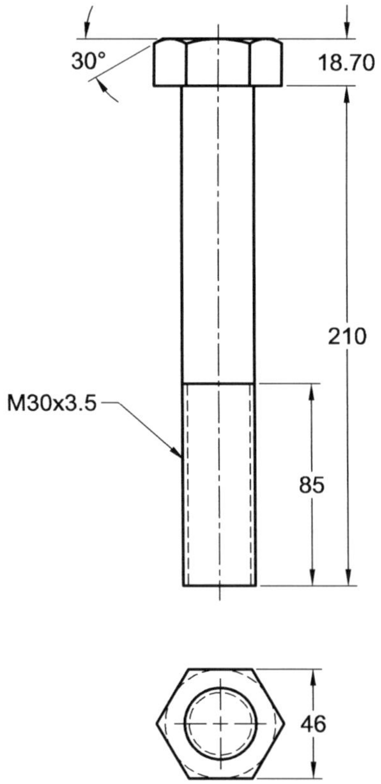

Figure 10-92 Dimensions of the Bolt

Answers to Self-Evaluation Test

1. Master representation, **2.** Graphics representation, **3. Distance**, **4. Centered**, **5.** BOM, **6.** T, **7.** T, **8.** F, **9.** T, **10.** T

Chapter 11

Generating, Editing, and Modifying the Drawing Views

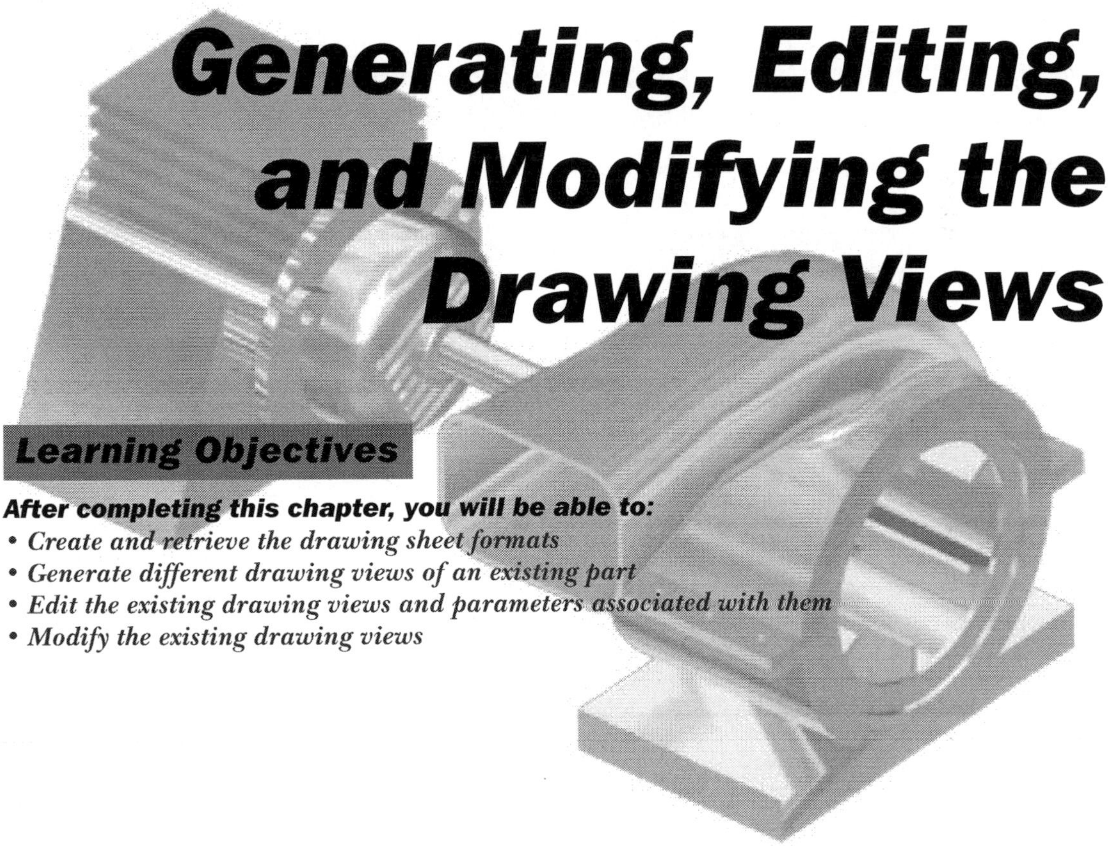

Learning Objectives

After completing this chapter, you will be able to:

- *Create and retrieve the drawing sheet formats*
- *Generate different drawing views of an existing part*
- *Edit the existing drawing views and parameters associated with them*
- *Modify the existing drawing views*

THE DRAWING MODE

In the earlier chapters, you learned about creating the parts in the **Part** mode and assembling different parts in the **Assembly** mode. In this chapter, you will learn to generate the drawing views of the parts and assemblies created earlier. Drawing views are generated in the **Drawing** mode. One of the major advantages of working with this software package is its bidirectional associative nature. This property ensures that any modifications made in the model in the **Part** mode are reflected in the drawing views of the model and vice versa. In Creo Parametric, there are two types of drafting methods: Interactive drafting and Generative drafting. In this chapter, you will learn Generative drafting.

To generate the drawing views, first you need to start a new file in the **Drawing** mode of Creo Parametric. To do so, choose the **New** button from the **Quick Access** toolbar to display the **New** dialog box. Select the **Drawing** radio button in the **New** dialog box, as shown in Figure 11-1.

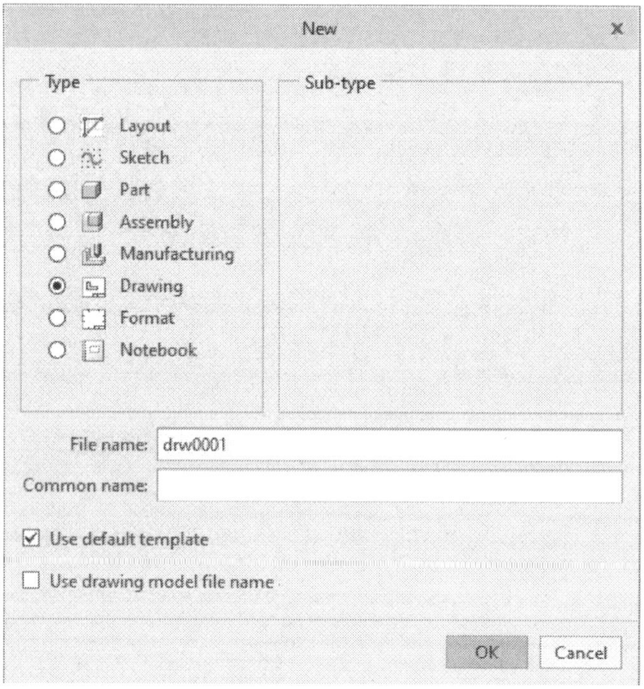

Figure 11-1 *The **New** dialog box*

Specify the name of the drawing in the **File name** edit box and then choose the **OK** button to display the **New Drawing** dialog box, as shown in Figure 11-2.

New Drawing Dialog Box

The **New Drawing** dialog box is used to specify the template that will be used while starting a new file in the **Drawing** mode. The options available in this dialog box are discussed next.

Default Model Area

The **Default Model** area is used to specify the name of the model whose drawing views you want to generate. You can specify the name of the model in the name edit box or select the model using the **Open** dialog box that will be displayed when you choose the **Browse** button. If a part or assembly file is already opened in the current session then the name of that model will be displayed by default in the edit box of the **Default Model** area.

Specify Template Area

The **Specify Template** area is used to specify whether you want to use an empty sheet, predefined formats, or the default template available in Creo Parametric. There are three radio buttons in this area. The **Use template** radio button is selected by default. When the **Empty with format** radio button is selected, the dialog box is modified, as shown in Figure 11-3. When you select the **Empty** radio button, the **New Drawing** dialog box is modified, as shown in Figure 11-4.

Orientation Area

The **Orientation** area will be available only when you select the **Empty** radio button from the **Specify Template** area. The buttons in this area are used to specify the orientation of the sheet. You can select a standard size sheet with a portrait or a landscape orientation using the **Portrait** or **Landscape** button. You can also specify a sheet with the user-defined size by choosing the **Variable** button.

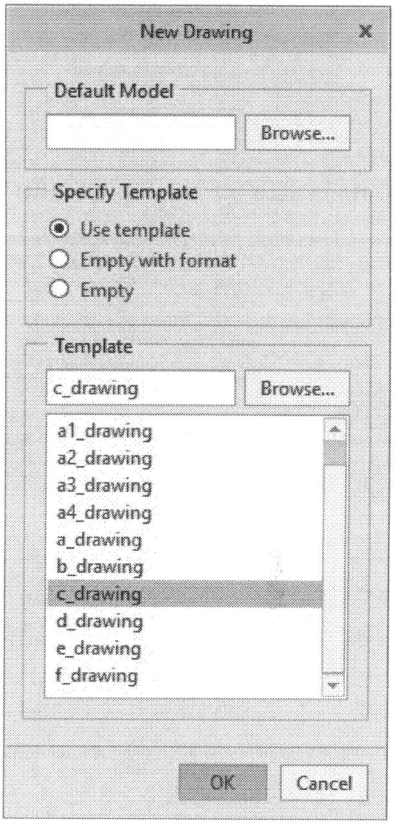

***Figure 11-2** The **New Drawing** dialog box with the **Use template** radio button selected*

Size Area

The options in the **Size** area are used to set the size and the units of the sheet. The **Size** area will be available only when you select the **Empty** radio button from the **Specify Template** area. The options in this area are discussed next.

Standard Size

The **Standard Size** drop-down list is used to select a drawing sheet of standard size. This drop-down list is available only when you choose the **Portrait** or **Landscape** button from the **Orientation** area.

Inches and Millimeters Radio Buttons

These radio buttons are selected to set the standards for the user-defined sheets. You can set the size of the sheet in inches or in millimeters. These buttons are available only when you choose the **Variable** button from the **Orientation** area.

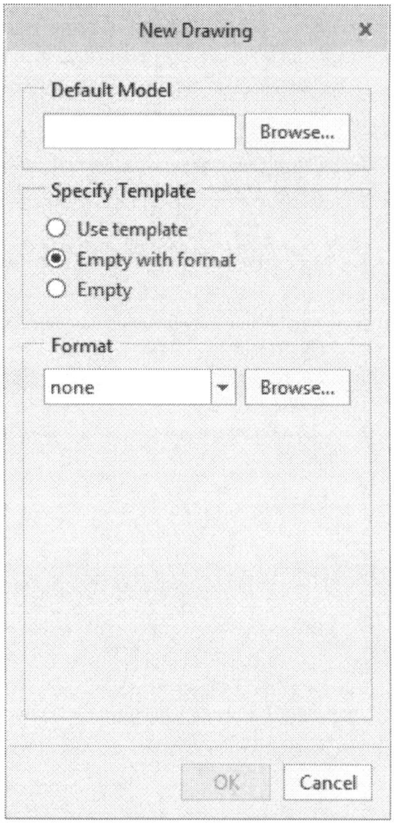

***Figure 11-3** The **New Drawing** dialog box with the **Empty with format** radio button selected*

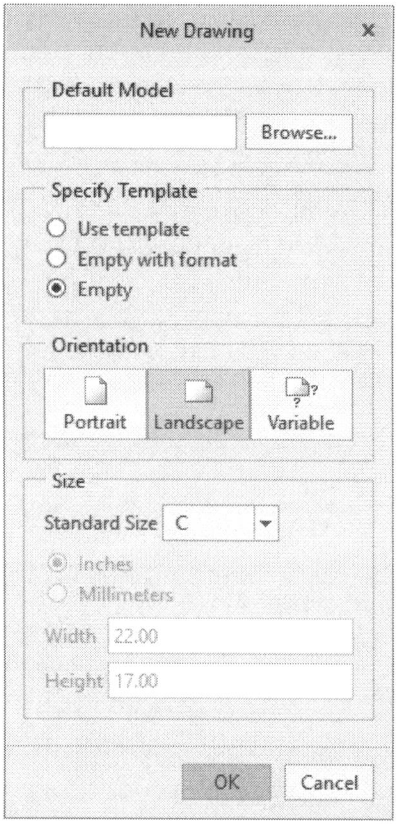

***Figure 11-4** The **New Drawing** dialog box with the **Empty** radio button selected*

Width and Height Edit Boxes

These edit boxes are used to specify the width and the height of the user-defined drawing sheets. These edit boxes are available only when you choose the **Variable** button from the **Orientation** area.

Format Area

The **Format** area will be available only when you select the **Empty with format** radio button from the **Specify Template** area. The options in this area are discussed next.

Format

The **Format** drop-down list is used to select the available formats.

Browse

The **Browse** button is chosen to display the **Open** dialog box for retrieving the drawing formats. By default, there are only eight standard system formats that can be retrieved. However, you can create your own user-defined formats that can be retrieved later.

Choose the **OK** button from the **New Drawing** dialog box to proceed to the **Drawing** mode; a drawing sheet of the specified size and orientation will be placed on which you can generate the drawing views. The Drawing user interface of Creo Parametric has been enhanced to simplify your tasks and make it easy to perform. Additionally, the **Ribbon** has been introduced to display all frequently used tools collected logically in various groups and tabs. If required, you can increase the drawing area by minimizing the **Ribbon**. To do so, right-click on a tab in the **Ribbon** and select **Minimize the Ribbon** check box.

GENERATING DRAWING VIEWS

In Creo Parametric, the first view that you need to generate is the general view. This view mostly acts as the parent view for the remaining views. Different methods used for generating various views are discussed next.

Generating the General View

Ribbon: Layout > Model Views > General View

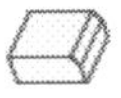
General View

The General view is the first view that is generated on the drawing sheet. This view can be the top, front, right, left, bottom, back, trimetric, isometric view, or any user-defined view of the model. To generate the general view, choose the **General View** tool from the **Model Views** group in the **Layout** tab of the **Ribbon**; the **Select Combined State** dialog box will be displayed, as shown in Figure 11-5. In this dialog box, if you select the **No Combined State** option from the Combined state names collector and then choose the **OK** button, you will be prompted to select a center point to place the drawing view in the drawing sheet. Select the center point; the preview of the drawing with default orientation will be displayed in the drawing sheet and also the **Drawing View** dialog box will be displayed, as shown in Figure 11-6.

***Figure 11-5** The **Select Combined State** dialog box*

If you select the **DEFAULT ALL** option from the Combined state names collector of the **Select Combined State** dialog box and choose the **OK** button, you will be prompted to select a center point to place the drawing view in the drawing sheet. Select the center point; preview of the drawing with the default orientation will be displayed in the drawing sheet and also the **Drawing View** dialog box will be displayed, refer to Figure 11-6.

Tip

*1. If you have not specified any model in the **Default Model** area of the **New Drawing** dialog box then choose **Layout > Model Views > General View** from the **Ribbon**; the **Open** dialog box will be displayed. You can select the model from the **Open** dialog box.*

*2. Creo Parametric automatically selects the model in the **Default Model** area of the **New Drawing** dialog box. This model is selected from the present session of Creo Parametric. If the model selected automatically is not the required model, then change the model by choosing the **Browse** button in the **New Drawing** dialog box.*

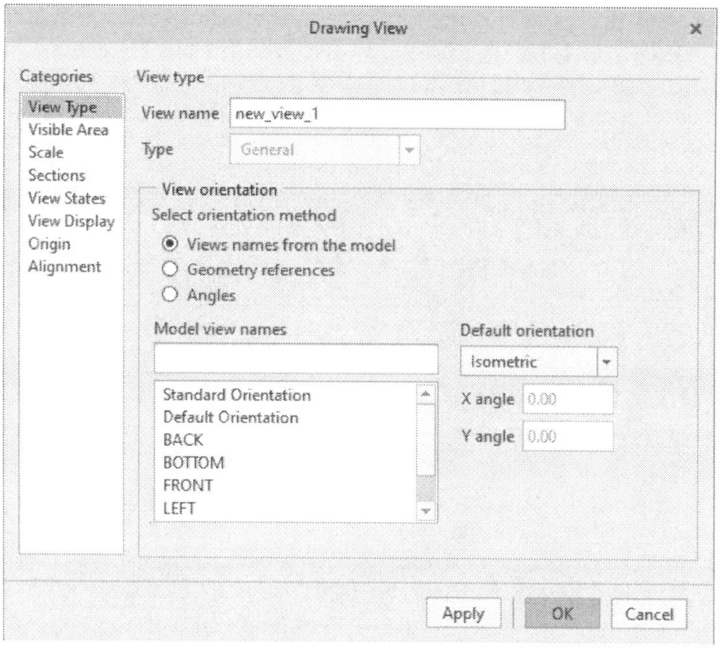

Figure 11-6 *The **Drawing View** dialog box*

Note

*If you do not want to be prompted for selecting a combination state, select the **Do not prompt for combined state** check box from the **Select Combined State** dialog box.*

The **Drawing View** dialog box contains the options that are required to generate a general view. The **Categories** list box in the dialog box lists all parameters that are required to generate a drawing view in Creo Parametric. The **View Type** option is selected by default in the **Categories** list box and its related options are displayed on the right side of the **Categories** list box. These parameters/options are discussed later in this chapter.

The procedure to generate a general view is given below:

1. Choose **Layout > Model Views > General View** from the **Ribbon**; you will be prompted to select a center point.

Note

*If the selected model is an assembly model, then you can create default exploded view of the assembly by choosing the **DEFAULT ALL** option from the **Select Combined State** dialog box which will be displayed on choosing the **General View** tool.*

2. As soon as you select the center point, the **Drawing View** dialog box will be displayed and the default view of the model will be displayed on the drawing sheet. In the **Type** drop-down list, the **General** option is selected by default and in the **View orientation** area, the **Views names from the model** radio button is selected by default. This radio button allows you to select the standard orientations of the model from the **Model view names** list box. If you select

the **Geometry references** radio button, you can select the planar faces or the datum planes to orient in a particular orientation. If you select the **Angles** radio button, the **Angle value** edit box and the **Rotation reference** drop-down list will be displayed below the **Reference Angle** area. You can change the orientation of the model by specifying the required angle value in the **Angle value** edit box.

3. Select the view that you want to generate as the general view from the list provided below the **Model view names** list box.

4. After selecting the required options, choose the **Apply** button and then choose the **OK** button from the **Drawing View** dialog box.

Tip

*To set the model orientation using the **Geometry references** radio button, select the view type from the **Reference 1** drop-down list and then from the model, select the face or the datum plane that you need to see from the selected view. Similarly, select an option from the **Reference 2** drop-down list and then select a datum plane or a face from the model. For example, if you select **Top** option from the **Reference 1** drop-down list, then from the model, you need to select the face or the datum plane that you need to keep in the top orientation in the drawing view.*

Generating the Projection View

Ribbon: Layout > Model Views > Projection View

The projection views are the orthographic views generated by projecting lines normal to the existing view, refer to Figure 11-7. Before generating a projected view, you need to make sure that at least one parent view is already present on the drawing sheet. To generate a projected view, choose **Layout > Model Views > Projection View** from the **Ribbon**. Now, move the cursor to a location where you need to place the view and specify a point on this location. Depending on the point selected to place the view, the resulting view will be a top, bottom, front, back, right-side, or left-side view.

The scale factor of these views will be the same as that of the parent view from which they are projected. If more than one view exists that can be the parent view of the projection view, then you will be prompted to select a projection parent view for the new view.

Generating the Detailed View

Ribbon: Layout > Model Views > Detailed View

Detailed views are used to provide the enlarged view of a particular portion of an existing view. To generate a detailed view, choose the **Detailed View** tool from the **Model Views** group of the **Layout** tab in the **Ribbon**; you will be prompted to specify the center point for detail on an existing view. Define a center point on the view whose detail needs to be generated; you will be prompted to sketch a spline without intersecting other splines to define an outline. Draw a closed spline to define the outline of the detail view and exit the drawing mode by clicking the middle mouse button. Next, you need to specify the placement point for placing the detail view. You can change the name, scale, or the reference point of the parent view and can also select the boundary type of the detailed view using the options from the **Drawing View** dialog box, which will be displayed when you double-click on the detailed view.

Figure 11-7 shows a drawing sheet with various drawing views.

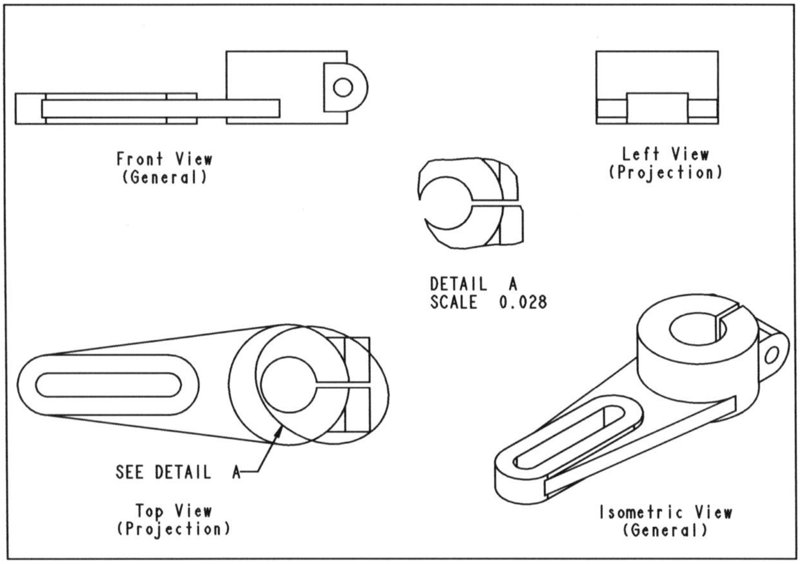

***Figure 11-7** Drawing sheet with various drawing views*

Generating the Auxiliary View

Ribbon: Layout > Model Views > Auxiliary View

The auxiliary views are generated by projecting normal lines from a specified edge, axis, or datum plane of an existing view. The view scale will be the same as that of the parent view. To generate an auxiliary view, choose the **Auxiliary View** tool from the **Model Views** group of **Layout** tab in the **Ribbon**; you will be prompted to select an edge, axis, or datum plane from the front surface of the main view. Select the edge or surface normal to which you need to place the generated view; a rectangular box will be attached to the cursor. Select a point on the drawing sheet to place the view on the drawing sheet. Figure 11-8 shows the edge to be selected as reference to generate the drawing view and the resulting auxiliary view.

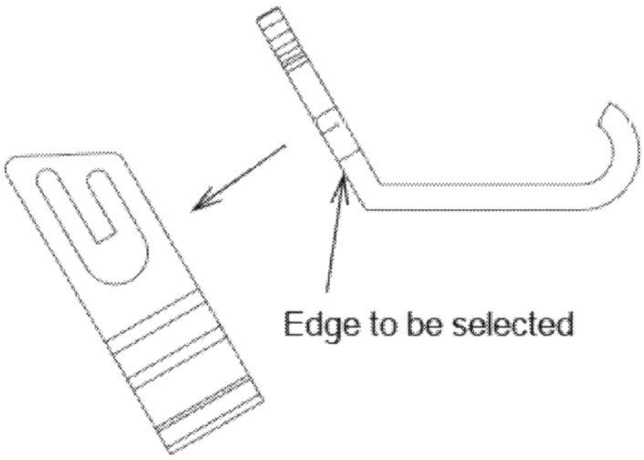

***Figure 11-8** Resulting auxiliary view*

Generating the Revolved Section View

Ribbon: Layout > Model Views > Revolved View

A revolved section view is the view that is generated from an existing view by revolving the section through an angle of 90 degrees about the cutting plane and then projecting it along the length. Remember that the cutting plane of the revolved section view is normal to the parent view. The procedure to generate a revolved section view is discussed next.

1. Choose the **Revolved View** tool from the **Model Views** group in the **Layout** tab of the **Ribbon**; you will be prompted to select a parent view for the revolved section.

2. Select the view from the drawing sheet that will be defined as the parent view for generating the revolved view.

3. Next, you need to select a center point on the drawing sheet to place the view. Select a point anywhere in the drawing sheet; the **Drawing View** dialog box will be displayed. The **Create New** cross-section option is selected by default from the **Cross-section** drop-down list in the **Revolved view properties** area of the **Drawing View** dialog box. The **XSEC CREATE** menu is also displayed on the lower right corner of the screen.

Note

*The **Create New** option is selected by default in the **Revolved view properties** area of the **Drawing View** dialog box, only if there is no existing revolved section present.*

4. Choose **Planar > Single > Done** from the **XSEC CREATE** menu; the message input window will be displayed prompting you to specify the name of the cross-section. Enter the name of the cross-section and accept it; the **SETUP PLANE** menu will be displayed and you will be prompted to select a planar surface or datum plane.

5. Select the plane along which the view will be sectioned, as shown in Figure 11-9. The resulting section view will be placed in-line with the section plane.

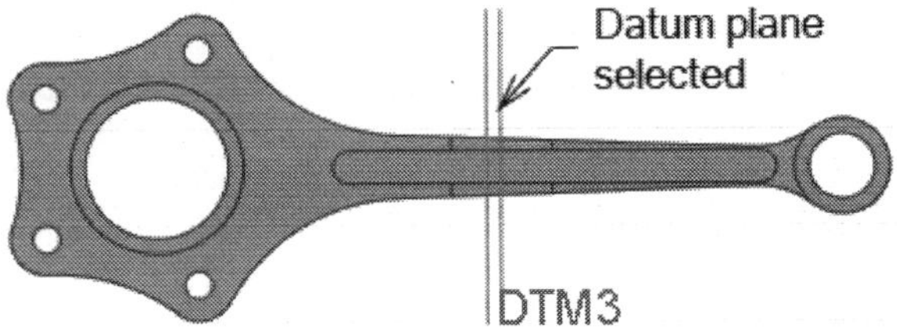

Figure 11-9 *Datum plane for the revolved section view*

6. Choose the **Apply** button and then the **OK** button to exit the **Drawing View** dialog box. The resulting sectioned view is shown in Figure 11-10.

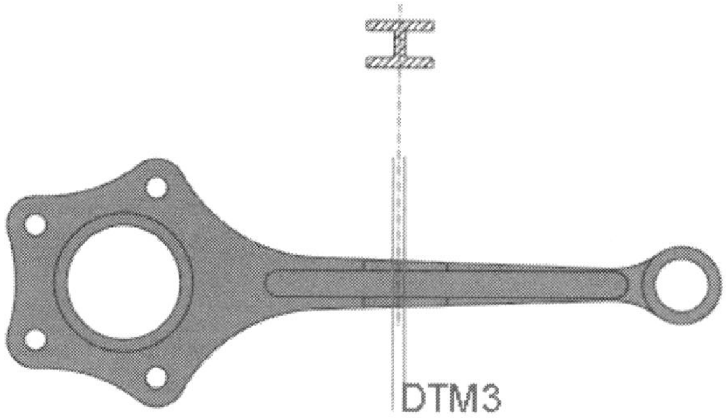

Figure 11-10 Resulting revolved section view

Note
Copying and aligning of a view will be discussed later in this chapter.

Drawing View Dialog Box Options

The remaining types of drawing views that you can generate in Creo Parametric require the options available in the **Drawing View** dialog box. Therefore, it is important for you to understand these options before proceeding further.

View Type Option

When you invoke the **Drawing View** dialog box, the **View Type** option is automatically selected in the **Categories** list box. The **View name** edit box is used to enter the name of the drawing view. The **View orientation** area lists the options to orient the drawing view. These options are the standard options of orienting the model.

Visible Area Option

The **Visible Area** option is used to set the display of the view. You can display a drawing view as a full view, a partial view, a half view, and a broken view. On choosing the **Visible Area** option, the **Drawing View** dialog box will be displayed, as shown in Figure 11-11. The options in the **View visibility** drop-down list of this dialog box are discussed next.

Full View

The **Full View** option is selected by default in the **View visibility** drop-down list. This option can be combined with any view type to generate a drawing view displaying the complete view.

Half View

The **Half View** option can be used on the projection, auxiliary, or general view to generate a drawing view that displays only the half view of the part. Generally, an existing view is selected to display the half view. For generating a half view, double-click on an existing view to display the **Drawing View** dialog box. In the **Categories** list box, choose the **Visible Area** option to invoke the **Visible area options** area and then select **Half View** from the **View visibility** drop-down list. On doing so, you will be prompted to select a reference plane that will be used to remove half of the drawing view. The reference plane can be a datum plane or a planar surface, and must be normal to the screen in the selected view. Select the reference plane; an arrow will be displayed attached to the selected plane. This arrow indicates the portion of the view to be removed. You can also flip the direction of the arrow by using the **Flips the side to keep** button available below the **Half view reference plane** collector in the dialog box. Generally, this type of view is generated for symmetric parts. Therefore, you can specify the type of symmetry line using the **Symmetry line standard** drop-down list. Figure 11-12 shows the reference plane to be selected and Figure 11-13 shows the resulting half view.

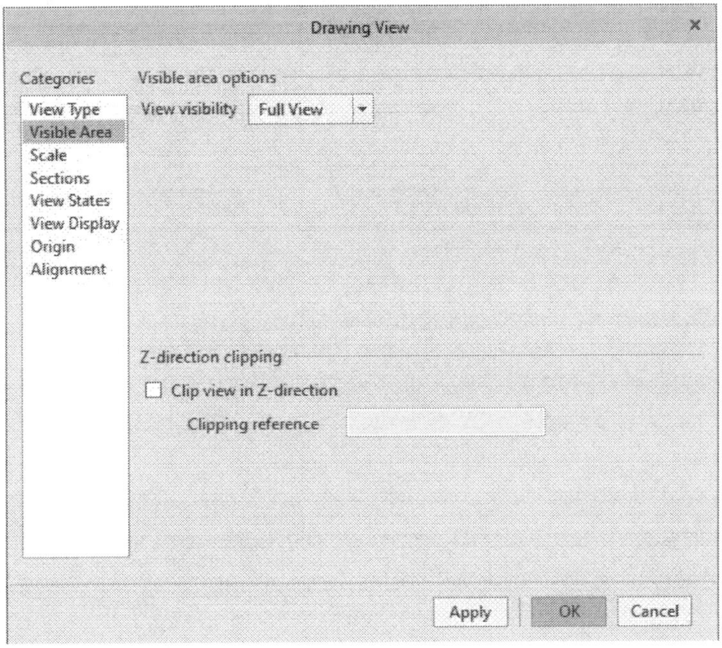

*Figure 11-11 The **Drawing View** dialog box with the **Visible Area** option chosen*

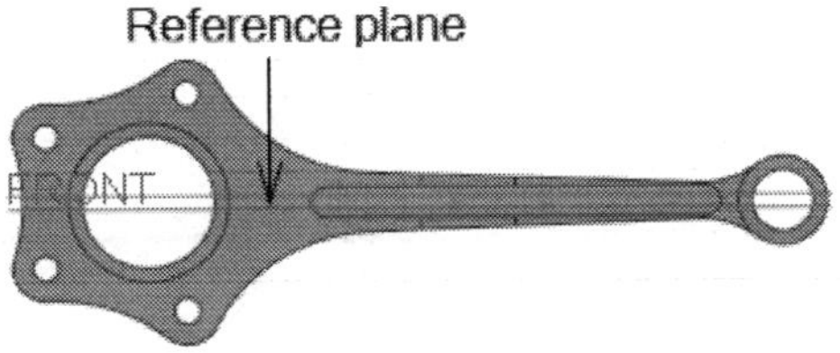

Figure 11-12 Reference plane to be selected to generate a half view

Figure 11-13 Resulting half view

Partial View

The **Partial View** option can be used on the projection, auxiliary, revolved, or general views to generate a view that displays a specified portion of the view. To convert an existing view to a partial view, double-click on that view. Choose the **Visible Area** option from the **Categories** list box and select the **Partial View** option from the **View visibility** drop-down list; you will be prompted to select a new reference point. Select a reference point on the selected view and you will be prompted to draw a spline that will define the outer boundary on current view, refer to Figure 11-14. Select the center point on the view and then draw the spline. Press the middle mouse button and exit the spline creation. Now, choose the **Apply** button from the

Drawing View dialog box. You will notice that only the area of the view inside the spline is retained and the rest of the portion of the drawing view is cropped, refer to Figure 11-15.

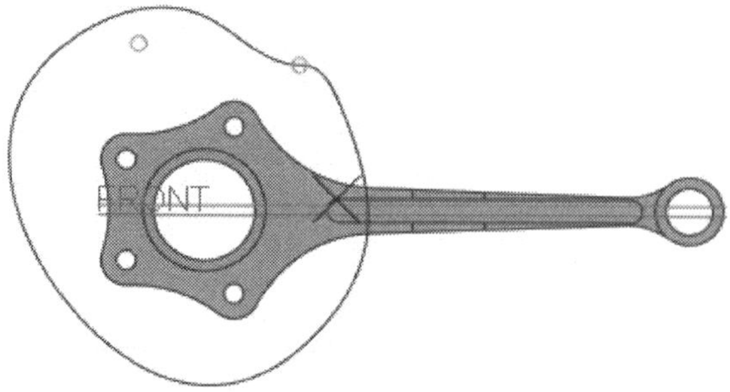

Figure 11-14 Spline drawn on the view

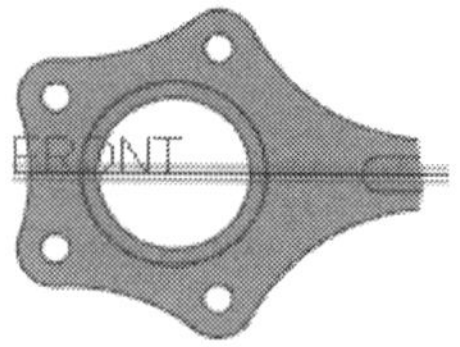

Figure 11-15 Cropped drawing view

Broken View

This type of view is used for the parts having a high length to width ratio. The **Broken View** option can be used on the projection or general view to generate a view that is broken along the horizontal or vertical direction using the horizontal or vertical line. To generate a broken view, select the **Broken View** option from the **View visibility** drop-down list. Choose the **Add break** button in the **Drawing View** dialog box to add break lines. Select an edge from the view and move the cursor vertically and then select a point on the screen to draw the first vertical break line. Move the cursor and select a point up to which you want to break the view, refer to Figure 11-16. You can also specify the style of the break line using the **Break Line Style** drop-down list available on the extreme right of the list box. Now, choose the **Apply** button from the **Drawing View** dialog box to create the resultant broken view and then choose the **OK** button to exit from it. Figure 11-17 shows the resultant broken view.

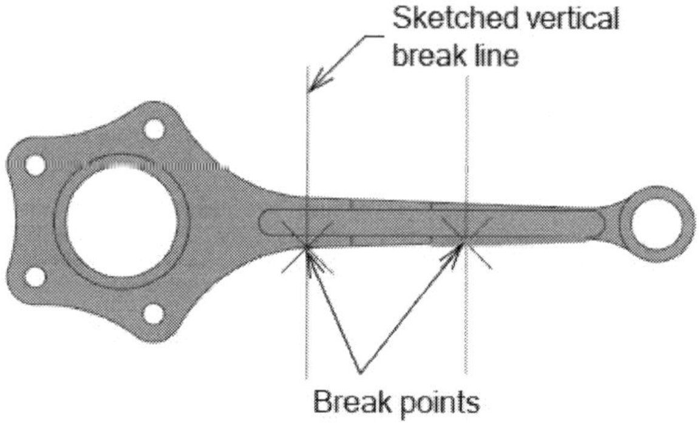

Figure 11-16 References to break the view

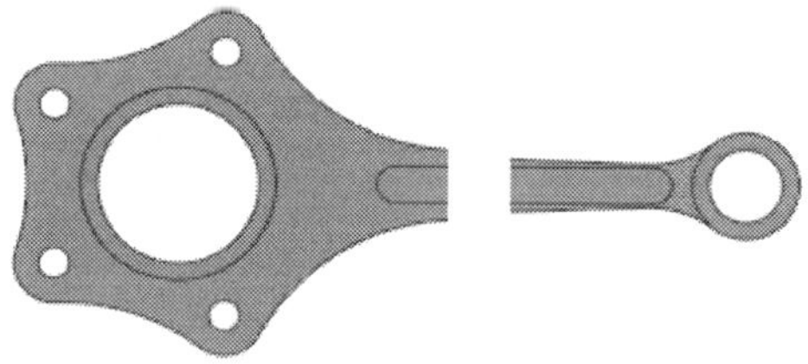

Figure 11-17 Resultant broken view

Figure 11-18 shows various views that have already been discussed in this section.

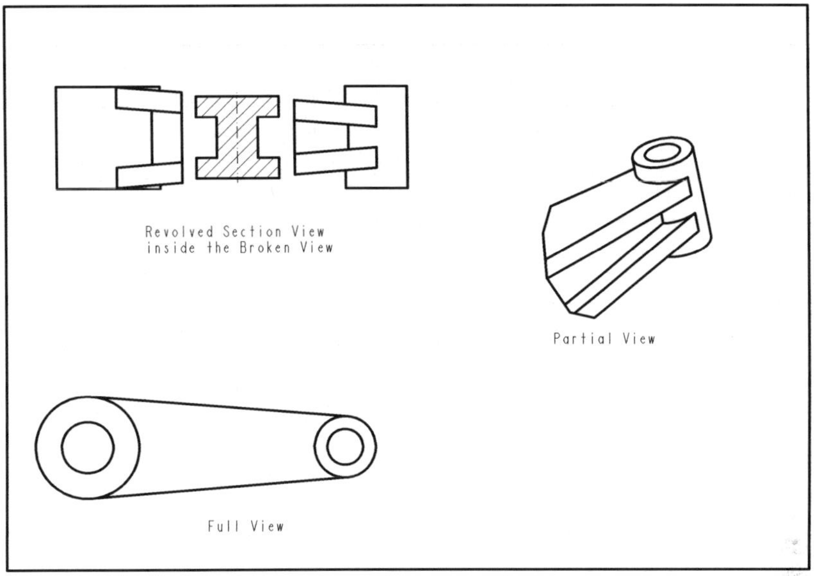

Figure 11-18 *Different drawing views*

Scale Option

When you select the **Scale** option, the **Scale and perspective options** area will be displayed on the right of the **Categories** list box. The **Default scale for sheet** radio button is selected by default. Therefore, all generated views are scaled with the default sheet scale. The second radio button is **Custom scale**. This radio button, when selected, allows you to enter a scale for the drawing view. When you modify the scale factor of a view, the view associated with this parent view is also scaled.

The third radio button is **Perspective**. This radio button is selected to generate a perspective view. After selecting the **Perspective** radio button, you can specify the distance from the eye and the diameter of the view circle. Figure 11-19 shows a perspective view of a model.

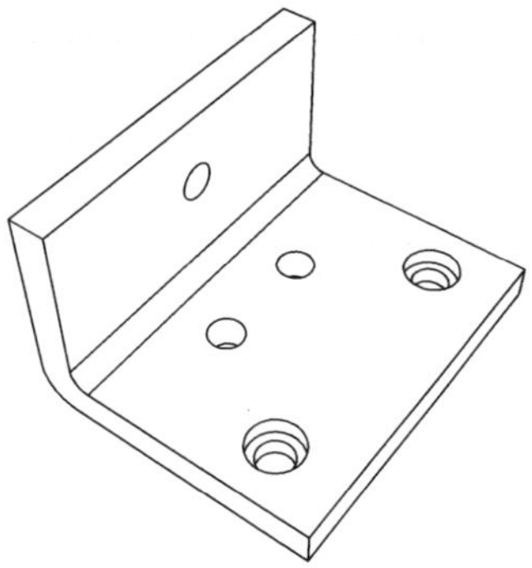

Figure 11-19 *A general perspective view*

Sections Option

The section views are generally used to display drawings of a part that is complicated from inside. As it is not possible to display the inside of the part using the conventional views, therefore these views are cut (sectioned) using a datum plane or a planar surface and the resulting section view is displayed. When you choose the **Sections** option from the **Categories** list box in the **Drawing View** dialog box, the **Section options** area will be displayed. In this area, the **No section** radio button is selected by default. Select the **2D cross-section** radio button and then choose the **Add cross-section to view** button to activate the options to create a section. The options in the **Section options** area and the methods used for creating the section plane are discussed next.

Full Section View

Consider a part that is cut throughout its length, width, or height and the cut portion is removed from the display. The remaining portion, when projected normal to the cutting plane, displays the full section view.

To generate a full section view, you first need to generate a projected view from one of the views existing on the drawing sheet. After generating the projected view, double-click on it to invoke the **Drawing View** dialog box. Choose the **Sections** option from the **Categories** list box. The options available for generating the section view are displayed on the right of the **Drawing View** dialog box. Select the **2D cross-section** radio button to define a section plane. Now, choose the **Add cross-section to view** button to add a section plane. The **Full** option in the **Sectioned Area** drop-down list allows you to create a full section view.

When you choose the **Add cross-section to view** button, the **XSEC CREATE** menu will be displayed. Choose **Planar > Single > Done** from the **XSEC CREATE** menu; you will be prompted to specify the name of the cross-section. Specify the name of the cross-section in the message input window. After choosing **OK** from the window, the **SETUP PLANE** menu will be displayed and you will be prompted to select a planar surface or a datum plane. Select the datum plane from the other view; the name of the section view will be displayed in the **Name** drop-down list. Scroll to the right in the dialog box and click once in the **Arrow Display** collector to invoke the selection mode and select the view to display the section view arrows. You can also flip the material side of the section view using the **Flip** button. Choose the **Apply** button from the **Drawing View** dialog box. Figure 11-20 shows the top and section views.

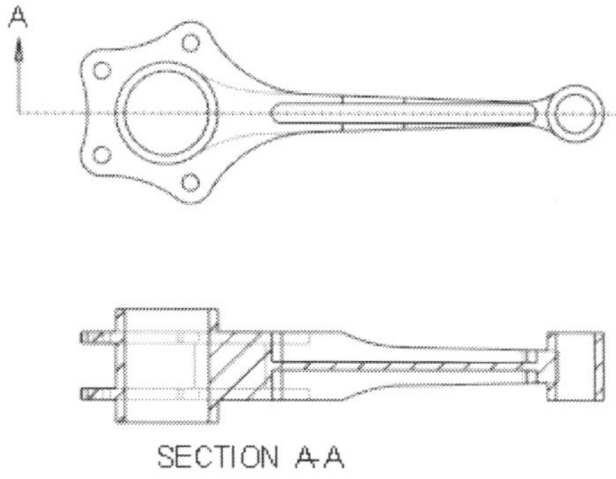

Figure 11-20 *Top and section views*

In this section, you learned to generate a full section view by defining a planar section plane. Now, you will learn to generate a full section view by defining an offset section plane. You need to make sure that before generating a section view by defining an offset section plane, the part file of the current model should be closed.

Invoke the **Drawing View** dialog box by double-clicking on an already generated projected view. Now, choose the **Sections** option from the **Categories** list box on the left of this dialog box. Select the **2D cross-section** radio button and choose the **Add cross-section to view** button to define the section plane. If there are some existing section planes, select the **Create New** option from the **Name** drop-down list; the **XSEC CREATE** menu will be displayed. Choose **Offset > Both Sides > Single > Done** from the **XSEC CREATE** menu; the message input window will be displayed. Specify the name of the cross-section and choose the **OK** button.

The model will be displayed in a sub-window on the left of the screen. The **SETUP SK PLN** menu will also be displayed and you need to select a sketching plane for drawing the sketch that represents an offset section plane. Select the sketching plane from the subwindow. Choose the **Okay** option from the **DIRECTION** submenu; the **SKET VIEW** submenu will be displayed. Choose the **Default** option from this submenu. Draw the sketch of the offset section using the sketch tools available in the **Sketch** menu of the subwindow, refer to Figure 11-21. After drawing the sketch, choose **Sketch > Done** from the menu bar to exit the sketcher environment.

Note
The sketch created for the section must be a continuous line.

Click once on the **Arrow Display** collector to invoke the selection mode and then select the view in which the section arrows will be displayed. Choose the **Apply** button from the **Drawing View** dialog box. Figure 11-21 shows the top view, offset sectioned front view, and sectioned general view of a part.

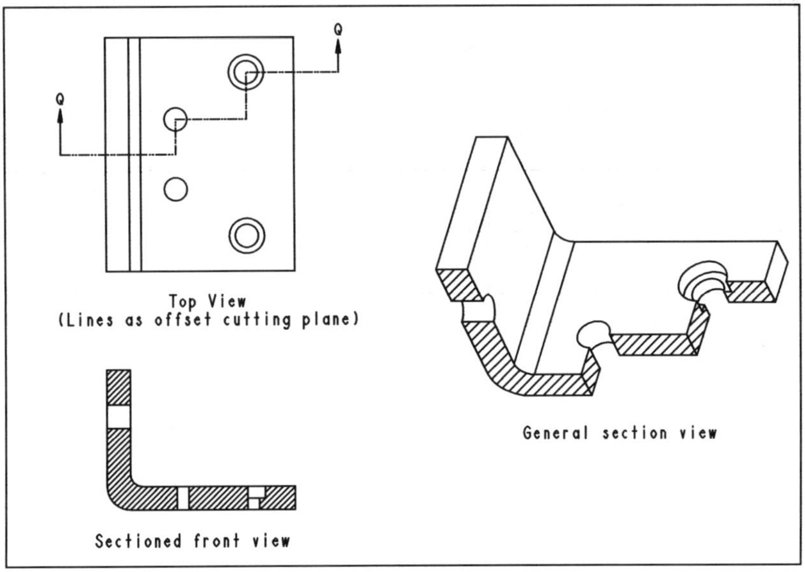

Figure 11-21 *The section views created using the offset section plane*

To generate an isometric section view, generate the general view and switch the orientation of the general view to isometric. Now, select the **Sections** option from the **Categories** list box and sketch a section plane using the sub-window as discussed earlier. Note that you cannot select a datum plane to define the section plane for an isometric section view. Figure 11-22 shows the top, full section front, and full section isometric views.

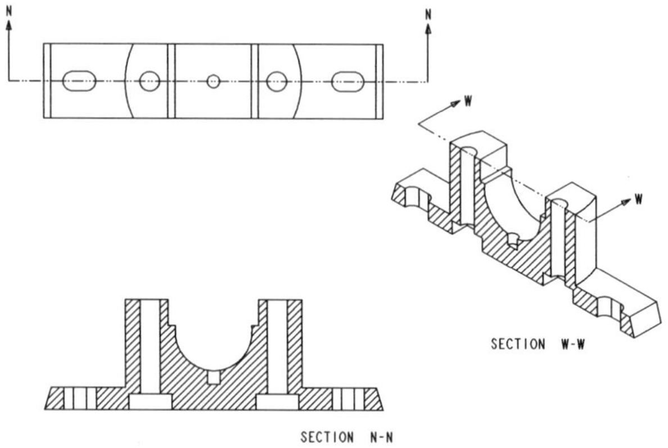

Figure 11-22 *Top view, full section front view, and full section isometric view*

Half Section View

Consider a part that is cut half way through the length, width, or height and the front cut portion (front quarter) is removed from the display. This part when projected is called the half section view. In this projected view, only half of the part will be displayed sectioned and the other half of the part will be displayed as it is. To generate the half section view, you need to select the **Half** option from the **Sectioned Area** drop-down list while specifying the parameters for sectioning the view. After selecting this option, you are prompted to select the reference plane for the half section creation. Select the plane from the drawing view; you will be prompted to pick the side to keep. Select a point on the side of the section view that you need to keep. Select the parent view where the section arrow will be placed and choose the **Apply** button from the **Drawing View** dialog box. Figure 11-23 shows the resulting half section view.

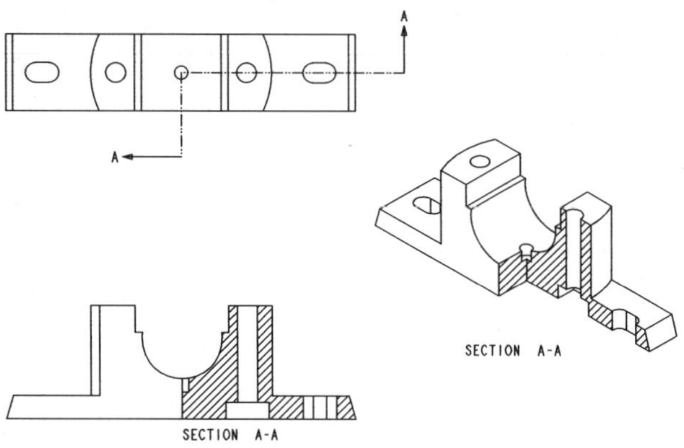

Figure 11-23 *Top view, half section front view, and half section isometric view*

Local Section View

The Local section view is used when you want to show a particular portion of the view in the section, and at the same time, not section the remaining view. The local section area is specified by drawing a spline around it. To generate the local section view, you will first generate the full section view, as discussed earlier. After generating the full section view, select the **Local** option from the **Sectioned Area** drop-down list. Next, select a point on the selected view and draw a spline that will define the area of the drawing that needs to be sectioned.

To draw a local section isometric view, you need to draw a section plane and then choose the **Local** option. Define a center point for the local section and sketch the spline. Figure 11-24 shows the local section front view and the local section isometric view.

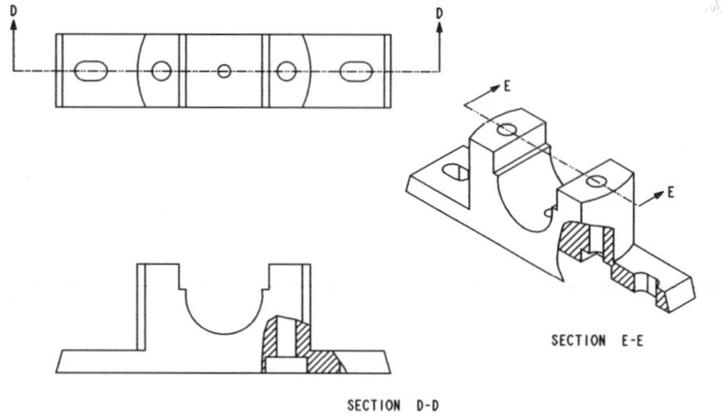

Figure 11-24 *Top view, local section front view, and local section isometric view*

Model edge visibility

Creo Parametric provides you two options for displaying the edges of the section drawing views. If the **Total** radio button is selected in the **Section options** area of the **Drawing View** dialog box, a section view is created that displays all visible edges of the section view, in addition to the section area. This option can be combined with the full, half, local, full (unfold) and full (aligned) views. Figure 11-25 shows the front and total sectioned left view.

If the **Area** radio button is selected, then only the sectioned area of the section view is displayed and no other edges of the view are displayed in the area cross-section view, refer to Figure 11-26. This option can be combined with the full, full (unfold) and full (aligned) views.

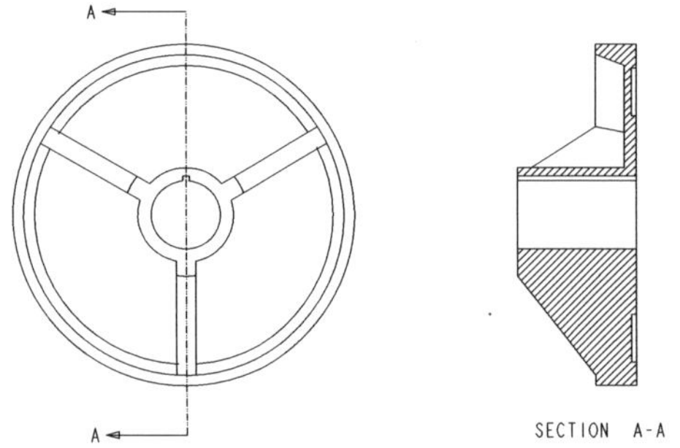

Figure 11-25 The front view and the total sectioned left view

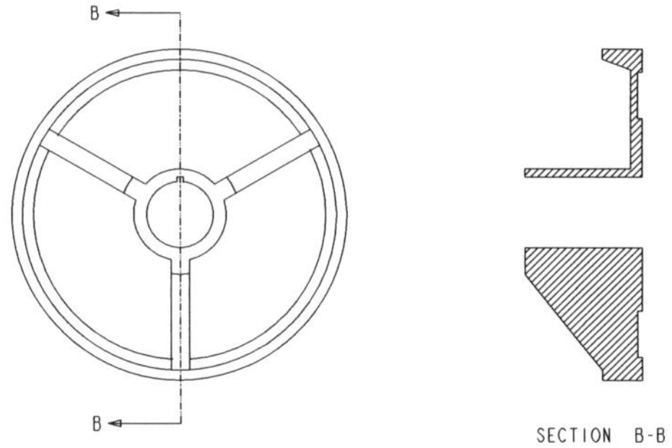

Figure 11-26 The front view and the area cross-section view

Full(Aligned)

This view is used to section those features that are created at a certain angle to the main section planes. You can align sections to straighten these features by revolving them about an axis that is normal to the parent view. Generally for this section view, the section plane is sketched. Remember that the axis about which the feature is straightened should lie on all cutting planes. This means that the lines that are used to sketch the section plane should pass through the axis about which the feature will be revolved. Figures 11-27 and 11-28 show these views when the **Area** radio button is selected. Figures 11-29 and 11-30 show these views when the **Total** radio button is selected.

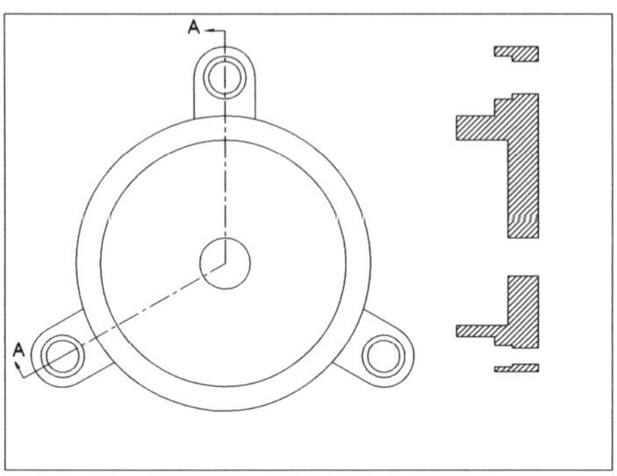

Figure 11-27 The area cross-section view with normal lines of projection

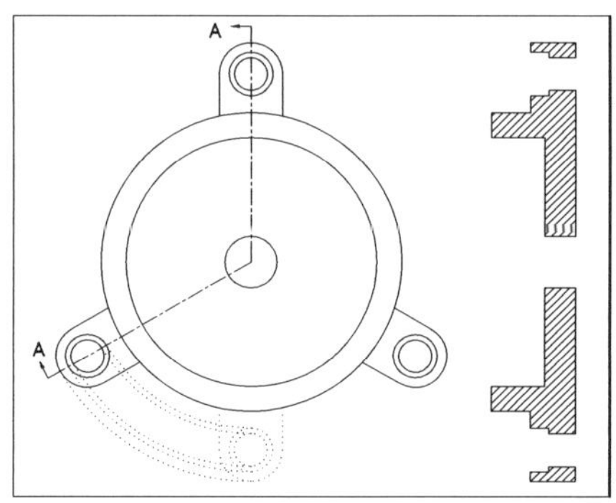

Figure 11-28 The area cross-section view with aligned lines of projection

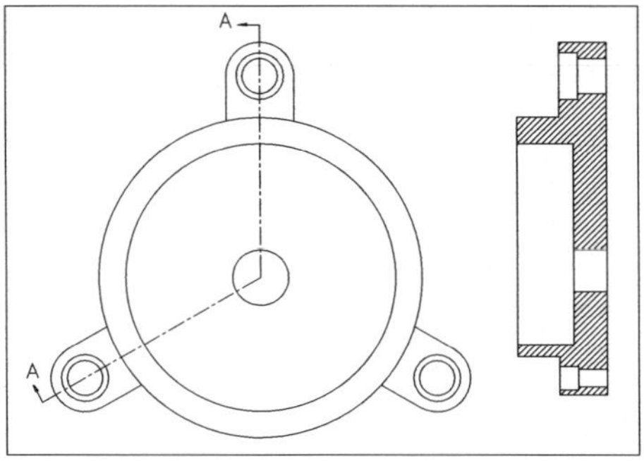

Figure 11-29 *The total cross-section view with normal lines of projection*

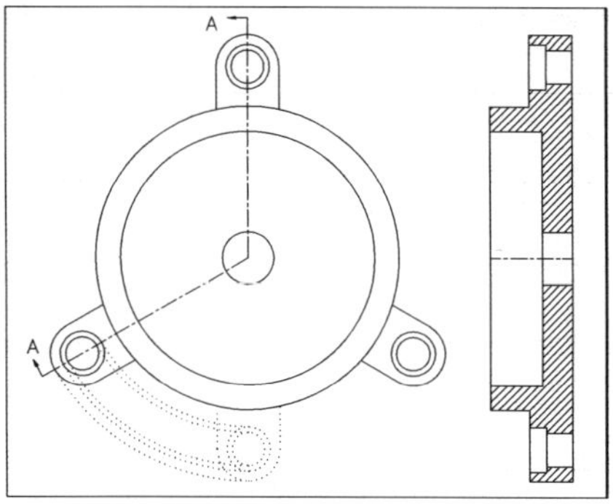

Figure 11-30 *The total cross-section view with aligned lines of projection*

Full(Unfold)

The **Full(Unfold)** option is used to generate the section view by unfolding the section surface of an offset section view. This type of view is only generated using a general view. To generate this type of view, first you need to generate a general view in the required orientation. You can place this general view anywhere on the drawing sheet. Now, choose the **Sections** option from the **Categories** list box on the left of the **Drawing View** dialog box. Now, define the section plane using the offset option choose the **Full(Unfold)** option from the **Sectioned Area** drop-down list and complete the creation of the full unfold section view. Figure 11-31 shows the model and the section plane. Figure 11-32 shows an offset section view and a full unfold section view.

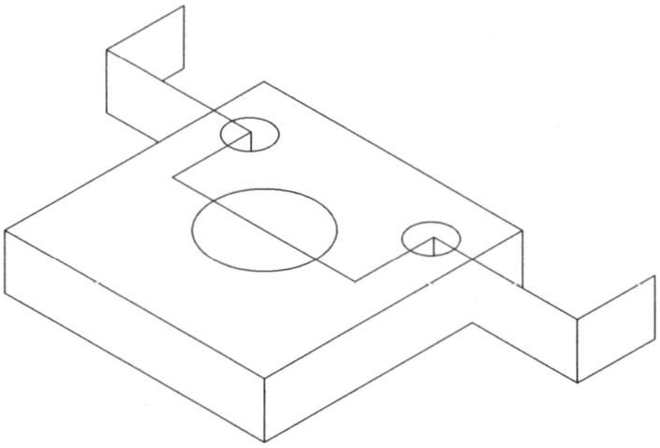

Figure 11-31 *The model and the section plane*

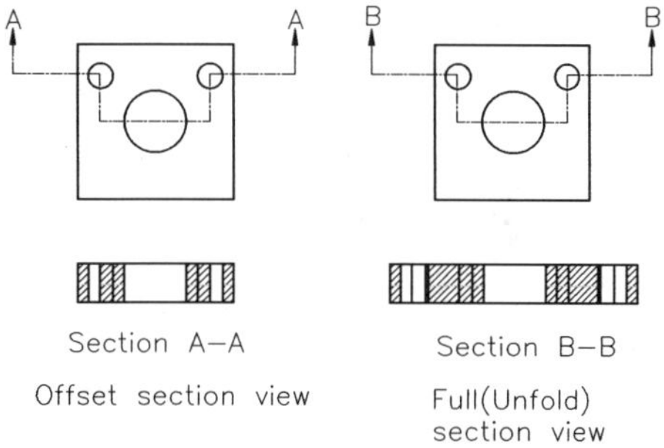

Figure 11-32 *Offset section view and full unfold section view*

Once you have understood the main options of the **Drawing View** dialog box, you can generate the copy and align the view. The method to do so is discussed next.

Generating the Copy and Align View

Ribbon: Layout > Model Views > Copy and Align View

The **Copy and Align View** tool is used to align views that are generated from an existing partial view. Therefore, to generate this type of drawing view, it is necessary that first a partial view is generated. You will learn more about the partial views later in this chapter.

The procedure to generate a copy and align view is discussed next.

1. To generate the copy and align view, choose the **Copy and Align View** tool from the **Model Views** group of the **Layout** tab in the **Ribbon**; you will be prompted to select an existing partial view to be aligned with. Once you have selected a partial view, you will be prompted to select a center point for the drawing view.

2. Select a point anywhere on the drawing sheet to place the view. The view is placed at the selected location. You will be prompted to specify a center point for detail on the current view and to sketch a spline to define the outline of the copy and align view.

3. Specify the center point of the spline and draw the spline. Press the middle mouse button to finish sketching the spline. The drawing view will be cropped along the spline drawn; you will be prompted to select a straight line (axis, segment, datum curve) alignment on the current view.

4. Select an edge from the current view, as shown in Figure 11-33, to align it with the partial view. The copied view is aligned with its parent partial view, as shown in Figure 11-34.

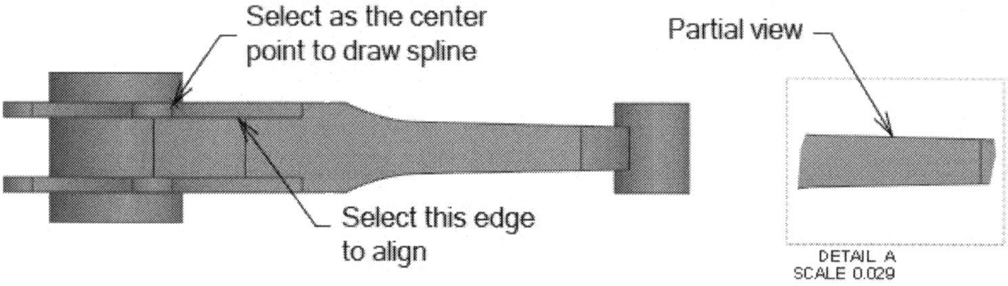

***Figure 11-33** Selecting an edge from the current view to align with the partial view*

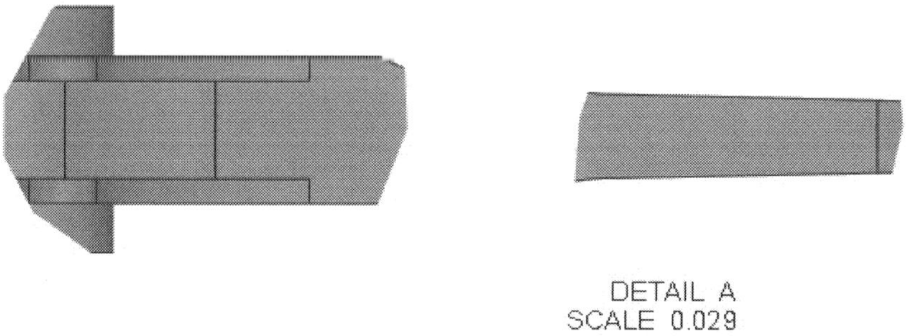

***Figure 11-34** Resulting copy and align view*

Generating the 3D Cross-Section View

The 3D cross-section allows you to display 3D sections in the drawing views. This type of sectioning may be required to show complex cross-sections of a model. Note that the 3D cross-sections are created using zones. Therefore, you first need to create zones and then use them to create 3D cross-sections. The procedure to create zones and 3D cross-sections is discussed next.

1. Open the part file of the model in the **Part** mode and choose **View Manager** from the **Manage Views** drop-down in the **Model Display** group of the **View** tab in the **Ribbon**.

2. Choose the **Sections** tab and then choose the **New** button; a flyout will be displayed, as shown in Figure 11-35. Next, choose the **Zone** option from the flyout to define a new cross-section in the model.

3. Enter the name for the new cross-section and press ENTER; a dialog box with the name that you have specified for the new cross-section will be displayed, refer to Figure 11-36.

Note

In the example discussed above, the name for the new 3D cross-section is given as ***Xsec0001****. Therefore, the dialog box shown in Figure 11-36 is named as* ***Xsec0001****.*

4. Now, you need to define the planes for the zone creation. Select the datum plane through which you want to create the zone.

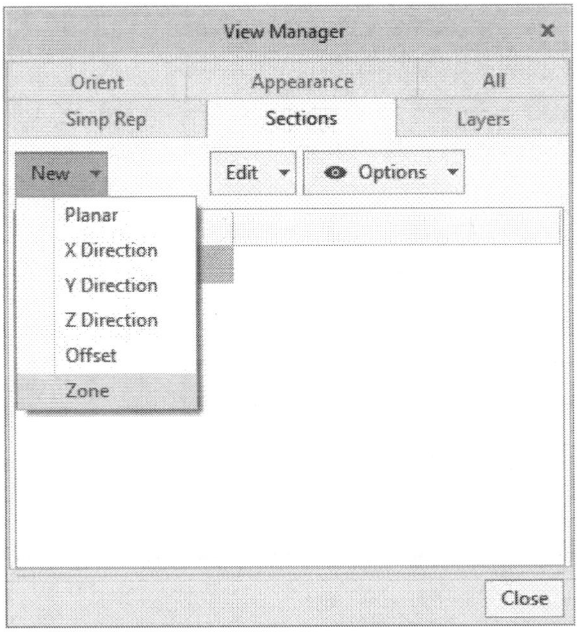

Figure 11-35 The flyout displayed

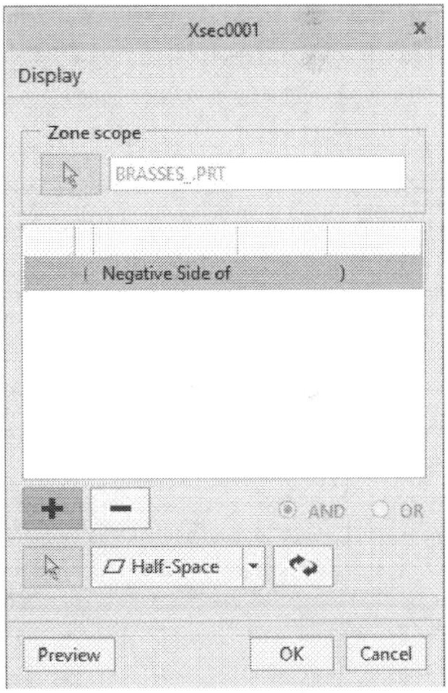

Figure 11-36 *The* ***Xsec0001*** *dialog box*

5. You can add multiple reference entities or datum planes for defining the zone by choosing the **Add a reference to the zone** button from the **Xsec0001** dialog box. You can also remove references specified for the zone by selecting the reference from the **zone scope** area and then choosing the **Remove a reference from the zone** button from the **Xsec0001** dialog box.

6. By default, the direction for creation of the zone is chosen as the negative side of the selected reference entity. You can change the orientation by choosing the **Change orientation** button from the **Xsec0001** dialog box.

7. After specifying all the references for the zone, choose the **OK** button to finish the zone creation and exit the **Xsec0001** dialog box. Next, choose the **Close** button from the **View Manager** dialog box.

Note

*If you want to make the 3D cross-section active in the drawing area, select it from the **View Manager** dialog box and right-click to invoke the shortcut menu. Choose the **Activate** option from the shortcut menu; the 3D cross-section will be activated in the drawing area.*

8. Start a new drawing file and insert a general view in the drawing sheet by choosing the **General View** tool from the **Model Views** group of the **Ribbon**.

9. Choose the **Sections** option from the **Categories** list of the **Drawing View** dialog box. Next, select the **3D cross-section** radio button from the **Section options** area; the 3D cross-section with the name **Xsec0001** will be displayed in the drop-down list next to this radio button.

10. The **Show X-Hatching** check box is selected by default in the **Section options** area. If you need to display the hatching lines, then select the **Show X-Hatching** check box and choose the **Apply** button from the **Drawing View** dialog box. The 3D cross-section will be displayed on the drawing sheet.

Figure 11-37 shows the trimetric view and the 3D cross-section of a part model inserted in an A4 drawing sheet.

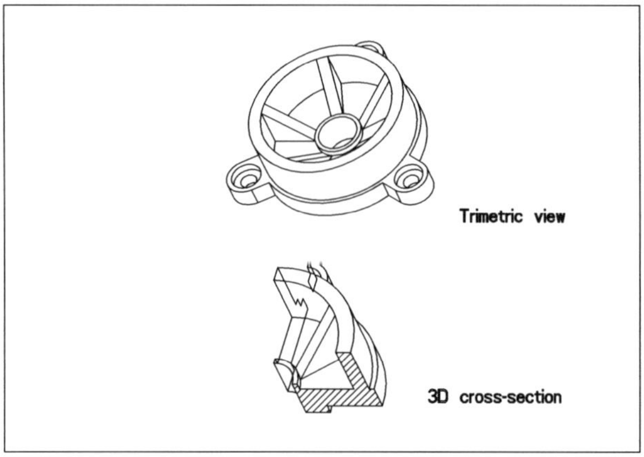

***Figure 11-37** The trimetric view and the 3D cross-section in an A4 drawing sheet*

Note

*You can also display the 3D cross-section in the **Shading** mode.*

EDITING THE DRAWING VIEWS

In Creo Parametric, you can edit a drawing view as well as the items in the drawing views. All options of modifying the drawing views or items related to it can be chosen from the shortcut menu that will be displayed when you right-click and hold the right mouse button after selecting the view or the item. However, these options can also be chosen from the groups available in the **Ribbon**. Creo Parametric allows you to perform the following types of editing operations on the drawing views.

Moving the Drawing View

When a view is generated, its movement is locked by default. To unlock the drawing view, choose **Lock View Movement** either from the shortcut menu that will be displayed on right-clicking and holding the right mouse button on the drawing view or directly from the **Drawing Tree**. Alternatively, choose the **Lock View Movement** toggle button from the **Document** group to unlock the movement of the views. When you select the drawing view, it will be displayed inside a boundary. To move the drawing view, select it and then drag it to the required location on the drawing sheet. Remember that if you select the view that has some child views, the child views will also move along with the parent view in order to maintain their alignment with the parent view. Also, the projected views can be moved only in the direction of projection.

Note

*The **General** and **Detailed** views can be moved to any new location because they are not the projections of any view.*

Erasing the Drawing View

The **Erase View** button available in the **Display** group of the **Layout** tab in the **Ribbon** is used to temporarily remove the selected drawing view from the sheet. However, the view still remains in the memory of the drawing and can be resumed at any point of time. As the view is not completely removed from the memory, you can also erase a view that has some child views associated to it and it will not affect the child view. Once a view is erased, a box will be displayed in place of the view displaying the name of the view. To resume the view, choose the **Resume View** button from the **Display** group of the **Layout** tab in the **Ribbon**. Alternatively, select the erased drawing view from the drawing or directly from the **Drawing Tree** and then right-click on it. Choose **Resume View** option from the shortcut menu displayed.

Note

When you erase a view, the leaders and dimensions that were attached with the view are also erased. When you resume an erased view, the leaders and dimension values are displayed again.

Deleting the Drawing View

To delete a drawing view from the drawing sheet, select the view from the drawing or from the **Drawing Tree** and hold down the right mouse button to invoke a shortcut menu. Choose the **Delete** option from the shortcut menu. Once the view is deleted, no information related to the deleted view remains in the memory of the drawing. Remember that if a view has some child views associated with it then it is also deleted. Before deletion, Creo Parametric informs you that views associated with this view will also be deleted. You can choose the **Yes** button to delete the view or choose the **No** button to continue with the view from the displayed dialog box.

Adding New Parts or Assemblies to the Current Drawing

Ribbon: Layout > Model Views > Drawing Models

To generate the drawing views, you can add more parts or assemblies, apart from the default part or assembly. To do so, choose **Layout > Model Views > Drawing Models** from the **Ribbon**; the **DWG MODELS** rollout will be displayed, as shown in Figure 11-38.

Next, choose the **Add Model** option from the **DWG MODELS** rollout to add a model to the current drawing. When you choose this option, the **Open** dialog box will be displayed. You can select the new model to be added using this dialog box. Remember that the latest model added will be the current model and the drawing views generated will be of this model. You can change the current model by choosing the **Set Model** option from the **DWG MODELS** rollout. Similarly, you can also delete a model from the current drawing. However, only the model that does not have any view generated from it can be deleted.

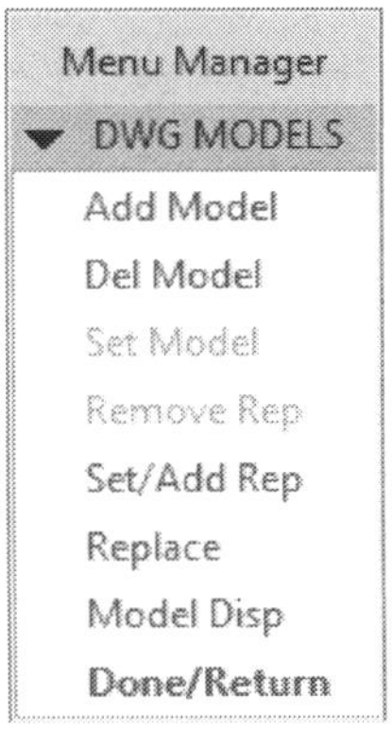

*Figure 11-38 The **DWG MODELS** rollout*

MODIFYING THE DRAWING VIEWS

You can also make the following modifications in the existing drawing views.

Changing the View Type

Select the drawing view that needs to be modified and hold the right mouse button to invoke the shortcut menu. Choose **Properties** from the shortcut menu; the **Drawing View** dialog box will be displayed, as shown in Figure 11-39. The **Drawing View** dialog box can also be invoked by double-clicking on the view to be modified. The **View Type** option in the **Categories** area is chosen by default. You can modify the view type by selecting any option from the list displayed under **Model view names** collector. Remember that only the general, projection, and auxiliary view types can be modified using this option.

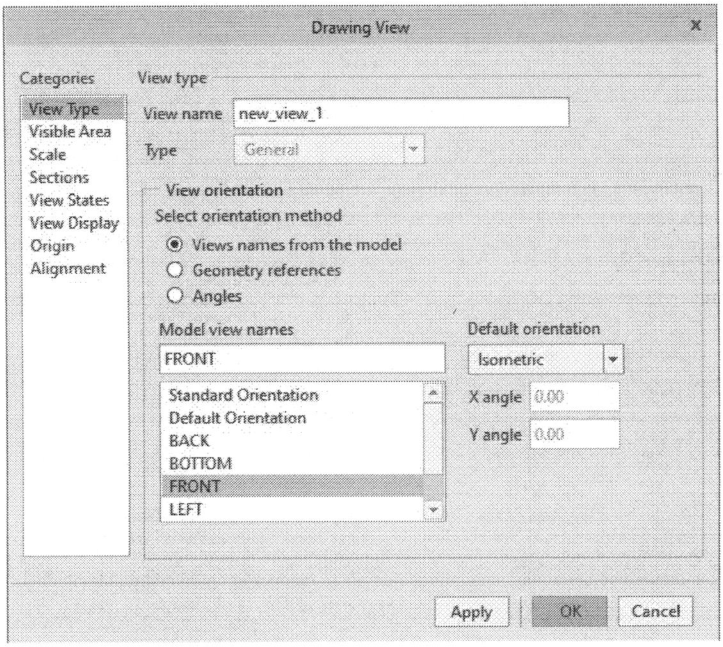

Figure 11-39 *The* ***Drawing View*** *dialog box*

Changing the View Scale

The scale of a view can be modified using the **Scale** option from the **Categories** list box of the **Drawing View** dialog box. You can modify the scale factor of the views that are generated using the general view and detailed view options.

Note

The scale factor represents the scaling of the drawing model. For example, if the scale is set to 0.25, Creo Parametric scales the drawing views to one-quarter (1/4) of the actual size of the model.

Reorienting the Views

Double click on the general view to be reoriented to display the **Drawing View** dialog box. You can use the options from this dialog box to change the orientation of the selected view. However, if some child view is associated with the general view, then the child view will also be reoriented accordingly.

Modifying the Cross-sections

You can flip the side of sectioning and replace, delete, or rename the sections in the views using the **Sections** option of the **Categories** list in the **Drawing View** dialog box.

Modifying Boundaries of Views

You can modify the boundaries of the detailed or partial views using the option in the **Drawing View** dialog box. To modify the boundary of a detailed view, double-click on it to invoke the

Drawing View dialog box. The selection mode is invoked by default in the **Reference point on parent view** selection area. Therefore, you can specify the center point for the boundary of the detailed view. To sketch the boundary of the detailed view again, click once in the **Spline boundary on parent view** selection area and draw the boundary on the parent view. To modify the boundary of a partial view, invoke the **Drawing View** dialog box and choose the **Visible Area** option from the **Categories** list box. Using the options available on the right of this dialog box, you can modify the boundary of the detailed view.

Adding or Removing the Cross-section Arrows

If you have not specified the cross-section arrows while generating the section views, you can specify them by selecting the section view and holding down the right mouse button to invoke the shortcut menu. Choose the **Add Arrows** option from the shortcut menu; you will be prompted to select the view where the arrows should be displayed. Select the view in which you need to display the arrows.

To remove arrows, select them and choose the **Delete** button from the right-click shortcut menu.

Modifying the Perspective Views

Double-click on the perspective view to display the **Drawing View** dialog box and choose the **Scale** option from the **Categories** list box. The options related to the perspective view are available. You can modify the eye point distance or the view diameter using these options.

MODIFYING OTHER PARAMETERS

Apart from modifying the drawing views, you can also modify other parameters related to the drawing views. For example, you can modify the size and the style of the text, scale factor of all drawing views, cross-section hatching, and so on. All this is done by selecting the item. When the item is highlighted, hold down the right mouse button to invoke the shortcut menu. You can select the options from the shortcut menu to modify that item.

You can modify any parameter associated with the drawing views. Depending on the item selected to modify, the options related to it vary. The options displayed in the shortcut menu vary from item to item.

Editing the Cross-section Hatching

Select the hatching from a drawing view. When the hatching turns green in color, hold down the right mouse button to invoke the shortcut menu. Choose the **Properties** option from the shortcut menu; the **MOD XHATCH** menu will be displayed, as shown in Figure 11-40.

The parameters related to cross-section hatching that can be modified are: the spacing of the hatching, angle of the hatching lines, offset value, and the line style of the hatching lines. There are also some standard hatch patterns that are available in Creo Parametric. You can retrieve these standard patterns by using the **Open** dialog box displayed upon choosing **Retrieve** from the **MOD XHATCH** menu.

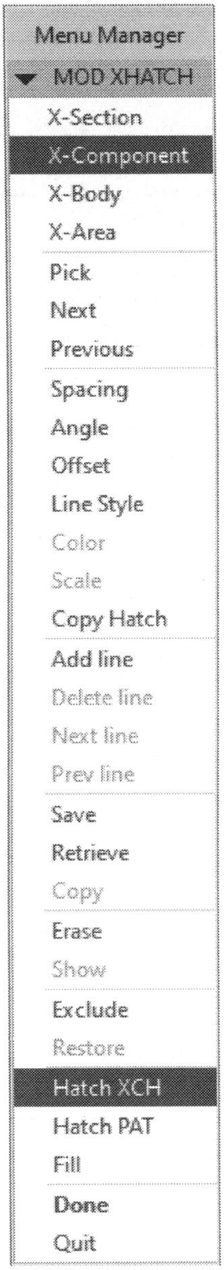

***Figure 11-40** The **MOD XHATCH** menu*

TUTORIALS

Tutorial 1

In this tutorial, you will generate the drawing views of the model created in Exercise 4 of Chapter 8, as shown in Figure 11-41. Select the A4 size drawing sheet and generate all drawing views shown in Figure 11-42. **(Expected time: 45 min)**

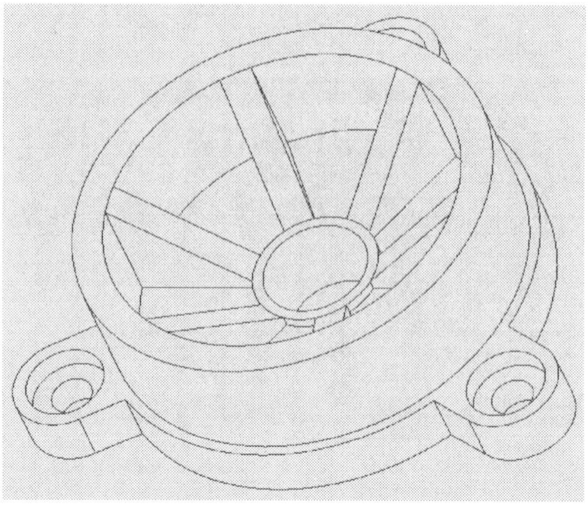

Figure 11-41 *Part for generating drawing views*

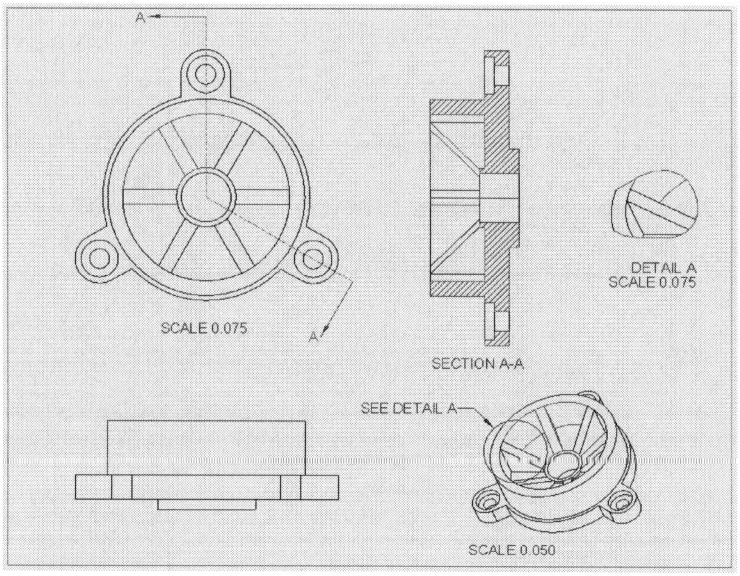

Figure 11-42 *The drawing views to be generated*

The following steps are required to complete this tutorial:

a. Start a new drawing file and select the required drawing sheet.
b. Generate the top view.
c. Generate the front view taking the top view as the parent view.
d. Generate the sectioned right view by defining a plane on the top view of the mode.
e. Generate the default 3D view of the model.
f. Generate the detail view from the 3D view.

Before you start generating drawing views, set *C:\Creo8.0\C11* as working directory. Copy the model file *c08_exr04.prt* from the *c08* folder and paste it in the folder named *C11*. The drawing generated in this tutorial will be saved in the *C11* directory with an extension *.drw*. The part file *(.prt)* and the drawing file *(.drw)* should lie in the same directory or folder.

Note

You cannot open a drawing file if the part file, from which the drawing view was generated, has been deleted. Also, if the part file is removed from the folder where its drawing file is placed or if you rename the part file after generating the drawing views, the drawing file will not open.

Starting a New Drawing File

To generate drawing views in Creo Parametric, you first need to start a new file in the **Drawing** mode.

1. Choose the **New** button from the **File** menu to display the **New** dialog box.

2. Select the **Drawing** radio button and then enter ***C11tut1*** as the name of the file.

3. Choose the **OK** button from the **New** dialog box to display the **New Drawing** dialog box.

4. Choose the **Browse** button and select *c08_exr04.prt* for generating the drawing views.

5. Select the **Empty** radio button from the **Specify Template** area.

6. Choose the **Landscape** button from the **Orientation** area, if it is not chosen by default, as shown in Figure 11-43.

7. Select **A4** from the **Standard Size** drop-down list in the **Size** area. Choose the **OK** button to proceed to the **Drawing** mode.

   An A4 size sheet with the landscape orientation is displayed in the drawing area.

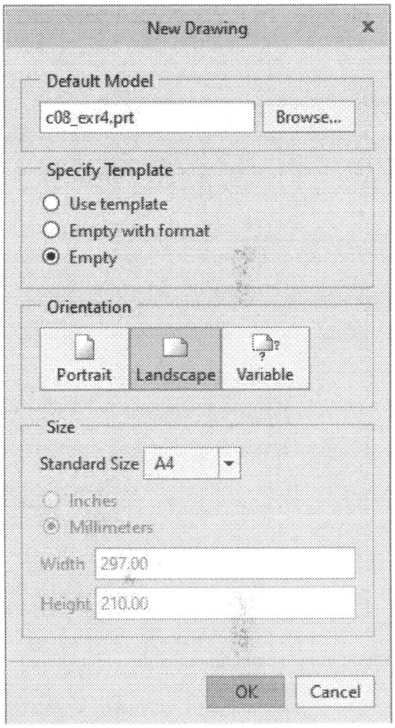

***Figure 11-43** The **New Drawing** dialog box*

Generating the Top View

You can generate any view as the first view on the drawing sheet. However, in this tutorial, you will generate the top view as the first view.

1. Choose **Layout > Model Views > General View** from the **Ribbon**; the **Select Combined State** dialog box is displayed. Choose the **OK** button from the dialog box; you are prompted to specify the center point for the drawing view.

2. Specify the center point for the first view, which is on the top left corner of the sheet, as shown in Figure 11-44. On specifying the center point, the **Drawing View** dialog

box is displayed along with the preview of the drawing view in the default orientation. In the **Model view names** list box, select the **TOP** option and choose the **Apply** button. The default view is oriented as the top view.

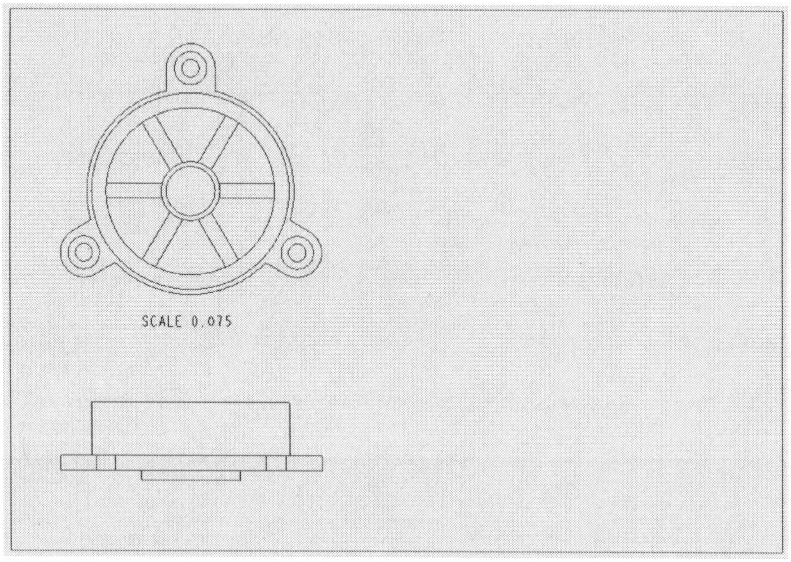

Figure 11-44 *The top and front views*

Note

Although you can move a view anywhere on the sheet, yet you should try to place the drawing view inside the sheet and also leave space for the other views to be generated.

3. Select the **Scale** option from the **Categories** list box in the **Drawing View** dialog box; the **Scale and perspective options** area is displayed. Select the **Custom scale** radio button and enter **0.075** as the scale factor for the view. Choose the **Apply** button; the scale of the view is modified.

4. Choose the **OK** button from the **Drawing View** dialog box to complete the placing of the top view.

5. Turn off the display of datum planes, axes, points, and coordinate system by choosing the respective buttons from the **Datum Display Filters** drop-down in the **Graphics** toolbar.

6. Change the model display to no hidden by choosing the **No Hidden** option from the **Display Style** drop-down. Next, choose the **Repaint** button from the **Graphics** toolbar to repaint the screen.

Tip

If the view that you have placed on the drawing sheet is not at the proper location, you need to move the drawing view. To do so, select the drawing view; the selected drawing view is enclosed in a green box. Next, right-click and hold the right mouse button to display a shortcut menu. Choose the ***Lock View Movement*** *option from the shortcut menu displayed and then press the left mouse button inside the box and drag it to place it at the required location.*

Generating the Front View

The front view of the model is generated from the top view, which is already placed on the drawing sheet.

1. Choose **Layout > Model Views > Projection View** from the **Ribbon**; you are prompted to specify a center point for the drawing view.

2. Specify the center point for the front view below the top view, as shown in Figure 11-44.

3. Click anywhere on the drawing sheet to clear the current selection set.

   If the front view of the model is not similar to the one shown in Figure 11-44, then you need to change the first angle projection to the third angle projection. To do so, choose **Prepare > Drawing Properties** from the **File** menu; the **Drawing Properties** dialog box is displayed. Choose the **change** button available next to **Detail Options** in this dialog box; the **Options** dialog box is displayed. Enter **projection_type** in the **Option** text box; the options in this dialog box are displayed in the **Value** drop-down list. Select **third_angle** from the drop-down list and then choose the **Add/Change** button. Next, choose the **OK** button to accept the changes and exit the dialog box.

> **Tip**
> *It is recommended that you use the* ***Repaint*** *button available in the* ***Graphics*** *toolbar to remove any temporary information in the drawing area and to refresh the screen.*

Generating the Section View

To generate a section view, a section must be defined on the model. You will sketch this section to generate the section view for this tutorial.

1. Choose **Layout > Model Views > Projection View** from the **Ribbon**; you are prompted to select a projection parent view.

2. Select the top view as the projection parent view; you are prompted to specify a center point for the drawing view.

3. Specify the center point on the right side of the top view to place the view.

4. Once the view is placed, double-click on it to display the **Drawing View** dialog box.

5. Choose the **Sections** option from the **Categories** list box in the **Drawing View** dialog box; the section options are invoked in the **Drawing View** dialog box.

6. Select the **2D cross-section** radio button and choose the **Add cross-section to view** button; the window area below it is activated. Select the **Create New** option from the **Name** drop-down list; the **XSEC CREATE** menu is displayed.

7. Choose **Offset > Both Sides > Single > Done** from the **XSEC CREATE** menu.

8. Specify the name of the cross-section as **A** in the message input window and press ENTER. Once you have specified the name of the cross-section, a separate window displaying the part appears. Also, you are prompted to select a sketching plane.

Note

*If the model in the **Part** mode is opened in another window, the subwindow will not appear and you have to manually change the window. Choose **View > Window > Windows** from the **Ribbon** and select the part file that is opened. Now, you can continue with the sketcher environment.*

9. Choose the **Plane Display** tool from the **Show** group of the **View** tab in the **Ribbon** from the original window to display datum planes. You may need to repaint the screen in the subwindow by choosing the **Repaint** option from the **View** menu to view the datum planes, if required.

10. Select the **TOP** datum plane from the subwindow. Choose the **Okay** option from the **DIRECTION** submenu; the **SKET VIEW** submenu is displayed.

11. Choose the **Bottom** option from the **SKET VIEW** submenu and select the **FRONT** datum plane from the subwindow.

12. Choose the **Line** tool from the **Line** flyout in the **Sketch** menu and sketch the lines, as shown in Figure 11-45. These lines define the section plane.

13. Align the lines and modify the angular dimension to **120**, as shown in Figure 11-45.

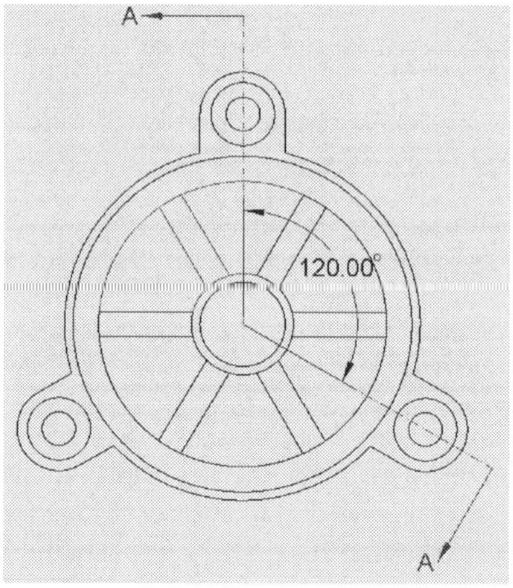

Figure 11-45 *Sketch for generating the total aligned section view*

14. Choose **Sketch > OK** from the menu bar of the subwindow.

15. Choose the **Full(Aligned)** option from the **Sectioned Area** drop-down list; you are prompted to select an axis.

16. Select the central axis of the model from the front drawing view. You may need to turn on the display of the central axis if its display is turned off.

17. Scroll the bar in the **Drawing View** dialog box to the right edge. Click in the field below the **Arrow Display** column; you are prompted to pick a view for arrows where the section is perpendicular. Select the top view to display the arrows.

18. Choose the **Apply** button from the **Drawing View** dialog box and then exit the dialog box; the section view is displayed.

19. Turn off the display of the datum planes and the datum axes.

Tip
If you do not want to display the section arrows in any view, you can use the middle mouse button (in case of three button mouse) to abort the creation of arrows.

Modifying the Hatching

You need to modify the spacing between the hatch lines to make the hatching dense. In this section, you will select the type of filter from the Filter drop-down list to select the hatch lines.

1. Select the **X-Section** filter from the Filter drop-down list available in the **Status Bar** below the drawing area. Select the hatching lines from the drawing sheet; the hatching lines turn green.

2. Press and hold the right mouse button to invoke the shortcut menu. Choose the **Properties** option from the shortcut menu; the **MOD XHATCH** menu is displayed. Alternatively, you can double-click on the hatching lines to invoke the **MOD XHATCH** menu.

3. In the **MOD XHATCH** menu, choose the **Spacing** option to display the **MODIFY MODE** submenu.

4. The spacing between the hatching lines needs to be reduced. To do so, choose the **Half** option and then choose the **Done** option in the **MOD XHATCH** menu. On doing so, the hatching appears more dense, refer to Figure 11-46.

5. Set the filter back to the **General** option using the Filter drop-down list.

Generating the General View

In this section, the General view needs to be generated to show the 3D view of the model, which is the trimetric view.

1. Choose the **General View** tool from the **Model Views** group in the **Layout** tab of the **Ribbon**; the **Select Combined State** dialog box is displayed. Choose the **OK** button from the dialog box; you are prompted to specify a center point for the placement of the view.

2. Specify the center point for placing the general view below the section view, as shown in Figure 11-46; the **Drawing View** dialog box is displayed.

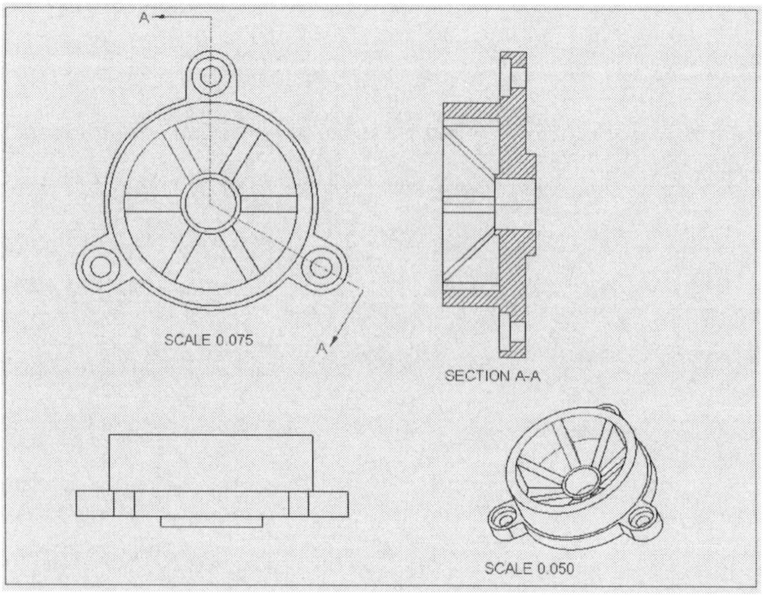

Figure 11-46 Different drawing views

3. Set the value of the scale for the new view to **0.05** using the **Scale** option from the **Drawing View** dialog box.

4. Choose the **Apply** button and then exit the dialog box.

Generating the Detail View

As mentioned earlier, the detail view is required to provide details of a particular portion of the drawing view. In this section, you need to give the details of one of the ribs of the model.

1. If the general view is selected (highlighted in green box), then you need to click once on the drawing sheet to exit the current selection set. Next, choose the **Detailed View** tool from the **Model Views** group of the **Layout** tab in the **Ribbon**; you are prompted to select a center point for detail on an existing view. This center point will be the center of the detailed view.

2. Select the center point for the detail on one of the ribs in the trimetric view, refer to Figure 11-47; you are prompted to sketch a spline about the center point that you selected.

3. Draw the spline and then press the middle mouse button to exit the **Spline** tool.

4. Select the center point for the placement of the drawing view. The detailed view is generated and scaled by default.

5. Double click on the detail view to invoke the **Drawing View** dialog box.

6. Select **Circle** from the **Boundary type on parent view** drop-down list.

7. Select the **Scale** option from the **Categories** list box. Set the value of the custom scale to **0.075**. Choose the **Apply** button and exit the dialog box.

   A note is also displayed with an arrow attached to the trimetric view. You may need to move the note to a suitable position by selecting it and then dragging the cursor. The final sheet after generating all views is shown in Figure 11-47. If you are unable to select the note, then choose the **Annotate** tab from the **Ribbon**. On doing so, you will be able to select it and move it.

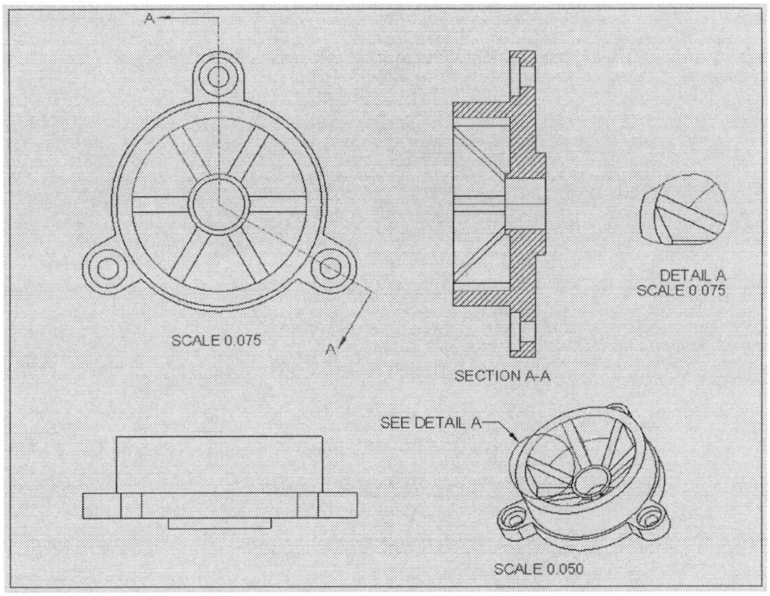

***Figure 11-47** The drawing views for Tutorial 1*

> **Tip**
> *If any view or the text on the drawing is overlapping or is not at the desired place on the sheet, select it and then move it by dragging the mouse.*

Saving the Drawing File

In this section, you need to save the drawing file that you have created.

1. Choose the **Save** button from the **Quick Access** toolbar; the **Save Object** dialog box is displayed with the name of the drawing file that you have entered earlier.

2. Choose the **OK** button to confirm the action.

Closing the Drawing File

After saving the drawing file, you need to close it.

1. Choose **File > Close** from the menu bar to close the drawing window.

Tutorial 2

In this tutorial, you will generate the drawing views of the part created in Tutorial 1 of Chapter 9. The part is shown in Figure 11-48. The drawing views that need to be generated are shown in Figure 11-49. **(Expected time: 45 min)**

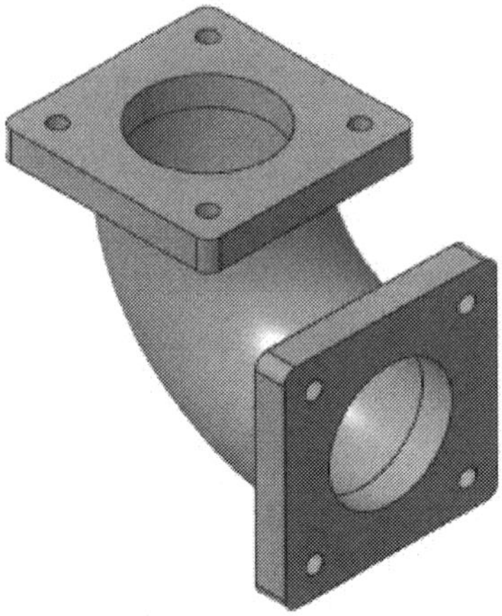

Figure 11-48 *Model for generating drawing views*

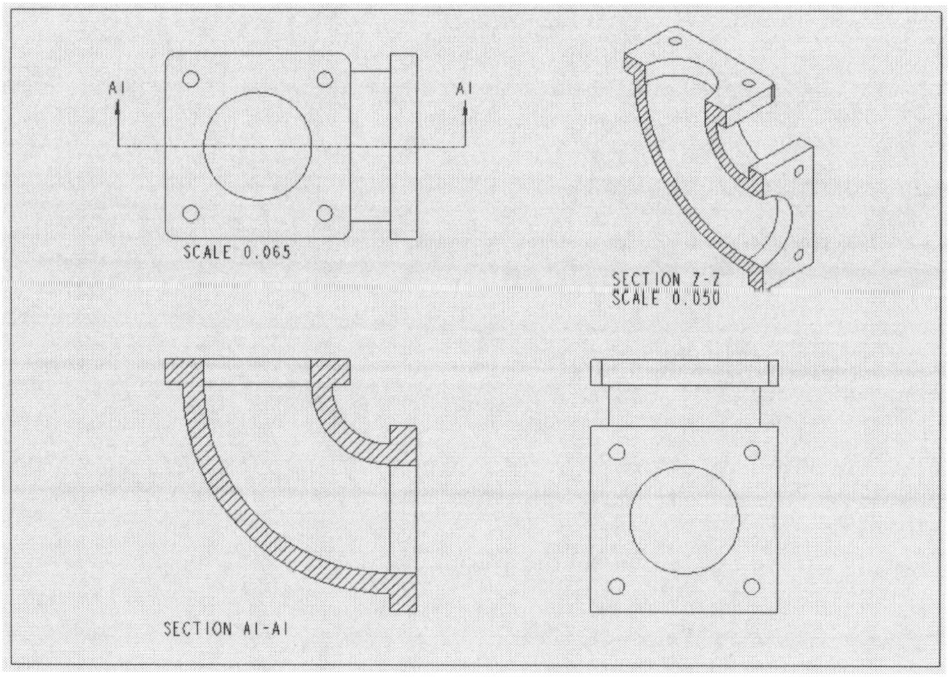

Figure 11-49 *Drawing views to be generated*

Before you start generating drawing views, copy the file *c09tut1.prt* from the *c09* folder in the current directory.

The following steps are required to complete this tutorial:

a. Start a new drawing file and select the size of the drawing sheet.
b. Generate the top view.
c. Generate the sectioned front view by defining the **FRONT** datum plane as the section plane.
d. Generate the right view of the sectioned front view.
e. Generate the isometric sectioned view. The section will be defined by drawing a line on the **TOP** datum plane.

Before you start generating the drawing views, set the working directory to *C:\Creo8.0\C11*. The part *(.prt)* file and the drawing *(.drw)* file should lie in the same directory or folder.

Starting a New Drawing File

To generate the drawing views, you first need to start a new drawing file.

1. Choose the **New** button from the **Data** group in the **Ribbon** to display the **New** dialog box.

2. Select the **Drawing** radio button and then enter **C11tut2** as the name of the file.

3. Choose the **OK** button from the **New** dialog box to display the **New Drawing** dialog box.

4. Choose the **Browse** button to select *C09tut1.prt* from the *C09* folder for generating drawing views and then choose the **Open** button.

5. Select the **Empty** radio button from the **Specify Template** area.

6. Choose the **Landscape** button from the **Orientation** area.

7. Select **A4** from the **Standard Size** drop-down list in the **Size** area. Choose the **OK** button from the **New Drawing** dialog box to proceed to the **Drawing** mode.

Generating the Top View

First, you need to generate the top view. All other views, except the sectioned isometric view, will be the child views of the top view. You need to generate the top view first because the required sectioned front view can be generated only from the top view. The right view can also be generated independently, but then this view will not help generate any other required view.

1. Choose the **General View** tool from the **Model Views** group in the **Ribbon**; the **Select Combined State** dialog box is displayed. Choose the **OK** button from this dialog box; you are prompted to specify a center point for the placement of the view.

2. Specify the center point for the placement of the top view close to the upper left corner of the drawing sheet; the **Drawing View** dialog is displayed.

3. Choose the **TOP** option from the **Model view names** list box and then choose the **Apply** button.

4. Choose the **Scale** option from the **Categories** list box. Next, choose the **Custom scale** radio button and enter **0.065** in the edit box.

5. Choose the **OK** button to exit the **Drawing View** dialog box.

   If necessary, move the view, refer to Figure 11-50, and then choose the **No Hidden** option from the **Display Style** drop-down in the **Graphics** toolbar. You may also need to repaint the screen.

Generating the Sectioned Front View

The sectioned front view of the model is generated from the top view. Before proceeding use the **Plane Display** button from the **Datum Display Filters** drop-down in the **Graphics** toolbar to turn on the display of datum planes and repaint the screen.

1. Choose the **Projection View** tool from the **Model Views** group of the **Layout** tab in the **Ribbon**; you are prompted to specify a center point for placement of the projection view.

2. Specify the center point for the placement of the front view below the top view, as shown in Figure 11-50.

3. Select the newly generated view and invoke the shortcut menu. Choose the **Properties** option to display the **Drawing View** dialog box.

4. Choose the **Sections** option from the **Categories** list box to display the section related options in the dialog box.

5. Select the **2D cross-section** radio button and choose the **Add cross-section to view** button; the window area below it gets activated. Select the **Create New** option from the **Name** drop-down list, if it is not selected; the **XSEC CREATE** menu is displayed.

6. Choose **Planar > Single > Done** from the **XSEC CREATE** menu.

7. Enter **A1** as the name of the cross-section in the message input window and then press ENTER to display the **SETUP PLANE** submenu; you are prompted to select a planar surface or a datum plane.

8. Select the **FRONT** datum plane (the plane that cuts the part horizontally through the center of the cylindrical feature in the top view) from the drawing area.

9. Scroll the bar in the **Drawing View** dialog box to the right. Click on the field below the **Arrow Display** column; you will be prompted to pick a view for arrows where the section is perpendicular. Select the top view to display the arrows.

10. Choose the **Apply** button and then exit the dialog box.

Modifying the Hatching

The offset distance between the hatching lines in the front sectioned view is large. Therefore you need to reduce the distance between the hatching lines.

1. Select the **X-Section** filter from the **Filter** drop-down list in the Status Bar. Select the hatching lines from the sectioned front view in the drawing sheet; the hatching lines turn green. Hold the right mouse button to invoke the shortcut menu. Choose the **Properties** option from the shortcut menu; the **MOD XHATCH** menu is displayed.

2. In the **MOD XHATCH** menu, choose the **Spacing** option to display the **MODIFY MODE** submenu.

   The spacing between the hatching lines has to be reduced.

3. Choose the **Half** option twice and then choose the **Done** option in the **MOD XHATCH** menu. Now, the hatching appears more dense.

4. Click once in the drawing area to remove the X-hatch from the current selection set. The sheet after placing these two views should look similar to the one shown in Figure 11-50.

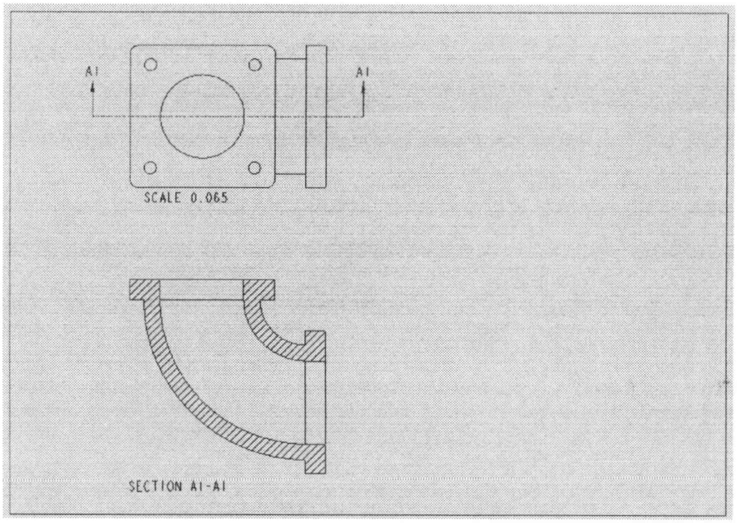

Figure 11-50 *Drawing sheet after generating the top view and the sectioned front view with the display of datum planes turned on*

5. Set the selection filter back to **General**.

Generating the Right View

The right view is the projection of the sectioned front view. Before proceeding further, turn off the display of the datum planes.

1. Choose **Layout > Model Views > Projection View** from the **Ribbon**; you are prompted to select the projection parent view.

2. Select the sectioned front view as the parent view.

3. Specify the center point for the placement of the drawing view on the right of the sectioned front view. The right view of the model is placed, as shown in Figure 11-51.

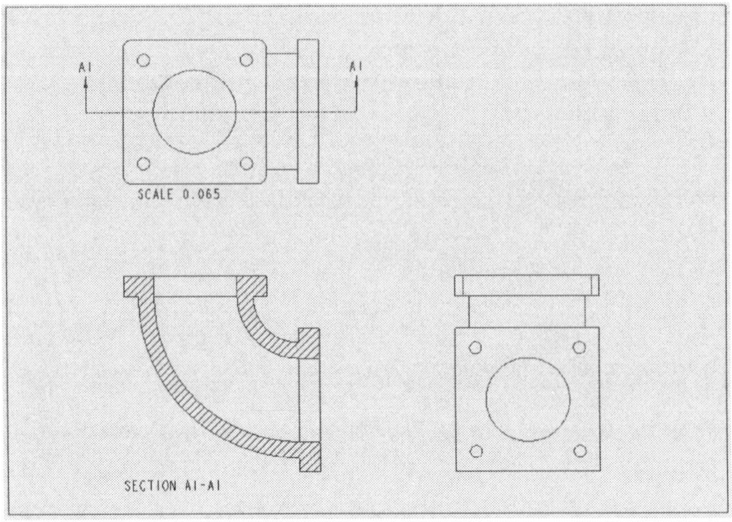

Figure 11-51 The drawing sheet after generating the top, sectioned front, and right views

Generating the Isometric Section View

The isometric section view is an independent view and it will be generated by using the **General** option.

1. Choose the **General View** tool from the **Model Views** group in the **Ribbon**; the **Select Combined State** dialog box is displayed. Choose the **OK** button from this dialog box; you are prompted to specify a center point for the placement of the view.

2. Specify the center point on the upper right corner of the drawing sheet for placing the view; the **Drawing View** dialog is displayed.

   The default view is a trimetric view but you need to display the isometric view. Therefore, you need to change the orientation.

3. Choose the **Isometric** option from the **Default orientation** drop-down list; the isometric view of the model is displayed on the drawing sheet.

4. Choose the **Scale** option from the **Categories** list box. Select the **Custom scale** radio button and enter **0.05** as the scale factor in the edit box displayed. Next, choose the **Apply** button.

5. Choose the **Sections** option from the **Categories** list box to display the section options in the dialog box.

6. Select the **2D cross-section** radio button and then choose the **Add cross-section to view** button; the collector below it gets activated. Select **Create New** from the name drop-down list; the **XSEC CREATE** menu manager is displayed.

7. Choose the **Offset > Both Sides > Single > Done** from the **XSEC CREATE** menu manager; the message input window is displayed.

8. Enter **Z** as the name of the section in the message input window and press ENTER; a separate window is displayed with the model.

9. Select the **TOP** datum plane from the subwindow. You may need to turn on the display of the datum planes. Choose the **Okay** option from the **DIRECTION** submenu; the **SKET VIEW** submenu is displayed.

10. Choose the **Right** option from the **SKET VIEW** submenu and select the **RIGHT** datum plane from the subwindow.

11. Choose **Sketch > Line > Line** from the menu bar and draw a line, as shown in Figure 11-52. This line creates a section plane.

12. Choose **Sketch > Constraint > Coincident** from the menu bar and align both ends of the line to the edges.

13. Choose **Sketch > OK** from the menu bar.

14. Choose the **Apply** button to place the view and then exit the dialog box.

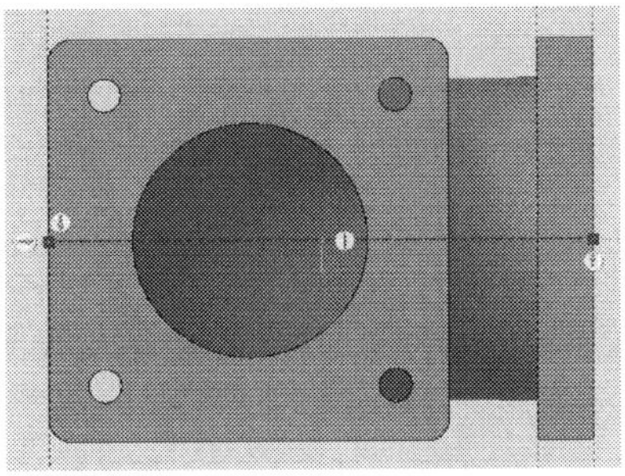

Figure 11-52 *The sketch for the section line with the constraint*

Note

If the drawing view placed on the sheet overlaps the boundary of the drawing sheet, then you can move the drawing view. To do so, select the drawing view; the selected drawing view is enclosed in a green box. Now, press the left mouse button inside the box and drag it to place it at the desired location.

Tip

*You can lock or unlock a drawing view using the **Lock View Movement** button from the **Document** group of the **Layout** tab.*

Modifying the Hatching

The spacing between the hatching lines in the sectioned isometric view is large. Therefore, you need to reduce the distance between them. Use the same procedure that was discussed earlier in this tutorial to modify the spacing between the hatch lines. The sheet after placing all views should look similar to the one shown in Figure 11-53.

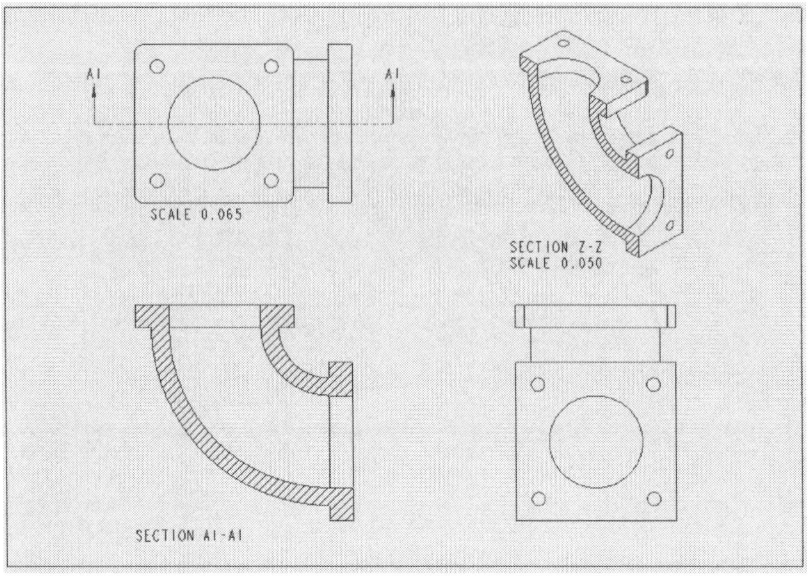

Figure 11-53 Different drawing views generated

Saving the Drawing File

In this section, you need to save the drawing file.

1. Choose the **Save** button from the **Quick Access** toolbar; the **Save Object** dialog box is displayed with the name of the drawing file that you have entered earlier.

2. Press ENTER to save the file.

Closing the Drawing File

After you have saved the drawing file, close the drawing file.

1. Choose **File > Close** from the menu bar; the drawing window is closed.

Self-Evaluation Test

Answer the following questions and then compare them to those given at the end of this chapter:

1. The __________ view is used when you want to show a particular portion of the view in a section and at the same time, do not want to section the remaining view.

2. The __________ option allows you to temporarily remove the selected drawing view from the sheet.

3. The __________ option is used to redisplay the drawing views that are erased using the **Erase View** option.

4. Using the __________ dialog box, you can reorient the general view.

5. The cross-section hatching on the sectioned portion can be modified using the __________ menu.

6. The bidirectional associative nature of a software package implies that when one file related to the part model is modified, the modifications are reflected in other related files as well. (T/F)

7. General view is the first view that is generated in the sheet. (T/F)

8. The **Full** section view is the most widely used section view. (T/F)

9. Section views are generally created for the models having features that are not clearly visible in the standard view. (T/F)

10. Broken views are used for the parts that have a high length to width ratio. (T/F)

Review Questions

Answer the following questions:

1. Which of the following options, when chosen, displays all the erased drawing views on the sheet?

   (a) **Delete** (b) **Resume View**
   (c) **Move** (d) **Erase**

2. Which of the following buttons in the **Orientation** group of the **View** tab is used to make all the objects in the drawing area visible on the screen?

   (a) **Zoom Out** (b) **Zoom In**
   (c) **Refit** (d) None of these

3. Which of the following options, when selected, displays only the area of the section view that is sectioned?

   (a) **Total** (b) **Area**
   (c) **Align** (d) None of these

4. Which of the following options is used to permanently remove a drawing view from the sheet?

   (a) **Resume** (b) **Erase**
   (c) **Delete** (d) **Move**

5. Which of the following options is used to modify the numeric value associated with the drawing views?

   (a) **Xhatching** (b) **Any Item**
   (c) **Value** (d) None of these

6. In Creo Parametric, you can change the view type of an existing view. (T/F)

7. In Creo Parametric, you can reorient only the general views. (T/F)

8. In Creo Parametric, you can flip the side of the cross-section views. (T/F)

9. The **Orientation** area in the **New Drawing** dialog box will be available only when you select the **Empty** radio button from the **Specify Template** area. (T/F)

10. The orientation of a model saved in the **Part** mode can be used to orient the drawing view in the **Drawing** mode. (T/F)

EXERCISE

Exercise 1

Generate the drawing views of the model created in Exercise 3 of Chapter 5 on an A4 size sheet. The model is shown in Figure 11-54. The drawing views to be generated are shown in Figure 11-55. **(Expected time: 45 min)**

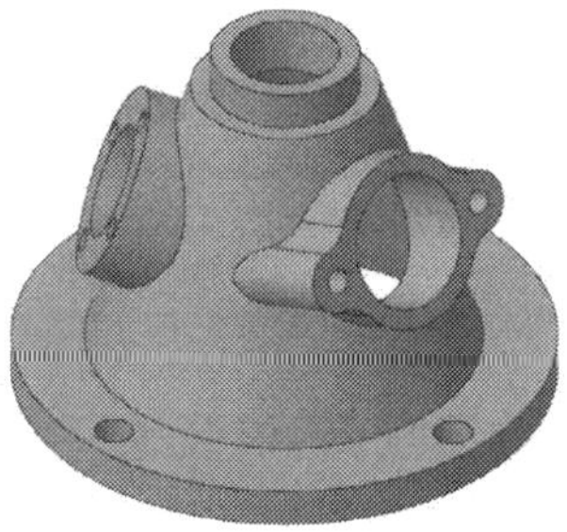

Figure 11-54 *Part for generating the drawing views*

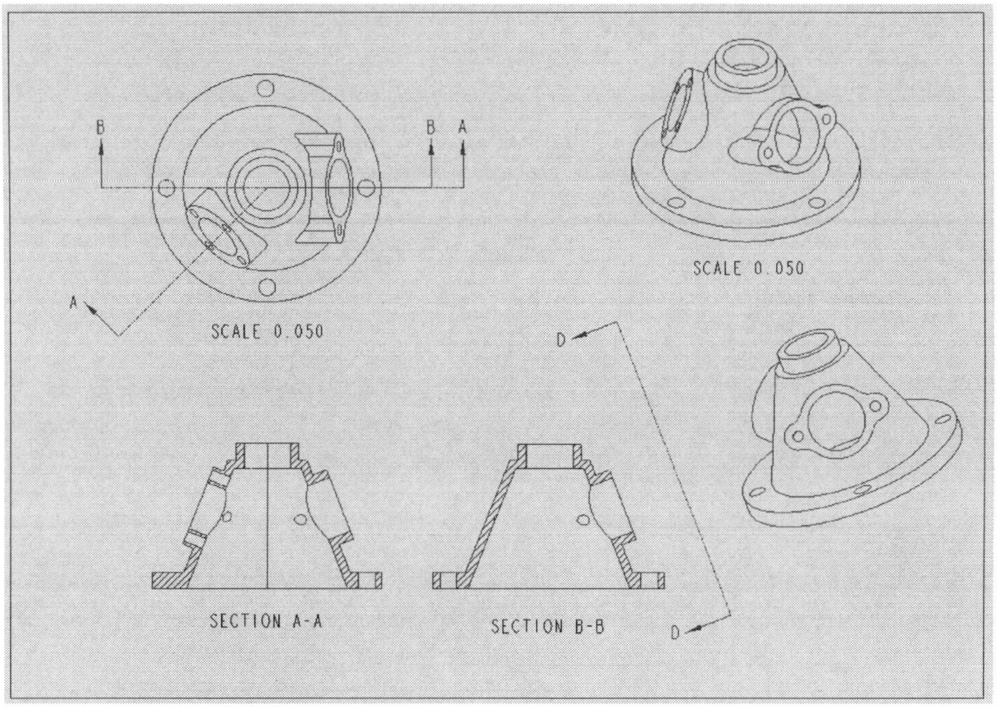

Figure 11-55 *Drawing views to be generated in Exercise 1*

Answers to Self-Evaluation Test

1. Local Section, **2.** Erase, **3.** Resume View, **4.** Drawing View, **5.** MOD XHATCH, **6.** T, **7.** T, **8.** T, **9.** T, **10.** T

Chapter 12

Dimensioning the Drawing Views

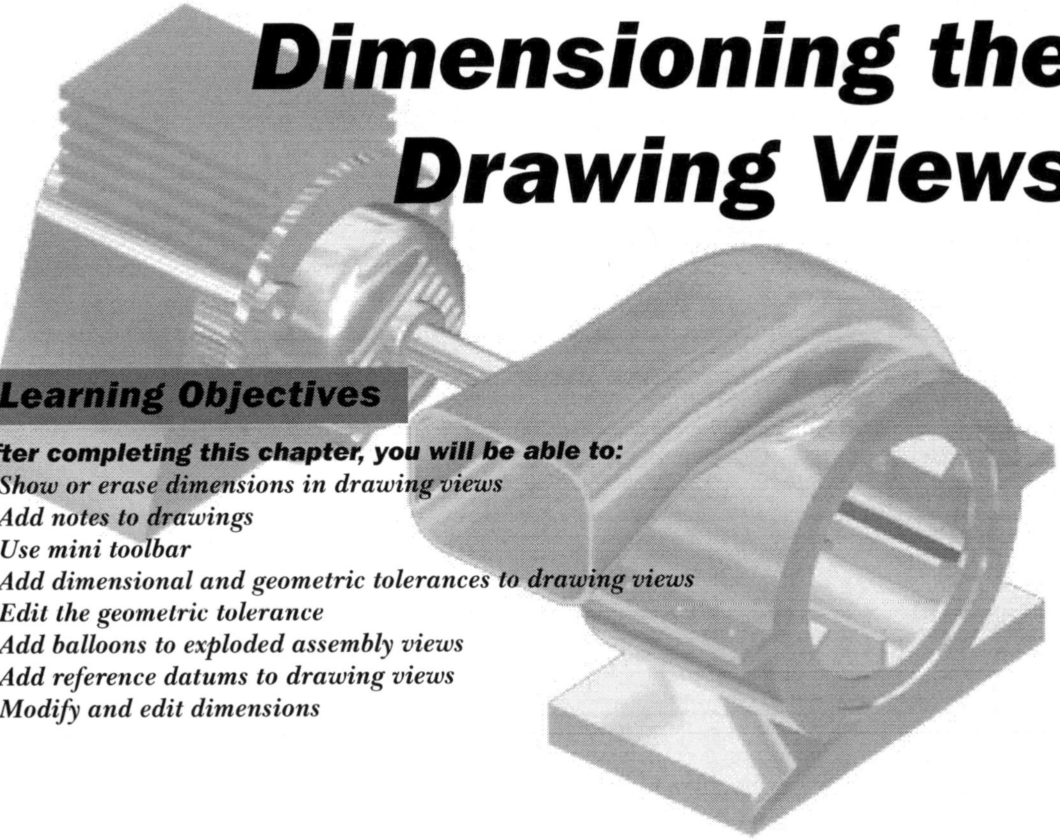

Learning Objectives

After completing this chapter, you will be able to:

- *Show or erase dimensions in drawing views*
- *Add notes to drawings*
- *Use mini toolbar*
- *Add dimensional and geometric tolerances to drawing views*
- *Edit the geometric tolerance*
- *Add balloons to exploded assembly views*
- *Add reference datums to drawing views*
- *Modify and edit dimensions*

DIMENSIONING THE DRAWING VIEWS

Once you have generated the drawing views, you need to generate dimensions, add notes, symbols, balloons, and so on in the drawing views. These dimensions are assigned to each entity of sketches in the model or the features associated with the model. These dimensions are associative in nature and they can be used to modify or drive the dimensions of a part; therefore these dimensions are called driving dimensions. Dimensions can be generated using the **Show Model Annotations** dialog box, refer to Figure 12-1. The options in the dialog box are discussed next.

Note

You can also create dimensions in the ***Drawing*** *mode, but these dimensions cannot drive the dimensions of the part. Therefore, the dimensions that you create in Creo Parametric are called driven dimensions.*

Show Model Annotations Dialog Box

Ribbon: Annotate > Annotations > Show Model Annotations

When you import a 3D model to a 2D drawing, by default, the dimensions and other information of the model are stored in the drawing but they remain invisible. You can let the required model information be shown or erased on a particular view by using the options in the **Show Model Annotations** dialog box. The visible items are referred to as shown and the items that are not visible are referred to as erased. The erased annotations are not permanently deleted from the model as they remain stored in the database. To show or erase model information, choose the **Show Model Annotations** tool from the **Annotations** group of the **Annotate** tab in the **Ribbon**; the **Show Model Annotations** dialog box will be displayed, as shown in Figure 12-1. This dialog box enables you to generate various parameters related to the model in the drawing view. There are six tabs in this dialog box. These are **Show the model dimensions**, **Show the model gtols**, **Show the model notes**, **Show the model surface finishes**, **Show the model symbols**, and **Show the model datums**. These tabs are discussed next.

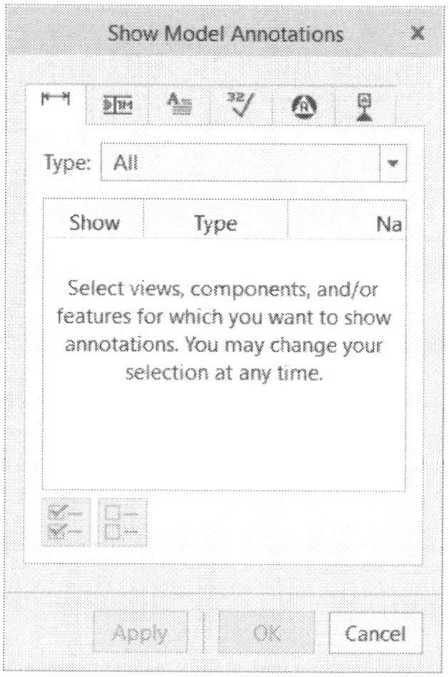

Figure 12-1 *The* ***Show Model Annotations*** *dialog box*

Show the model dimensions Tab

The **Show the model dimensions** tab is chosen by default in the **Show Model Annotations** dialog box. The options in this tab enable you to generate parametric dimensions in the drawing views. The bidirectional associative nature of the Creo Parametric enables you to modify the model by modifying the dimensions from the drawing or the 3D environment.

Type Drop-down List

This drop-down list contains options that can be used to specify which dimension you want to be displayed in the drawing. These options act as filters to display the dimensions. The options displayed in this drop-down are **All**, **Driving dimension annotation elements**,

All driving dimensions, **Strong driving dimensions**, **Driven dimensions**, **Reference dimensions**, and **Ordinate dimensions**. By default, **All** is selected.

List box

The list box consists of the dimensions to be displayed, their names, and a check box on the left of each dimension. You need to select a check box to display the corresponding dimension value in the selected drawing view. By default, all check boxes are cleared. If you want all the dimensions to be displayed in that view, then choose the **Select All** button. If you want to erase all the dimensions displayed in the view, then you need to choose the **Erase All** button. These two buttons are displayed below the list box.

Show the model gtols Tab

On choosing this tab, a list box will be displayed with a list of the tolerances applied to the dimensions in the part or the assembly environment. In this case, you can select the options to show or erase the geometric tolerances as you did in the case of the **Show the model dimensions** tab.

Show the model notes Tab

Choose this tab to show or erase the notes that were created in the **Part** mode. If any change is made to a note in the **Drawing** mode, the note in the **Part** mode is automatically modified. Similarly, if you modify this note in the **Part** mode, the modifications are reflected automatically in the **Drawing** mode.

Show the model surface finishes Tab

Choose this tab to show or erase the surface finish symbols associated with the part shown in the drawing views.

Show the model symbols Tab

Choose this tab to show or erase any symbol associated with the model.

Show the model datums Tab

Choose this tab to show or erase the critical measurement locations on the drawing view.

After selecting the required parameters, choose the **Apply** button to accept the settings and continue the selection process. Choose the **OK** button to accept the settings and exit the dialog box or choose the **Close** button to close the dialog box and cancel the settings performed.

Figure 12-2 shows some dimensioned drawing views.

The tools in the **Annotations** group of the **Annotate** tab of the **Ribbon** are discussed next.

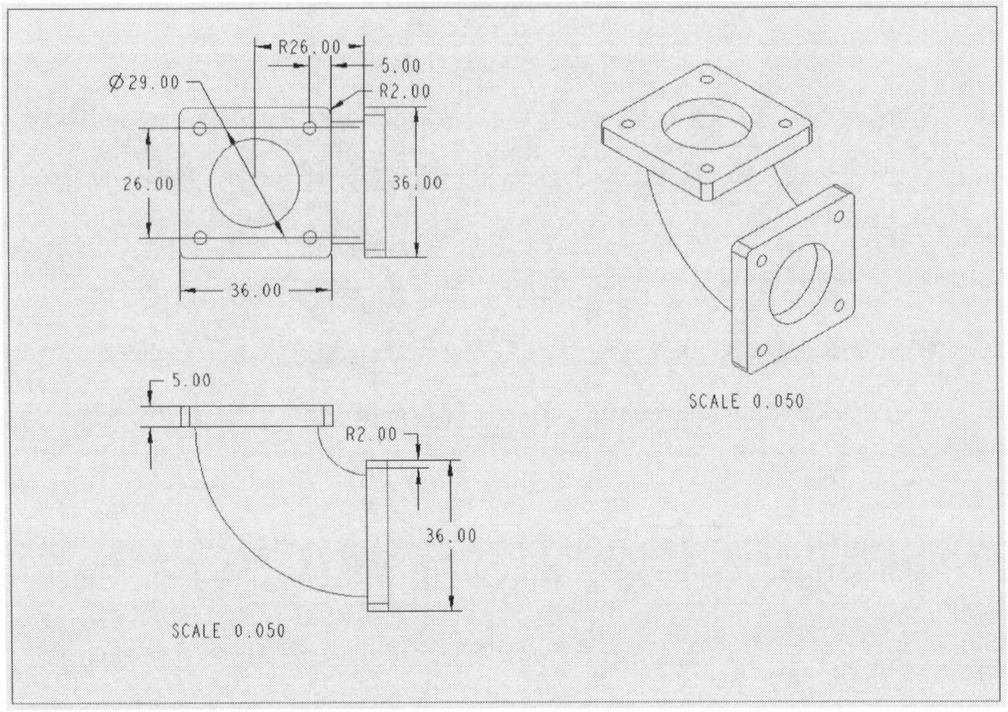

Figure 12-2 Dimensioned drawing views

Dimension

Ribbon: Annotate > Annotations > Dimension

The **Dimension** tool is used to create dimensions with new references in the drawing views. To create dimensions, choose the **Dimension** tool; the **Select Reference** dialog box will be displayed. You can select the required type of reference button from the dialog box to dimension an entity.

In Creo Parametric 8.0, defining dimensions is faster, easier, more visual, and intuitive. Typical creation of a radial dimension provides for up to four flip states that represent different witness line and arrow configurations.

In Creo Parametric, the flipping of dimension text and arrow provides better visibility, depending on the position of the cursor during placement of a dimension. As a result, the desired visibility of text is provided faster and with fewer clicks. When creating the dimension, move the cursor to different locations to see an update of the dimension text and arrow positions.

Tip

*You can also flip the dimension arrow by choosing the **Flip Arrows** tools from the mini toolbar.*

Surface Finish

Ribbon: Annotate > Annotations > Surface Finish

The **Surface Finish** tool is used to create the surface finish symbols in the drawing views. To create symbols, choose the **Surface Finish** tool; the **Open** dialog box will be displayed. Select the required symbol name from this dialog box and choose the **Open** button; the **Surface Finish** dialog box will be displayed along with the preview of the symbol and you will be prompted to place the symbol. Select the entity to which the surface finish symbol is to be assigned. You can also specify the roughness height in the **roughness_height** text box of the **Variable Text** tab. You can change the selection by again choosing a symbol from the **Browse** button in the **Definition** area.

Reference Dimension

Ribbon: Annotate > Annotations drop-down > Reference Dimension

The **Reference Dimension** tool is chosen to generate reference dimensions in the drawing views.

Jog

Ribbon: Annotate > Edit > Jog

The **Jog** tool is used to create a jog. On choosing the **Jog** tool, you will be prompted to select a dimension or leader type to create a jog. Select the required entity to create a jog.

MINI TOOLBAR

In Creo Parametric 8.0, when you click on 2D annotations and other drawing objects such as drawing views and tables, a context-specific mini toolbar is displayed. This mini toolbar is useful for faster and more intuitive command access.

Figures 12-3 and 12-4 show the mini toolbars displayed on clicking radial dimension and geometric tolerance, respectively.

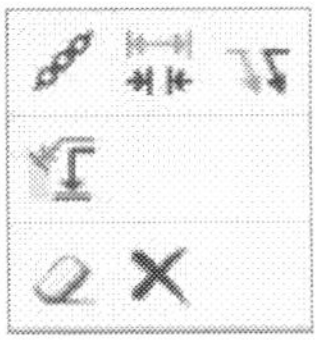

Figure 12-3 The mini toolbar displayed for radial dimension

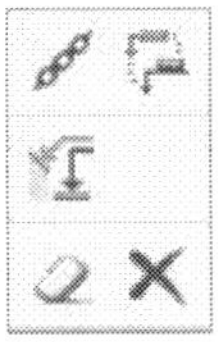

Figure 12-4 The mini toolbar displayed for geometric tolerance

ADDING NOTES TO THE DRAWING

Ribbon: Annotate > Annotations > Note

The notes added to a model in the **Part** mode can be easily displayed in the drawing views by using the **Show Model Annotations** dialog box. However, if you have not added any note to the model in the **Part** mode and want to add it in the drawing views, then you have to create them. To create various type of notes, choose **Annotate > Annotations > Note** down arrow from the

Ribbon; a flyout will be displayed, as shown in Figure 12-5. In this flyout, different types of options to create note are available. Choose the required type of note from the flyout. For example, if you choose the **Unattached Note** option from the flyout, the **Select Point** dialog box will be displayed. Choose the required button from the **Select Point** dialog box. Next, by using the left mouse button, specify a point on the drawing where you want to place the note; the **Format** tab will be added to the **Ribbon** and the text window will be displayed in the drawing area. Now, choose the required style and symbol from the **Style** and **Text** groups in the **Format** tab. Next, click twice outside the text box to finish the note. The note will be placed at the specified point. Now, you can drag it to any required location.

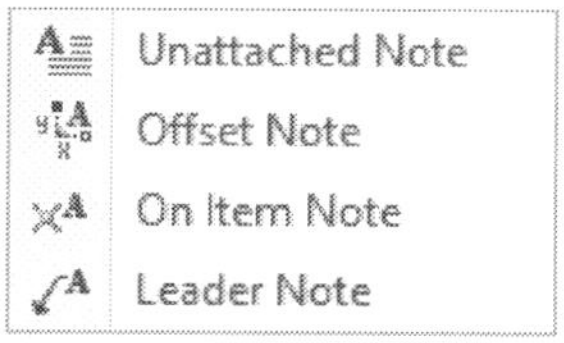

***Figure 12-5** The flyout with various note type options*

ADDING TOLERANCES TO THE DRAWING VIEWS

You can add geometric tolerances in the dimensions. Tolerance is defined as the difference between the maximum and minimum variations in the dimensions of the selected component. It is almost impossible to manufacture a component to the exact dimensions. In such cases, the tolerance value is added to dimensions to make sure that some variation that occurs during manufacturing can be taken care of. However, when you actually send a part for manufacturing, there are other parameters along with the dimension tolerances that may vary and require some tolerances. Depending upon these factors, the tolerances are divided into two types: dimensional tolerances and geometric tolerances.

Note

*By default, the **Tolerance** option in the **Dimension** dashboard is disabled. To enable the **Tolerance** option in the drawing mode, choose the **Drawing Properties** option from the **Prepare** flyout of the **File** menu; the **Drawing Properties** dialog box will be displayed. In this dialog box, click on the **change** option adjacent to the **Detail Options** row; the **Options** dialog box will be displayed. Choose the **Find** button at the bottom left corner of the dialog box to invoke the **Find Option** dialog box. Next, type **tol_display** in the **Type keyword** edit box and choose the **Find now** button. Select **tol display** in the **Choose option** list and then choose the **yes** option in the **Set value** drop-down list. Next, choose the **Add/Change** button and then choose the **Close** button to close the **Find Option** dialog box. To close the **Options** dialog box, choose the **Close** button.*

Dimensional Tolerances

These are the variation in the standard dimensional values. You can select a dimension and modify the type of tolerance from the **Dimension** dashboard. The **Dimension** dashboard is discussed later in this chapter. Figure 12-6 shows the drawing view having dimensions with the tolerance.

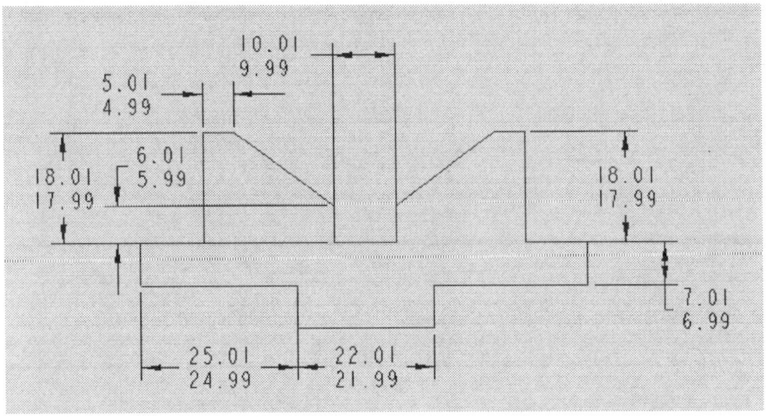

Figure 12-6 Drawing view showing dimensions with tolerances

Geometric Tolerances

Ribbon: Annotate > Annotations > Geometric Tolerance

When you send a component for manufacturing, you have to provide various other parameters along with the dimensions and dimensional tolerances. These parameters can be the geometric condition, surface profile, material condition, and so on. All these parameters are defined using the geometric tolerances. The geometric tolerances (will be called gtol henceforth) can be added to a drawing using the **Geometric Tolerance** dashboard, shown in Figure 12-7. To display this dashboard, choose **Annotate > Annotations > Geometric Tolerance** from the **Ribbon**; *gtol* will be attached to the cursor. Click to specify the location where you need to place the *gtol*. The options in the **Geometric Tolerance** dashboard are discussed next.

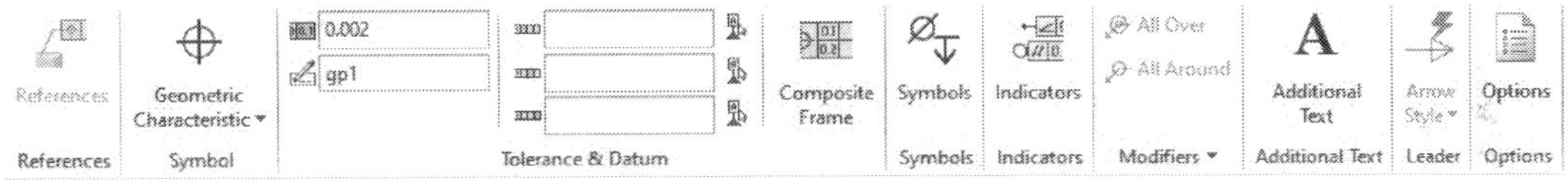

*Figure 12-7 The **Geometric Tolerance** dashboard*

References Group

The tools in the **References** group are accessible only in the **Part** mode.

Symbol Group

The tools in the **Symbol** group allows you to specify the geometric characteristic symbol for the selected entity. You can specify the geometric characteristics as straightness, flatness, circularity, cylindricity, profile of a line, profile of a surface, angularity, perpendicularity, parallelism, position, concentricity, symmetry, runout, or total runout.

Tolerance & Datum Group

The tools in the **Tolerance & Datum** group allow you to control the gtol values used for the selected entity.

The Tolerance value edit box is used to specify tolerance value. Additionally, you can include symbols or modifiers from the **Symbols** group that will appear along with the tolerance value. This tolerance value appears in the second compartment of the gtol frame.

By default, the Name edit box displays the name of the gtol generated by the system. You can use the gtol name within callouts to display the complete gtol frame and its value within other annotations.

The datum reference text boxes allows you to specify the primary, secondary, and tertiary datum references. These references can be single or compound datum references along with their modifiers. Contents of these three datum references appear as three compartments of the selected gtol frame.

The **Composite Frame** tool is used to invoke the **Composite Frame** slide-down panel. In the slide-down panel, you can define primary, secondary and tertiary datum planes for different tolerance values. You can add, delete, shift up and down, or modify the rows in the slide-down panel.

When you select the **Share datum references** check box, the primary, secondary, and tertiary reference values of the first row are used for all the other rows, and you cannot modify these three datum references for other rows. Also the reference selection button gets disabled for the rows other than the first row.

Indicators Group

The **Indicators** tool in the **Indicators** group allows you to control the tolerance applied to the selected geometry. Choose the **Indicators** button to invoke the **Indicators Frame** slide-down panel. You can specify indicator types (Direction feature, Collection plane, Intersection plane, and Orientation plane), symbols, and datum reference for the selected gtol frame. The specified indicators will be displayed after the gtol frame according to the requirements of ISO 1101-2012.

Each row of the **Indicators Frame** slide-down panel allows you to specify a single indicator that is displayed together with the gtol frame. Using the buttons available in the panel, you can add, delete, shift up and down, or modify the row representing individual indicators.

For indicator types such as direction feature, collection plane, and orientation plane, you can specify the symbol as angularity, perpendicularity, or parallelism. Similarly, for intersection plane, you can specify the symbol as perpendicularity, parallelism, or including.

Modifiers Group

The tools in this group allow you to specify the modifier for the selected gtol frame. Remember that you can specify only one modifier for the selected geometric symbol.

Additional Text Group

Choose the **Additional Text** tool in the **Additional Text** group to invoke a slide-down panel. The text boxes in this panel allow you to specify additional text to be displayed above, below, on the left, and on the right of the frame of the selected gtol frame.

Leader Group

The **Arrow Style** drop-down list in this group is used to specify the arrow style of the leader displayed for the selected gtol frame.

Options Group

The options available in the slide-down panel of the **Options** group are used to control the position of the gtol frame driven by model. Note that, the **Options** tool in the **Options** group will be enabled only when you have selected a geometric tolerance created in the **Part** mode of Creo Parametric. Select the **Driven by model** check box in the **Placement position** area to set the placement position of the selected gtol frame driven by the model geometry.

The **References driven by model** and **Attach point(s) driven by model** check boxes under the **Attachment** area are used to specify the attachment point of the selected gtol frame.

The check boxes in the **Designation** area are only available in the **Part** mode. Choose the **Control Characteristics** tool to designate the annotation elements of the selected gtol frames.

Symbols Group

Choose the **Symbols** tool in the **Symbols** group to invoke the **Symbols** slide-down panel. The options in this panel are used to add symbols to the edit boxes available in the **Tolerance & Datum** group.

Figure 12-8 shows the drawing view with reference datums and gtol frame.

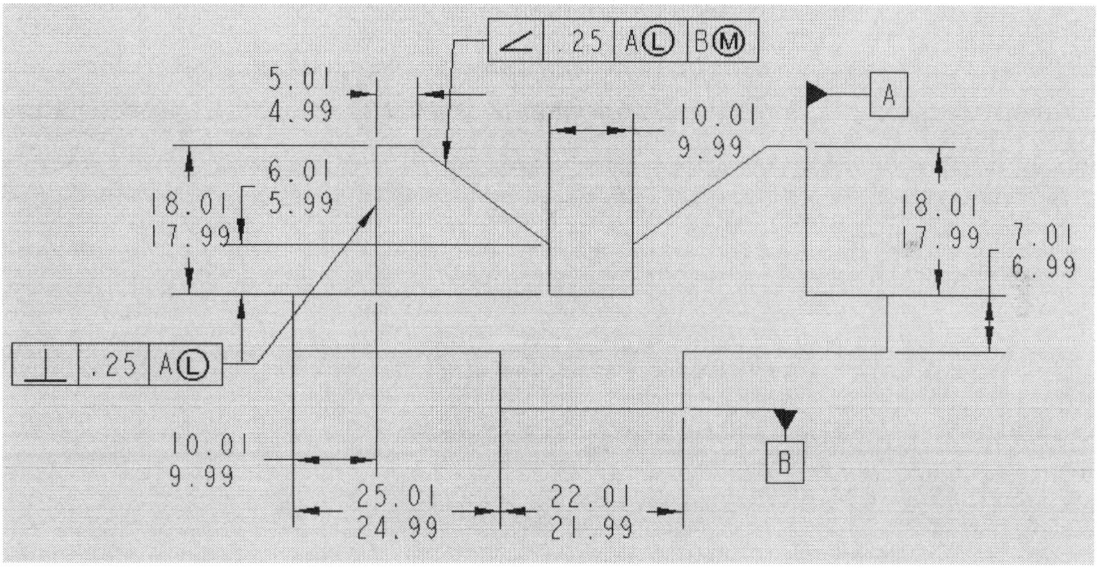

***Figure 12-8** The drawing view with reference datums and gtol frame*

EDITING THE GEOMETRIC TOLERANCES

The gtol frame added to the drawing views can be edited using the options of the **Geometric Tolerance** dashboard. To edit a gtol frame, select it from the drawing area; the **Geometric Tolerance** dashboard will be displayed in the drawing area where you can modify the attributes of the selected gtol frame.

ADDING BALLOONS TO THE ASSEMBLY VIEWS

Ribbon: Annotate > Annotations drop-down > Balloon Note

The **Balloon Note** tool is used to create balloons associated with the assembly in the drawing views. To add a balloon to the assembly drawing views, choose the **Annotate > Annotations > Balloon Note** from the **Ribbon**; the **NOTE TYPES** menu will be displayed. In this menu, choose the type of leader to be used, if required. Next, choose the **Make Note** option from this menu; the **Select Point** dialog box will be displayed. Choose the required button from the **Select Point** dialog box. Next, specify a point on the drawing where you want to place the balloon by using the left mouse button. Enter the text in the **Enter NOTE** edit box and choose the required symbol from the **Text Symbol** window displayed at the bottom. Choose the **Accept value** button to accept the text. On doing so, you will again be prompted to enter a note in the **Enter NOTE** edit box. Now, choose the **Cancel input** button; the balloon will be placed at the specified point, refer to Figure 12-9.

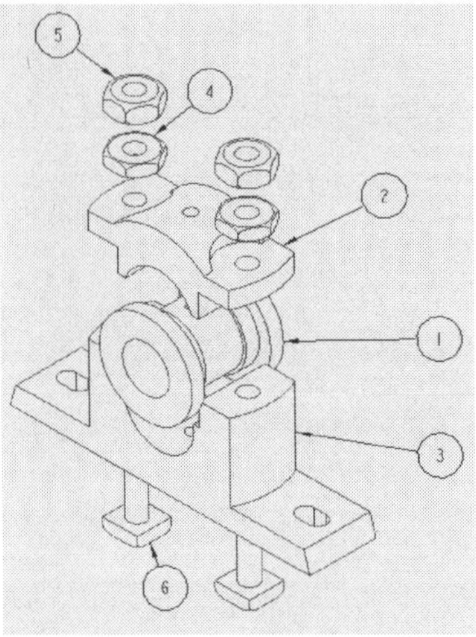

Figure 12-9 *Drawing view with balloons*

Tip

1. The notes or balloons can be modified and edited using the options in the shortcut menu displayed on selecting and right-clicking on the corresponding note or balloon.

*2. To generate automatic balloons in assembly drawings, you can use the tools available in the **Create Balloons** drop-down of the **Balloons** group in the **Table** tab. Note that automatic balloons can be created only when you have inserted a BOM table in the drawing.*

ADDING DATUM FEATURE SYMBOL TO THE DRAWING VIEWS

To specify geometric tolerance on a geometry, you must specify datum references. In the drawing mode of Creo Parametric, the **Datum Feature Symbol** tool is used to specify the location of datum references. To specify a datum reference, choose the **Datum Feature Symbol** tool in the **Annotations** group of the **Annotate** tab; a datum feature symbol will get attached to the cursor and you will be prompted to select an edge, an entity, a dimension, a gtol, a dimension witness line, a datum point, a draft datum point, a curve, a point on a surface, a vertex, a cosmetic sketched entity, an entity, a vertex, or a section entity. As you click on the desired entity; the datum feature symbol gets attached to the selected entity and you can place the datum feature symbol by clicking the middle mouse button. As you place the datum feature symbol in the drawing area; the **Datum Feature** dashboard gets activated automatically, as shown in Figure 12-10.

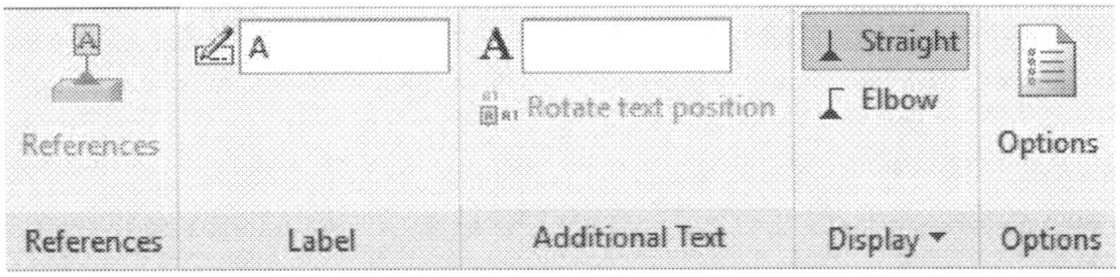

*Figure 12-10 The **Datum Feature** dashboard*

The **References** tool in the **References** group is only accessible in the Part mode. You can use the Label edit box in the **Label** group to specify label for the datum feature that will appear in the datum feature symbol frame. You can use a combination of one or two capital alphabets from A to Z (except I, O, and Q) or can also use a combination of an alphabet and a number to specify the label of the datum feature symbol.

If the label entered in the Label edit box is invalid a message will pop up informing that the label entered is not a standard label syntax.

The tools in the **Additional Text** group enable you to specify additional text that appears along with the instance of the datum feature symbol. You can specify any alpha-numeric string with no limitations as the additional text. The label for the datum feature symbol and the additional text that you specify appears horizontally aligned to the screen. You can choose the **Rotate text position** option to change the position of the additional text around the frame of the datum feature symbol.

You can choose the tools available in the **Display** group to control the display of the leader with which the datum feature symbol is attached in a geometry. You can select the **Straight** or **Elbow** option to specify the appearance of the leader. You can also select the **Display per ASME Y14.5m-1982** check box in the **Display** drop-down so that the datum feature symbol appears according to the ASME standard. Note that, the **Straight** and **Elbow** options are not available after you select this check box.

Figure 12-11 shows the datum feature symbols added to the drawing view to specify the geometric characteristics of the given model.

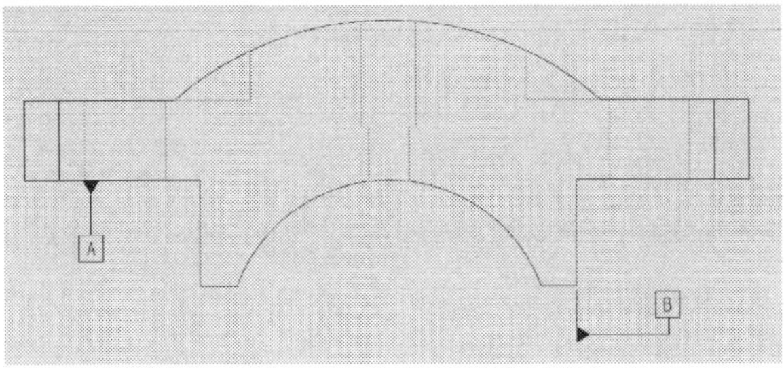

Figure 12-11 Datum reference symbols added to a drawing view

MODIFYING AND EDITING DIMENSIONS

Creo Parametric allows you to edit and modify the dimensions assigned to the drawing views. There are various methods of editing and modifying the driving dimensions. Before these methods are discussed, it is important for you to learn the terminology used in these methods. Figure 12-12 shows the terminology that is used in the Drawing mode of Creo Parametric.

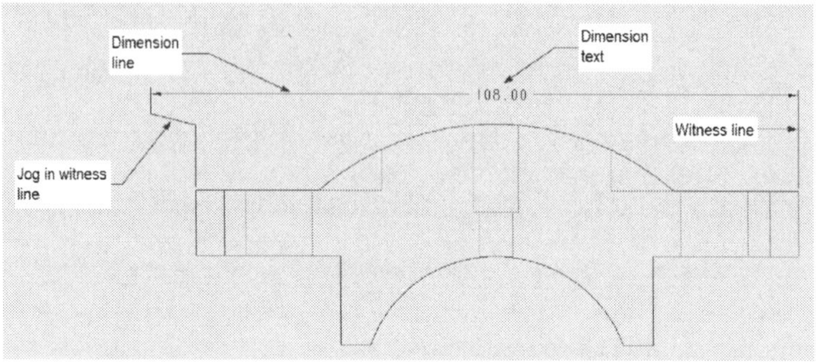

Figure 12-12 *Parameters associated with dimensions*

Modifying the Dimensions Using the Dimension Dashboard

You can modify the properties of the dimensions in a drawing using the **Dimension** dashboard. The **Dimension** dashboard is automatically displayed above the drawing area when you place a newly created dimension or select an existing dimension, refer to Figure 12-13. The options available in this dashboard are discussed next.

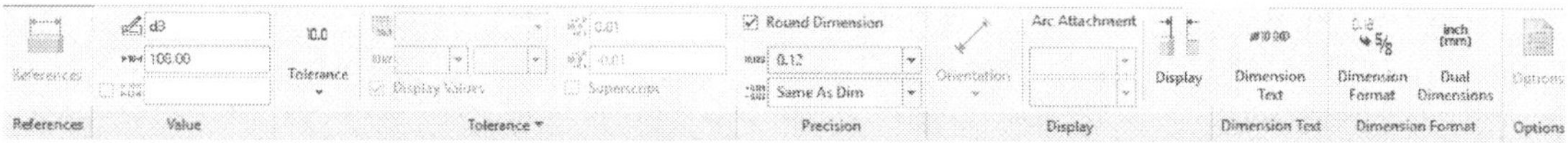

Figure 12-13 *The **Dimension** dashboard*

References Group

This tool is available only in the **Part** mode. The **References** tool in the **References** group allows you to control references that are used to create the selected dimension. In the **Part** mode, click the **References** option; the **References** dialog box will be displayed listing details of the references that were used to create the selected dimension.

Value Group

The tools available in the **Value** group allow you to control the dimension name and its value. The Name edit box displays the symbolic name of the selected driven dimensions. You can modify the symbolic name of the dimension using this edit box. The Nominal Value edit box displays the nominal value of the selected dimension. You can modify the nominal value if you select single or multiple driving dimensions.

The Override check box enables you to override the nominal value of the selected driven dimension. For driving dimensions, the Override Value check box will be disabled. If you select a single dimension and specify a new value in the Override Values edit box, the override value will appear as a dimension value in the graphics area.

Tolerance Group

The options available in the **Tolerance** group allow you to control the tolerances applied to the selected dimension. Choose the **Tolerance** tool to display the tolerance modes that you can specify for the selected dimension. This tool is disabled by default and the procedure to enable this option is discussed earlier in this chapter.

The Display Values check box is available only when you have selected **ISO** as the **Tolerance** in the **Drawing Properties** dialog box. The Upper and Lower tolerance edit boxes are available only when the tolerance mode is set to Limits, Plus-Minus, and Symmetric. Also the **Superscript** check box is enabled only when **Symmetric** is selected as the tolerance mode.

You can select the **Tolerance Value Only** check box available in the expanded **Tolerance** group to display the tolerance values instead of the dimension value for the selected dimension.

Precision Group

The tools available in the **Precision** group allows you to define the precision of the dimension and tolerance values of the selected dimension. The **Round Dimension** check box is selected by default. As a result, the dimension value after precision will get round off.

Display Group

The options available in the **Display** group are used to control the display of the selected dimension. The **Orientation** tool in this group allows you to specify the orientation of the selected dimension. This tool will be activated only when you select a dimension. Depending on the type of the selected dimension, a list of corresponding orientation options appears in the drop-downs available in this group.

The options in the **Arc Attachment** area allow you to set the location of the attachment points for the linear dimension defined with an arc or circle as its references. Choose the **Display** tool to invoke a slide-down panel. You can use the options in this slide-down panel to specify the text orientation and configuration for the selected dimension.

Dimension Text Group

The **Dimension text** tool available in the **Dimension Text** group allows you to control the display of the dimension text. Choose the **Dimension Text** tool to invoke the slide-down panel that has the options to modify the dimension text, specify the prefix and suffix, and insert symbols with the dimension text. You can select more than one dimension at a time and all the selected dimensions are updated with the modified dimension text, along with the prefix and suffix.

Dimension Format Group

The **Dimension Format** tool available in the **Dimension Format** group allows you to specify the display of dimensions in decimal or fractional format. You can also specify units for the angular dimensions using the **Angular Dimension Units** drop-down list.

The **Dual Dimensions** tool in the **Dimension Format** group is used to specify the placement of primary and secondary dimensions. By default, the secondary dimensions are not displayed in the drawing. To enable the representation of dual dimensions, choose **File > Prepare > Drawing Properties** to invoke the **Drawing properties** dialog box. Next, click the **change** option adjacent to the **Detail Options** row to display the options dialog box. Enter **dual_dimensioning** in the **Option** edit box and press ENTER. Next, select desired configuration for the dual dimensions in the **Value** drop-down list and choose the **Add/Change** button and then choose the **Apply** button to turn on the display of dual dimensions in the drawing. After specifying the settings for the dual dimensions, choose the close button to close the dialog boxes.

Options Group

The options available in this group have already been discussed earlier in this chapter.

Cleaning Up the Dimensions

Ribbon: Annotate > Edit > Cleanup Dimensions

Generally, dimensions displayed in the drawing views are scattered and improperly placed. This, in turn, makes the drawing views look very confusing. Creo Parametric helps you to clear this confusion by using two methods. The first method is to place all dimensions manually at the appropriate location in the view. The second and more convenient method is to place these dimensions in a proper order by choosing the **Cleanup Dimensions** tool from the **Edit** group in the **Annotate** tab of the **Ribbon**. On doing so, the **Clean Dimensions** dialog box will be displayed and you will be prompted to select dimensions or individual views to clean. After you select the dimension to clean, press the middle mouse button; the options in this dialog box will be enabled, as shown in Figure 12-14. Note that a dimension is already selected to show all available options. These options are discussed next.

Placement Tab

The options in this tab of the **Clean Dimensions** dialog box are used to place the dimensions in the required location. The distance between the two dimensions and the distance between the view boundary and the dimension are set under this tab. The options in this tab are discussed next.

Space out dimensions

This check box is used to maintain the distance between two dimensions. Most of the other options in this tab are available only when you select this check box.

Offset

The **Offset** edit box is used to specify the offset distance between the first dimension and the view boundary.

Increment

The **Increment** edit box is used to specify the offset distance between two dimension lines.

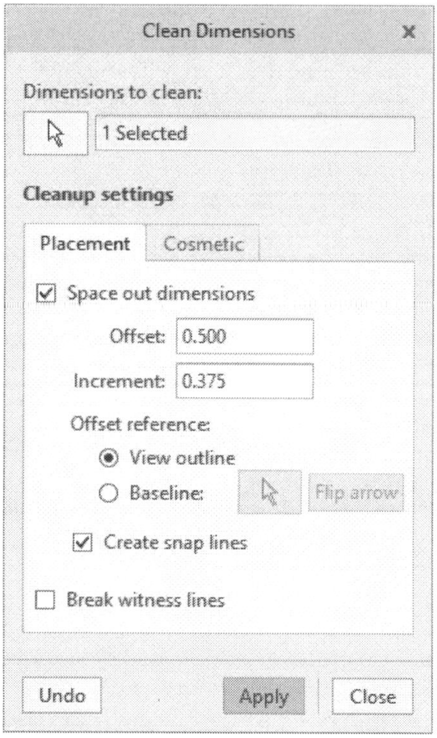

*Figure 12-14 The **Clean Dimensions** dialog box with the **Placement** tab chosen*

Offset reference Area

Options in this area are used to define the reference for offset. These options are discussed next.

View outline: The **View outline** radio button is selected to specify the offset distances taking the outline of the view as reference.

Baseline: The **Baseline** radio button is selected to specify the offset distance taking a selected flat edge, datum plane, snap line, detail axis line, or view border as the reference. When you invoke this option, you will be prompted to select either of the above mentioned entities as the reference.

Create Snap Lines

The **Create snap lines** check box is selected to create the snap lines, which can be used to clean the dimensions. The snap lines will be created at the distances specified in the **Offset** and the **Increment** edit boxes. If this check box is cleared, the snap lines will not be created.

Break witness lines

The **Break witness lines** check box is selected to break the witness lines if some dimensions overlap each other.

Cosmetic Tab

The options in the **Cosmetic** tab of the **Clean Dimensions** dialog box are used for determining the location of the dimension value with respect to the dimension line, refer to Figure 12-15. The arrows and the orientation of the dimension text can also be controlled using this tab. The options in this tab are discussed next.

Flip Arrows

The **Flip arrows** check box is selected to flip the arrows of the dimensions that are not easily adjusted inside the dimension lines.

Center Text

The **Center text** check box is selected to place the dimensions at the center of the dimension lines.

Horizontal and Vertical Buttons

These buttons are used to specify the placement of the horizontal or vertical dimension text outside the dimension line if they do not fit.

Figures 12-16 and 12-17 show the drawing views before and after using the **Clean Dimensions** dialog box.

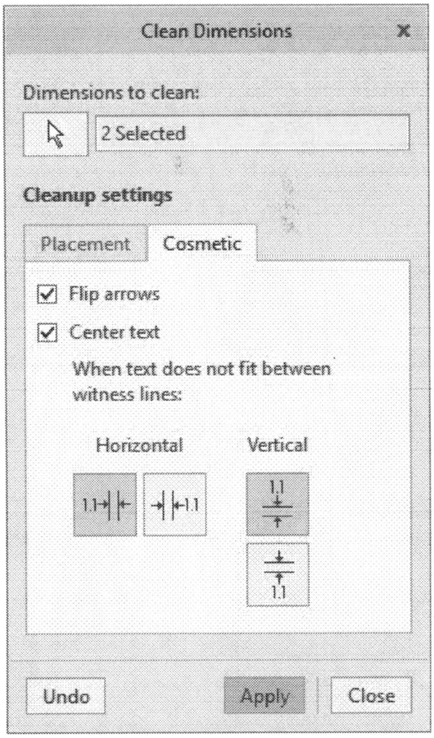

*Figure 12-15 The **Clean Dimensions** dialog box with the **Cosmetic** tab chosen*

Note

The system does not show a reference datum plane in a drawing view unless it is perpendicular to the drawing area.

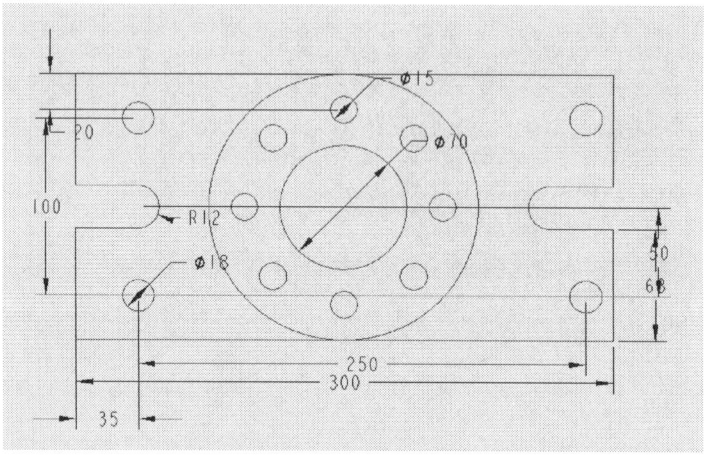

Figure 12-16 *Drawing view with dimensions before cleaning*

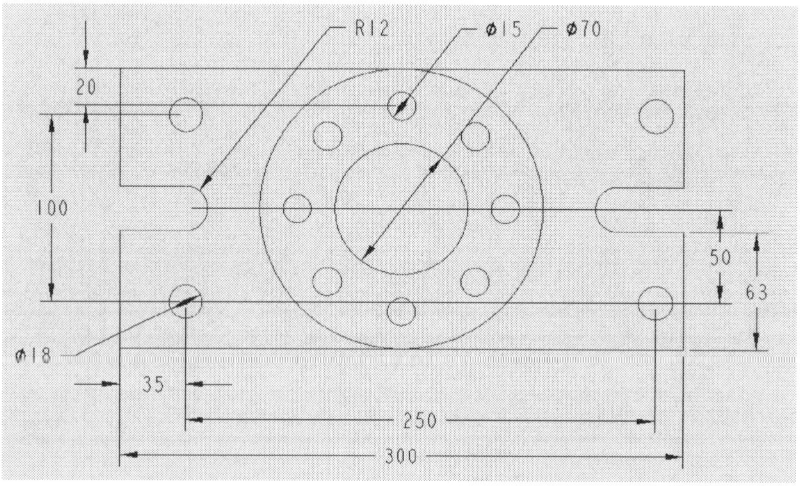

Figure 12-17 *Drawing view with dimensions after cleaning*

Note

*The **Create snap lines** check box is cleared for Figure 12-17.*

TUTORIALS

Tutorial 1

In this tutorial, you will generate drawing views of the part shown in Figure 12-18. The views to be generated are shown in Figure 12-19. **(Expected time: 45 min)**

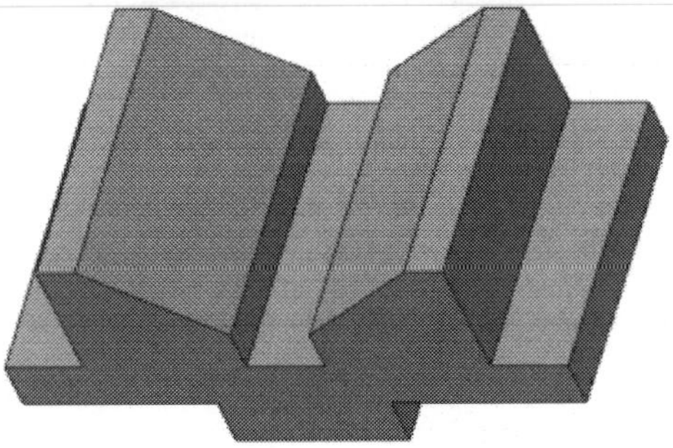

***Figure 12-18** Model for generating drawing views*

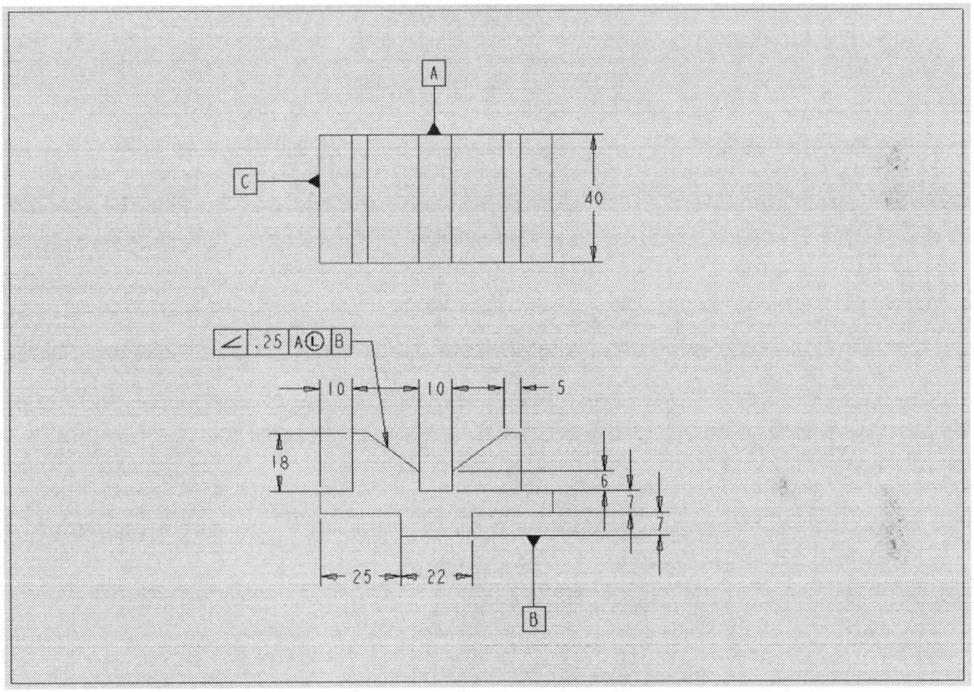

***Figure 12-19** Drawing views to be created along with the dimensions*

The following steps are required to complete this tutorial:

a. Create the model in the **Part** mode using the given dimensions and save the file.
b. Open a new drawing file in the **Drawing** mode and generate the top and front views of the model on the drawing sheet.
c. Generate the dimensions in the views and clean them using the **Clean Dimensions** dialog box.
d. Add geometric tolerance to the required entities.

Before you start creating the model and the drawing views, set the working directory to *C:\Creo-8.0\C12*. Make sure that the *C12* folder exists inside the *Creo-8.0* folder.

Creating the Model

First, create the given model in the **Part** mode. Its views will be generated later.

1. Create a new **Part** file and name it as *C12tut1*.

2. Create the model in the **Part** mode and save this model.

Starting a New Drawing File

To generate drawing views, you need to start a new drawing file.

1. Choose the **New** button from the **File** menu; the **New** dialog box is displayed.

2. Select the **Drawing** radio button from the **Type** area and enter the name of the drawing as *C12tut1* in the **File name** edit box.

3. Choose the **OK** button from the **New** dialog box; the **New Drawing** dialog box is displayed.

4. If the name of the model is not displayed in the **Default Model** area, choose the **Browse** button and select the model. Select the **Empty** radio button from the **Specify Template** area, the **Landscape** button from the **Orientation** area, and the **A4** option from the **Standard Size** drop-down list.

5. Choose the **OK** button from the **New Drawing** dialog box to enter the **Drawing** mode. The **Drawing** mode is invoked and a new A4 size drawing sheet is displayed.

Generating the Drawing Views

You need to generate the top and front views of the model.

1. Choose the **General View** tool from the **Model Views** group of the **Layout** tab in the **Ribbon**; the **Select Combined State** dialog box is displayed. Choose the **OK** button from the dialog box.

2. Now, click in the sheet to specify the center point for the drawing view, refer to Figure 12-20. When you specify the point, the **Drawing View** dialog box is displayed.

3. Choose the **TOP** option from the **Model view names** list in the **View orientation** area.

4. Choose the **Scale** option from the **Categories** area and select the **Custom scale** radio button, edit the scale factor for the view to **0.06**.

5. Choose the **Apply** button and choose the **OK** button. The default view is oriented as the top view and scaled to the specified scale factor.

6. Choose the **No Hidden** option from the **Graphics** toolbar and repaint the screen using the **Repaint** tool from the **Graphics** toolbar. The top view is shown in Figure 12-20.

7. Choose the **Projection View** tool from the **Model Views** group of the **Layout** tab in the **Ribbon**. Make sure that the angle of projection is in the third angle projection, else you will not get the desired result.

8. Specify the center point for the drawing view by clicking below the top view, as shown in Figure 12-20.

   The sheet after generating the two views is shown in Figure 12-20.

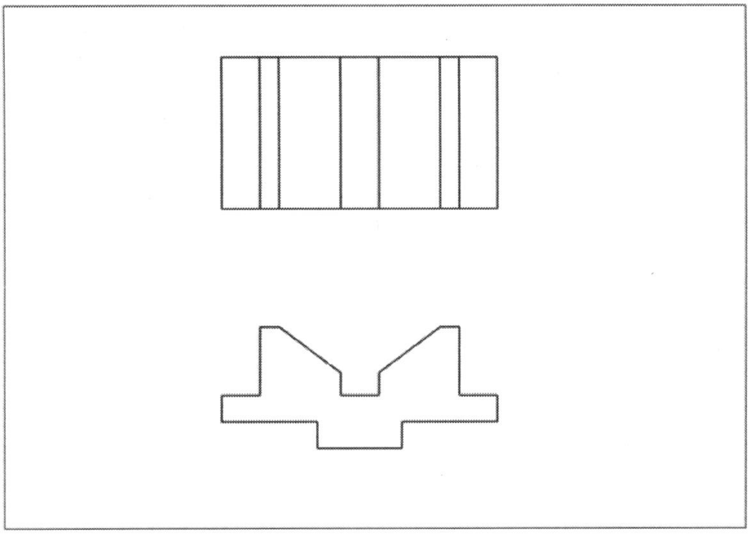

Figure 12-20 *Top and front views of the model*

Placing Datum Feature Symbol

1. Choose the **Datum Feature Symbol** tool from the **Annotations** group of the **Annotate** tab; the datum feature symbol gets attached to the cursor, refer to Figure 12-19.

2. Click on the top edge of the top view and drag the cursor to the desired point and press the middle mouse button to place the symbol in the drawing. The datum feature symbol is automatically named as **A** because this is the primary datum feature of the model.

3. Repeat steps 1 and 2 to place secondary and tertiary datum feature symbol B and C, respectively.

Dimensioning the Drawing Views

1. Choose **Annotate > Annotations > Show Model Annotations** from the **Ribbon**; the **Show Model Annotations** dialog box is displayed with the **Show the model dimensions** tab chosen by default. Also, you are prompted to select a view to display its annotations.

2. Select the front view (the view at the bottom); all dimensions of this view are listed in the dialog box.

3. Choose the **Select All** button available below the list box; all dimensions are displayed as strong dimensions in the selected view.

4. Choose the **Apply** button to accept and continue the settings.

5. Now, select the top view from the drawing area; its dimension is listed in the **Show Model Annotations** dialog box.

6. Select the check box on the left of the dimensions listed; the dimension of the selected item is displayed in the drawing area as a strong dimension.

7. Choose the **OK** button from the dialog box to accept the settings and exit the dialog box.

Placing the Dimensions in Order

The dimensions displayed in the drawing views are scattered and improperly placed. You need to place the dimensions in proper order using the **Clean Dimensions** dialog box.

1. Choose **Annotate > Edit > Cleanup Dimensions** from the **Ribbon** to display the **Clean Dimensions** dialog box; you are prompted to select the dimensions to be cleaned.

2. Select all dimensions from both the views by using the window selection method; the selected dimensions are highlighted. Press the middle mouse button to activate the options in the **Placement** tab.

3. Enter **0.5** in the **Offset** and **Increment** edit boxes.

4. Clear the **Create Snap Lines** check box. Choose the **Apply** button and then choose the **Close** button to exit the **Clean Dimensions** dialog box.

5. After cleaning the dimensions, the drawing views should look similar to the ones shown in Figure 12-21. Select and move the dimensions if necessary. You can also increase the text size of the dimensions if required.

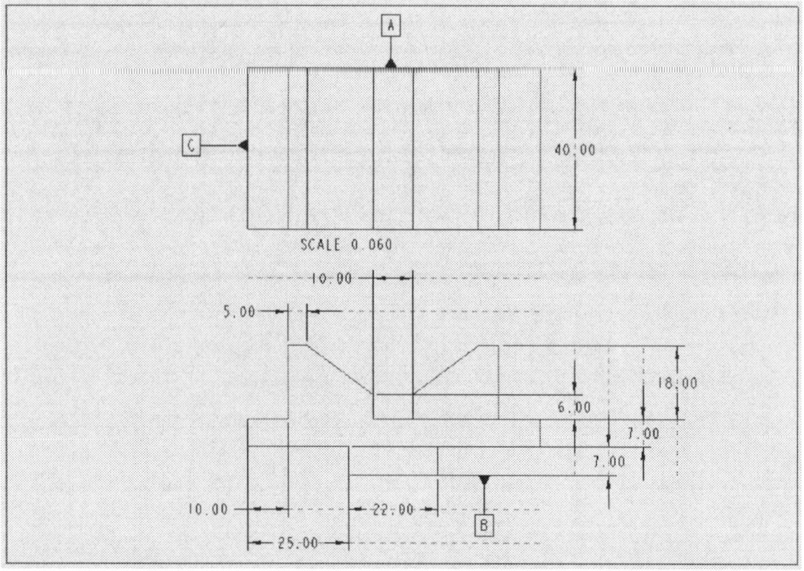

***Figure 12-21** Top and front views of the model with dimensions and datum planes*

Displaying the Geometric Tolerance

1. Choose **Annotate > Annotations > Geometric Tolerance** from the **Ribbon**; the gtol frame gets attached to the cursor and you are prompted to select edge entities, dimensions, gtols, notes, dimension, witness lines, coordinate systems axis, center axis endpoints.

2. Select the left-inclined edge from the front view and then click the middle mouse button at the desired place in the drawing area to place the gtol frame. As you place the gtol in the drawing view; the **Geometric Tolerance** dashboard gets activated.

3. Choose the **Angularity** option from the **Geometric Characteristic** drop-down list and type **0.25** in the Tolerance value edit box.

4. Specify the primary datum reference for the gtol by choosing the **Select the datum reference from the model** button from the **Tolerance & Datum** group; the **Select** dialog box is displayed. Select the **A** datum feature symbol from the drawing area and choose the **OK** button from the **Select** dialog box; the name of the datum feature symbol is displayed in the gtol frame.

5. Click in the edit box of the primary datum reference and choose the **Symbols** tool to invoke the symbols gallery. Select the least material condition symbol from the gallery; the symbol is inserted in the gtol frame.

6. Similarly select the secondary datum reference using the button available adjacent to the Secondary datum reference edit box or you can type the name of secondary reference in the edit box. Remember that the name of the datum reference is always in capital letters.

   The drawing views after adding the geometric tolerances should look similar to the ones shown in Figure 12-22. You need to move the note for the geometric tolerances to a location shown in Figure 12-23 by selecting and dragging it. Similarly, arrange all dimensions and datums on the drawing sheet, as shown in Figure 12-23.

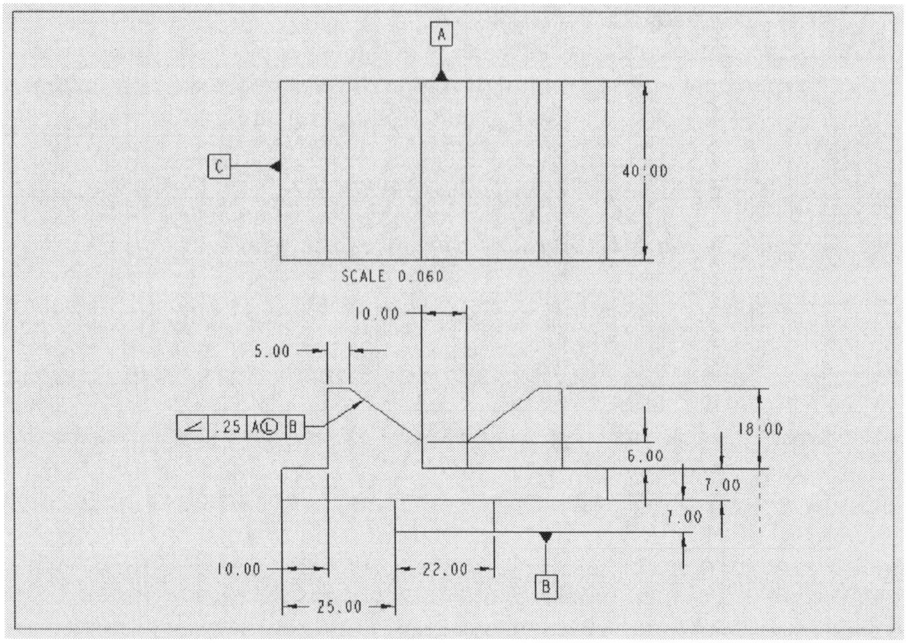

Figure 12-22 Top and front views of the model with dimensions, datum planes, and geometric tolerances

Saving the Drawing File

Now, you need to save the drawing file that you have created.

1. Choose the **Save** button from the **Quick Access** toolbar; the **Save Object** dialog box is displayed with the name of the drawing file that you had entered earlier. Press ENTER to confirm the saving of the file.

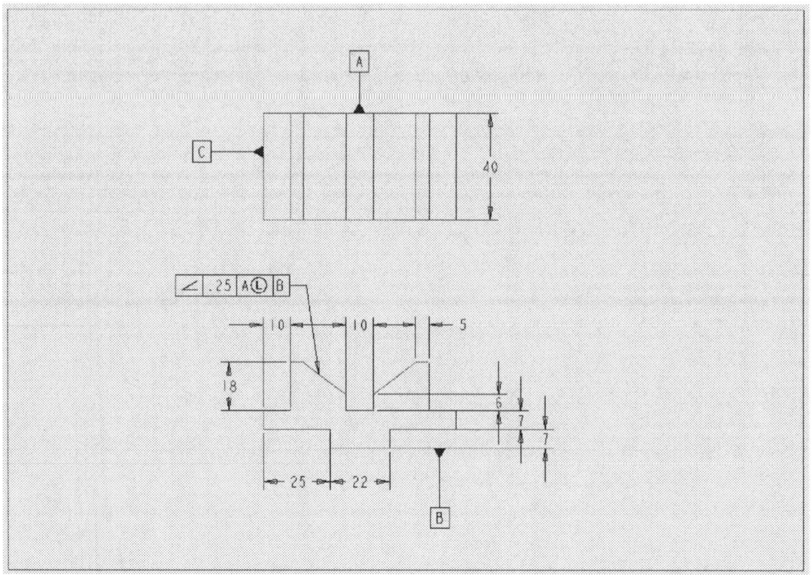

Figure 12-23 The required drawing views after arranging all dimensions and datums

Tutorial 2

In this tutorial, you will generate the drawing views of the model shown in Figure 12-24. The dimensioned drawing views are shown in Figure 12-25. **(Expected time: 45 min)**

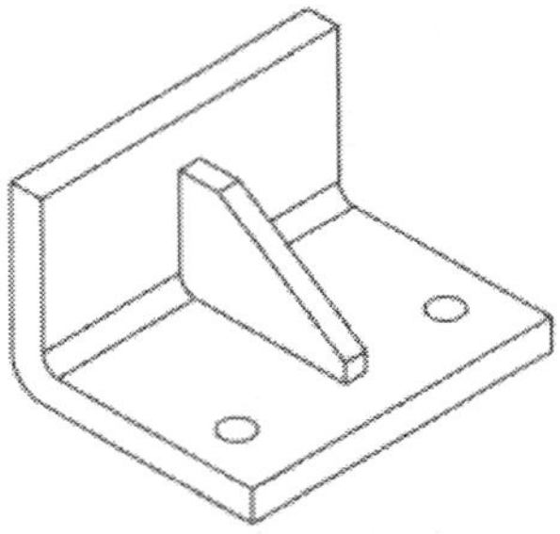

Figure 12-24 *Model for generating drawing views*

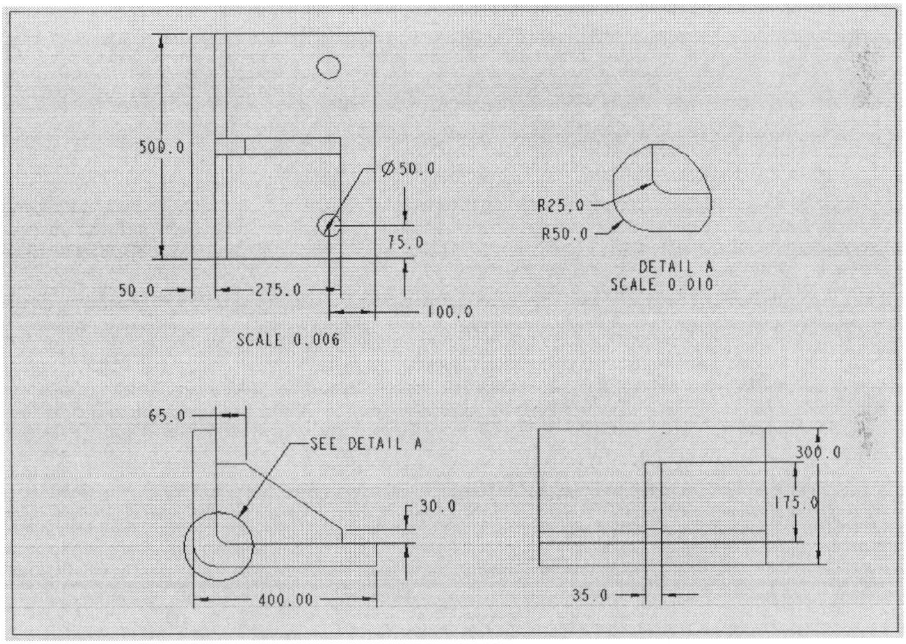

Figure 12-25 *Drawing views to be generated in Tutorial 2*

The following steps are required to complete this tutorial:

a. Create the model in the **Part** mode using the given dimensions and save the file.
b. Open a new drawing file in the **Drawing** mode and generate the top view, front, right, and detailed views of the model on the drawing sheet.
c. Dimension the views and clean the dimensions using the **Clean Dimensions** dialog box.

Make sure the working directory is set to *C:\Creo-8.0\C12*.

Creating the Model

First, in the **Part** mode, you need to create the model whose drawing views are to be generated.

1. Create the model in the **Part** mode using the given dimensions.

2. Save this model created in the **Part** mode with the name *C12tut2*.

Starting a New Drawing File

To generate the drawing views, you first need to start a new drawing file.

1. Choose the **New** button from the **File** menu; the **New** dialog box is displayed.

2. Select the **Drawing** radio button from the **Type** area and enter the name of the drawing as *C12tut2* in the **File name** edit box.

3. Choose the **OK** button from the **New** dialog box; the **New Drawing** dialog box is displayed. In this dialog box, the name of the model is displayed in the **Default Model** area.

4. If the name of the model is not displayed in the **Default Model** area, choose the **Browse** button and select the model.

5. Select the **Empty** radio button from the **Specify Template** area, choose the **Landscape** button from the **Orientation** area, and select the **A4** option from the **Standard Size** drop-down list.

6. Choose the **OK** button from the **New Drawing** dialog box; you enter the **Drawing** mode and a new drawing sheet of A4 size is displayed on the screen.

Creating the Drawing Views

First, the top view of the model needs to be generated and then the front view will be generated by projecting it from the top view. The right view will be projected from the front view. The detail view can be placed on the drawing sheet by specifying a scale for it.

1. Choose **Layout > Model Views > General View** from the **Ribbon**; the **Select Combined State** dialog box is displayed. Choose the **OK** button from the dialog box.

2. Now, click on the top-left side of the sheet to specify the center point for the drawing view; the **Drawing View** dialog box is displayed.

3. Choose the **TOP** option from the **Model view names** list in the **View orientation** area.

4. Choose the **Apply** button; the default view is oriented as the top view.

5. Choose the **Scale** option in the **Categories** display area and specify the scale factor for the view as **0.006** in the **Custom scale** edit box. Exit the dialog box and then choose the **No Hidden** option from the **Display Style** drop-down in the **Graphics** toolbar. Repaint the screen using the **Repaint** tool from the **Graphics** toolbar. For the top view, refer to Figure 12-26.

   The front view of the model needs to be generated from the top view.

6. Choose **Layout > Model Views > Projection View** tool from the **Ribbon**. Next, specify the center point for the drawing view which is below the top view, refer to Figure 12-26. Make sure that the angle of projection is the third angle projection else the desired projection will not be displayed.

7. Similarly, place the right and detailed views on the drawing sheet, refer to Figure 12-26.

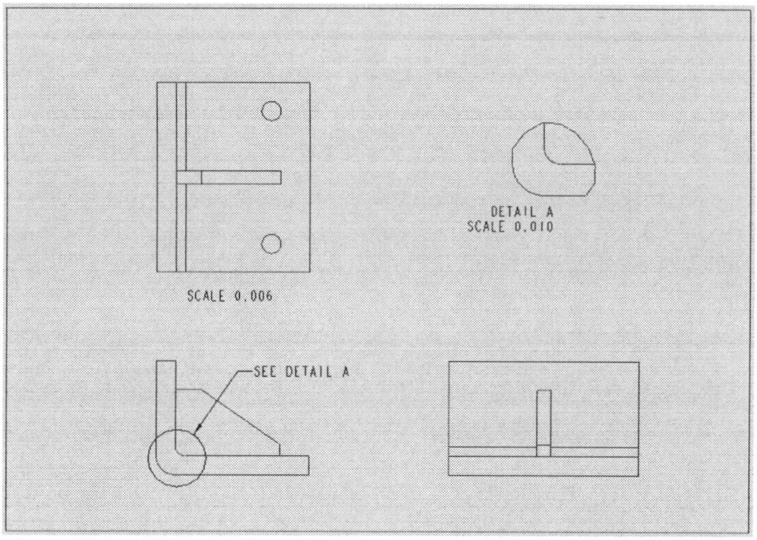

***Figure 12-26** Drawing views before adding dimensions*

Dimensioning the Drawing Views

1. Choose the **Show Model Annotations** tool from the **Annotations** group of the **Annotate** tab in the **Ribbon**; the **Show Model Annotations** dialog box is displayed with the **Show the model dimensions** tab chosen by default. Also, you are prompted to select a view to display its annotations.

2. Move the cursor toward the round feature in the detailed view and click when the round feature gets highlighted; the dimensions of the radius of round feature are listed in the dialog box.

3. Choose the dimensions of the round feature listed in the dialog box. All the radial dimensions are displayed in the selected view as strong dimensions.

4. Choose the **Apply** button to accept the settings and continue.

5. Now, select all the three views; all dimensions of the three views are listed in the dialog box.

6. Choose the **Select All** button; all dimensions are displayed on the view as strong dimensions.

7. Choose the **OK** button to accept the settings and exit the dialog box.

Placing the Dimensions in Order

The dimensions displayed in the drawing views are scattered and improperly placed. You need to use the **Clean Dimensions** dialog box to place the dimensions in order.

1. Choose the **Cleanup Dimensions** tool from the **Edit** group of the **Annotate** tab in the **Ribbon** to display the **Clean Dimensions** dialog box; you are prompted to select the dimensions to clean.

2. Select all dimensions from the top and front views by dragging a rectangle around them. Press the middle mouse button so that the options in the **Clean Dimensions** dialog box are displayed.

   When you select the dimensions, the radius dimensions in the drawing views are not selected. This means that the radius dimensions will remain unaffected by the values that you will enter in the **Clean Dimensions** dialog box.

3. Enter **0.4** in the **Offset** edit box and **0.3** in the **Increment** edit box.

4. Clear the **Create snap lines** check box. Choose the **Apply** button and then choose the **Close** button to exit the **Clean Dimensions** dialog box.

   After you clean the dimensions, you will notice that the dimensions are not placed in the required order. So, you need to manually place the dimensions in proper order.

5. Select the dimension and drag it to the desired location on the drawing sheet.

   Some dimensions are repeated. For example, the diameter of the hole feature is displayed twice in the drawing view. So, you need to erase the repeated and not required dimensions.

6. Select the dimension and hold down the right mouse button on the repeated dimensions to display a shortcut menu. Choose the **Erase** option from the shortcut menu.

   In some cases, the dimensions are displayed in the views in which you do not want them to be displayed. In such cases, you can switch those dimensions to the other views. To switch the dimension to other view, select it and hold the right mouse button; a shortcut menu is displayed. Choose the **Move to View** option from it.

   After manually placing the dimensions, the drawing views should look similar to the views shown in Figure 12-27.

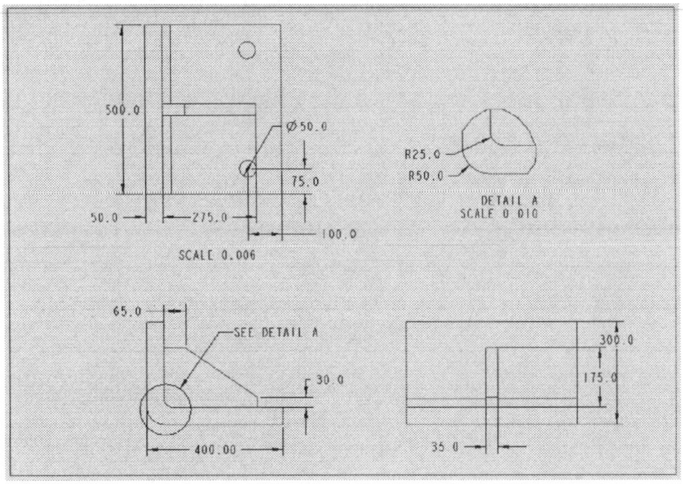

Figure 12-27 The required drawing views

Saving the Drawing File

1. Choose the **Save** button from the **Quick Access** toolbar; the **Save Object** dialog box is displayed with the name of the drawing file specified earlier.

2. Press ENTER to confirm the saving of the file.

Self-Evaluation Test

Answer the following questions and then compare them to those given at the end of this chapter:

1. The __________ dialog box is used to display dimensions in drawing views.

2. The dimension value can be modified by choosing the __________ option from the shortcut menu displayed on selecting the dimension and then right-clicking on it.

3. The __________ dialog box is used to clean dimensions in the drawing views.

4. The configuration file can be opened by choosing __________ from the **File** menu.

5. The **Decimal** and __________ radio buttons are available in the **Format** area of the **Dimension Properties** dialog box.

6. The reference datum planes are used to display geometric tolerances in a drawing view. (T/F)

7. If you modify a dimension value in the **Drawing** mode and then regenerate the drawing view, the modification will also be applied to the model in the **Part** mode. (T/F)

8. The notes created on a model in the **Part** mode can also be displayed in a drawing view in the **Drawing** mode. (T/F)

9. When you select a drawing view to display dimensions, all dimensions are placed in proper order in the drawing view. (T/F)

10. Most of the editing operations in a drawing view can be done using the shortcut menu that is displayed when you hold the right mouse button on the item to be modified. (T/F)

Review Questions

Answer the following questions:

1. Which of the following dialog boxes is used to display dimensions in a drawing view?

   (a) **Dimension Properties** (b) **Clean Dimensions**
   (c) **Show / Erase** (d) None of these

2. Which of the following dialog boxes is used to modify the properties of the text in a drawing view?

   (a) **Dimension Properties** (b) **Clean Dimensions**
   (c) **Show / Erase** (d) None of these

3. Which of the following dialog boxes is used to display the geometric tolerances in a drawing view?

   (a) **Geometric Tolerance** (b) **Clean Dimensions**
   (c) **Show / Erase** (d) None of these

4. Which of the following options in the shortcut menu is used to switch a dimension from one view to another?

   (a) **Erase** (b) **Flip Arrows**
   (c) **Move Item to View** (d) None of these

5. Which of the following radio buttons in the **Type** area needs to be selected in the **New** dialog box to open a new drawing file?

   (a) **Assembly** (b) **Sketch**
   (c) **Part** (d) None of these

6. The dimensions erased using the **Erase** option of the shortcut menu are also deleted from the model in the **Part** mode. (T/F)

7. Notes or balloons can be modified and edited using the options in the shortcut menu displayed on right-clicking. (T/F)

8. The system does not show a reference datum plane in a drawing view unless it is perpendicular to the drawing area. (T/F)

9. Generally, dimensions displayed in the drawing views are scattered and improperly placed. (T/F)

10. You cannot move text dynamically on a drawing sheet. (T/F)

EXERCISE

Exercise 1

In this exercise, you will generate the drawing views of the model created in Exercise 2 of Chapter 8, as shown in Figure 12-28. Add dimensions to the views, as shown in the Figure 12-29.

(Expected time: 45 min)

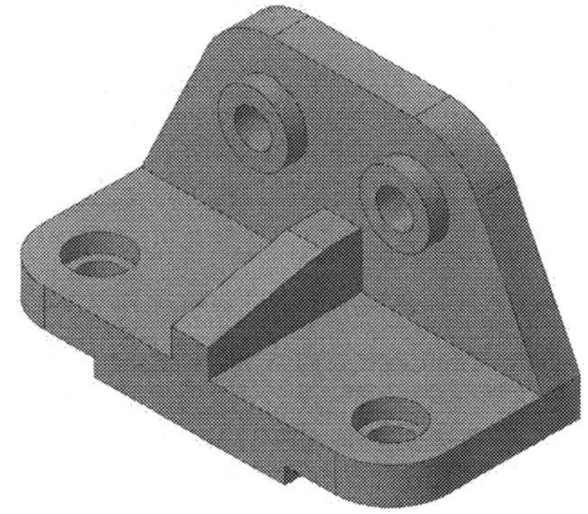

Figure 12-28 *Isometric view of the model*

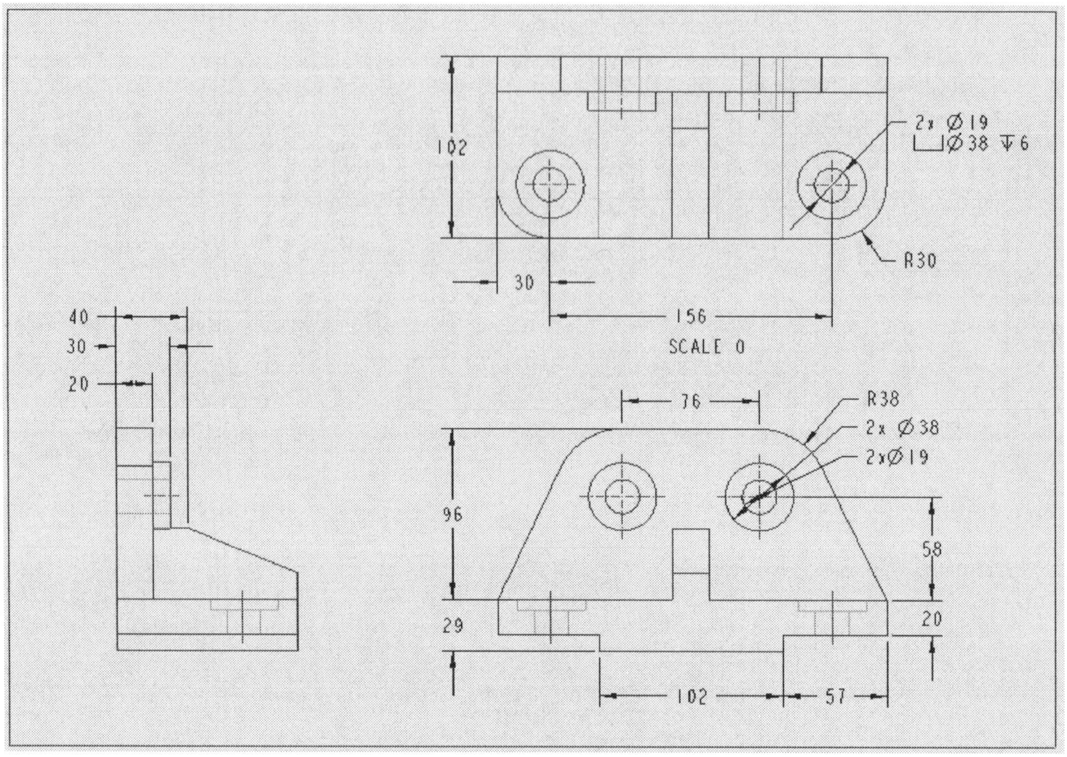

***Figure 12-29** Orthographic views of the solid model*

Note

*In the given drawing views, to generate the center lines go to **Annoaate > Annotations > Show Model Annotations > Show** the model datums.*

Answers to Self-Evaluation Test

1. **Show / Erase**, **2.** **Modify Nominal Value**, **3.** **Clean Dimensions**, **4.** **Options**, **5.** **Fractional**, **6.** T, **7.** T, **8.** T, **9.** F, **10.** T

Chapter 13

Other Drawing Options

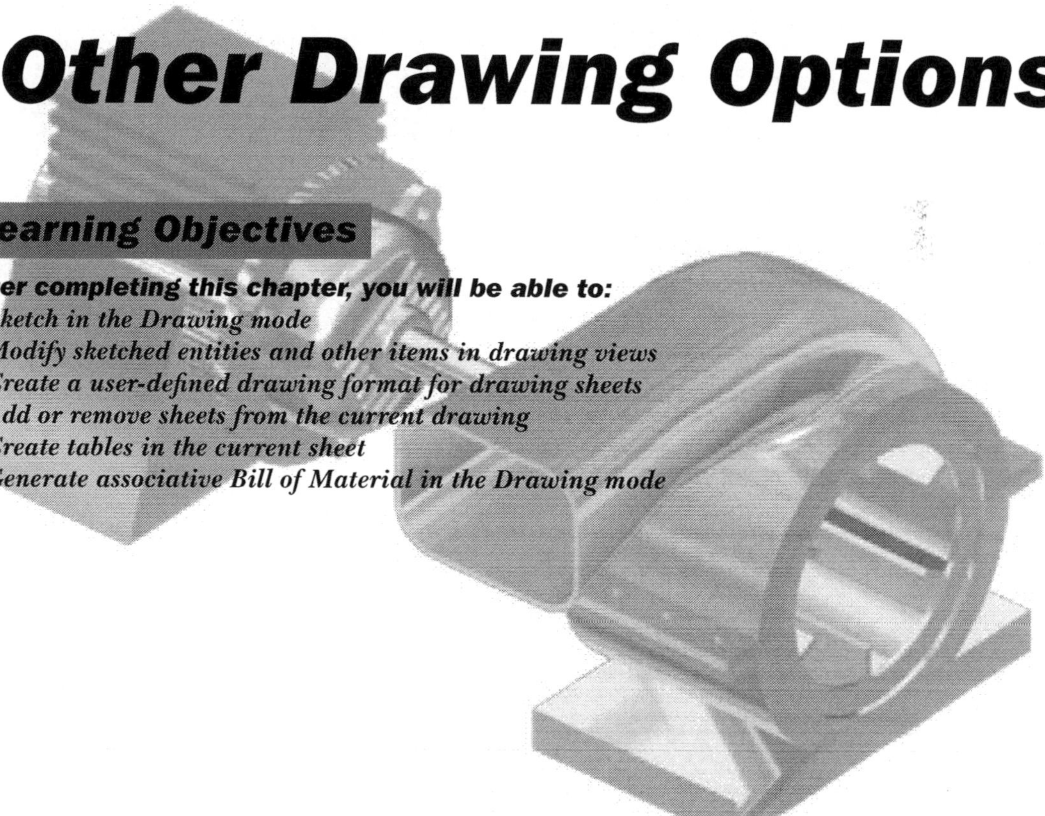

Learning Objectives

After completing this chapter, you will be able to:

- *Sketch in the Drawing mode*
- *Modify sketched entities and other items in drawing views*
- *Create a user-defined drawing format for drawing sheets*
- *Add or remove sheets from the current drawing*
- *Create tables in the current sheet*
- *Generate associative Bill of Material in the Drawing mode*

SKETCHING IN THE DRAWING MODE

Sketching is one of the most important tools available in the **Drawing** mode. Sketching in the **Drawing** mode is called drafting. As discussed earlier, there are two types of drafting techniques in Creo Parametric: Generative drafting and Interactive drafting. Any item on the drawing sheet that is not generated from a model is called a draft entity or a draft item. Drafting is extensively used for creating user-defined formats, drawing tables, and drawing title blocks in the formats. Sketching in the **Drawing** mode is almost similar to that in the other modes of Creo Parametric. Sketching can be done by using the tools in the **Sketch** tab of the **Ribbon**. The tools used to sketch in the **Drawing** mode are discussed next.

Entity Select

The **Entity Select** tool is used to switch to the selection mode. This tool is available in the **Settings** group of the **Sketch** tab in the **Ribbon**. You can select the drawing views, sketched entities, dimensions, notes, and so on from the drawing by simply clicking on them with the left mouse button. This tool is extensively used for moving the drawing views or the items in the drawing view. You can select the entity and drag using the left mouse button to move it.

Tip
To delete an entity or an item from a drawing view, you can select the item and then press the DELETE key. You can also select the entity and choose the ***Delete*** *option from the shortcut menu that is displayed when you hold the right mouse button.*

Line

Ribbon: Sketch > Sketching > Line Drop-down

The tools in the **Line** drop-down available in the **Sketch** tab are used to draw a line/arc or chain of lines or arcs. The **Line/Arc Chain** tool is used to create a line/arc or chain of lines/arcs by selecting two points in the drawing area, the **Single** tool is used to create the individual line, and the **Tangent** tool is used to create a tangent line between two entities. To create a line/arc or chain of line and arcs, choose **Sketch > Sketching > Line drop-down > Line/Arc Chain** from the **Ribbon**; the dimension mini popup toolbar will be displayed, as shown in Figure 13-1. The options in this popup toolbar are discussed next.

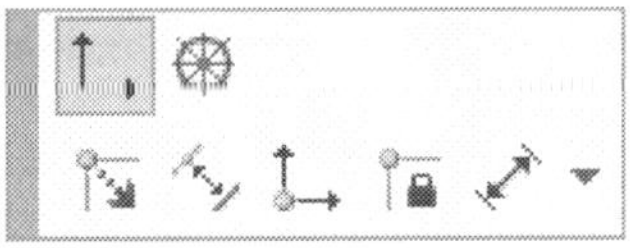

Figure 13-1 *The dimension mini popup toolbar*

Define geometry position using Cartesian dimensions

This button is used to generate the reference point of the line by using the cartesian coordinate system.

Define geometry position using Polar dimensions

This button is used to to generate the reference point of the line by using the polar coordinate system.

Move the dimension base point to another location

By using this button, you can move the base point, which is the origin, to another location.

Changes the reference of the current dimension

By using this button, you can change the reference of the current dimension.

Move the dimension base point to sketch origin

By using this button, you can move the dimension base point to the origin.

Locks the base point position

By using this button, you can lock the position of base point.

Change the dimension orientation Drop-down List

The options available in this drop-down list are used to force the dimension orientation manually. There are three options in this drop-down list: **Change the dimension orientation to be slanted**, **Change the dimension orientation to be horizontal**, and **Change the dimension orientation to be vertical**.

Circle

Ribbon: Sketch > Sketching > Circle Drop-down List

The tools in the **Circle** drop-down list are used to create circles. The **Center and Point**, **2 Point**, **3 Point**, **3 Tangent**, and **Concentric** tools are used to draw circles in the drawing mode. The working of these tools is same as those discussed in Sketch mode.

Arc

Ribbon: Sketch > Sketching > Arc Drop-down List

The tools available in the **Arc** drop-down list are **3-Point / Tangent**, **End Center and Ends**, **Tangent**, and **Concentric**. These tools are used to draw arcs in the drawing mode.

The other geometric entities that can be drawn by using the tools in the **Ribbon** are splines, construction circles and lines, ellipses, and points. You can also create a fillet and a chamfer between two entities. You can also use the edges of the model and also create offsets.

Snapping Guides

Ribbon: Sketch > Settings > Snapping guide Drop-down List

The **Snapping Guides** drop-down list is used to apply snapping preferences. The option available in drop-down list, as shown in Figure 13-2, are discussed next.

Horizontal/Vertical

When this option is selected, the cursor automatically snaps to horizontal line while sketching.

Vertical

When this option is selected, the cursor automatically snaps to vertical line while sketching.

Parallel

When this option is selected, the cursor automatically snaps to parallel line while sketching.

Papendicular
When this option is selected, the cursor automatically snaps to perpendicular line while sketching.

Equal
When this option is selected, the cursor automatically snaps to lines with equal length while sketching.

Midpoint
When this option is selected, the cursor automatically snaps to midpoint of line while sketching.

Tangent
When this option is selected, the cursor automatically snaps tangency while sketching.

Coincident
When this option is selected, the cursor automatically snaps to other sketch element to coincide while sketching.

Diagonal
When this option is selected, the cursor automatically snaps to diagonal points while sketching.

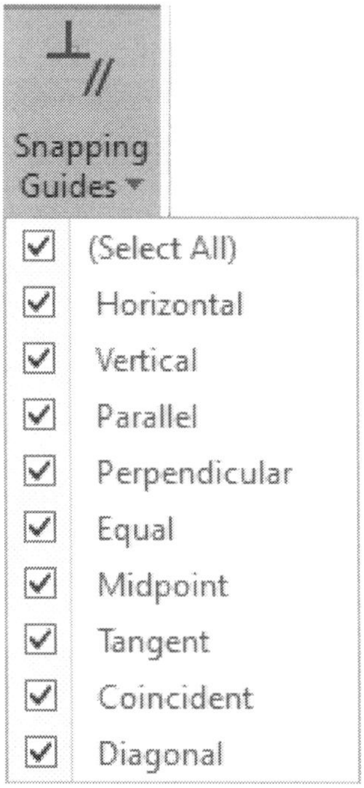

*Figure 13-2 The **Snapping Guides** drop-down list*

Snapping Setting

Ribbon: Sketch > Settings > Snapping Setting drop-down list

The options available in this drop-down list are used to control the snapping of cursor while sketching. These are discussed next.

Snap to Model Geometry
When this check box is selected, the sketched entities automatically snap to the model geometry.

Instant Snapping to Geometry
When this check box is selected, the cursor automatically snaps to the points on the the model geometry.

Snap to Grid
When this check box is selected, the cursor automatically snaps to the intersection points of the grid.

MODIFYING SKETCHED ENTITIES

There are various options to modify draft entities. The operationsor modifications that can be applied on the draft entities are discussed next.

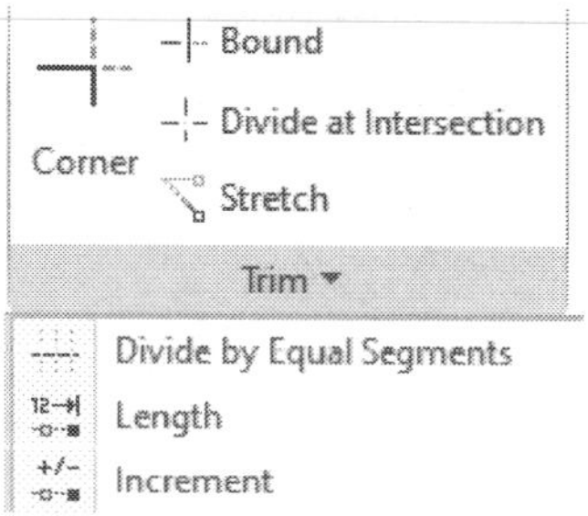

__Figure 13-3__ The tools in the __Trim__ group of the __Ribbon__

Trimming the Draft Entities

The sketched or draft entities can be trimmed using the tools available in the **Trim** group of the **Sketch** tab in the **Ribbon**. The tools in the **Trim** group are discussed next, refer to Figure 13-3.

Divide at Intersection

Ribbon: Sketch > Trim > Divide at Intersection

The **Divide at Intersection** tool is used to divide two selected entities at the point of intersection. When you choose this tool, you are prompted to select two entities. Select the two entities to divide them at the intersection point. If a warning message is displayed while selecting the entities, then choose the **OK** button from it. The message informs you that the command will break all the parametric sketching references.

Divide by Equal Segments

Ribbon: Sketch > Trim > Divide by Equal Segments

This tool is in the extended **Trim** group and is used to divide the draft entities into segments of equal length. To do so, choose the **Divide by Equal Segments** tool and then select the entities to divide. After selecting the entity or entities, press the middle mouse button; the Message Input Window will be displayed and you will be prompted to enter the number of equal segments. Enter the number of equal segments into which you need to divide the entity and press ENTER.

Corner

Ribbon: Sketch > Trim > Corner

This tool is used to trim the two selected entities so that a corner is formed after trimming. Make sure you select the entities using the CTRL key.

Bound

Ribbon: Sketch > Trim > Bound

This tool is used for deleting or extending a portion of an entity by defining a bounding entity.

Length

Ribbon: Sketch > Trim > Length

This tool is available in the expanded **Trim** group. Using this tool, you can trim or extend an entity up to a specified length.

Increment

Ribbon: Sketch > Trim > Increment

This tool is available in the expanded **Trim** group. It is used to extend or shorten an entity in specified increments. To extend the entity, enter a positive increment value and to shorten the entity, enter a negative increment value.

Stretch

Ribbon: Sketch > Trim > Stretch

The **Stretch** tool is used to stretch a sketched entity. The vertices of the sketched entity that are included inside the selection window will be moved from their original location and stretched, and the remaining vertices will remain stationary.

Transforming the Draft Entities

The tools in the **Edit** group of the **Sketch** tab of the **Ribbon** are used to translate or move the draft entities. The tools in the **Edit** group are shown in Figure 13-4 and are discussed next.

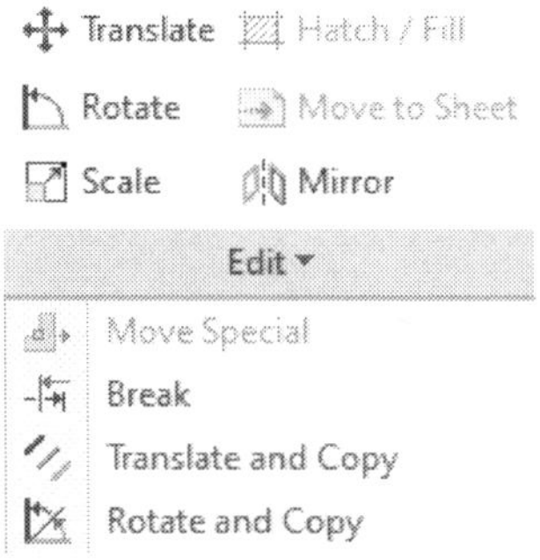

*Figure 13-4 The tools displayed in the **Edit** group*

Translate

Ribbon: Sketch > Edit > Translate

The **Translate** tool is used to move the selected entity or entities on the drawing sheet. To translate the entities, invoke this tool and select the required entities. Next, press the middle mouse button to confirm the selection; the **GET VECTOR** menu along with the **Select Point** dialog box will be displayed, as shown in Figure 13-5 and Figure 13-6. The options in the **GET VECTOR** menu are used to specify the direction of translation.

*Figure 13-5 The **GET VECTOR** menu*

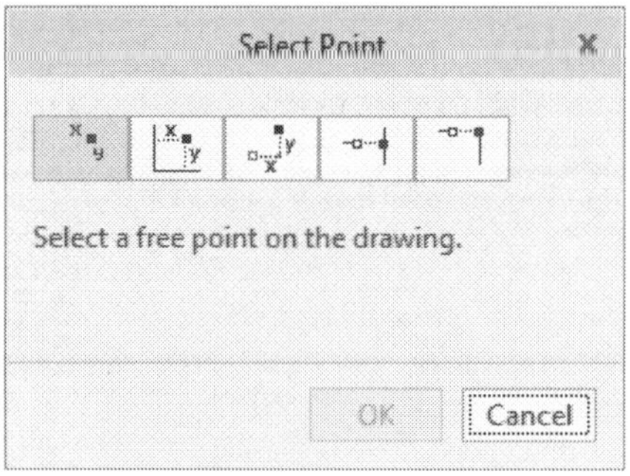

*Figure 13-6 The **Select Point** dialog box*

Translate and Copy

Ribbon: Sketch > Edit > Translate and Copy

The **Translate and Copy** tool is available in the expanded **Edit** group. This tool is used to translate the selected entity and at the same time make a copy of the entity. After selecting the entities, press the middle mouse button; the **GET VECTOR** menu and the **Select Point** dialog box will be displayed. After defining the translation distance and direction, you need to specify the number of copies you want to create.

Rotate

Ribbon: Sketch > Edit > Rotate

The **Rotate** tool is used to rotate the selected entity. On choosing this tool, you will be prompted to select the entity that you want to rotate. After selecting the entity or entities, press the middle mouse button; you will be prompted to select the center point for rotation. Select the center point; the Message Input Window will be displayed and you will be prompted to enter the rotation angle in the counterclockwise direction. The default direction of rotation will be counterclockwise. However, you can rotate the selected entity in the clockwise direction by specifying a negative rotation angle.

Rotate and Copy

Ribbon: Sketch > Edit > Rotate and Copy

The **Rotate and Copy** tool is available in the expanded **Edit** group and is used to create a pattern of the selected draft entity. Choose this tool, select an entity, and press the middle mouse button; the **Select Point** dialog box will be displayed and you will be prompted to select the center point of rotation. Select the center point of rotation in the drawing; the Message Input Window will be displayed and you will be prompted to enter the rotation angle in counterclockwise direction. Enter the rotation angle and press the middle mouse button to accept the entered rotation angle value. Now, enter the number of copies in the Message Input Window; the rotated copies of the selected entity will be placed on the drawing sheet.

Scale

Ribbon: Sketch > Edit > Scale

The **Scale** tool is used to scale the selected entity. To do so, select the entity to scale and press the middle mouse button; the **Select Point** dialog box will be displayed and you will be prompted to select the origin point for scaling. Select the origin point; the Message Input Window will be displayed. The new scale factor can be specified in the Message Input Window. A scale factor greater than 1 will increase the size of the selected entity and a scale factor less than 1 will reduce the size of the selected entity.

Mirror

Ribbon: Sketch > Edit > Mirror

The **Mirror** tool is available in the **Edit** group. This tool is used to mirror the selected entities using an existing sketched line. To do so, choose the **Mirror** tool from the **Edit** group of the **Ribbon**. On choosing this tool, you will be prompted to select a 2D entity. Select the 2D entity and choose the **OK** button from the **Select** dialog box; you will be prompted to select a line. Select the line about which you want to create the mirrored copy.

Grouping Entities

Ribbon: Sketch > Group > Draft Group

A draft entity can be grouped with a note, geometric tolerance (gtol), dimension, or with another draft entity. In Creo Parametric, the note, gtol, and dimensions are called detail items. After grouping the draft entities, the operation applied on any one of them is also applied to the grouped entity. Choose **Sketch > Group > Draft Group** from the **Ribbon** to form a group among draft entities like lines, arcs, circles, and so on. The operation performed on any one of the grouped entities is performed on all the entities forming that group. When you choose this option, the **DRAFT GROUP** menu will be displayed, as shown in Figure 13-7. Use the required option from this menu to create the draft group.

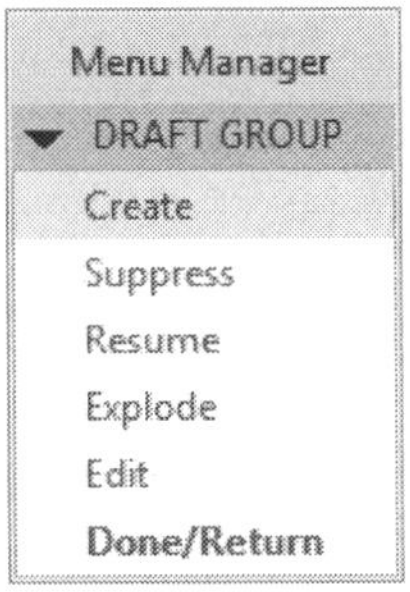

*Figure 13-7 The **DRAFT GROUP** menu*

Relate

Ribbon: Sketch > Group > Relate View drop-down

The draft entities can be related to the view such that the draft entities move or re-scale along with the related views. The options to relate can be invoked by choosing **Sketch > Group > Relate View** drop-down from the **Ribbon**. On doing so, a drop-down menu will be displayed, as shown in Figure 13-8. Note that this drop-down will be activated only when you select the drawing view. The tools in this menu are discussed next.

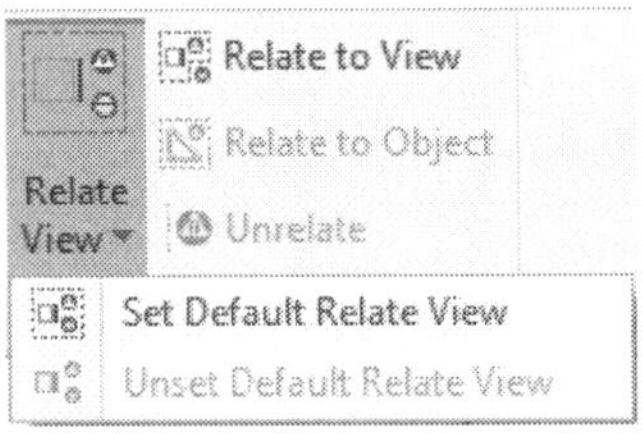

*Figure 13-8 The **Relate View** drop-down*

Set Default Relate View

Ribbon: Sketch > Group > Relate View drop-down > Set Default Relate View

This tool is available only when the drawing view is selected and is used to set the draft view as the current view. The draft entities that are drawn after setting a default draft view are grouped with the current draft view. Now, if you move the draft view, then the grouped entities will also move with it.

Note

When you move a view that forms a group with draft entities, the draft entities also move. But when you move the entities, the draft view does not move.

Unset Default Relate View

Ribbon: Sketch > Group > Relate View drop-down > Unset Default Relate View

This tool is used to unset the current draft view and is activated only when a current draft view exists.

Relate to View

Ribbon: Sketch > Group > Relate to View

This tool is used to relate a view with a draft entity or entities. This tool is activated in the **Group** group only when a draft entity is selected. On choosing this tool, you will be prompted to select a view to which you want to relate the draft entity. Select the drawing view from the drawing sheet.

Relate to Object

Ribbon: Sketch > Group > Relate to Object

This tool is used to relate notes, dimensions, gtols, and other objects to a drawing view. This tool will be activated in the **Group** group only when an object is selected. On choosing this tool, you will be prompted to select dim, dim arrow, gtol, note, symbol, axis, or datum point that will serve as the host.

Unrelate

Ribbon: Sketch > Group > Unrelate

This tool is used to unrelate the previously related objects. This tool is activated in the **Group** group only when a related object is selected.

Drafting by Using the Drawing Views

You can use the generated drawing views to create an entity or the whole drawing view. To use a generated drawing view, choose the down arrow at right of the **Edge** button in the **Sketching** group of the **Sketch** tab in the **Ribbon**; a drop-down will be displayed with two tools: **Use Edge** and **Offset Edge**. These tools are discussed next.

Offset Edge

Ribbon: Sketch > Sketching > Edge drop-down > Offset Edge

This tool is used to offset the selected edges to the specified offset distance. When you choose this tool, the **OFFSET OPER** menu will be displayed, as shown in Figure 13-9. There are two options in this menu: **Single Ent** and **Ent Chain**. The **Single Ent** option is used to select a single entity from the drawing view, whereas the **Ent Chain** option is used to select a chain of entities from the drawing view. After choosing any one of the options from the **OFFSET OPER** menu, if you move the cursor over the drawing view, each edge of the model in the drawing view will be highlighted in green color. After selecting each edge of the model, you need to define the offset distance.

Figure 13-9 *The* ***OFFSET OPER*** *menu*

Use Edge

Ribbon: Sketch > Sketching > Edge drop-down > Use Edge

The **Use Edge** tool is selected to draw a copy of the selected edge in the drawing view. This tool will be available only when you have at least one drawing view in the current drawing sheet and if it is selected.

Converting a Generated View to a Draft View

Ribbon: Layout > Edit > Convert to Draft Group

To convert a generated view into a draft view, select the view to convert and choose the **Convert to Draft Group** tool from the **Edit** group of the **Layout** tab in the **Ribbon**; the **Confirm** dialog box will be displayed with a message that the selected view(s) will be broken up into draft entities. Choose the **Yes** button from this dialog box. After converting a generated view to a draft view, all objects in the view will be converted to individual entities.

Note
*The **Convert to Draft Group** tool is activated in the **Edit** group only when the display style of the view is **Hidden Line**, **No Hidden**, or **Wireframe**.*

USER-DEFINED DRAWING FORMATS

Creo Parametric provides you with some standard drawing formats for generating drawing views. These standard formats have standard sheet sizes, tables, and title blocks. However, sometimes you may need to create a user-defined drawing format that is specifically designed as per your requirements, including sheet size, tables, and title block. Choose the **New** button from the **File** menu to display the **New** dialog box. In this dialog box, select the **Format** radio button from the **Type** area and then specify the name of the format in the **File name** edit box to create the user-defined format. When you choose the **OK** button, the **New Format** dialog box will be displayed, as shown in Figure 13-10.

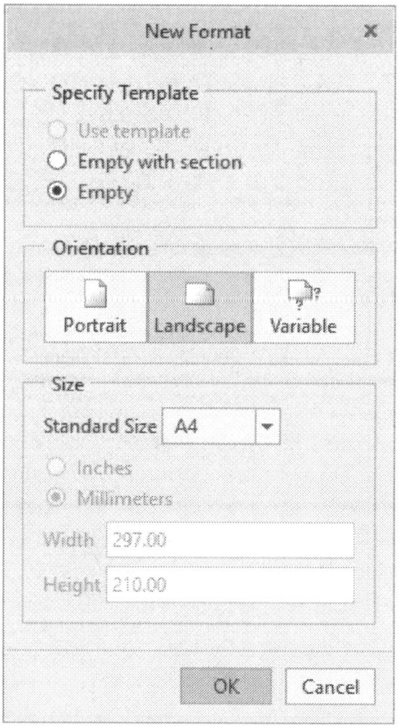

***Figure 13-10** The **New Format** dialog box*

You can set the size and orientation of the format sheet by using this dialog box. After setting the parameters, choose the **OK** button from this dialog box to proceed to the **Format** mode; the **Format** mode will be invoked.

You can create the desired entities in this format by using the **Sketch** tab of the **Ribbon**. You can also add the text to the format by using the **Note** button in the **Annotations** group of the **Annotate** tab of the **Ribbon**. Figure 13-11 shows a user-defined format created using the A4 size sheet. This format consists of the user-defined title block. Figure 13-12 shows the drawing views generated on the user-defined format.

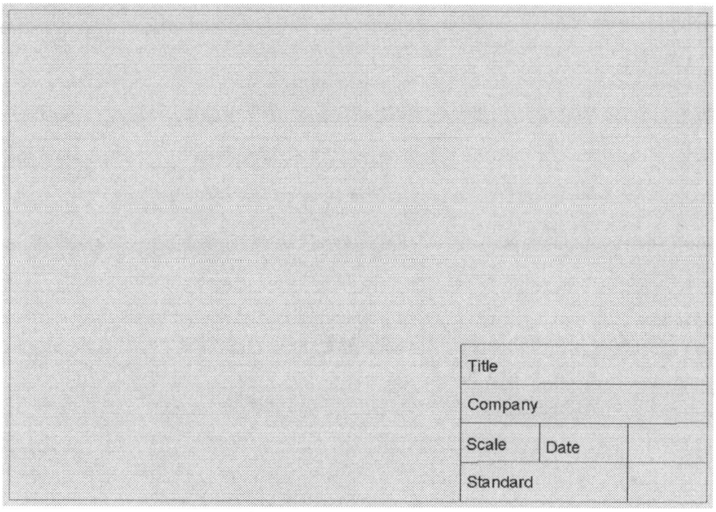

Figure 13-11 A user-defined format

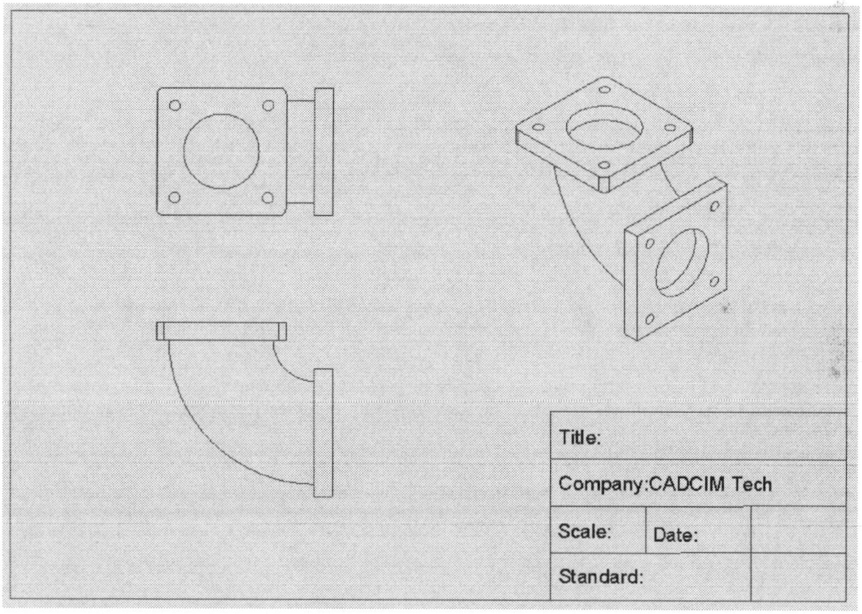

Figure 13-12 Drawing views generated on the user-defined format

RETRIEVING THE USER-DEFINED FORMATS IN THE DRAWINGS

Once you have created a user-defined format, you can retrieve and use it in the **Drawing** mode as a sheet for generating the drawing views. To retrieve a user-defined format, select the **Empty with format** radio button and choose the **Browse** button from the **New Drawing** dialog box to invoke the **Open** dialog box. In this dialog box, a folder named **System Formats** will be displayed with some predefined formats. You can retrieve these predefined formats or browse to the location where you have saved a user-defined format created earlier.

Note

*You can add or replace formats in a drawing by using the options in the **Sheet Setup** dialog box that will be displayed when you hold the right mouse button on the drawing sheet and choose the **Sheet Setup** option. You can add another sheet and then select the format from this dialog box.*

Tip

The user-defined format that you create is saved in the .frm format. The location of this format will be the working directory that was selected at the time when the format was created.

ADDING AND REMOVING SHEETS IN THE DRAWING

Ribbon: Layout > Document > New Sheet

To add sheets to the current drawing, choose the **New Sheet** tool from the **Document** group of the **Layout** tab in the **Ribbon**; a new sheet will be displayed and added to the current drawing sheet. Also, the sheet number will be displayed in the **Sheets** bar at the bottom left of the screen. You can also click on the (+) sign displayed at the bottom left of the screen to add new sheet to the current drawing. You can generate various views of a model on multiple sheets. All these drawing views on different sheets are contained in a single drawing file.

To remove a sheet from a drawing, right-click on the required **Sheet** tab available at the bottom of the drawing area to invoke a shortcut menu. Choose the **Delete** option from the shortcut menu to delete that particular sheet; the **Delete Selected Sheets Confirmation** message box will be displayed. Choose the **Yes** button from the message box to confirm the deletion of the current sheet.

CREATING TABLES IN THE DRAWING MODE

Ribbon: Table > Table drop-down

You can create any kind of tabular representation in the **Drawing** mode easily by using the tools in the **Table** tab of the **Ribbon**, as shown in Figure 13-13. The tools in this tab are discussed next.

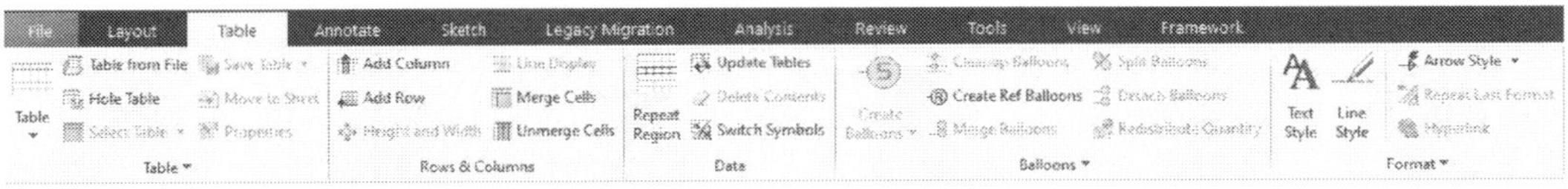

*Figure 13-13 The **Table** tab of the **Ribbon***

Table Tool

To create a table, choose the **Table** tool from the **Table** group in the **Table** tab of the **Ribbon**; the **Table** drop-down will be displayed, as shown in Figure 13-14. Move the cursor over required boxes displayed in the drop-down; the number of selected rows and columns of the table will be highlighted on the top-left corner of the drop-down. Click on the position where you get the required number of rows and columns; a rectangle will be attached to the cursor and you will be prompted to define the location. Click on the desired location to insert the table. You can also create the table by choosing the **Insert Table** tool in the same drop-down. On doing so, the **Insert Table** dialog box will be displayed, as shown in Figure 13-15. In this dialog box, there are four areas which are discussed next.

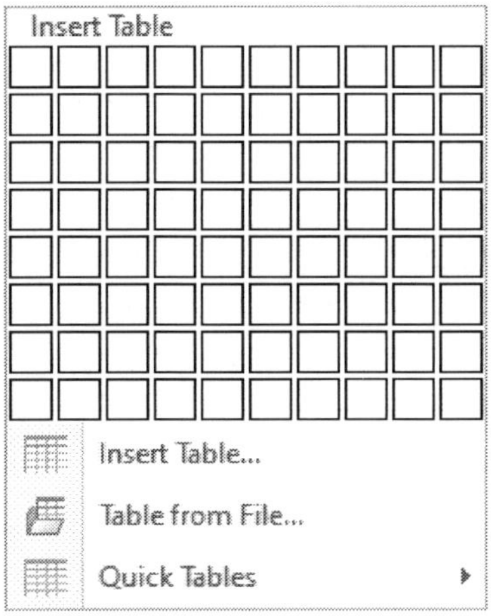

*Figure 13-14 The **Table** drop-down*

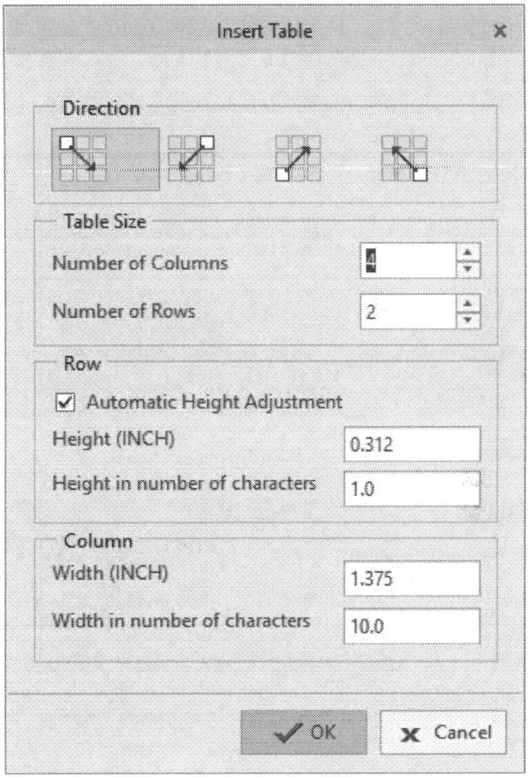

*Figure 13-15 The **Insert Table** dialog box*

Direction

In this area, there are four buttons to define the direction of table growth. The direction of table growth is the one in which the new rows and columns will be added later.

Table Size

In this area, there are two spinners to control the table size. The **Number of Columns** spinner is used to set the number of columns in the table. The **Number of Rows** spinner is used to set the number of rows in the table.

Row

In this area, there are two edit boxes and a check box to control the height of rows. Select the **Automatic Height Adjustment** check box to automatically change the height of row to adjust the text written in the row. If you want to set the height manually, then enter the value of height in inch in the **Height (INCH)** edit box or enter the value of height in number of characters in the **Height in number of characters** edit box. The height of one character is 7 mm by default.

Column

In this area, there are two edit boxes which are used to control the width of the columns. To change the width of columns, you need to enter the value in inch in the **Width (INCH)** edit box. You can also enter the width value in number of character in the **Width in number of characters** edit box. The width of one character is 5.95 mm by default.

Table from File Option

To insert a table that already exists in the *.tbl* or *.csv* file format, choose **Table > Table > Table from File** from the **Ribbon**; the **Open** dialog box will be displayed. Browse and select a *.tbl* or *.csv* file from this dialog box to open it.

Add Column

The **Add Column** button is used to add a column to an existing table. To do so, choose **Table > Rows & Columns > Add Column** from the **Ribbon**; you will be prompted to select a point inside the existing table to add a column. Specify a point on the column line; a column gets added to the table.

Tip

*To save a table with the .tbl file format, choose **Table > Table > Save Table drop-down > Save as Table** from the **Ribbon**. The **Save Drawing Table** option is activated only when existing table is selected in the drawing sheet.*

Add Row

The **Add Row** button is used to add a row to an existing table. When you choose this button from the **Rows & Columns** group of the **Table** tab of the **Ribbon**, you will be prompted to select a point inside the existing table to add a row.

The following are the editing operations that can be performed on the tables:

1. To move a table, select it and then drag it by holding it from one of the points that are highlighted on the table.

2. To enter text in any of the cells of the table, you need to double-click inside the cell; the **Format** tab will be added to the **Ribbon**.

3. To delete a table from the drawing, click anywhere on the table and choose the **Select Table** option from the **Select Table** drop-down in the **Table** group of the **Table** tab in the **Ribbon**. Press and hold the right mouse button, and then choose the **Delete** button from the shortcut menu displayed. You can also use the DELETE key after selecting the required items.

   To delete an individual row/column, move the cursor over the boundary of the required row/column; the entire row/column will be highlighted in green color. Click when the entities are highlighted and then choose the DELETE key.

> **Tip**
> *In the **Drawing** mode, most of the modifications and editing can be done by using the shortcut menu that will be displayed when you select the item to modify and then hold the right mouse button. The options in this menu vary and depend on the item selected.*
>
> *It is recommended that you use the options in the **Table** tab of the **Ribbon** to create the title blocks in the formats. The text can easily be added to the title block by double-clicking in the cell.*

GENERATING THE BOM AND BALLOONS IN DRAWINGS

The Bill of Material (BOM) is a representation of the components and their parameters that are used in the assembly. In the **Drawing** mode, the BOM is generated and is associative in nature. Therefore, any modification in the assembly, such as addition or removal of components, is automatically reflected in the BOM.

The procedure to generate the BOM and Balloons in the **Drawing** mode is discussed next.

Consider a case in which you have started a new drawing sheet of standard size A with a user-defined format. Now, follow the steps discussed next to generate the BOM and balloons.

Creating a Table for the BOM

Before proceeding to generate the BOM, first you need to create a table for it. The items to be listed in the BOM will be displayed in this table.

1. Choose **Table > Table drop-down > Insert Table** from the **Ribbon** to display the **Insert Table** dialog box.
2. Set the **Number of Columns** spinner to **3** and **Number of Rows** spinner to **2** in the **Table Size** area.
3. Clear the **Automatic Height Adjustment** check box in the **Row** area.
4. Set the height of rows as **0.5** in the **Height (INCH)** spinner in the **Row** area.
5. Set the width of column as **0.5** in the **Width (INCH)** spinner in the **Column** area.
6. Choose the **OK** button to exit; the **Select Point** dialog box will be displayed and you will be prompted to locate the origin of the table.
7. Click at the desired location to place the table; the table gets inserted in the drawing.
8. Select a cell in the second row and choose the **Height and Width** tool from the **Rows & Columns** group of the **Table** tab; the **Height and Width** dialog box is displayed.
9. Enter the value of width as **3** in the **Width(drawing units)** edit box in the **Column** area of the dialog box and then choose the **OK** button.

Setting the Repeat Region

Ribbon: Table > Data > Repeat Region

After creating the table for the BOM, you need to define the repeat region in this table. Repeat regions are smart cells that can expand depending on the amount of data inserted in the cell. The second row will be defined as the repeat region.

1. Choose **Table > Data > Repeat Region** from the **Ribbon**; the **TBL REGIONS** menu will be displayed.

2. Choose the **Add** option from the **TBL REGIONS** menu and the **Simple** option from **REGION TYPE** menu; you will be prompted to locate the corners of the region.

3. Select the first cell of the second row and then the third cell of the second row to define the region. Next, press the middle mouse button twice to exit the menu.

Creating the Column Headers

After creating the table and setting the repeat region, you need to define the column headings in the table.

1. Double-click in the first cell of the first row; the **Format** tab will be added in the **Ribbon**. Enter **Index** and click twice outside the table to exit.

   Similarly, enter **Part Name** and **Qty** in the second and the third cells.

Assigning the Report Symbols in the Repeat Region

The information in the cells that are defined as repeat region is determined by the text that is written in the form of report symbols. These report symbols associate the information in the cells with the assembly file directly.

1. Double-click in the first cell of the second row; the **Report Symbol** dialog box will be displayed. Select **rpt** from the dialog box; the **rpt** options are displayed in the dialog box.

2. Select the **index** option from the dialog box.

3. Double-click in the second cell of the second row; the **Report Symbol** dialog box will be displayed. Select **asm** from the dialog box; the **asm** options are displayed.

4. Select the **mbr** option from the dialog box and then select the **name** option from the dialog boxes that appear. This will ensure that the names of the parts in the assembly are displayed in this column. The number of rows will be automatically modified depending on the number of parts in the assembly.

5. Double-click in the third cell of the second row; the **Report Symbol** dialog box will be displayed. Select **rpt** from the dialog box; the **rpt** options are displayed. Select the **qty** option from the dialog box.

Note

In Creo Parametric 8.0, ***&mdl.name, &mdl.type, &mdl.User Defined*** *are the report parameters added for repeated regions in detailed drawings. With these parameters, you can directly reference and display predefined or user-defined model parameters.*

Generating the Exploded Drawing View

Ribbon: Layout > Model Views > General View

After setting all the options, you need to generate the isometric view of the exploded state of assembly. Therefore, you should have created an exploded state of assembly. To generate the exploded drawing view, follow the steps listed next.

1. Choose **Layout > Model Views > General View** from the **Ribbon**; the **Select Combined State** dialog box will be displayed. Choose the **OK** button from the dialog box and you will be prompted to select the center point for the drawing view.
2. Select the point on the drawing sheet to place the view; the **Drawing View** dialog box will be displayed. Select the **Isometric** option from the **Default orientation** drop-down list in the **Drawing View** dialog box. Choose the **Apply** button. Specify the scale factor, if required.
3. Choose the **View States** option from the **Categories** list box. Select the **Explode components in view** check box. Select the name of the exploded state from the **Assembly explode state** drop-down list and then choose the **Apply** button.
4. Close the **Drawing View** dialog box.

Note that the BOM automatically appears on the table by extending the repeat region. Also, note that the **Qty** column is empty. If the BOM is not displayed, choose **Table > Data > Update Tables** from the **Ribbon**.

Setting the No Repeat Option

Ribbon: Table > Data > Repeat Region

If an assembly has more than one quantity of a component, the components are repeated in the BOM. Therefore, you need to set the option to omit the repeated instances in the BOM.

1. Choose **Table > Data > Repeat Region** from the **Ribbon**; the **TBL REGIONS** menu will be displayed.
2. Choose the **Attributes** option from the menu; you will be prompted to select a region.
3. Move the cursor over the table below the first row. All rows, except the first row, are highlighted. The highlighted portion is called region. Select the region; the **REGION ATTR** submenu will be displayed. Choose the **No Duplicates** option and then choose the **Done/Return** button.
4. Press the middle mouse button twice to exit the menu.
5. Choose **Table > Data > Update Tables** from the **Ribbon**, if the information is not updated in the table.

Generating Balloons

Ribbon: Table > Balloons > Create Balloons drop-down > Create Balloons - All

Balloons are generated by using the repeat region. Before generating balloons, you need to set the repeat region for the generation of balloons.

1. Choose the **Create Balloons - All** tool from the **Create Balloons** drop-down available in the **Balloons** group of the **Table** tab in the **Ribbon**; balloons are attached with all the parts of the assembly in the exploded view. The balloons also contain the index number so that the parts in the assembly can be referred to as BOM.

2. Drag the balloons to place them appropriately in the drawing sheet. You can select a balloon and then hold the right mouse button to invoke the shortcut menu. From this menu, you can select the required options to modify the attachment points of balloons.

TUTORIALS

Tutorial 1

In this tutorial, you will create a format of size A and add the title block to the format. Then you will retrieve the format in the **Drawing** mode and create the table for BOM. Next, you will generate the exploded isometric view of the **Pedestal Bearing** assembly created in Tutorial 2 of Chapter 10. Also, you will add balloons to the drawing view, as shown in Figure 13-16.

(Expected time: 45 min)

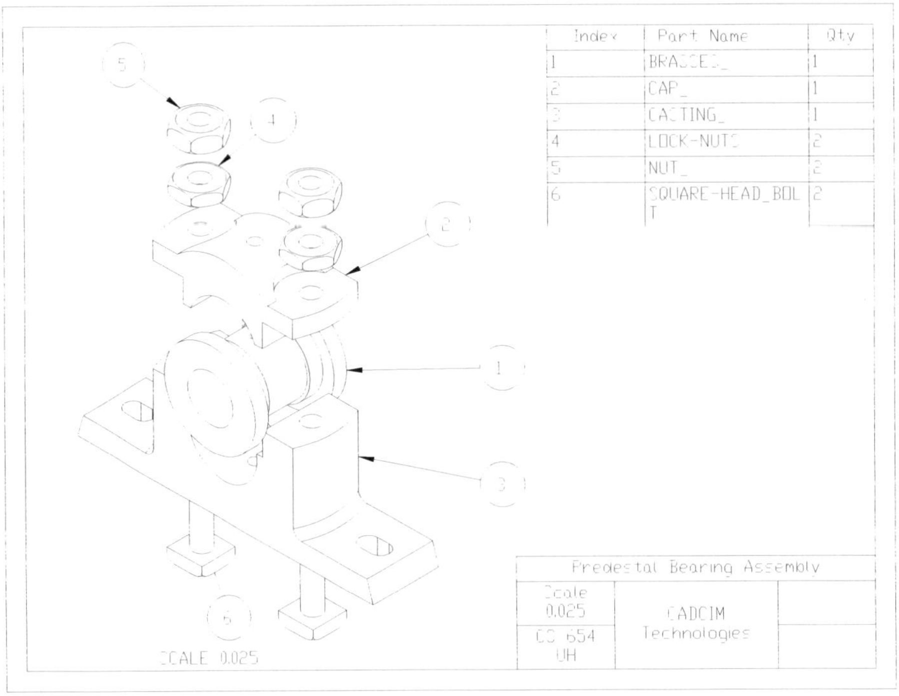

Figure 13-16 *The drawing view of the assembly showing the BOM and balloons*

The following steps are required to complete this tutorial:

a. Start a new file in the **Format** mode. Create the format of the drawing and then add the title block in the format.
b. Save the format file and then close it.
c. Start a new drawing file in the **Drawing** mode. Select **Pedestal Bearing** as the model and retrieve the format that you have created.
d. Create the table and enter the headers.
e. Define the repeat region and then assign the report symbols.
f. Generate the exploded isometric view of the assembly, .
g. Add balloons to the drawing.

Copy the *Pedestal Bearing* folder from the *c10* folder to the *c13* folder. As this is the first tutorial of the chapter, therefore you need to create the *c13* folder inside the *Creo-8.0* folder. Set the working directory to *C:/Creo-8.0/c13/Pedestal Bearing*.

Starting the Format File

As evident from Figure 13-16, the exploded view of the Pedestal Bearing is generated on a drawing sheet with a customized format. Therefore, you need to create a format before generating the drawing view. This format will be retrieved later.

1. Choose the **New** button from the **Quick Access** toolbar; the **New** dialog box is displayed.

2. Select the **Format** radio button from the **Type** area in the **New** dialog box and enter the name as **Format1** in the **File name** edit box. Choose the **OK** button from the **New** dialog box; the **New Format** dialog box is displayed.

3. In this dialog box, the **Empty** radio button in the **Specify Template** area is selected by default and the **Landscape** button in the **Orientation** area is also chosen by default. If not so, you need to set them that ways.

4. Select **A** from the **Standard Size** drop-down list in the **Size** area and then choose the **OK** button; the **Format** mode is invoked. Note that a sheet of size **A** is displayed on the screen. This is evident from the text displayed below the sheet on the screen.

Creating the Format

1. Choose **Sketch > Sketching > Edge > Offset Edge** from the **Ribbon**; the **OFFSET OPER** menu is displayed. Choose the **Ent Chain** option from this menu and then select all the four border lines of the format by drawing a window around them. Press the middle mouse button after the selection has been made; the Message Input Window is displayed and also, an arrow pointing outward is displayed on the format boundary. This arrow displays the direction of the offset. To offset the lines in the opposite direction, specify a negative offset value.

2. Enter **-0.25** in the Message Input Window and press ENTER. Next, press the middle mouse button twice to exit the **OFFSET OPER** menu.

3. Choose **Table > Table > Table** drop-down **> Insert Table** from the **Ribbon**; the **Insert Table** dialog box is displayed.

4. Choose **Table growth direction: leftward and ascending** button from the **Direction** area to set the direction of table growth.

5. Set the value to **3** in both the **Number of Columns** and **Number of Rows** spinners in the **Table Size** area.

6. Enter **0.5** as height in the **Height (INCH)** edit box and **1.2** as width in the **Width (INCH)** edit box.

7. Choose the **OK** button from the **Insert Table** dialog box; the **Select Point** dialog box is displayed and you are prompted to select a point to place the table. Also, the preview of the table gets attached to the cursor.

8. Click on the desired location in the drawing to place the table, refer to Figure 13-17.

9. Select any cell of the second column and then choose the **Height and Width** tool from the **Rows & Columns** group in the **Table** tab; the **Height and Width** dialog box is displayed.

10. Enter the value **2** in the **Width (drawing units)** edit box in the **Column** area and then choose the **OK** button; the title block created will be similar to the one shown in Figure 13-17. Note that the title block created is not the required one. Therefore, you need to modify it.

Figure 13-17 *Format with the title block*

11. Select the entire first row and then choose the **Merge Cells** button from the **Rows & Columns** group.

12. Similarly, select the second and third cells in the second column and choose the **Merge Cells** button.

The table after merging rows and columns should look similar to the one shown in Figure 13-18.

13. You need to set the alignment of the text that will be later entered in the cells of the table. To do so, double-click on any of these cells; the **Format** tab is added.

14. Click on the **Text Style** option from the **Style** drop-down of the **Format** tab; the **Text Style** dialog box is displayed.

15. In the **Note/Dimension** area of this dialog box, select the **Middle** option from the **Vertical** drop-down list and select the **Center** option from the **Horizontal** drop-down list. Exit the **Text Style** dialog box by choosing the **OK** button.

16. Now, the text is aligned vertically to the middle of the cell and horizontally to the center of the cell. Similarly, set the alignment in all the cells of the table individually.

Figure 13-18 *Modified title block*

Saving the Format File

You need to save the format file that you have created so that it can be used as a template in the **Drawing** mode. Using this template, you will generate the exploded drawing view of the Pedestal Bearing. The file will be stored in the *.frm* file format.

1. Choose the **Save** button from the **Quick Access** toolbar; the **Save Object** dialog box is displayed with the name of the file that you entered earlier. Press ENTER.

2. Close the current window by choosing the **Close** button from the **Quick Access** toolbar.

Starting a New Drawing File

You need to start a new drawing file to generate the exploded drawing view of the Pedestal Bearing.

1. Choose the **New** button from the **File** menu to display the **New** dialog box. In this dialog box, select the **Drawing** radio button and specify the name of the drawing as *c13tut1*. Choose the **OK** button; the **New Drawing** dialog box is displayed.

2. In the **New Drawing** dialog box, choose the **Browse** button from the **Default Model** area and select the assembly file named *pedestalbearing.asm*.

3. Select the **Empty with format** radio button from the **Specify Template** area in the **New Drawing** dialog box.

4. Select **Format1** from the **Format** drop-down list. If **Format1** is not available in the drop-down list, choose the **Browse** button to locate **Format1**. Then, choose the **OK** button to start a new drawing file with the selected format.

Creating the Table for BOM

1. Choose **Table > Table > Table** drop-down **> Insert Table** from the **Ribbon**; the **Insert Table** dialog box is displayed.

2. Choose the **Table growth direction: leftward and descending** button from the **Direction** area.

3. Set the value to **3** in the **Number of Columns** spinner and **2** in the **Number of Rows** spinner in the **Table Size** area of the dialog box. Now, you need to enter the dimensions of the BOM table.

4. Enter **0.5** as height in the **Height (INCH)** edit box and **1.2** as width in the **Width (INCH)** edit box.

5. Choose the **OK** button from the **Insert Table** dialog box; the **Select Point** dialog box is displayed and you are prompted to select a point to place the table.

6. Click on the desired location to place the table; refer to Figure 13-21.

7. Select any cell of the second column and then choose the **Height and Width** tool from the **Rows & Columns** group in the **Table** tab; the **Height and Width** dialog box is displayed.

8. Enter the value **3** in the **Width (drawing units)** edit box in the **Column** area and then choose the **OK** button.

9. Similarly, select any cell of the third column and change its width to **0.8**.

Setting the Repeat Region

Repeat regions are the smart cells in a table that can expand depending on the amount of data inserted in a cell. You need to define the second row as the repeat region.

1. Choose the **Repeat Region** tool from the **Data** group of the **Table** tab in the **Ribbon**; the **TBL REGIONS** menu is displayed.

2. Choose the **Add** option from the **TBL REGIONS** menu; you are prompted to locate the corners of the region.

3. Select the first cell of the second row and then select the third cell of the second row to locate the corners of the region. Press the middle mouse button twice to exit the menu.

Creating the Column Headers

Now, you need to create headings in the table.

1. Double-click in the first cell of the first row; the **Format** tab is added to the **Ribbon.** Enter **Index** and click outside twice to exit.

2. Similarly, enter **Part Name** and **Qty** in the second and third cells, respectively.

Assigning the Report Symbols in the Repeat Region

The information in the cells that are defined as repeat region is determined by the text written in the form of report symbols. These report symbols associate the information in the cells with the assembly file directly.

1. Double-click in the first cell of the second row; the **Report Symbol** dialog box is displayed, as shown in Figure 13-19. Select **rpt** from this dialog box; the **rpt** options are displayed, as shown in Figure 13-20.

*Figure 13-19 The **Report Symbol** dialog box*

*Figure 13-20 The **Report Symbol** dialog box with the **rpt** options*

2. Select the **index** option from the dialog box; **rpt.index** is added to the cell.

3. Double-click in the second cell of the second row; the **Report Symbol** dialog box is displayed. Select **asm** from this dialog box; the **asm** options are displayed.

4. Select the **mbr** option from the dialog box; the **mbr** options are displayed.

5. Select the **name** option from the dialog box; **asm.mbr.name** is added to the cell.

6. Double-click in the third cell of the second row; the **Report Symbol** dialog box is displayed. Select **rpt** from this dialog box; the **rpt** options are displayed.

7. Select the **qty** option from the dialog box.

If the text selected to enter in the cells is overlapping the other text or is extending beyond the cell, ignore it. This is because the BOM will be created in the repeat region where this text exists and after the BOM is created, the text of the BOM will fit inside the cells. The drawing sheet after creating the column headers and entering the report symbols is shown in Figure 13-21.

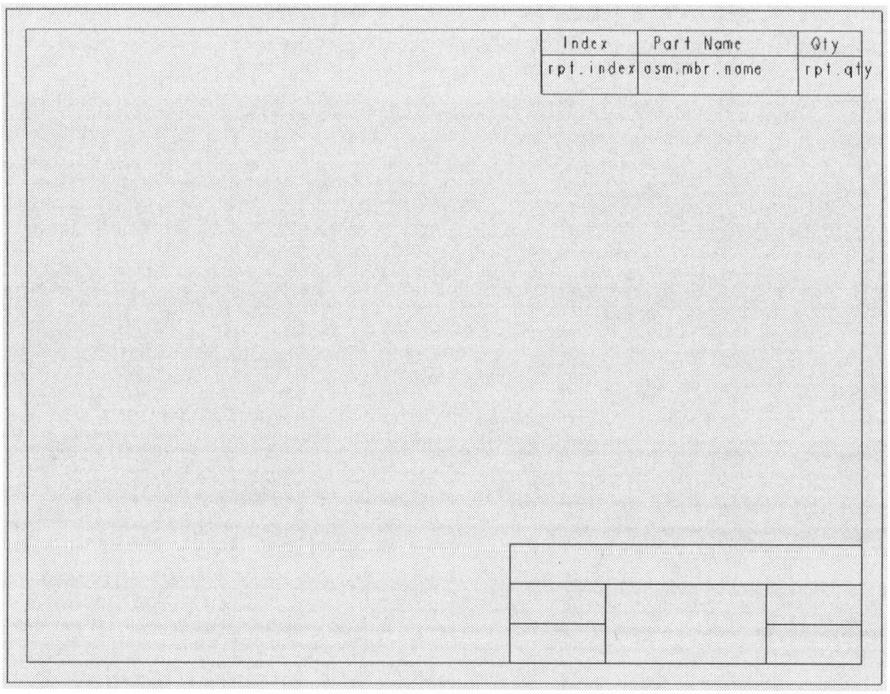

Figure 13-21 *Drawing sheet after adding the report symbols*

Generating the Exploded Drawing View

In Tutorial 2 of Chapter 10, you created the exploded view of the Pedestal Bearing in the **Assembly** mode. The name of the exploded view that you created was **EXP1**. As a result, the exploded view of the Pedestal Bearing is integrated with the assembly file that you copied from the *c10* folder. Now, you will use this exploded state to generate the exploded drawing view.

1. Choose the **General View** tool from the **Model Views** group in the **Layout** tab of the **Ribbon**; the **Select Combined State** dialog box is displayed.

2. Select the **Do not prompt for combined State** check box and then choose the **OK** button to exit the dialog box; you are prompted to select the center point for the drawing view.

3. Select a point on the left of the drawing sheet; the **Drawing View** dialog box is displayed. Select the **Isometric** option from the **Default Orientation** drop-down list in the **Drawing View** dialog box and then choose the **Apply** button.

4. Choose the **Scale** option from the **Categories** list box. Next, select the **Custom scale** radio button from the **Scale and perspective options** area. Enter the value **.025** in the edit box and then choose the **Apply** button.

5. Select the **View States** option from the **Categories** list box. Next, select the **Explode components in view** check box and select the **EXP1** from the **Assembly explode state** drop-down list. Next, choose the **Apply** button.

6. Close the **Drawing View** dialog box to complete the placement of the required view. If the model is not placed properly inside the boundaries of the drawing sheet, unlock it and then drag the drawing view and place it, as shown in Figure 13-22. Note that the BOM automatically appears in the table by extending the repeat region and the **Qty** column is empty.

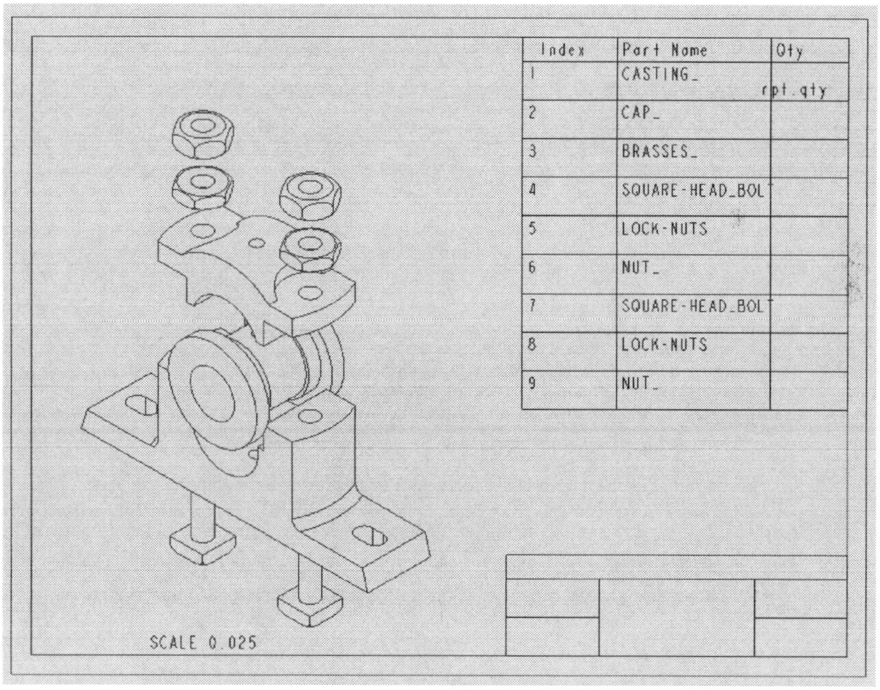

Figure 13-22 *Exploded isometric view of the assembly*

Note

1. If the BOM is not displayed, then choose ***Table > Data > Update Tables*** *from the* ***Ribbon****.*

2. You may need to select the ***No Hidden*** *option from the* ***Display Style*** *drop-down list from the* ***Graphics*** *toolbar to change the view.*

Setting the No Repeat Option

As some components are repeated in the assembly, they are also repeated in the BOM. The BOM needs to be set such that no part name is repeated.

1. Choose the **Repeat Region** tool from the **Data** group of the **Table** tab in the **Ribbon**; the **TBL REGIONS** menu is displayed.

2. Choose the **Attributes** option from the menu; you are prompted to select a region.

3. Move the cursor over the table below the first row. All rows, except the first row, are highlighted in orange. The highlighted portion is called region. Select the region; the **REGION ATTR** submenu is displayed. Choose the **No Duplicates** option and then choose the **Done/Return** button from the submenu.

4. Press the middle mouse button twice and exit the menu; the drawing sheet appears, as shown in Figure 13-23.

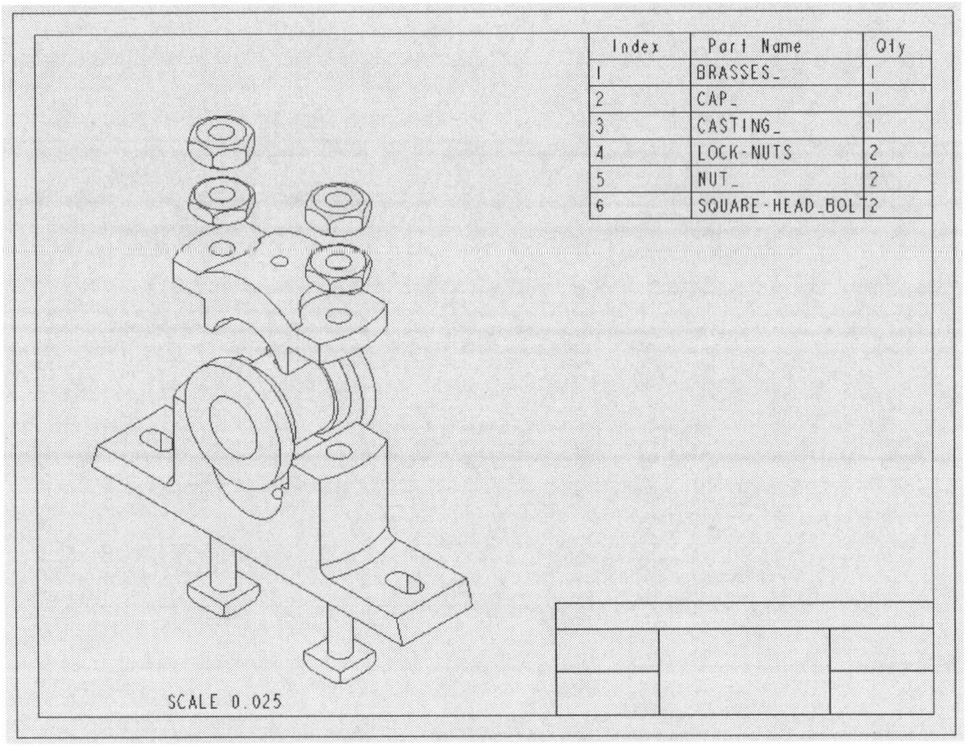

Figure 13-23 *The drawing view showing BOM with no duplicates*

5. If the information is not updated in the table, choose the **Update Tables** tool from the **Data** group of the **Table** tab from the **Ribbon** to update the information.

Generating Balloons

Balloons are generated by using the repeat region. Before generating the balloons, you need to set the repeat region for generating balloon.

1. Choose the **Create Balloons - All** option from the **Create Balloons** drop-down list available in the **Balloons** group of the **Table** tab in the **Ribbon**; balloons are attached with all the parts in the exploded view. The balloons contain the index number so that the parts in the assembly can be referred to as BOM.

2. Drag the balloons to place them appropriately in the drawing sheet, as shown in Figure 13-24. You can select a balloon and then hold the right mouse button to invoke the shortcut menu. From this menu, you can select the required options to modify the attachment points of balloons.

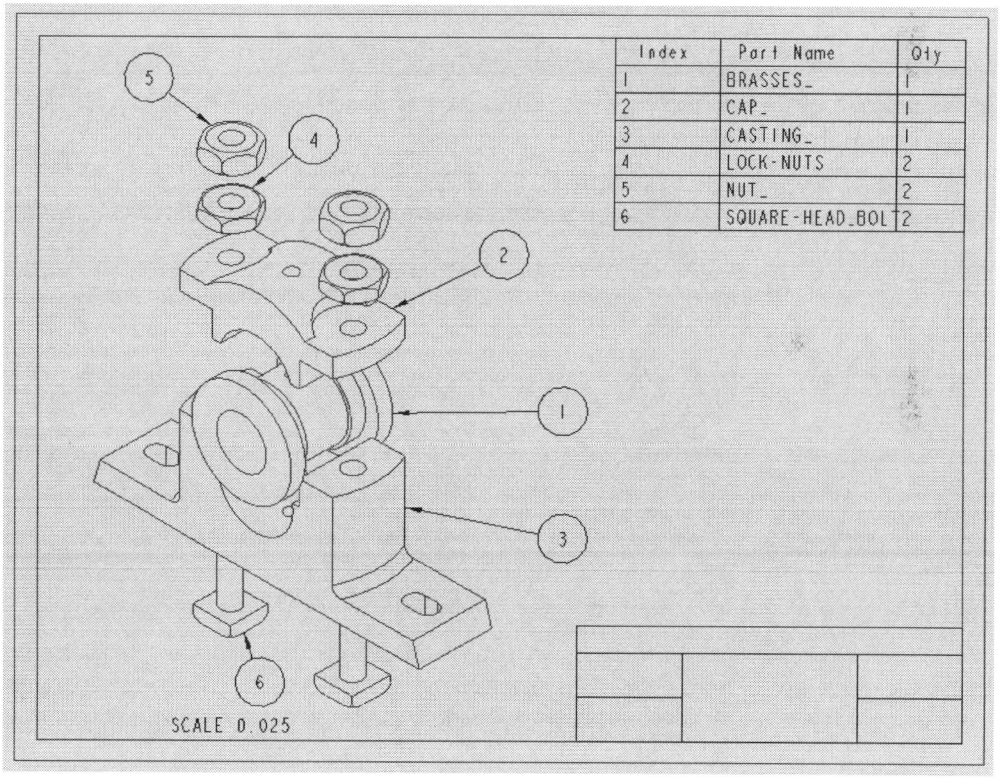

Figure 13-24 The drawing view of the assembly showing the BOM and balloons

Entering Text in the Title Block

The text alignment in various cells of the title block was defined while creating the format. As a result, when you enter the text, it will automatically be aligned to the middle left.

1. Double-click on the top cell in the table; the **Format** tab is added to the **Ribbon**.

2. Now, enter **Pedestal Bearing Assembly** in the cell. Next, choose the **Text Style** tool from the **Style** group drop-down of the **Format** tab; the **Text Style** dialog box is displayed. Now, select the **Default** check box at the right of the **Height** property in the character area and choose the **OK** button.

3. Double-click on the cell that is in the first column and the second row; the **Format** tab is added to the **Ribbon**.

4. Now, enter **Scale** and then press ENTER. Now, enter **0.025** and click outside twice to exit; the text Scale: 0.025 is entered in the two lines in the title block.

5. Similarly, enter the text in all the remaining cells one by one, as shown in Figure 13-25.

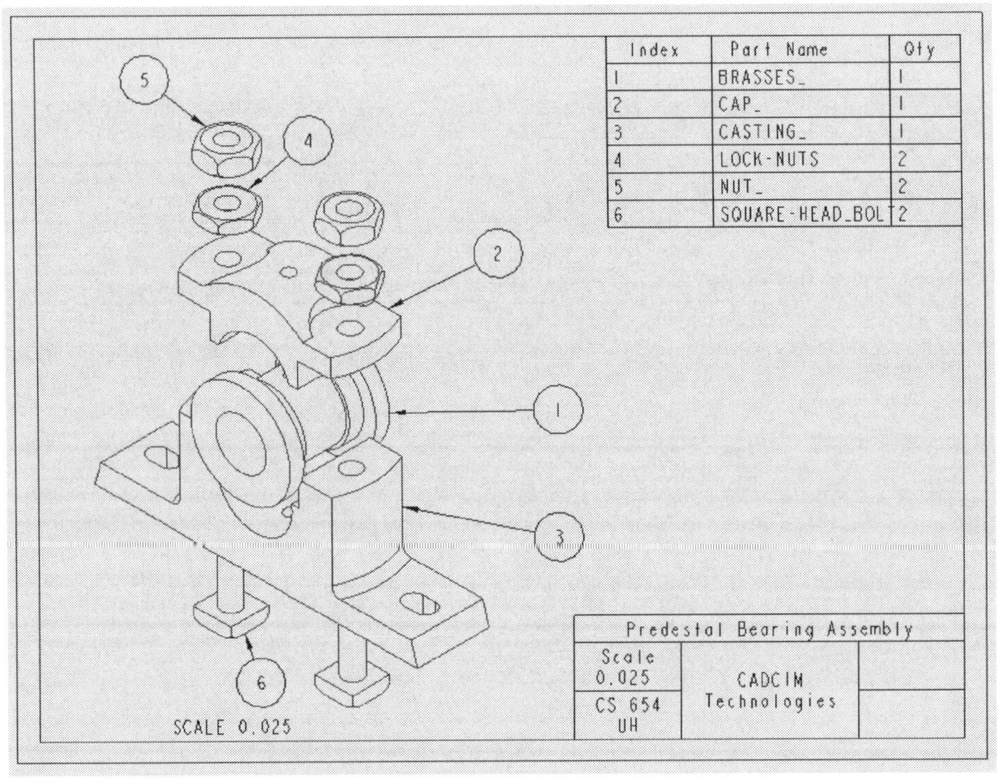

Figure 13-25 The drawing view of the assembly showing the title block, balloons, and the BOM

Saving the Drawing File

1. Choose the **Save** button from the **Quick Access** toolbar to save the drawing file; the **Save Object** dialog box is displayed. Now, save the file by pressing ENTER.

Closing the Window

The drawing file has been saved and now you can exit the **Drawing** mode.

1. Choose **File > Close** from the menu bar to close the current session.

Note

The occurrence of components in BOM depends on the order in which they were assembled in the assembly. In other words, the component placed first will be placed first in the BOM.

Tutorial 2

In this tutorial, you will create a format of size A and add the title block in the format. You will retrieve the format in the **Drawing** mode and generate the exploded isometric view of the Shock Assembly created in Tutorial 1 of Chapter 10. Also, you will add the associative Bill of Material and the Balloons to the drawing view, as shown in Figure 13-26.

(Expected time: 45 min)

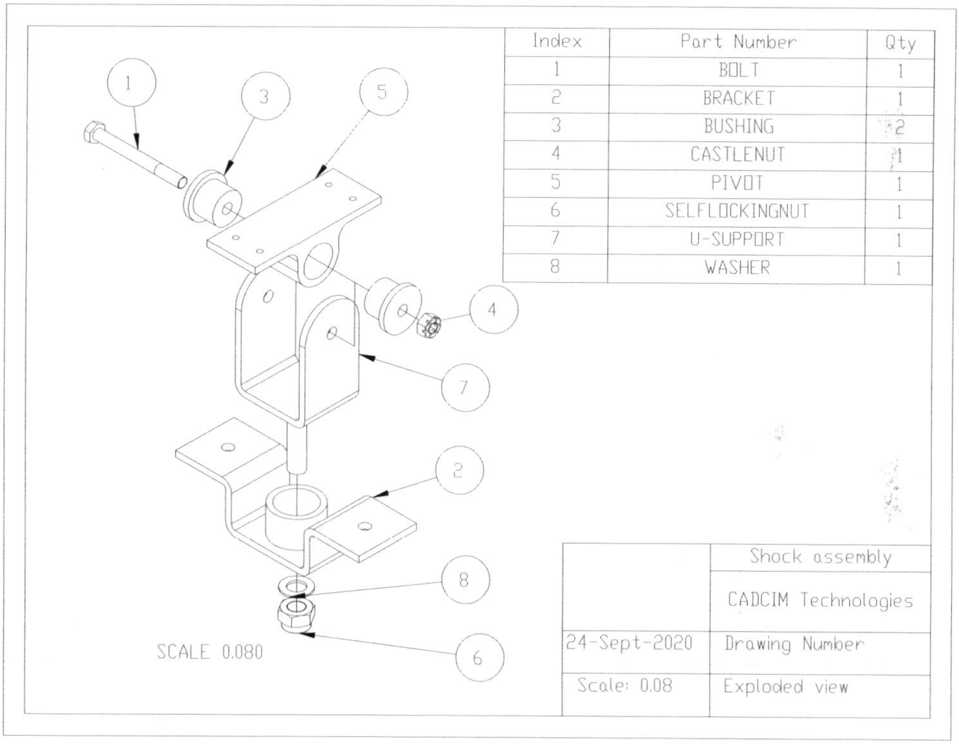

__Figure 13-26__ The drawing view of the exploded assembly with BOM

The following steps are required to complete this tutorial:

a. Start a new file in the **Format** mode. Create the format of the drawing and add the title block in the format.

b. Save the format file and then close it.

c. Start a new drawing file in the **Drawing** mode, select *shockassembly.asm* as the model, and retrieve the format created earlier.

d. Create the table for BOM and enter the headers.

e. Define the repeat region and then assign the report symbols, see.
f. Generate the exploded isometric view of the assembly and generate balloons in the drawing view.
g. Add text in the title block.

Copy the *Shock Assembly* folder from the *c10* folder to the *c13* folder and set this folder as the working directory.

Starting the Format File

As mentioned in the tutorial description, you need to create a format of size A that will be retrieved later to create the drawing view of the Shock Assembly.

1. Choose the **New** button from the **Quick Access** toolbar to display the **New** dialog box.

2. Select the **Format** radio button from the **Type** area in the **New** dialog box and name the file as *Format2*. Choose the **OK** button from the **New** dialog box; the **New Format** dialog box is displayed.

3. Select the **Empty** radio button from the **Specify Template** area and the **Landscape** button from the **Orientation** area of the **New Format** dialog box, if they are not already selected.

4. Select **A** from the **Standard Size** drop-down list in the **Size** area and then choose the **OK** button to invoke the **Format** mode. Note that a sheet of size A is displayed on the screen. This is evident from the text displayed below the sheet on the screen.

Creating the Format

1. Choose **Sketch > Sketching> Edge > Offset Edge** from the **Ribbon**; the **OFFSET OPER** menu is displayed. Choose the **Ent Chain** option and then select all the four border lines of the format by drawing a window around them. Press the middle mouse button after the selection has been made; an arrow is displayed pointing outward. This arrow displays the direction of the offset. Since you need to offset the lines in the opposite direction, you will specify a negative offset distance.

2. Enter **-0.25** in the Message Input Window and press ENTER. Press the middle mouse button twice to exit the **OFFSET OPER** menu.

3. Choose **Table > Table** drop-down **> Insert Table** from the **Ribbon**; the **Insert Table** dialog box is displayed.

4. Choose the **Table growth direction: leftward and ascending** button from the **Direction** area.

5. Set the value to **2** in the **Number of Columns** spinner and **4** in the **Number of Rows** spinner in the **Table Size** area.

6. Enter **0.5** as height in the **Height (INCH)** edit box and **1.2** as width in the **Width (INCH)** edit box.

7. Choose the **OK** button from the **Insert Table** dialog box; the **Select Point** dialog box is displayed and you are prompted to select a point to place the table.

8. Click on the required location to place the table, refer to Figure 13-27; the table is placed at the specified place.

9. Select any cell of the second column and then choose the **Height and Width** tool from the **Rows & Columns** group in the **Table** tab; the **Height and Width** dialog box is displayed.

10. Enter the value **3.2** in the **Width (drawing units)** edit box in the **Column** area and then choose the **OK** button.

11. Select any cell in the second row and choose the **Height and Width** tool from the **Rows & Columns** group in the **Table** tab; the **Height and Width** dialog box is displayed.

12. Enter the value **1** in the **Height (drawing units)** edit box in the **Row** area of the dialog box and then choose the **OK** button. Note that you may need to clear the **Automatic height adjustment** check box to enter the **Height (drawing units)**.

13. Select any cell in the first row and choose the **Height and Width** tool from the **Rows & Columns** group in the **Table** tab; the **Height and Width** dialog box is displayed.

14. Now, clear the **Automatic height adjustment** check box in the **Row** area. Enter the value **0.7** in the **Height (drawing units)** edit box in the **Row** area and then choose the **OK** button.

    The title block created will be similar to the one shown in Figure 13-27. However, this title block is not the required one. Notice the difference between the title block shown in Figure 13-27 and the one shown in Figure 13-28. So, you need to modify the title block that you have created such that it is similar to the one shown in Figure 13-28.

Figure 13-27 *Format with the title block*

15. Select a cell in the first row and a cell in the second row of the first column to merge them. Then, choose the **Merge Cells** tool from the **Rows & Columns** group in the **Table** tab of the **Ribbon**; the cells are merged.

    The table after merging the rows and columns should look similar to the one shown in Figure 13-28. You need to set the alignment of the text that will be entered in the cells of the table later.

Figure 13-28 *Modified title block*

16. Double-click in a cell to invoke the **Format** tab and click on the **Text Style** option from the **Style** drop-down of the **Format** tab; the **Text Style** dialog box is displayed.

17. In the **Note/Dimension** area of this dialog box, select the **Middle** option from the **Vertical** drop-down list and the **Center** option from the **Horizontal** drop-down list. Next, choose the **OK** button to exit the dialog box.

    Now, the text is aligned vertically to the middle of the cell and horizontally to the center of the cell. Similarly, set the alignment in all the cells of the table individually.

Saving the Format File

You need to save the format file that you have created so that it can be used as a template in the **Drawing** mode. The file will be stored in the *.frm* file format.

1. Choose the **Save** button from the **Quick Access** toolbar; the **Save Object** dialog box is displayed with the name of the file entered earlier. Next, press ENTER to save it.

2. Close the current window by choosing **File > Close** from the menu bar.

Starting a New Drawing File

You need to start a new drawing file for generating the exploded drawing view of the Shock Assembly and for generating the BOM.

1. Choose the **New** button to display the **New** dialog box. Select the **Drawing** radio button from the dialog box, specify the name of the drawing as **c13tut2**, and then choose the **OK** button; the **New Drawing** dialog box is displayed.

2. Choose the **Browse** button from the **Default Model** area and select the assembly file named *shockassembly.asm*.

3. Select the **Empty with format** radio button in the **Specify Template** area of the **New Drawing** dialog box.

4. Select **Format2** from the **Format** drop-down list. If **Format2** is not available in the drop-down list, choose the **Browse** button to locate **Format2**. Next, choose the **OK** button to exit the **New Drawing** dialog box.

Creating the Table for BOM

1. Choose **Table > Table** drop-down **> Insert Table** from the **Ribbon**; the **Insert Table** dialog box is displayed.

2. Choose **Table growth direction: leftward and descending** button from the **Direction** area.

3. Set the value to **3** in the **Number of Columns** spinner and **2** in the **Number of Rows** spinner in the **Table Size** area.

4. Enter **0.5** as height in the **Height (INCH)** edit box and **1.2** as width in the **Width (INCH)** edit box.

5. Choose the **OK** button from the **Insert Table** dialog box; the **Select Point** dialog box is displayed and you are prompted to select a point to place the table.

6. Select a point close to the upper right corner of the inner rectangle; the table is placed at the specified point.

7. Select any cell in the second column and then choose the **Height and Width** tool from the **Rows & Columns** group in the **Table** tab; the **Height and Width** dialog box is displayed.

8. Enter the value **3** in the **Width (drawing units)** edit box in the **Column** area and then choose the **OK** button. Similarly, change the width of the third column to **0.8**.

Setting the Repeat Region

Repeat regions are the smart cells in a table that can expand depending on the amount of data inserted in the cell. The second row will be defined as the repeat region.

1. Choose **Repeat Region** from the **Data** group of the **Table** tab in the **Ribbon**; the **TBL REGIONS** menu is displayed.

2. Choose the **Add** option from the **TBL REGIONS** menu; you are prompted to locate the corners of the region.

3. Select the first cell of the second row and the third cell of the second row to define the region. Press the middle mouse button twice to exit the menu.

Creating the Column Headers

Next, you need to create headings in the table.

1. Double-click in the first cell of the first row; the **Format** tab is added to the **Ribbon.** Enter **Index** in the cell and click outside twice to exit.

2. Similarly, enter **Part Name** and **Qty** in the second and third cells, respectively.

Entering the Report Symbols in the Repeat Region

As mentioned earlier, the information in the cells that are defined as the repeat region is determined by the text written in the form of report symbols. These report symbols associate the information in the cells with the assembly file directly.

1. Double-click in the first cell of the second row; the **Report Symbol** dialog box is displayed. Select **rpt** from this dialog box; the **rpt** options are displayed.

2. Select the **index** option from this dialog box.

3. Next, double-click in the second cell of the second row; the **Report Symbol** dialog box is displayed. Select **asm** from this dialog box; the **asm** options are displayed.

4. Select the **mbr** option from this dialog box and then select the **name** option from the options that appear.

5. Double-click in the third cell of the second row; the **Report Symbol** dialog box is displayed. Select **rpt** from this dialog box; the **rpt** options are displayed.

6. Select the **qty** option from this dialog box. The drawing sheet after adding the report symbols is shown in Figure 13-29.

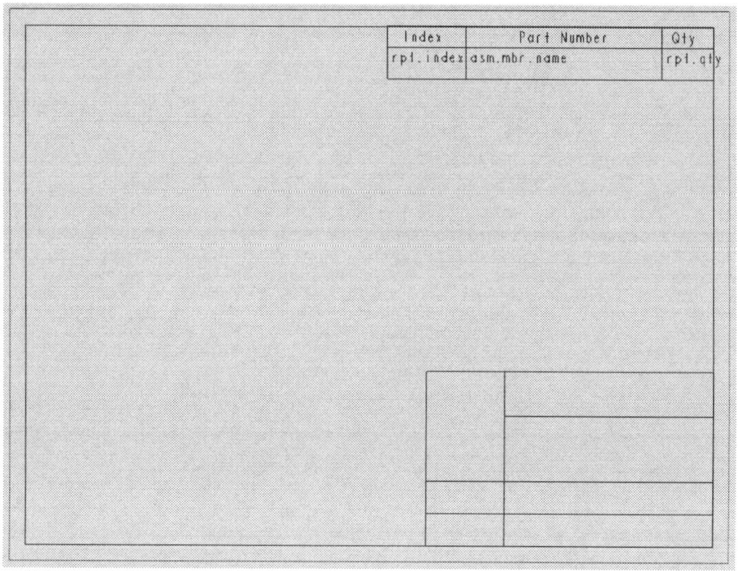

Figure 13-29 Drawing sheet after adding the report symbols

Generating the Drawing View

In Tutorial 1 of Chapter 10, you created the exploded view of the Shock Assembly in the **Assembly** mode. The name of the exploded view that you have specified was **EXP1**. As a result, the exploded view of the Shock assembly has been integrated with the assembly file that you copied from the *c10* folder. Now, you will use this exploded state to generate the exploded drawing view in the **Drawing** mode.

1. Choose the **General** tool from the **Model Views** group of the **Layout** tab in the **Ribbon**; the **Select Combined State** dialog box is displayed.

2. Select the **Do not prompt for Combined State** check box and then choose the **OK** button; the dialog box is closed and you are prompted to select the center point for the drawing view.

3. Select a point on the left of the drawing sheet; the **Drawing View** dialog box is displayed. Select the **Isometric** option from the **Default orientation** drop-down list in the **Drawing View** dialog box and then choose the **Apply** button.

4. Select the **Scale** option from the **Categories** list box. Also, select the **Custom scale** radio button from the **Scale and perspective options** area. Enter **.003** in the edit box and then choose the **Apply** button.

5. Select the **View States** option from the **Categories** list box. Select the **Explode components in view** check box. Select the **EXP1** option from the **Assembly explode state** drop-down list and then choose the **Apply** button.

6. Choose the **Close** button from the **Drawing View** dialog box; the exploded isometric view of the assembly is displayed.

If the model is not placed properly inside the boundary of the drawing sheet, unlock it and then drag the drawing view and place it, as shown in Figure 13-30. If the BOM is not displayed, choose **Table > Data > Update Tables** from the **Ribbon** to display it.

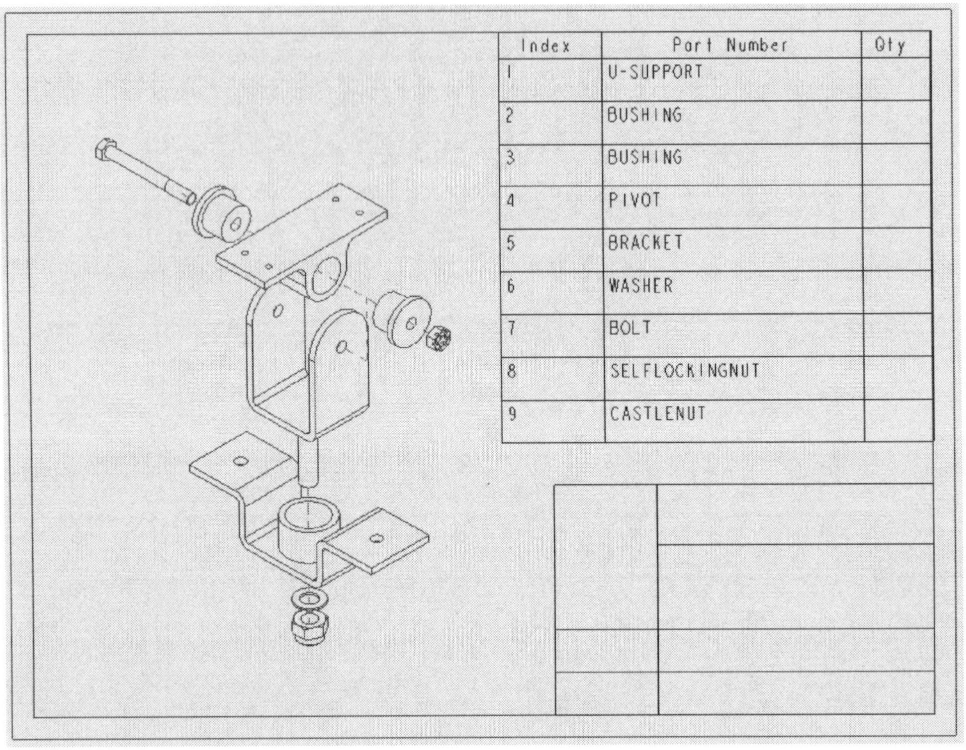

Figure 13-30 *Exploded isometric view of the assembly*

Setting the No Duplicates Option

The Shock assembly consists of two instances of Bushing. Therefore, in BOM, it is also repeated. But, the BOM needs to be set such that no part name is repeated.

1. Choose **Repeat Region** from the **Data** group of the **Table** tab in the **Ribbon**; the **TBL REGIONS** menu is displayed.

2. Choose the **Attributes** option from the menu; you are prompted to select a region.

3. Move the cursor on the table below the first row; all the rows, except the first row, are highlighted in green. Select the region; the **REGION ATTR** submenu is displayed. Choose the **No Duplicates** option and then choose the **Done/Return** option.

4. Press the middle mouse button twice to exit the menu. The drawing sheet appears, as shown in Figure 13-31.

5. If the information is not updated in the table, choose **Table > Data > Update Tables** from the **Ribbon** to update it.

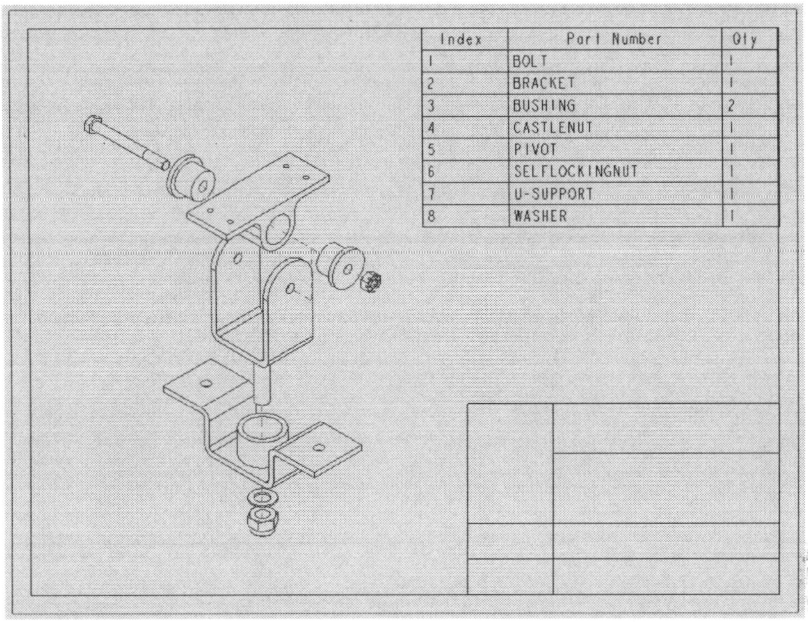

Figure 13-31 *The drawing view of the assembly showing the BOM and balloons*

Generating Balloons

Balloons are generated by using the repeat region. Before generating the balloons, you need to set the repeat region for the balloon generation.

1. Choose the **Create Balloons - All** option from the **Create Balloons** drop-down list available in the **Balloons** group of the **Table** tab of the **Ribbon**; balloons get attached with all the parts of assembly in the exploded view. The balloons contain the index number so that the parts in the assembly can be referred to as BOM.

2. Drag the balloons to place them appropriately on the drawing sheet, refer to Figure 13-31. You can modify the attachment points of a balloon by selecting the balloon and right-clicking. Next, choose the required option from the shortcut menu displayed to modify the attachment points.

Entering Text in the Title Block

1. Double-click in the cell formed by merging the first row and the second column; the **Format** tab is added to the **Ribbon**.

2. Enter **Shock assembly** in the cell. Next, choose the **Text Style** tool from the **Style** group drop-down of the **Format** tab; the dialog box is displayed. Now, select the **Default** check box at the right of the **Height** property and choose the **OK** button.

3. Now, select the cell from the second row and the second column. Enter **CADCIM Technologies** in the cell and click outside twice to exit.

4. Similarly, enter the text in all the remaining cells one by one, as shown in Figure 13-32.

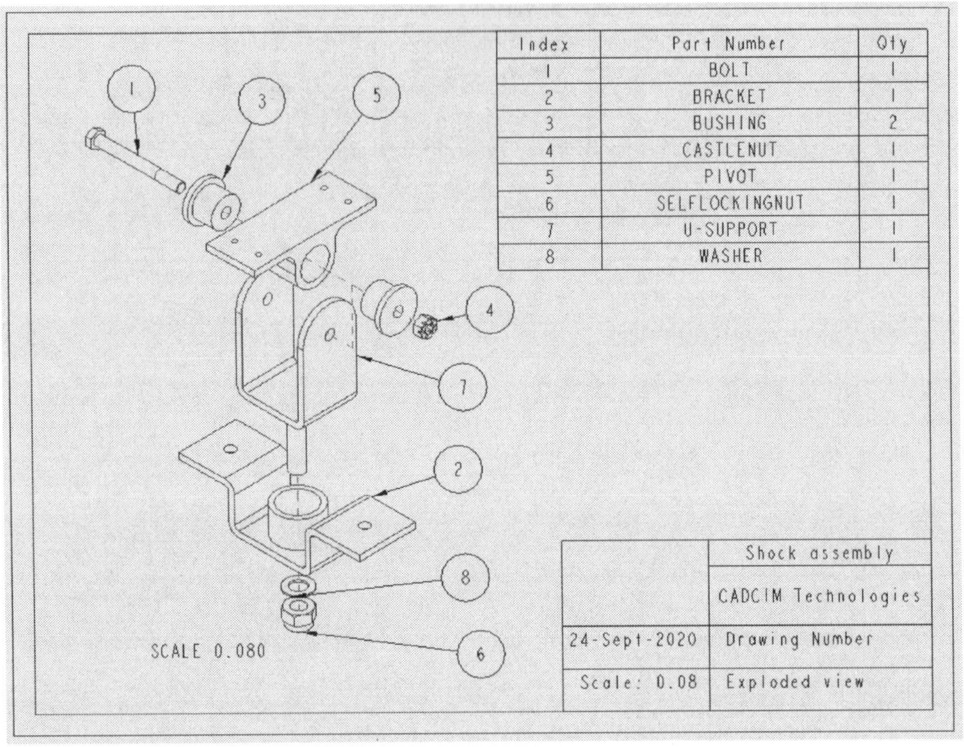

Figure 13-32 *The drawing view of the assembly showing the title block, balloons, and the BOM*

Saving the Drawing File

1. Choose the **Save** button from the **Quick Access** toolbar to save the drawing file; the **Save Object** dialog box is displayed. Press ENTER to save the file.

Closing the Window

The drawing file has been saved and now you can exit the **Drawing** mode.

1. Choose **File > Close** from the menu bar to close the window.

Self-Evaluation Test

Answer the following questions and then compare them to those given at the end of this chapter:

1. The __________ button of the **Balloons** group in the **Table** tab of the **Ribbon** is used to generate balloons.

2. The __________ tab is added, when you double-click on the cell of a table.

3. The __________ option is chosen to avoid repetition of components in BOM.

4. The __________ key can also be used to delete a draft entity.

5. The __________ button of the **Table** group in the **Table** tab of the **Ribbon** is used to draw tables in the **Drawing** mode.

6. When you move a view that forms a group with draft entities, the draft entities also move. (T/F)

7. Creo Parametric provides you with some standard drawing formats for generating drawing views. (T/F)

8. Creo Parametric allows you to create the user-defined formats. (T/F)

9. The BOM generated by using the repeat region is associative. (T/F)

10. One drawing file has only one drawing sheet. (T/F)

Review Questions

Answer the following questions:

1. Which of the following modes in Creo Parametric is used to create an associative BOM?

   (a) **Part** (b) **Format**
   (c) **Drawing** (d) **Sketch**

2. Which of the following dialog boxes is displayed on double-clicking in the repeat region cell?

   (a) **Drawing View** (b) **Format**
   (c) **Report Symbol** (d) **TBL REGIONS**

3. Which of the following buttons enables you to draw geometric entities continuously?

   (a) **Line** (b) **Circle**
   (c) **Chain** (d) **Arc**

4. Which of the following radio buttons is selected in the **New** dialog box to create a user-defined format for the drawing sheet?

   (a) **Part** (b) **Format**
   (c) **Drawing** (d) **Sketch**

5. Which of the following menus in the menu bar has the **Close** option to close the current window?

   (a) **Info** (b) **Window**
   (c) **View** (d) **Model**

6. You can open more than one window in Creo Parametric. (T/F)

7. It is not possible to draw splines in the **Drawing** mode. (T/F)

8. There are two types of drafting in Creo Parametric: Generative drafting and Interactive drafting. (T/F)

9. You can import a text file into the **Drawing** mode. (T/F)

10. The **Draft Group** option in the cascading menu is used to add or remove notes, surface symbols, or gtols from the selected dimension. (T/F)

EXERCISE

Exercise 1

Create a format of size A and then retrieve it in the **Drawing** mode to place the Right view, and an exploded isometric view of the **Cross Head** assembly created in Exercise 1 of Chapter 10. Add the associative assembly BOM and the balloons to the drawing view, as shown in Figure 13-33. **(Expected time: 45 min)**

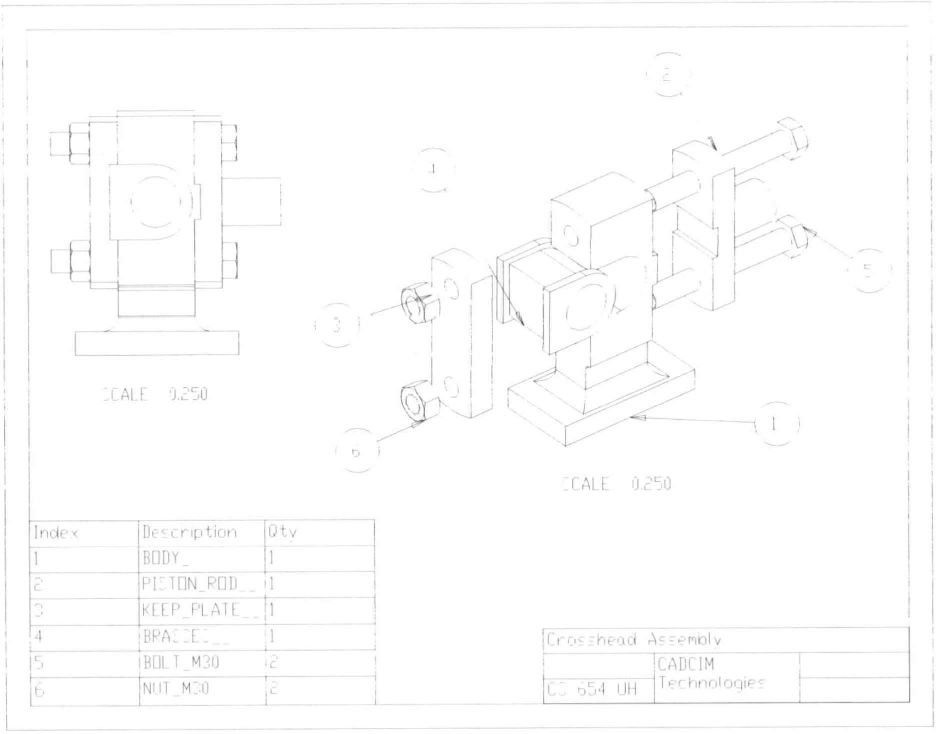

Figure 13-33 Drawing for Exercise 1

Answers to Self-Evaluation Test

1. **Create Balloons**, **2.** **Format**, **3.** **No Duplicates**, **4.** DELETE, **5.** **Table**, **6.** T, **7.** T, **8.** T, **9.** T, **10.** F

This page is intentionally left blank

Index

R

S

T

U

V

W

X

Other Publications by CADCIM Technologies

The following is the list of some of the publications by CADCIM Technologies. Please visit *www.cadcim.com* for the complete listing.

AutoCAD Textbooks

- AutoCAD 2022: A Problem-Solving Approach, Basic and Intermediate, 28th Edition
- AutoCAD 2021: A Problem-Solving Approach, Basic and Intermediate, 27th Edition
- AutoCAD 2020: A Problem-Solving Approach, Basic and Intermediate, 26th Edition
- Advanced AutoCAD 2021: A Problem-Solving Approach, 3D and Advanced, 25th Edition
- Advanced AutoCAD 2018: A Problem-Solving Approach, 3D and Advanced, 24th Edition

Autodesk Inventor Textbooks

- Autodesk Inventor Professional 2022 for Designers, 22nd Edition
- Autodesk Inventor Professional 2021 for Designers, 21st Edition
- Autodesk Inventor Professional 2020 for Designers, 20th Edition

AutoCAD MEP Textbooks

- AutoCAD MEP 2020 for Designers, 5th Edition
- AutoCAD MEP 2018 for Designers, 4th Edition
- AutoCAD MEP 2016 for Designers, 3rd Edition

AutoCAD Plant 3D Textbooks

- AutoCAD Plant 3D 2021 for Designers, 6th Edition
- AutoCAD Plant 3D 2020 for Designers, 5th Edition
- AutoCAD Plant 3D 2018 for Designers, 4th Edition

Autodesk Fusion 360 Textbook

- Autodesk Fusion 360: A Tutorial Approach, 2nd Edition

Solid Edge Textbooks

- Solid Edge 2021 for Designers, 18th Edition
- Solid Edge 2020 for Designers, 17th Edition

NX Textbooks

- Siemens NX 2021 for Designers,14th Edition
- Siemens NX 2020 for Designers,13th Edition
- Siemens NX 2019 for Designers,12th Edition
- NX 12.0 for Designers, 11th Edition

NX Mold Textbook

- Mold Design Using NX 11.0: A Tutorial Approach

NX Nastran Textbook

- NX Nastran 9.0 for Designers

SOLIDWORKS Textbooks
- SOLIDWORKS 2021 for Designers, 19th Edition
- SOLIDWORKS 2020 for Designers, 18th Edition
- SOLIDWORKS 2019 for Designers, 17th Edition
- Learning SOLIDWORKS 2019: A Project Based Approach
- SOLIDWORKS 2020: A Tutorial Approach

SOLIDWORKS Simulation Textbooks
- SOLIDWORKS Simulation 2018: A Tutorial Approach
- SOLIDWORKS Simulation 2016: A Tutorial Approach

CATIA Textbooks
- CATIA V5-6R2020 for Designers, 18th Edition
- CATIA V5-6R2019 for Designers, 17th Edition

Creo Parametric Textbooks
- Creo Parametric 7.0 for Designers, 7th Edition
- Creo Parametric 6.0 for Designers, 6th Edition

ANSYS Textbooks
- ANSYS Workbench 2019 R2: A Tutorial Approach
- ANSYS Workbench 14.0: A Tutorial Approach

Creo Direct Textbook
- Creo Direct 2.0 and Beyond for Designers

Autodesk Alias Textbooks
- Learning Autodesk Alias Design 2016, 5th Edition
- Learning Autodesk Alias Design 2015, 4th Edition

AutoCAD LT Textbooks
- AutoCAD LT 2022 for Designers, 14th Edition
- AutoCAD LT 2020 for Designers, 13th Edition

EdgeCAM Textbooks
- EdgeCAM 11.0 for Manufacturers
- EdgeCAM 10.0 for Manufacturers

Autodesk Revit MEP Textbooks
- Exploring Autodesk Revit 2021 for MEP, 7th Edition
- Exploring Autodesk Revit 2019 for MEP, 6th Edition

AutoCAD Civil 3D Textbooks
- Exploring AutoCAD Civil 3D 2020, 10th Edition
- Exploring AutoCAD Civil 3D 2019, 9th Edition
- Exploring AutoCAD Civil 3D 2018, 8th Edition

AutoCAD Map 3D Textbooks
- Exploring AutoCAD Map 3D 2018, 8th Edition
- Exploring AutoCAD Map 3D 2017, 7th Edition

RISA-3D Textbook
- Exploring RISA-3D 14.0

Autodesk Navisworks Textbooks
- Exploring Autodesk Navisworks 2021, 8th Edition
- Exploring Autodesk Navisworks 2020, 7th Edition
- Exploring Autodesk Navisworks 2019, 6th Edition

AutoCAD Raster Design Textbooks
- Exploring AutoCAD Raster Design 2017
- Exploring AutoCAD Raster Design 2016

Bentley STAAD.Pro Textbooks
- Exploring Bentley STAAD.Pro CONNECT Edition, V22, 4th Edition
- Exploring Bentley STAAD.Pro V8i (SELECTseries 6)

Autodesk 3ds Max Design Textbooks
- Autodesk 3ds Max 2021 for Beginners : A Tutorial Approach, 21st Edition
- Autodesk 3ds Max 2020 for Beginners : A Tutorial Approach, 20th Edition
- Autodesk 3ds Max 2019 for Beginners : A Tutorial Approach, 19th Edition
- Autodesk 3ds Max Design 2015: A Tutorial Approach, 15th Edition

Autodesk 3ds Max Textbooks
- Autodesk 3ds Max 2021: A Comprehensive Guide, 21th Edition
- Autodesk 3ds Max 2020: A Comprehensive Guide, 20th Edition
- Autodesk 3ds Max 2019: A Comprehensive Guide, 19th Edition
- Autodesk 3ds Max 2019 for Beginners: A Tutorial Approach, 19th Edition

Autodesk Maya Textbooks
- Autodesk Maya 2022: A Comprehensive Guide, 13th Edition
- Autodesk Maya 2020: A Comprehensive Guide, 12th Edition
- Character Animation: A Tutorial Approach

Pixologic ZBrush Textbooks
- Pixologic ZBrush 2021: A Comprehensive Guide, 7th Edition
- Pixologic ZBrush 2020: A Comprehensive Guide, 6th Edition
- Pixologic ZBrush 2018: A Comprehensive Guide, 5th Edition
- Pixologic ZBrush 4R8: A Comprehensive Guide, 4th Edition

Fusion Textbooks
- Blackmagic Design Fusion 7 Studio: A Tutorial Approach, 3rd Edition
- The eyeon Fusion 6.3: A Tutorial Approach

Flash Textbooks

- Adobe Flash Professional CC : A Tutorial Approach, 3rd Edition
- Adobe Flash Professional CC: A Tutorial Approach
- Adobe Flash Professional CS6: A Tutorial Approach

MAXON CINEMA 4D Textbooks

- MAXON CINEMA 4D R20 Studio: A Tutorial Approach, 7th Edition
- MAXON CINEMA 4D R19 Studio: A Tutorial Approach, 6th Edition
- MAXON CINEMA 4D R18 Studio: A Tutorial Approach, 5th Edition

Computer Programming Textbooks

- Introducing PHP 7/MySQL
- Introduction to C++ programming, 2nd Edition
- Learning Oracle 12c - A PL/SQL Approach
- Learning ASP.NET AJAX
- Introduction to Java Programming, 2nd Edition
- Learning Visual Basic.NET 2008

Oracle Primavera Textbooks

- Exploring Oracle Primavera P6 Professional 18, 3rd Edition
- Exploring Oracle Primavera P6 v8.4

AutoCAD Textbooks Authored by Prof. Sham Tickoo and Published by Autodesk Press

- AutoCAD: A Problem-Solving Approach: 2013 and Beyond
- AutoCAD 2012: A Problem-Solving Approach
- AutoCAD 2011: A Problem-Solving Approach
- AutoCAD 2010: A Problem-Solving Approach
- Customizing AutoCAD 2010
- AutoCAD 2009: A Problem-Solving Approach

Coming Soon from CADCIM Technologies

- Flow Simulation Using SOLIDWORKS 2021

Made in the USA
Columbia, SC
21 January 2023

75599395R00424